Royal Institution of Naval Architects

International Maritime Conference

Pacific 2008

AA005662

January 29-31, 2008
Sydney, Australia

Printed from e-media with permission by:

Curran Associates, Inc.
57 Morehouse Lane
Red Hook, NY 12571
www.proceedings.com

ISBN: 978-1-60560-459-6

Some format issues inherent in the e-media version may also appear in this print version.

Royal Institution of Naval Architects
10 Upper Belgrave Street
London, SW1X 8BQ, UK

Royal Institution of Naval Architects

International Maritime Conference
2008

TABLE OF CONTENTS

A Combinatorial Algorithm for Motion Planning in an Autonomous Unmanned Undersea Vehicle Equipped with Forward-Looking Sonar 1
S. Chan

Gas Turbine Power and Propulsion Options for the Modern Warship 22
D.J. Bricknell, R. Partridge

How Do You Ensure Safety Excellence in Your Operations 32
P. Davison

Meeting the Maritime Manpower Challenge The Nautical Profession 36
D. Bendall, L.A. Holder

Human Systems Integration: How Does it Fit Into Warship Design? 43
P. Elischer, M.R. Grech

Seakeeping Behaviour of a Damaged Warship 53
T. Andrewartha, G. Thomas, S. Cannon

Conversion MV "DRONNING INGRID" to M.V."AFRICA MERCY" 65
A. Budd

Underway Replenishment Ship Designs 74
D.J. Bricknell, P.E. Vedlog

Coastal Protection Vessels – Recent Developments 84
D.J. Bricknell, C. Fjeldstad

Law and an Interdisciplinary Approach to Maritime Affairs 94
J. Allsop

Marine Surveyors Meeting the Demands of the Next Decade 104
A.R. Brink

The Graduated Application of Survivability Modelling in the Warship Design Process 112
G. MacDonald, P. Horstmann

An Effective Corvette Platform Training System 144
H. Kucharzewski, R. Wiedemann, M. Chang

Generalized Form Ray Tracer (GFRT) for Use Within the Measure of Total Integrated System Survivability (MOTISS) Program 153
P. Morrisseau, J.A.W. Sajdak

Cost Savings for Warships Using the Reconfigurable Hull Form Concept 163
M. Renilson

It's Time to Restore the Role of Shipping in the Australian Transport and Logistics Industry 169
P. Crumlin

Incat Builds for Japan 198
C. Clifford

Effect of Tuned Mass Absorbers on Large Structures 204
J. Marco, I. MacGillivray

Progress in Simulation-Based Design 215
K. Fach, V. Bertram

Preliminary Frigate Design using a Multi-objective Evolutionary Algorithm 223
W. Smith, T. Ray

Australian's Future Maritime Air Warfare Destroyer Capability – Balancing National Interests 233
W. King

Port State Control 235
L. Emmett

Determination of Ship Grounding and Accident Scenarios of a Geographical Area Using Risk Analysis 249
A.R.M. Sidek, B.G Prusty,T. Ray

The Study of the Effective Education Technique of Ship Management System and the Present Conditions of the Japanese Fleet Management 259
H. Kido, K. Sakaguchi

Robust Ship Designs - Balancing Uncertainty - Impacts of Future Trends 265
T. Ray, W. Smith

Geometric Non-Linear Analysis of Stiffened Structures: A Review 275
R.E. Ojeda, B.G. Prusty, N. Lawrence

A Guide to the Assessment of Vessel Wash within Sheltered Waterways 285
G.J. Macfarlane, G. Cox

Hydrodynamic Design of High Speed Multi-Hulls and Motion-Control Devices Using a Panel Method 296
D. Sadovnikov

Naval Ship Assurance – A New Approach 306
J. Buckley, P. James, B. Cuckson, S. Marshall

Submarine Power and Propulsion - Trends and Opportunities 315
J. Buckingham, C. Hodge, T. Hardy

Minor Warship Roles - How Technology is Leading to a New Vessel Type 329
A. Kimber, W. Giles

Energy Scavenging Unmanned Ocean Vehicles Affordable Persistent Ocean Surveillance 339
P. Kilbourn

The Use of Warship Vulnerability Analysis Tools As a Measurement of Blast and Ballistic Protection Requirements 350
M. Buckland

Reduction of Roll Motion of a Surfaced Submarine in Beam Seas 360
A. Davies, D. Ranmuthugala, J. El-Atm, Y. Tso

Advanced CFD Simulation Of Submarine Dynamics 370
K.H. Lee, C. Pregnalato

Seakeeping Operator Guidance 379
J. Colwell, D. Stredulinsky

The Changing Face of International Marine Casualty Investigation, New Ways for Modern Times 389
P. Foley

Novel Antifouling Surfaces Mimicked from Nature 397
A. Scardino

Explorer AUV Missions in Coastal Newfoundland 408
R. Lewis, S. Adams, N. Bose, J. Anderson

Modelling 100,000 Years of Tropical Cyclone Waves 415
T.A. Hardy, L. Mason, J. McConochie

Hydrofoils Applied to Canting-keel Yachts 425
R.S. Milne, P.J. Helmore

Update on van Oortmerssen's Resistance Prediction 437
P.J. Helmore

Amphibious Ships 451
P. Brown

Use of Slocum Glider Technology in Coastal and Marine Applications 460
V. V. Anokhin, J. R. Romero, R. Steedman

Coming to Terms with the New SOLAS Probabilistic Sub-Division and Damage Stability Provisions 470
R. Gehlin

Integrated Waste Management in the Maritime Environment 484
K. Harvey

Alternative Shock Trial Techniques for the RAN Air Warfare Destroyer 494
W. Reid, L. Deyong, P. Elischer

A Trial Sensor Network for the Armidale Class Patrol Boat 504
C.P. Gardiner, P. Vincent, A. Wilson, D. Ellery, T. Armstrong

Predicting Reduction of Tankers' Hull Girder Section Modulii 515
G. Wang, L.D. Ivanov, A.K. Seah

Improvements and Updates to Intact Stability Standards Applied to Domestic Vessels in Australia 522
M. Flapan

Integrated Ship Survivability for the Royal Australian Navy 533
L. de Yong

SEA4000: The Australian Air Warfare Destroyer Program – Combat Systems Challenges and Opportunities 543
J. Davis

MAN Diesel High Power Density Engines and After-Sales Support in Naval and Commercial Applications 544
R. Losch

KSS – Efficient Catamaran Build Process in Composites 577
D. Kelsall

Meeting Future Military Capability with a Network Centric Architectural Framework 583
A. Cameron, P. Williams

Optimal Naval Warship Design For Fabrication And Maintenance 595
G.W. Sloan

Flammability and Fire Resistance Issues of Composite Structures for Ships 609
Z. Mathys, I. Grabovac, N.S. John, S. Russo

SEA4000: The Australian Air Warfare Destroyer Program – Shipbuilding Challenges and Opportunities 619
J. Carr

The Application of Classification to High Performance Craft 620
T. Ingram, D. Novak

Residual Strength Variation of a Welded Structural Component due to a Pore 631
A. Ghosh

Classification of Naval Support Ships 640
B. Simpson

The Tenix Marine LHD Ship Build Program 655
K. Malpress

Innovative Gear Solutions in Naval Propulsion Systems 661
F. Hoppe

DAD Helicopter-Ship Aid for Maximising Helicopter and Ship Operational Capability, Improving Safety and Usage Monitoring 671
D. Howe

Advances in Gyro-Stabilisation of Vessel Roll Motion 682
T. Perez, P. Steinmann

Delegate List 693

Author Index

WELCOME

On behalf of the Organizing Committee, the Royal Institution of Naval Architects, the Institute of Marine Engineering, Science and Technology and Engineers Australia, it is my pleasure to invite you to the Pacific 2008 International Maritime Conference. The Conference is being held in association with the RAN Sea Power Conference as part of the Pacific 2008 Maritime Congress.

Building on the success of Pacific 2006, our Program Committee has arranged an excellent programmed around the theme of the Conference — Meeting the Maritime Challenges. A wide range of topics will be addressed by outstanding speakers from Australia and overseas. Many papers will focus on issues relating to the major recently approved RAN projects but others will cover matters as diverse as weather prediction and the design of modern yachts. There will be something for everyone in the maritime industries and in 2008 the IMC is joined by the American, Australian, British, Canadian and Dutch (ABCD) Working Group on Human Performance at Sea with specialized papers on the challenges faced by man in the maritime environment to be presented on one day of the Conference.

I look forward to welcoming you to the Pacific 2008 International Maritime Conference.

JOHN C. JEREMY
Conference Chairman

A Combinatorial Algorithm for Motion Planning in an Autonomous Unmanned Undersea Vehicle Equipped with Forward-Looking Sonar

Steve Chan
stevechan@internode.on.net

Abstract This paper presents the conceptual design of integrated vision and motion planning capability of an Undersea Unmanned Vehicle (UUV). Vision is achieved through Forward Looking Sonar, which is responsible for producing dynamic snap shots of the vehicle's work environment, S, populated with obstacles. Motion planning is through auto-execution of a combinatorial algorithm, previously developed for robotic motion planning. A program, performing Delaunay triangulation of S, is executed at the completion of every snap shot. The obstacle-avoiding path is either a special sub-complex of the resulting CW complex. At this stage, motion execution of UUV is treated similar to a robot and the work environment is considered as a semi-algebraic set of a configuration space. Strictly speaking, such a conceptual modeling is valid for a time period of the order of R/V, where R is the vision range of the sonar and V is the velocity of UUV. This means that over any time interval [t_1, t_2], successive sub-intervals exist over which the path-planning problem of UUV is similar to that of a robotic system.

The paper will also present a broader solution for a system of UUVs communicating with the Ownship through a Mobile Service Station (MSS), such that the communicating UUVs act as sensor elements of a network while the MSS acts as a concentrator. The paper will also address the integrity of the communication link between MSS and UUVs using the standard transactional read-write locks. We also outline the backbone design of an undersea acoustic channel that pays attention to multipath spread and Doppler spread.

SECTION 1 INTRODUCTION

This paper aims at presenting a conceptual design of a system of communicating autonomous unmanned undersea vehicles. The purpose of the study is to explore the idea of using such a system as a realization of a networked system of sensors. Issues discussed include

- Vision and guidance of the vehicle
- Communication issues from the vehicles to the Ownship and possibly among themselves. The paper also speculates on endowing each of such vehicles with a certain degree of computing capability.

The vehicle is equipped with an array of forward-looking sonar [FLS]. At a given time t, this array captures the positions of the obstacles in a finite domain [Zimmerman 2004]. The intention is to use these coordinates as input, the vehicle will plan a collision free path through S. The vehicle must be able to do this autonomously. This is achieved by having a program performing a Delaunay triangulation of S using that what is captured by the FLS at the moment [http://www.cgal.org/]. The execution of the program produces a collision free path for the vehicle to traverse. At this new position in S, the FLS performs another capture and the process is repeated. Here we are assuming that at any time, the number of obstacles populating S is finite.

The triangulation process means that the continuum S is replaced by a collection of tetrahedral. The precise meaning of the phrase "replaced by a collection of tetrahedral" can be made precise in computational geometry. The obvious interest in such a study lies in the case when the obstacles are mines. The important property of a Delaunay tetrahedron is that its interior (away from the vertices) is free of obstacles/mines. Hence if the only requirement is collision free, any path through the interior without hitting the vertices will do. There are certain subtle points we have not addressed. One example is what happens if three mines are collinear? Another is what if S is not convex? Variations of this algorithm can handle additional requirements such as shortest distance traversal or shortest time traversal. More recently developed techniques in performing Delaunay triangulation also allow moving obstacles to be tackled [Eguchi 2001].

It is highly desirable that we should endow the vehicles with capability of uploading information either directly to the Ownship or to a concentrator of information, which we call MSS. Since then, the fleet will function as a networked system of sensors.

Hence we demand there should be underwater wireless channels for acoustic data transmission through which the vehicles can upload their tracking data. With this requirement, comes a host of difficult problems. Some of these are concurrency control, multipath spread, Doppler spread, noise, bit rate yield, parameter estimation etc. Most of these problems have been extensively discussed in the literature [Bessios 1996]. In a later section, we will tackle the problem of concurrency control from the point of view of mobile computing drawing on some of our previous works [Chan 1998, 1999]. The fundamental reason for stressing the importance of concurrency control in such a system is we do not want incomplete data and each uploading/reporting must be completed before the next can begin. This is why concurrency control is also called serializability.

Towards the end of the paper, we outline a simple channel model showing how do multipath spread and Doppler spread can affect bit yields.

This paper is organized as follows. In the next section, we discuss various issues regarding to the use of Delaunay triangulation in a real time fashion. In the third section, we show in detail serializability and epsilon serializability can also be achieved through proper use of lock systems. This enables the integrity of the information uploaded and is the first step towards building a networked system of sensors through the use of communicating UUV. The other harder problems related to channel integrity and signal reliability will only be touched upon.

SECTION 2: PATH-PLANNING-ALGORITHM

Historically path panning has been most intensively studied in robotics. One of the most outstanding contributions to this topic has been made by Canny [Canny 1988]. There are also other important seminal works [Schwartz 1987], [Schwartz 1982] and [Schwartz 1984].

Robot motion planning deals with the following problem: Given a robotic system B consisting of a collection of rigid bodies and having a total of r degrees of freedom which is free to move in a 2 or 3 dimensional space V admitting a collection of obstacles whose geometry is known to B, it is required to find a continuous collision-free path for B. This path serves to connect an initial placement of B to a desired target placement. Satisfactory solution of this problem leading to algorithms that can be realistically implemented as output would constitute a direct attack on non-heuristic robotics and would be tremendously useful to precision robotics

In his 1987 ACM award winning thesis, Canny [Canny 1988] gives the problem a most unambiguous definition and provides, with extremely elegant tools from differential topology, an algorithm for computing such a collision free path. His approach is as follows.

A work environment S is a semi-algebraic set [Whitney 1957] in the configuration space of B. To find a collision-free path in S that will join two desired placements of B is the same as to compute a one-dimensional subset R of S such that every connected component of S contains a single connected component of R.

This condition is called the "road map" condition and an R satisfying this condition is called a "road map" of S. Hence the motion planning problem will be solved if once S is given, then one such R can be computed. Canny produces an algorithm, which computes R and also the curves linking the placements to R. The algorithm is recursive and has a run-time complexity given by

$$T_R = O\{n^r (\log n\,(2d)^{O(r)} \log^3 w + (2dr)^{O(r^*r)} \log^2 w)\}$$

Here r is the number of degrees of freedom of B, n is the number of constraining equations and inequalities that define S, d is the degree of the constraining polynomials and w is the magnitude of the coefficient of the constraint polynomials.

This expression has run-time exponents, which are linear in the number of variables. In practice the number of degrees of freedom of a robotic system and hence the number of variables is a small constant. For a system with a small number of degrees of freedom, say between three and six, the Canny algorithm can be implemented. However, from a real time system point of view, this algorithm is actually useless because of the fact that the obstacles are not static and the number of them is not predictable.

In an effort to reduce the run-time complexity of the Canny algorithm, Chan and Lai design a combinatorial algorithm for robot motion planning [Chan 1998]. The basic technique employed is algebraic topology of semi-algebraic sets [Hironaka 1957]

To understand this construction, a bit of background in combinatorial topology is in order. This is the reason why the algorithm is described as a combinatorial algorithm. A standard textbook on this subject is [Spanier 1966].

The starting point is to consider a triangulation (K, f) of S. Let L be a subcomplex of K. Then the pair (|K|, |L|) is a relative subcomplex with a 1-skeleton $(|K|, |L|)^1 = |K^1 \cup L|$. By the corollary in [Spanier 1966], the pair $(|K|, |K^1 \cup L|)$ is 1-connected. In other words, the required "road map" is $|K^1 \cup L|$. The result of this alternative construction is expressed as

Algorithm (Combinatorial)

Let S be a work environment mathematically expressed as a semi-algebraic set. Let (K, f) be a triangulation of S. Then, a "road map" in the sense of Canny may be constructed as the 1-skeleton $|K^1 \cup L|$ of the relative CW complex (|K|, |L|), where L is a subcomplex of K.

This is the starting point of our adaptation of motion planning for robotic systems to UUV. Before we carry on with the details, three observations are in order. First, the work environment S of a robot is required to be semi-algebraic, i.e. it is defined by algebraic inequalities. This is usually satisfied. Second, a semi-algebraic set is convex. Third, the combinatorial algorithm is based on combinatorial topology and therefore, is different from the construction by Canny, which is based on differential topology. In the result of [Chan 1998], the starting point is a triangulation of S, while in [Canny 1988], the starting point is to construct a stratification of S. One may ask, is it always possible to triangulate a semi-algebraic set? The answer is yes. This is a deep result of [Hironaka 1975].

Before we switch over to UUV, let us look at further fine details / implications of this combinatorial algorithm. We will

- Construct the road map for a simple S
- Use integer programming techniques and graph theoretic techniques to determine optimal paths.

Example

Let ABCD and EFGH be rectangles with vertices enumerated in a counterclockwise direction and with EFGH inside ABCD and E facing A as shown in the following diagram,

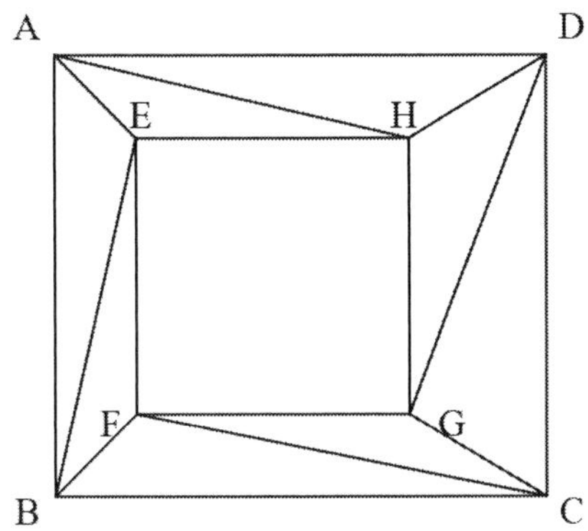

Figure 1 Road map of a 2-dimensional environment

Let the work environment, S be the space between the two rectangles. Then S is stratified by A, B, C, D, E, F, G, H as 0-dimensional manifolds, the open intervals AB, BC, CD, DA, EF, FG, GH, HE as 1-dimensional manifolds and the S-∂S as a 2-dimensional manifold. A triangulation of S is as shown in figure 1.

Let K be the complex consisting of

- A, B, C, D, E, F, G, H as 0-cells,
- AB, BC, CD, DA, EF, FG, GH, HE, AE, EB, FB, FC, GC, GD, HD, HA as 1-cells,
- Triangles AEB, BEF, BFC, CFG, CGD, DGH, DHA, HAE as 2-cells.

Let L be the subcomplex consisting of A, B, C, D, E, F, G, H, AB, BC, CD, DA, EF, FG, and GH, HE, AE. The 1-dimensional skeleton, K^1 is obtained from the above description of K with all the triangles left out. The road map is $|K^1 \cup L|$. This means that if P and Q are two points in the work place S, an obstacle-avoiding path can be obtained in the following way: Go from P to the nearest 0-cell ($\in$ L) and then travel along the 1-cells until you come to the 0-cell that is nearest to Q. Complete the journey by following the connecting curve from that 0-cell to Q.

Shortest Path Determination

Consider S being the space inside a torus

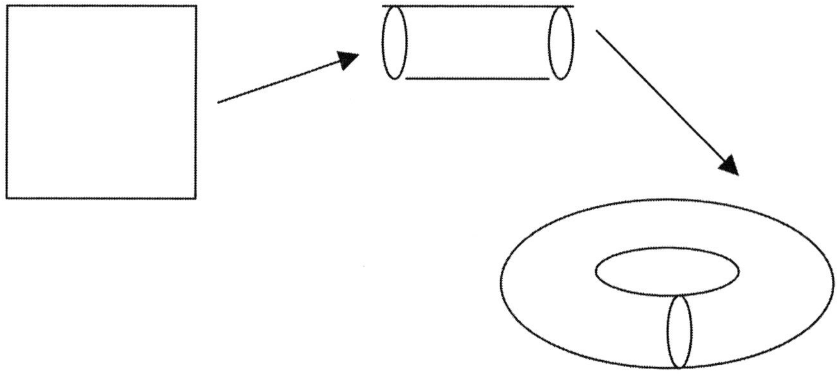

Figure 2: The work environment (S) inside a torus

In this case, the triangulation goes as follows:

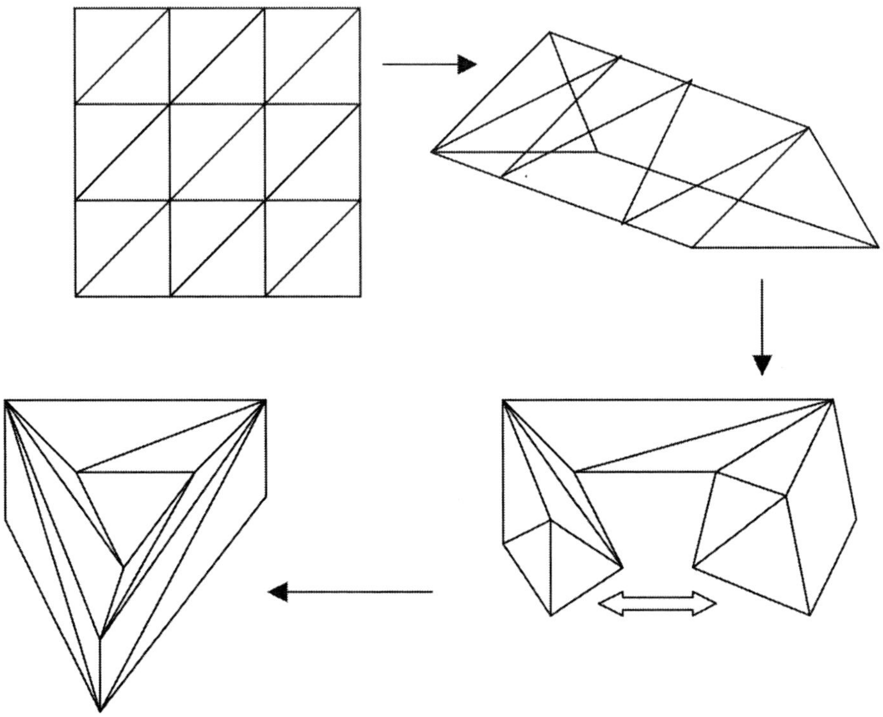

Figure 3: Triangulation of a torus

When the torus is "opened" up, we have the following

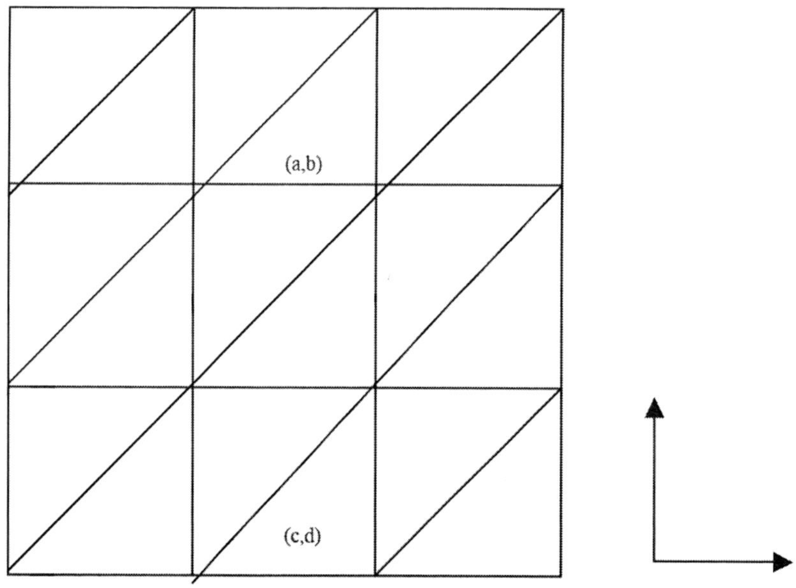

Figure 4: The work place of a torus

The shortest path analysis then goes as follows.

Let P be the point with coordinates (a, b) and Q be the point with coordinates (c,d). To go from P to Q along the shortest path will require the following steps:

- First go from P(a, b) to the nearest 0-cell, let it be called P'.
- Travel from P' to the 0-cell that is nearest to Q. Let this 0-cell be called Q'.

The length of the path from P' and Q' will be along the 1-cells and therefore will have a length given by an expression of the form $j_1L + j_2 (\sqrt{2}) L$, with j_1 and j_2 being integers ≥ 0 and L being the length of a grid line. If PP'+QQ' > d (P, Q), then set $j_1 = j_2 = 0$. Here d (P, Q) stands for the Euclidean distance between P and Q. Otherwise, the shortest path will be determined as the least upper bound of all expressions of the form $j_1L + j_2(\sqrt{2})L$ subject to fixed end point conditions. In this way we are led to a problem in integer programming [Kaufmann 1977]. Because on a torus, opposite sides of the square are identified, therefore, this integer -programming programming problem has to be solved over a finite field.

Clearly, with the motion-planning algorithm formulated in a combinatorial way, it is possible to pursue further other optimality constraints.

Now let us return to our main pursuit to adapt the combinatorial algorithm to path determination for an undersea UUV. Again, we make the observation that the algorithm calls for performing a triangulation on the semi-algebraic set S. There is no specified requirement on the nature of the triangulation. Hence we will choose a special triangulation that will further simplify the life of the UUV. We assume that the 3D FLS captures a the coordinates of a group of obstacles populating in a domain S. This domain S is the work environment of the UUV. There is no guarantee that this S is semi-algebraic of even convex also there is no guarantee that there are no collinearly situated obstacles. This assumption constitutes the limitation of our analysis.

How do we improve the combinatorial algorithm so that we can apply it to real time applications? This is one of the main purposes of this work. First, we note that in a real time environment, the programming coming out from the algorithm has to be repeatedly executed. This definitely calls for a much-improved run-time complexity compared to that given by [Canny 1988]. As the basic thesis of having to deal with semi-algebraic set is fixed, we can only change the strategy in triangulating the semi-algebraic set S. In this paper, we propose that the "semi-algebraic set S" captured by the FLS should be triangulated according to Delaunay, i.e. we use Delaunay triangulation [Preparata 1985], [Guibas 1985 a], [Guibas 1985 b], [Devillers 1997]. In particular, the improved construction suggested in [Devillers 1997] is especially appropriate. Among other desirable features, the construction is optimal on worst-case data and has a complexity of $O(n \log n)$ which is a far cry from the exponential complexity of the Canny algorithm. Secondly, the construction as explained in [Guibas 1985 b] and [Devillers 1997] is randomized and is therefore capable of handling obstacles in motion. This is a distinct advantage, which can handle only static obstacles. In undersea environments, obstacles are always in motion.

We will now try to understand more on this last point. In the two-dimensional case, a Delaunay triangulation of S is a covering of S by Delaunay triangles. Similarly, in the three-dimensional case, a Delaunay triangulation of S is a covering of S by Delaunay tetrahedrons. These objects have the following property

a) let p, q, r be obstacles in a 2D S, the triangle with p, q, r as vertices is said to be Delaunay if there is on other obstacle inside the circum-circle of pqr. Edges of a Delaunay triangle are described as legal edges or Delaunay edge.
b) Let p, q, r, s be obstacles in a 3D S, the tetrahedron with p, q, r and s as vertices is said to be Delaunay if there is no other obstacle inside the circum-sphere of p, q, r and s. The same description applies to the edges of the tetrahedron.

Our interest is obviously in the three dimensional case. Provided that we can treat the output from the FLS as a semi-algebraic set, then by virtual of [Hironaka 1957], triangulation of the set is always possible. Hironaka does not prescribe how to do the triangulation. This is how Delaunay triangulation comes in.

1. The triangulation is a covering of S, therefore, both the initial and the final placements of the UUV have to be in one of the tetrahedral. This is the geometric meaning of being a cover.
2. The interior of any tetrahedron is obstacle free. In particular, the obstacles can be mines and therefore, the UUV cannot sail through them. As long as the UUV is navigating through the interiors of the tetrahedral (not hitting

any vertices), its traversal is collision free. This is actually a strong form of the combinatorial algorithm announced in [Chan 1998]

3. What happens when an obstacle arrives at the scene or when one flows away. When that happens, some of the edges of the triangulation will cease to be legal. The triangulation will not be Delaunay any more and some tetrahedral will now contain obstacles. It is imperative that the algorithm be called into action to perform a local repair. This means that only those edges that do not satisfy the condition of being Delaunay need to be changed. This drastically reduces the computing overhead.
4. We will do a few steps to illustrate how we handle new obstacles.

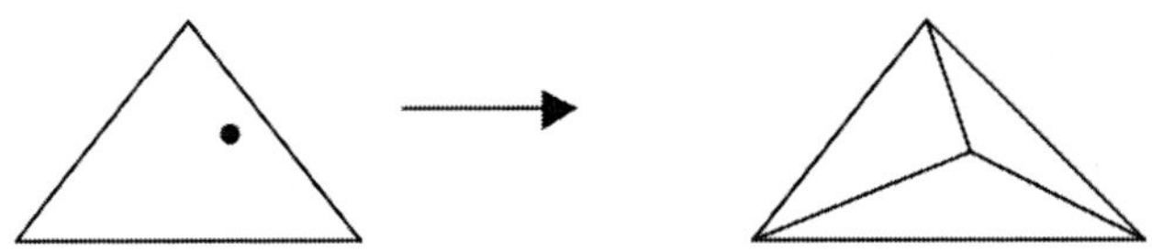

Figure 5: Triangulation changes for inserting a new obstacle

The dot in the left diagram of Figure 5 denotes a new obstacle. The triangulation is now non-Delaunay as the interior of the tetrahedron is now not void. What we do now is to add three new edges and have them tested for being Delaunay. If they are all Delaunay, then the repair is done. Otherwise, we perform an operation called edge-flip [Guibas 1985 b] until every edge of the triangulation is Delaunay. This treatment is first performed by Guibas, Knuth and Sharir [Guibas 1985 b] and since then, detailed examples have been worked out in computer graphics lecture notes.

SECTION 3: UNDERSEA WIRELESS DATA TRANSMISSION

In this section we envisage that the Ownship sends off a fleet of autonomous unmanned vehicles. To realize the vision that these vehicles will perform the role of networked sensors, they have to have the capability of communicating with each other. Thus we are looking into the realm of wireless hydro-acoustic data transmission. This is a vast engineering domain of profound complexity. To outline a realistic design in the space of one paper is just beyond anybody's endeavor. The purpose of this section is just to outline the hierarchy of problems that one has to address. At the end of the section, we will produce an example showing how to estimate output bit rates for a given set of input engineering conditions. In our view, the hierarchy comprises of the following problems,

- Concurrency control

- Multi-path symbol interference resolution
- Doppler spread
- Data bit estimate.

3.1 Concurrency control

In hard-wired environments, the problem of concurrency control is well settled. Solutions are expressed in terms of the various locking theorems. Gray's treatise [Gray 1993] provides a comprehensive treatment of the topic.

The relevance of concurrency control can be understood as follows. Among the fleet of vehicles, there is one, which has the proxy of the Ownship. This vehicle may be the Ownship itself and plays the role of mobile service station (MSS) in mobile computing. Members of the fleet are reporting their tracking output to the MSS. It is imperative that there is not to be any partial data reported and that all the tracking reports to the MSS are serialized without needing to impose on the fleet members as to when to report. This requirement alongside with others says that the transmission of tracking data to the MSS has to be transaction in nature.

Precisely, this means that the MSS has to be running as a transaction processing system. It functions as a concentrator of distributed data for reliability, availability and performance. In it, there needs to be implemented fault-tolerant storage for fault-tolerant processes between itself and fleet members. More importantly, the MSS of the fleet has to ensure that all tracking reports from fleet members satisfy the well-known ACID properties enforced in transaction processing. In brief, these ACID properties are,

- Atomicity: A reporting induces changes to the state of the reporting agent. These change either all happen or none happens. These changes include database changes, messages and actions on transducers. In other words, a partial report has always to be rolled back.
- Consistency: A report must be a correct transformation of the state of the reporting vehicle. The actions taken as a fleet do not violate any integrity constraints associated with the state.
- Isolation: Even though reports execute concurrently, it appears to each report R, others are executed before R or after R, but not both.
- Durability: Once a report completes successfully, it changes to the state 'survive failures'.

The isolation requirement is also called concurrency control requirement or serializability requirement. In a hardwired environment, the general implementation technique is via the locking theorem as explained in [Gray 1993]. In the following, we will outline how may the locking theorem, or a variation of it, be extended to a wireless mobile environment. *So baring the complexities introduced by new carrier (sound wave instead of electromagnetic wave) and medium (undersea instead of air), it is actually feasible to have a version of the locking theorem for wireless acoustical data transmission.*

Speculatively, if each fleet member is endowed with computation capability and its action upon its environment be allowed to be determined by its own experience, then it is at least partially qualified as a mobile agent. At the formal level, a mobile agent operates flexibly and rationally in a variety of environmental circumstances with its given perceptual and effectual equipment. Such a visionary structure of the fleet can only be realized in the distant future.

3.2 Serializability in Mobile Environment

As we have said above, a UUV should upload the tracking data via a transaction. What then is a transaction in a mobile environment? The purpose of this section is to make precise this notion and to announce a couple theorems, which will ensure the integrity of the upload. The proofs of these theorems may be found in [Chan 1999], [Chen 2004]..

Normally as in [Gray 1993], we will use x, y, ... to denote data objects and use the symbol "<<" to denote chronological ordering relation. We will also use the letter "r" for the operation "read", "w" for the operation "write", "a" for "abort" and "c" for "commit". Now let T be a subset of $\{r[x], w[x] \mid x \text{ is a data item}\} \cup \{a, c\}$, then the following statements are obvious;

a) $a \in T$, iff $c \notin T$;
b) if any operation z is c or a (whichever is in T), for any other operation $p \in T$, then $p << z$;
c) if $\{r[x], w[x]\} \in T$, then either $r[x] << w[x]$ or $w[x] << r[x]$, i.e. conflicting operations are ordered.

In mobile environments, it is assumed that each proxy (could be served by the MSS in small systems) while the Ownship holds a master clock against which all local clocks synchronize [Levi 1990] With this in mind, we can now understand a transaction in a mobile environment in the following way.

A transaction T is a function from a totally ordered set **T**(local time of a proxy) to the set **ZETA** = OP × DATA ∪ SYM inducing a partial ordering relation << in **ZETA** defined by $p << q$ iff $T(t_1) = p$ and $T(t_2) = q$ such that $t_1 < t_2$, where < is the ordering relation in **T**. The sets OP, DATA and SYM have the following meaning:
OP = {r, w}, i.e. the fleet member UUV does only either "read" or "write" in uploading its tracking data;
DATA = set of data objects;
SYM = {contact-proxy, page –invalidation, sleep, involuntary-sleep, wakeup, commit, abort, move}
Because of obvious limitation, we do not go into detail explanation for each of these symbols. Interested readers are referred to [Chan, 1999], [Chen 2004] or [Gray 1993].

A typical sequence of operations of a UUV during uploading will be as follows:

Contact-proxy -> r[x] -> w[x] ->contact-proxy -> page-invalidation -> ... -> contact-proxy -> sleep -> wakeup -> ... ->commit

In this sequence, the proxy (held by MSS first acquires the appropriate set of locks for the UUV. Once a write is executed, the proxy is asked to broadcast a report of invalidation to proclaim the existence of a new version of x. In the above example, the UUV, after serving a number of operations of T, decides to go to sleep, i.e. temporarily withholds action. This is dome by first contacting the proxy, then flushing all its dirty pages (to the MSS) and releasing all the locks held. Upon waking up, the UUV goes through a process similar to start-up, i.e. there will be fetches of data objects and lock requests through the proxy. The remaining operations are then executed under the covering of the locks. Finally, the transaction is committed through delayed writes.

How do we enforce concurrency control and how do we detect concurrency control violations? These are the problems we need to solve using the preliminary notions we develop above. We will first define the notion of transaction histories and equivalence of transaction histories. Then, using the artifact called serializability graph (SG) introduced by Bernstein [Bernstein 1987], in a hardwired environment, Gray was the first to prove that concurrency control is achieved if there is no cycle (worm hole) in the SG. This no-cycle condition is later found to have to be modified if it is used to pick out violation of serializabilityi n a mobile environment [Chan 1999], [Chen 2004]. The modification may be stated as follows. In the classical hardwired case, cycles are not allowed, but in the mobile case, the correct statement is cycles with self intersections are not allowed provided that all the notions for serializability are defined and match the corresponding notions in the hardwired case. Subsequently, it is found that the no-cycle condition is too strict and we can actually relax it a bit and we can still achieve the type of scenario discussed in [Gray 1993]. This is the notion of epsilon serializability [Pu 1990], [Chen 1999], [Chan 1999]. This notion is especially useful if the data is rapidly changing. Speculatively speaking, epsilon serializability may be a very appropriate constraint to enforce on a fleet of UUV uploading tracking data.

Due to the limitation of space, we do not reproduce the mobile versions of the no-cycle theorems as proofs may be found in the literature. The rest of the section will be devoted to the setting up of the no-cycle condition leading to the statements of the theorems in the mobile environment.

It must be stressed that the availability of these theorems does not solve (nor implement) the problem of doing undersea acoustic data transmission. These theorems only guarantee that such a vision is possible. Up to solving numerous practical problems [Bessio 1996] (and refernces contained therein), one day we may yet have a workable networked sensor system.

Now we introduce the notions of transition histories and no-cycle conditions.

Classically, let T = {T_1,..., T_n}be a set of transactions where each is associated with an ordering relation $<<_j$. The complete history H over T is defined to be a partial order $<_H$ such that

a) $H = \bigcup_{j=1}^{n} T_j$

b) $<_H = \bigcup_{j=1}^{n} <<_j$

c) For any two conflicting operations p and q, either p $<_H$ q or q $<_H$ p.

In a hardwired environment, two histories H1 and H2 are said to be equivalent if I) they are defined over the same set of transactions; ii) they order conflicting operations of non-aborted transactions in the same way, i.e. for any conflicting operations p_j and q_k belonging to transactions T_j and T_k, $p_j <_{H1} q_k \Leftrightarrow$ (H1 replaced by H2). In mobile environments, the history H represents a merging of the operations of a set of transactions according to a master clock held by Ownship. Any two of these merges are deemed equivalent if they are defined over the same set of transactions and preserve their relative orders.

Now we go onto introducing SG. In general a transaction may be made up of blocks of operations separated by either voluntary sleep or involuntary sleep. This is the case when the UUV may decide to take a break from its uploading activity for whatever reason. Thus, in the following discussion, we may represent a transaction T_i = {$T_{i\alpha}$}; the second index being the block index. A history is said to be serial if it runs one block of a transaction at a time. Thus a serial history is one that is of the following form: $[T_{i\alpha}, T_{j\beta}, ...]\ (\beta > \alpha\ for\ i = j)$

The notion of SG is then defined as follows. Given a history over the set {T_j}, the SG for H, denoted by SG(H), is a directed graph whose edges are blocks of transactions in {T_j}that are committed and whose edges are all $T_{j\alpha} \rightarrow T_{k\beta}$ (j is different from k), such that one of the operations of $T_{j\alpha}$ precedes and conflicts with one of the operations of $T_{k\beta}$ for some α,β. Furthermore, we identify blocks that are in the same transaction with the same value of the first index. The following diagram is taken from [Gray 1993] to illustrate the idea.

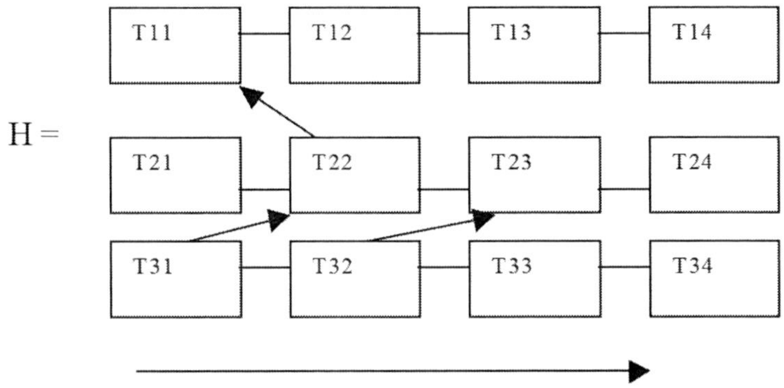

Figure 5: A serial history (The arrow indicates direction of local time.)

Consider now the following history

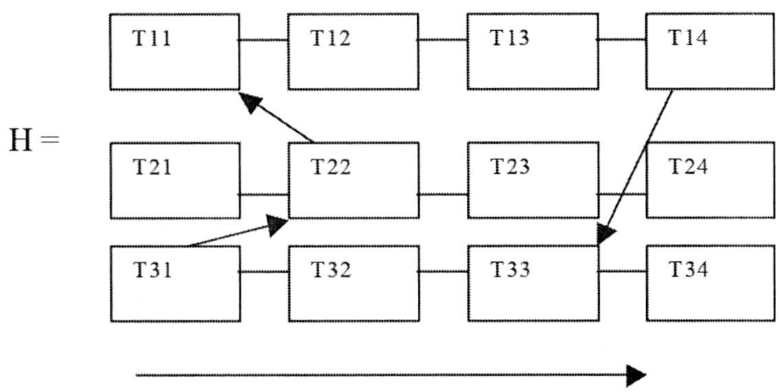

Figure 6: A history of wormhole SG in mobile environment

As can be easily realized, this latter history gives rise to a SG with the following cycle.

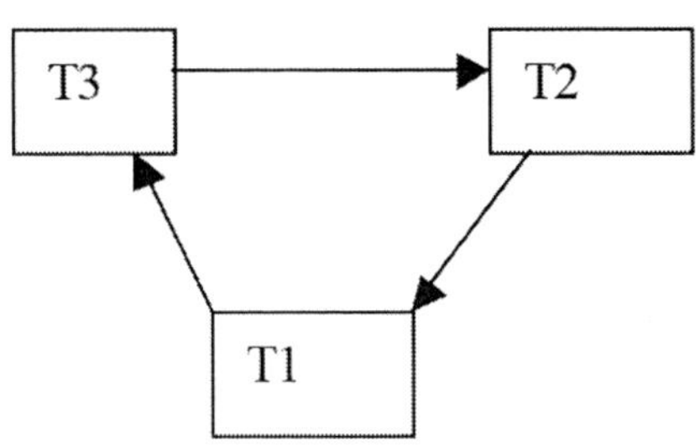

Figure 7: A wormhole (cycle) SG in mobile environment

Figure 7 is an example of a wormhole graph not allowed in classical hardwired environments. It is allowed in mobile environments because there is clearly no violation of causality with respect to local time. Because of the presence of this parameter, transactions are not point-like objects (as they are in the classical case) but are not extended objects,

Consider now the following history

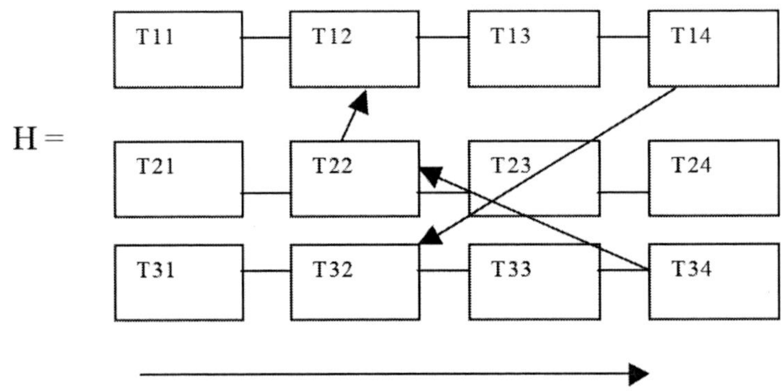

Figure 8: A history with self-intersection wormhole in mobile environment

This history gives rise to the following SG(H)

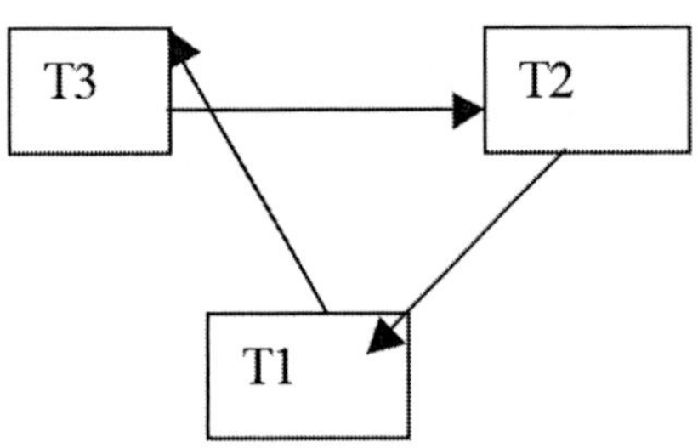

Figure 9: A SG with self – interesting cycle in mobile environment

The history in Figure 8 is of the following form

$$H = T_{34} \parallel T_{22} \parallel T_{12} \parallel \cdots \parallel T_{14} \parallel T_{32} \parallel \cdots$$

The violation of causality is obvious from the following segment

$$|| T_{34} || \cdots || T_{32} || \cdots$$

We can now state the serializability theorem for mobile environment.

Theorem: A history is serial if and only if there is no wormhole with self-intersection in the serialization graph SG.

The proof can be found in [Chan 1999] and in [Chen 2004]

How can we ensure this happen? This is answered by the next theorem known as the locking theorem for mobile environment. But first a few definitions before the theorem is stated.

A transaction block is well formed if each READ, WRITE and UNLOCK action is covered by a corresponding lock, and all locks are released by the end of the transaction block. A transaction block is two-phase if it has a growing phase during which it only acquires locks, then a shrinking phase in which it only releases locks. A history is legal if it does not grant conflicting locks to two different transaction blocks at the same time.

Theorem: If all transaction blocks are well formed and two-phase, then any legal history is serial.

This is called the locking theorem in mobile environment.

3.2 Example of an underwater acoustic channel

In this section, we describe an underwater acoustic channel due to Zakharov and Kodanev [Zakharov 2000]. The rationale for describing such a work is to highlight the process that once the computing platforms in the UUV, MSS and Ownship resolve all the problems pertaining to concurrency control, the next group of problems to tackle will be those handled by this model.

An OFDM transmitted signal may be represented as

$$s(t) = A\sum_{k=0}^{N-1} \cos[\omega_k t + \phi(k)] \qquad (3.2.1)$$

Here N = number of frequency components, A = amplitude, $f_k = f_c - F/2 + k/T_s$, f_c is the central frequency, T_s is the symbol duration and $F = N/T_s$ is the frequency bandwidth of the transmitted signal, $\phi(k)$ is the amplitude modulation of the frequency component. Let Q be the number of frequencies by which simultaneous data transmission is carried out. Then the overall data transmission consists of a number of OFDM symbols. The data rate is F/Q.

The next thing to do is to set up a channel model which expresses the transmitted signal x(t) in terms of s(t) and Doppler parameters. Once this is done, there are estimation theory steps leading to the estimate of the data bit d_m.

Briefly, the reasoning goes as follows. Approximate the impulse response by Dirac delta function $\delta(t)$ leading to a simplified expression of the received signal

$$x(t) = \sum_{l=1}^{L} s(t - \tau_l)\exp(2i\pi f_l t) \qquad (3.2.2)$$

Here, x(t) is the received signal, L is the number of macro-rays with Doppler parameters η_l and delay τ_l and $f_l = f_c(1/\eta_l - 1)$.

With the channel model in place, the sequence of operations to be performed consists of

a) End-front signal processing;
b) Estimate of delays and Doppler parameters ($Z_l(k)$);
c) Estimate of transfer function of the macro-rays ($\hat{H}_l(\omega)$);
d) Diversity combining and decoding

The output is the spectral sequence $Y_l(k) = Z^*_l(k)\ \hat{H}_l(\omega_k)M_2(k)$, where the last factor provides randomization of the transmitted signal.

Finally, the combining of Q frequency diversity signals is performed

$$w_l(m) = \sum_{i=0}^{Q-1} \mathrm{Re}(Y_l(m + iN/Q)) \quad m = 0,...,N/Q \ \ -1 \qquad (3.2.3)$$

Here m is the index of the data bit within an OFDM symbol. The multipath-Doppler combining is performed by

$$w(m) = \sum_{l=1}^{L} w_l(m)$$

The estimate of the data bit d_m is

$$\hat{d}_m = sign\{w(m)\} \qquad (3.2.4)$$

This is the output of the acoustic channel. The foregoing process draws the attention of the designers as to where the problems lie.

SECTION 4: CONCLUSIONS

In this paper, we have outlined a conceptual design of a system of communicating autonomous unmanned undersea vehicles. Our vision is such a system can ultimately be used as a networked system of sensors. We have touched upon the following problems:

1. Each UUV has to be equipped with a FLS to provide vision.
2. Each UUV has to be equipped with a computing platform running in real-time a Delaunay triangulation program based on the strong form of the combinatorial algorithm.
3. Other than running motion-planning software, the computing platform of the UUV has also to run transaction-processing software uploading the tracking result to a information concentrator, the MSS.
4. Between the MSS and any UUV of the fleet, there has to be an undersea acoustic channel capable of resolving multipath spread and Doppler spread via orthogonal frequency division multiplexing.

REFERENCES

[Zimmerman 2004]: Zimmermann, J. Z., *A 3D, Forward-Looking, Phased Array, Obstacle Avoidance Sonar for Underwater Vehicles* www.farsounder.com/news/NewsArticle.php/

[Eguchi 2001]: Eguchi, G. *Delaunay Triangulation,* Section 6.838, Computational Geometry, October 2001

[Bessios 1996]: Bessios, A. G., Caimi, F. M., *High-Rate Data Communications on Underwater Acoustic Communications, Framework at the Physical Layer,* MPE Volume 2, pp 449-485, 1996.

[Chan 1998]: Chan, S., Chen, J., *Transaction Processing in a Mobile Environment.* Presented at the International Conference on Theoretical Computer Science, Hong Kong, April, 1988.

[Chan 1999]: Chan, S., Chen, J., *A Mobility Concurrent Support Scheme for Rapidly Changing Data,* Proceeding of the 17th IASTED International Conference on Applied Informatics, Innsbruck, Austria, 1999.

Chan, S., Chen, J., *Serializability and Epsilon serializability in a Mobile Environment,* Proceeding of the 17th IASTED International Conference on Applied Informatics, Innsbruck, Austria, 1999.

[Canny 1988]: Canny, J. F., *The Complexity of Robot Motion Planning,* MIT Press, 1988.

[Schwartz 1987]: Schwartz, J., Yap, C. K., *Algorithmic and Geometric Aspects of Robots,* Lawrence Erlbaum, 1987.

[Schwartz 1982]: Schwartz, J., Sharir, *On the 'Piano Mover's Problem, II, General Techniques for Computing Topological Properties of Real Algebraic Manifolds,* Computer Science Department, New York University Report, 1982.

[Hopcroft 1984]: Hopcroft, J., Schwartz, J., Sharir, M., *On the Complexity of Motion Planning for Independent Objects: PSPACE-Harness of 'Warehousemen's Problem',* International Journal for Robotics Research, volume 3, 76, 1984

[Whitney 1957]: Whitney, H., *Elementary Structure of Real Algebraic Varieties,* Annals of Mathematics, Volume 3, 66, 1957.

[Chan 1998]: Chan, S., Lai, K. F., *Combinatorial Approach to Robot Motion Planning,* Presented at the International Conference on Computer Science at Johannesburg, 1998.

[Spannier 1966]: Spannier, E., *Algebraic Topology,* McGraw Hill, 1966.

[Hironaka 1957]: Hironaka, H., *Stratification,* Proceedings of Symposia in Pure Mathematics, Volume 29, 165, 1957.

[Kaufmann 1977]: Kaufmann, A., Henry-Labordere, A., *Integer and Mixed Programming: Theory and Applications,* Academic Press, 1977.

[Guibas 1985 a]: Guibas, L., Stolfi, J., *Primitive for the manipulation of general subdivisions and computation of Voronoi diagram,* ACM Transaction on Graphics, Volume 4, 74 – 123, 1985.

[Guibas 1985 b]: Guibas, L., Knuth, D., Sharir, M., *Randomized Incremental Construction of Delaunay and Voronoi diagrams,* Algorithmica Volume 7, 381 – 413, 1985.

[Devillers 1997]: Devillers, O., *Improved Incrementa; Randomized Delaunay Triangulation,* INRIA Research Report, 1997.

[Gray 1993]: Gray, J. and Reuter, A., *Transaction Processing: Concepts and Techniques,* Morgan Kaufmann Publishers, Inc,, 1993

[Chen 2004]: Chen, J., *Data and Knowledge Transaction in Mobile Environments,* PhD Thesis, University of Western Sydney, 2004.

[Levi 1990]: Levi, S., Agrawala, A., *Real Time Systems Design,* MaGraw-Hill, 1990.

[Bernstein 1987]: Bernstein, P. A., Hadzilacos, V., Goodman, N., *Concurrency Control and Recovery in Database Systems,* Addison-Wesley, 1987.

[Zakharov 2000]: Zakharov, Y. V., Kodanev, V.P., *Multipath-Doppler Diversity of OFDM signals in a Underwater Acoustics Channel,* Proceedings of 2000 IEEE International Conference on Acoustics, Speech and Signal Processing, Volume 5, 2941-2944, 2000.

GAS TURBINE POWER AND PROPULSION OPTIONS FOR THE MODERN WARSHIP

David J Bricknell
Director Systems, Rolls-Royce Naval Marine, UK
Richard Partridge
Naval System Integration Manager, Future Projects, Rolls-Royce Naval Marine, UK

SUMMARY

Modern naval combatants appear to be dividing into three distinct ship types: ocean capable patrol-craft, globally deployable combatants, and the developing new class of fast littoral combatant. Each ship type places different and significant demands on the power and propulsion system. This paper describes the technical and operational issues driving the demands on the power and propulsion system including ship speed, un-refuelled endurance, fuel type, on-board maintenance and the new electric mission-systems. Gas turbine powered mechanical-geared systems, gas turbine alternator powered integrated full-electric systems and hybrid mechanical/electrical systems are all examined and their contribution to the success of the system is discussed.

1. INTRODUCTION

The **globally deployable combatant** is a 'blue-water' capable ship operating throughout the world's oceans. It provides area air defence with large numbers of Vertically Launched (VL) missiles, large multi-function radar, and gunfire support. It carries special forces, one or two medium/large helicopters, and deploys Anti-Submarine Warfare ASW assets through unmanned vehicles. The role of the ship requires it to have a long un-refuelled endurance together with a good sustained speed for accompanying carrier battle groups. This requires the capability to loiter for prolonged periods and benefits from a good sprint speed. The modern ship still has a substantial complement (about 200 reduced from the more usual 350 or so from a generation ago) and spends a considerable time away from base port, requiring systems and equipment to be reliable and maintainable with or without an accompanying infrastructure. Current and future ships fulfilling this requirement are around 8,000t/150m (in a conventional displacement-type hull) with an all gas turbine power and propulsion system driving twin propellers for a ship speed in excess of 30 knots. Examples of ships in this class are the US DDG51, Japan's Kongo and South Korea's KDXIII. European ships offering lower overall capability in terms of speed as well as missiles are the Royal Navy's Type 45, Franco-Italian Horizon and FREMM, the Dutch LCF, Spanish F100 and German F124. It is no surprise, perhaps, that many of the designs arose from the nine-nation NATO NFR90 programme.

The **Ocean Capable Patrol Craft** is a 'green-water' vessel patrolling near to the base country with capability to operate off the continental shelf out to say 1000nm or so from the shoreline.

Current and future ships fulfilling this role are around 2,000t/90m Offshore Patrol Vessel OPV (or Corvette) with minimal crew number (~60 complement) and a shorter un-refuelled endurance of about 2,500nm at 12-15 knots. The OCPC is generally a more complex and capable ship than a purely Exclusive Economic Zone (EEZ) protection vessel and its role is more than just fisheries and minerals protection. The ships tend to be faster (25 –27 knots) than the typical 20-knot EEZ OPV but slower than the Global Combatant. This class of vessel has tended to be all-diesel powered, primarily for initial cost reasons but also because speed aspirations are limited.

The developing new class of **fast combatant** currently focuses around broadly two types of craft: the first is the Swedish Visby-type own-littorals combatant and the second is the trans-oceanic littoral combatant like the US Littoral Combat Ship. Both types of vessel are designed to operate close to shore either in one's own littorals or in those of another country. The vessel characteristics are high-speed, shallow draft, waterjet propulsion, and gas turbine/diesel combinations.

2. GLOBAL COMBATANT

The globally deployable combatant needs long un-refuelled endurance and a good sprint speed both for coastline dominance and keeping up with large carrier battle group **[1]**. The ship will spend time throughout the speed range.

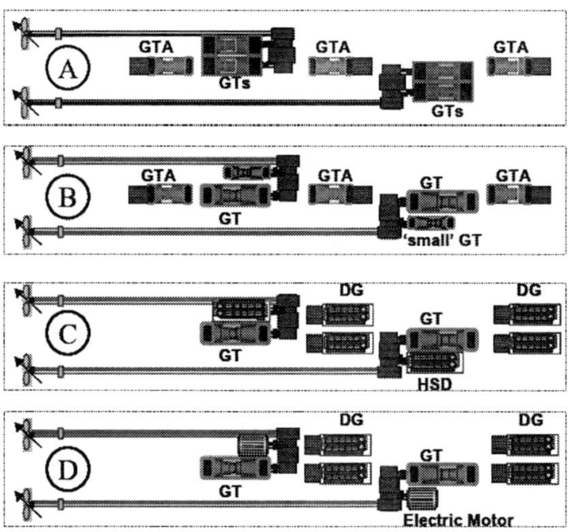

Figure 1 –Propulsion system options for Global Combatant: COGAG system - four medium size gas turbines sharing the duty equally and combining for a high maximum speed (A); an alternative all gas turbine COGOG - two large and two small gas turbine, each optimised for the sprint and endurance speed ranges (B); the very fuel efficient twin large GT and twin cruise high-speed diesel CODOG system (C); and the hybrid electrical- mechanical CODLOG system (D).

This Global Combatant's operational requirement will demand of the power and propulsion system a fuel-efficient, high-power, low at-sea-maintenance/upkeep system and supporting equipments. Current and near-future designs for this class of ship are over 8,000t with sprint speeds of 32 knots or more. The ships are twin propeller-driven and require installed propulsive power generally in excess of 60MW. Older designs have an un-refuelled range of about 4,500nm at 18 knots whilst future requirements are for 6000nm or more at a similar or greater speed. A ship's service and mission system electrical load of about 3 - 4 MWe is demanded once the ship is at sea and combat ready.

Mechanical drive power and propulsion system options are shown in Figure 1. Today's most commonly installed system adopts four medium power marine gas turbines coupled in a COGAG arrangement (total propulsion of 72MW) supported by three small gas turbine alternators to provide electrical load (3x2.5MW). The system delivers low-at-sea maintenance by virtue of its all gas turbine prime-mover selection but it falls short in delivering long un-refuelled endurance or in providing any electrical power headroom for any future adoption of electric mission systems.

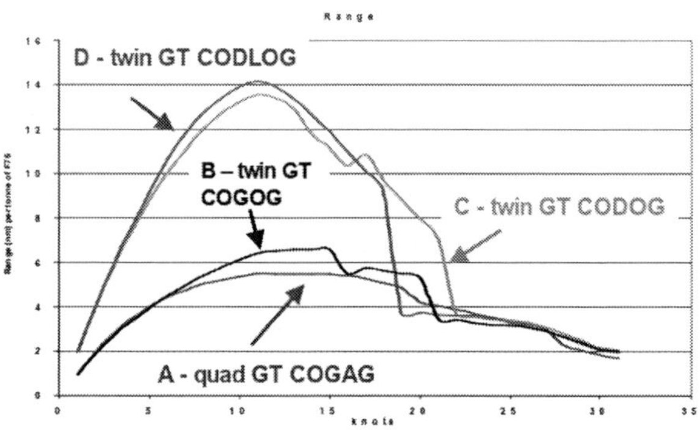

Figure 2 – Un-refuelled endurance comparison for the Global Combatant's potential power and propulsion systems showing the significant endurance improvements that can be achieved by moving to an optimised cruise/boost propulsion system

Delivering improved un-refuelled endurance with such a system is relatively straightforward now that more powerful marine gas turbines are available. Figure 2 shows the nautical miles per tonne of fuel throughout the ship's operational speed range for four systems. The first is the current quad 18MW GT and three 2.5MW ship service generators; the next three replace the four 18MW GTs with two 36MW GTs thereby maintaining the ship's maximum speed but releasing the space available for dedicated cruise engine system. The choice of cruise system is then between smaller GTs, high-speed diesels or hybrid electric drive (either powered by diesel generators or by gas turbine alternators); the aim for each system being to deliver a cruise speed of at least 18 knots. Figure 3 shows comparative initial costs and weights.

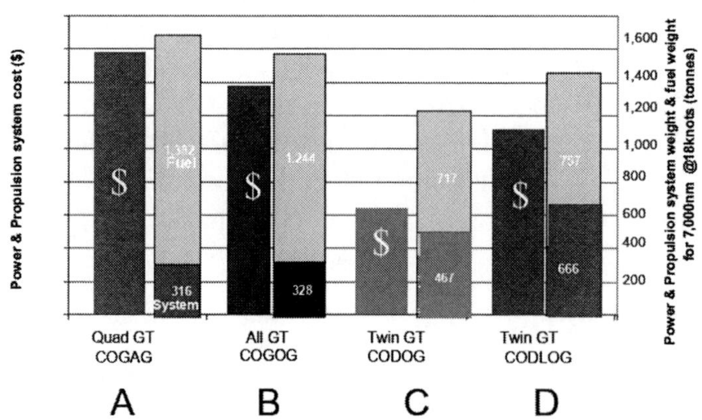

Figure 3 – Initial Cost and weight comparison for the Global Combatant's Power and Propulsion Systems.

An all-electric power and propulsion is not included here as the power density required for the Global Combatant's 32 knots is more than one and a half times that currently being achieved by today's technology **[2]**.

For the hybrid system, and in order to gain sufficient benefit from integrating the power and propulsions systems, selecting the ship speed at which one transitions from electric drive to mechanical drive is important. At 10 knots the Global Combatant requires about 1.5MW whilst at 18 knots this is around 10MW, after allowing for system losses. To arrange the ship for efficient operation at 18 knots the hybrid electric drive would require four x 4.5MW DGs for ships power and propulsion, allowing for some redundancy in the generator system. Figure 4 illustrates a hybrid-electric Global Combatant.

Figure 4 – An illustration of a hybrid-electric Global Combatant.

3. OCEAN CAPABLE PATROL CRAFT

The Ocean Capable Patrol Craft OCPC extends littoral capability into 'local' green-water. Good sea-keeping and a powerful combat system are required but within a relatively compact and inexpensive platform with a limited complement. This has tended to lead to a new and very populous class of ship sometimes described as Corvettes (or even Frigates) but just as often described as Offshore Patrol Craft (OPVs) although they clearly have a role above and beyond protection of a country's Exclusive Economic Zone (EEZ).

Currently this class of ship are designed at around 90m with a displacement of a little less than 2,000t and a ship's maximum speed from 25knots to a sprint speed of 28 or maybe 30knots. This ship length hits the main resistance 'trough' at about 22 knots and the main resistance 'hump' at about 30 knots (Froude number of 0.31 to 0.54) and generally these ships are designed for service speeds nearer to 25 knots. Achieving 25 knots will require a propulsion power of about 18MW whereas achieving 30knots will require around 30 MW.

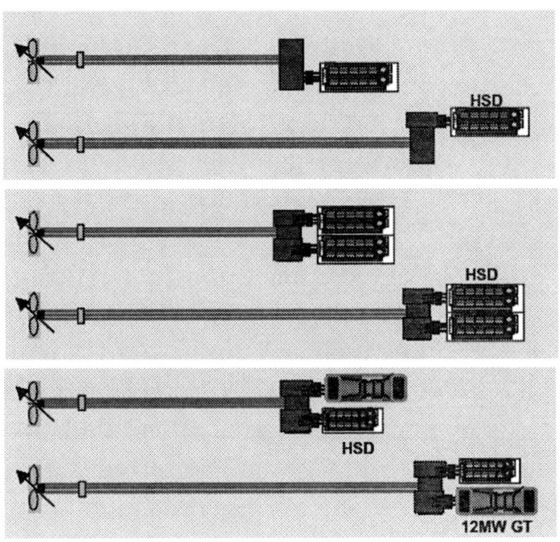

Figure 5 – Power and Propulsion Options for Ocean Capable Patrol Craft showing the 'standard' Twin Diesel (8 – 16MW) and quad-diesel CODAD system (16 – 32MW), and a 'new' CODAG 32MW system utilising the new 20V high-power higher speed diesels and a mid-power gas turbine with a combining gearbox (lower).

With a maximum ship speed below 35 knots, propellers are normally adopted thereby locating the engine room near amidships due to shaft rake necessary to achieve a clearance between hull and propeller tip, as illustrated by Figure 6.

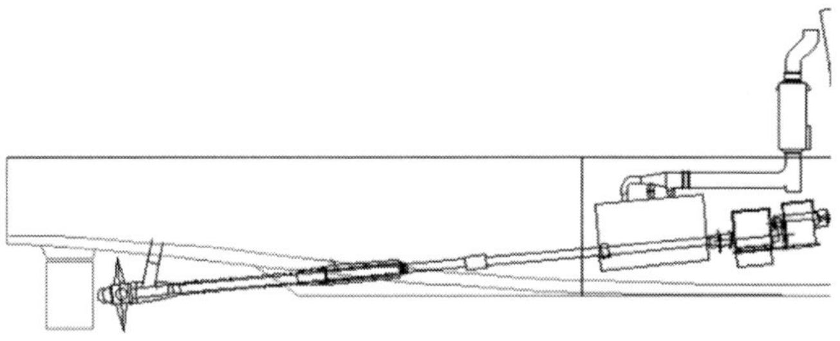

Figure 6 – Illustrative 'Shaft-Rake' impact in conventional combatant

Quad high-speed diesel (HSD) CODAD configurations have been normal for this ship type for some time but now recent developments with higher power 1000rpm engines have encouraged more twin diesel system configurations. High-speed diesels (HSDs) are adopted due to the power density and the low overall power required for the relatively low ship speed. Increased on-board maintenance effort of HSDs may well be acceptable due to the proximity to the ship's permanent shore-base. Twin diesel solutions may need to add a low power loiter drive to achieve acceptable low ship speed performance (<7 knots) and to reduce the impact on diesel maintenance due to extended low power running.

Whilst replacing HSDs in a quad-diesel CODAD arrangement on a like-for-like basis with gas turbines is likely to reduce on-board maintenance effort, it will also increase ship's first cost and significantly impact on un-refuelled endurance; few benefits can be gained from the characteristics of modern marine gas turbines in this configuration. However, it may be that the latest generation of HSD, with very high levels of power density, will complement the modern marine gas turbine in a sympathetically designed system arrangement and will once again open up the possibility of gas turbines in this ship type.

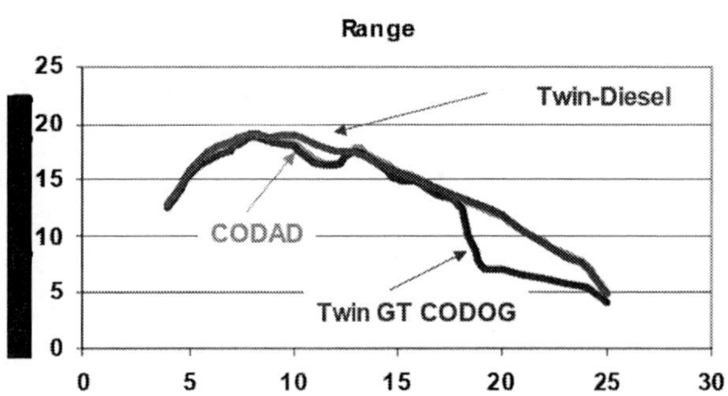

Figure 7 – Un-refuelled range comparison for an Ocean Capable Patrol Craft with all HSDs, all medium size GTs and a combination of HSDs and GTs

Very compact and fuel-efficient high-speed diesels (@ 1800 - 2100rpm) are now available up to 4.3MW and at this power the OCPC is able to achieve an endurance speed of around 20 knots. More general acceptance of multi-speed combining gears allows twin gas turbines at 12 MW each to be added to the small high-speed diesels thereby delivering maximum powers of around 30MW. Such a system delivers very low installation impact with high un-refuelled endurance at 20 knots and below, at a competitive initial cost albeit it still is situated near amidships. System options are shown in Figure 5, fuel usage at Figure 7 and initial weight and cost comparisons for the higher power systems are shown in at Figure 8.

As can be seen from figure 07, for a 30 knot plus OCPC, excellent endurance and low fuel burn can be achieved by both the CODAD and CODAG systems but the CODAG system shows a significant improvement in system weight and an improvement in combined fuel and system weight with a lower initial cost: the lower overall ship weight means less ship power which potentially leads to lower overall fuel consumption.

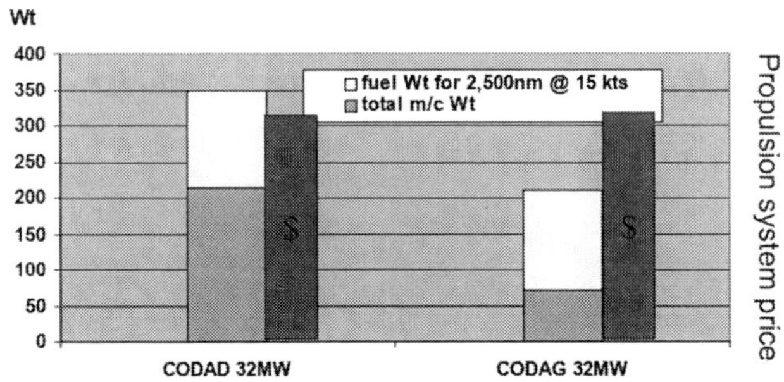

Figure 8 – Initial System Weight and Cost comparison for Ocean Capable Patrol Craft

As with the Global Combatant the mid-size gas turbine at about 12MW would be required to deliver good sprint-speed fuel consumption within a compact and cost effective package.

However in addition to the growth of the OCPC class of ship there is a also a trend toward faster vessels more attuned to operating in near-to home littorals whilst growing toward the size necessary for green-water operation and an aviation capability. Both the Swedish Visby and the Omani Baynunnah class at about 70m and capable of speeds greater than 35 knots are demonstrating the characteristics required for 21st Century warfare; both are waterjet powered and Visby is gas-turbine powered.

4. FAST COMBATANTS

Whilst fast vessels are not a wholly new breed of ships, the adoption of waterjets and gas turbines has enabled a new class of littoral vessels to be developed. In many respects the first of this new breed was the Swedish 'Visby' class. This is a relatively small vessel (approx 70m), designed to be deployed in a country's own littorals. High sprint speed, shallow draft, extended low-speed running, and low noise and low magnetic signature are key characteristics but green-water capability, organic aviation and very high un-refuelled endurance are not attainable in a ship of this size. The US Littoral Combat Ship has been developed in order to operate in other countries' littorals. It is shallow draft, fast and agile but it is also designed to be deployable trans-ocean (from one green-water area to another). It has multi-mission capability due to the modular payload and requires transportable infrastructure to deploy and support it.

Figure 9 – Illustrative Fast Littoral Combatant developed to enable the comparison of Power and Propulsion options for a green-water fast combatant.

This section examines the type of vessel necessary to dominate a country's own littorals but coupled with an ocean going capability for domination of the 'green-water' and including a credible defence against helicopter launched anti-ship missiles. Defence of one's own country whilst providing significant force-projection in near waters, particularly to deploy in key shipping choke-points, requires vessels with similar characteristics as Visby but in this case these must be coupled to higher un-refuelled endurance and the capability to deploy in the green-water battle-space. These additional characteristics, coupled with the ability to deploy an aviation capability (including helicopter and unmanned vehicles) as well as special forces deployment, demanding manoeuvring at high-speeds, will require a minimum hull size of around 100m.

Power and Propulsion system options for a fast green-water combatant have a considerable impact on the ship's size and general arrangements when high maximum speed is coupled with long endurance. For the purposes of power and propulsion system options comparison a target green-water ship displacement of 2,500t has been used and the fuel load and mission systems were kept common. The fast littoral craft design developed is illustrated in Figure 9.

Propulsion options considered are shown in the following four figures. Figures 10 shows a single Gas Turbine and Twin Diesel CODAG configuration, ,

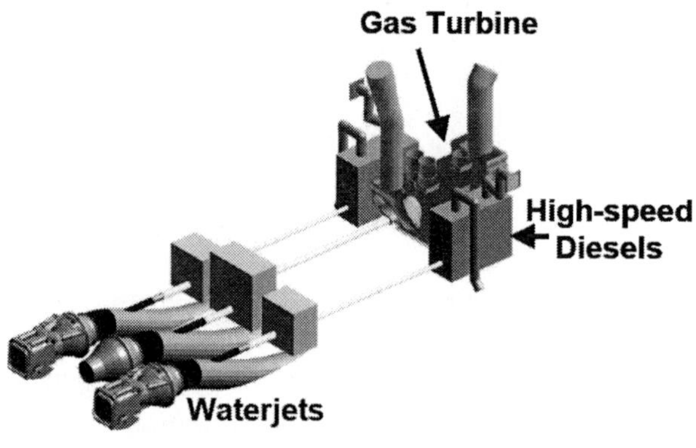

Figure 10 - Single 40MW Gas Turbine CODAG and twin 5.5MW high-speed diesels producing 51MW. The single gas turbine arrangement and twin diesels can be mounted in the same compartment thereby reducing the impact on the size of the ship.

Figure 11 shows a twin Gas Turbine CODOG arrangement

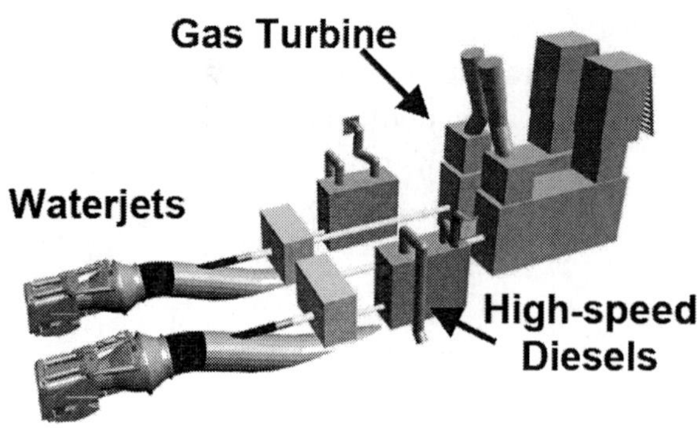

Figure 11 - Twin 25MW Gas Turbine CODAG producing 61MW. Twin GTs are mounted in different compartments from the diesels.

Figure 12 an innovative Visby-style quad mid-size Gas Turbine and twin small Diesel configuration - this arrangement leads to the smallest and lightest ship design with excellent maximum speed and endurance.

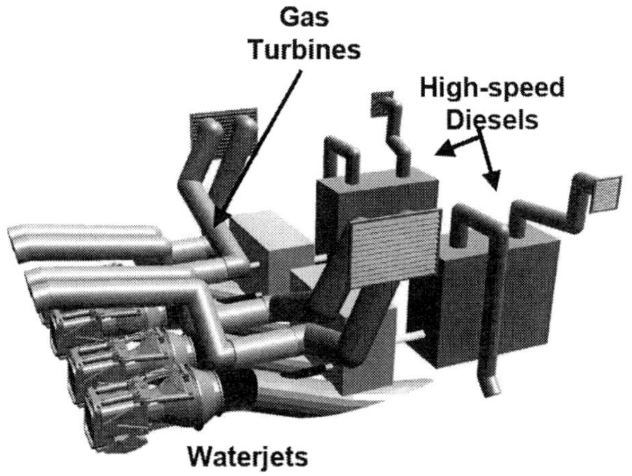

Figure 12 - 'Visby-style' CODOG system with four 12.75MW cold-end drive gas turbines delivering 51MW in a CODOG arrangement with two 5.5MW high-speed diesels..

Figure 13 All-electric propulsion. This has a significant impact on the ship's general arrangement, structural and machinery weight and on unrefuelled endurance because of the significantly larger ship demanded by the space and weight required.

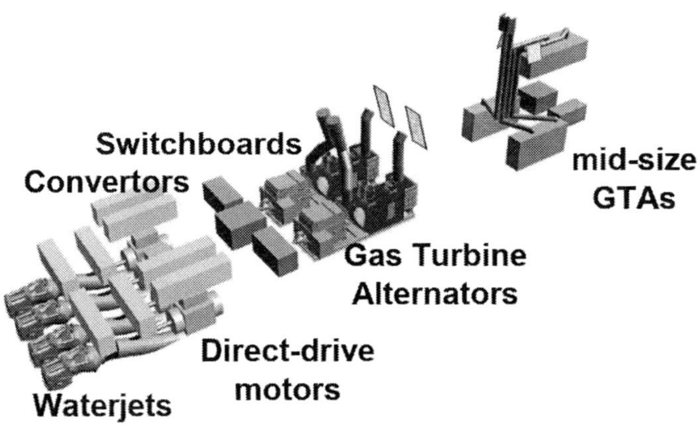

Figure 13 - Integrated Full-Electric Power and Propulsion arrangement with 80MW propulsion power.

A ship and system weight comparison is shown in Figure 14. The Visby-Style propulsion system can be seen as the lightest in weight and also has the lightest ship structural weight. The all-electric power and propulsion system is the heaviest by quite some margin as well as incurring the heaviest structural weight. The choice and availability of a particular power and size of gas turbine can have a significant impact in overall ship size and displacement.

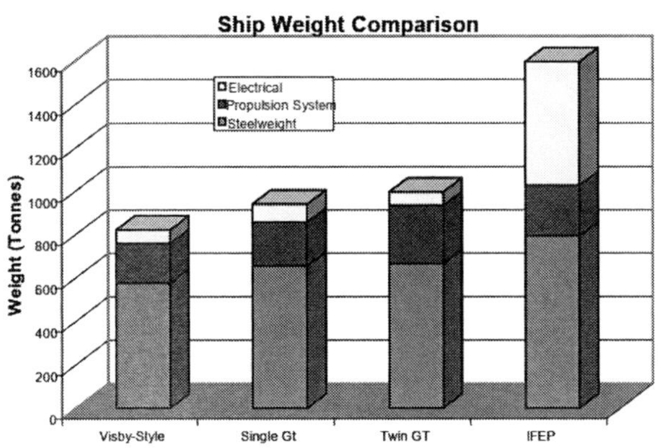

Figure 14 Ship and system weight comparison. Accompanying the substantial IFEP electrical system weight is also an increase in ship's structural weight.

A scale comparison of the three fast ship types can be seen in Figure 15 showing the 10m increase in ship length when going from Visby-style to conventional GT and an further 10m again when adopting IFEP.

Figure 15 – Comparison of the 'same mission' fast littoral ship when comparing different power and propulsion options. The quad lightweight, cold-end drive Gas Turbine/Diesel combination delivers an overall smaller and lighter ship with long unrefuelled endurance and high maximum speed

The Visby-style arrangement gains considerable benefit from the cold-end drive arrangements and exhausting over the stern allowing a smaller ship overall and by trading off the poorer fuel consumption of a smaller Gas Turbine against the smaller and lighter ship and its lower power requirement. The IFEP ship with Gas Turbine Alternators demands a considerably larger ship to cope with the size of the electrical motors and the size and weight of the generation and transmission systems leading to a considerably larger ship. Range is significantly curtailed due to the poor overall transmission efficiency.

Between the more conventional single and twin GT arrangements, the single GT gains benefit from the narrower beam of the ship allowing for a higher maximum speed. Table 1 shows the power and propulsion system comparison at a ship level.

Table 1 Power and Propulsion system analysis for the Fast Corvette.

	Twin GT	Single GT	Visby-Style	IFEP
Length OA m	116.3	116.3	**105.6**	126.4
Beam m	15.5	15.0	**15.0**	16.5
Displ deep t	2865	2722	**2508**	3776
Displ light t	2261	2240	**1942**	2786
Speed knts	46.8	44.4	**48**	48
Mission length days	20.6	20.6	**24.8**	14

The 'small' cold-end drive gas turbine is assumed fitted as 'un-packaged' as on Visby and other similar ships. Power density is very high with a unit weight less than 2 tonnes. No such engine is currently available for such a system, but the basis exists for a marine version of a current regional jet aero-gas turbine.

The Visby-Style arrangement also benefits from excellent fuel consumption throughout the ship speed range as is illustrated in figure 16.

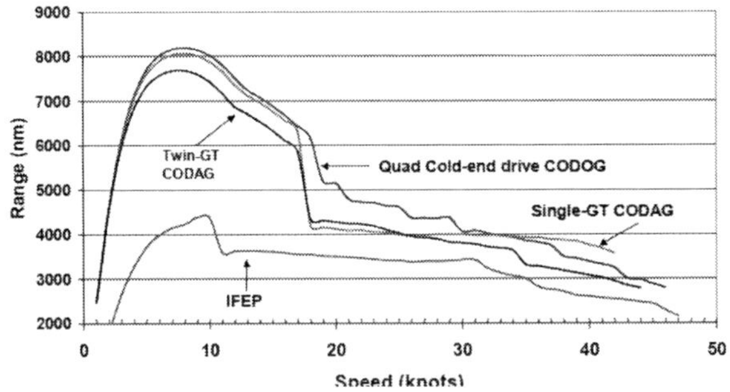

Figure 16 Endurance calculations for the different power and propulsion options

Characteristics of the gas turbine alternator suitable for adoption in an IFEP version of the green-water fast combatant are dominated by the space available in a ship of this size: a key driver for such a GTA is its overall length which must not be much over 10m including package and alternator if it is to fit between an acceptable damage compartment length. **[3]**

5. CONCLUSIONS

Whether future naval combatants follow the conventional speed route or turn to faster designs, gas turbines would appear to have a secure future. High-power and high-power density are the distinguishing features for the selection of the power system once ship speeds increase, and the ship arrangement should be adapted to maximise the Gas Turbine's benefits.

Modern propulsion systems for large global combatants are likely to adopt a twin-gas turbine system, in conjunction with propulsion diesels or an electric drive system designed specifically for cruise operation. Whilst retaining the same top ship speed as the legacy quad Gas Turbine arrangement, these modern systems offer a greatly enhanced range or endurance capability and potentially lower initial and operational cost.

Existing high-speed diesels are well suited for conventional OCPC-type vessels; their modus operandi means that multiple-GT arrangements are unsuitable. However, new and very compact high-speed diesels, in combination with a modern circa 12MW GT, may offer an attractive alternative, in terms of combining low installation impact and high endurance at competitive cost as well as improving vessel sprint speed.

Fast combatants with agility and shallow draft are a rapidly developing class of ship. Power and propulsion arrangements for these green-water combatants have a considerable impact on ship size. IEP systems demand a significantly larger ship to accommodate the plant and range is substantially curtailed due to the relatively poor overall transmission efficiency. However, a 'Visby-style' CODAG arrangement, using a new small cold-end drive marine gas turbine, may offer significant programme and operational benefits.

REFERENCES

[1] Bricknell, David J and Skarda, Robert - Power Growth in Naval Combatants, DTA Asia Feb 2005

[2] Bricknell, David J and Partridge, Richard – Power and Propulsion for the Modern Global Combatant; Pacific 2006

[3] Casson, Wood, Bricknell, Daffey, Partridge – Power and Propulsion for the new global combatant; RINA Warship 2006

How do you ensure Safety Excellence in your operations

P. Davison

DNV Maritime Solutions, Singapore

Abstract

A recent study has revealed an increase in accident frequency in shipping the last few years. Simultaneously, the maritime stakeholders' tolerance for poor safety decreases. This development has caused more and more shipping companies to put safety at the top of their agenda. Success or failure of these initiatives is dependent upon their ability to handle human and organizational factors. This article first describes the key safety challenges for the maritime industry. Secondly, the article outlines ours perspective on how to successfully address and improve human and organisational factors in shipping. It is argued that shipping has a lot to learn from other industries like, for instance, offshore and aviation to reach the goal of safety excellence.

Keywords

Safety; safety culture; Safety performance; Human and organizational factors; Change-management

1. Introduction

Historically shipping has used references as seamanship, seaworthiness and compliance to describe the safety standard of a vessel (Soma, 2004). History has also demonstrated that these parameters have restricted potentials when it comes to accident prevention (Kristiansen, 1999). *Safety Excellence* is a reference to available safety best-practices that easily can be applied in shipping irrespective of regulative requirements.

2. The shipowners challenge

In a historic perspective, shipping is known for its poor focus on safety. However, significant improvements have been achieved. On average the total loss ratio is reduced by 2 - 3% every year (Soma, 2003). Despite this development there has been a tendency in recent years that the frequency of serious accidents (partial losses) is increasing (DNV, 2007).

Although the frequency of total losses is declining, the consequences of such accidents have been, and still are, increasing. The public now have a zero tolerance for accidents and environmental damages. Today the cargo owners consolidate into fewer and larger companies. These companies are more vulnerable both in terms of disturbances to logistic chains and loss of reputation. As a result most stakeholders (i.e. the public, cargo owners, regulators etc) demand high transparency on safety issues. Port states, underwriters, vetting and media are more frequently addressing safety issues in general, but especially after accidents with severe consequences.

Given that the frequency of accidents is reduced, how can then safety be a business challenge for shipowners? The key reason for this is that the increased consequence of accidents is not outweighed by reduced accident frequency. Hence, the risk faced by shipping companies may actually be higher than ever. The tendency for increasing frequency only makes risks even higher. Risk considerations is a fundamental component of sound and sustainable business management, and the increasing safety risks poses a challenge to shipowners.

An additional challenge is that shipping in fact is so safe today that a significant share of ships sail through their life time without any serious accidents. Therefore, shipping companies that hold on to their reactive approach of making safety improvements solely based on incident information will have little information to assure that they manage the increasing risks effectively. The key to reducing the risks shipping companies face is to address and improve the complex interrelations between human and organisational factors.

3. Focus on human and organisational factors

A century of research into accidents has shown that most incidents involve the human element (Soma, 2003). We also know that humans by nature make mistakes, errors and shortcuts. In the airline industry this has been an accepted fact for years, but not so much in the maritime industry until just recently. Research has shown that an airline pilot makes an average of 4.9 errors per hour, but still this industry is among the safest in the world. That implies that any organisation handle errors when they occur, not only prevent them. Therefore we need to focus on how to develop organisations and systems that do not allow errors to escalate into incidents. This implies a holistic view on human an organisational factors. Therefore the hypothesis of this study is that shipping has a lot to learn from other industries in improvement of human and organisational factors through a holistic view on safety.

3.1 A holistic view on safety

The core of a holistic view is to address a wide range of organisational dimensions having an effect on safety and being fundamental in creating a learning safety organisation (Rasmussen, 1997; Perrow, 1999; Sagan 2004). The key organisational dimensions are:

(1) Strategy and Goals
(2) Processes
(3) People and Competence
(4) Roles and Responsibilities
(5) Tools

The five organisational dimensions point to arenas where improvement potentials can be identified and achieved. There are interactions between these dimensions that are critical for safety, and that is the reason why a holistic view on safety is of utmost importance (Rasmussen, 1997). Generally, all the five dimensions should be addressed, but for a specific organisation the improvement potentials may be more obvious and important within some of the dimensions.

3.2 What can we learn from other industries?

Other industries like offshore, space, nuclear energy, aviation and process industry are typically at least a decade ahead of shipping when it comes to safety. This is important as research has proved that there common characteristics for safe organisations across industries (LaPorte, 1991; Hollnagel, 2004; Reason, 1997; Rasmusen, 1997; Weick, 2001; Soma, 2004). These characteristics are summarised as:

Strategy and Goals
Clear and well-formed strategy and no conflicting goals is the cornerstone of all safety work, and a safety strategy is a prerequisite for success in achieving safety beyond compliance.

A safety philosophy is worth nothing without unequivocal commitment from all management tiers, top management included (Rundmo, 2003). Safety must be an integral part of the corporate strategy and an important organisational goal. It must also be recognised that safety as an organisational goal can come into conflict with other corporate goals (e.g. to be a profitable business), and in reaching the corporate goals, some level of risk will always have to be accepted. The objective of safety work is to eliminate unacceptable risk.

Processes
There is always a set of key organisational processes that create a drive towards improved safety. If these safety processes fail, safety work and continuous improvement will be suboptimal. These processes need to be more "intelligent" than the systems' ability to produce errors and accidents.

In many shipping organisations the safety processes are not functioning optimally, and hence there is a considerable improvement potential to be gained if this process is improved and strengthened. Well-functioning safety process should be both reactive and proactive. The process should be reactive in the sense that measures are implemented as a learned response to near-misses and accidents, and proactive in the sense that findings of risk assessments onboard and onshore, suggestions from crews, efforts to improve attitude and competence etc result in a continuous strengthening of the safety barriers.

People and Competence
Addressing people and competence are of utmost importance in safety work and is the key in achieving lasting improvement in safety performance. In accident free organizations people have an in-depth understanding of the shipboard operations combined with a paranoia for failures.

Attitude is an aspect of this dimension where considerable improvement potential can be achieved. Correct attitudes act as "lubrication" to safety-critical processes, and behaviour is strongly rooted in attitudes. Attitudes are therefore a key dimension to address in realising improvement potentials. It is however very difficult to change attitudes directly. Therefore it is important to focus on behaviour. When you do things correctly you will over time modify attitudes addressing the behaviour.

Roles and Responsibilities
Accident free organisations are characterised by safety commitment from the top. This commitment is shared by the whole organisation and fuelled by extremely efficient communication. Because it is acknowledged that errors do occur, it is important to focus on organisational redundancy. Organisational redundancy implies that the operators have the possibility to reveal and correct errors through overlapping responsibilities, competencies and possibilities to monitor each other.

Another characteristic is flexible organisations. This implies that the organizations adapt to what the situation requires. Addressing and understanding the culture of an organisation is therefore of utmost importance in understanding and realising improvement potentials.

Tools
Tools (and technology) are put in place to support important processes, people and competence issues, as well as organisational roles and responsibilities.

Tools and technology should be understood to cover a wide range of measures. In a shipping safety perspective, tools and technology would include risk assessment tools, incident investigation tools, crew resource management, IT systems for ship-to-shore communications etc.

4. Safety Excellence program

The objective of a program to reach *"Safety Excellence"* is to obtain a sustainable positive impact on safety performance through a holistic view on safety. To do so there is a need for both knowledge about safety and knowledge on how to make successful changes. A key is to access best practices and to effectively implement them in the shipping organisation. Without ownership and significant tailoring the implementation will fail. Therefore it is important to follow a specific work method by first identifying improvement potentials, defining appropriate solutions before entering the implementation phase. Through the phases some key areas are of special importance:

- **Benchmarking of leading indicators:** How is the company performing on key organisational factors?
- **Safety best practices:** What are the best practices both within shipping and other industries regarding safety?
- **Implementation of best-practices:** How to implement new safety practices in a specific shipping company?

Most international shipping companies today collect information about incidents and accidents on its vessels and are able to draw lessons learned from these events. Such organisational learning is of course of great importance, but we commonly see a need to improve how this learning is communicated and used with organisation in order to achieve real improvement. Most companies do not put the required resources into implementation of lessons learned. Again, successful implementation is the key to achieving this.

4.1 Three phase approach

Too often safety improvement initiatives arise from insufficient or biased information; an incident, an external requirement, a few observations or an idea. Typically the initiative is materialised in a procedure everybody is expected to comply with after an e-mail distribution. And over and over again, naturally, the procedures cause more frustration than effect, if it is recognised at all.

A good way of structuring this program is to apply a three phase approach. First identify improvement potentials, then design solutions and at last conduct a sustainable implementation of the solutions. The order and quality of the phases is important.

4.2 Improvement identification

Phase 1 is designed to provide answer to three key questions:

(1) Is the Client OK with regard to safety?
(2) What challenges and problems do they have?
(3) What are the prioritised improvement areas?

The real benefits of the program are realized in phase 2 and 3. Phase 1 will provide the required information to structure a Phase 2 (solution development) project. Since all client organisations have a unique set of safety challenges and problems, the Phase 2 project will need to be customized based on the findings from Phase 1.

4.3 Solution development

The second phase is less rigid than Phase 1, as there may be a range of unique challenges and problems that the specific company should prioritise. The scope for a Phase 2 will therefore need to be customized based on the Phase 1 results.

The design of the solutions should include descriptions of the desired behaviour or practice (the "to be" situation) and how this is preformed today (the "as is" situation). The approach used in bridging the gap between the "as is" and the "to be" situation is what solution development is all about. The implementation plan should detail the activities required in actually bridging this gap.

4.4 Implementation and follow up

When the appropriate solutions targeting the prioritised improvement potentials have been developed, together with an implementation plan for these solutions, it is about to realise the improvements through implementation (phase3). The implementation process will normally be done in three stages:

Pilot: Implementation on a few vessels or onshore departments (depending on the solution in question) where the implementation concept is tested, evaluated and adjusted accordingly
Roll-out: When the pilot stage is finalized a full-blown roll-out is conducted. A roll-out should in some cases also be performed stepwise to assure the quality of the implementation.
Completion: When approaching full implementation according to the implementation plan, it is important to assure that the "to be" situation is achieved and that the improvements are sustainable.

To make a successful implementation it is necessary to monitor performance of the project and the benefits it creates. Such benefit tracking will be described in more detail later in this document.

5. Case – "Concerned CEO"

DNV has several examples of the "Concerned CEO" scenario. In the specific case briefly presented here the CEO had joined the company six months earlier. Even though the company had not yet experienced any serious accidents the CEO realised that if something happened, the company could potentially loose business. The project was carried out by DNV in close cooperation with the Client. All three phases were performed in succession, each phase being tailored to the findings and results of the previous phase.

A phase 1 was initiated with the following actions: Review of strategy / mission statements, review of safety performance data, questionnaire survey to all crews, interviews of key shore and sea personnel, workshops with key personnel. The objective of this was to gather and structure the existing knowledge within the organisation and to determine the status of key safety processes in order to develop the "as is" situation. The conclusions were:

- **Is the safety OK?**: No, huge variance and increasing frequency trends.
- **What challenges?:** Strategy: Low cost not quality. Processes: learning process failed. People: new / inexperienced crews. Responsibilities: Unclear / not taken. Tools: Training, procedures
- **Prioritised improvement areas:** Safety department revitalisation, Management safety awareness. Communication, Teamwork, Leadership both ashore onboard and ship/shore.

The phase 2 (solution development) was initiated by a series of risk assessment and root cause analysis workshops of key shipboard operations (Soma, 2006) to explore the findings in phase 1 and increase insight and awareness of the participants. Based on this a "to be" situation was structured. Key solutions areas were:

- Shore management alignment and awareness
- Adjustment of KPIs ashore and onboard
- Improved induction and familiarisation training.
- General safety training though seminars and CBTs

The various solution areas are currently being implemented. The general safety and awareness training is in the rollout phase while the remaining solutions are in the pilot phase. Already several quick wins were observed.

- The communication improved
- The trust in management improved
- Captains became committed
- Near miss reporting boomed
- Lost Time Incident decreased
- Number of risk assessments boomed
- TMSA score improved from 1 to almost 4
- Improved reputation in the market

6. Conclusion

This paper has presented why shipping has a need for extensive safety improvements in the coming years. Shipping is behind other industries. That means that we can learn and improve safety quite easily. The hypothesis was that an improvement project taking a holistic view on human and organisational factors would have a significant impact on safety performance. This has been demonstrated through a safety excellence program where best-practices were transferred to shipping from other industries.

Secondary benefits have also been identified. It will be better in the eyes of the stakeholders such as society, shareholders, insurance, class, finance, yards, authorities and not the least the employees. This is essential to retain human resources in a global maritime labour market increasingly characterised by a lack of competent personnel and competition between the shipping companies.

7. References

DNV (2007) ; DNV Press release presented at : http://www.dnv.com/press/

Hollnagel, E. (2004); *"Barriers And Accident Prevention"* Ashgate publishing Limited, UK

Kristiansen, S., Soma, T. (1999), *"Effectiveness of Safety Measures in Maritime Operations"*. 16th. ESReDA, Oslo.

LaPorte, T., Consolini, P. (1991), *"Working in practice but not in theory: Theoretical challenges of "High -Reliability Organisations""*, Journal of Public Adm, Research & Theory, 1(1), 19-47

Norton, K. (1996), *"The balanced Scorecard; translating strategy into action"*, Harvard business school press, USA,

Perrow, C. (1999), Normal accidents - living with high-risk technologies,(2. edition) Basic Books, N.Y

Rasmussen, J. (1997), *"Risk management in a dynamic society : A modelling problem"*, Safety science Vol. 27, No 2/3, pp. 183-213, Elsevier science Ltd, Great Britain

Reason, J. (1997), *"Managing the Risk of Organisational Accidents"*, Ashgate, Hampshire, UK

Rundmo T., Hale A. (2003) *"Managers' attitudes towards safety and accident prevention"*, Safety Science 41, pp557-574

Sagan, (2004), *"Learning from normal accidents"*, Organization & Environment, Vol. 17 No. 1

Soma, T. (2003), *"What are the causes of ship accidents?"*; 2nd Int. Conf. Maritime transport, Barcelona,

Soma, T., (2004), *"Blue-Chip or Sub-Standard?"*, Doctoral thesis at NTNU IMT 2004-8.

Soma, T., Rafn C. (2006), *"Using risk assessment workshops in accident investigation"*, RINA conference on Learning from Marine Incidents 3.

Weick, K., Sutcliffe, K., (2001), *"Managing the Unexpected"*, John Willey & Sons

Meeting the Maritime Manpower Challenge
The Nautical Profession

By David Bendall,
Maritrade / Videotel Marine International, Australasia
Email: David@maritrade.com.au
and Leonard Holder,
Past President Nautical Institute, Chairman Videotel Marine International,
Email: lenholder@lenholder.freeserve.co.uk

NOTE: This is a personal perspective from two long-serving members of the Nautical Institute and does not necessarily reflect the policy or priorities of its Council and wider membership)

GOOD NEWS NOW – DANGER AHEAD

Goods news travels fast and all of us at recent conferences, in many sectors of the shipping industry, have been enjoying the good news about 1000 plus new ships coming into service each year for the next few years and the increased income they will generate.

At the same time, the world economy has been changing the lifestyles of many of the young people who in past years might have been expected to seek jobs as seafarers. More Ships + Fewer Seafarers = A Major Problem. With some exceptions, the shipping industry does not have a great track record for careful planning for its future manpower requirements or of careful nurturing and corporate responsibility for the people who are serving at present. Seafarers – particularly experienced shipmasters and chief engineers cannot be instantly produced. It takes time and in the interim, they will become a rarer breed than they are today. Sea service will become a "sellers' market" and key staff will be able to demand higher salaries and better conditions. Already, there are stories of officers being "poached" with offers of higher salaries. So the sooner the industry starts planning the better.

Professional bodies have a role to play. We will review past successes and limitations of Nautical Institute initiatives in setting, monitoring and improving professional standards including, publications, courses, diplomas, Seaways, seminars, participation in quality assurance etc

THE NAUTICAL INSTITUTE

Those that have read Michael Plumridge's excellent history of the Institute "A Nautical Odyssey"[1] will understand that, in the early 1970's the seafaring world was changing. Some of the best old established companies, which had recruited, trained, employed, mentored, guided and developed the careers of senior sea staff for their whole working lives were being replaced by a much more money-orientated system of almost "casual" employment. The main bastions against unfair treatment were the

[1] 2003 Plumridge M J "A Nautical Odyssey" A history of the Nautical Institute's first 30 years. Published by the Nautical Institute

Trades Unions, but they were weakened by the flagging out of fleets beyond their sphere of influence.

On 26th April 1969 a meeting was held in Liverpool at which it was decided to ask deck officers and masters whether they would welcome the formation of a nautical professional body to "raise professional standards". The members of the profession welcomed the idea. The shipowners through the British Shipping Federation and the Trades Unions objected to its formation – seeing it as another player in an already crowded industry, needlessly complicating their dealings.

The maritime world owes a lot to Captain Malins of the Marine Society who put forward the concept of a nautical professional body in 1969, Captains Frank Main and Robert Smith who grasped the idea and were determined to make it work, and all the hundreds of other masters and officers who put their weight behind it and have spent many years of their own time making it what it is today. Another person who deserves special mention is Julian Parker, the first Executive Secretary, who steered the energy of volunteers into productivity and progress.

Working with Shipowners and Trades Unions

Nearly 40 years on, shipowners and trades unions recognise the value of a good professional body. The Nautical Institute has established a niche for itself promoting professional education, sharing and circulating up-to-date information and publications, offering advice to members, to Governments and others, encouraging continuing career development and many other activities that support the raising of professional standards at sea.

The Wider Maritime Sphere

Through its members, it has a major presence in seafaring, ports, pilotage, surveying, management, maritime law and insurance and not least maritime education where it influences the education and training of future seafarers.

NAUTICAL KNOWLEDGE AND EXPERIENCE

There are two major sources of nautical expertise

1. Knowledge of principles, regulations, theory and underlying concepts which is the province of nautical college lecturers, and
2. Practical experience, which rests with senior deck officers and masters at sea.

The problem, which we all face for the future, is to pass that expertise on to the next generation. What part can the Nautical Institute play?

Working with the International Maritime Organization

Since the Institute was formed, some people who have been delighted with its progress have been the Secretaries General of IMO. The regulations, codes and resolutions which flow from IMO will be effective only if the Governments implementing them have the will to do so, and also the means. The Nautical Institute helps to bridge the gap between a line in a syllabus or regulation and its application in the workplace.

Publications

The Nautical Institute, through its publications, has helped IMO inform and develop programmes in developing countries. Mr William O'Neil was and is very supportive and has written forewords for a number of NI publications. With headquarters within a few yards of one another, continuing a close working relationship is vital for the future. The current Secretary General, Mr Mitropoulos shares with the Nautical Institute the desire to raise standards in shipping.

Seaways and Electronic Media

The Journal "Seaways" is the lifeblood of an organisation which has members scattered across the world in hundred of countries and at sea on all the world's oceans. The Institute has been lucky to have contributors who have provided relevant and high quality reports and articles and editors who have been skilled in editing and presenting them.

In more recent years the gap between branches and members has been reduced through more instant means of communication such as email and the Internet. As more ships get access to broadband, this must improve still further. Good communications help to bind the profession together.

Nautical Colleges

Early meetings of the Institute were often held in colleges, where serving officers were studying and staff members were keen to keep up to date and close to current practices at sea. There have always been strong personal links with the World Maritime University, where the late Professor Zade was a great enthusiast. Continuing professional development for college lecturers can take many forms, seeking higher qualifications, carrying out research, returning to serve at sea, etc In today's cost-conscious, poorly paid, short-staffed colleges these ideas might be regarded as "over the top", but the least any lecturer can do is to keep up-to-date and in touch by being a member, reading the Journal "Seaways" and mixing with seagoing colleagues who are at the forefront of operations in industry. The lecturers owe it to their students to prepare them for the future.

Formal links, where the local Nautical Institute branch nominates a member or members of the Governing Body or Advisory Board of a college have also been tried but tend to be less effective than informal personal involvement of members.

Learning from Mistakes: Working with Insurers

Everyone learns from their mistakes, but it would be a safer and better world, if we all learned from everyone's mistakes. Through the pages of the journal, through conference proceedings and in particular from MARS (Marine Accident Reporting Scheme) the Institute has brought to light the lessons learned from casualty investigations and the lessons learned by the brave few who have shared the near-miss situations with others, so that all may learn. The Institute has a mutually beneficial relationship with marine insurance and P&I staff, each side learning from the other.

Working with Accountants and Lawyers

Two other professions that seafarers need to communicate with are Accountants and Lawyers. Understanding accountants is relatively easy – if you cannot put a monetary value on it and count it, they think it is worthless. So much for loyalty, ethics and a sense of professional pride. Dealing with lawyers is more complex. Anyone who has had to provide evidence or give statements in court will realise that lawyers think in a different way from the rest of us. Most legal battles are just that – battles. They are confrontational and the ammunition is evidence, so you need to make sure to "keep your powder dry" and not give the opposition easy targets. The "Mariner's Role in Collecting Evidence" has helped the innocent seafarers understand their possible roles. The Institute has shared many conferences with lawyers and the working relationship improves all the time.

WORKING WITH GOVERNMENTS

The governing bodies of many professions actually establish the curriculum and training programmes and examine candidates for qualifications. With many thousands of cadets and officers in the world fleet, this was not something that could be contemplated by a small group of founding members. A system of government examinations and certification already existed and levels of membership were based upon that in the first instance.

Setting Professional Examinations?

Despite the fact that governments set only minimum standards and the Institute wants to raise standards above the minimum, the statutory certificates remain the criterion for membership. The opening paragraph of this paper indicates that there are going to be problems in meeting minimum standards in the future, so perhaps the best the Institute can do is to work with governments in establishing and implementing proper minimum standards at the basic level, and use other ways to raise standards above that level through supplementary courses, publications, lectures, Continuing Professional Development etc. One local example of cooperation is the fact that the Nautical Institute Square Rig Sailing Ship Certificate is required for the issue of a sailing ship endorsement to Australian Certificates under Marine Orders Part 3.

WORKING WITH OTHER PROFESSIONAL BODIES

The Marine Engineers

From the early days many people thought that the Deck and Engineering Officers should have one professional body. In the 1970s, the Institute of Marine Engineers was keen to align itself with other graduate branches of engineering and was not really representative of the body of experienced non-graduate seagoing engineer officers. Working relationships with the Institute of Marine Engineers have always been good, and in its new guise of the Institute of Marine Engineering, Science and Technology (IMarEST) they are even better. The recent NI AGM saw the signing of a formal agreement between the two bodies.

Naval Architects

This is another graduate profession, with a distinguished history of academic excellence and innovation. Relationships between ship design and operation are always better when the designer understands the practical operation of the ship he or she is designing. Most seafarers will be aware of design faults, most naval architects will already be working on their next commission and may not want to hear about them. It will be more difficult in a world where ships are owned in one country, designed in another, built elsewhere and operated by a manager in another. For the future, the lines of communication must be kept open leading to better understanding between designers and users. In this the Classification Societies have an important role to play, and the Nautical Institute Lloyds Register "Alert" programme is a superb example. To ensure continuing and strengthening co-operation in the future the Nautical Institute and the Royal Institution of Naval Architects (RINA) are about to sign a formal agreement (Memorandum of Understanding) similar to that with IMarEST. This is part of the strategy to develop a single, inter-disciplinary, professional voice for the industry.

Looking to the local scene in Australia, the November meeting about the salvage of the "Pasha Bulker" indicates that the local branches of RINA, INMAREST, CMMA and NI are well on this road

THE BRANCH NETWORK

One of the greatest assets of the Nautical Institute and a key to the future is the Branch Network. Being able to meet with like-minded people and exchange professional views and information in a local as well as an international context is vital if the organisation is to be a truly global body. Having started in the UK and British Commonwealth, it is not surprising that some people used to think of it as a "British Club". It might have developed that way, seeking royal patronage and looking inward, but it didn't. The expanding of branches into the USA, Greece, The Philippines, Ukraine, etc set the trend for the future. There are now too many branches to mention them all here.

The South East Australia Branch, "your Local Branch" has quarterly meetings in February, May, August and November and has a strong working relationship with shipping companies, port authorities and the Australian Maritime Safety Authority www.nisea.org

Some countries, such as Japan, already had strong shipmasters' and officers' associations and others such as Singapore decided in the early days to "go it alone". Relationships with these countries have always been positive and some things, such as publications, have been widely shared but it would be better if the profession spoke with one professional voice internationally and shared resources – everyone would benefit.

WORKING WITH COMPANIES

Videotel Marine International was formed in 1973, about the same time as the Nautical Institute was developing. Videotel shares many of the Institute's aims. Both are based in London, with representation overseas, both work closely with IMO in trying to raise standards in colleges and training centres across the world. Both are interested in professional training at sea, the Nautical Institute through senior members mentoring trainees and young members and Videotel with the provision of high quality training materials and courses on board thousands of ships.

A Corporate Affiliate Scheme provides organisations with the opportunity to become closely associated with Institute. It has many benefits, including access to knowledge, contacts and resources through the Institute's network of over 40 branches worldwide and members in over 110 countries. Videotel was pleased to become one of the first corporate affiliates.

Working in Quality Assurance and Accreditation

Whilst it is important that the Institute works closely with other players in industry, it is even more important that it keeps its independence as an acknowledged and unbiased centre of reference for quality standards in the industry. The Nautical

Institute is already established in monitoring dynamic positioning and oil cleanup courses. More recently it explored with SIGTTO the monitoring of training and competence standards for gas carriers, and is now an accreditator on behalf of SIGTTO. The Nautical Institute also accredits a growing list of other specialist courses.

POSSIBILITIES

Every five years the Institute formulates a strategic plan, the most recent, initiated by Dr Phil Anderson as President, covers the years 2006 to 2010, ensuring that the Institute addresses the real issues which concern the shipping industry, the naval service and our profession and to ensure that the Institute continues to provide the kind of service required by its members.[2]

Becoming a Major Training Centre?

The Institute has for many years acted as a distance-learning centre, adding diploma schemes to its authoritative books. Run by volunteers on a very low budget, they have not had the wide influence they would have had, if they had been commercially managed mainstream courses. Perhaps the best way to influence and raise standards is to monitor and accredit the courses provided by others. Nevertheless, the Institute's courses are now being recognised as foundation qualifications by some academia and sector bodies such as harbour and pilotage authorities.

KEY FACTORS FOR THE FUTURE

Increased Membership

The Institute will only provide more and better services if more people share the ideal of raising standards and become members. Compared with the benefits of membership, the cost is minimal, and should be an accepted part of being a recognised member of the profession.

More Branches in Centres of Manpower Supply

When the Philippine Branch was inaugurated, it was immediately evident that there was a huge thirst for nautical knowledge in the colleges and manning agencies in Manila and elsewhere. The same thirst for knowledge is evident in other developing centres of manpower supply. Forming local branches with experienced seafarers and organising the supply (and if necessary translation) of Nautical Institute textbooks is an ideal way to quench that thirst.

Publications, Conferences, "Seaways", Electronic Media etc

The future should build upon past successes with publications and conferences, often working alongside other bodies with similar interests in achieving high standards. In particular further links with the Institute of Marine Engineering, Science and Technology (IMarEST) and the Royal Institution of Naval Architects (RINA) will be explored.

---

[2] Nautical Institute website: www.nautinst.org

Continuing Cooperation with Other Industry Bodies

There are no longer any industry bodies standing on the sidelines criticising the formation of the Institute. It is seen as 100% useful and positive in its efforts. It is still an organisation of individuals with the ideal of raising standards in the profession, and will work with all those who wish to join them in achieving that goal.

LH DB 19/11/2007 Bendall Holder paper for Pacific 2008.doc

Human Systems Integration: How does it fit into Warship Design?

Paul Elischer and Michelle R. Grech
Maritime Platforms Division
Defence Science and Technology Organisation
Australia

paul.elischer@dsto.defence.gov.au
michelle.grech@dsto.defence.gov.au

ABSTRACT
This paper provides an overview of Human Systems Integration (HSI) and its application in the capability development lifecycle process. Effective HSI techniques must be developed in order to support challenges facing the maritime industry today, one of the most crucial ones being reduced manning. The Defence Science and Technology Organisation (DSTO, Maritimes Platform Division) actively supports the RAN in the application of HSI techniques in a number of projects, and this presentation outlines some of these activities. Issues discussed include such factors as human factors engineering, system safety, habitability, training and human performance in a maritime environment.

HUMAN SYSTEMS INTEGRATION: WHAT IS IT?

Human-Systems Integration (HSI) is aimed at incorporating human factors and systems engineering tools and methods into the development and acquisition of systems and equipment to improve system performance and safety throughout the whole capability life cycle (3). Figure 1 provides an overview of the manner in which the human element fits within the capability of a military environment, highlighting its large-scale influence. HSI covers a broad spectrum of issues and these are covered in the various domain headers discussed below.

HSI places a great deal of emphasis on systems integration which includes the interfaces between the physical and human systems (both operating and maintenance personnel). Unfortunately, the effective integration of the human system has in most cases been inadequately considered and sometimes completely neglected. This has led to systems that are difficult to operate, expensive to run, hard to provide training for or just unsafe. An evaluation undertaken by the US on four newly acquired army systems identified a number of problems directly attributed to the fact that HSI was not seriously or adequately considered (14). The problems included:

- human factors issues not addressed;
- doctrine or concepts incomplete or ill-suited to the soldier;
- manpower requirements were underestimated;
- skills and abilities required were undetermined or underestimated;
- training was untested;
- training devices were unavailable.

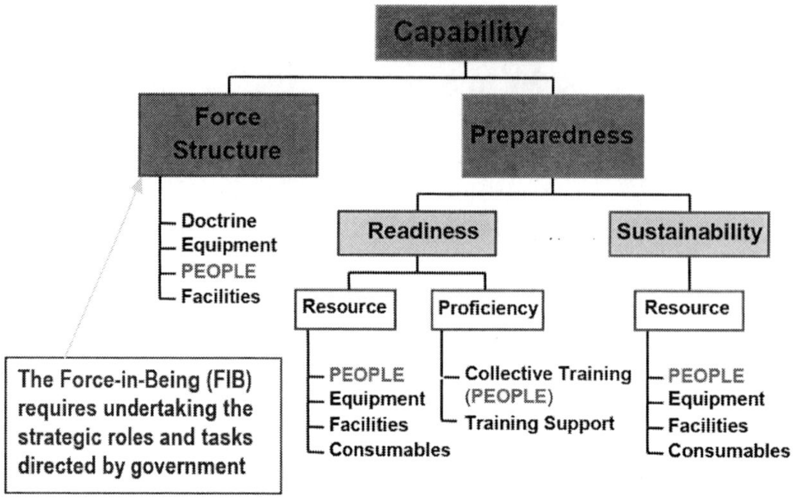

Figure 1: The human element in the capability life cycle

The literature cites a number of definitions describing HSI, all with the common theme of considering the human element throughout the capability life cycle. One definition cites that '*human system integration is the implementation of an interdisciplinary technical and management process for integrating human considerations within and across all system and platform elements of warship design and throughout its capability life cycle*' (11). Ideally it should be initiated early in the acquisition process to ensure that the design and operational requirements are cognisant of human capabilities and limitations. These should include, for example, such aspects as cognitive, physical and sensory skills required for training in order to achieve the optimum outcomes in terms of operational requirements and mission effectiveness. HSI ensures that the following fundamental question is asked early and answered in the acquisition life cycle process '*can these operators and maintainers, using this equipment, under these conditions, with this training, accomplish their mission*' (16).

HSI DOMAINS

HSI is divided into a framework incorporating a number of domains dealing with varying human factor and system engineering issues. The number of domains cited in the literature varies from a core of five, covering human factors engineering, crewing, personnel, training and system safety with additional domains, such as human survivability, habitability, human survivability, etc. being included to ensure focus on other specific areas of interest. A review carried by DSTO on HSI in the Australian Defence Force (ADF) (6) recommends that initially the ADF should use the core domains and expand these as the need arises. Following careful consideration, two additional domains, 'habitability' and 'organisational and social', have been added to address maritime issues. The maritime HSI domains together with their respective areas of human factors and systems engineering are summarised in Table 1. Table 1 also includes areas involvement for each domain, highlighting the elements with which DSTO is currently involved. A brief description of each domain is provided below:

Human factors engineering: The comprehensive integration of human capabilities and limitations into system definition, design, development, and evaluation to promote effective human machine integration.

Crewing: The availability, number of men and women, military and civilian, required and available to effectively operate, maintain, sustain and provide training for systems.

Personnel: The human knowledge, skills, abilities, and cognitive and physical capabilities and characteristics required to train for, operate, maintain, and support a system in peacetime, contingency operations and conflicts.

Training: The education, instruction, on the job training and resources i.e. tools, devices, simulators and procedures required to provide [Navy] personnel with requisite knowledge, skills, and abilities to properly operate, maintain, and support [Navy] systems.

System Safety: The identification of risk factors in naval systems and the physical environment that can increase opportunities for system caused injury or death and the subsequent implementation of system design characteristics that serve to minimize the potential for mishaps causing death or injury to personnel operating, maintaining, and supporting the system.

Habitability: The characteristics of systems, facilities, personnel services, and living conditions that result in a high level of personnel morale, quality of life, health and comfort adequate to sustain maximum personnel effectiveness, support mission performance and avoid personnel recruitment and retention problems.

Organisational and Social: This domain is a relatively recent consideration and as such, it is still evolving. Its inclusion is in response to the organisational complexity that is a key feature of providing Networked Enabled Capability (NEC). These issues have always been present on platforms but NEC operations have increased their importance for effective operations. The issues include: shared situational awareness, cultural issues, trust and information sharing, 'Soft' interoperability, alternative organisational.

Table 1: Domains proposed for the Royal Australian Navy HSI Process

Human Factors Engineering	Crewing	Personnel	Training	System Safety	Habitability	Social and Organisational
Operator roles, functions and tasks Human-System Interface Situational awareness Workload Workspace Environment	Crewing Workload Crewing concepts Force Availability Phasing Recruitment Retention Skill Mix/Special Skills	Physical Personnel Factors Cognitive Personnel Factors Previous experience/ training Human-Human Interaction Structure Workforce Planning Occupational Standards	Legacy Transfer Type Availability Training concepts and strategy Simulation Operational tempo	Error sources Accident and safety/health hazard voidance User behaviour Total System risk reduction /mitigation Reliability and redundancy Threat environment Casualty equipment Medical injury/systems Fatigue and Stress	Quality of life and work Environment limits and controls Noise Vibration Toxicity Electrical Anthropometry Personnel services	Shared situational awareness Cultural issues Social Factors Trust and information sharing "Soft interoperability Alternative organizational configurations

- Not currently being addressed
- Maritime Platforms Division (Project)
- Maritime Platforms Division (other)
- Other DSTO Divisions

THE CASE FOR HSI

The literature is in agreement that proper application of HSI would certainly avoid some of the issues listed above and would result in significant life-cycle cost savings. In the preparation of a business case for the Canadian Forces it was cited (3) that personnel costs account for some 40% of expenditures and incorporates the highest cost outlay in the budget. Such high costs could be avoided or reduced by tighter matching of selection and training of personnel with the system task performance requirements. Studies (3) conducted in the 1980s showed that 20% of female candidates entering the equipment operator trade and another 10% entering the non-trades drivers course were failing during training. Research to examine the reason for such a high failure rate in the female population showed that these rejection rates could be substantially improved by simply modifying the physical size and strength of the selection standards required for these trades.

HSI can also avoid costs associated with 'rejects' in the operation of equipment. For example: accurate data on body sizes can be used to optimize the number of items in different size ranges of clothing or equipment and avoid wastage; usability evaluations can provide features to equipment that is misused or unused; training needs analysis and human factors engineering can avoid the costs of training to compensate for designs that are difficult to use.

HSI PROCESS: WHERE ARE WE GLOBALLY

Australia's Maritime HSI effort is still at the foundation stage when compared with what is being done in other developed countries such as the USA, UK and Canada. Formal HSI programs are already in use by the USA UK and Canada. Australia is currently in the process of implementing HSI policies, training, programs and guidance documents via Materiel Requirement Sets 'Human Machine Interface' (1), Human Systems Integration Framework Implementation Program (Defence OHS Program Office) (2) and Human Systems Integration for the Australian Defence Force (6). Whilst the approach to HSI may differ across countries, the overall philosophy remains the same, which is incorporating human factors issues into the development and acquisition of systems and equipment to improve military performance throughout the whole capability life cycle. The next sections provide a brief overview of some successful HSI initiatives undertaken in the USA, UK and Canada. Finally, a brief description will be provided of HSI initiatives currently being undertaken in Australia.

USA HSI Process

The USA has some of the most mature HSI processes and programmes in place. The US Department of Defense, for example, follow HSI instruction guidelines (DoD 5000.2) (17) for new acquisitions, upgrades etc. These guidelines mandate the initiation of a HSI program from the onset of design and/or procurement in order to optimize total system performance, minimize cost of ownership and ensure that the build will accommodate the user population that will be operating maintaining and supporting the system. The thrust of the instruction is to ensure that all aspects of the platform and equipment systems incorporate effective human systems interfaces, achieve the required level of human performance and address PERSTEMPO/OPTEMPO issues commensurate with attaining the requisite skill base.

From a maritime perspective, in 2002, the Naval Sea Systems Command announced the creation of the Human Systems Integration Directorate, NAVSEA 03 (10). Its role was to examine the entire Navy acquisition process from the human-performance standpoint and to review the business-as-usual approach of the Navy to sustaining operational readiness. This directorate now serves as NAVSEA's single point of contact with US Navy stakeholders including, the Chief of Naval Operations, Commander Fleet Forces Command, Office of

Naval Research, other systems commands, Naval Education and Training Command, and other Navy, Defence Department, and joint offices for HSI-and human-performance-related activities.

To support the HSI approach the US Navy has developed a Systems Engineering, Acquisition and Personnel Integration (SEAPRINT) program (11). The US Navy now focuses on a sailor centred approach which is designed to address the impact of advanced technology and automated systems. In this regard, SEAPRINT, shown in Figure 2, was developed to apply the HSI process throughout the systems engineering process across all domains. SEAPRINT identifies tools and processes to aid in the decision making process commensurate with different types and size of systems. It addresses aspects such as early initiation of HSI, identification of issues and planing analysis, document/crosswalk HSI requirements, making HSI a factor in source selection, executing integrated technical processes, conducting trade-offs and HSI milestone assessments. SEAPRINT is an iterative program designed to explore the impact of manpower, personnel and training solutions and their inter relation with modern technological systems and concepts being employed in modern warship design.

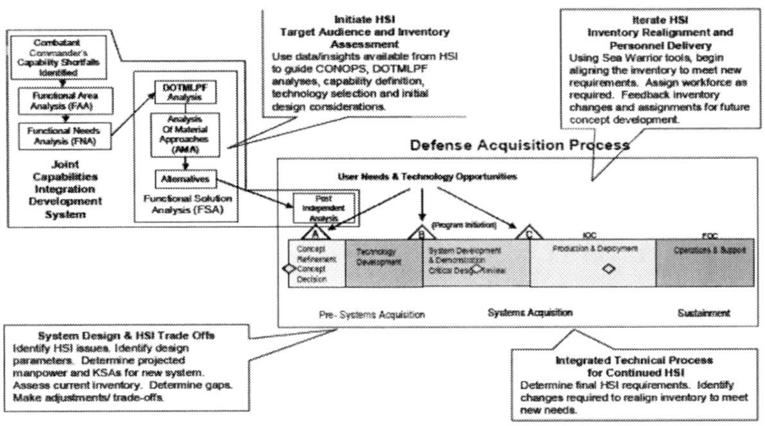

Figure 2: The Systems Engineering Process with Human Systems Integration (11)

UK HSI (HFI) Process

The UK Ministry of Defence (MOD) has adopted the SMART Acquisition program as a long-term initiative to acquire defence capability. This was implemented as a follow-up to a strategic defence review in the late 1990s and embodies the Human Factors Integration (HFI) tri-service policy that was mandated in 1998 by the UK MOD for all acquisition projects (4). Due to factors such as technology and capability improvements, alternative crewing strategies, reduced crew numbers and through life costs, like for like replacement is no longer an option. Essentially the SMART acquisition process is designed to 'acquire defence capability faster, cheaper, better and more effectively integrated' (4). SMART acquisition generated a fundamental change in the procurement approach and evaluates how a capability will integrate with other capabilities to achieve optimum effect. It requires a more flexible, imaginative approach in which issues are addressed on their merits rather than by reference to what has been done. To complement the SMART project the UK, MOD developed a comprehensive guide 'Human Factors Integration (HFI) Practical Guide for Integrated Project Teams (IPT)' (15) which provides the necessary guidance tools and methods to implement the HFI approach. Figure 3 demonstrates the manner in which the HFI management works within a SMART procurement project.

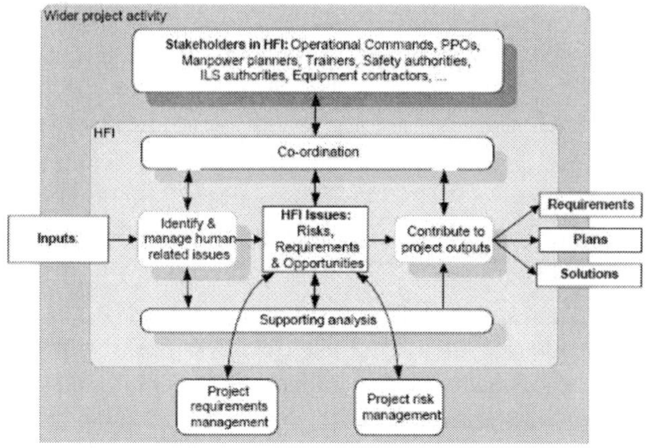

Figure 3: Management of HFI within the Smart Procurement Project (4)

Canadian HSI Process

The requirement to include HSI in the Canadian Department of National Defence (DND) acquisition projects has increased over the last few years (5). As a result, HSI is now included in the template for the Statement of Operational Requirement (SOR) within DND. The approach is very structured and includes the HSI plan, system definition and staff characteristics. It also includes identification of HSI risks, HSI process management, definition of project scenarios and measures, requirements and specifications for HSI domains, evaluation of HSI aspects, verification, validation and management and HSI handover and monitoring. Figure 4 illustrates the model incorporating HSI used by the department of National Security in the materiel acquisition and support process (MA&S). The MA&S process is conducted within the context of the life-cycle management system (LCMS), and the defence planning and management (DP&M) process when larger acquisition projects are conducted. The MA&S life-cycle combines the LCMS stages and the DP&M process phases. The process includes identification, options analysis, definition, implementation, in-service; and disposal. The HSI process across the MA&S life-cycle is outlined in Figure 4 below.

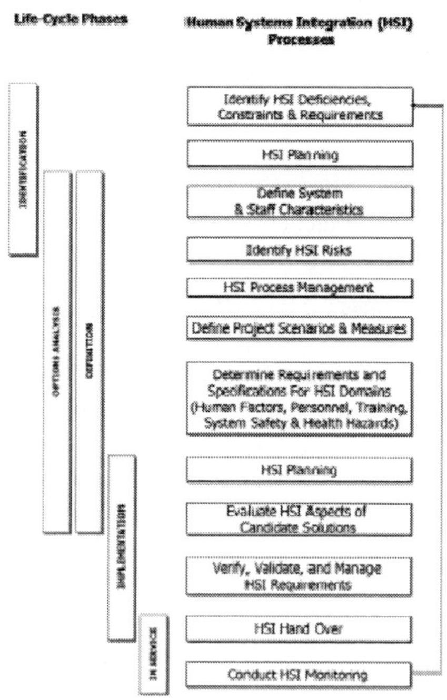

Figure 4: The HSI Process in the MA&S Model (5).

Australian HSI Process

ADF doctrine advocates that its capability is its people irrespective of the platform (sea, land or air), and the quality and well being of its people dictates the effectiveness of the ADF and the level of capability it can deliver. An issue of concern, which was highlighted in the Kinnaird Review (9), is the lack of human factors consideration within the Defence Capability Lifecycle (DCL) processes. Current attempts to address the impact of the human factors in acquisition have largely been ad hoc and retrospective in nature. Further attention is drawn to this issue whenever a number of human factors problems arise during capability operational lifecycle. The fixes or follow-on studies to resolve such issues have resulted in major cost outlay for defence that could have been substantially reduced if HSI considerations were incorporated at the initial stages of the DCL. Nevertheless, this is certainly changing, with increasing defence awareness and emphasis on research on the implications and ramifications of omitting the inclusion of a formal HSI program into the design and or acquisition process. The implications of not seriously adopting HSI include; life cycle costs, personnel issues (retention and recruitment) system safety and mission effectiveness.

HSI and System Safety

In March 2006, the Defence OHS Committee (DOHSC) directed Defence OHS Program Office (DOHSPO) to assess where Defence is in regard to 'Safe Design' principles within the DCL (8). The subsequent Safe Design Review (SDR) analysed processes and practices involved in major, minor and rapid acquisition programs conducted throughout the DCL. The methodology involved consultation with service and group representatives and a review of relevant documentation to identify Safe Design related processes, practices and opportunities for improvement in materiel acquisition. This culminated in the DOHSC endorsement of a phased implementation of a HSI framework with the aim of improving the identification, elimination and management of hazards throughout all phases of the DCL. The SDR identified a number of shortfalls in the current ADF system safety approach. These included amongst others:

- lack of identified authoritative sources for the provision of Safe Design and HSI Subject Matter Expertise (SME);
- lack of HSI advice and assistance across DCL activities;
- HSI not identified as a critical component of, or considered beneficial to, system design and acceptance into service considerations (8).

On a positive note, the SDR does indicate that there is room for improvement in the application of 'Safe Design' principles in all phases of the DCL. Some of the key issues identified are the need for improvement in the areas of Safe Design and HSI.

Use of HSI Principles in the ADF

The findings of the SDR, wholly or in part, have been and are currently being used within the ADF. Its implementation has been driven by a number of factors including: legacy issues, where deficiencies in an existing platform have been identified and require rectification etc. and for supporting new acquisitions for Navy by providing informed advice on the technical risks, addressing issues described in the HSI domains discussed previously. The following provides a snapshot of some the areas that have been addresses by DSTO.

New Acquisitions: HSI was included as part of the overall risk assessment for new acquisitions which included the Air Warfare Destroyer (SEA 4000 Project) and the Landing Helicopter Dock (LHD), (JP2048 Project). In addition, HSI has been incorporated as part of the development of the science and technology plan. This is particularly relevant for Australia from a social and organizational aspect as Australia mainly acquires systems and equipment from overseas with the result that the design reflects the social and cultural characteristics of the design nation.

Crewing, Personnel and Training: Minimum crewing philosophies commensurate with operational requirements, the implementation of alternative crewing strategies, advances in available technologies and OPTEMPO/PERSTEMPO issues have placed increasing demands on crew, support personnel and training requirements. Optimal crewing is not minimal crewing, hence the need for HSI. These issues have already compounded on recruitment and retention, leading to workload issues (both physical and cognitive), higher stress and fatigue, which ultimately effects crew safety and mission effectiveness. Essentially what the ADF is attempting to achieve in this area is best described by the model shown in Figure 5.

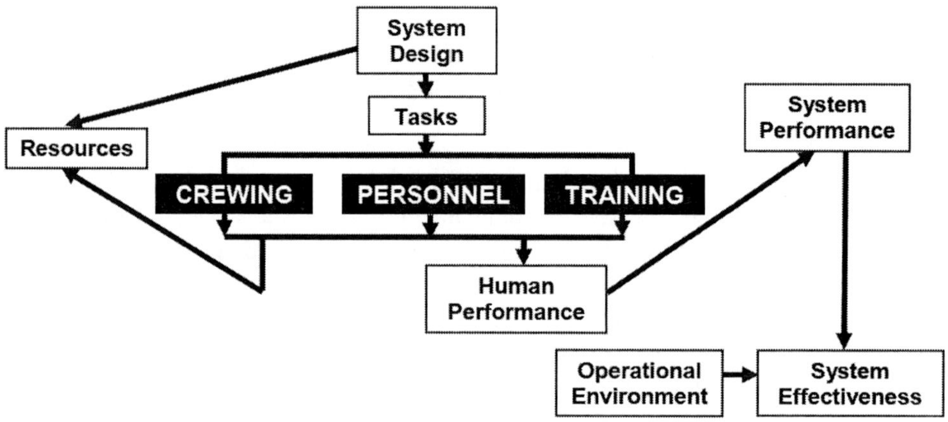

Figure 5: Crewing, personnel and training, HSI model.

DSTO has undertaken a number of projects in this area, examining human performance at sea and developing workforce planning and crew modelling tools. Some of this work is described briefly in the next sections.

Human Performance at Sea: These studies comprise both subjective and objective measures to identify and quantify the major contributors to fatigue in the maritime environment and the effect of fatigue on crew safety, operational performance and mission effectiveness. Evidence increasingly indicates that fatigue contributes to serious incidents in the workplace and mitigating strategies need to be identified and put in place. Human performance studies conducted include:

- assessment of human error in maritime operations;
- examination of situation awareness and new technology;
- examination of the effect of workload, stress, fitness, nutrition, and ship motion on fatigue, maritime accident investigations; and
- assessment of motion induced fatigue, whole body vibrations and injury due to slamming of fast craft transits (wave impacts).

Workforce planning and crewing modelling: The following comprise current modelling and simulation tools used by DSTO in the area of crewing and sustainment.

- ITHINK, POWERSIM. ARENA: These are time based simulations where feedback from processes is assessed on the overall performance of the system. The results can be optimized and assessed for risk.
- NaMOS: The model addresses the question of how do you allocate 18 crews to 12 boats (patrol boats), while operating under the constraints such as PERSTEMPO, crew leave/training, boat maintenance, schedules, peak usage periods and contractual requirements. It produces a fleet availability schedule and crew assignments.
- SENMA: This addresses the question, "given what we have, can and how do we meet our goals?" People are recruited, spend a certain amount of time in each rank and are then promoted.
- CGT: The complement generation tool addresses optimized manning of Naval Platforms. It applies a "Top Down" approach i.e. functional decomposition of a platform's mission requirements, allocation of the functions to machine or human elements and determines the total demand for personnel on the platform.
- ANUBIS DEMAND MODEL and the SUCCESSION MODEL: ANUBIS addresses forward planning demand and used in conjunction with SENMA (supply model). It asks the question, are the number of positions sustainable? The SUCCESSION MODEL is similar to ANUBIS but there is a much greater focus on the individual and it records age, time in level, time in employment, pay etc.

CONCLUSION: LESSON LEARNT AND THE WAY FORWARD

This paper has highlighted the importance of applying and incorporating HSI principles throughout the capability life cycle in a planned and structured way. This includes new designs and procurement requirements and the development and acquisition of systems and equipment to improve military performance. The scale of the system, its required capability and the resources needed to support that capability should never be underestimated.

HSI should be applied from the onset and not be regarded as an issue after the event. Retrospectively, not applying HSI principles can be a costly process not only in upgrading equipment or improving design but also as a result of potential injury to personnel and the subsequent medical treatment that may be required. It may also impact on potential litigation where chronic injuries are concerned.

HSI also impacts on recruitment and retention issues. Low crew morale due to any number of human factor issues, i.e. working conditions, environmental stressors and unrealistic work demands etc. can adversely influence career decisions. Considerable time, effort and monies are expended in training personnel to attain the skill sets required to operate a system and this could be a wasted if personnel leave the organization earlier than expected.

In terms of the way forward, much progress has been made in the recognition and implementation of HSI in the procurement capability life cycle of naval assets. DSTO will be involved in the technical risk assessment processes for all new acquisitions and this certainly includes HSI considerations. Within this context, the implementation of HSI for combat systems is well underway in both planning and research. There is also endorsement by the DOHSC for the phased implementation of a HSI framework to improve the identification, elimination and management of hazards throughout all phases of the defence capability lifecycle. In addition, guidance documentation is being prepared for the integration of human needs and requirements into the capability lifecycle.

A crucial element is that the HSI approach is strongly supported and endorsed by Navy as evidenced by the increasing emphasis on the human element in the DSTO research programs for navy.

REFERENCES

[1] Australian Defence Force (2006). Maritime Materiel Requirements Set: DEF (AUST) 5000–VOL 06 Combat System Requirements PT 02 SEC 03–ISS 01 Draft, 24 Nov 2006.

[2] Australian Defence OHS Program Office (2007). Human Systems Integration Framework Implementation Project (Draft Version 0), February 2007, Unpublished.

[3] Beevis, D. (1999). Human Systems Integration in Systems Acquisition & Development. Toronto ON: H/HFCS DCIEM.

[4] Booher, H. R. (2003). Handbook of Human Systems Integration. U.S.: Wiley Interscience.

[5] Canadian DND repository for information on Human Systems Integration (HSI) Web site http://www.drdc-rddc.gc.ca/researchtech/projects/hsi/process/index_e.html

[6] Cockshell, S. and Hanna, S. (2006). Human Systems Integration for the Australian Defence Force: Integration of Human needs and Requirements into the Capability Life Cycle. Endiburgh: DSTO-CR-2006-0209.

[7] Dennis, J. F. (2007). Human Systems Integration Electronic Systems Laboratory. PhD Thesis: Georgia Tech Research Institute, US.

[8] Fleming, P.J. (2007). Human Systems Integration Framework Project (HSIFIP): Proposed Phase 1 Scope of Work. Canberra: Defence Occupational Health Safety Compensation Branch.

[9] Kinnaird, M. (2003) Defence Procurement Review 2003

[10] Maxwell, G., Robert Bost, J. (2004). "The Navy Must Put People First". *United States Naval Institute Proceedings*: Mar 2004. 130(3); 90-93.

[11] McGovern, N. J. (unknown) Human Systems Integration in the US Navy: Policy, Process and Practice.

[12] Greenley, M. P. and, (1999) Human Systems Integration in Warship Design: The "Way Ahead" for Human Factors Engineering Tools. *Greenley & Associates Incorporated.* PWGSC Contract No. W7711-8-7464.

[13] Pearce M.L. (1994). Liveware Integration.(Final Report No. AC/243 (Panel 8) TR/18). Brussels: NATO Defence Research Group.

[14] Promisel, D.M., Hartel, C.R., Kaplan, J.D. and Marcus, A. (1985). Reverse Engineering: Human Factors, Manpower, Personnel and Training in the Weapon System Acquisition Process. (Report ARI-TR-659) Alexandria, Virginia: Army Research Institute for the Behavioral and Social Sciences.

[15] UK Ministry of Defence, (2007). Human Factors Integration (HFI) Practical Guidance for IPT's. London: Ministry of Defence © Crown Copyright.

[16] Vallard, I. (1999). Human System Integration Overview. DRDB, Director of Science and Technology-Human Performance,

[17] US DoD Defense Instruction 5000.2, Defense Acquisition management Policies and Procedures, Part 7 Section B.

Seakeeping Behaviour of a Damaged Warship

T Andrewartha[1], G Thomas[2] and S Cannon[1]
[1]Maritime Platforms Division, Defence Science and Technology Organisation, Melbourne, Victoria, Australia
tristan.andrewartha@dsto.defence.gov.au
[2]Australian Maritime College, Launceston, Tasmania, Australia
g.thomas@amc.edu.au

ABSTRACT

In the unfortunate event that a naval vessel is damaged, decisions must be made regarding the reduction in the vessel's capability and what damage control procedures can be initiated to both restore this capability and minimise further damage. In order to make an informed decision, it is vital that Commanding Officers (COs) have access to reliable information regarding the vessel's structural integrity, stability and seakeeping performance. A reduction in any of these parameters may have a significant influence on the operability of the vessel.

A preliminary program of work has been conducted to investigate the change in ship motions due to an angle of list induced on the vessel after sustaining damage. This study incorporates both experimental and numerical simulations of a listed vessel at rest in a seaway. During the experimental program several parameters were investigated to ascertain their influence on the vessel's motions. These parameters include initial list angle, direction of list and wave conditions.

The results from this study show that for a vessel in beam seas there is no significant effect on vessel motions due to an angle of list. However, in head seas the introduction of an angle of list has a marked effect on the amplitude of roll motions. The example operator guidance generated shows that numerical modelling can be applied to the analysis of the operability of a vessel against set limiting criteria. This information and methodology could be incorporated into a rapid damage assessment tool along with sea-load predictions to assist COs in their decision making process in the event of an emergency.

INTRODUCTION

Recent incidents have occurred where a warship has sustained damage following an unexpected event. Whether the damage was sustained by offensive actions, such as that inflicted upon the USS Cole, or via accidents, such as the grounding of HMS Nottingham, the crew need to respond quickly to manage the situation. Within the commercial sector, organisations such as Lloyds Register of Shipping offer a Ship Emergency Response Service (SERS) that provides advice in situations similar to these. The objective is to save the ship and prevent loss of cargo or pollution of the sea. In the case of a warship, tools are being developed to provide onboard assistance to the crew following an incident. DSTO is currently developing a Rapid Assessment Tool for demonstrating the potential of these tools within the Royal Australian Navy (RAN). This paper reports on work which contributes towards the development of this tool, and in particular investigates the seakeeping behaviour of a ship which has an angle of list following damage.

DAMAGE SCENARIO

The terms Float-Move-Fight are often used to describe the extent of damage sustained by a warship. Following a damage incident the CO must rapidly assess the vessel's situation with respect to these three terms. If the ship is unable to float the situation is one of safely evacuating the crew. If the ship is able to float the CO must determine whether the vessel is able to move within the seaway. Depending upon the mission and the extent of damage this decision will be whether to move the ship to a safe haven for repairs or to continue the mission. Alternatively, if the damage is minor and the damage control processes have secured the ship the CO must decide whether systems onboard are operational or not. A typical scenario might be that as a result of damage the ship has sustained a list and it is the influence of this list on the operability of systems that needs to be determined.

A numerical and experimental analysis has been performed to assess the seakeeping characteristics of a vessel that has an angle of list after a damage event. Using *FD-Waveload* [1], a frequency domain, three-dimensional panel method seakeeping code, a series of vessel conditions were examined. In addition, model tests were conducted to assess the applicability of the code for predicting motions when the vessel is in a damaged state. The application of the numerical simulations is demonstrated through the development of operator guidance for a sample onboard system.

PREDICTION OF DAMAGED SHIP MOTIONS

The motions of a vessel can be modelled using several methods depending on the resources available and the output required. The simplest method is frequency-domain linear strip theory. Increasing in complexity, 3-D panel method codes in the frequency-domain allow greater flexibility in the types of scenarios that can be modelled and have the capacity to model more complicated shapes including catamarans. Some codes such as *FD-Waveload* have the capacity to model the motions of asymmetrical bodies. More complicated and computationally demanding still are time-domain seakeeping codes which also generally have the capacity to model non-linear effects and time dependent characteristics.

The simulation of intact ship motions in waves is based on a numerical solution to the six simultaneous equations of motions [2] given in a generic form in eqn 1. In this equation, M_{jk} are the components of the generalised added mass matrix, A_{jk} are the added mass coefficients, B_{jk} are the damping coefficients, C_{jk} are the restoring coefficients and F_j are the exciting force (j = 1..3) and moment (j = 4..6). The accuracy of the simulation is dependent on the accuracy of the determination of the hydrodynamic forces and moments [3].

$$\sum_{k=1}^{6} \left[\left(M_{jk} + A_{jk} \right) \ddot{\eta}_k + B_{jk} \dot{\eta}_k + C_{jk} \eta_k \right] = F_j e^{i\omega t} \text{ for } j = 1..6 \qquad \text{eqn. 1}$$

Examining the damaged ship scenario, some studies have been conducted using time-domain codes to compute motions. These have been coupled to computational fluid dynamics (CFD) codes to model the ingress/egress of water and free surface effects in parallel with the motions [3,4,5]. By modelling such a condition, the number of variables requiring consideration greatly increases. Since the vessel is taking on water, trim, list and displacement are all changing and the traditional equations of motion for a floating body are no longer valid due to constantly changing boundary conditions. When using coupled time-domain motion and CFD codes, the equations of motion are solved for each time step using the computed external forces. Not only does this increase the complexity of the model, but it also significantly increases the computational power and time required.

In order to simplify this very complex problem, the current study assumes that damage control procedures have been affected, and a new equilibrium condition with constant angles of list and trim, and draft is established. By making this assumption, it is possible to predict the motions of this vessel using a frequency-domain 3-D panel method code. The primary advantage of this approach is that each scenario is significantly quicker to assess than a time-domain simulation, which lends itself to the generation of operational guidance.

NUMERICAL PREDICTIONS

By examining a vessel with an angle of list, a degree of asymmetry is introduced into the problem. A 3-D panel method is more capable than a traditional strip theory seakeeping code of modelling the asymmetrical submerged surface and therefore *FD-Waveload* was utilised for this study. *FD-Waveload* is a frequency-domain, panel method seakeeping code which is based on the zero-speed Green function with a forward speed correction. Full details of the methodology can be found in [1].

A series of conditions was analysed using *FD-Waveload* including a variety of headings and angles of list. A preliminary study was first conducted to ensure that the results were not dependent on the hull form definition. This included a number of mesh sensitivity studies in both upright and listed cases to ensure consistent results across different mesh densities. Once it was established that there were no significant differences, meshes were generated for several different vessel attitudes including upright, and listed to 15 and 20 degrees to starboard.

MODEL EXPERIMENTS

A series of experiments was undertaken to assess the accuracy of the numerical predictions. These were conducted in the Australian Maritime College's (AMC) model test basin which forms part of the Australian Maritime Hydrodynamics Research Centre (AMHRC). This collaborative research organisation was established in late 2002 as part of the Australian Commonwealth Government's Major National Research Facilities Program by the AMC, DSTO and the University of Tasmania. The model test basin is 35 m in length, 12 m wide with a water depth of up to 1.0 m. Waves are generated by a multi-element piston-type wave-maker.

The vessel used throughout this investigation was a typical frigate (Figure 1). The principal particulars of the model and equivalent full scale vessel are given in Table 1.

The tests were conducted for both beam and head sea conditions for a series of different angles of list (Table 2). Throughout this testing program, the model was restrained using a bridle system similar to that used in past experimental investigations [6] (Figure 1). This system used weighted cords attached at the bow and stern of the model, ensuring that it was free to heave, pitch and roll and allowed partial freedom in surge, sway and yaw. The model's ballast was configured so that it could be readily moved to generate the required angle of list while attempting to maintain a constant roll radius of gyration.

Table 1 Vessel Principal Particulars

		Model 1:70	Full Scale
LBP (m)		1.578	110.5
B (m)		0.198	13.9
T (m)		0.064	4.45
Δ		10.682 kg	3600 tonnes
LCG fwd midships (m)		-0.061	-4.27
KG (m)		0.064	5.81
R44 (%B)	0 deg list	37.85	
	15 deg list	37.74	
	20 deg list	38.07	

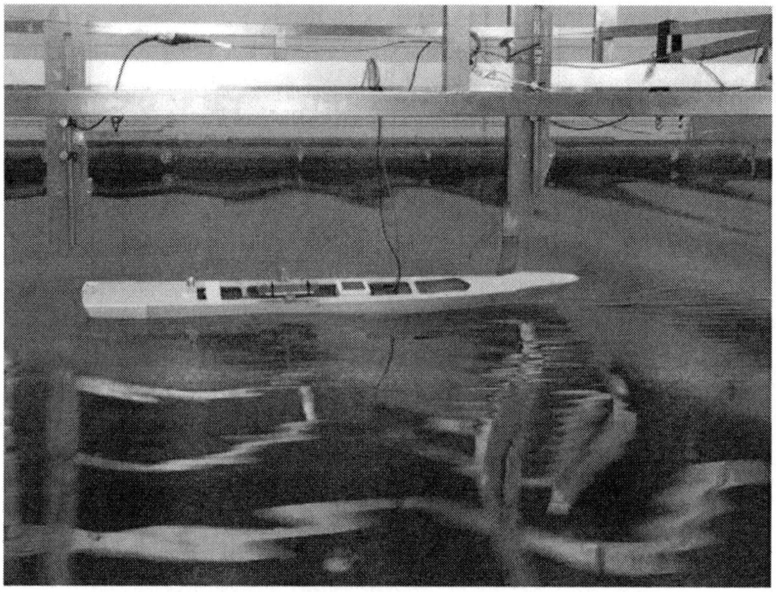

Figure 1 Frigate model restrained using bridle system

Table 2 Experimental Conditions

Condition Number	Vessel Heading (deg)	Initial Angle of List (deg)	Initial List Direction
1	90	0	-
2	90	15	Away from wave
3	90	15	Towards wave
4	180	15	-
5	180	20	-

Measurements were taken using a Crossbow NAV420 sensor, which is capable of measuring motions in six degrees of freedom, and a series of linear voltage displacement transducers (LVDTs). Roll decay tests were also conducted to determine both the natural roll period and the roll damping as a function of initial list angle. In this test the model was heeled over and then released. A time trace was taken of the rolling motions. This was then used to determine the roll damping coefficient and natural roll frequency.

RESULTS AND DISCUSSION

The results of the experiments were produced in the form of non-dimensionalised response amplitude operators (RAOs), allowing ready comparison between conditions. In order to correctly represent the listed vessels, the numerical mesh was offset such that the global origin occurred in the same *y*-plane as the listed centre of gravity. As a result, the rolling motions were modelled as relative angles both experimentally and numerically.

Figure 2a shows the roll motions of the frigate in beam seas for the two angles of list tested. Examining this RAO, there is little difference between the rolling motions of an upright vessel compared to a vessel listed to 15 degrees. This suggests that the rolling motions are not significantly affected by introducing an angle of list. Moreover, there is little discernable difference between listing the vessel towards or away from the waves.

D'Este and Contento [6] found that a listed vessel will exhibit different peak roll amplitudes to an upright vessel which is in contrast to the current investigations. This may be attributed to the variation in the submerged hull forms of the vessels. The previous study was based on a Ro-Ro, with a high block coefficient and well-defined bilge radius whereas the current study was based on a frigate hull form with a lower block coefficient (Figure 4). By listing the Ro-Ro, the underwater shape would be significantly altered, most likely affecting the roll damping, while the rounded form of the frigate would have less effect.

The heave results in beam seas (Figure 2b) show some degree of dependence on the angle of list. When the frigate was listed away from the wave, the heave motions between 0.3 and 1.2 rad/s were greater than for an upright vessel. When it was listed towards the wave, heave motions in lower frequency waves were found to be lower than for an upright vessel; however this trend is reversed for intermediate wave frequencies.

As expected, insignificant pitch motion was measured in beam seas (Figure 2c). The small pitch motion was due to the asymmetric fore and aft distribution of volume in the frigate hull form.

Considering Figure 3a, there was a clear relationship between angle of list and roll motion in head seas. This is most likely due to the submerged shape of the frigate. While an upright vessel is generally symmetrical about the centreline, the listed vessel is asymmetrical below the waterline. As a result, the forces acting on the hull are no longer symmetrical, resulting in a shift in the force balance which generates the rolling motion. Assuming that the upright vessel will not roll in head seas, comparing this with the 15 degree and 20 degree lists shows a distinct increase in rolling motions with increasing angle of list.

In head seas the experimental heave results (Figure 3b) show that increasing the list angle will for the most part not change the heave motions with the exception of a secondary peak which occurs at around 0.7 rad/s. This secondary peak increased with a larger angle of list. The angle of list had very little effect on the pitch motions in head seas (Figure 3c). This however is expected given that the longitudinal distribution of volume did not change greatly by listing the vessel.

When comparing the head seas and beam seas conditions it was found that the beam sea roll RAO is higher in magnitude for the list angles investigated. Therefore, the expected rolling motions would be less in the head seas condition. For example, with a list of 15 degrees, the roll RAO reaches double the magnitude in beam seas compared with head seas. This suggests that from an operational perspective it would be beneficial for a listed vessel to maintain a head seas condition, if possible, to minimise the rolling motions.

Examining the correlation between the numerical and experimental results, overall *FD-Waveload* predicted the motions of a listed vessel quite well. The roll RAO results in beam seas (Figure 2a) show a small increase in magnitude for an increase in angle of list, while little difference was visible between the 90 degree and 270 degree conditions. Comparing these with the experiments, the correlation was very good in terms of trends. There was however a slight shift in the resonant frequency, which affects comparisons for an individual frequency, particularly for those less than the resonant frequency. It is also interesting to note the difference in the peak area between the experimental and numerical results. The experimental roll RAO features a narrower peak region and will therefore have a smaller area under the curve. This will affect the results when sea spectra are applied.

In beam seas the comparison between the heave experimental and numerical results is less satisfactory than for roll and pitch. The trends derived from the experiments are however replicated to a degree in the numerical results (Figure 2b). For the condition with the frigate listed away from the waves, an intermediate peak is visible in the numerical results at approximately 0.7 rad/s. This appears to follow the peak in the experimental results. Similarly, a trough is evident in the experimental results in the listed toward the waves condition which is also replicated in the numerical results. While these trends are both identified in the numerical results, they are not as extreme as was visible in the experimental results. This will most likely also result in quite different motions with the application of sea spectra to determine the heave in irregular seas.

Little can be deduced from the experimental-numerical pitch comparison since they were of negligible magnitude (Figure 2c). However, a slight peak is clearly visible in both the numerical and experimental results.

Experimentally the roll magnitude increased with increasing angle of list (Figure 3a) in head seas. Similarly in the numerical results, a marked increase in roll magnitude was noted. As with the beam seas results, the resonant peak occurred at a lower frequency for the numerical results. In order to increase the clarity of the numerical trends, a series of extra list angles were modelled including 5, 10 and 25 degrees. These extra RAOs reinforce the theory that for an increase in list, an increase in roll magnitude can be expected in head seas. As with in beam seas, there is a clear difference between the area under the experimental and numerical head seas roll RAOs. This will be magnified when sea spectra are applied in order to extract the motions in irregular seas.

The numerical heave RAOs in head seas (Figure 3b) generally follow the experimental trends showing little difference between the two list angles. The main difference between these results is that the secondary peak is not visible in the numerical results. The numerical pitch results generally follow the trends visible in the experimental results but with very little difference in pitch motions between the different list angles (Figure 3c).

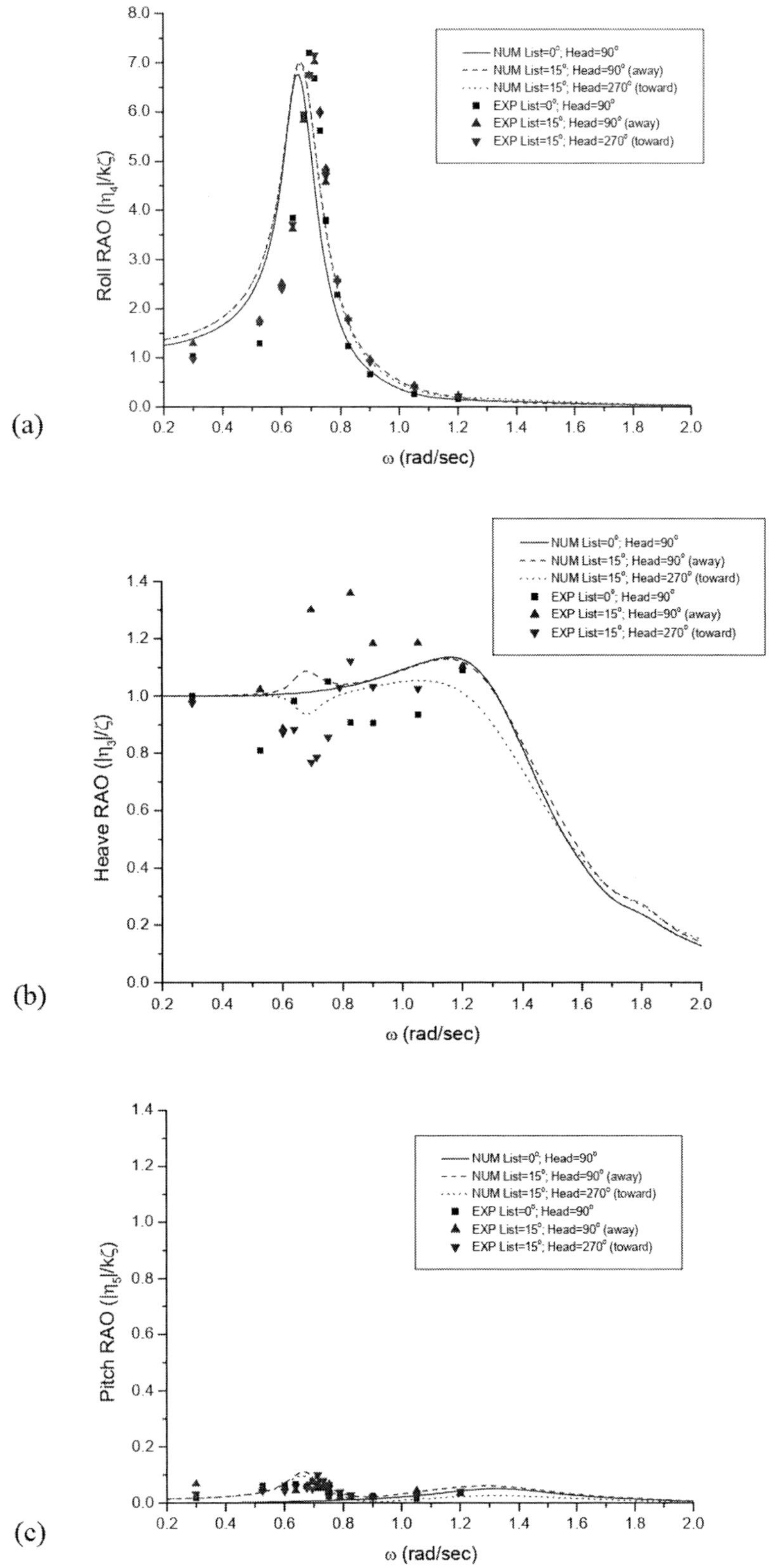

Figure 2 Beam seas RAOs; a: roll; b: heave; c: pitch

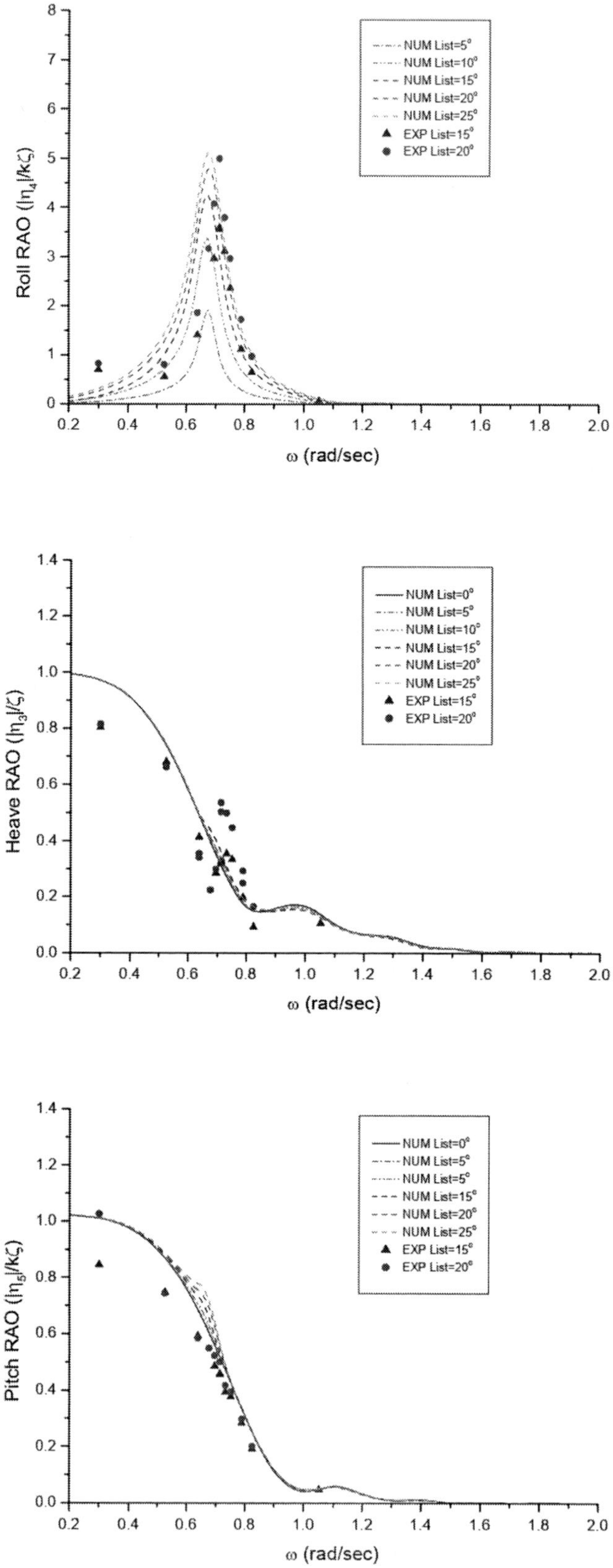

(a)

(b)

(c)

Figure 3 Head seas RAOs; a: roll; b: heave; c: pitch

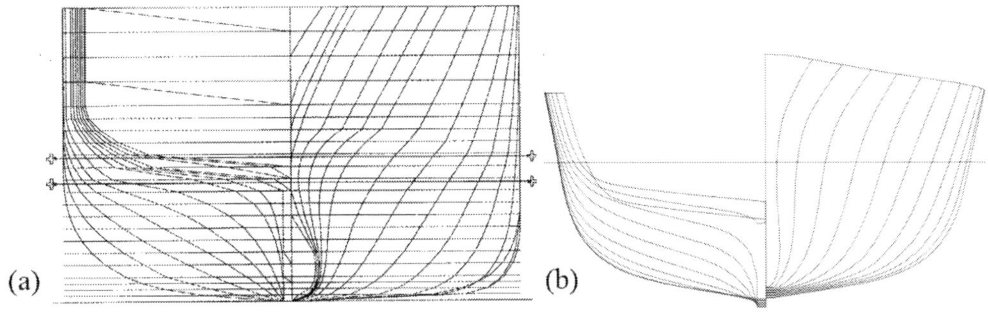

Figure 4 a: Ro-Ro body plan [6]; b: Frigate body plan

A roll decay test was also conducted on the frigate for the different angles of list. From the resulting time traces the natural roll period was found to remain constant at approximately 1.059 seconds model scale (8.37 seconds full scale). The roll damping coefficient however reduced with increasing angle of list (Figure 5). This may be attributed to the emergence of a bilge keel as the angle of list was increased and may explain the under prediction of roll motions in head seas for large angles of list.

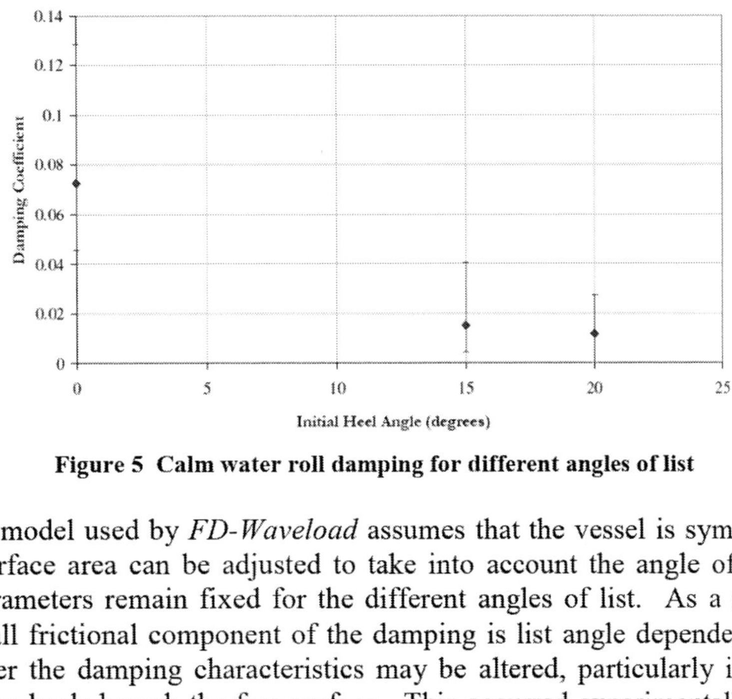

Figure 5 Calm water roll damping for different angles of list

The damping model used by *FD-Waveload* assumes that the vessel is symmetrical. Only the wetted surface area can be adjusted to take into account the angle of list, while the remaining parameters remain fixed for the different angles of list. As a result, only the relatively small frictional component of the damping is list angle dependent. By heeling the vessel over the damping characteristics may be altered, particularly in the instances where the bilge keels breach the free-surface. This occurred experimentally on a number of occasions particularly at large angles of list. In order to establish if this was the main contributing factor to the under-prediction, it would be necessary to modify the damping model to allow the asymmetry to be modelled; this capability is currently unavailable in *FD-Waveload*.

Overall, the results have shown that for the tested frigate hull form in beam seas, an angle of list will not significantly affect the roll motions. In head seas however, introducing an angle of list introduces an asymmetrical submerged form and will result in significant changes to rolling motions for the angles investigated. Generally, rolling will increase with angle of list. *FD-Waveload* is shown to give reasonable overall predictions of the motions of a listed vessel, both in head and beam seas. When considering individual frequencies however, there are some significant differences requiring further investigation.

EFFECT OF LIST ON SYSTEM CAPABILITY

A methodology for assessing the effect that the permanent list angle has on the operability of a particular ship system is demonstrated using the results obtained from *FD-Waveload*. By modelling the motions for a series of headings, speeds, sea states and angles of list, the performance of a damaged vessel can be compared to ship system limiting criteria. In this case study, the maximum significant roll amplitude that the sample on-board system can operate was used as a criterion. This limiting criterion was a roll angle of 17.6 degrees. In order to obtain the significant roll motion, the static list angle was added to the relative roll angle output from *FD-Waveload*. For example, for an angle of list of 15 degrees, the maximum allowable significant roll amplitude of the vessel would be only 2.6 degrees in the direction of the permanent list, whereas the maximum roll angle in the opposing direction could be 32.6 degrees from the equilibrium position.

Figure 6 shows two polar plots detailing the performance, against the limiting criteria, for different operating conditions (i.e. different angles of list). Figure 6a is for the upright condition, while Figure 6b is for a list of 10 degrees to starboard. Both plots are for operations in mean sea state 6, based on the ITTC two parameter sea spectrum. In these plots, the green region represents significant roll motions of up to 17.6 degrees and is allocated a pass rating accordingly. If the desired operating speed/heading combination falls in a red region, the significant roll motions could be expected to be greater than 17.6 degrees.

This is only a simplistic study and a full operability calculation could be performed as per the Materiel Requirement Set [7]. This would take into consideration the many factors contributing to the operability of a system, while the current study has merely considered the effect of an angle list on the capability of the system in terms of one limiting criterion in a given sea state.

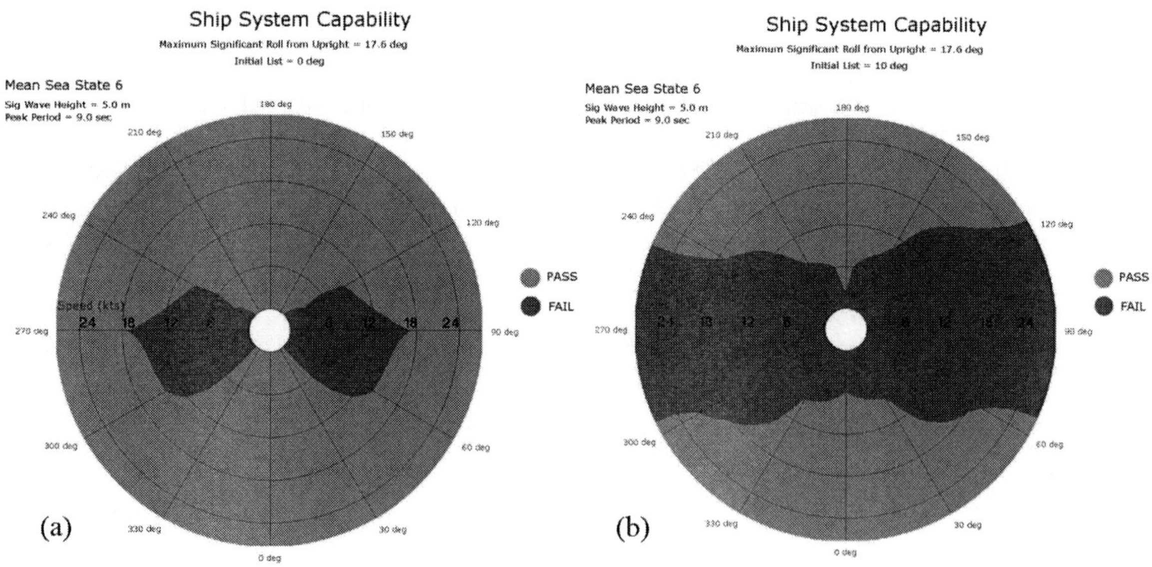

Figure 6 Frigate Ship System Capability - a: Initial List = 0 deg; b: Initial List = 10 deg

By comparing the red areas in Figure 6a and 6b, it is clear that by introducing an initial angle of list the system onboard the ship has a more limited range of operability. It is also important to note the asymmetry in this plot; with an initial list of 10 degrees to starboard, the motions are expected to be slightly larger with seas coming from the starboard side as indicated by the larger area of red.

Figure 7 shows the change in capability over a period of time for a damage event. The scenario discussed above forms part of the overall capability assessment. The capability of the ship system was determined by allocating a pass/fail rating to each point of the speed/heading matrix. By then comparing the total number of passes with the number of points in the matrix, the capability percentage was determined, in this particular case the operability of the ship system.

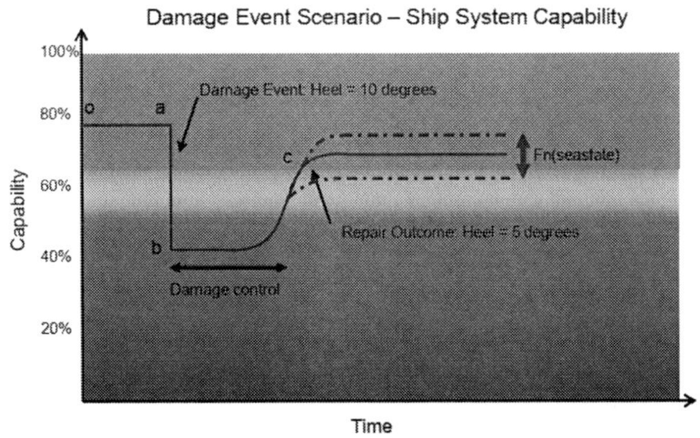

Figure 7 Effect of list on Ship System Capability

Considering Figure 7, based on the performance of an upright frigate in sea state 6, the system operability is 77 %. A damage event occurs at point *a* resulting in the vessel listing to 10 degrees (*b*). In this condition the system capability drops to 42%. At this point, damage control procedures are put in place, such as cross flooding, which reduces the list to 5 degrees (*c*). The resulting operability level is dependent on sea state, but in sea state 6, this capability is now around 69%, a marked improvement from the operability whilst the vessel had a 10 degrees list. If it was not possible to correct the list, the reduction in system performance would place significant operational restrictions on the vessel.

This case study has only considered the effect of the angle of list on the performance of a particular system onboard a vessel. There are many other factors which should be considered including water ingress and the associated dynamic effects, as well as any damage to equipment and systems. A future aim is to generate an operator guidance tool to provide COs with more accurate data to make decisions on operational capability in a damage situation.

CONCLUSIONS

An experimental and numerical investigation has been conducted to examine the effect of an angle of list on the motions of a vessel. Experiments were first conducted to examine this effect, and provide valuable validation data for comparison with the results of the numerical code *FD-Waveload*. This work has led to the development of an example methodology for assessing the performance of a damaged vessel against set limiting criteria through the use of a case study.

The results of this study have shown that *FD-Waveload* has the capacity to predict the motions of a vessel with an angle of list with some limitations regarding prediction of responses at individual frequencies. Through the use of a case study, the results of this investigation have been applied to generate a concept for predicting the effect of list on the capability of an onboard ship system on a frigate. This methodology can be extended to examine any of a range of other limiting criteria applicable to naval surface platforms.

Along with the consideration of sinkage, free surface and sloshing, it may be prudent to extend this study to different types of hull-forms in support of the theory that the hull form body section affects the resulting motions with changing angle of list. Extending this study to take these factors into account will most probably result in the need for time-domain analysis to accurately model the fluid flooding in and out of the damaged compartment.

ACKNOWLEDGEMENTS

The authors would like to acknowledge Tim Lilienthal for his contribution to the experimental component of this study.

REFERENCES

1. Martec Ltd. (2006), *FD-Waveload 2006A – Theory Manual,* Martec Limited, Nova Scotia, Canada.

2. Salvesen, N., Tuck, E.O., Faltinsen, O. (1970), *Ship Motions and Sea Loads*, Society of Naval Architects and Marine Engineers, Transactions, pp 251-287.

3. Laskowski, A., Jankowski, J. (2005), *Simulation of Damaged Ship Motion in Irregular Waves*, 10th Naval Platform Technology Seminar, Singapore, pp 235-243.

4. Umeda, N., Kamo, T., Ikeda, Y. (2004), *Some Remarks on Theoretical Modelling of Damaged Stability*, Marine Technology, Vol. 41, No. 1, pp 45-49.

5. Woodburn, P., Gallagher, P., Letizia, L. (2002), *Fundamentals of Damaged Ship Survivability*, Royal Institution of Naval Architects, Transactions, Vol. 144, pp 19-39.

6. D'Este, F., Contento, G. (2000), *On the Mathematical Modelling of Roll Motion of a Ship with a List*, International Journal of Offshore and Polar Engineering, Vol. 10, No. 4, pp 316-320.

7. Department of Defence (2003), *DEF(AUST)5000 ADF Maritime Materiel Requirement Set, Volume 3, Part 6 – Seakeeping*, Navy Systems Branch, Navy Platform Systems.

8. Papanikolaou, D., Zaraphonitis, G., Spanos, D., Boulougouris, E., Eliopoulou, E. (2000), *Investigation into the Capsizing of Damaged Ro-Ro Passenger Ships in Waves*, Proceedings, 7th International Conference on Stability of Ships and Ocean Vehicles (STAB2000), February 2000, Tasmania, Australia.

9. Schindler, M. (1998), *Damage Stability Tests of a Model*, Schiff & Hafen, Vol. 50, No. 5, Seehafen Verlag, pp 28-36.

Conversion MV "DRONNING INGRID" to M.V."AFRICA MERCY"

Alan Budd
Mercy Ships Australia
alan.budd@mercyships.org

Before moving into the substance of my presentation, I would like to give you some background as to why we, as a charity, would want to convert a train ferry to a hospital ship.

Mercy Ships is an international, interdenominational Christian humanitarian organisation taking hope and healing by ships, to the world's poorest nations.

This grew from a vision of our founder, Don Stephens, in 1964 when he and a group of young people survived a hurricane in the Caribbean. In 1978 he purchased the MV Victoria from the breakers yard in Genoa, and with a team of volunteers "resurrected" the vessel over the next four years, to realize his vision.

Since 1978, Mercy Ships, a global charity, has operated a fleet of hospital ships in developing nations. Mercy Ships takes hope and healing to the poor and needy, by mobilizing people and resources worldwide.

During this time Mercy Ships have operated four ships, Anastasis, Good Samaritan/ Island Mercy, Caribbean Mercy and now the Africa Mercy. We also have two land bases in Sierra Leone, New Steps for Polio and enforced amputation victims and the ACFC for VVF sufferers. The organization is represented by fourteen offices throughout the world, with the International Operations Centre located in Garden Valley Texas. The United Kingdom office is situated in Stevenage, and Australian office in Caloundra, Queensland.

Working in partnership with local communities, Churches, government and non-governmental agencies in selected countries, the goal of Mercy Ships is to be a locally relevant catalyst for transformational development.

I might mention here that over 90% of our Mercy Ships people are volunteers who pay their own travel expenses to and from the ship, crew fees whilst living and working on the ship, and all personal insurances. The current crew fee is USD150 per week for short term volunteers.

To show the work that is done on a typical outreach, I will run a short video of an outreach team working in Sierra Leone from our Anastasis. From this it will be seen the advantages of using ships in this type of operation. 9 minute video "Give a Little", prepared by Colleen Conley.

In order to expand the 30-year history of serving over 90 of the poorest countries in the world, Mercy Ships' vision is to rationalise our operating capacity to bring hope and healing to nations low in the UN Human Development Index (HDI) in Africa and ultimately to other deserving countries.

Given the ageing of the Anastasis and the Caribbean Mercy, both having been built in 1950, Mercy Ships leaders in the mid-90's commenced the search for a "new" ship to augment its operational capacity. In 1999, the search ended with the purchase of the Dronning Ingrid, a 16,572 tonne GRT Danish train ferry (built by the Danish Government in 1980 at the cost of US$150 million) that had been made redundant by the completion of a tunnel and bridges.

Ms. Anne Gloag from Scotland purchased this ship for £Stg 4 million in the name of the Balcraig Foundation as a gift to Mercy Ships.
This ship was renamed the Africa Mercy to serve as the fourth and largest hospital ship in the Mercy Ships fleet.

The two most significant changes required to be made were to provide;
1 a complete "World" class hospital, and
2 suitable accommodations to house all those involved in the operation of the ship, the hospital, and for the numerous outreach services.

Special attention was therefore required for the following aspects to suit her for her unique service;-
The construction and equipping of her new hospital and the associated spaces,
The construction of a variety new accommodation and activity spaces,
Revised bunker spaces, new fresh water tanks, conversion of various tanks to cold stores and other uses,
New steel works,
The installation of two new 2.6MW alternators, each of twelve tonnes, and their associated electrical works,
Twenty vehicles to be carried on the upper deck,
New 150 man lifeboats and other lifesaving equipment,
Sixteen twenty foot loaded containers in a new hold,
New crane with plinth, and a swimming pool which is yet to be constructed on the upper deck".

The original contract for the conversion of the vessel was with Cammell Laird who unfortunately went into receivership some 18 months into the contract. This cost our project two years in time and GBP1million in money.

The AFM has now been refitted in the shipyard of A&P Tyne, in Newcastle on Tyne, UK. In conjunction with the above work was that of continuous maintenance and survey work.
On completion Africa Mercy became the world's largest non-government hospital ship with six state-of-the-art operating theatres, 78 hospital beds and a total complement of 474 which includes 70 children's berths in family suites to serve primarily in the poorer nations in Africa.
In addition to providing a wide range of healthcare services, the AFM facilitates the delivery of integrated community development services in many forms with far-reaching impact in improving the quality of life for hundreds of thousands of African people.

Let's look at a brief history of the Africa Mercy in her life as the Dronning Ingrid, prior to purchase for the Mercy Ship fleet. She was classed Bureau Veritas + I 3/3 E, Passenger Ferry + Machinery, Deep Sea - limited to the Baltic & Danish Territorial Waters. This restriction was imposed at the request of the Danish owners, DSB now Scandlines.

Her principal dimensions are; Length overall 152 meters, 498.68 ft, Maximum breadth 23.7 meters, 77.76ft, Maximum draft 6 meters, 19.68ft, and Gross Registered tonnage 16,572 Tons.
The railway system of Denmark, by geographical necessity, has utilized a fleet of passenger and train ferries over many years, and is still operating a quite substantial fleet in that country's waterways. In particular, the main inter-city line from Copenhagen to Odense necessitated the crossing the of the Store Baelt also known as the Great Belt, a twenty kilometre wide stretch of water separating Jutland and Zealand. In the late nineteen seventies the government decided to modernise part of their fleet with three new ships. These were to be the Dronning Ingrid, a train ferry, and sister ships the Prins Joachim and the Kronprins Frederick. The Dronning Ingrid was built at Helsingor Vaerft, Denmark, at a cost of US$150,000,000, and entered service in 1980 on the Store Baelt service.

The service requirements for this vessel were for the fast and safe transit of this waterway, quick turn around and fulfilling the necessities of the passengers during this passage. Being of a short duration, some eighty minutes, did not require the provision of cabin accommodation for passengers. However sufficient cabins were provided for 48 crew.
The bulk of service for passengers was the catering facility, which was sufficient to provide cafeteria seating for up to nine hundred people on each crossing. The serving areas, along the centreline of the ship, were divided into port and starboard sides to allow one side to operate while the other was serviced on the return journey. Dining was in cafeteria and restaurant areas on 5 and 6 decks adjacent to the servery areas.
Shops, lounges, and toilet facilities were provided for the further comfort of the 2200 day passengers.
The train deck consisted of four rail tracks running the length of the vessel, separated by a centre trunk structure. Into this centre trunk, and attached to the ship sides were incorporated the passenger "platforms" and access to the ship's passenger's areas via stairways and lifts, forward and aft of the centre of the vessel. The requirement for the fast movement of passengers from the trains to the ship's facilities and return necessitated an efficient "passenger proof" system. This was provided for by the concept of a "Train Square", a "two-storey" space off which the shops and fast food bars were accessed. Not unlike a city underground railway station. With the completion of the fixed link, consisting of tunnels and bridges, across the Great Belt in 1998, the Dronning Ingrid was no longer required.

Although the Dronning Ingrid did not meet the immediate requirements of Mercy Ships, it was evident that a suitable conversion would produce a highly efficient vessel capable of maintaining the facilities to the high standards required by our organisation.

She was propelled by 6, B&W Alpha V16 Diesel engines, type 16 U 28LU each of 3120 kW at 775 rpm, totalling 18,720 KW, driving 2 controllable pitch propellers, to give a service speed of 19 knots. Fuel consumption of the main engines was 98t/24 hr 200secR/MDO, but is now 25t/24hrs at a speed of 13 knots using just one engine on each shaft. It could be 2 or 3 on either side, and for manoeuvring we run all 4 so that we can use the shaft alternators to supply power for the bow thrusters.
The main engines are now configured two to each shaft port and starboard. Each main engine is clutch connected to the main gear box driving each shaft, which also drives a Brown Boveri shaft generator of 865kVA at 1000rpm. The No. 1 engines have been disconnected from the gearboxes, each having front end power take off couplings to drive the, 2.6 Mega watt alternators for harbour power supply.

Calculations had shown the existing four alternators, to be insufficient for the anticipated maximum hospital/hotel loading of 2.6 Mega Watts, and thus required the new configuration. In service it has been found that full load in port is around 1.6MW so there is plenty of spare capacity

Two new ABB Automation 3281KVA 750rpm, 2.6 Mega Watt alternators are clutch coupled to the forward end of number one main engine port and starboard. The changed operating conditions of these engines were reviewed by B&W and found to require no changes. These engines are situated in number one main engine room now called the harbour generator room, and were originally attached by clutch to the main gearbox in number two main engine room by an extended shaft passing through a watertight bulkhead at frame 72. These alternators were lowered into position, through holes cut in the train deck.

The major electrical work was carried out by Peto (North East) Ltd. Main electrical current is 380V, 50hz 3ph. This is reduced to 220V single phase in the accommodation. Forward of the number four main engine compartment is the auxiliary engine room, housing four alternators. These are original engines, Frichs diesels typeA8.185CUS, each of 736kVA at 1000 rpm with a fuel consumption of some 3t/day/gen at full load, so with three Frichs running in port consumption was about 9t/day, and now with one B&W running it is about 8t/day MDO,

Positioned directly above these engines, and adjacent to the main control room on deck two is the electrical control room and main switchboard rooms which required extensive modification to accommodate the new alternator distribution.

The main switchboard room is slightly aft of the engines – an additional switchboard was added in what used to be the Electricians workshop to receive the power from the two new alternators – this switchboard is connected to the existing switchboard for distribution

The Main and auxiliary engines are cooled by fresh water systems, the high temperature at 60 degrees and the low temperature at 30 degrees. Both these closed fresh water systems are cooled through plate type salt water cooled heater exchangers.

Because of the limited operational range in her normal life, Dronning Ingrid carried comparatively little fuel, 276 cubic meters; hence the fuel tank capacity was found to be insufficient for Africa Mercy's International voyages and long outreaches.

The existing four of the existing eight dry tanks between frames 4 and 42 were converted to fuel tanks for the carriage of Medium Diesel Oil, fitted with the appropriate pipe work and gauges and integrated into the existing fuel system. The total fuel oil capacity is now 1,087 cubic meters an increase of some 811 cubic meters. This now gives an at sea endurance of some 30 days or 9,360 nautical miles… or a total of 125 days on station in port.

Bunkering stations located forward were closed down, and all bunkering connections are now through watertight door access to modified positions port and starboard aft, in way of frame 10.

Air conditioning and ventilation requirements for the proposed area of operations indicated the existing arrangements were inadequate and a major upgrade found necessary. Two Carrier Transicold 30HXC310-PH3 Chillers, each with a cooling capacity of 1043 kw are installed on deck 4 forward. These provide chilled water to air handling units on decks 4 and 7. Air conditioning machine rooms are situated, 5 rooms on deck 4, two of which, located P and S aft, are to provide for the hospital deck, and 3 existing rooms on deck 7. All existing fan coil units have been enlarged to suit the new conditions. The refrigerant is R134A. The range of climatic conditions assumed for sizing the plant is:

	Ambient	Indoor
Summer (primary)	+36 Deg C 80% RH	+24 Deg C(+0-2 Deg C) 50 to 60% RH
Summer (secondary)	+30 Deg C 90% RH	+22Deg C(+0-2Deg C) 50 to 60% RH
Sea Water	+32 Deg C	
Winter	-5 Deg C	+22 Deg C(+2-0 Deg C)

As this vessel can and will serve in other parts of the world as well as Africa, provision has been made for adequate environmental comfort for all on board.
Since taking up operation in West Africa in May of last year, the installation has performed to specification, although we haven't reached worst case loading yet – it has been relatively mild in Liberia.

The hot water system is existing - there are three means of heating the domestic hot water, electric calorifiers, a small plate heat exchanger that uses waste heat from the engine 60 deg jacket water system or we have two oil fired "boilers". These latter don't actually produce steam but rather hot water previously mainly used for heating of accommodation and keeping plant warm in winter months.
It was estimated that 80 – 100mt per day of fresh water would be the normal consumption, with the hospital functioning. The actual consumption is about 60 tonnes per day. We are not quite at full capacity yet but encouraged that most of the time we are lower consumption than Anastasis for a comparable number of people on board. Total FW tank capacity is now 950m^3.
The vessel will normally load water from shore facilities and, where necessary pass it through a deep bed fine filter followed by chlorination, followed by activated carbon filters to then remove the excess chlorine. Developments in the generation and treatment of fresh water from harbour water are being followed closely with regard to future possible installation in this vessel.

With a maximum resident population of some 474 persons on board, sewage treatment is of vital importance. Servac was the contractor for upgrading the existing vacuum collection type sewage system, which was processed ashore and sufficient for the original transient trade. The additional sewage treatment plant consists of two Hamworthy Super Trident TS 20 plants. This involved the fitting of additional vacuum pumps and the installation of considerable amounts of stainless steel piping throughout the vessel. Both grey and black water is handled by this system. The new storage tank has been constructed with a capacity of six cubic meters This system can hold up to 635m3 of sewage if necessary. The calculated sewage daily production was some 50t/day. This plant is capable of meeting all known Legislative requirements for operation at sea or in port under tropical conditions. It has the capacity to cope with the full ship's complement and all the hospitals requirements. This installation is located in auxiliary engine room number two.

For the disposal of waste, a Team Tech Incinerator has been installed in auxiliary engine room number 5(auxiliary generator room), now called auxiliary generator room. This unit has a capacity of 400 litres, is oil fired and has an operating temperature of 1200 degrees C. The efficiency of this incinerator is sufficient to dispose of all hospital waste ensuring no possible contamination from disposed materials from either hospital or domestic sources. (we wouldn't have the capacity for all domestic waste)

In considering what is arguably the most important of the changes, let's now focus on the hospital;

By the insertion of a new deck, now designated deck 4, at 2.5 meters above the train deck, this full deck forward of frame 25 became available to accommodate 196 persons, below which is deck 3, the rail deck, became our hospital. The actual rail deck, provided sufficient volume for six operating rooms on the starboard side, consisting of two Ophthalmic, one General, one Gynaecological, one Orthopaedic and one Maxillo-Facial. Directly astern of the operating area are the Pharmacy, CT Scan, X-ray, Laboratory and associated equipment areas. The port side, from frame 25 to frame 129 accommodates a seventy eight bed hospital ward which includes recovery, high dependency and intensive care facilities. This area is partitioned into designated wards.

The patient waiting area is located in the area at the fwd end of the container space at the hospital level. Forward from frame 129 to 175, berthage capacity for ninety medical staff, and a classroom from 175 to 198 on that deck. Adjacent to and included in the boundaries of the operating rooms and wards are consulting and treatment rooms.

The air to the Operating Theatres area passes through a special HEPA (highly effective particle extraction) filtration system to ensure the cleanest air possible and have balanced flow. The HVAC System overall including the Hospital is fairly simple, no high tech controls. Medical air and Oxygen is processed on board and is protected by emergency dump valves. Oxygen is produced in a dedicated facility located in a 20 foot container on Deck 8. Compressed Air and Oxygen is distributed throughout the hospital through a built in piping system with specialized quick connect fittings in the Operating Theatres, Labs and Wards.

In the event of a fire or other emergency, compressed air and oxygen in receivers located on Dk 8, can be dumped to atmosphere and the complete system depressurized from a valve station on the open deck.

Ship side access to the hospital is via openings in way of frames 20 to 25 on number 4 deck port and starboard. This hospital is capable of delivering surgery and treatment of world standard in areas of real poverty.

Together with alterations to decks 2,3,4,5,6 and 7, the total accommodation is now 474 persons berthed in cabins of various configurations from two dormitory style cabins – one 8 berth and one 10 berth the rest are either 4, 3, 2, 1 or family suites to single self contained cabins, and to 26 family suites holding up to eight persons. Not included in this number are the 78 patients in the hospital ward on deck 3.

Because of the number of long term volunteer staff in family groups on board, there is a need for schooling facilities for their children, which is not normally encountered in "conventional" ships. The school/classroom on deck 3 is not part of the school for children but rather for teaching various things – could even be technical training. Spaces are provided on deck 3 between frame 175 and frame 197, also deck 6, between frame 28 and frame 45, and on deck 7 between frames 30 and 49 for school class rooms ranging from nursery to University entrance level.

Some 420 tonnes of steel were added to the ship during the conversion, mainly for decks, bulkheads and accommodation.

The bow and stern doors being no longer required were seal welded in the closed position, and full bulkheads fitted at frames 25 and 175, and a new collision bulkhead at frame 198. All this work has been carried out to the requirements and survey of Bureau Veritas.
The wheelhouse required various changes and these were, basically that we changed most of the electronics including VHF, Radars, Echo sounder, speed log, etc. Also added a GMDSS station and VDR– all supplied by Sperry and installed by our own crew
Local manufacturer Solar Solve Marine, were generous in donating 15 Solarsafe roller screens to the wheelhouse windows.
The bridge mounted B&W Alpha engine control equipment was refurbished, as were all communication, navigation and alarm systems. A new machinery monitoring alarm system by Lyngso was added to replace the original system.
An electronic chart system was donated by Transas, another example of the generosity of people when they know the work of Mercy Ships.

The after docking bridge, on deck seven, was not required in the vessel's new role, so was removed, and a new accommodation block was manufactured by Cammell Laird and added where the aft bridge had been – it consists of two decks – 6 and 7 and a total of 9 family cabins, 4 on deck 6 and 5 on deck 7.

Aft from frame 25 to the stern door, there is sufficient capacity for the stowage of sixteen, twenty foot containers, separated by the centre casing. Access to these holds is provided by hatches, one each side, cut athwartship, of sufficient opening to allow the lifting of one container at a time. The openings are provided with coamings and steel covers of sufficient strength to carry two containers on the covers. Into both covers are mounted a plug hatch of sufficient dimensions to allow access for pallet size cargo slings.
These containers will carry mainly stores for the hospital and other ship services – there are also containers with tools for the construction teams and vehicle maintenance. The aft crane has a lifting capacity of 23 tonnes, at 19 metres. This is an electro-hydraulic crane, mounted on deck No 5, and stiffened to deck 4. The arrangement is such that vehicles mounted on the upper level of the boat deck can be easily lowered to either side or over the stern of the vessel. The capacity is such that all containers will be easily handled.

It will be noted that two, one hundred and fifty person life boats are situated on the upper deck mid ships. These SEL10.5 boats were donated by a Swiss family trust at a cost of GBP91,500! Handling of these will be by two, gravity multiple pivot davits, type GLP.MP170, manufactured by Global Davits.
Forward of these are two, fifty person lifeboats handled by Wellin gravity davits. This gives a boat capacity of two hundred per side.
Additional to these will be 6, 35 man rafts which will be davit launched with an existing Wellin raft davit. The same davit (new one supplied by NED Deck) will also launch the new Rescue Boats— one on either side. The rafts will provide another two hundred and ten places. (the existing carousels unfortunately had to be removed as they were not strong enough to carry the new rafts with A packs in them – the original rafts were fitted with B packs for coastal work).

Main fire zones have been changed and they are noted on the new GA to accommodate new cabin arrangement changes and to satisfy SOLAS regulations for passenger ships.
The engine room fire protection is a CO2 fixed smothering system supplied by Cosalt. It consists of 20 x 45kg bottles and some 62 nozzles; this was a replacement for the Halon

system which was originally fitted. A high fog mist system is fitted in way of the oil burning equipment in the machinery spaces in compliance with the latest regulations. The rest of the vessel is fitted throughout with a sprinkler system and a smoke detecting system supplied by Autronica of Norway.

A fwd rudder and steering gear was fitted to this vessel. This allowed the precise positioning of the vessel into the loading facilities in her former role. This gear was controlled from the after docking bridge it could also be controlled from the forward bridge – in actual fact in operation they always docked bow first – they used the aft bridge for reversing out). In her new role there is sufficient manoeuvrability available with the twin bow thrusters and existing stern rudders, and twin screws. There being no further need for this bow rudder, it has been permanently and securely locked in line.
With the vessel spending so much time alongside the wharf, the use of the self polishing and other high performance anti fouling paints is not warranted. Her sister ships, Anastasis and Caribbean Mercy, were using International Interclene which were found to be satisfactory. We find the grass that tends to grow on the outboard side falls off pretty quickly once underway – the Interclene works really well but we do usually try to scrape the props before departure using our own divers

The major contractor for this project was A&P Tyne, and the major subcontractor responsible for the outfitting was MSC (Europe).

The vessel is now Classed Passenger Ship Ice II SDS x hull x mach, Bureau Veritus and registered in Malta.
Her registered owners are Africa Mercy Malta Ltd, a division of Mercy ships International, and the registered office is Mercury House, 125 Old Mint Street, Valetta, Malta VLT 13. . Operators are Mercy Ships based in Garden Valley, Texas

The magnitude and cost of this project is obviously quite substantial, approximately GBP32m. When one considers that all funds are donated, the generosity of people throughout the world is truly awesome.
As opposed to normal commercial shipping, Mercy Ships is financed by donations, and progress depends on the inflow of donations. I emphasize there is no Government assistance for the work of the Mercy Ships fleet. The tasks being performed by our fund raisers, and the response by corporations, charity funds, church groups and individuals is as I have said, awesome.
Anastasis has now been paid off as has the Caribbean Mercy, and Africa Mercy will have more than twice the throughput of these two vessels combined. It is envisaged that Africa Mercy will touch the lives of some one million people each year. This is a worthy contribution to a country where it was recently stated that 87 million people would face calamity in the coming year.

During the eight years of conversion many volunteers came from many parts of the world to work on the vessel, in the engine room, on deck, in administration, in a host of other situations, and in doing so donated many skills and much valuable time toward the rebuilding of Africa Mercy. Many more volunteers are needed now and into the future to operate AFRICA MERCY in her many roles, as she takes hope and healing to the needy of AFRICA.

In conclusion I commend this project to you for your support in whatever way you feel compelled...

Feel free to spread the word of what can be and is being done by Mercy Ships, and its volunteers, with a ship that was redundant, and is now giving new life to many hundred of thousands of suffering people in some of the world's poorest countries.

www.mercyships.org.au

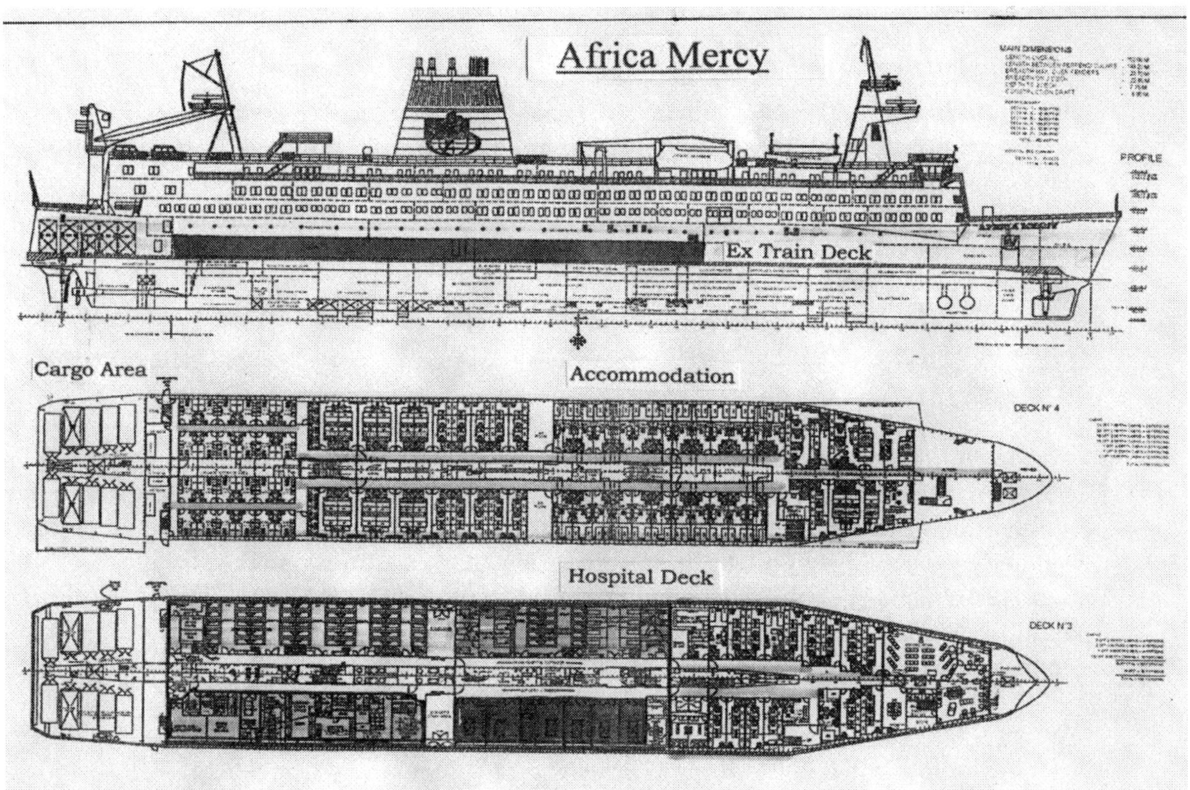

Major changes to Africa Mercy

Underway replenishment ship designs

David J Bricknell
Director, Systems and Product Strategy, Rolls-Royce Naval, UK
Per-Egil Vedlog
Design Manager, Rolls-Royce AS, Norway

SUMMARY

Rolls-Royce has recently completed a number of designs of varying deadweight capacity for Underway Replenishment vessels capable of 20 to 27 knots. The designs utilise commercially developed fully proven and efficient hull-forms from successful Rolls-Royce Ro-Ro, Ro-Pax or Containership designs. The ship design provides for an identified incremental upgrade of capability from build through the service life of the vessel. These upgrade paths allow for tailoring of the solution to the individual customers requirements. Acquisition costs can be kept low as the starting point is a commercial standard vessel. The addition of military capability such as aviation facilities, self-protection systems, additional dual capacity replenishment rigs and the potential for future upgrade has been designed into the ship to reduce the risk of obsolescence and to ensure the customer gets exactly what they need, when they need it. There is a high degree of confidence in the design performance due to an already proven hull-form, detailed structural definition, sea-keeping analysis, systems selection and general arrangement drawings. Economic analysis has been conducted on a range of propulsion systems to ensure that value for money can be demonstrated through life. The paper discusses the requirement analysis, hull-form selection, design development and analysis that has provided a high confidence, low risk, adaptable and upgradeable design.

1. INTRODUCTION

The world's changing geo-politics have seen an upsurge in warships being deployed worldwide in multinational operations. Nations involved in these often fast-developing crises are seeking to put early-entry forces into theatre to rapidly stabilise the situation through their presence or to take some form of military action. Sustaining this presence demands long-term support to those forces, and that in turn requires re-supply. Getting into theatre rapidly also demands either pre-positioning or support along the route for fuel to allow a rapid transit. For all these reasons many nations are looking to acquire or to improve their Underway Replenishment (UnRep) capability.

Attracting the attention of Governments, Navies and acquisition authorities are the recent changes in International Maritime Organisation regulation in particular MARPOL with its amendment of 2003 **[1]** terminating category one tankers (>20,000dwt crude, fuel, HFO or lub oil as cargo and not complying with post 1980 regulations) without double hulls in 2005 and Cat 2 and 3 tankers (>20,000dwt complying with post 1980 regulations or 5,000 – 20,000 dwt) without double hulls from 2010 – hence the rather accelerated acquisition timescales facing many navies. Figure 1 illustrates the tanker gap challenge.

Currently two ways of providing new MARPOL compliant capability are generally considered: converting ships taken up from trade and or building new ships.

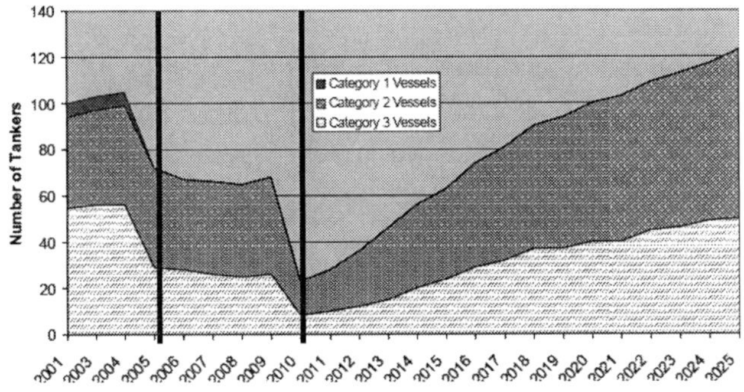

Figure 01 – The Tanker Gap from a MarPol Perspective

2. ADAPTING COMMERCIALLY TRADING SHIPS

Taking ships up from trade can be a very cost effective and readily available solution to an immediate requirement. The success of such a venture depends upon what the 'Naval' role of the vessel will be and how close the donor vessel's existing capability is to the requirement. Some good examples are forward repair ships, where very similar requirements for both Navy and the Offshore industry exist, and with Ro-Ro/Logistics vessels, where transport of vehicles and vehicles carrying logistics from A-to-B is a similar role for the Navy and Merchant marine. These vessels are not required to operate 'up-threat' and hence can rely wholly upon commercial construction rules including Flag-state safety regulations.

Adapting commercially trading ships for Underway Replenishment is also often considered. This is normally aimed at product tankers to deliver fuel whilst at sea. A recent example, HMAS Sirius, illustrates some of the benefits and some of the compromises inherent in such a project.

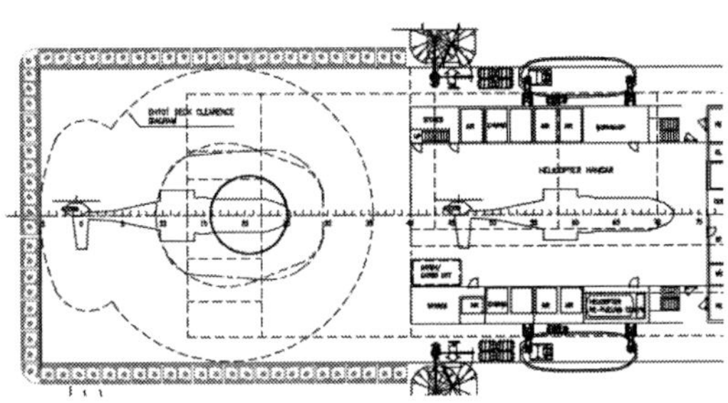

Figure 02 – Aviation arrangements

Navies' requirements for Underway Replenishment vary but essentially they require a capability to provide clean fuel, lubricants, dry stores, and sometimes ammunition, to the underway fleet. Replenishment ships are also required to provide valuable aviation assets both to VERTREP personnel and stores to other ships, and to maintain the rotary wing assets of the ships own helicopters and those of the larger fleet. Fuels and lubricants, including ships own fuels, can vary from about 9,000dwt to 25,000dwt – a typical mix of liquid cargo for the larger vessel might be 15,000cu.m of Marine Gas Oil, 7,000cu.m of aviation fuel 1,000cu.m of fresh water and 300cu.m of lub oil. Often added to this is a dry cargo and provisions requirement of about 5,000cu.m and maybe some container stowage. At sea delivery of stores and liquids is achieved by Replenishment-At-Sea/Fuelling-At-Sea RAS/FAS equipment.

The aviation capability will require around 55m of ship length for a large helicopter single-spot flight deck and a hangar capable of a (second line) maintenance and helo stowage. This is shown in figure 02. A twin-spot in-line medium-size helicopter and side-by-side hangar will fit within a similar length.

In order to keep up with a task group the ship will need to be able to maintain task group speed which is 18 knots, and would be more effective and able to move around the group if it has a speed of 20 knots. For some navies a speed in excess of 27 knots would be required. Add the aviation team to a ship's crew capable of operating around the clock and delivering fuel and stores at sea to two ships simultaneously and the ship accommodation requirement will be around 90 to 110 berths. The ship hull needs to be efficient as well as dynamically stable. The directional stability is required to enable the ship to take position some 30m from the ship it is replenishing and maintain that course.

The operational profile of a typical replenishment ship is also rather different from that of a commercially trading products tanker **[2]**. The replenishment ship spends considerably more time at slow to medium ship speeds than the A-to-B commercial vessel. This requires a much more flexible propulsion system capable of sustained operation at speeds other than the single principal design point of a commercial vessel.

In terms of carrying fuel oil a critical characteristic, and the single most important requirement driving much replenishment ship acquisition activity currently, is MARPOL regulations requiring ships carrying fuel oil to have double-hulls. The International Maritime Organisation's Maritime Pollution (MARPOL) convention covers prevention of pollution of the marine environment by ships but provides an exemption for naval vessels. However given that Replenishment vessels operate away from home waters, the potential environmental impact, and the potentially limited access that could apply to an 'exempted' vessel is leading many navies to speed up acquisition of compliant vessels **[3]**. Given that MARPOL was introduced in 1973 and updated in 1992 plenty of time existed for the fleet to be updated by the required date of 2015 (Cat 2 and 3 tankers). However in 2003 IMO brought forward the cut-off for operation of Cat 1 tankers to 2005 (many Unrep ships would be subject to this) and Cat 2 and 3 tankers to 2010 – hence the accelerated acquisition timescales facing many navies.

But double hulls are only part of the fuel oil issue. Naval ships run on F76, a distillate fuel similar to the commercially available Marine Gas Oil (MGO). Unlike MGO, F76 has often been stored for long periods of time before use on the naval ship. Contamination with particulates, water and microbial growth means that 'dirty' fuel will be transferred unless significant and continuous cleaning is undertaken underway by the delivering ship: the frigate Captain will be distinctly unhappy to receive dirty fuel which will then overwhelm his own fuel polishing system **[4]**.

Delivering fuel and stores at sea is limited by sea conditions. A dominant factor is the sea state under which operations can be undertaken; ship length is important as pitch is often a limiting factor **[5]**. Delivering fuel in SS7 and stores in SS6 will require a ship of at least 200m whereas ships of 170m or 150m will be at a Sea State or two disadvantage due to ship motion.

Features most associated with naval vessels such as shock tolerance are not generally demanded but the replenishment ship's radiated noise could well become an issue during the replenishment activity. If operating within a radius of the operational fleet some minimum CBRN and a two compartment damage stability criteria is the minimum most countries would accept.

Selecting the right type of donor ship to be converted therefore needs careful consideration. A commercial products tanker provides a maximum speed of 15knots, no aviation capability, no Replenishment-at-Sea capability and very limited dry stores. The crew size on a vessel would normally be in the mid-twenties. A typical products tanker is illustrated in Figure 03.

Figure 03 - Typical products tanker – showing the limited space available for aviation arrangements and the minimal size superstructure accommodation block

In order to provide the additional capabilities necessary for an effective replenishment ship a significant amount of work would need to be undertaken unless compromises are accepted on operational capability. A flight deck and hangar would need to be added. Putting the flight deck in front of the bridge due to the aft mounted accommodation structure provides a very poor flight deck position especially as additional accommodation would need to be added and the RAS/FAS rigs will prove a flying hazard. Landing on top of tanks full of fuel is unlikely to pass a safety case and hence flying may not even be allowed. Adding the flight deck aft will require a significant cantilevered structure over the transom and would still not allow anything more than a flying spot with poor access for VERTREP stores.

The replenishment fuelling rigs will need to be added, together with their additional winches, and generator capacity, about 1.25MW per rig is required. Accommodation will also need to be added for a further 30 or more people, and a further 30 if an aviation complement is included. Then there are lifeboats, larger galley and stores, communal spaces, messes, dining rooms, garbage disposal, waste disposal, air-conditioning systems and sewage treatment. Adding additional bulkheads to comply with two-compartment damage stability is unlikely to be worth considering. Similarly resiliently mounting machinery to reduce noise signature is unlikely to be undertaken.

A tanker hull-form is bluff bowed and has a very high block coefficient to enable it to carry plenty of cargo in its length at a relatively low speed. This however means that inflows into the propellers and rudders is poor, efficiency is low and noise relatively high. Most commercial tankers have poor directional stability and tend to course-weave and hence will not provide a very effective ship to provide stores to another ship in close proximity. Even a substantially modified tanker will not provide the speed performance or capabilities required for an underway replenishment sip unless operational requirements are significantly compromised.

The Royal Australian Navy ship *HMAS Sirius* provides a recent and very successful example of such a project. **[3]**. An underway replenishment capability was achieved by adapting the double-hulled commercial products tanker, *MV Delos*. The Commonwealth acquired *Delos* in June of 2004 for $52m (US$35.7m) and time chartered until August 2005 for $8.22m. After undergoing modifications ($63m) for underway replenishment, improved accommodation and sewage treatment plant upgrade, command and control facilities, the installation of a solids cargo container deck, and having a flight deck fitted, she was renamed *HMAS Sirius* and delivered in September 2006. *Sirius* carries around 30,000cu.m of fuel and about 5,000cu.m of aviation fuel. She has accommodation for 60, can replenish ships at sea by day and night, is capable of replenishing two ships simultaneously, and has transfer points for fuel, water and stores. Sirius achieves speeds of 15 knots with a single directly-driven shaft, powered by a hard-mounted 6 cylinder slow-speed engine.

Although considered a success in terms of a project which is underbudget and ahead of timescale,it should be noted that the original Commonwealth Defence requirement **[6]** identified that the current vessel (*Westralia*) because of its maximum speed of 16 knots was not able to fully integrate with any task group that required flexibility to operate at higher speeds. It also had limited facilities and the inability to land on an aft flight deck meaning a reduced aviation capability that degraded afloat support effectiveness. The capability guidance also stated that ships within the maritime operations support capability should have a maximum speed in excess of 18 knots **[10]**. Environmental expediency and a rapidly approaching MARPOL deadline modified the operational requirements and has delivered an environmentally sustainable ship with rather similar characteristics to the one it replaced.

3. NAVAL AUXILIARIES USING THE COMMERCIAL SUPPLY CHAIN

The Authors' company has significant experience with the commercial supply chain having delivered designs from which over 750 ships have been constructed including complex offshore vessels, products tankers, Ro-Ros, Ro-Pax and containership vessels from shipyards world-wide.

A solution to the MARPOL/Military Capability dilemma is to utilise an existing commercial design capability to design a replenishment ship based wholly on commercial Class and Flag rules but having in mind the naval capability requirement. The design is based on the most appropriate hull form from the current portfolio. For a 20knot plus design this will most likely be either a Ro-Ro, Ro-Pax or Container Ship design. Figure 04 illustrates an Unrep ship based on a bespoke containership design.

Figure 04 – Replenishment ship design based on bespoke Containership

This hull form is then adapted at the design phase to include the liquids and solids storage, the aviation facilities - including flight deck and hangar, the replenishment system, and the additional accommodation and support infrastructure. The ship design draws on the experience of a range of ship designs including products tankers and Ro-Pax.

Keeping wholly to a commercial specification, Class Commercial Rules, and Flag regulations and building and delivering the ship, in a manner that the Commercial Shipbuilding supply chain is comfortable with, can generate a significant capability at an acceptable price. The key advantage of these ships is that they are purpose-designed for replenishment-at-Sea with dedicated stores handling areas and equipment, aviation facilities to provide fleet helicopter support and maintenance, and the ability to provide solids or liquid stores quickly and efficiently. This approach enables a navy to adopt a commercial ship design with a basic level of replenishment capability but incorporates spaces where additional capability can be added; margins are set for the full capability vessel. The adoption of commercial standards is not a hindrance for a ship that is designed for robustness with redundancy in power and propulsion to provide a survivable platform. Survivability can be quickly and cost-effectively made under incremental upgrades planned in at the design stage.

4. THE FAMILY OF ROLLS-ROYCE UNREP SHIPS

	14,000 dwt - 20 knot	14,000 dwt - 27 knot	25,000 dwt - 20 knot
Main Dimensions			
Length Over All	173.20 m	188.20 m	223.40 m
Length Between P.P.	161.70 m	176.70 m	210.00 m
Breadth mld.	26.60 m	26.60 m	28.80 m
Depth 3rd.deck	13.30 m	13.30 m	13.30 m
Draught Summer	9.00 m	9.00 m	9.00 m
Scantling Draught	9.30 m	9.30 m	9.50 m
Performance			
Service Speed	20 knots	27 knots	20 knots
Deadweight	14000 tonnes	14000 tonnes	25,000 tonnes
Ship Capacity			
Marine Diesel Oil	2300 m^3	3900 cu.m	3800 cu.m
Fresh water	1000 m^3	1000 cu.m	1000 cu.m
Water ballast	7000 m^3	7000 cu.m	11400 cu.m
Cargo Capacity			
Marine Diesel Oil	8000 m^3	8000 m^3	15000 m^3
Fresh water	1000 m^3	1000 m^3	1000 m^3
Avcat	2000 m^3	2000 m^3	7000 m^3
Lub oil	1000 m^3	1000 m^3	1000 m^3
Dry Cargo	4100 m^3	4100 m^3	4100 m^3
Provisions	1000 m3	1000 m3	1000 m3
Containers	17 FEU or 34 TEU	17 FEU or 34 TEU	21 FEU or 42 TEU
Class. Notation	+1A1 , DG-P , DEICE , OPP-F , HELDK-SHF , F-AMC , LCS-DC , E0 , RP , Clean Design		

Table 1 - UnRep ship family characteristics

The authors' company has recently completed a number of UnRep ship designs of varying deadweight capacity and for ship speeds from 18 knots up to 27knots, depending upon customers' requirements. The designs utilise commercially developed, fully proven and efficient hullforms from successful Rolls-Royce Ro-Ro merchant ship designs. The ships have been designed to provide minimum initial acquisition costs with an incremental upgrade through the service life of the vessel. These upgrade paths allow for tailoring of the solution to any customer's specific requirements. The addition of military capability such as aviation facilities, self-protection systems, additional dual capacity replenishment rigs and the potential for future upgrade has been designed into the ship to reduce the risk of future platform obsolescence.

The two variants of the 14,000dwt surface combatant support ships and the 25,000dwt carrier support ship have the characteristics described in Table 1.

A detailed structural definition, seakeeping analysis, systems selection and general arrangement drawings have all been completed, as has an economic analysis on a range of propulsion systems.

Figure 05 - MV Olympic Champion

The Ro-Ro MV Olympic Champion (Figure 05) is designed by Rolls-Royce. It is a 204m long, 27.5-knot (service speed) ferry, two of which are currently operating in the Mediterranean. This hull-form is well proven in service and has been extensively tank tested. It is the base for the family of Underway Replenishment ships described here.

The hull-form resistance for these options was estimated using MARINTEK EmPower, which uses a database of similar geometries and resistance tests to provide a derived resistance curve with high confidence. The hull-form is very efficient when compared to others in the class **[5]**. This being mainly due to the constant evolution of merchant ship designs due to strong competitive pressures and rapid ship class builds. The structural design of the ship follows commercial practices and includes a double-hull to meet IMO MARPOL regulations. The structure has been designed to comply with major IACS Class rules.

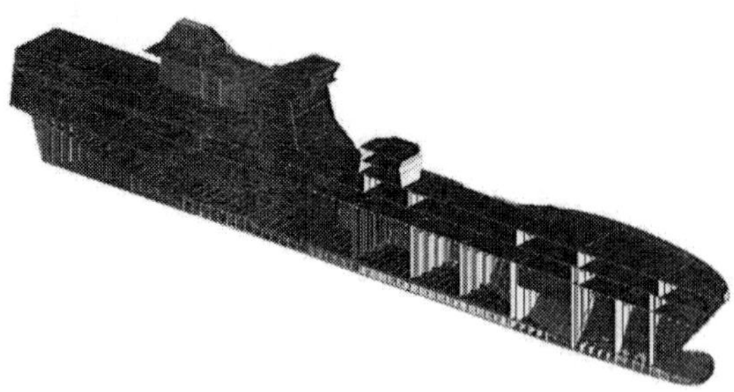

Figure 06 - Underway Replenishment Ship NAPA Global Steel Model

A NAPA steel model illustrated in Figure 06 was constructed and analysed to provide confidence that the steel weight was correct. Accurate steelweight estimates significantly contribute to the subsequent accuracy of stability and powering predictions.

5. GENERAL ARRANGEMENTS

The illustrations and general arrangement drawing included in the paper refer predominantly to the 25,000dwt vessel but the ships are from a family and hence share many common features, capacity of F76 and of AVCAT being the principal differences. The 20 knot 25,000dwt is shown in 3-D rendered form in figures 07.

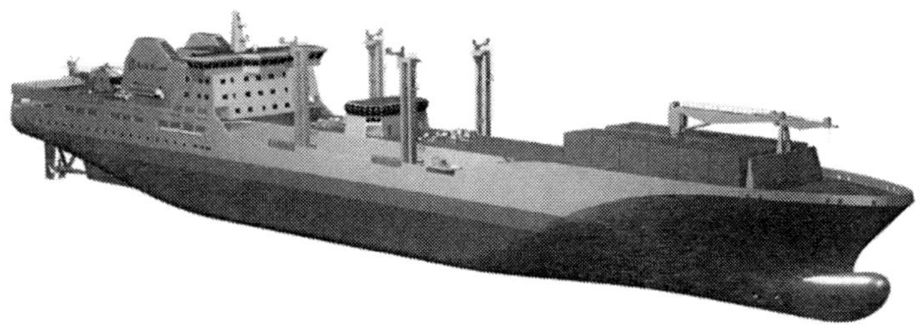

Figure 07 - Rendered 3-D model – 25,000dwt Vessel

The tank deck is enclosed to reduce the effects of corrosion on the pipe systems and hence reduce the maintenance burden. Atmospheric ventilation of this space will be controlled and monitored. The flightdeck is designed to withstand the crash-on-deck loads and normal wheel loads of the Merlin helicopter. The flightdeck is illustrated in Figure 08

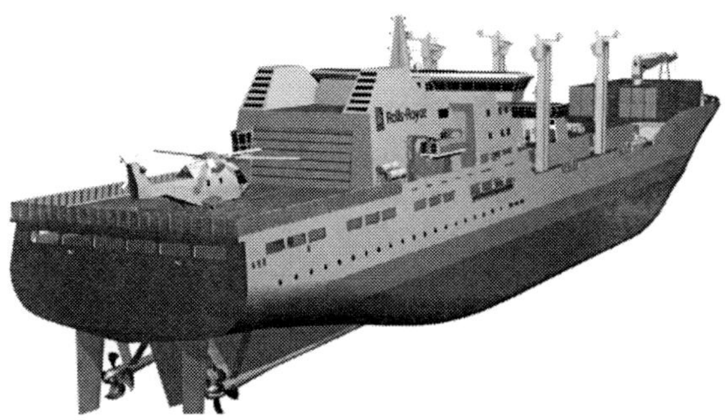

Figure 08 - After view of 14,000dwt Ship showing the helicopter and flight deck

The upper deck is dominated by the Replenishment-At-Sea and Fuelling-At-Sea (RAS/FAS) systems.

The whole arrangement of the ship aims to provide the most effective reception, movement, treatment, storage and delivery of fuels, liquids and dry stores. The ship also includes a substantial aviation facility to enable operation and maintenance of the ship's helicopter, as well as the capability to undertake maintenance of fleet aviation assets for ships who do not have the necessary space to do this maintenance themselves.

The clearway enables rapid access of dry stores to the dual-purpose rigs as well as the between deck storerooms. Figure 09 shows the ship's profile and upper deck arrangements. To comply with maritime legislation all accommodation is aft of the bulkhead at the front end of the superstructure. In front of this bulkhead are all the cargo areas, especially those for the fuel. The accommodation is designed to support a combined navy and civilian complement. The naval cabins are located around the inside of the aft accommodation area and the civilian crew cabins are to merchant accommodation standards, with natural light around the outside of the area. All cabins are en-suite in this example. Within the aft area are the frozen, cold and dry provisions storerooms with access to the stores deck and flightdeck via a lift. The lift will be cleared for carriage of munitions from the below waterline magazine forward, to either the stores deck or for loading onto the helicopter. These are illustrated in Figure 10.

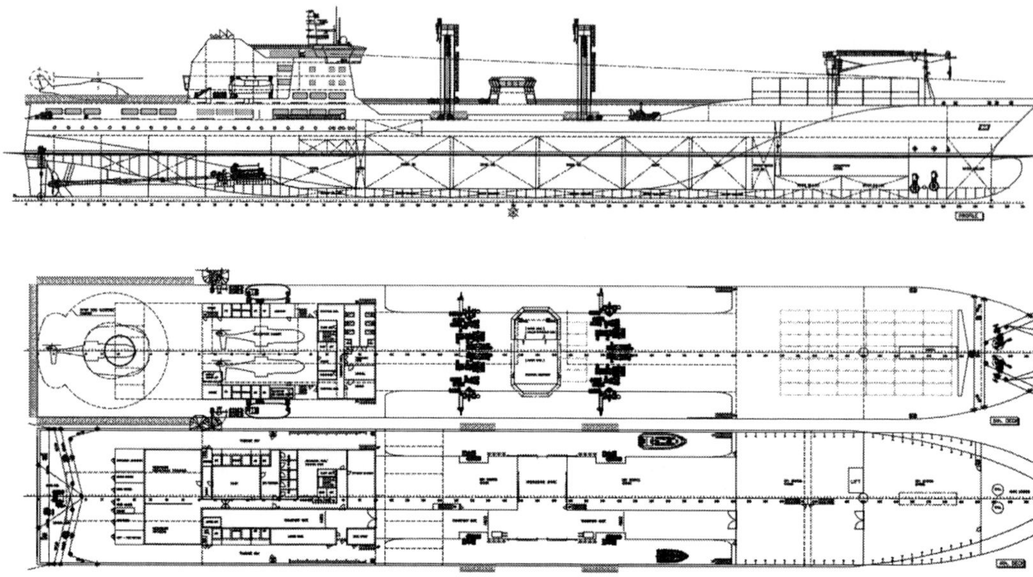

Figure 09 - GA of upper deck arrangements

The citadel for the ship covers the volume aft of the forward bulkhead of the superstructure. There are two cleansing stations, one servicing the foredeck and the other the quarterdeck and flightdeck. Both are however accessible from all parts of the upperdeck to ensure there is no wasted decontamination effort. The ship is designed For-But-Not-With (FBNW) a pre-wet system for full Chemical Biological Radiological and Nuclear (CBRN) capability.

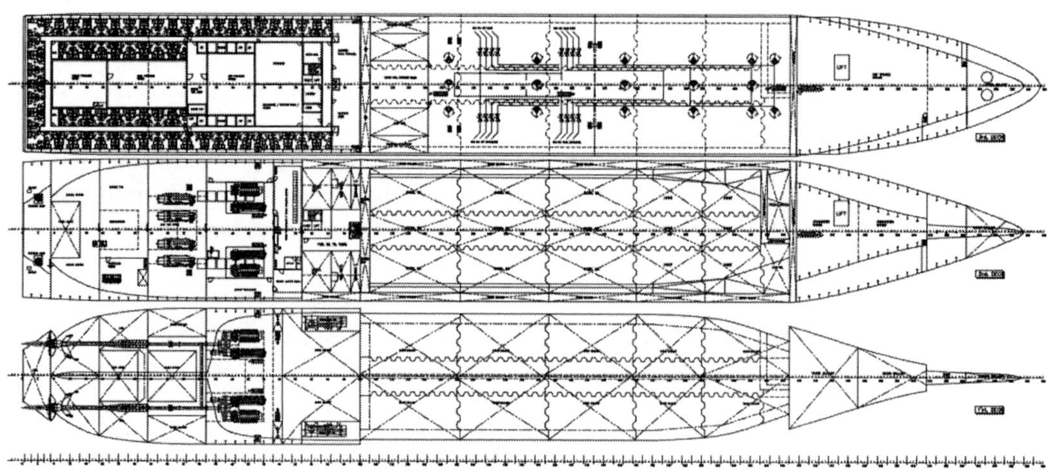

Figure 10 - Accommodation and tank arrangements

The aviation facilities for the ship include a flightdeck that provides space for large helicopters such as a Chinook or EH101/Merlin. The hanger has been designed to accommodate two folded Merlin helicopters and to facilitate maintenance, including main rotorhead and gearbox removal and replacement. There is space around the hanger to embark a modularised support system for the helicopters as well as workshop, stores, office and planning space for the ship's embarked flight. The flightdeck is fitted for a fixed fire-fighting foam installation as well as the necessary systems for night vision goggle and reduced visibility operations.

The fuel tanks have nitrogen inert gas systems to minimise fire risks as well as comprehensive fuel treatment systems to enable fuel to be cleaned prior to delivery to the receiving vessels. The tanks are also split athwartships into three sections with the centre tanks receiving the cleaned fuel before it is discharged to the receiving ship; this ensures that the stability is retained. A full damaged stability analysis has been conducted including asymmetric damage cases to ensure that IMO and UK MOD stability standards have been met.

The cargo will be delivered from the tanks using commercial high-reliability deepwell pumps, running at a rate of up to 600m^3 per hour per tank. This will enable fuel to be pumped via twin hoses on two rigs at up to 2400m^3 per hour, providing rapid fuelling of large units such as Carriers and Assault ships. Alternatively, one rig per side can be used for fuel and the other for dry stores. At least one stern reel can be fitted to enable over the stern fuelling-at-sea operations to other ships, including those without RAS receiving moveable highpoints, such as patrol boats and corvettes.

6. CONCLUSIONS

The Rolls-Royce family of Underway Replenishment ships maintains the commercial basis of the ship designs as far as possible including maximising the producability using flat panels and corrugated bulkheads wherever possible. The addition of military features and standards only where necessary supports the design objective of keeping the costs as low as possible whilst maximising the mission capability. The result is an optimal balance between commercial and military technology to provide an affordable yet capable solution.

7. REFERENCES

1 MARPOL 73/78 Reg 13F Annex 1, 2001 amendment and 2003 amendment 13G

2 Kimber, Andy; Vik, AM; Future Naval Tankers – Bridging the environmental gap – the cost effective solution, WMTC 2006

3. The Auditor-General Audit Report No. 20 2006-07 Performance Audit – Purchase, Chartering and Modification of the New Fleet Oiler, Department of Defence, Defence Materiel Organisation. Commonwealth of Australia ISSN 1036-7632, ISBN 0 642 80939

4. NES 320 Requirements for the Design and Installation of Fuel Systems in HM ships.

5. Bricknell, David; Skarda, Robert; Vedlog, Per-Egil, A Family of Underway Replenishment Ships, 'WMTC' March 2004

6. Defence Media Release: HMAS Sirius – New Fleet Oiler Delivered under budget and ahead of schedule Tuesday 30 January 2007

COASTAL PROTECTION VESSELS – RECENT DEVELOPMENTS

David J Bricknell
Director - Systems, Rolls-Royce Naval, UK
Cato Fjeldstad
Sales Manager, Rolls-Royce AS, Norway

SUMMARY

This paper describes the development of a new class of Coastal Protection Vessels derived from a well-proven range of offshore standby vessels. These ships are flexible multi-role ships capable of providing emergency towing, firefighting and pollution control as well as general law enforcement, task-force co-ordination and survivor facilities. Sustained speed and long endurance in rough weather combined with a low crew-size, aviation facilities and fast rescue boats all contribute to a flexible government response capability complementary to the current classes of Customs Cutters, Research Vessels and Offshore Patrol Vessels. A range of Coastal Protection Vessels have now been delivered to a number of customers worldwide with more ships currently under construction. This paper provides an update to that given at Pacific 2006.

1. INTRODUCTION

Until 1967 a Nation's jurisdiction over its adjacent sea effectively was limited to 3 miles, generally, although somewhat ambitiously, considered to have been derived from a 19th Century view of the range of a canon ball. In 1967 the 3-mile limit was replaced by a 12-mile limit (measured from the low waterline) and the water lying within this limit is considered the sovereign territory of the controlling nation who then has rights over national security and customs affairs in the same manner as on land. Foreign ships however have 'rights of innocent passage' provided they are flying their nation's flags and are 'exposed' i.e. no submarines.

The Exclusive Economic Zone, or EEZ, is an area beyond and adjacent to the territorial sea under in which the rights and jurisdiction of the coastal state and the rights and freedoms of other states are governed. The EEZ is regulated by the 1982 United Nations Convention on the Law of the Sea **[1]** that came into effect in 1994. Part V of this convention describes the rights of the coastal state in the EEZ as sovereign rights for exploring and exploiting, and conserving and managing the natural resource of the waters superjacent to the seabed and of the seabed and its subsoil. Jurisdiction is provided for the establishment and use of artificial islands, installations and structures; marine scientific research; and for the protection and preservation of the marine environment. The EEZ extends up to 200 nautical miles from the low water line.

Part VI of the UN Convention concerns the Continental Shelf where it extends beyond the distance of the EEZ i.e. beyond 200 miles. In 2007 Ireland became the first nation to receive from the 'UN Commission on the Limits of the Continental Shelf' confirmation that it could extend its jurisdiction to its South West beyond the 200 miles of its EEZ **[2]**. The rules and requirements to be met by Part VI are complex but once a state achieves agreement from the UN Commission it has the rights to exploit its natural resources located on or under the seabed, including oil, gas and minerals. Nations have the possibility of claiming continental shelf up to 350miles from their low water mark. In Ireland's case the additional territory amounts to some 39,000 square kilometres. In Russia's claim, not yet agreed, it could amount to more than 1.2 million sq. km based on the arctic sea floor being a continuation of the Siberian continental shelf [3].

The 12-mile 'sovereign territory' and the 200-mile EEZ have led governments to respond by procuring assets capable of performing customs, coastguard and general policing and constabulary activities. Nations carry out their responsibilities through a wide and often diverse range of organisations including fisheries protection services, customs services and navies using a wide range of assets, both air and sea based to do it. With an eye on the current global terrorist threat, the security of coastal populations, and marine pollution, some nations are restricting access to their EEZs and in many cases this is approved by IMO or other regional or global organisations **[4]**. In order to successfully implement this many nations are consolidating their various authorities under one co-ordinating command or integrating them into a single Authority. The Australian Joint Offshore Protection Command, established in 2005, is an example.

For the sea-based assets, four vessel types are used mostly. These are:

- Customs Cutter – this vessel has to implement a nation's sovereign territory responsibilities within the 12-mile limit - they tend to be relatively small 10-40m with limited range and no organic aviation.

- Research/surveillance vessel - these vessels conduct the marine scientific research defined by the convention – they are larger than customs vessels (generally over 40m) are quiet in operation with good seakeeping and long endurance for patrol in the 200nm zone. They are not required to intervene where disputes occur or laws are being broken and hence are unarmed and because they have no pursuit role they have a relatively low maximum speed and again no organic aviation.

- Offshore Patrol Vessel (OPV) – these vessels are normally operated by the country's navy and because it delivers law enforcement (of the Convention's rights) within the EEZ the vessels tend to be (lightly) armed and of a warlike appearance. OPVs will tend to be larger than research vessels due to the requirement to land a helicopter (and sometime to provide organic aviation). The Flight deck length can be a significant ship size driver and ship motion is generally reduced with a longer ships. OPVs are also required to pursue ships that break the law within the EEZ and as such tend to be of a higher speed than the types of ships that are likely to be transgressing the law. In the case of the EEZ this is usually fishing vessels. Sustained speeds of 15 or 16 knots in high-sea states with sprint speeds of 20 or 21 knots are generally considered sufficient. Damage to the OPV can be expected particularly during boarding or in worst cases being rammed and so these vessels are designed to meet naval stability standards and are designed to be more robust than a naval combatant.

- Ocean Capable Patrol Craft (OCPC) – with security concerns having increased significantly recently and the rights to engage in naval activities in other countries EEZs remaining a controversial topic: different interpretations of the use of weapons, surveillance and hydrographic activities are held by different countries. This had led to a new class of Patrol Craft more closely resembling a naval corvette with significantly greater armament and higher pursuit speeds.

Patrol, surveillance and policing of the continental shelf where it extends beyond the 200 mile zone is currently an unknown although it is a candidate for the OCPC. However significant interest is being shown in the arctic area by all the countries bordering it and this looks to be leading to a new class of 'Ice-Capable Patrol Craft'.

None of the above ship types however adequately covers a subject that is increasingly of concern to nation states in and around their coastline, territorial waters, EEZ and continental shelves: the containment or prevention of marine accidents and minimising the after effects of any incident. In order to implement these 'disaster prevention and recovery roles' a different class of vessel is required - one which includes the capabilities such as high bollard-pull for towing of stricken vessels, fire-fighting, and pollution control as well as disaster relief in the case of a significant number of casualties. Such vessels are, as a secondary capability, often capable of delivering other roles including general law-enforcement and fisheries patrol as well. This paper describes a range of Coastal Protection Vessels (CPV) developed to meet this growing requirement.

2. MARITIME ACCIDENTS

Marine accidents within, and outside of a country's EEZ have the potential to cause considerable damage to the marine life and to the coastal eco-system as well as to passengers and crew aboard any stricken vessel or vessels. Constant vigilance coupled with the ability to respond rapidly, ideally before such an emergency becomes a disaster, requires a new class of vessels designed specifically for the task and sufficiently multi-capable to be able to deal with a variety of potential disasters. Figure 01 shows the principal trade routes of marine traffic around the world and against this it is little surprise that accidents are concentrated off the coasts of particular countries.

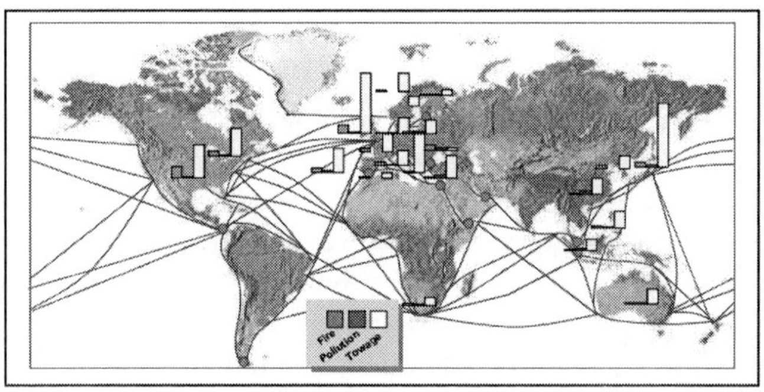

Figure 01 – Principal Trade Routes and Maritime Accidents by type shown by country or region. ***Data is IMO published data for 2001 to 2003 inclusive.***

Analysis of these accidents shows that most require emergency towage. Of course timely intervention by a suitably equipped towing vessel can prevent an emergency becoming a crisis but fire on the stricken vessel and any pollution from that vessel can have quite a disproportionate impact on the local marine ecosystem and any dependant industries. Both fire and pollution incidents identified in the figure leave little doubt that this is an issue that needs addressing.

2. CHARACTERISTICS OF A COASTAL PROTECTION VESSEL

The characteristics required of such emergency response vessels are those developed following recent marine accidents. These include: emergency towing capability (bollard pull); fire-fighting; pollution control – oil spill containment and skimming and processing capability; and survivor facilities for people recovered from the sea or from the stricken ship. Ship speed is important for timely response to the emergency but the required hull characteristics for a fast vessel are not always in line with the characteristics required for high bollard pull or for pollution control. Vessels speeds of 18 to about 20 knots are seen as achievable.

Emergency Towing

Most Offshore Patrol Vessels are specified as being required to tow a vessel of similar size in a moderate sea state. This requirement has little impact on the vessel's propeller characteristics required for maximum speed but equally will make little impact on a stricken oil tanker which is likely to be many times greater in displacement. High bollard pull, required by Coastal Protection Vessels, is the delivery of thrust at low speed and influences both the propeller diameter, any nozzle and the hull-lines aft in order to achieve good water inflow conditions. Figure 02 shows emergency towing of the stricken containership *Napoli* by a coordinated effort of UK and French CPVs.

Figure 02 – MV Napoli under tow by a UT515 supported by a UT719T off the coast of Devon, UK.

Fire-Fighting

A significant number of marine accidents are due to fire. In order to address fire at sea the Coastal Protection Vessel will need to deliver water onto the stricken vessel from alongside or from standing-off some distance. Figure 03 shows a UT 512 with fire monitors in action.

Figure 03 – UT512 deploying significant firefighting assets.

Three standards of Firefighting capabilities are defined by Class Societies These are:

'FiFi I' vessels are equipped with at least two monitors that, in total, have a discharge rate of 2400 m^3/hr. The monitors are able to throw water to a height of 45 metres and to a minimum of 120 metres distance.

'FiFi II' vessels are equipped with three or four monitors that, in total, are able to deliver 7200 m^3/hr a distance of 150m and to a height of 70m.

'FiFi III' vessels have three or four monitors capable of delivering 9600 m^3/hr a distance of 150m and to a height of 70m.

Pollution Control Operations

In the case of stricken vessels the oil spill source will be known but often oil spills will arise either before the vessel is known to be in danger or from some other cause and in these cases the detection of the pollution from airborne assets becomes important. Once oil is spilled, mechanical containment by booms, usually deployed by small boats, and mechanical recovery or skimming provides the means of containment and recovery. Effective coordination of the pollution incident means that any Coastal Protection Vessel will need to be equipped with extensive Command, Control and Communications capability.

3. COASTAL PROTECTION VESSELS

Coastal Protection Vessels have been delivered into a number of key geographical areas. These are owned and operated on a country-by-country basis. Ownership of the assets vary with some privately owned and operated on behalf of a coastguard agency and others owned and operated directly by coastguard agencies and even by navies. Despite the effects of marine accidents respecting no political boundaries there appears to be only limited signs of shared resources but this may come in the future.

Emergency Towing Vessels

The United Kingdom's Emergency Towing Vessel (ETV) scheme aims to protect the country from marine pollution arising from drifting disabled ships. *Anglian Princess and Anglian Prince,* operated by Klyne Tugs are UT719-T designs with a bollard pull of about 180 tonnes. Built at the Yantai Raffles yard in China, they are now on long-term charter to the UK's Maritime and Coastguard Agency. Anglican Princess is based in Dover and is shown in figure 04.

Fig 04 Anglian Princess UT719-T

The design is a development of the basic UT719 platform. To date, seven vessels have been built as multi-functional tug supply vessels, with a further three on order. Two more have been built under the UT719-R label as field safety standby vessels for the North Sea.

The Klyne Tugs UT719-T is optimised for towing and anchor handling but also has a substantial supply capacity and a FiFi 1 firefighting outfit. It is 67.4m long, has a beam of 15.5m and a design draught of 5.2m. Built to Lloyd's Register class +100A1+LMC, UMS, FiFi-1, it is powered by two medium speed engines providing 6,000kW. Maneuverability is high with two independent high lift rudders and two 600kW tunnel thrusters at the bow and a single 900kW unit at the stern.

Fig 05 - UT-738 'R' Class *Pacific Responder*

In Australia the government finalised in August 2006, a long term charter of Swire Pacific's UT738 'Pacific Responder' to be based in Cairns, to provide emergency towing services covering the Northern Great Barrier Reef and the Torres Straight region. It is 64.3m loa with an 80t bollard pull and has been modified for its new ETV role. The R class is shown in figure 05.

Pollution Control Vessels

Two powerful towing and pollution prevention vessels have been built at the Zamakona shipyard for the Spanish marine safety agency SASEMAR (see figure 06). These vessels are a derivative of the UT722L offshore design and have a main engine power of 16,000kW giving a bollard pull in excess of 220 tonnes. Several types of oil skimmer system are fitted, in conjunction with a very large capacity for recovered oil: 1,700m^3. In addition to a large towing winch there is a foredeck winch and fenders for escort work.

Fig 06 - UT-722L Emergency Towing and Pollution Control Vessel for Spain showing the oil boom extended.

Multi-Purpose Coastal Protection Vessels

Multi-Purpose Coastal Protection Vessels are designed to undertake a variety of coastguard and EEZ management roles including standby and rescue, firefighting, salvage and also general law enforcement and fishery control.

Four countries have recently acquired multi-purpose Coastal Protection Vessels. France has acquired the UT515, Norway the UT 512, Iceland the UT 512L, and India the UT517. These vessels share similar characteristics:

France. Type-numbered UT515, the 80m-long vessels will combine towing, salvage, coastguard and safety standby capability. An installed power of 16,000kW (4x4,000kW diesels) will provide a bollard pull of about 200 tonnes (tanker assistance up to approx 300,000 dwt) and a speed of 19.5 knots. The ship is designed to ICE C and includes Fi-Fi 2 with a 500m^3 oil recovery tank. Accommodation is provided for 25-35 with two fast rescue craft launched by heave compensated davits. The ship is shown in Figure 07.

Fig 07 – UT-515 Coastal Protection Vessel for France combined very high bollard pull with a high ship speed.

The Les Abeilles company in Groupe Bourbon operate these ships on a long-term charter to the French Navy. The ships were built at Myklebust Verft, one of the Kleven Maritime yards in Norway.

Norway. The UT512 design for Norway *K/V Harstad* is from the same family as the UT515 and has been acquired by Remøy Shipping for charter to the Norwegian Coastguard for a comparable coast protection role. The vessel was built at Aker Søvikness on the West coast of Norway.

K/V Harstad (see Figure 08) is designed and fitted for emergency towing of tankers up to about 200,000dwt and has a full outfit of oil-spill control equipment with a tank capacity for recovered oil of more than 1,000m³. It has twin 4,000kW diesels enabling a bollard pull of about 110tonnes and a ship speed of about 18.5 knots (19.5knots optional) to allow the vessel to reach the scene of an accident quickly. As its roles will include patrolling, the ship will have fast boarding/rescue boats, a gun on the foredeck and a comprehensive civil and military communications system. Like other Norwegian coastguard vessels the crew of 26 will be a mix of civilian and military personnel. The 83m long vessel is, on a long-term charter to Kystvakten.

Fig 08 – K/V Harstad UT512 recently accepted by the Norwegian Coastguard as a multi-purpose Coastguard Vessel. Ship's armament has not yet been fitted.

Operating along the full length of Norway's coastline and throughout the country's exclusive economic zone will involve much time spent in the Barents Sea, so ice 1B class has been specified, along with anti-icing measures such as heated shelters for the two MOB/boarding boats.

For towing and emergency work this particular UT512 design includes a towing winch, a reinforced pushbow, Fi-Fi 1 fire-fighting systems, a hospital and an extensive range of ancillary equipment including line-throwing gear, a harpoon system for attaching tow lines, oil spill booms and skimmers and a $1{,}000m^3$ of tankage for recovered oil.

KV Harstad has also been adapted to launch and recover the Nato Submarine Rescue System NSRS. Figure 09 shows the ship deploying the NSRS. The NSRS is a system for rescuing personnel from damaged and sunk submarines. The system is air transportable and can be mounted on a pre-prepared vessel like *Harstad* in 1 – 2 days.

Fig 09 - UT512 KV Harstad deploying the Nato Submarine Rescue Systems NSRS.

Iceland: Iceland have recently ordered a UT512L version to undertake tanker assistance (up to approximately 200,000dwt), towing/salvage, fire-fighting, oil recovery, and general law enforcement. The vessel at 93.65m is about 10m longer than the standard UT512. Twin 4500kW diesel engines deliver the 100t of bollard pull and a ship speed of 19.5knots. Oil recovery tanks are 640m^3 and two 1200 m^3 monitors provide the Fi-Fi 1 capability. An artist's impression is shown in figure 10.

Fig 10 - UT512L Icelandic Coastguard Protection Vessel

India: A predicted 100 per cent increase in oil demand in India over the next 20 years will radically increase tanker traffic in India's EEZ. The Indian Coast Guard is building three vessels at the ABG shipyard in Surat to counteract potential marine incidents. The UT-517s (figure 11) are 94m-long vessels designed to carry a medium-sized helicopter and be equipped with oil booms and skimmers, the recovered oil, recovery tankage of 500m^3, either being held in tanks on board or transferred to inflatable barges that can be towed astern.

Fig 11 - UT517 Indian Coastguard's Pollution Control Vessel

The ship is capable of a speed greater than 20 knots with power provided by an innovative hybrid electric propulsion system with two 3,000kW diesels and twin 2,300kW electric motors; this system being ideally suited to the very varied operating profile of such a ship. Apart from the pollution control role, the vessels will undertake many other EEZ related duties, including surveillance and law enforcement, anti-smuggling, fishery protection, SAR, data collection and assistance with salvage and fire fighting (two 1200m^3 monitors).

4. ICE-CAPABLE COASTAL PROTECTION VESSEL.

With changing climate and an increasing interest in the extended EEZ allowed by part VI of UNCLOS, northern navies have acquired or are showing considerable interest in ice-capable Patrol Craft. The Arctic Ocean ice is melting, the North-West passage will soon be navigable for substantial parts of the year, and the whole region is rich in minerals, oil and gas.

Fig 12 - *KV Svalbard*, Ice-capable Patrol Vessel

Russia is preparing a significant claim for the Arctic continental shelf, Norway has built and is operating the *Svalbard* ice-capable patrol vessel, Greenland (Denmark) has an ice-capable patrol vessel under construction, and Canada has an urgent requirement for six to eight ice-capable Arctic Patrol Vessels.

Norway: *KV Svalbard*, (see figure 12) is an ice-capable Patrol Vessel constructed at the Langsten yard in Norway. The vessel is about 6500t and 104m long and Polar 10 ice notation (capable of operating in multi-year ice-floes where accidental ramming is expected). Propulsion is by diesel podded electric with an installed generator capacity of 14MW delivering 17.5knots and a bollard pull sufficient to tow a 100,000dwt vessel. Crew size is 48.

Greenland: is acquiring a 71m ice-capable Fisheries Inspection Vessel (figure 13). The ship is capable of operating in areas of 0.5m unbroken ice and is classed to ICE-1A with ICE-1A* for hull structure. Crew accommodation is provided for 45 persons in a gas tight citadel.

Fig 13 - NVC-810 Greenland Fisheries InspectionVessel

5. COASTAL PROTECTION VESSEL WITH SURVIVOR CAPABILITY

With the memory of recent coastal disasters in mind a multi-capable, multi-purpose Coastal Protection Vessel has been developed from the latest North Sea Multi-Field Standby Vessel the UT527 (see figure 14).

Fig 14 - UT527 Multi-field Standby Vessel on trials

In its Commercial role the vessel has been designed specifically for oil recovery, rescue, towing and firefighting but what differentiates the vessel from other CPVs is the extensive survivor facilities on-board and the time the ship can remain available at sea. The ship has been designed to enable it to remain at sea for three years without return: it changes its crew at sea, and it re-stores and replenishes at sea. To enable such an extensive operational period the ships systems, including its power and propulsion system, are designed both to be low maintenance but also to be easily maintainable at sea as well as graceful in its degradation should unplanned failures occur.

Although primarily a multi-field standby vessel, it can be readily adapted for a wide range of Coastal Protection duties. Built to Det Norsk Veritas class +1A1, EO, FIFI I/II it is capable of over 20 knots and can launch and berth its own rescue/work boat, used for towing oil booms in the stern. Oil recovery systems meet DnV oil recovery and NOFO specifications. Figure 15 shows the UT527 in firefighting mode.

Fig 15 - UT527 deploying her fire-fighting capability

In its Coastal Protection role the vessel's strengths are in fire fighting, pollution control, emergency towing (up to 130t bollard pull) and survivor recovery (up to 320 survivors) as well as implementing essential Command and Control of situations. General law enforcement, and fisheries patrol are additional capabilities. Long endurance or time-on-station (20,000nm at 16 knots), a maximum speed of greater than 20 knots, the ability to maintain station in high-sea-states, a stern well for launching fast boats, a medium size helicopter and space for survivors or embarked military forces identifies this ship as potential for both Disaster Relief activities or for Maritime Counter Terrorism activities. The UT527 as a CPV is illustrated in figure 16.

Fig 16 UT527 Multi-role, multi-capable Coastal Protection Vessel

5. ACKNOWLEDGEMENTS

The authors would like to acknowledge the innovative design work undertaken by the UT-design team at Rolls-Royce AS in developing this new class of ship.

6. REFERENCES

1. United Nations Convention on the Law at Sea, Third, 1982
2. Ireland can extend territorial waters, Irish Times, 07/04/2007
3. The battle for the next energy frontier; Yenikeyeff S M, and Krysiek, T F – Oxford Energy Comment August 2007
4. The disappearing right to navigational freedom in the exclusive economic zone; van Dyke, Jon M. 2005 Elsevier

Law and an interdisciplinary approach to maritime affairs

The Honourable Justice James Allsop
Federal Court of Australia
Justice.Allsop@fedcourt.gov.au

May I thank Professor Malek Pourzanjani for graciously agreeing to deliver this paper for me. I hope the views which I express (which are my personal views) reflect his too. But that is for him to say.

Captain Michael Bozier asked me to write a paper for this conference.

Why would he do that? This is a conference, the civilian side of which is concerned with topics such as naval architecture, shipbuilding, navigation and other technical aspects of the shipping industry. Of what possible interest is a lawyer, let alone a judge, to participants here?

If these rhetorical questions strike some chord, perhaps reinforced by the memory of some collision with the legal system in the past, I would like to try and persuade you that the present **relative** lack of interdisciplinary communication between lawyers and the shipping industry is not healthy for the maritime community. I should begin by saying that there are some lawyers with a deep and effective relationship with the maritime community. My thesis is that there should be more; and that the practical and theoretical communities should commune more, and enjoy doing so.

Michael Bozier and I first met some years ago in a case about rusted steel coils that had been carried from Kaohsiung in Taiwan to Sydney on a ship called *Arawa Bay.* He patiently and lucidly explained to me the workings of the McGregor hatch system and various other important aspects of the care and management of such a tramp ship in the carriage of that cargo.

We met again in later maritime legal adventures.

I was a barrister who attended to maritime briefs, along with other briefs concerning a reasonably wide range of commercial activity. A small selection of the topics covered over the years are a patent case about conveyor belts, a patent extension case about calcium antagonist anti-anginal drugs, cases about trademarks, insurance disputes about the operation of the primary insurance and reinsurance markets, cases about mine accidents, about roof bolts in coal mines, and countless other subjects.

None of this is either remarkable or special. Lawyers, including barristers, spend their lives delving into minute detail about narrowly focused subjects in order to understand the affairs of clients and to examine and cross-examine witnesses. Over time, they develop an eclectic, indeed ludicrously random, treasure chest of knowledge about countless topics.

For barristers and solicitors who attend, either from time to time, or more regularly, to the affairs of shipping clients, there is a constant struggle to piece together loose threads and blobs of knowledge to place the client's affairs in context and give useful advice.

When I went to the Federal Court bench in 2001, I joined a group of colleagues who attended to the maritime work of the Court. The New South Wales Registry had a "panel" of judges (of about 9) who hear maritime cases. Some had experience, as many Australians do, of sailing. With some particular and notable exceptions, few had any rigorous or coherent knowledge or understanding of the maritime or shipping industry. All were fine lawyers, deeply experienced in commercial law, industrial law and common law. But only a few had experience in shipping cases.

I spoke to Michael Bozier. He enthusiastically agreed with a suggestion that we begin a programme of education of the judges. My late colleague, Justice Richard Cooper from Queensland, was an enthusiastic proponent of the same task.

What did we do? First, we sat the judges down and Michael Bozier gave two superbly prepared day-long seminars accompanied by detailed notes. They were hugely well received. Judges began to appreciate how much of the activity of shipping and maritime affairs of which they were ignorant. (This is hardly surprising – Australia, after all, as our national anthem should say, is a "*land girt by beach*".)

At this point a fundamental question should be asked: Why should such structural knowledge about the shipping industry be important to lawyers and judges?

The answer is that an appreciation of such matters is important to deal efficiently and knowledgeably with the affairs of his or her client, or with the disputes brought by litigants before him or her. It is a matter not only of despatch and expedition, but also accuracy, reliability and confidence. There are at least two elements to this: cost and the quality of legal thinking.

First, there is cost. To explain this kind of background to a judge, or an advising lawyer, in the context of a law case or in obtaining advice, takes time. Time and expert assistance costs money.

Secondly, and more importantly, the development of rules and legal principle is not some detached ethereal process. Rules grow out of honest common sense in the practical affairs of commercial men and women. One cannot, as an adviser or rule maker, begin to appreciate what ought to be the rule until one understands what is the practical reality of how business and affairs are conducted. This is especially so in respect of a body of law (the law maritime) which has its origins in the wisdom of the civil law, the customs and usages of the sea and nations who use the sea, and the commerce and intercourse of nations. Whilst each nation's maritime law can be said to be domestic or municipal, the roots of all maritime law are necessarily international, growing from common usages of international commerce and international activity. There is a deep, elemental practicality to maritime affairs, which tends to demand a similar practicality and fairness in the law. To ascertain, elucidate and explain that law requires a familiarity with the activity itself.

This education of the judges who were interested in maritime work continued in 2003 and 2004 with lectures on salvage, collision and the shipping industry from other scholars, former mariners, and participants in the industry such as Professor Edgar Gold, Captain Ken Ross and Dr Sarah Derrington.

In 2005, the process began to take a more formal structure. In that year, pursuant to a decision of all the judges of the Court, the Chief Justice nominated thirteen judges,

Australia wide, to attend to maritime disputes in the Court. The organisation was such that there would be judges in each city: two in Brisbane, four in Sydney, three in Melbourne, two in Adelaide and two in Perth, to attend to Admiralty and maritime cases. All these are port cities. All have arrest work, involving, from time to time, the onerous responsibility of the custody of valuable, working merchant ships. All have a volume of *in personam* maritime disputes, principally cargo claims, but from time to time other matters such as charterparty, salvage and collision matters and other like disputes. These judges deal with the matters arising in their own Registries at first instance, and comprise the pool of judges who deal with maritime appeals across the country.

This structure was chosen to give the qualities of both specialisation and depth, by having a sufficient number of judges on the Court to ensure that the Court would never be "caught short" in the provision of suitable judges for maritime cases. In years gone by, another system had prevailed, which is one more frequently used in courts where there is a far more significant volume of maritime work: that is to have a single Admiralty judge who would do all first instance work. The difficulties created by this in a country with numerous ports, a wide geographic spread of work and the need for a knowledgeable appeal court are clear. Also, when such a judge leaves the Court there can be a dearth of knowledge and experience.

The Court has built on Michael Bozier's early work and carries on regular education days for its maritime judges drawing on scholars, shipping men and women and self education. The education programme in 2007 has seen lectures by the judges of the Court to each other by video-link and in person on various topics including a review of Australian Admiralty practice; an outline review of overseas Admiralty practice; the onus of proof in cargo claims and the relationship between articles 3 and 4 of the Hague-Visby Rules; marine insurance; the international and domestic regimes for oil pollution liability; charterparties; ship registration and the role of the flag; the role of the IMO; an outline of the UNCLOS; an outline of port state control; and the concept of a safe port. In 2008, a series of 8 lectures are planned on the legal regimes for the carriage of goods by sea.

The Court is seeking to develop a scholarly educational exchange with the Australian Maritime College ("AMC") and the Shanghai Maritime University and to build on its relations with the Maritime and Shipping Law Unit of the University of Queensland ("MASLU").

The developments have, in effect, led to the organisation within the Court of what can legitimately said to be an operating maritime court, supported and maintained by a programme of judicial education. This is underpinned by skilled Marshals and Registrars, who deal with administrative work in arrests and with mediations. Many of the Registrars have experience in maritime affairs and specialised training in maritime law. The arrangement is also underpinned by a lower court of federal judges in the Federal Magistrates Court, which has nominated specialised maritime judges.

During my consultations with Michael Bozier, we realised that there was no course of study in Australia specifically designed for people such as lawyers, brokers and others who wish to enter the maritime industry shore-side, to understand the basic elements of shipping, such as the nature and structure of ships, ship stability, forces on a ship, cargo handling, tug work, ship handling and propulsion. Lawyers who spend some time acting for their clients begin to learn about these things. This is often achieved, however, in a random way.

Such a course should be a recognised facet of the education of the non-seafaring maritime community.

If one doubts this need for familiarity with practice, turn the question around. Can anyone really contemplate a master of an ocean-going commercial vessel not having sufficient basic legal understanding of aspects of public international law, IMO instruments, laws governing security and safety, commercial law insofar as it touches the operation of bills of lading, insurance and charterparties, the rules international and relevant domestic about the crew's rights and the notions underlying powers of arrest in Admiralty jurisdiction? It is, of course, vital for ship masters, agents, brokers, insurers and others in the industry all to have a grounding in legal principle.

It is equally vital for maritime lawyers to have a coherent basis to understand ships and shipping. This is not for the purpose of becoming naval architects or brokers or shipmasters (any more than the master's knowledge is likely to equip him to be a practising shipping lawyer); but it is absolutely essential to understand the basic elements of the affairs being dealt with.

I am not intending to be critical of those in the profession, but to a degree this has been lost. For instance, today the Australian Transport Safety Bureau ("ATSB") conducts regular enquiries about events – groundings, collisions and the like. These inquiries are generally done without lawyers. The information obtained in such inquiries is withheld from the public, unless the ATSB decides to release it. It cannot be subpoenaed for legal proceedings. This privacy and confidentiality is for the express purpose of enhancing safety and enhancing systems to prevent occurrence and re-occurrence of accidents. The inquiry is not to find fault and allocate blame, but to find out what happened and how the system can be improved. Thus, people involved in such events are encouraged to be frank, without fear of retribution, rather than to be defensive and self-protective. This was a profound and deliberate policy. By and large, it has been a success. But it has also meant that generations of lawyers (7 years is a generation in the law) have ceased to be educated in shipping matters because of the absence of Boards of Marine Inquiry.

This may be no bad thing in terms of the policy sought to be developed by the ATSB. I am not suggesting a change of ATSB's policy. Rather, I only wish to highlight the end of a previously important source of education about shipping affairs.

A lack of detailed knowledge in lawyers of the conduct of maritime affairs will affect the Australian maritime community. How?

One example is skilled dispute resolution.

Disputes are part of commerce. They are an incident of the very activity itself.

Maritime dispute resolution has, especially in the last thirty to forty years, been conducted primarily in arbitration. There are exceptions to this. Much cargo claim work still occurs in the courts. But this, by and large, is a result of statutory policy. Skilled non-legal commercial men and women and skilled lawyers in a small number of well-established centres have monopolised the arbitration work. London, in particular, through its great concentration of maritime skill, has caused vast amounts of foreign currency to flow into England by the provision of these services.

Yet, as can be seen from this conference, the level of skill in Australia in maritime affairs matches most countries in the world.

We have a vast amount of sea-borne trade in and out of this country.

We have a fine body of maritime colleges, led by the AMC.

We have legal and court system which is, by world standards, first class.

Geographic diversity and distance, however, make coherent organisation and development of pools of work with a critical mass difficult. In part because of this, interdisciplinary exchange is vital to the development of professional skill in those undertaking maritime activity and those charged with the responsibility of understanding and applying the law in relation to maritime activity.

Interdisciplinary exchange in other areas is common. Patent lawyers and patent judges meet regularly with scientists. Medico-legal societies cross-fertilise the ideas of the medical profession and lawyers.

Maritime affairs, founded as they are on international business and activity, invite close liaison and discussion among scholars, lawyers and business people.

From time to time, when this does happen, one is reminded of the huge benefit gained from it. I recently had the privilege of chairing a hypothetical put on by the Company of Master Mariners of Australia in Brisbane. It took the *Zim Mexico* incident from the United States, where a master of a ship was gaoled after there had been a fatality caused by his ship hitting a shore-side crane, knocking workmen from the crane. A number of members of the maritime community and lawyers participated, as members of a panel, to work through the events and give an insight into how the problem might be addressed in Australia. It was illuminating to see real exchange at this level between lawyers and commercial men and women.

One of the problems in dealing with maritime disputes is the difficulty of obtaining and retaining the relevant evidence. This is often brought about by those in charge of the dispute (very often the lawyers) not understanding enough about the industry to understand what evidence is vital, where that evidence is and how relevant information is to be obtained and maintained. This is vital in circumstances where ships, masters and crew travel the world and there is often a limited amount of time to identify and obtain the relevant evidence. Once again, a greater general familiarity with the operation of the activities would enhance the lawyer's skill and capacity.

Recently, there has been established an important institution under the auspices of the Australian Centre for International Commercial Arbitration (ACICA) and with the support of the Commonwealth Attorney-General: the Australian Maritime and Transport Arbitration Commission (AMATAC). It has been formed to facilitate and develop arbitral and mediation dispute resolution of domestic and international maritime and transport disputes in Australia. It is an organisation primarily run, or intended to be run, by commercial men and women. It will include lawyers in its number. But it is not a lawyer's organisation. It will be a forum for education and for dispute resolution. Together with the existing well-known bodies such as the Nautical Institute, the Company of Master Mariners, the Maritime Law Association of Australia and New Zealand and other maritime professional bodies, AMATAC may provide a focus both for interdisciplinary exchange and dispute resolution.

As I said, this country sends away very large amounts of money in dispute resolution costs. That is partly based on inertia; it is partly based on the skill of those other centres which provide those services.

Australia has a significant proportion of world sea-borne trade coming into and out of this country. A greater emphasis on interdisciplinary exchange is vital. I would urge that interested lawyers be welcomed, in some fashion, into maritime societies. Educate them. Make them understand your business. Require them to come to you and to understand your industry. If you do this, you are more likely to obtain a lawyer or a judge who does not have to be told, at least quite at the same length, what your business is and what it is that you were trying to do in the case at hand.

There is a growing recognition in Australia of the need for maritime skills to be maintained. The dearth of sea-going professionals in pilotage, port management and marine surveying has been well-discussed and is well-documented. The same real skill shortage (in the sense of lawyers who understand the maritime industry) may arise in the important service aspect of the community in the legal profession.

The problem is being addressed. In terms of dispute resolution, the reorganisation of the Federal Court in its Admiralty and maritime jurisdiction and the setting up of AMATAC are two important steps that have occurred. They need to be reinforced and built upon.

Secondly, the maritime teaching institutions are beginning to co-ordinate to better use their depth of skill. In this respect, MASLU and the AMC are two of the outstanding maritime institutions in the country. Together, they are combining legal and practical education in a scholarly and thorough way.

The AMC has taken a vitally important step in the creation of a maritime policy unit. This grew out of an industry seminar in February 2007, which was well attended by people from all parts of the maritime community. One resolution carried unanimously was to establish in the AMC a policy unit to provide in-depth analysis for all aspects of the industry, in particular in the area of the development of national policy.

There is room for the development of a thoroughly developed, specifically designed course for lawyers to understand ships and the shipping industry. Sydney Technical College is to begin such a course. I hope it prospers.

All these things are important.

Much more is capable of being done.

Given their importance to the Australian economy, those who comprise the maritime community should demand and expect the legal community to be skilled and knowledgeable not only, as they are, in the law, but also in the affairs of the maritime industry. This demand is one that should not just be made on private lawyers, but also on the court system and governmental system as well.

Ultimately, the Australian maritime community needs a legal profession and dispute resolution system sufficiently skilled to be capable of winning the confidence of the Australian and international maritime communities. The development in this region of the commercial shipping enterprise of so many countries (China, Malaysia, Korea, Japan, Vietnam, the Philippines, India, Cambodia and Indonesia) requires that Australia participate as a knowledgeable and skilled performer and repository of skill in maritime affairs. An essential component of that is a skilled legal profession and dispute resolution system. The sophisticated legal skills are here and have been for decades. What must be enhanced is the interdisciplinary knowledge and communication between sectors of the industry and the legal community.

This will enable Australia to participate with the growing legal centres of the regions: Singapore, Shanghai, Hong Kong and Tokyo and elsewhere, in providing skilled legal, arbitral and court assistance to the region. It is in the interests of the entire Australian maritime community to see this occur.

Captain Allen R. Brink
President
MIIMS FNI MRIN FCMS HCMM

Slide 1

Marine Surveyors
Meeting the Demands of the Next Decade

What does the marine industry expect from Marine Surveyors in the next ten years?

How will the growing manpower demands of the marine industry impact on the marine surveying profession in the next ten years?

How can the marine industry be sure that marine surveyors that are available are of the correct standard?

Slide 2

What does the marine industry expect from marine surveyors in the next ten years?

Confidence in the ability of the surveyor

Maintenance of standards of ability
A competent service
A professional attitude
Accountability

Slide 3

What does the Industry expect?
From

1. Marine Surveyors
Suitably qualified and Highly Motivated
Technologically up to date
Experienced in their area of speciality
Members of a suitable Professional Institute
Properly identified
Appropriately insured

2. Professional Organisations
Properly managed Institutes
With strong membership criteria
Having suitable training programmes
And structured career paths for new recruits
Continual Professional Development Standards (CPD)
World-wide membership providing acknowledged standards for all members

Slide 4

How will the growing manpower demands of the marine industry impact on the marine surveying profession in the next ten years?

What is Going to be required?

World Fleet Developments 1980 – 2006

1980 - 600 Million DWT

2000 - 720 Million DWT

2006 - 960 Million DW

This could mean that by 2016 the total could be 1280 Million DWT

Ship Building in 2005	
Oil Tankers	**518**
Dry Bulk Carriers	**264**
Containers	**491**
General cargo	**221**
Total Orders =	**1569**

Total Cargo carrying Ships = 50,214 **Figures courtesy of World Fleet Statistics 2006**

Slide 5

World Trade

90%

of Global Trade is carried by sea

THIS FIGURE IS NOT GOING TO DECREASE!

Slide 6

Vessel Surveys Include
Oil Tankers
Gas Tankers (LPG/LNG)
Chemical Carriers
Bulk Carriers
Dry Cargo Carriers
General Cargo Carriers
Container ships
Refrigerated Cargo Carriers
RO-RO Cargo Carriers
RO-RO Passenger Carriers
Cruise Liners
Fishing Vessels
Coastal Vessels
Offshore Units

Slide 7

Other Surveying Requirement

SISM/ISPS and Port & State Control
Small and Leisure Craft
Mega and Super YachtsOffshore
Mobile Units and Oil Rigs
Marine Accidents and
Casualty Investigations
Engineering Surveys

Slide 8

Leisure Craft

Yachts (All sizes)
Motor boats (All sizes)
Speed Boats – Ribs – Small Fishing Vessels
Jet Skis – Inland Waterways Vessels
Dinghies – Cruisers – Commercial Vessels
Safety Boats - Inflatables – Sail Boards

Slide 9

What Surveying Manpower Will be Required in the Future

Classification Surveyors
Accident Investigators
Cargo Surveyors
Port & State Control Officers
Engineering Surveyors
ISM/ISPS Consultants
Small Craft Surveyors
Hulls & Machinery Surveyors
Offshore/Mobile Units/SPSO Surveyors

Slide 10

Where Have Marine Surveyors Come From?

Historical Fact

In previous years most cargo, hulls and machinery and engineering surveyors have been: Retired Master Mariners or Retired Chief Engineers - Along with Some 1st and 2nd Officers and other Marine Engineer Specialist

These professional seafarers have put their experience to good use during the years after they have come ashore and leading upto their retirement to supplement their pensionsWith the advent of better pay, fewer seafarers and longer service, fewer seafarers require a second career, this has meant that the experience gained by these very professional People is being lost.

Slide 11

UK Career Opportunities for Former Seafarers (2006)

Table shows career opportunities in UK Only.

Slide 12

What Marine Surveying Skills Are Required ?

For Classification Societies - All areas (But primarily QA skills)
ISM, Port & State & Specific surveys etc – 14 Areas IdentifiedHulls, Machinery and Engineering surveys – 26 Areas Identified
Cargo Surveys - 13 Areas identified
Yachts & Small Craft and all equipment – 16 Areas Identified
Offshore, and Submersible vehicles – 5 Areas Identified

Total number of Skill Areas identified = 74

Slide 13

Marine Surveyors Age Profile for 2006

Courtesy of the International Institute of Marine Surveying

400							
350 to 399					351		
300 to 349							
250 to 299				260			
200 to 249							
150 to 199							
100 to 149							
50 to 99		51	95				
1 to 49	12					25	4
Ages	20	30	40	50	60	70	80

Slide 14

Career Opportunities?

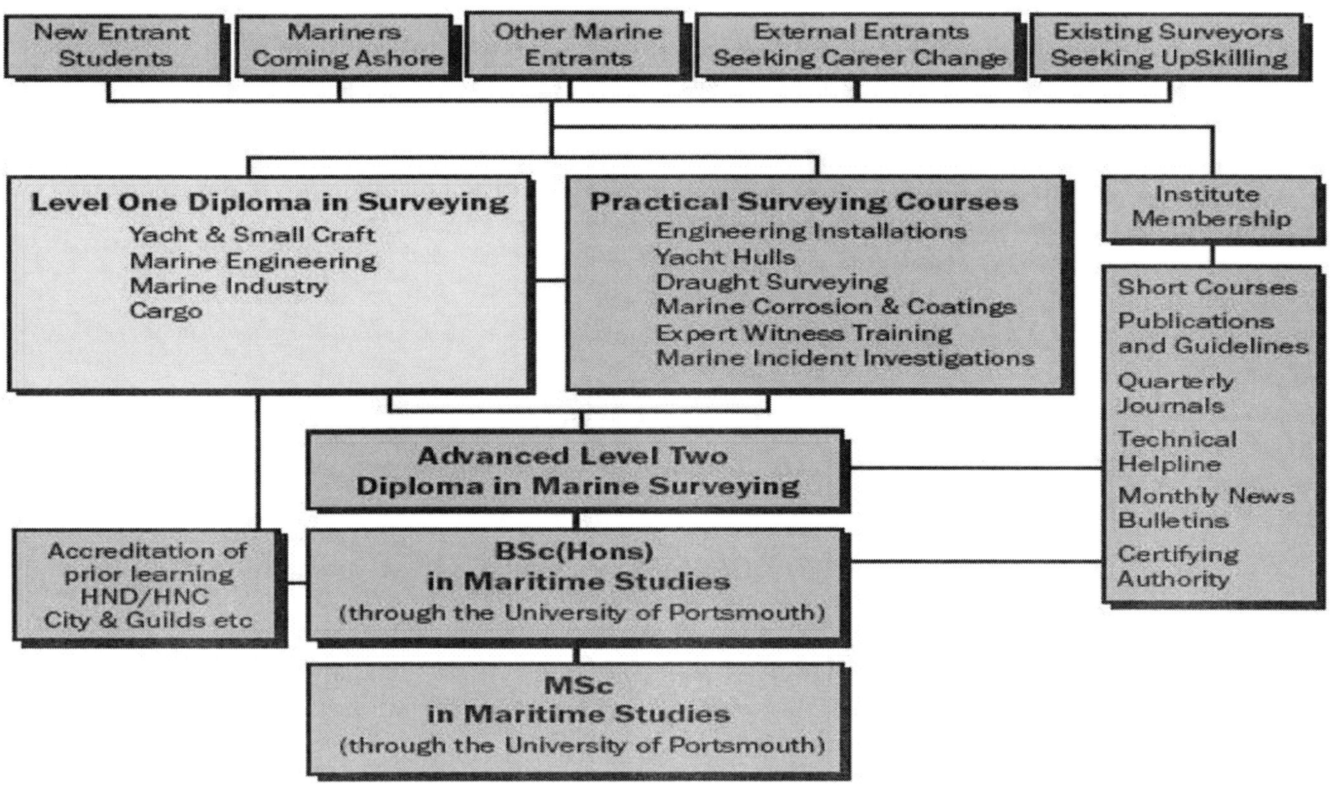

Slide 15

How can the marine industry be sure that marine surveyors that are available are able to do the job?

Marine Surveying
The Role of the Institutes

Robust Membership Criteria
Strong Training Programmes
Continual Monitoring of Standards
Maintenance of valid Membership Registers
Act as the voice of Marine Surveyors world wide
Provide Members with Conference/Seminar and Training Course opportunities
Ensure members remain up to date through Personal Development (CPD)

Slide 16

Innovations for the Next Decade

International Register of all marine surveyors – Verifiable through the Internet
Educational opportunities for all surveyors through Educational Establishments in their own countries
Skill Training for all surveyors though local training facilities
Inventive methods of recruitment to increase Institute membership to provide a worldwide network of members who the Industry can call upon to provide the service they require

Slide 17

Conclusion

The coming decade presents the Industry with manpower requirements never seen before.

The Industry is demanding qualified, highly motivated, specifically trained marine surveyors

The International Institute of Marine Surveying (IIMS) will listen to and work with the marine industry to provide suitable surveyors that can meet the demands that the next decade will bring

The IIMS will seek recognition from the IMO for the work of all marine surveyors and encourage the enrolment of all surveyors onto a new International Register

The Graduated Application of Survivability Modelling in the Warship Design Process

Author/Presenter: G MacDonald (gmacdonald@bmtdsl.com.au)
BMT Defence Services (Australia), Melbourne, Australia
co Author/co Presenter: P Horstmann
QinetiQ, UK

1. Introduction – what will be covered (cost benefit & survivability)

2. Why is it important
Cost Capability Trade off –

Survivability – Warships are not built for peace, they are designed to go in harms way operating in a manmade hostile environment with the intent of achieving mission objectives. Survivability is the measure of the capability of a naval ship to avoid and/or withstand the hostile environment in which it is operating while performing that mission and as such is a key metric of the systems capability.
The three main elements of survivability are

- Susceptibility, the ability to avoid detection;
- Vulnerability, the ability to withstand a hit (wepon effects);
- Recoverability, the ability to reconfigure and restore the damaged system after a hit.

ie. Survivability = f (Susceptability, Vulnerability, Recoverability)

The ability of a warship to complete its mission after an attack depends on its susceptibility (ability to avoid being hit), vulnerability (level of damage when hit) and recoverability (ability to recover from a given level of damage). Survivability is the combination of these three elements. This paper describes the importance of considering all the aspects of survivability at the earliest stage in the design and then managing it throughout the design process. Methods which can be used to assess survivability and measures which can be taken to reduce vulnerability are discussed. An insight into how recoverability data can be used to optimise competing warship designs is presented along with an example of how susceptibility data can be combined with vulnerability data to provide a comprehensive analysis package.

An essential element of ship design is the integrated measurement of the ship's Susceptibility and Vulnerability to specific weapons threats, as well as its ability to recover from attack by those weapons. Alion's advanced survivability efforts include generating platform specific survivability requirements and simulating and analyzing the effective performance capability of the ship against weapons threats. "Susceptibility Measures" are defined as the predicted probability of a ship to avoid and/or defeat and attack by a weapon threat. "Vulnerability Measures" are the predicted probabilities of ships meeting selected levels of mission performance after sustaining a hit. "Recoverability Measures" are the probability distribution of the time required to stop the spread of damage and to return to a preselected/optimum capability after such a hit. Integrating and analyzing these factors supports the development of the Survivability requirements of the ship, its systems and components, and the development of a more survivable ship design.

Alion's experts are highly qualified and experienced in defining Survivability Requirements and all aspects of Susceptibility, Vulnerability and Recoverability efforts in ship design, including:

- Selection of threat weapons and operational mission threat environments
- Analysis of Ship Signatures and Design Features to reduce Susceptibility
- Analysis of the effects of weapons on a wide range of structural materials
- Design, instrumentation and management of large scale weapons effects testing, including prediction of results and post test analysis
- Design of armoring and other protective features
- Physics-based modeling of explosive/fragment effects on structure using advanced Finite Element and Compression Fluid Dynamics simulations and Alion's Measure of Total Integrated Ship Survivability (MOTISS) Program
- Stochastic modeling of weapons effects on complex ship designs
- Discrete event physics based modeling of ship progressive fire/flooding/damage
- Discrete event modeling of damage control and ship recovery
- Identification of root causes of design systems failures and development of alternative design recommendations

Generation of Requirements for Survivability/Performance/Detail Contractual Specifications

3. Who should be interested and when in the through life program

4. How would they use the data

5. How is it done

a. Cost Benefit method

b. Cost benefit using survivability ie S=f(S,V & R)

In naval ship design, vulnerability reduction has not been given the same priority as susceptibility reduction. This has lead to an unbalance in operational requirements. However, when maximizing survivability is the objective, the current thinking is that there must be a balance between susceptibility and vulnerability reduction. Threshold requirements based on acceptable levels of both vulnerability and susceptibility must be established. However designers must be given sufficient latitude to improve overall survivability by balancing, or trading off susceptibility and vulnerability.

c. Metrics of S&V

d. How do we establish the values against the metrics ie analysis

e. Examples

6. Where are we now

7. Where to in the future
8.

In an ideal world we would be able to afford the capability we desire however xxxx

Irrespective of the stage in the life cycle of a naval platform from initial concept development through to in-service engineering change proposals, design trade-offs occur in balancing the objectives of delivering the most cost effective solution to the specific capability being pursued during which the opportunities to optimise the performance, cost and schedule characteristics of the associated activity. In an ideal world this would be achieved through a structured evaluation and comparison of a range of potential solutions against defined objectives and constraints.

1.2 Trade-offs take place in a cascading format with progressively more granularity being available at each succesive level from compliance with defence policy and national requirements, through trades that take place in the delivery of one or more capabilities, to those that take place within the boundary of an individual programme.

It is within the individual program that this paper concentrates and the trade off between capabilities competeing for the program dollar, in this case between the elements that constitute a vessels survivability. This is illustrated in the diagram below:

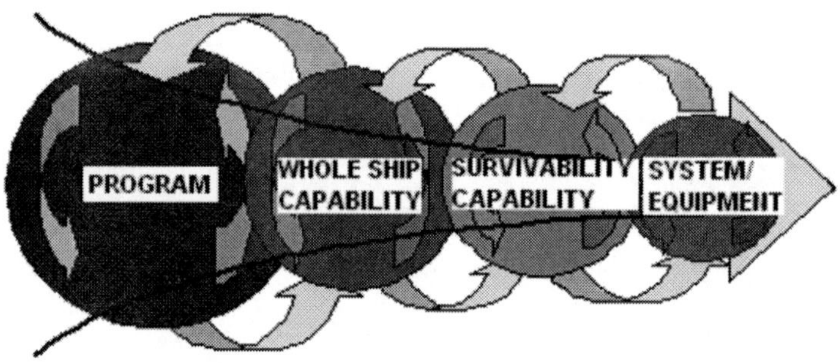

1.3 Conducting trade-offs is also necessary to ensure that an acceptable level of capability is available to the front-line user that is both affordable and deliverable within an agreed timescale. These Trade-offs must achieve balance between Military Capability (MC), Operational Analysis (OA), adjustments to the Equipment budget, developments in technology and risk reduction activities.

1.5 Best practice would be as follows:

For a specific programme (ie a new capability or an in-service platform) the Operational Concept Document should specify in qualitative and quantitative terms the Survivabilty metrics required to achieve the military capability being acquired, this would be derived from consideration of the threat environment and desired probability of mission success;

The Survivability metrics, traceable back to the OCD would be read down into the Functional and Performance Specification (with increased granularity as appropriate) highlighting impact/constraint on specific functions and providing guidance in way of DID's detailing how compliance with these requirements are to be demonstrated to the client at each stage of the acquisition program;

The Test Concept Document would provide guidance as to how the Survivability performance will be measured for Acceptance into Naval Service and through life.

The Survivability metrics and methodology for evaluation of compliance together form the benchmark against which analysis of trade-offs should be conducted.

In setting the survivability metrics it is essential that the client perform some initial trade-offs focussed on adjusting requirements, the budget and timescale in the light of capability and risk considerations such that the bar isn't set too high with resulting industry proposals providing unpleasnt program surprises.

The FPS DID should contain guidance on the Survivability trade space within the capability and the potential areas where trade-offs are encouraged. Where industry would like to consider studies outside this boundary, the client would be responsible for considering the merit of such proposals. Whenever trade-off decisions are taken, they would be supported by appropriate evidence and the rationale recorded in programme documentation.

5. Capability Trade-Offs

5.1 Within Solution Trade-Offs can take place throughout the life of the platform however the objective and impact of any trade's varies depending on where the programme is in the Cycle.

5.2 Requiremnt Definition Phase.

The early stages of a programme will be focussed on ensuring that it is correctly scoped in terms of its user requirements, budget and time envelope. This will be an iterative process to ensure that the user's aspiration for effective capability is correctly balanced against the resource available and a realistic assessment of the time to acquire a solution system. The client should specify the Key Survivability metrics that define the capability banded with Threshold and Objective Measures of Effectiveness (MoEs) and Measures of Performance (MoPs). Cost and schedule should be developed using standard forecasting techniques and provided at 10%, 50% and 90% confidence levels. Each dimension should take appropriate account of risk.
At this stage trade-offs will take place in the 3 dimensions of Cost Schedule and Capability to explore the envelope such that the client can issue a request for proposals/tender from industry.

5.4 Design/synthesis stage

In service stage

a. During this phase industry, in developing their potential solutions to the programme SRD. From this point onwards, it will be possible to adopt a more formal process for conducting trade-offs in order to ensure that the solution proposed at MG has been optimised within an acceptable PCT envelope but with trade space in all 3 dimensions. To achieve this for performance, the Specification will be set between the Threshold and Objective MoP for each Survivability Metric. Cost and time trade space will be based on 3 point estimates in accordance with IAB Guidance.

b. Construction In the Demonstration and Manufacture (D&M) Phases, trade-offs will continue to be conducted across the DLODs in conjunction with the selected contractor. If it is considered impossible to remain within the PCT envelope agreed by the Approving Authority (AA) by conducting further trade-offs, then it will be necessary to seek a change in the envelope's parameters.

The Trade Space

As previously indicated the acquisition balance of investment decision for a specifc capability requires exploration of the Cost Schedule, capability trade space with the objective of achieveing best value for money.

The increasing cost of ownership of military capability is driving acquisition practice towards acquiring the best equipment, within considerable attention to cost and risk while achieving a balance of performance and schedule) with considerable constraint for systems without regard for their cost. We should develop only affordable and achievable. Cost As an Independent Variable (CAIV) is an acquisition management tool that we can use to achieve that result. It is a strategy that entails setting aggressive, yet realistic cost objectives when acquiring defense systems and managing the achievement of these objectives. Cost objectives must balance mission needs against projected out year resources and, taking into account existing technology, likely maturation of new technologies and anticipated process improvements in both DoD and industry. As we determine system performance, schedule and cost objectives (based on cost-performance and cost schedule trade-offs), the acquisition process will make cost more of a constraint and less of a variable, while nonetheless obtaining the needed military capability of the system.
The CAIV is the acquisition management tool that considers thresholds and objective parameters. The objective of CAIV is to ensure that we field an affordable system by linking cost, to the performance and schedule called out in the ORD. Furthermore, the objective performance parameter should show a significant or measurable improvement over the threshold. Ideally, the IPT will have this type of information available on at least the KPPs and as many other parameters a possible. The challenge facing the IPT is determining what performance the Operating Forces really needs, when they need the capability, and where costs can reasonably be avoided.

Although much discussion of CAIV is centered on new systems, CAIV principles are applicable throughout a system's life cycle.

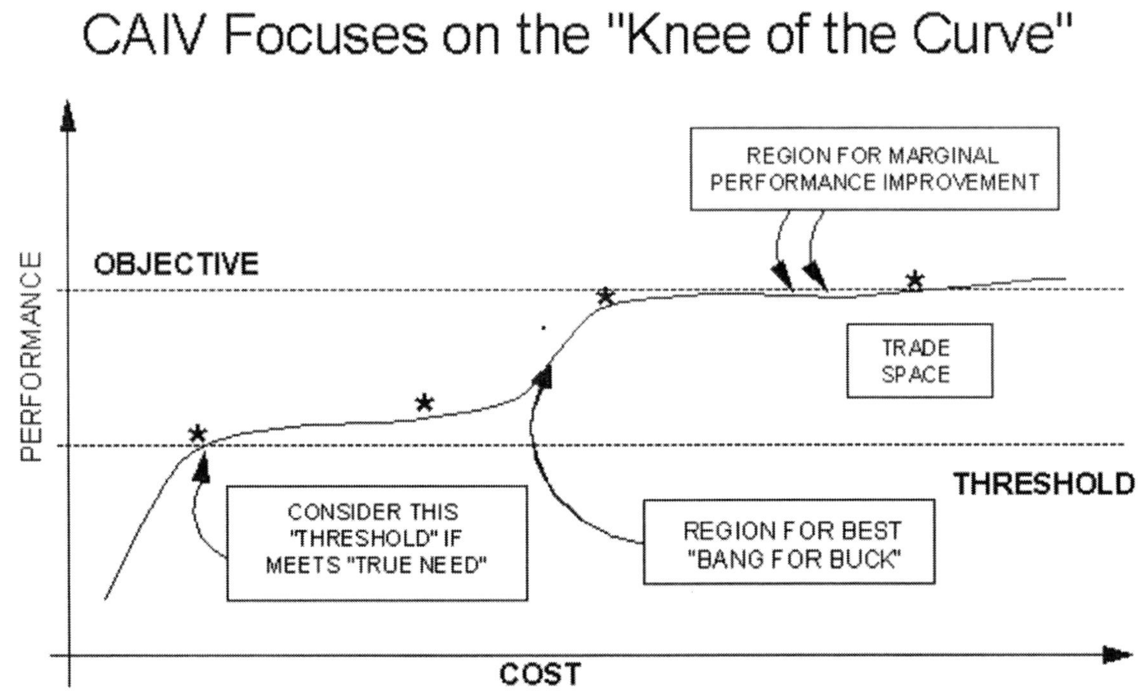

2. CAIV Defined. Cost As an Independent Variable is an acquisition strategy focusing on cost-performance trade-offs in setting program goals. This strategy formalizes the process for cost performance trade-offs and engages the warfighter, the developer, and the supporter to facilitate meaningful trade-offs to arrive at an affordable balance among performance and schedule. These trade-offs will enable the warfighter to make choices that will provide the best performance from the system within available resources.

3. Conceptual Approach. A key tenant of the CAIV approach is a far stronger user role through participation in setting and adjusting program goals throughout the program particularly in the cost-performance tradeoff process. CAIV processes identify the most cost effective solution within the performance floor (threshold) and the upper cost ceiling. CAIV fosters development of a design that is both affordable and satisfactory to the end user. As the end user, the Operating Forces are principally represented by the RO. Knowing that MCOTEA will conduct a thorough and independent evaluation of the system further ensures that the system will meet the user's requirements. Working within that context, a sample process for achieving the objective of CAIV includes:

- Setting realistic but aggressive cost objectives early in each acquisition program to ensure cost is considered equal in importance to performance and schedule considerations. Cost objectives are normally established in the APBA and updated for each phase of an acquisition program. Once cost objectives are established, cost becomes more of a constant or constraint than a variable,

- Fostering early and continuous user participation in setting and adjusting program goals,

• Encouraging identification of the "trade space" (i.e., cost gradient with respect to performance) around the cost objective,

• Requiring developers to address cost throughout the life cycle process by devising appropriate metrics for tracking progress in setting and achieving cost objectives,

• Managing risks to achieve cost, schedule and performance objectives,

• Motivating government and industry managers to achieve program objectives, and

• Putting in place for fielded systems additional incentives to reduce operating and support costs.

4. CAIV in ORD Development. CAIV is based on the principle that the best time to reduce life cycle cost is early in the acquisition process. The introduction of CAIV into the ORD development process has two parts:

• Determining which parameters should be included in the ORD and

• Determining the values that should be associated with the parameters that are included.

For the first, the key issue is to ensure that the parameters included in the ORD are those needed to fulfill the needs of the Operating Forces and are met without "gold plating" the proposed material solution. In the examples listed later, the inclusion of a bayonet lug on the combat shotgun could be considered as a parameter that was in excess of what the FMF needed. Ensuring that excess parameters are not included is neither easy nor subject to an easy-to-follow set of rules. For the second, determining the values that should be associated with included parameters is the subject of cost-performance trade-off analysis. The initial cost-performance trade-off analyses should be conducted before ORD parameters and general acquisition approach (e.g., non-developmental item) are finalized. After the MNS is approved and a successful MS O decision, thorough analyses of alternative material concepts and their cost implications are conducted. These analyses normally take the form of an CA or TEA. In addition to shaping the ORD, these analyses provide advice to the MDA at MS I regarding whether a new program is warranted. CAIV tenets should be explicitly considered in the CA "Scope of Analysis" (tailored accordingly in the case of an TEA). Thus, the first CAIV analysis inherent in the development of the CA/TEA can be extracted directly or lead naturally to more robust cost-performance and cost-schedule CAIV analyses.

The cost-performance tradeoff analyses explore the relationship between performance parameters and their associated costs. These trade-offs will help identify performance enhancers, potential superfluous requirements and cost drivers. They will also help identify potential program cost, schedule and performance risks. These analyses should consider what is technologically feasible to meet the user requirements and support establishment of aggressive (but achievable) cost, schedule, and performance thresholds, and reasonable objectives. The IPT (with the RO taking the lead) will use this information to help them arrive at threshold requirements in the ORD. Note: Cost-performance tradeoffs are not intended to arbitrarily "lower the height of the bar" or "dumb down" the true needs of the user. However, requirements must not be established in a "vacuum," that is without regard to cost and schedule. Otherwise, the Operating Forces may not receive anything because the apparent solution is not technologically feasible, is cost prohibitive, or will take an inordinate amount of time to develop. The PO builds on this information to develop the program's acquisition strategy, and negotiate specific performance parameters, cost objectives and schedule information in the APBA at each milestone. Once system performance and objective cost are agreed to through trades, cost is then a constraint and less of a variable. Changes in the program can occur, as long as the threshold values in the ORD and APBA are achievable. Changes which result in breaching approved ORD/APBA threshold values must be presented to the approval authorities for decision.

The application of CAIV to developing ORDs does not have to be as complicated as that shown above. The information contained in Figure D-1 may not be reasonably available on many parameters. The examples discussed below do not contain this level of detail. However, they do illustrate how simple CAIV principles were applied to programs to help make the program more affordable, refine the ORD and still meet the real needs of the FMF.

EXAMPLE -- Joint Combat Shotgun

A market survey revealed that several commercial shotguns met 90% of the ORD requirements. The question facing the acquisition community was, "Is meeting 90 percent of the requirement good enough?" In this case, three requirements were not met. The commercial shotguns did not have a bayonet lug, a front sight post, or a dust cover. Cost information from the market survey further revealed that relaxing these requirements would result in a savings of 32 to 56 percent over developing a brand new shotgun that could satisfy all the "requirements." Mandating that all requirements be met would also result in not meeting Initial Operational Capability (IOC) date. Thus, relaxing the requirement would result in a significant cost savings and timely delivery to the FMF weapons that met the most significant performance parameters. After much discussion, the requirements were relaxed via an amendment to the ORD -- a CAIV success story.

NOTE: In the above example, the bayonet lug requirement was considered by some to be superfluous -- not a true need of the FMF. CAIV analysis exposed the superfluous requirement.

EXAMPLE -- LW155 (for illustration only)

The LW155 is replacing the M198 155mm howitzer. The range requirement is 30 kilometers. During developmental testing the prototype demonstrated no more than 29.8 kilometers. Engineers estimated the costs associated with engineering changes to the gun in order to achieve the required range at >$2.0 million. MCCDC user representatives queried the FMF concerning the range requirement and the associated cost. The FMF did not feel the additional 200 meters was worth the cost and recommended relaxing the threshold range requirement.

5. Conclusions. There is no magic to applying CAIV. The IPT must apply a great deal of common sense in trying to determine if meeting a specific requirement is worth the cost. And it must do this early on as well as throughout the life of the ORD. The IPT is encouraged to discuss the results of the cost-performance trade-offs with the FMF and their respective chains-of-command. The IPT should also be on guard to not "take the easy way out" by hastily relaxing the requirement to make it easy on the material developer (PO and contractor). However, we should first determine if Commercial Items (COTs) or Non-Developmental Item (NDI) items can meet the needs of the FMF. We can field COTs/NDI solutions faster and cheaper than new development programs. Also keep in mind that cost objectives must be not only

realistic but also aggressive. Ultimately, the material solution described in the ORD must above all meet the needs of the FMF.
NOTE: This annex focused on cost performance trade-offs. Similarly, consideration should also be given to cost schedule trade-offs. Finally, a pre-planned product improvement is an excellent way to defer desired performance until it becomes technologically affordable.

Relationship Between C ost or Price and Technical Streng th
Defence acquisition focuses on achieving a "best valu e" source selection. B est value m eans that th e Gove rnmen t will
perform a cost/technical trade-off analysis such that business judgement will be exercised in selecting the
most advantageous alternative to the Government, considering both the technical merit and costs of
proposals. The determination of best value will be made by comparing the differences in the value of
performance capability factors with the differences in the costs proposed. The Government will make this assessment through the development of
trade-off analyses and other analytic studies that involve the assessment of benefits of superior
performance capability features - for example, economic benefits clearly attributable to increased
productivity; probability of successful contract performance; unique and innovative approaches or
capabilities - versus added costs.

1. Introduction

1.1 What is The Solution Trade-Off Process (ST-OP)? (See the 'ST-OP Process' graphic at Appendix 1)

ST-OP is an engineering (of performance, cost and time characteristics) process that is applied iteratively in the design process in order to optimise a proposed solution to a capability need. Its application may have a reactive impact on the definition of the capability need and associated approval limits.

In this context, 'optimisation' is considered to be:

- the eradication of non-compliances with Approval Limits,
- the eradication of non-compliances with capability requirements and constraints,
- the ensuring of affordability,
- the maximisation of Value For Money (VFM),

within acceptable levels of risk and with sufficient confidence of achievability.

The four steps comprising any one iteration are:

1. Analyse the (proposed) Solution.
2. Evaluate the Analysis.
3. Incorporate Changes.
4. Record the Analysis, Evaluation and Changes.

A single iteration, or a set of iterations, may be needed to achieve a specific objective.

Assessment Phase: ST-OP should be applied to each option being considered in the latter part of the Phase, after the option has been (initially) developed for appraisal, but before COEIA is conducted, to ensure fair appraisal.

3. Entry Criteria

A set of iterations of ST-OP should not commence until the following entry criteria have all been met:

1. The objectives of the trade study analysis have been established

2. The trade-off objectives are consistent with each other.

3. The requirements are nominally complete, and is associated with a CONEMP and scenarios. Requirements are quantified, with effectiveness envelopes. All user requirements / capability constraints are prioritised, and the authorities for permission to trade at each level of priority are understood (this is particularly important for interoperability and DLOD trade-offs across capability boundaries).

4. **Approval and Not-To-Exceed Limits** are known (to include both affordability and COO).

5. The limit of acceptability (e.g. 50%, 90% confidence) for **level of risk** / uncertainty is agreed with the DEC / IPTL.

6. Relevant **SRDs** are identified, and if not nominally complete are at least populated to identify the CONUSE, and satisfy all KURs and priority 1 user requirements / capability constraints. Requirements are quantified. All system requirements / constraints are prioritised, and the authority for permission to trade at each level of priority are understood (again, particularly important for interoperability trade-offs across solution boundaries).

7. The **definition of each discrete proposed solution** to be optimised:

    - is uniquely identified.
    - is under version control at least, change and configuration management at best.
    - embraces all systems contributing to the satisfaction of the subject URD that require investment, including their enabling systems (which must include the complete support solution).
    - includes critical interoperability with other systems.
    - embraces the contribution from all DLOD.
    - includes the identification of critical technologies.
    - includes the identification of significant assumptions.

- includes a through-life time-line, which identifies desired ISD / OSD, and which discriminates between any individual acquisition increments.
- includes the definition of the content of each acquisition increment.
- is developed to a level of detail and maturity that is appropriate to the current point in the acquisition life-cycle.
- Is at a level of detail sufficient to enable credible performance, cost, schedule, and risk analysis, including determination of three-point estimates for time and cost, and uncertainty of performance / effectiveness predictions.

4. Exit Criteria

A set of iterations of ST-OP should cease when an appropriate mix of the following exit criteria have been met:

- The proposed solution demonstrably excludes elements that do not deliver VFM.

- The proposed solution satisfies the URD (which may itself have been subject to trade-offs).

- Diminishing returns per iteration indicate that further optimisation effort at the current time is inappropriate. The proposed solution can be deemed optimised.

- The window of opportunity (usually time and budget) for optimisation is closed. The proposed solution can be deemed to be the best that can be determined within the resources available.

- In the Assessment Phase, the proposed solution lies within the PCT envelope, at an acceptable limit of risk / uncertainty.

- In the Demonstration and Manufacture Phases, the proposed solution lies within the capability / cost / time approval limits at an acceptable limit of risk / uncertainty, or it is decided to make a re-submission to the Approving Authority.

5. ST-OP Step 1 - Analyse the Proposed Solution

5.1 Related Guidance

The guidance for this Step should be read and implemented in conjunction with the following AMS topics:

- Planning and Scheduling.
- Cost of Ownership (COO).
- Modelling, Simulation & Synthetic Environments (MS&SE).
- Risk Management.
- 3 Point Estimating.
- Technology Management.

5.2 Introduction

The purpose of Step 1 is to enable determination of the extent to which the proposed solution, as-input, is acceptable. It can also serve as:

- a sensitivity analysis performed to determine which characteristics of the solution are (or are not) capability / cost / schedule drivers.
- an impact analysis for tentative decisions taken in a preceding iteration, after which decisions can be ratified or re-considered.

The Step is essentially mechanistic 'number-crunching' in nature. The definition of the solution must remain unchanged throughout the Step, though more clarifying detail can be added as found necessary to enable credible modelling and analysis.

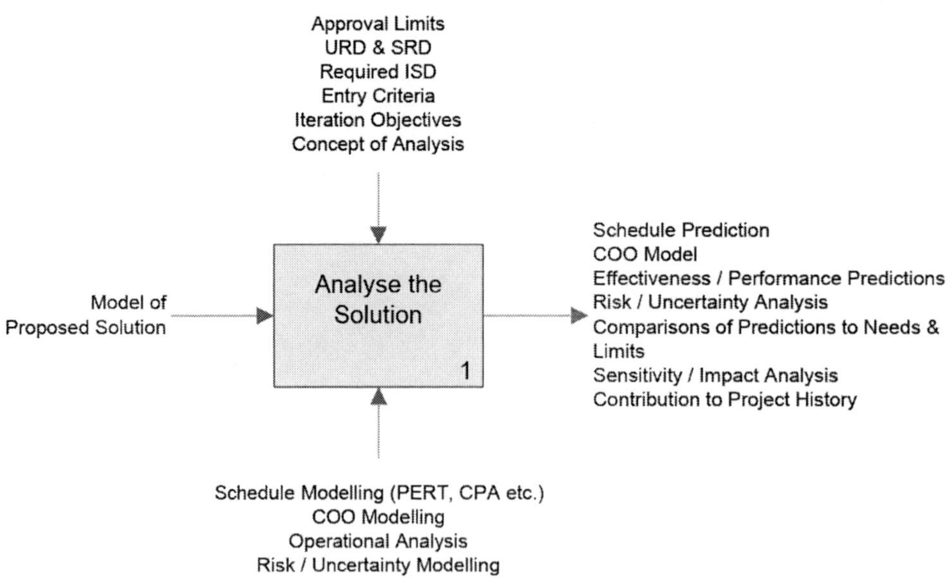

5.3 Method

Start with a check to ensure the Entry Criteria are met. Then -

- **Schedule** - Construct or refresh the through-life schedule for the solution. Use PERT, CPA, Critical Chain Management, or a similar technique to identify critical and near-critical paths, and associated floats. Ensure that the schedule is sensitive to typical dependencies and critical paths such as –

    - the Acquisition and Acceptance Strategies in respect of how the acquisition is to be incremented,
    - the chain from user / operator / maintainer competency definition through training needs analysis et. al. to delivery of trained personnel prior to entry into service of the new equipment,
    - the chain from definition of availability / sustainability requirements through to the delivery of the support solution prior to entry into service of the new equipment,
    - the chain from determination of the verification / validation criteria, through the evolution of the test capability, to the generation and evaluation of evidence.
    - technology maturity.

  Be aware that infrastructure development can be a particularly long-lead item.

- **Cost** - Then construct or refresh the EP / STP and COO models for the solution and its through-life schedule. The COO model should be parametric or data-driven so that it can be used for cost engineering. The cost / work breakdown structure should reflect its use as an engineering, not budgeting, tool.

- **Effectiveness / Performance** - Obtain / refresh predicted-performance data for the solution. The measures of performance (MoP) should be as identified in the SRD. Data, preferable derived from simulation, statistical analysis, trials etc. must be accompanied by its provenance. Predictions must include the exchange of services across capability / system boundaries.

  Feed the predicted-performance data into OA in order to predict the level of effectiveness deliverable by the solution.

- **Risk / Uncertainty** - Identify and analyse the major risks and uncertainties. Input the conclusions to the derivation of three-point estimates for schedule and cost. Ensure that performance / effectiveness predictions also quantify uncertainty.

- **Comparison** - Compare the results of the above analyses with the Approval Limits, URD, SRD, and required ISD. Identify where the comparison is acceptable (green), marginal (amber) or unacceptable (red).

  Also compare the results with those of the preceding iteration (if applicable) in order to identify and quantify progress through the iterations toward achieving the optimisation objectives. This comparison constitutes an 'impact analysis' of the changes introduced at the preceding Step 3. It can also serve as a sensitivity analysis.

- **Output** –
    - the results of the analysis for evaluation.

1. INTRODUCTION

Warships represent a significant investment for any country. It follows that, regardless of whether they are being used to defend national boundaries or project power, that they are as survivable as possible. In times of war, warships can expect to come under attack e.g. HMS Sheffield and HMS Coventry. Even in times of relative peace warships are still likely to be targeted, e.g. the USS Stark and the USS Cole. It is imperative that the investment a warship represents is capable of withstanding these attacks and is able to

continue fighting. This paper describes how the three aspects of warship survivability: susceptibility, vulnerability and recoverability can be assessed together to optimise survivability. This can be represented by considering figure 1 which shows a diagram of capability (e.g. a ship's ability to provide air defence) versus time .

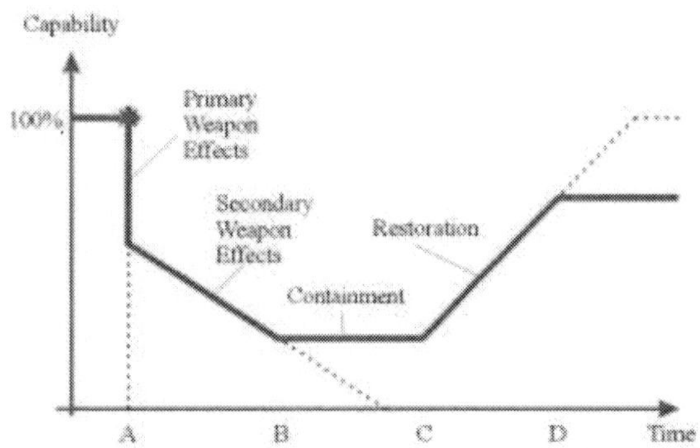

Figure 1: Capability versus time

Everything prior to point A in figure 1, where the ship is fully capable, is a measure of the ships susceptibility to attack. This relates to the ships signature, its hard/soft kill systems and the sophistication of the attack At point A the ship is successfully attacked and there is an immediate drop off in capability due to the primary weapon effects, e.g. blast and fragmentation. Secondary effects such as flooding and fire could result in a further loss of capability (point A to B). This could, of course, drop to zero if the ship is sunk. The journey from point A to B is vulnerability, which is a measure of a ship's in

built resistance to weapon effects. The containment and restoration phases, point B to point D, cover the ship's ability to recover from the damage. However, in reality the recoverability process starts somewhere between point A and point B when the crew initiate damage control and fire fighting activities. The amount of capability that can be maintained during an attack, or restored afterwards, is a good measure of the design of a warship.

2. SUSCEPTABILITY

In the simplest terms susceptibility is a measure of a ship's ability to avoid being hit. There are a number of ways to reduce susceptibility.

2.1 AVOID DANGER

The most obvious way to reduce susceptibility is simply to avoid any hazardous areas. If this option is to be entertained then the notion of having a warship in the first place should also be considered.

2.2 SIGNATURE

If a ship is going to venture into a potential hostile environment then it is advisable to reduce the ship's signature as much as possible. A warship's radar signature is of key importance and this paper will culminate in the description of a method to link the output from the interaction between a ships radar cross section (RCS) and missile seeker logic to the SURVIVE vulnerability assessment code. Spectre is a RCS prediction code, developed by QinetiQ to model the RCS of large, complex targets. It uses a high frequency approximation to Maxwell's equations based on physical optics and the method of equivalent currents. The code has been operational for over fifteen years and has been thoroughly validated against measured data. It is capable of producing a variety of outputs, including down and cross range profiles, which are suitable for feeding into simulated seeker algorithms. Other aspects of signature that need to be considered to
reduce susceptibility are:

- Noise: The quieter something is , the less likely it is to be detected and targeted. QinetiQ have developed systems such as HVME to monitor a platform's self noise so that measures can be taken to minimise emissions.
- Magnetic: Threats such as mines can fuse off a platform's magnetic signature. Codes, such as TMSS, have been developed to consider mine activation. TMSS has now been linked to the vulnerability code SURVIVE to provide a comprehensive assessment capability.
- Infra Red (IR): It is difficult to fully remove the heat generated within warships but the IR signature can be minimised. Work is underway to link prediction codes such as SIREX to SURVIVE.
- EM: A warships communications can often give away it's position. The issue needs to be managed, particularly given the propensity of combined operations, net centric warfare and the increasing desire for full theatre picture compilation.
- Visual: The oldest form of stealth and one that may increase in prominence given the increase in the asymmetric threat.

2.3 HARD/SOFT KILL

The ability to decoy incoming weapons or physically shoot them down also falls under susceptibility. Options range from chaff launchers to electronic jamming. Given the asymmetric threat, increasing emphases is being placed on hard kill measures such close in weapon systems (e.g. Goalkeeper).

3. VULNERABILITY

3.1 INCREASED SUSCEPTABILITY

Despite having sophisticated defensive systems warships are still vulnerable to attack, as incidents like the USS Cole have shown. Indeed the increasing asymmetric threat highlights the vulnerability of expensive assets to relatively unsophisticated attack. Even with more conventional attacks there are factors that restrict the ability of a warship to fully utilise its defensive systems. These include:

- Complex rules of engagement which may require the positive identification of threats.
- Increased littoral operations, and hence reduced sensor performance.

Both of the examples generally lead to reduced reaction times and hence a greater chance of taking damage. Operations in the littoral also open up ships to a greater range of threats such as shells and anti-tank weapons.

3.2 WEAPON EFFECTS

If the defensive systems have been overwhelmed, got around or simply not worked then the warships needs to have some inbuilt resistance to weapon effects. For an above water attack the first damage mechanism to be dealt with is usually the **kinetic energy** of the
incoming weapon. This could be anything from small calibre ammunition to supersonic anti-ship missiles. Although some smaller weapons may not have much penetrative power, some warship systems may be vulnerable, particularly if they have poorly protected equipment and/or a lack of redundancy. Detonating munitions will produce high velocity **fragmentation** (figure 2) that will penetrate through the ship's structure and damage equipment (and hence systems). The fragments can be pre-formed or natural.

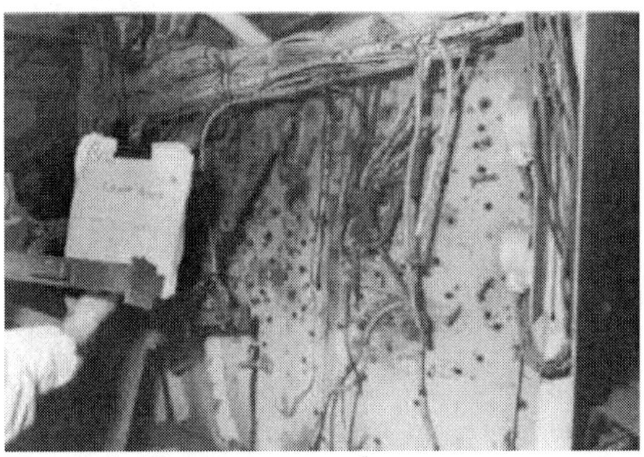

Figure 2: Fragmentation damage

Internal and external detonations will produce an impulsive **bl ast** loading onto the ship. This is also capable of damaging structure and equipments. Internal detonations will be more devastating (figure 3) as the build of gaseous products will produce a longer duration quasi-static load which can spread the internal damage over a large area

Figure 3: Blast damage

Underwater detonations have two primary effects. Firstly a **shock** wave will emanate from the explosion. The equipment on the ship will be excited by a high frequency low displacement motion. This loading will be attenuated depending on where the equipment is within the vessel. Secondly, due to the expansion of gaseous products of the detonation, a bubble will be formed. Once the bubble reaches maximum radius, it will collapse in on itself due to hydrostatic forces. As the pressure in the collapsing bubble increases it will reach a point where it re-expands, and so on. This oscillation may induce **whipping** on the ship. This global loading can be severe enough to break the back of the ship if the detonation is in the correct place (or wrong place depending on your point of view)! Whipping takes place over a longer time scale than shock and will reload equipment with a low frequency high displacement motion. Secondary effects such as flooding, fire and residual strength are covered in section 4.

3.3 VULNERABILITY ASSESSMENT

In order to reduce the vulnerability of warships the UK MoD, through QinetiQ, developed computer codes to assess the extent of the problem. The initial development produced above water codes such as REVS and underwater codes such as SSVUL and SUBVUL. These codes are all encompassed in the modern code SURVIVE. The primary aims of SURVIVE and the analysts who operate it, are:

- To measure the vulnerability of a warship.
- To explain why the vulnerability exists.
- To suggest measures to reduce the vulnerability.

SURVIVE works by building a representation of the warships structural layout and internal systems, figure 4.

Figure 4: SURVIVE equipment layout

As well as storing the physical locations of equipment and their properties a logical definition of how the system operates is also required. A weapon model is defined detailing, amongst other things:

- . Explosive type and quantity.
- . Terminal velocity
- ..Physical properties.
- . Fragmentation pattern.

To highlight vulnerable aspects of the ship most vulnerability analyses attack the ship using a 2D linear grid, as shown in figure 5. The red areas below represent vulnerable areas for a particular system.

Figure 5: SURVIVE 2D linear grid

When the data from susceptibility studies is combined with vulnerability information, overall survivability conclusions can be drawn. Data from the Spectre RCS prediction code has been linked with SURVIVE. This allows the vulnerability analysts to consider the areas of the ship where particular weapons are most likely to hit. Figure 6 shows an incoming missile vector attacking a Leander class frigate. This is covered in more detail in section 5

Figure 6: Leander Missile attack vector(from Spectre)

Vulnerability assessment can also provide casualty assessment and various type of hazard and safety case analysis. These range from the magazine layout and protection strategies to the effect on surrounding infrastructure from detonations in harbour.

3.4 VULNERABILITY REDUCT ION

Of paramount importance for any warship is not just to have a survivability policy but the will to implement and enforce it. Vulnerability reduction measures have the greatest effect if they are designed in from the outset. Vulnerability assessment should not be considered as an audit function. Vulnerability engineers have showed this to great effect
by working with the project teams and designers of the T45 and CVF. QinetiQ has even developed a tool, "SURVIVE LITE", to allow designers to assess vulnerability at the early stages of design. This is fully compatible with SURVIVE and makes the transition from the early concept stage through to the detailed design stage straightforward.
The main point, is too manage vulnerability throughout the design process. There is little point having a well thought out policy for the primary equipment on the ship and ruining that good work by adding, for example, a vulnerable electrical distribution system.
It is important to use the data from any vulnerability assessment to inform the vulnerability reduction process. For most above water attack scenarios the warship is likely to remain afloat, as has been seen in numerous incidents. However, it is also likely that the primary weapon systems will be inoperable, rendering the ship impotent. The following measures can be taken to help reduce vulnerability:

- Employ a good zoning/water-tight sub division policy.
- Provide as much redundancy within key systems as possible. Where there is redundancy, separate it. Figure 7 shows two components, only one of which is required to provide a capability (e.g. a generator). However, in the top picture they are too close together and are both killed in the same attack. The lower picture shows how a relatively minor move of the equipment results in the capability being preserved.

Figure 7: Separate redundant components

Figure 7: Separate redundant components

- Group critical components together. If this doesn't happen then an attack anywhere on the ship will have a chance of knocking out key systems.
- If key equipments cannot be duplicated then protect them.
- Protect key spaces. Investigate armour schemes and blast resistant structure/doors. Even the compartment layout can help.
- Use shock mounts or rafts to isolate equipment from the effects of shock.
- Use signature management to reduce vulnerability.

4. RECOVERABILITY

Recoverability is specifically concerned with restoring a platform's capability against time and is defined as: 'The ability to recover capability, partially or fully, within a specified timescale and to sustain that recovered capability for a given period'.

4.1 RECOVERABILITY MANAGEMENT

The UK MOD Recoverability Capability Working Group (RvCWG) was created to manage recoverability issues. The group is formed from key stakeholders including representation from equipment capability managers, Fleet, training, research, equipment support and future platform organizations. This level of representation strives to ensure coherence in the MOD and provides wider endorsement of programmes and policy statements generated by various sub-groups. The group has only recently been established and is maturing in its understanding of recoverability detail, but the inclusion
of future platforms has been essential. Inclusion at an early stage is helping to ensure that new vessel projects are approaching recoverability in a similar manner as defined by the RvCWG, whilst providing valuable feedback on the advice and guidance given.
A programme of work to define recoverability by specific and measurable metrics is underway. The ability to define and quantify recoverability is the corner stone to the entire capability management process. The first milestone reached in this process was the identification of seven key pillars, together with associated subfunctions (figure 8). Obviously, any attempts at modelling recoverability would have to contain the pillars and sub-functions

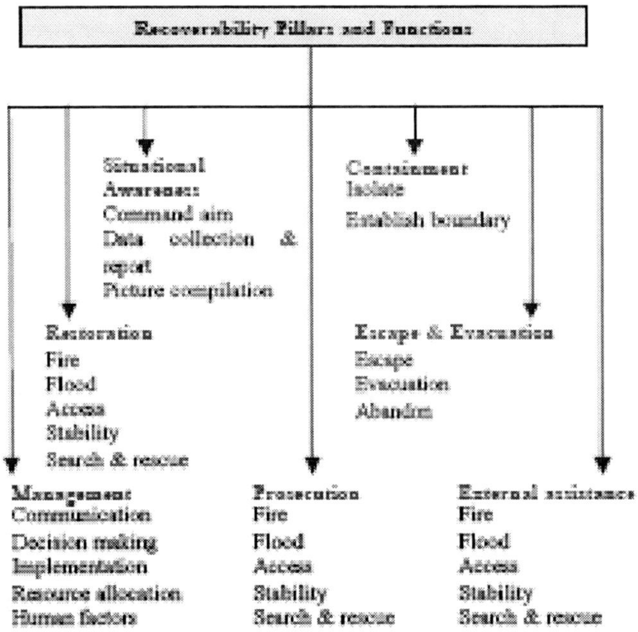

Figure 8: Recoverability Pillars

4.2 RECOVERABILITY REQUIREMENTS

The RvCWG agreed that acceptance of recoverability targets will be via design review, scenario analysis and/or recoverability modelling. To fulfil the latter, the MESH IPT contracted QinetiQ to extend its vulnerability assessment code, SURVIVE, to include a recoverability modelling capability. This capability is planned to be completed in 2007 and some of the requirements it will meet are listed below:

- The ability to conduct quantifiable and measurable manpower/equipment trade off's against platform recoverability.
- Assess the recoverability effects of inserting new technology both portable and fixed.
- Identify bottlenecks, critical paths and vulnerabilities in the damage control and fire fighting (DCFF) activities performed by platforms.
- Assess the recoverability of a platform within different environments including action-state, peacetime cruising and alongside.
- The ability to assess a specific platform in terms of recoverability relative to a known established benchmark.
- Respond to the requirements of the command aim and react to any revisions made during a typical platform recoverability process.
- Model the key secondary damage mechanisms e.g. fire, smoke, flooding, system malfunction, casualties and blocked access
- Evaluate and characterise the broad range of recoverability activities, technologies and systems assessing their impact on overall platform survivability as well as the interfaces.
- Develop a wide range of realistic operating scenarios (flying, RAS, littoral ops etc) to fully assess the performance of a platform to restore capability.

- Evaluation of proposed ship designs, includingmanning, against recoverability (survivability) requirements and agreed acceptance criteria.
- . The ability to select a post damage time frame and assess the current status of the platform in terms of operational capability, manpower allocation and equipment status.
- The ability to assess and monitor movement of personnel throughout a platform.

4.3 SECONDARY DAMAGE MECHANISM

In order to fully assess recoverability secondary effects such as fire and flooding had to be upgraded within SURVIVE. The fire algorithm had to address the insulation materials
used on board ships and the fire load provided by the compartments on the ship. The effect of crew actions on the spread on the fire was then implemented. The flooding algorithms were modified to considered pumping actions by the crew and self flooding during fire fighting. A large angle stability algorithm has now been included within SURVIVE.

4.4 CREW MOVEMENT

On of the primary developments for recoverability modelling within SURVIVE was the addition of a capability to move crew and equipment around the vessel. The crew are tracked individually at each timestep using a network model to calculate available routes through the ship. An example of a complete node network is shown in figure 9. It should be noted that not all routes are available at all times. For example, a smoke - logged
passageway will only be available to crew using breathing apparatus.

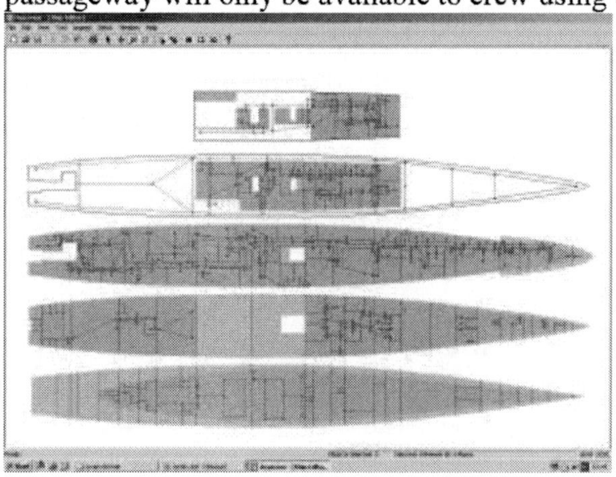

Figure 9: Node network within SURVIVE

4.5 DC&FF AND EQUIPMENT REPAIR

The DC&FF and equipment repair covers the following:

- DC&FF methods and equipment specification.
- . Diagnostic procedures and times.
- . Equipment failure criteria and t imes to repair (where possible).

- Spares availability.
- System reconfiguration via the network model.

Figure 10 depicts the layout of DC&FF equipment together with other equipment shown in green. It should be noted that although the DC&FF equipment are shown as icons, this is purely for clarity. Within SURVIVE, they are modelled in the same manner as other equipment and are subject to the same damage effects and system dependencies. For example, before a firehose can be used, SURVIVE checks that there is no damage to the hose itself and that the HPSW system can supply the hydrant it is connected to.

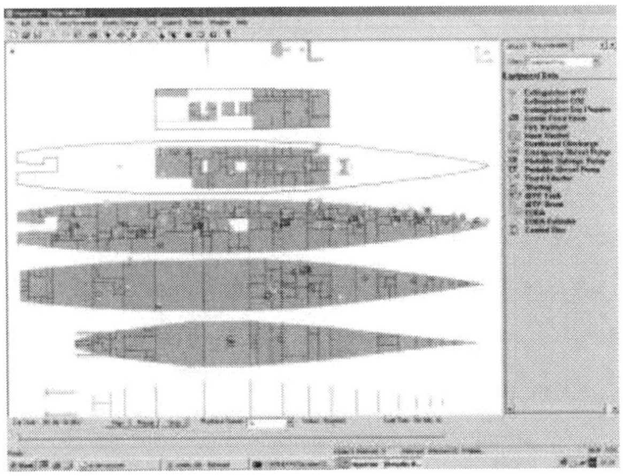

Figure 10: DCFF equipment within SURVIVE

4.6 AUTOMATIC RECONFIGURATION

Another network model will be used to emulate automatic system reconfiguration post-damage (e.g. chilled water). This same model will also calculate total load requirements for electrical and fluid systems. Where a shortfall in generating capacity is encountered, the control module will decide which equipments are shut down on a priority basis. This capability will help optimise the layout of reconfigurable systems.

4.7 SIMULATION CONTROL

The recoverability analysis within SURVIVE is controlled by a module which is responsible for the allocation of manpower, and other resources, with a view to meeting the command aim at all times. Where this is not possible, such as after an attack, it will do the following:

- Compile a picture of the event as per real-life. e.g. manual surveys of compartments, sensor data (if deemed to be operational).
- Prioritise DC&FF and warfare activity.
- Make decis ions based on it's perception of the damage - not necessarily the real picture.

4.8 RECOVERABILITY OUTPUT

A recoverability simulation will be capable of generating vast quantities of very detailed data. It is therefore necessary to reduce the data to enable the user to interpret the results more easily. The screen shots contained within this paper give some idea of potential
output. Additionally, data can viewed as animations through time, graphs of capability through time or even single physical parameters such as temperature versus time for a single compartment. Figures 11 and 12 depict a vessel before and after a hit from a missile. The resulting damage to equipment, and therefore systems, is calculated by SURVIVE and displayed as follows. Firstly, equipment coloured green are undamaged. Note, although an equipment is shown as undamaged it may still require the restoration of a service(s). This is explained later in the system view. Secondly, equipment coloured red are damaged beyond repair (complete replacement required). Finally, those equipments coloured amber are damaged in so far as they either require a small repair or resetting in the case of trips. The right hand viewing pane in Figures 11 and 12 display the ships functions, services and other systems and even individual components.

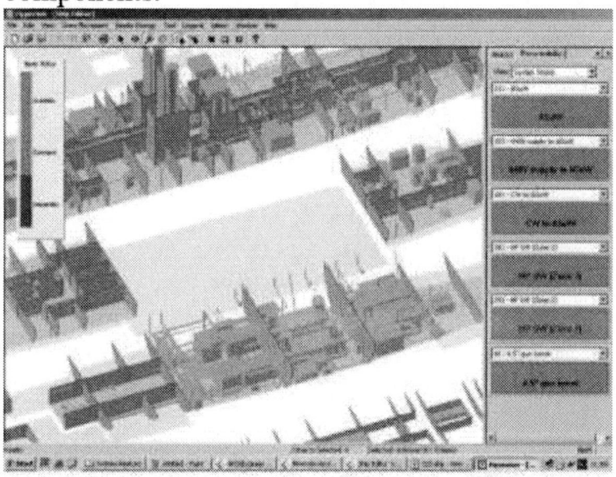

Figure 11: Equipment, systems & functions pre-damage

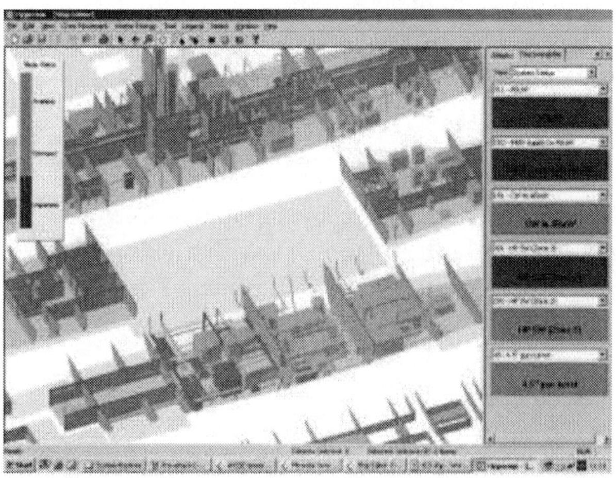

Figure 12: Equipment, systems & functions post-damage

Figure 13 shows a simplistic breakdown of the AAW function. In order for the AAW function to perform it requires all of the components with the exception of the

electrical supply which has an alternative. Figure 14 shows sample output at 4 different times
through the simulation, using the same traffic -light colour codes as before. Pre-damage indicates that everything is functioning correctly. One second after the hit we can see that the AAW function is no longer available. Clearly, this is due to the lack of both an electrical supply and hydraulic supply within the system chain. At the service detail level we can see that the normal electrical supply has failed but the alternative supply is not yet connected. After 5 minutes the alternative electrical supply has been connected but the
AAW function is still not available due to the lack of hydraulics. This has been corrected after 2 hours and AAW is clearly functioning even though the normal electrical supply is damaged.

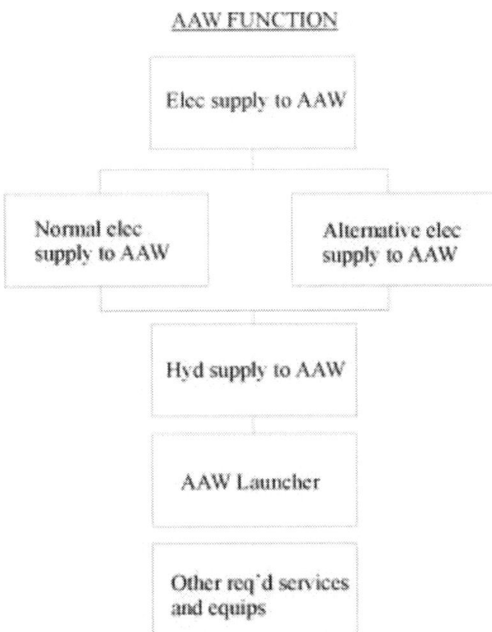

Figure 13: Simplified AAW Function

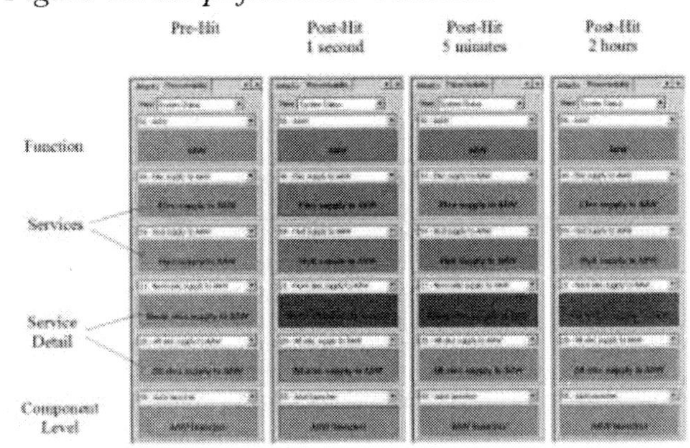

Figure 14: AAW Function breakdown through time

It will be possible in SURVIVE to assess competing design options to optimise platform recoverability.

5. SURVIVABILITY

Warship susceptibility and vulnerability have historically been studied separately, although both are related. Operational analysis would typically consider first the probability that a ship is hit and then (independently) the probability that particular capabilities are destroyed. Although this is a reasonable first approximation, it neglects the impact of ship signature on hit location and hence the probability of a system being damaged. Recoverability assessments have been conducted through analysis of past events and best practice. In an attempt to address this, QinetiQ has carried out an initial investigation into the feasibility of assessing the interactions between signatures and vulnerability (and recoverability when it becomes available). This has linked the outputs from QinetiQ's radar signature prediction tool Spectre to the vulnerability assessment tool SURVIVE, via a number of representative missile seeker algorithms. The impact of signatures on survivability was then examined for a simple test case.

5.1 ISIM

As described previously a typical SURVIVE assessment will involve simulating a series of missile attacks, arranged across a uniform grid. This is an effective technique for identifying where the vulnerable regions of a ship are, but does not take into account the biasing of hit locations caused by radar signature. To assess the feasibility of including these effects, a simple harness for holding different possible missile seeker algorithms was developed, known as the Integrated Survivability Interface Module (ISIM). This takes the output from Spectre, applies the selected algorithm and then produces
a non-uniform attack grid for use in SURVIVE. ISIM was intended to demonstrate the feasibility of linking signature data from Spectre into SURVIVE rather than provide a complete solution. As a result, the seeker algorithms used were simple generic ones, rather than detailed representations of specific missiles. Three different approaches were considered: hotspot tracking, range-gated centroid tracking and monopulse tracking.
These are described below.

5.2 HOTSPOT TRACKING

This algorithm uses cross-range RCS profile data generated by Spectre. The missile's aim point is assumed to be the centre of the cross-range cell with the largest RCS. The algorithm assumes that the missile's range gate covers the entire ship, which is appropriate for some older designs of missile.

5.3 RANGE-GATED CENTROID TRACKING

To demonstrate a seeker algorithm that involved range gating, the range-gated centroid tracker was introduced. This exploits two dimensional RCS data of the kind shown in Figure 15. The algorithm selects the aim point via a two stage process. Firstly, the row of cells with the greatest RCS is selected. This represents centring the range gate on the peak of RCS in downrange. Then the centroid of RCS for this row of cells is calculated. This is used to define the change in missile direction required. It should be noted that, when there are several large scatterers of similar RCS, the centroid may not correspond to any individual hotspot, but may fall between them.

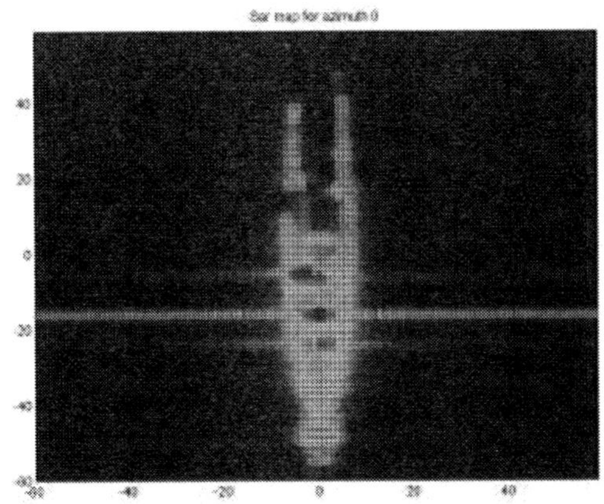

Figure 15: Spectre generated SAR image of HMS Scylla

5.4 MONOPULSE TRACKING

Monopuls e tracking is a more complicated approach commonly used in western anti-ship missile systems. It works by illuminating the target with multiple beams and using the difference between the signals received from each beam to steer the missile. The precise aim point selected will depend on how the monopulse beams and target RCS interact as the missile flies towards the target.

5.5 LEANDER TEST CASE

To assess the impact of the different aim point selection algorithms, a series of basic tests were carried out, using the former Royal Navy Leander class vessel HMS Scylla (Figure 16).

Figure 16: Spectre CAD model of HMS Sylla

To test the impact of ISIM generated aim points, two generic systems were added to an existing SURVIVE model of a Leander class frigate: propulsion (figure 17) and anti-air warfare (AAW, figure 18). These were chosen for demonstration purposes only and it must be noted that a normal SURVIVE model would be much more detailed than this.

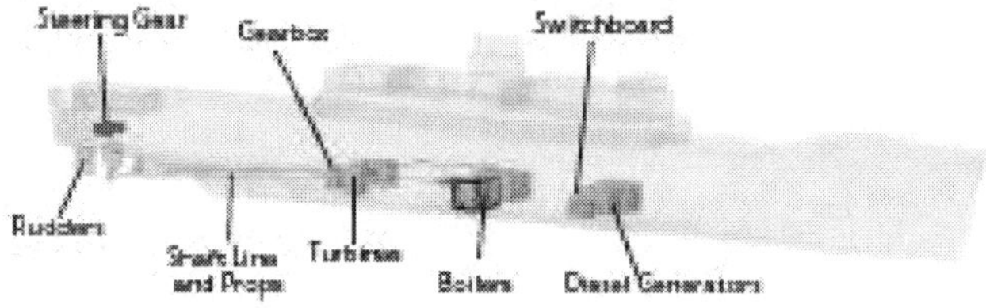

Figure 17: Generic propulsion system

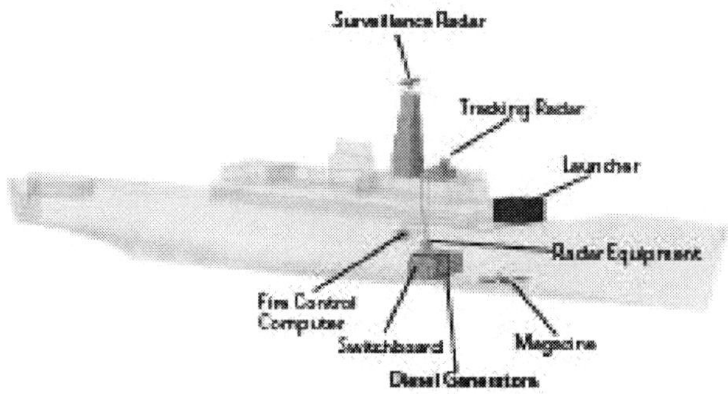

Figure 18: Generic AAW System

5.6 TEST CASE RESULTS

To assess the effects of hotspot tracking, the RCS of a simple CAD model of HMS Scylla was calculated using Spectre over narrow angular regions about 0, 45, 90, 135 and 180 azimuth. The generic demonstration missile was assumed to be 10m above sea level and operating at 10GHz, VV polaris ation. In each case, the resulting aim point was within 10m of the ship's centre. The range-gated centroid algorithm was tested at 45 intervals along the starboard side from 0 (bow on) to 180. In each case, the algorithm resulted in an aim point on a peak of RCS, figure 19.

Figure 19: range-gated-centroid tracking (HMS Scylla)

The complexity of the monopulse tracking algorithm prevented its inclusion in the prototype version of ISIM. However, some simulated fly-ins were carried out against randomly generated ship-like targets. These showed considerable variation of the aim-point with range, as illustrated in figure 20.

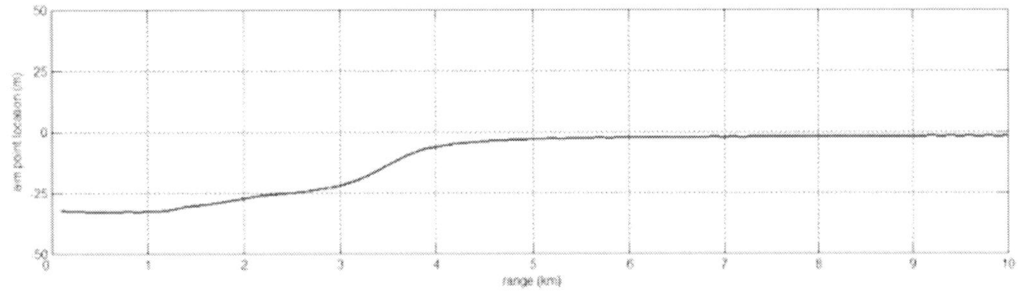

Figure 20: variation of aim point with range for the generic complex target

In the case of this generic target, the aim point varies from about -5m cross-range at 10km range to about -32m in cross-range at 1km (figure 20). This approximately corresponds to the location of the largest source of RCS on the target, the same result as the simpler hotspot tracker would have given. To test whether this was always the case, 20 further runs were carried out using monopulse tracking against targets with random RCS distributions and analysed manually. Of these, 16 (80%) hit the target within 1m of the peak of RCS, with the remaining 4 (20%) hitting secondary peaks. A further set of 1000 runs was also carried out and analysed automatically. Of these, 80.2% had final
aim points within 1m of the maximum value of RCS in cross-range.
These results suggest that, although hotspot tracking is a reasonable first approximation to missile behaviour, introducing a monopulse tracking algorithm into ISIM would significantly improve the accuracy of the results. The convergence of the statistical results for large numbers of runs suggest another possibility: that it may be possible to develop simple rules describing missile behaviour on the basis of more detailed simulation. These could then be used to incorporate the interactions between signatures and vulnerability into higher level models used for operational analysis.
To assess the impact of seeker algorithms on vulnerability, the hit-points generated by the range-gated centroid tracker (figure 22) were fed into SURVIVE and the results compared with those from a uniform attack grid (figure 21). In each diagram, the dots represent the attack locations and the colours represent the probability of damaging a given system.

Colour	Chance of disabling system
Grey	None
Blue	1 – 20%
Green	21-40%
Yellow	41-60%
Pink	61-80%
Red	81-100%

Table 1 – colour scheme used in attack grid diagrams

Figure 21: Standard 2D attack grid

The grid spacing for the 2D run used a standard fidelity setting, with 3m spacing between shots in the x direction and 2.5m spacing in the z direction (to hit each deck). Sensitivity runs have shown that this is the optimum spacing for assessing frigate sized vessels. As can be seen from figure 22, the hit distribution from ISIM is very different. Hits tend to cluster around the main superstructure region, with none being registered aft of that and relatively few forward. These hits were also caused by the missile locking on to the superstructure but, due to the slight dive, hitting the forward section instead.

Figure 22: Attack grid from ISIM

Table 2 shows the difference in system vulnerability given by SURVIVE for this example. The concentrated superstructure hits predicted by the ISIM approach have reduced the vulnerability of the propulsion system by approximately one third compared to the linear grid, but have slightly increased the vulnerability of the AAW system. This suggests that including signatures and seeker algorithms into overall vulnerability assessments will significantly affect the estimated overall survivability.

	Propulsion		AAW	
Grid type	% hits	% vuln	% hits	% vuln
Uniform 2D	36%	29%	57%	51%
ISIM	32%	20%	56%	56%

Table 2 – impact of the ISIM grid on system vulnerability

6. CONCLUSIONS

It is imperative that warships are protected and that measures are taken to maximise their survivability. Survivability can be optimised by addressing it at the earliest stage in the design. This survivability should be managed throughout the design process to ensure that the addition of service systems does not compromise the initial survivability. Tools are available to assess all aspects of survivability. These tools should be used to
optimise the design of warships. The ISIM model has been developed to demonstrate the
feasibility of integrating signature data and survivability calculations. ISIM uses the signature data generated by Spectre to determine missile aim points, which are then

used to define attacks in SURVIVE. This capability can be used to improve the accuracy of existing vulnerability calculations and also to model the effects of novel technologies on overall survivability. Tests carried out with the generic missile models on a basic ship model suggest that the radar signature could have a significant impact upon the overall survivability of the vessel, by biasing the hits towards certain parts of the structure.

The initial tests carried out on simple monopulse trackers suggest that it may be possible to establish simple statistical rules to describe the output of detailed analyses. This would allow the interactions between signatures and survivability to be included in higher level OA models.

AN EFFECTIVE CORVETTE PLATFORM TRAINING SYSTEM

Dr. Harro Kucharzewski
MarineSoft GmbH/Managing Director
info@marinesoft.de

Robert Wiedemann
MTU Friedrichshafen GmbH/Senior Manager for Naval Project
robert.wiedemann@mtu-online.de

Matthew Chang
MarineSoft Pte Ltd/Project /Business Development Manager
matthew@marinesoft.com.sg

ABSTRACT

This paper will present an On Board Training System (OBTS) for navy corvettes. Machinery operation simulation enables operators to interact with the ship's engine and automation system as it can be done in real. The project mainly focuses on a realistic simulation of the interaction between the ship's main engines and the onboard automation system both delivered by MTU. In order to achieve a most realistic training environment the overall goal was to design a training system that behaves and reacts close to reality. Embedded into the onboard training simulation is an interface to MTU's Monitoring and Control System MCS-5. Herewith the training simulation is stimulated by real monitoring data and gives back a realistic output to the MCS for training purposes. Beyond the intrinsic value of OBTS training extended training forms should be made available. Such kind of an extended OBTS includes also features for the Land Based Training (LBTS). This configuration enables trainees to perform in different kinds of training like guided system training, SPOT ON training and Team Resource Management.

The paper presents an essential strategy to form such streamlined OBTS and LBTS training for platform system operation incorporating the origin automation system. Thus it allows to assign, exercise and assess daily onboard duties as realistic as possible.

1. INTRODUCTION

Meeting today's and tomorrow's training needs has always been a challenge for navies worldwide. With limited natural resources and limited time for training purposes, there exists a growing need to constantly find new and innovative ways to maximise training outcome and efficiency. One way this can be achieved is the use of state-of-the-art technology as a training opportunity for enhancing mariner's performance.

One way to sharpen the profile, knowledge and skills of a mariner is the usage of intelligent simulators and simulations, which are not only foster technical skills, but also are able to react and behave in military terms.

With the increasing sophistication of modern ship systems and rising training costs, navies and marine industries around the world have to explore new ways for training in order to optimize costs.

PC based Platform Specific Simulation running on COTS hardware is actually a widely used option for delivering sophisticated training content embedded in a learning and training environment.

Distinguished by many experiences and contributions in navy training projects our intention behind Simulation Based Training was always to define firstly the objectives and requirements of the (simulation based) training process as a whole and taken from that to create a training concept. These formulated concepts are the heart of the simulation based training environment. These concepts determine essential functionalities of the simulation and not vice versa the simulation as a training tool allows determined training activities without to give instructors any influence using their own training methodology.

This approach was also applied in defining learning and training requirements for MTU's simulation based training philosophy for a Multipurpose Corvette.
Beside a Land based training system there was also demanded an opportunity for training on board.

2. THE IDEA OF *DISTRIBUTED* TRAINING SERVER FOR *CONTROL AND OPERATION SYSTEMS*

What's the problem today?

Identifying the trend that usually test simulators are compiled to on board training systems this question comes up: Are these so called on board *training* systems really dedicated to the current training needs and requirements of a Navy?
Obviously the answer must be "no".
Today's On Board Training Systems (OBTS) usually are lacking sufficient quality in sophisticated training technology as it is implemented satisfactorily in CBT systems or advanced land based training simulators (LBTS).

All training carried out at OBTS needs the skills of an experienced instructor. But with no sufficient training technology even for the most skilful instructor it is difficult to reach training results from OBTS covering all necessary knowledge and skills. Thus long term training outcome that can be generated herewith is seldom at the level desired.

3. REQUIREMENTS FOR MEANINGFUL OBTS

An extended OBTS particularly designed for training purposes is supposed to include

- a high degree of fidelity
- real time system behaviour,
- technical functionality
- and training content

whereas all features are dedicated to the framework of training procedures and tasks.

This may support the learning and training process optimally during all stages of training. Extended systems considering this approach are the best guarantee for a high training outcome.

Technical requirements

In order to achieve a most realistic training environment the overall goal was to design a training system that behaves and reacts close to reality. Therefore the technical concept is based on an embedded training system. Generally the technical simulation serves as shell for using parts of the real on board automation system. The technical solution focussed mainly on two crucial aspects:

- the operational simulation of the MTU engines and the propulsion for Multipurpose Corvette
- and an interface between this simulation and the real onboard automation system MCS-5 delivered by MTU.

Due to the original data visualization given by the native automation system and its close interaction with the technical simulation a highly realistic training environment can be established on board.

Training requirements

Instead of only simulation functions provided by an OBTS a meaningful or extended OBTS should offer at least the same training stages as a LBTS:

- Training Planning
- Training Preparation
- Briefing
- Exercise Supervision
- Assessment
- Debriefing
- Training Administration for instructor guided or instructorless kind of training.

During the last years there was developed an essential strategy to form such stream-lined OBTS and LBTS training concept for platform system operation. It allows to assign, exercise and assess one's daily duties onboard.

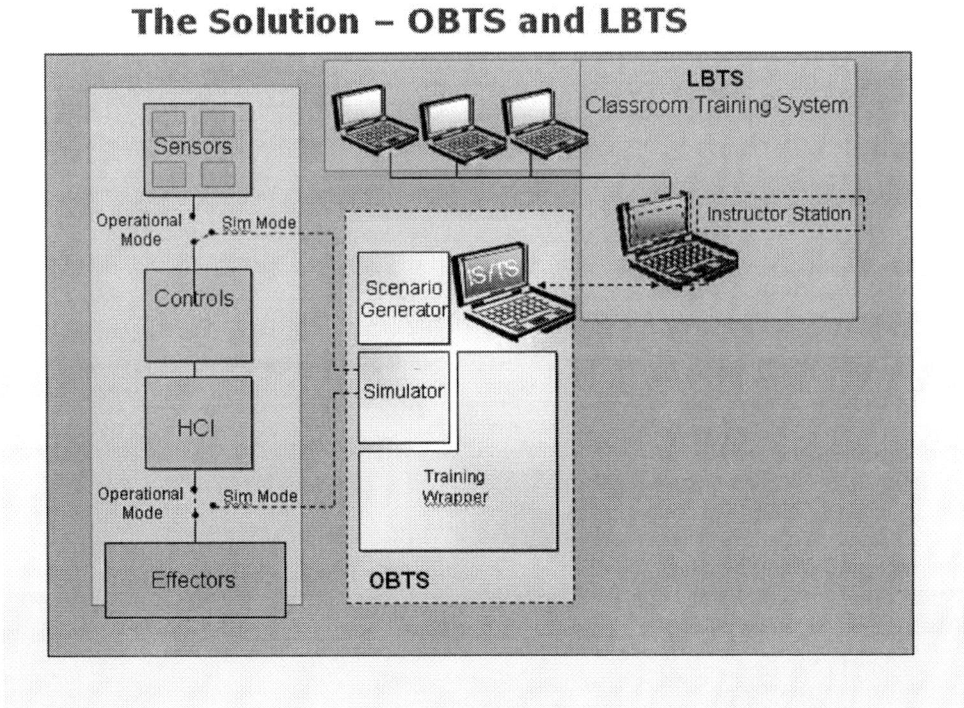

Exceptional training functionality can be achieved by the utilization of specifically designed training components from a land based training network. Thus the instructor/trainee on board can be supported with regard to all main training matters:

- Familiarization
- Basic Operation
- Trouble Shooting
- Team Operation

4. THE IMPLEMENTATION

Multipurpose Missile Corvette

Length:	72m
Tonnage:	840 to
Engines.	4 x MTU 595
Gearbox:	1 x Renk combined DAD & 2 x Red. Gearbox
Propulsor:	3 x KaMeWa Waterjet

Automation System

- **Propulsion Automation**
  Remote Control System RCS-5 DAD for propulsion plant with waterjet.
- **Electrical Power Management System**
- **Shipboard Automation**
  with MCS-5 for Bilge, Door and Ventilation system
- **Fire Detecting System**
- **Equipment Health Monitoring EHM**
- **Uninterruptible Power Supply UPS**
- **Closed Circuit Television System CCTV**
- **Engine Telegraph System MTS-5**
- **Onboard & Land base Team Training System**
- **Battle Damage Control System BDCS**

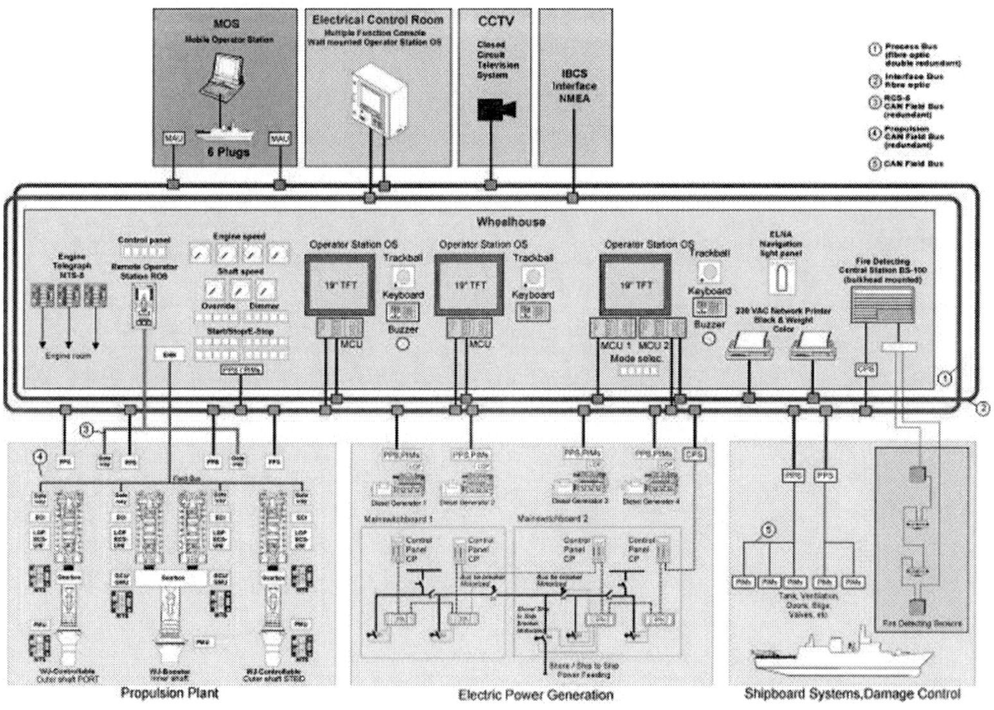

5. THE PC-BASED LBTS

The LBTS is PC-based and consists of an instructor station (IS) and trainee stations (TS) with enhanced training capabilities like self training, assessment and knowledge management. Training and self-training can be effectively carried out at the land based Trainee Stations, and can be continued at the original consoles on board by the OBTS. The TS provides the Human Computer Interface (HCI) of the Integrated Control System (ICS) and emulates local controls and panels of subsystems relevant to the operation in daily routine and emergencies.

The implemented Simulation Based Training functionality (SBT) allows to carry out interactively all essential operational procedures at any time independent of the real use of the ICS consoles. The user at the trainee station will be presented the same information as it is displayed to the operators at the local controls of the ICS system. In this respect computer-assisted instruction could replace the presence of an instructor.

Furthermore to the training of troubleshooting procedures the trainees can deepen into maintenance activities.

Every new application requires an introductory session to be able to use it effectively. Although the simulator program is build with an intuitive interface, it is important to allow first time trainees of this training tool to be accustomed to the simulator environment and pickup essential navigation skill before induction of any formal simulation exercises.

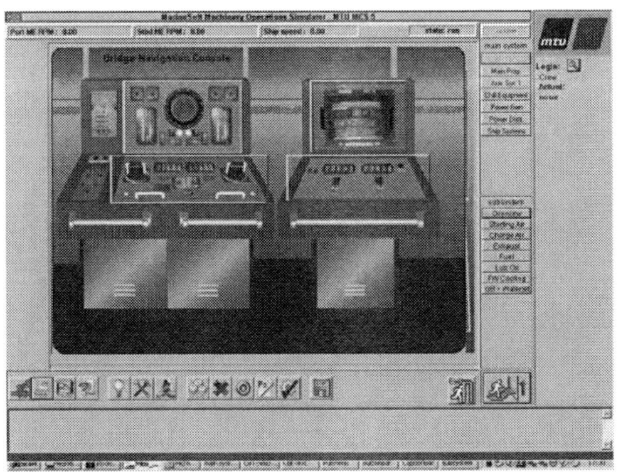

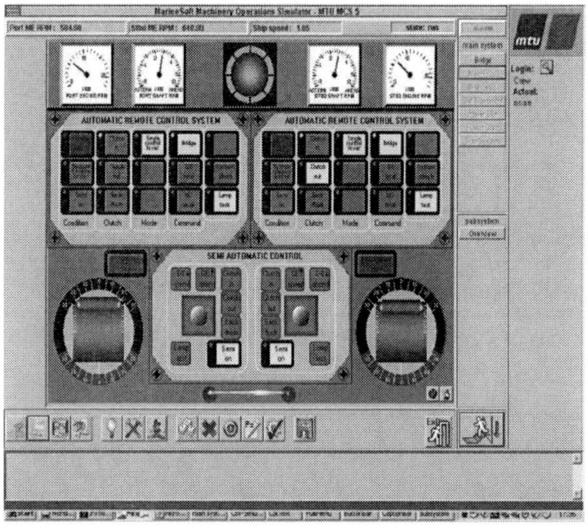

Simulated Bridge console

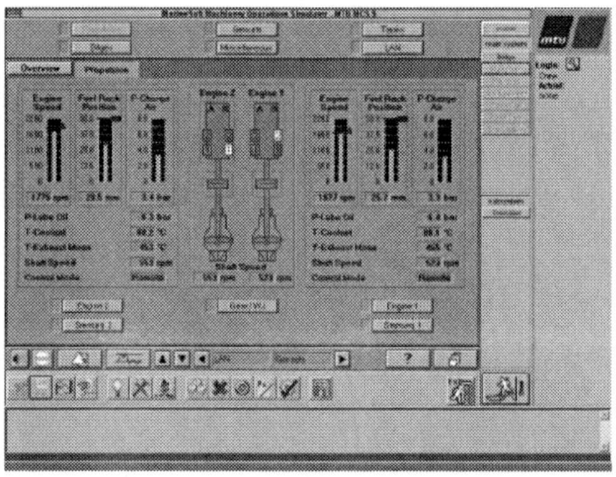

Original MCS 5 mimic integrated into simulation

6 SPOT ON TRAINING

The **SPOT-ON-Training** (Self **p**ace **o**n **t**ime **o**n **n**eed) environment is mainly dedicated to the purpose of self-learning and self-instruction. Arising from technologically driven trends, e-learning will gain more and more importance in simulation based training methodology.
Considering this fact Spot On training promotes not only institute-based self-paced learning ashore but also on board on-the-job training activities with the option for supervision from ashore (advanced distributed learning).
This kind of training has diverse features to structure self-paced learning; trainees can watch a demonstration, take prescribed exercises, and take tests to access their competency level.

It is usable for the purpose of:

- Consolidation of lessons, the practice of exercises and preparation for examinations within the institute-based training environment,
- Familiarisation with new machinery systems and equipment,
- Coaching, refreshing, and the drill emergency procedures
- Assessment of competency before promotion, etc. within an on-the-job training.

However it should be emphasized that the most satisfactory results will be achieved by using a combination of self-learning and instruction by an experienced tutor.

7 TEAM RESOURCE MANAGEMENT (TRM) TRAINING

TRM Training embodies the idea of co-operation between individuals using a common simulation platform.
This can be realised with custom-built exercises based on the simulator capability. In this concept, the operation of the simulated system is shared among the members of the group. Each member will inherit a particular set of system like the fuel oil system or cooling water system to operate. The entire group is inter-operating on a single platform. Each execution in the program will have interdependent effects on the system operational stability. The individual actions have positive or negative impact on the group efforts and influence the exercise outcome.

The success of the exercises centres on teamwork among team members as well as resource management skill of the team leader. This concept alone can meet a large spectrum of training needs, as it is required beyond the training of technical abilities.

Because of the complexity of system interoperation and a certain degree of human operation error, the system needs a period of time for balancing itself or arising out of a system imbalance before an alarm or failure occurs. This is the reason why this kind of training is deeming to be very challenging.

Team Resource Management Training is an invaluable tool especially in the areas of fostering team effort and communication as well as management and leadership skills of team leader and therefore an essential part in the assessment of upper classes.

Author's Biographies

Dr.-Ing. Harro Kucharzewski is the Managing Director of MarineSoft Entwicklungs- und Logistikgesellschaft mbH since 1990 with key qualification in ship automation and IT-based Simulator Training Development. He studied automation and thermodynamics at Technical University Odessa/Ukraine where he graduated as qualified engineer in 1969. Afterwards, he was working at the ship yard Rostock as manager at the development and research department at the former Warnowwerft Warnemuende (now Aker Ship Yard Warnemuende). From 1977 to 1992, he was teaching automation and ship operation at the Maritime Academy Warnemuende/Wustrow. He took the doctoral degree (Dr.-Ing.) in the subject of usage of adaptive controllers on ships from the Maritime Academy of Warnemuende/Wustrow in 1982.

Mr Robert Wiedemann is the Senior Manager for Naval Projects in Marine Automation Systems of MTU Friedrichshafen GmbH.
He is responsible for the application engineering of control and monitoring systems for complex propulsion systems, electrical power management systems, ship auxiliary systems, BDCS Battle Damage Control Systems, OBTS On Board Training System and LBTS Land Based Training Systems for military and paramilitary vessels as well as for the Sales Support for the Region Asia Pacific.

Mr Matthew Chang is the Project/Business Development Manager (SEA) of MarineSoft Pte Ltd. He joins the company after his 6-year stint with RSN in 1999. In 2001-2003 he was the project manager for the PV and LST platform system emulators. In parallel to his project management and fiduciary roles, he is also actively involved in the business development aspect for MarineSoft GmbH in SEA region. His specialise fields are in the area of applied information technology in training and marine systems.

GENERALIZED FORM RAY TRACER (GFRT) FOR USE WITHIN THE MEASURE OF TOTAL INTEGRATED SYSTEM SURVIVABILITY (MOTISS) PROGRAM

Morrisseau, P.F.
Alion Science and Technology
pmorrisseau@alionscience.com

Sajdak, J.A.W
Alion Science and Technology
jsajdak@alionscience.com

ABSTRACT

Survivability is a probabilistic indication of a system's ability to obtain or maintain some prescribed designated capability. In the area of military systems, or of large commercial structures, it is difficult to assess survivability to real world events without conducting tests of those events on the actual system. Due to the high cost of these tests, modeling and simulation is of great importance to understanding the behavior of a system to a real world threats. The Measure Of Total Integrated System Survivability (MOTISS) is a modeling and simulation tool developed to provide system designers and engineers a tool which allows rapid survivability evaluations of real world threats in a cost efficient and comprehensive manner.

Within MOTISS the motions and behavior of many physical properties such as blast wave propagation, explosively formed projectile flight, fragmentation or projectile motion need to be evaluated. The numerical representation of such phenomena hinges upon an accurate ($\pm\sim 10[cm]$) and rapid ($\sim 10[ms]$) path, or ray tracer. This paper discusses the operation and functionality of the Generalized Form Ray Tracer, (GFRT) sub-routine developed for the MOTISS program as applied to blast location and spread, threat/fragment/ballistic tracking, and shaped charge jet tracking.

INTRODUCTION

The Measure Of Total Integrated System Survivability (MOTISS) is both a process and a program. It is a process by which a system of system's (such as a naval or commercial ship, a building or an armored vehicle) overall survivability is assessed probabilistically in the event of one or more threats. MOTISS uses first principle physics coupled with empirical data and tests to provide a rapid first order solution to a threat event many thousands of times, with each test slightly varying the input parameters in order to capture the chaotic nature of the random unknowns. This provides the ability to determine, probabilistically, what would, will, or did happen in a similar real world threat event. MOTISS allows the system designer, engineer or operator to locate and rank weaknesses quickly in order to allow for corrective action before those weaknesses are exploited in a real world scenario.

However, even utilizing first order physics routines, when analyzing an event such as a threat weapon detonation and its effects on a system, the routines tend to be computationally intensive, thus disallowing rapid evaluation and returning the designer or engineer to 'waiting for answers'. One of the leading contributors to slow computation time within modeling and simulation is particle tracking and contact determination. It is this iterative position, velocity and acceleration update along with the contact search algorithms that so slows finite element applications. And while increasing processing power often mitigates some of this computational cost, the prohibitive expense of acquiring and maintaining such "super-computers" is often beyond the means of most engineers. Thus to enable MOTISS to provide the capability of "rapid" evaluation to designers and engineers an arbitrary time limit of 10 micro-seconds per particle for flight and contact determination using a standard desktop processor (Intel 2.16GHz) was subjectively prescribed for the MOTISS particle or ray tracer. The 10ms/particle allows nearly 6000 particles to be tracked within one minute or 360,000 to be tracked within one hour. Furthermore, the MOTISS particle tracer was also required to enable use in tracking the path of three dimensional spherical blast waves and linear shaped charge jets as well as standard ballistics and explosive fragments. Thus the MOTISS particle tracer, in order to efficiently track and determine contact occurrences within the modeled system geometry by each of these disparate phenomena, was extended to a *Generalized Form Ray Tracer* (GFRT).

Within this paper the assumptions, operation and functionality of the GFRT sub-routine is discussed as applied within MOTISS for ballistic, fragment, jet and blast tracing and contact.

GFRT THEORY DEVELOPMENT

The simplest way to track a particle, given an initial position, direction, drag coefficient, and velocity, is to iterate through its position time history and at every step compare its position to the geometric coordinates of every other object of the model. This method however is very computationally expensive. First, at every position increment the particles current position must be compared with each possible contact object, and if not directly limited in the search direction, the possible contact entities could be in the many hundreds of thousands. Second, if the increment is too large (i.e. the time step long enough that the particles velocity results in large position change), then contacts may be missed, effectively having the particle in front of an object at one time and behind it at the next. In this way, the particle may never be close enough to the possible contact entity to fit within its limited search grid. Though it is possible to overcome both these problems by falling back to "bucket" or "voxelized" search patterns and variable time stepping based on the particle velocity and geometric constraints such as is used within many finite element, discrete event, or engineering physics applications; these generalized tracking and contact methods would prevent the MOTISS program from meeting the time requirement of 10ms.

In order to achieve the time limit of 10ms/particle for flight and contact determination, a new method had to be developed to track geometric relationships more efficiently within

MOTISS. Further, MOTISS required more capability than the simple tracking of a single particle. For this reason the developers of MOTISS parted from standard particle flight and contact algorithm methodologies (to be returned to later) and examined the fundamentals of ray tracing theory.

Ray tracing theory was developed primarily for computer rendering graphics with respect to lighting, shadowing, and texturing (Martynov, 2000). These principles too have similarity within blast modeling and particle flight. The idea of "visibility" from the center of the blast to that which it may potentially come into contact established ray tracing as the MOTISS algorithm of choice to accomplish blast, fragmentation, jet and ballistic tracking. In simplified terms, ray tracing limits the contact search algorithm to only those objects, or target points, which have a clear line of sight (or ray) between the object itself and the original source of the loading function. Only after contact and penetration then does an update of the source and new rays be performed. Further, if the directionality of the source is known, only a single ray between the source and a contact entity exists, thus contact is automatically defined. In this manner, a ray tracing algorithm can establish which structure is "visible" to a blast wave, and thus subject to blast loading, by using rays from the object to the source of the detonation and jetting, fragmentation and ballistics can establish the contacted structure by utilizing both the directionality ray and that of the objects return rays to the projectiles source position. **Error! Reference source not found.** illustrates is the ray-tracing method.

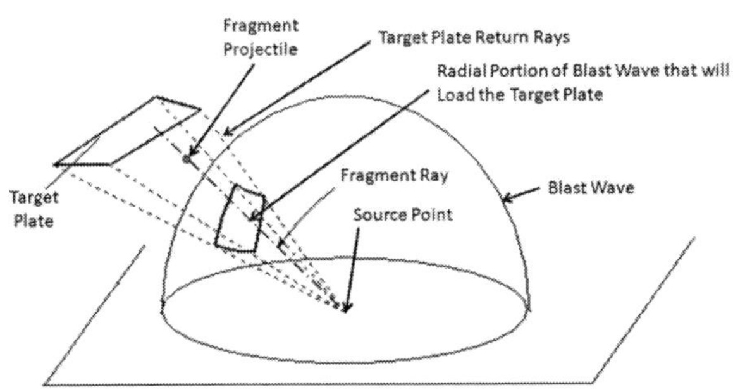

Figure 1 – Ray Tracer approach for determining "viewable" field for contact

Ray tracing does however come with three draw-backs in its application. The first is that ray tracing requires that the target object remains stationary to the source over the time duration required between the loading entity (blast wave, jet, ballistic or fragment) leaving the source and making contact with the object. While in highly dynamic system environments this may be a problem, the temporal assumptions within MOTISS, wherein the system response is treated in quasi-static time increments of seconds and loading is performed impulsively, negates the concern within this application.

The second draw-back of ray tracing results from its founding principles in computer rendering graphics and lighting. Light, having the benefit of traveling in waves, is not influenced by gravity (neglecting of course astronomical distances and cosmological effects), and thus ray theory does not permit parabolic motion. While this is sufficient for the applications of blast wave and linear jet tracing, ballistic and fragment motion is inherently parabolic. However, due to the short distances traveled by any fragment or ballistic within MOTISS (less than 100 meters) and the typical initial projectile velocities of 700 to 2000 meters per second, then the greatest induced error is approximately 10 cm. Since the modeling fidelity required by MOTISS is only at the decimeter scale, the induced parabolic error, additionally at the decimeter scale, may too be neglected.

The final draw-back in the application of ray tracing is in the time required to conduct the ray determinations, sorting and selection of all components and structure return rays. For this, we return to standard particle flight and contact analysis once the preliminary ray trace has limited the search. Using simple particle path tracking even when following straight lines, tracking large amounts of particles in a complex geometry can compound computation time. To streamline the contact portion of the algorithm, the GFRT takes advantage of voxelized nature of the model inherent in MOTISS. The Axis Aligned Bounding Box (AABB) formulation is a rectilinear mapping of the form of the system being modeled into boxes defined by a minimum and a maximum point with all sides aligned with an axis. The compartments or damage zones (DZ) and the vital components (VC) of the system are modeled in this formulation. Each DZ is bounded by structural elements or plates with specific material properties and each VC is assigned to a particular DZ. By understanding the relationship between the DZ's, VC's and plate elements, the tracking and contact algorithms can be made more efficient. Because the boundaries of each DZ are aligned with an axis and made up of contiguous plates, and each VC is associated with a particular DZ, the geometry search domain can be reduced based on the particle's present DZ location. If the particle is external to the model, the search domain can be reduced to only the DZ's and VC's which make contact with the environment. By reducing the amount of the geometry comparison loops by one or more orders of magnitude (i.e. 7568 plate elements reduced to 12 for a particular DZ) the computational speed also increases by orders of magnitude.

FUNCTIONAL SCHEMATIC AND FLOW

There are four different routines built on the GFRT theory; these are Ballistic, Frag, Jetting, and Blast. Ballistic is responsible for tracking the threat itself, be it a cruise missile, torpedo, bullet, etc. Frag generates and tracks fragmentation if present. If the threat is a shaped charge, Jetting tracks the resulting liquid jet, and finally, Blast is for analyzing the blast wave of detonating threats.

The Frag algorithm utilizing the GFRT theory is shown in figures 1-3. The general flow of the code throughout the program is governed by the fragment or projectile current position along the ray. Within the function, the modules are broken down in the Sub-Functions and Descriptions section.

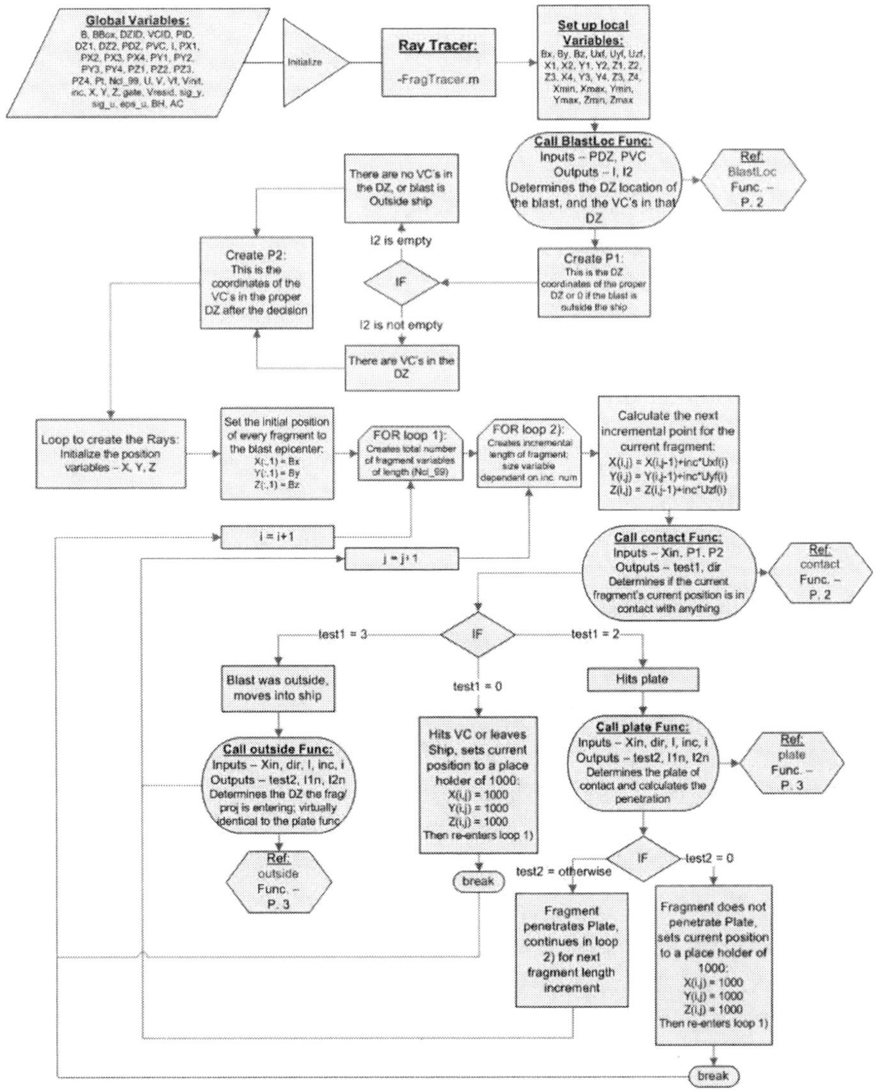

Figure 2 – Functional flow chart of the GFRT as applied to fragments

SUB-FUNCTIONS AND DESCRIPTIONS

The GFRT is broken into four sub-functions. The sub-functions are:

1. **BlastLoc** – At the onset of the threat-type event, determines the epicenter of the explosion or weapon firing. Inputs P1, and P2, are the X, Y, and Z min and max coordinates of all damage zones and vital components respectively; Outputs, I and I2 are the DZID associated with the blast and the VCID associated with the DZ respectively.

2. **contact** – Determines the state of each fragment or projectile as it progresses through space relative to the structure. It tests the location and hit state of the fragment or projectile for each iteration. Inputs, P1 and P2, like in BlastLoc, are the coordinates of the DZ's and VC's but are now specific to the index I and I2; so, P1(I) and P2(I2). Xin is the current position vector of the frag/proj, in the form [X Y Z]. Output, test is the indicator of the fragment's or projectile's contact

possibilities. If test = 0, the frag/proj hits VC or leaves ship, if test = 2, the frag/proj hits a plate, if test = 3, the frag/proj was outside the ship and is entering the ship, if test = 1, the frag/proj hits nothing and continues to the next incremental iteration.

3. **plate** – Locates the plate or plates involved in frag/proj contact. Determines the state of penetration, and the residual velocity or the penetration ratio if no penetration. Inputs, Xin is the current position of the frag/proj, dir is the direction indicator used to calculate which DZ boundary is being hit, thereby which plate is being contacted along with its unit normal vector for calculating the obliquity angle, I is DZID of the DZ that the frag/proj is currently in, inc is the increment size, i is the current frag/proj number of length 1-Ncl_99. Output, test initially determines if the frag/proj leaves the ship; if test = 0, the frag leaves the ship and the current analysis ends. Test is also set based upon the output of the penetrate function described in 4. I1 is the index of the next DZ entered if the plate is penetrated. If I1 is zero frag/proj leaves the ship, its analysis will end, and the next analysis begins. I2 is the VC index associated with the new DZ. ThePlate is the plate ID of the plate being contacted.

4. **penetrate** – Evaluation of the penetration of a ballistic through a target (structural, plate, armor or armor assembly). Inputs, Xin current frag/proj coordinate, ThePlate is the plate ID of contact as determined by the subroutine plate, I is the DZID of the current DZ, inc is the increment size, i is the current frag/proj number. Output, pass indicates whether or not the plate was penetrated, if pass = 1, the plate was penetrated and the fragment continues with a residual velocity, Vresid; if pass = 0, the plate stopped the frag/proj, the penetration ratio is calculated, and the next fragment calculation is started.

EXAMPLE CASES

Two platforms are used for example cases for verifying the fragment and projectile generation code; the first platform is the STAB 1, and the second platform is a notional DZ/VC configuration. First, a fragment generation test and a blast spread example will be observed using the STAB 1; then the plate contact analysis will be examined using the notional DZ/VC configuration.

TEST 1: FRAGMENTATION DISTRIBUTION ANALYSIS

For Fragment distribution analysis, the STAB 1 vessel is used to simplify the geometry in order to aid the visualization of the fragments. The Blast occurs within the ship, and spreads throughout. The focus of this test is to observe the randomized distribution of the fragmentation. The physics of the fragmentation follows the summarized fragmentation theory presented on the next three pages as determined by the TM5-1300 (Army Manual No. 5-1300, 1990.) The values for the important variables are provided in the table below.

Table 1 – Test 1 Variable Definition

Variable	Value	Definition
Bx	-72 [m]	X-component of the blast location
By	5.05 [m]	Y-component of the blast location
Bz	6 [m]	Z-component of the blast location
Uxw	-0.704	X-component of the warhead unit vector
Uyw	0.55	Y-component of the warhead unit vector
Uzw	0.45	Z-component of the warhead unit vector
EqTNT	90 [kg]	Equivalent TNT mass
MuFb	10 [deg]	Forward casing body frag. Mean Azimuth angle
MuAb	150 [deg]	Aft casing body frag. Mean Azimuth angle
SigFb	25 [deg]	Forward casing body fragment std. dev. to mean azimuth angle
SigAb	25 [deg]	Aft casing body fragment std. dev. to mean azimuth angle
Mub	18	Mean number of warhead casing bands
Sigb	6	Std. dev. of warhead casing band number

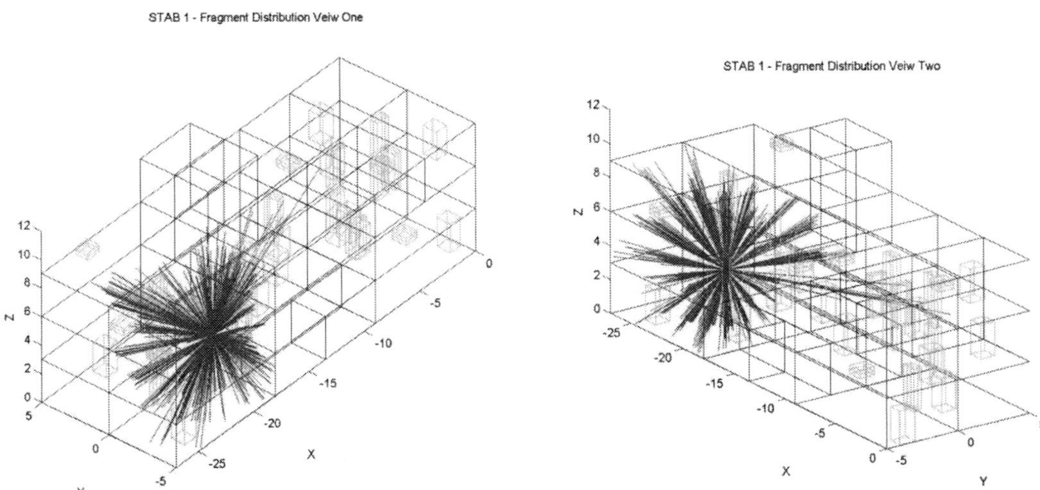

Figure 3 – STAB 1 Fragment distribution

The left hand side of the figure above shows the fragment distribution viewed from aligned perpendicularly to the warhead's unit vector. It can be seen the split of the cylinder from top to bottom. There are 901 fragments generated as found by the calculation of Ncl_99, which is the number of fragments to a 0.99 confidence level. The right view is aligned with the unit vector axis of the warhead. From this view, the banding experienced by the warhead case is evident. The quantity of bands is randomized based on the mean and standard deviation of the banded fragment parameters. For this analysis, the mean number of bands is 18, with a standard deviation of 6. The figure above depicts the mean of 18 bands for this analysis.

TEST2: SINGLE FRAGMENT TRACKING TEST

For this test, one fragment from a distribution will be examined and tracked through its path until it either exits the platform, or hits a VC. Penetration analysis is not examined here. The variable definitions important to the test are mentioned below.

Table 2 – Test 2 Variable Definitions

Variable	Value	Variable	Value
Bx	-72	Uyw	0.55
By	5.05	Uzw	0.45
Bz	6	DZ	27
Uxw	-0.704	Frag #	1

The complete path for every increment of Fragment #1 is listed in the table below. The information provided in the table is incremental X, Y, and Z locations, the increment that contact is present, the plate the fragment passes through as found by inspecting the geometry, the direction of the damage zone the fragment is leaving (forward, aft, port, starboard, ceiling or deck), and the plate the code calculated into the variable ThePlate.

Table 3 – Total Incremental Step for Fragment #1 in the Notional Ship and the plate contact analysis therein

Increment	X	Y	Z	Contact	The Plate by Inspection	Case - dir	Code Result
17	-73.867	5.9606	7.2024	CONTACT	By inspection the plate is 192	Case 5: Ceil	Plate 192
53	-78.067	8.0094	9.9078	CONTACT	By inspection the plate is 624	Case 5: Ceil	Plate 624
86	-81.918	9.8875	12.388	CONTACT	By inspection the plate is 854	Case 3: Port	Plate 854 Exit Ship

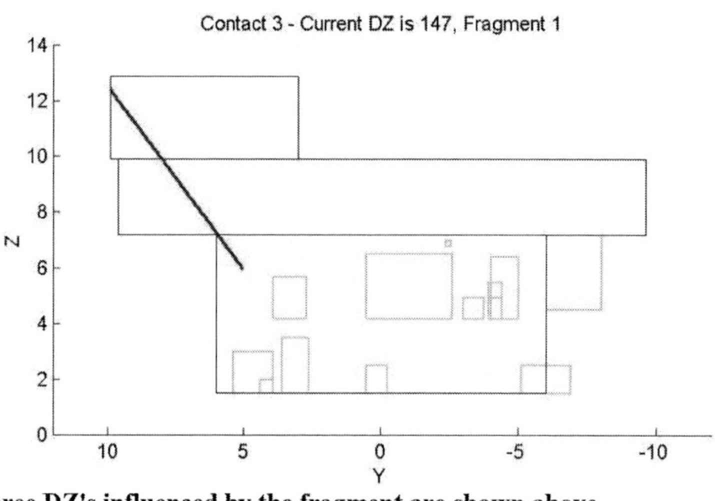

Figure 4 – The three DZ's influenced by the fragment are shown above

TEST 3: EXTERNAL AIREX BLAST SPREAD EXAMPLE

To exemplify the spread of an impulsive blast wave front, an external blast progression is shown on the STAB 1 for clarity. The blast wave geometry is solved using the solid angle sphere rendering algorithm (Balázs, 1997.) At any point of upon contact, an object will be influenced by the blast determined by the "visibility" of the object by the epicenter of the blast. In other words, what percentage of the overall object is shadowed or not shadowed by another object.

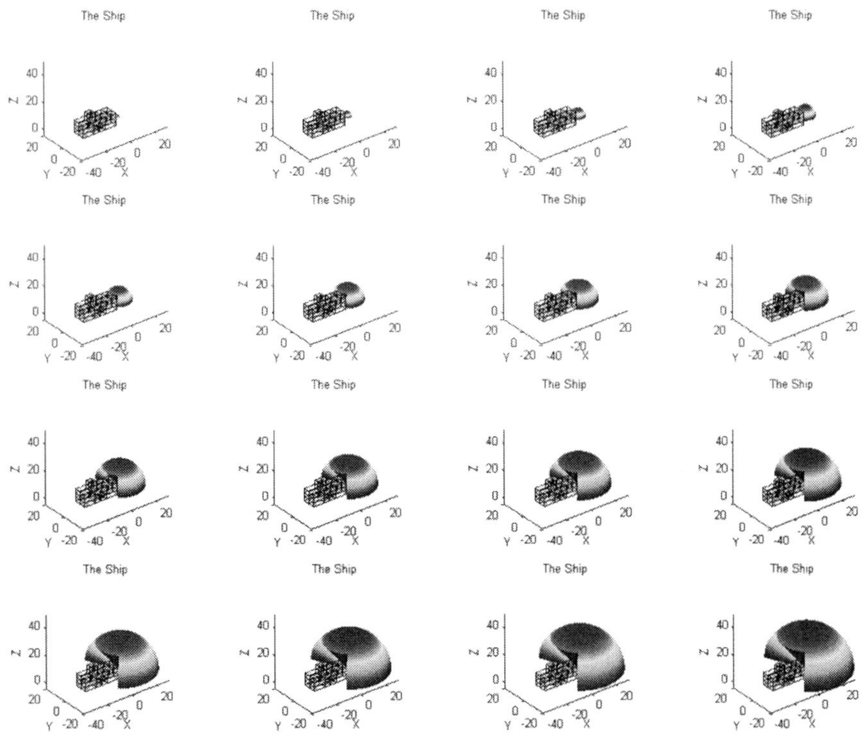

Figure 5 – Example Blast Wave Progression

Shown here is illustrated the assumption that the blast wave will not cross the Air/Water interface. Since the blast is an AirEx blast, the shock wave is stopped at the water line.

CONCLUSION

For the single fragment test, this particular fragment, originating in DZ 27, progressed through two more damage zones, and contacted three plates. Observing the geometry of the path line of the fragment, the plates observed matched the plates that the code generated. The three damage zones and the path line for the fragment shown in the figures above depict the progression of this fragment. The path can be visually verified as well as being geometrically verified through examination of the coordinates. This test shows that the plate function is operating correctly. The single fragment test took 0.000325[s] or 0.3[ms] to complete 86 increments of 15[cm] far below the 10[ms] goal for each fragment path.

The last test examines the external blast in air for the tracing of the blast wave front. The example illustrates the assumptions stated above. The blast wave front is eliminated by the surface of the water at the water line of the barge. Also, as the wave front encounters the ship, the azimuth and elevation solid angle are reduced from the shock wave influence. The propagation of the complete blast wave took 0.03[s] in 276 radial increments.

The computation time of the weapons effects module has been drastically reduced by the GFRT. Before its implementation, the threat modeling and the calculation of its effects took many minutes to complete; it now takes less than one second with the same degree of accuracy. By reducing the computation time for the weapons effects portion of MOTISS, it allows for the addition of higher fidelity and more refined calculations to the other portions of MOTISS.

REFERENCES

1. Popescu, L.M. "A Geometry Modelling System for Ray Tracing or Particle Transport Monte Carlo Simulation" Journal of Computer Physics Communications; 11 October 2002, *PACS:* 02.50N, 29.27.EG, 41.85.Ja, 87.53.Wz

2. Martynov, A, Mount D, "A Ray Tracing Algorithm (Sphere Rendering)" Independent Research Project; University of Maryland, Dept. of Computer Science, May 2000

3. Balázs, C, "Derivation of the Rendering Equation" for Central European Seminar on Computer Graphics for Students, 15 April 1997

4. Army Manual No. 5-1300 (TM5-1300), Navy Manual No. NAVFAC P-397, Air Force Regulation No. 88-22, "Structures to Resist the Effects of Accidental Explosions", Headquarters, Departments of the Army, the Navy, and the Air Force, Washington DC, 19 November 1990.

Cost savings for warships using the reconfigurable hull form concept

Martin Renilson, Australian Maritime College

Summary

Many warships are required to operate over two distinct speed ranges. This gives the naval architect quite a challenge when it comes to optimising the design.

It is widely accepted that a single set of prime movers will not be able to efficiently cope with the vastly differing power requirements for these two speed ranges, and hence warships are often fitted with two sets of prime movers for this purpose.

On the other hand, it is also obvious that a single hull form can not be optimised for these two distinct speed ranges. Earlier work (Renilson *et al,* 2006, 2007) demonstrated possible modifications to the hull that could be made to improve efficiency at the two distinct speeds. The most promising modifications identified from this work were changes in the stern shape, and the propulsion type. This paper focuses on these modifications, and discusses possible practical arrangements to achieve the changes necessary, along with estimated fuel savings as a function of vessel speed profile.

It is concluded that considerable fuel savings are possible, and that this concept warrants further consideration in order to reduce the through life cost of warship operation.

1 Introduction

Many warships are required to operate efficiently at two different speed zones: low speed cruise; and high speed sprint.

Historically, a single optimised hull form would be developed, with a balance between the different roles, based on the requirement set and on the operations envisaged. In many cases this would use two different prime mover configurations, one for the low speed and one for the high speed.

This paper discusses the possible advantages of a ship with a reconfigurable hull form, allowing optimisation for each of the two different speeds. This work built on a previous study (Renilson, *et al*, 2006, 2007) which demonstrated various improvements in operational efficiency.

A 93m Offshore Patrol Vessel (OPV), capable of a top speed of 40 knots, but with a normal cruising speed of only 15 knots, was used as the test case to demonstrate the concept.

2 Approach

The reconfigurable features that were considered are:

- reconfigurable stern shape;
- transom flap;
- propulsion; and
- prime movers.

It was assumed that the hull design for such a craft would be for the 40 knot operating regime, and that this would result in a non-optimum hull shape at the 15 knot cruise. An existing five metre long fine form model, which has a wide transom suitable for waterjet propulsion was used, as discussed in Renilson *et al* 2007.

A transom flap was designed for this hull form for 40 knot operation, and this was considered to be the standard vessel for the purpose of this work.

For the low speed operation an extension was fitted to negate the transom drag. A suitable extension was designed for the existing fine form, however because this has flat aft buttocks the extension was required to have a length of 14m, which was longer than desirable to achieve zero transom immersion whilst limiting the amount of separation that occurs.

A hull form designed for this purpose could have steeper buttocks aft, with an adjustable transom flap, resulting in a shorter stern extension.

It was assumed that for the low speed the reconfigurable vessel would be propelled using a single azimuthing thruster, the waterjet inlets would be covered to reduce the drag, and some of the water in the waterjet ducts evacuated using a pump (Renilson, *et al*, 2007).

Resistance experiments were conducted using the model with and without the flap and stern extension, and at three displacements to determine the influence of increasing the displacement due to the additional mass required for the reconfiguration. The principal particulars for the model and ship are given in Table 2.1.

	Model scale value	**Full size value**
Overall length (m)	5.318	98.9
Waterline length (m)	4.980	92.6
Displacement (t)	0.269	1,774

Table 2.1 Principal particulars of standard vessel

Propulsion coefficients for the thruster and the waterjet were obtained from existing data and used to convert the effective power (measured in the Ship Tank) to the required brake power.

3 Reconfiguration

The reconfiguration concept adopted is shown in figure 3.1. As can be seen, the stern extension is out of the water for the high speed operation, resulting in an immersed transom and exposure of the waterjet outlets. With this concept the stern extension is hinged at the top edge of the transom. The stern extension could then be rotated into position as shown in Figure 3.1. The thruster unit would not itself have to be retractable, as it would automatically come out of the water for the high speed operation as the stern rotates.

This has the advantage of it being easier to incorporate a larger diameter propeller (as the thruster is not retractable) which will give greater efficiency at low speeds. It will also result in lower RPM being required, and hence lower acoustic signature which may be very important to the low speed operating regime.

It may also be possible to make use of the additional deck space, for example during helicopter operations, as these would not be carried out at high speed.

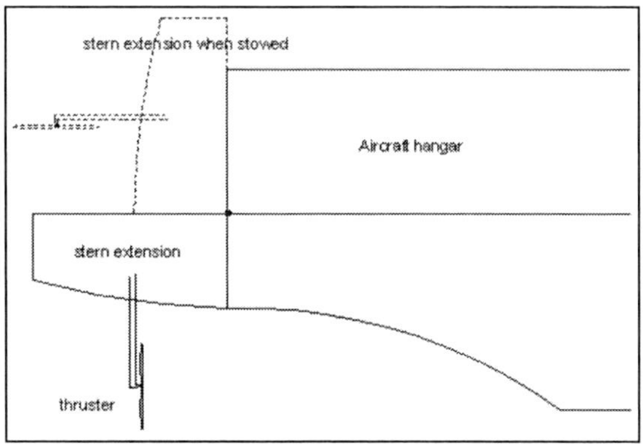

Figure 3.1: Simple schematic of hinge-able stern concept

The transom flap would be deployed for all high speed operations, and retracted during low speed operations. This is likely to have a small cost and mass impact which were not considered at this stage.

4 Mass estimates for reconfiguration

Inclusion of the stern extension and its associated machinery in a ship design would result in a mass penalty, which in turn will increase displacement and resistance. As discussed in Renilson *et al* 2007, the mass of the stern extension will depend on its length. It will also depend on whether the stern extension is to be considered structural, or not. As it will normally be deployed at low speeds, and hence in rough weather, it will need to cope with a degree of wave loading, however unless it is required to support a helicopter (as noted could be possible in section 3) it could be constructed much more lightly than normal hull structure.

There will be a reduction in mass due to the need for a smaller Diesel engine for low speed operation, the reduction in fuel required, and the evacuation of the waterjet ducts.

To obtain the total effect of the reconfiguration on the displacement of the vessel it would be necessary to carry out a full concept design study, which is beyond the scope of the present work. However, it is estimated that the net increase in displacement could be of the order of 250t.

5 Required power

The power required for both speeds with and without the extension and transom flap was obtained from model experiments conducted in the Ship Tank at QinetiQ, Haslar, which are fully reported in Renilson *et al* 2007.

The model experiments were conducted at three displacements to determine the effect of the mass of reconfiguration on the power required.

Propulsive efficiency for the propeller at 15 knots was estimated to be 0.65, and propulsive efficiency for the waterjet over the range were obtained from Svensson, 1994 and Svensson *et al* 1998.

The resultant required brake powers for the base case displacement (excluding the mass of the extension) are given in Table 5.1.

Speed (knots)	**Low speed configuration (kW)**	**High speed configuration (kW)**
15	1,400	2,600
40	-	35,000

Table 5.1: Required brake power for base case displacement (without stern extension))

6 Potential annual fuel saving

The annual fuel saving will depend on a range of factors including: percentage of time spent at each of the two different speeds; total annual deployment; specific fuel consumption of the prime movers; additional mass required for the configuration; and fuel cost.

For this study a specific fuel consumption of 190g/KWhr for the Diesel, and 225g/KWhr for the gas turbine were used, as discussed in Renilson *et al* 2007, and a fuel cost of A$1,000 per tonne was assumed.

An annual deployment of 200 days was assumed.

The annual fuel savings for the reconfigurable vessel, compared to the standard vessel, are presented as functions of the percentage of time operating at 15 knots in Figure 6.1 for the case where there is no change in mass due to the reconfiguration, and where the reconfiguration increases the vessel mass by 250t.

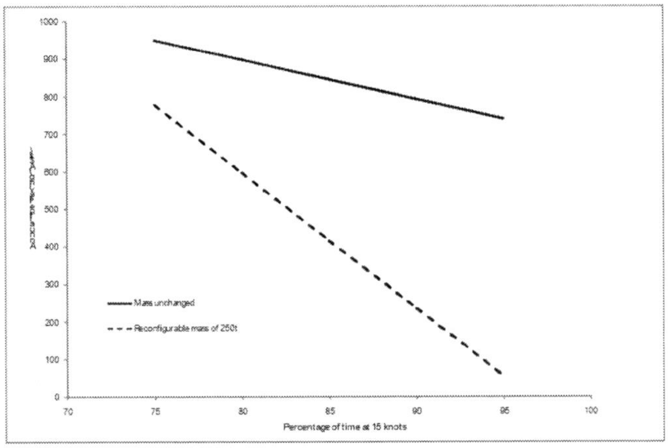

Figure 6.1: Annual fuel saving for an annual deployment of 200 days

As can be seen from Figure 6.1, this means that if an OPV of this size spends 85% of its time at 15 knots it would have an expected annual fuel saving of approximately A$400k, assuming a fuel cost of A$1,000 per tonne. This extrapolates to A$8M over a 20 year life.

Clearly, if the annual deployment were greater than 200 days then the savings would also be greater. Conversely, if the additional mass were greater than 250t then the savings would be less.

If fuel prices increase then the savings will directly increase.

7 Concluding comments

In order to investigate the benefits of a reconfigurable hull form, this concept was applied to a 93 m OPV with an assumed operating profile of 15 knots for 85% of the time and 40 knots for the remainder of the time over an annual deployment of 200 days.

This resulted in an annual fuel saving of the order of A$400k at a cost of A$1,000 per tonne. This is sensitive to the additional mass of the reconfiguration concept, as this causes an increased power requirement at 40 knots.

This concept could be applied to vessels of other sizes which require the ability to operate efficiently in two distinct speed regimes, including motor yachts.

8 Notation

LCG	Longitudinal centre of gravity
LWL	Waterline length
OPV	Offshore Patrol Vessel
P_B	Brake power
P_D	Delivered power

Δ	Displacement
ρ	Water density

9 Acknowledgements

The initial work was funded and carried out as part of the Weapon and Platform Effectors Domain of the MoD Research Programme. This was conducted when the author was employed by QinetiQ, and the author would like to acknowledge the assistance given to him by his colleagues there. In particular, he would like to thank: Owen Modral, Karl Randle, and Jim Taylor who worked with him on the initial studies; Chris Ross and Bob Scrace for their advice on ship design and hydrodynamics aspects respectively; and Chris Richardsen for his considerable input in overseeing the experimental activities.

10 References

RENILSON, M.R., MODRAL, O, and TAYLOR, J.W., 2007, *An investigation into the benefits of reconfigurable hull forms* International Shipbuilding Progress, 2007.

RENILSON, M.R., RANDLE, K. and TAYLOR, J.W., 2006, *A preliminary investigation into two speed operation of ships*, International Journal of Maritime Engineering, Vol 148, Part A2, 2006.

SVENSSON, R., 1994, *Waterjet propulsion – Experience from high powered installations,* Paper 3, RINA International Symposium on Waterjet Propulsion.

SVENSSON, R., BRIZZOLARA, S., and GROSSI, L., 1998, *Trial result including wake measurements from the world's largest waterjet installation,* Paper 12 RINA International Conference on Waterjet Propulsion.

SPEECH TO THE PACIFIC 2008 INTERNATIONAL MARITIME CONFERENCE

SYDNEY CONVENTION CENTRE

THURSDAY 31 JANUARY 2008

IT'S TIME TO RESTORE THE ROLE OF SHIPPING IN THE AUSTRALIAN TRANSPORT AND LOGISTICS INDUSTRY

BY PADDY CRUMLIN
NATIONAL SECRETARY, MARITIME UNION OF AUSTRALIA (MUA) AND VICE CHAIR, INTERNATIONAL TRANSPORT WORKERS FEDERATION (ITF)

SPEECH TO THE PACIFIC 2008 INTERNATIONAL MARITIME CONFERENCE

SYDNEY CONVENTION CENTRE - THURSDAY 31 JANUARY 2008

IT'S TIME TO RESTORE THE ROLE OF SHIPPING IN THE AUSTRALIAN TRANSPORT AND LOGISTICS INDUSTRY

PADDY CRUMLIN
NATIONAL SECRETARY, MARITIME UNION OF AUSTRALIA (MUA) AND VICE CHAIR, INTERNATIONAL TRANSPORT WORKERS FEDERATION (ITF)

Good morning

Judging by the subject matter of this Conference, and even of this session, it seems that my presentation did not fit neatly into the Conference themes. That is a bit of a surprise really because there is nothing particularly radical or unconventional about my subject matter. This probably reflects the fact that despite the significance of shipping to the Australian economy, it has not over the past decade received much policy or public attention. That situation will change under the Labor Government.

I hope also that my presentation will go some way to changing that situation.

For some reason, the reverence with which truckies are held in the Australian folklore and the fascination that so many have at various stages of their lives with trains, does not seem to exist in relation to ships, at least not in Australia. This hasn't always been the case in Australia, and it is quite different in many other parts of the world, particularly in island nations like ours.

Perhaps it's my profession as a trade union leader speaking about such matters which may seem unconventional and out of place – or at least may run against the conventional wisdom of what engages trade union leaders. Given the vilification that trade unionists received during the last Federal election campaign, one could be excused for thinking I was going to front up here today with a sawn off piece of anchor chain and create havoc.

The fact is that we, the Maritime Union of Australia, have never gone away, so we don't need to come back. We have been out there engaging, discussing, promoting and working in the Australian shipping industry despite the political cycles and despite failed policy actions specifically designed to crush the union. We might have long memories, but there's no place for holding grudges. The actions of Governments who attack the right of trade unions to exist and play a legitimate role in representing the interests and aspirations of workers in a democratic society reflects poorly on those Governments and their cohorts, not on the trade unions.

Our values remain steadfast, intact and legitimate. I hope in some small way to demonstrate why that is the case in the presentation I make today.

We will continue to promote shipping in general, and particularly a domestic shipping industry, because there are good public policy reasons for doing so.

Just consider these facts about, firstly, international shipping:

- Over 90% of world trade is carried by the international shipping industry. Without shipping, the import/export of affordable food and goods would not be possible - half the world would starve and the other half would freeze!

- There are around 50,000 merchant ships trading internationally, transporting every kind of cargo. The world fleet is registered in over 150 nations, and crewed by over a million seafarers of virtually every nationality.

- The United Nations Conference on Trade and Development (UNCTAD) estimates that the operation of merchant ships contributes about US$380 billion in freight rates within the global economy, equivalent to about 5% of total world trade.

- Over the last four decades total seaborne trade estimates have nearly quadrupled, from less than 6 thousand billion tonne-miles in 1965 to over 27 thousand billion tonne-miles in 2004.

- Between 1980 and 1999, the value of world trade grew at 12% per year, whilst total freight costs during this period increased by only 7%, demonstrating the falling unit costs of marine transportation.

    - Freight costs continue to represent, on average, less than 6% of the import value (or shelf price) of consumer goods. I will return to this issue.

And second, consider these facts about Australian shipping and the Australian maritime sector in general:

- Approximately 99 per cent of Australia's merchandise trade is moved by sea.

- Australia has the fifth largest shipping task in the world, in terms of tonnage and kilometres travelled.

- Our international maritime trade has increased significantly over the past decade, increasing from 408 million tonnes in 1994/95 to 681 million tonnes in 2004/05.

- Over the next twenty years, containerised trade is forecast to almost triple from 5.2 million Twenty-Foot Equivalent Units (teus) in 2004-05 to 14.9 million teus in 2024-25.

- Total non-containerised trade is forecast to more than double over the next twenty years, from 0.6 billion tonnes in 2004-05 to 1.4 billion tonnes in 2024-25.

- About 30 thousand visits were made by all ships at all Australian ports in 2004-05. The number of ship visits is forecast to nearly double over the next 20 years to around 54,800 in 2024-25.

- In the LNG trade, the quiet achiever among our bulk commodity trades, the Australian Petroleum, Production and Exploration Association (APPEA) has estimated that an extra 20-30 LNG tankers will be required to service new Australian LNG projects over the next 10 years (there are currently 11-12 LNG tankers in the trade now).

- We have also witnessed the interest of global capital in our ports. In 2006, both longstanding and principal stevedoring companies in Australia - P&O Ports and Patrick Corporation - were bought out:

  - P&O Ports was bought by Dubai Ports World, which catapulted it to the 3rd largest Global Network Terminal operator, while Patrick Corporation was bought by Toll Holdings, since split to form Asciano – the major Australian freight logistics company.

- In 2007, the Qld Government announced that Hutchison Port Holdings, one of the top 5 Global Network Terminal Operators, was the preferred tenderer for new wharfage at the Port of Brisbane.

I would ask that you contrast those revealing facts, which demonstrate the importance of shipping to the global economy and to Australia, with the state of the Australian shipping industry, which has been in virtual free fall over the past decade or so:

- There has been a decline from 66 to 45 Australian registered trading vessels:
  - a loss of 21 ships or a decline of 31.8%;

- There has been an increase in foreign registration of Australian ships, from 5 to 9:
  - an increase of 4 or an 80% increase in foreign registration;

- There has been a decline from 71 to 54 in the total number of Australian trading ships:
  - a decrease of 17 or 23.9%; and

- There has been a corresponding decline in deadweight tonnage (DWT – broadly, the measure of the weight that a vessel can carry) from 3.28 million DWT to 2.14 million DWT:
  - this represents a decline of 34.7% in DWT capacity of Australian shipping.

These statistics are graphically represented in Figures 1 and 2, derived using data from different sources, but which demonstrate the same downward spiral of Australian shipping.

Figure 1: The Australian Coastal Fleet – Number of Vessels and Total Deadweight – 1995-2006

FIGURE 37: THE AUSTRALIAN COASTAL FLEET - NUMBER OF VESSELS AND TOTAL DWT, 1995-2006

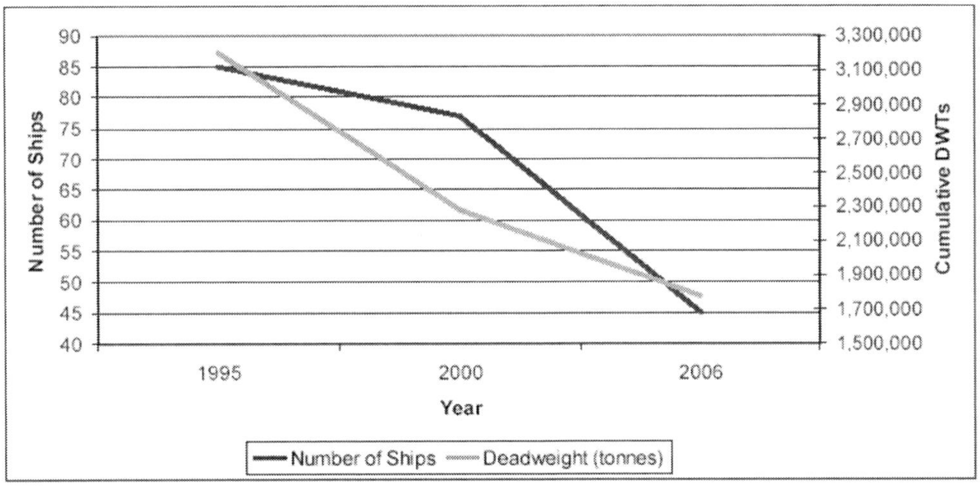

Source: BTRE 2006a, and Meyrick and Associates estimates

Source: International and Domestic Shipping and Ports Study, Meyrick and Associates/Booz Allen Hamilton, May 2007

Figure 2: The decline of Australian shipping

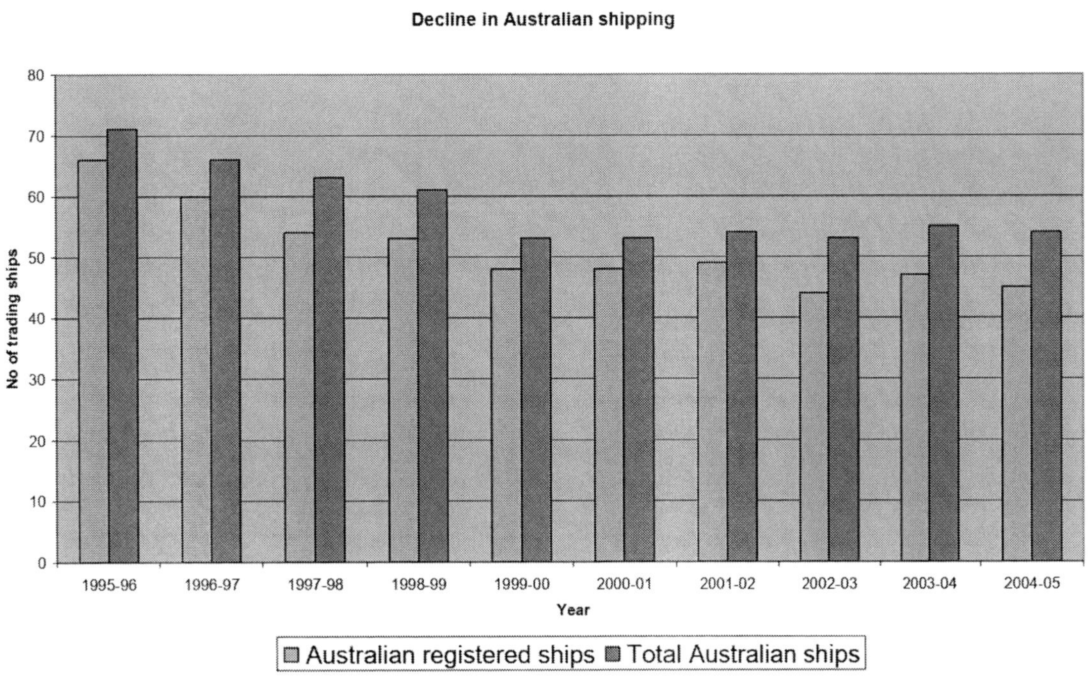

Source: *Australian Maritime Transport 2005*, a report for the Australian Shipowners Association prepared by the Apelbaum Consulting Group, March 2006

While it is true that the restructuring of the industry led to some loss of tonnage in the early 1990s, the really serious decline of the industry occurred under the Howard years.

On coming to power in 1996, the Howard Government took a number of deliberate steps aimed at strangling investment in Australian shipping.

It terminated the PAYE rebate scheme (a scheme that assisted seafarers in the intentional seafarer labour market); it terminated the shipping Capital Grants scheme (to have been applicable to vessels delivered on or before 30 June 1997) as at 30 June 1996 (though subsequently extended to 30 June 1997); and it terminated the accelerated depreciation provisions that supported investment in Australian shipping.

Despite a strong industry campaign arguing for a modest package of reforms to support Australian shipping, no positive initiatives occurred during the Howard years, except the Government initiated a number of reviews.

For example, in March 1999, the Report of Shipping Reform Group was handed to Government, and in May 1999 the Shipping Reform Working Group Report was also handed to Government. Still there was no Government action, except the then Transport Minister, John Anderson, made the now infamous statement that Australia was a nation of shippers, not a shipping nation.

Neither of these reports has ever been made public. There is widespread acceptance that they were not made public because they recommended the revitalisation of Australian shipping.

Out of frustration the Australian Shipowners Association established the Independent Review of Australian Shipping (IRAS) co-chaired by former Ministers for Transport the Hon John Sharp and the Hon Peter Morris. It reported in December 2001. Even that bipartisan report in support of the industry stimulated no action by the Howard Government.

It was not until 2004 that the then Transport Minister John Anderson set out what in fact is a non policy where he indicated that the key planks of Australian shipping policy are:

- That Australia's international trade should continue to rely on international shipping – which means foreign ships, and in many cases Flag of Convenience (FOC) ships; and

- A continuing reliance of the Coastal Trading Permit System to ensure Australian shippers can access foreign shipping – in contravention of the cabotage principles in the *Navigation Act 1912*, where permits were only meant to apply in limited circumstances to ensure that freight was not held up (permits were meant to be the exception, not the norm and are only meant to be issued when there is no suitable Australian vessel available).

The MUA is not opposed to the use of coasting trade permits. We are however opposed to the abuse of the permit system. The system has been blatantly abused through a variety of Howard Government sanctioned procedures, such as:

- Late applications for permits. Shippers and/or their agents are making application for a permit just before a cargo is required to be shipped, having discreetly ascertained that a foreign vessel is available to transport their cargo, thus making it almost impossible for an Australian vessel to be available.

- Manipulation of cargo sizes to avoid the use of a licensed ship. Another means by which shippers are obtaining permits is by stating a cargo size on the permit application that they know the licensed vessel cannot meet, or would only be able to meet with more than one voyage, making it uneconomical for the licensed ship to take the cargo.

- Freight price differentials, rendering the licensed ship unsuitable. A disturbing development is that the Department of Transport and Regional Services has apparently accepted the shippers' argument that if the freight price quoted by the licensed ship is higher than that quoted by the foreign ship then the licensed ship is unsuitable. Even if the foreign shipping lines seek to maximise freight revenue from Australian coastal voyages where a permit is issued, this aspect of their operation is still "cream" on the overall international voyage, (i.e. it is freight they did not factor into freight rate quotes for the principal voyage) and in any case freight rates can be adjusted to always undercut the Australian licensed vessel.

We have termed these practices contrived unavailability or contrived unsuitability.

The administration of the coastal shipping permit system has permitted foreign vessels, many of which are substandard Flag of Convenience vessels, sometimes with forged documentation, invariably carrying crews which have not been appropriately background checked and which may or may not be complying with the International Labour Organisations International Ship and Port Security Code (ISPS Code), to ply the Australian coast.

A vessel is considered as sailing under a Flag of Convenience where the beneficial ownership and control of a vessel is found to lie elsewhere than in the country of the flag the vessel is flying. There are some notorious registries with less than best practice commitment to safety, security, and labour standards.

In contrast to the Howard Government approach to Australian shipping, the UK Government has established bi-partisan support for its shipping industry, while the NZ Government has set out a positive reform agenda for NZ coastal shipping. The British Government adopted positive support measures such as a tonnage tax and fiscal initiatives to encourage a return of vessels to the red Ensign. The policies have had considerable success in dealing with the critical shortage of properly qualified officers worldwide and the growing negative impact that has on national economic and regulatory management and developments.

Over the past decade, Australian shipping policy failed to acknowledge the massive impact a maritime labour and skill shortage will have on port and maritime regulation, pilotage, survey and classification and related areas of maritime overview into the future. Without a viable and sustainable merchant marine we will not be able to deal with the demands placed on us by our reliance on shipping to deliver 99% of our nation's trade. Unless shipping policy is seriously addressed in the period ahead, it will be one of the great failures of this generation of policy makers and will see this country rely almost totally on migrating in those skills to meet contemporary requirements.

Despite this regressive state of affairs, I believe we are on the cusp of a significant change that will begin to restore an important role for shipping in the national freight and logistics industry.

Leaving aside the election of a Labor Government late in 2007, some of the other factors that have led me to this conclusion are:

- The calls by a number of respected industry leaders for a greater role for shipping in the national freight task. For example:

  - On 3 May 2007, the maritime press reported on a statement by the Executive Director of the Association of Ports and Marine Authorities, David Anderson, that the Australian transport chain should prepare for the re-emergence of coastal shipping as a formidable alternative to road and rail.

  - On 23 May 2007 the Executive Director of the Australian Logistics Council, Hal Morris, is reported as saying that coastal shipping was missing from the national transport policy agenda in favour of rail and road reforms, and that the Australian Logistics Council is concerned at the lack of attention paid at a Commonwealth level to the development of coastal shipping as an alternate transport mode.

  - The regular and positive interventions into the transport policy debate by the Australian Shipowners Association.

- The positive statements made by political leaders across the political spectrum. For example:

  - On 22 May 2007, the then Shadow Minister for Transport, Martin Ferguson, issued a statement saying that Labor is committed to a strong, efficient domestic shipping industry that offers fair pay and working conditions to Australian workers and the responsible use of single and continuous voyage permits. The statement went on to say that in Government Labor will:

    > "*undertake a review of coastal shipping policy as requested by the shipping industry to improve the efficacy of the current cabotage regime – with the objectives of sustaining a viable and efficient domestic shipping industry, decent wages and conditions for working Australians, responsible use of single and continuous voyage permits, and safety and security in our ports and shipping lanes.*"

  - On 24 May 2007, the then Parliamentary Secretary with responsibility for Maritime Transport, De-Anne Kelly, is reported to have indicated to the Regional Ports in Focus 2007 Conference that:

    > "*the Federal Government was looking at a means of supporting the growth of Australia's coastal shipping industry and that the Government wanted to see coastal shipping take a greater role in the freight task over the next 20 years.*"

  - On 1 August 2007, Labor issued a major policy statement on shipping entitled Labor's Vision for Shipping in a National Intermodal Transport System.

- Support from Parliamentary Committees of Inquiry for a stronger role for shipping in the national freight task. For example:

  - The House of Representatives Transport and Regional Services Committee report entitled *The Great Freight Task: Is Australia's transport network up to the challenge?* Of 13 August 2007, arising from its inquiry into the Integration of Road and Rail Interface with Ports over 2006-2007 concluded that:

    - Moving more freight by sea may be an option to alleviate some of the growing pressure on land transport networks.

    - Australia is necessarily reliant on international shipping for its import and export needs. However, the coastal shipping option for transporting freight between Australian cities is currently overshadowed by road and rail.

    - The European experience suggests that coastal shipping has significant potential to curb anticipated increases in heavy vehicle road traffic, rebalance modal shares, bypass land bottlenecks and provide a sustainable transport option.

    - The role of coastal shipping should not be overlooked when examining existing transport network operations and prospects for future freight efficiencies.

    - The Australian coastal shipping industry does face a number of operational challenges, including a decline in the number of Australian-registered vessels, an ageing fleet, cabotage, potential skills shortage, industrial issues and foreign competition.

    - It is arguable that these, and related requirements, have been valuable in developing a domestic industry with quality, reliability and safety records that are high by world standards. These standards are in sharp contrast to the condition of some foreign vessels operating on the Australian coast.

    - Ultimately, the significant growth in freight demand may require optimising the use of both Australian registered and foreign vessel freight capacities. However, the continued and necessary role for coastal shipping of freight, combined with the levels of domestic freight being moved by foreign shipping lines, strongly suggest to the Committee an opportunity to foster the national shipping industry for domestic freight movements. When examining the viability of the coastal shipping option, the Government will need to consider what, if any, protection or support the domestic shipping industry warrants.

    - Even a small modal shift in favour of domestic shipping should reduce transport sector energy consumption and emissions.

    - Coastal shipping's potential lies in transporting less time critical freight. It represents an environmentally beneficial and cost effective alternative to rail and road modes, for bulk cargo shipped over long distances. Sea transport does not require the same infrastructure investment or maintenance. A clearer government framework for the industry would help to combat perceptions that act as a barrier to investment.

- That the coastal shipping industry warrants examination by the Australian Government. This consideration should include whether changes are required to ensure legislative arrangements are commercially appropriate, and consistent with measures applicable to investment and taxation of road and rail modes.

- If Australia is to ensure it is in a position to meet the challenge of the growing national freight task, it must engage all transport sectors in logistics planning. With almost a quarter of the freight task predicted to be moved by sea in 2020, it is essential to take into account the coastal shipping industry and its capacity to share the freight task, as part of a comprehensive national transport strategy.

- The signs emanating from the Transport bureaucracies with responsibility for shipping policy, signalling that there is a growing interest in shipping as an important component of the freight task, after years of policy inaction. For example:

  - In July 2007, the Australian Maritime Group (a forum of maritime officials that reports to the heads of transport departments), under the auspices of the Australian Transport Council, comprising all the Transport Ministers, released a report prepared for it by Meyrick and Associates entitled *International and Domestic Shipping and Ports Study*.

    - The fact that the study was commissioned in the first place represents a sea change in thinking in the policy echelons of the Commonwealth and State transport departments. Further, the study essentially made the case for a revival of an Australian coastal shipping industry, and pointed to the types of policy options that might be considered by Government if shipping is to play a greater role in the national freight task.

- A renewed interest in the shipping policy debate in the maritime and mainstream media. For example:

  - In June 2007, at the time of the grounding of the *Pasha Bulker* at Nobbys Beach in Newcastle, the maritime press gave some attention to the role of Flag of Convenience (FOC) shipping and the significance of the form of commercial shipping contracts, particularly the dominance of Free on Board (FOB) shipping terms in Australia's bulk commodity export market.

  - On 19 May 2007, shipping policy made it onto the front page of the Australian, arising from an MUA website media release commenting on the outcomes of the ALP Conference. This was followed by a second front page article on 21 May, as well as an Editorial and further coverage on 22 May. The Financial Review gave shipping policy some attention on 21 May 2007. Such attention is unprecedented in recent Australian history.

- The analysis in the Australian newspaper during October 2007 of the role of shipping and freight rates in the context of the BHP Billiton call for a new international pricing mechanism for iron ore contracts.

Over this same period the MUA commissioned two important reports on Australian domestic shipping aimed at lifting the quality of the policy debate.

The first was a report prepared by the National Institute of Economic and Industry Research entitled *Australian Coastal Shipping: its Future Role* (unpublished) of June 2007. That report noted that:

> *"Coastal shipping carries a little under a third of Australian domestic freight, measured in tonne-kilometres. As a freight carrier, it therefore ranks with road and rail transport, each of which, similarly, is responsible for about a third of the Australian freight transport task (BTRE 2006b). However, coastal shipping performs its task with commendable efficiency. Shipping as a whole, of which coastal shipping is but part, employs no more than 3 per cent of the transport workforce, incurs no more than 3 per cent of transport industry fatalities, and is responsible for around 2 per cent of transport-sector greenhouse gas emissions. These are but small costs for nearly a third of the load."*

The report also concluded that:

> *"....container shipping is strongly competitive with land transport on the long haul between the East and West coasts of Australia. On the thousand-kilometre runs between the east coast capitals shipping is not currently competitive, but any one of the following events could easily precipitate a change.*
>
> - *If international liner ships ceased calling at all Australian ports, but instead selected one or two, coastal shipping would be in a position to gain some of the resulting local trade, particularly if low-cost transfer between liner and coastal vessels could be arranged. The resulting increase in service frequency would then improve its competitiveness overall.*
>
> - *The competitive position of coastal shipping would be improved if the ports reviewed their charging practices, removing monopoly elements and shifting elements of cost recovery from coastal shipping (which operates in a highly competitive market) to international shipping.*
>
> - *The competitive position of coastal shipping would likewise be improved if cost recovery for competing land transport were increased. Though rail is most directly competitive on the East-West run, its freight rates are set to compete with the market leader, road transport. The Productivity Commission has recently found that B-doubles are subsidised, these being the vehicles most likely to compete with rail and shipping. Independent calculations ......indicate that there is plentiful scope for increased cost recovery from road vehicles engaged in*

interstate trade, not only from B-doubles but from long-distance trucking in general. Increased cost recovery from trucks would then permit increased recovery from rail and improve the competitive position of shipping. However, this report does not advocate a move to full cost recovery for land transport infrastructure; rather, it advocates investment in shipping to level the playing field.

- *Finally, any increase in fuel prices, whether due to 'peak oil' or to greenhouse concerns, would advantage shipping (and rail) over road transport."*

The report identified a number of policy options which include the following:

- *"Better integration of transport, including shipping, into bulk manufacturing and commodity processing supply chains. Bulk transport so integrated should be eligible for support on the same conditions as the manufacturing operations to which it relates. This would include not only shipping but ports, road and rail.*

- *Provision of capital relief for shipping investment, so that the Australian register again becomes tax-competitive. The capital relief could include registration reform, accelerated depreciation (as was successful in the past), a switch to a tonnage tax (as is fast becoming world maritime standard practice) and measures to reduce the risks of shipping investment – integration into manufacturing supply chains, retention of cabotage and level playing field policies being important elements.*

- *The artificial cost disadvantages of shipping relative to land transport should be addressed by increasing cost recovery for land transport line-haul infrastructure (particularly road) and by reducing port costs loaded onto coastal shipping.*

- *Cabotage has in the past been conditional on the coastal shipping industry giving good service. The industry has met past targets for productivity improvement, and can do so again if its risk environment is managed so that investment is justifiable. Being made cost-effective through productivity improvements, cabotage will then yield spin-off benefits including improved security, improved backup for the navy, reduced environmental cost and a supply of skilled personnel to the maritime industries generally."*

The second report was prepared for the MUA by the Australia Institute, entitled *Climate Change and Australian Coastal Shipping,* published in October 2007. The essential messages that can be drawn from the Australia Institute report are that:

- Of all the domestic freight transport modes, coastal shipping is the least energy and emission intensive – shipping has positive greenhouse credentials.

- Shipping is at least 10 times more energy efficient that rigid trucks and about 5 times more energy efficient than articulated trucks.

- Any shift of freight from the land transport modes to shipping produces positive greenhouse benefits. Transport accounts for 15% of Australia's emissions (80 Mt of the 522 Mt of Australia's carbon emissions), with freight transport contributing about 1/3 of transport emissions (approx 30 Mt [37Mt on a full fuel cycle basis (in 2005)]). Transport emissions are growing at nearly 30% per annum, indicating that transport emissions must be reduced if Australia is to reduce its overall emissions:

  - coastal shipping contributes just 4% of freight transport emissions, compared to 84% by road transport.

- In contestable freight corridors such as the East-West corridor, there is potential to move more than 200 container shiploads (of 1700 TEUs) off the road and onto ships – that equates to reducing 26,637 truck movements (222 trucks) annually off the E-W highways (86,569 truck movements (594 trucks) if the N-S corridor is included), with substantial reductions in greenhouse emissions from the freight transport sector.

- Australian ships are more efficient and cleaner than the international fleet due to higher shipping standards in Australia – despite the fact that the Australian fleet is on average 5 years older than the international fleet (the average age of the Australian fleet is 17.5 years, indicating it is nearing the end of its commercial life).

- The age of the Australian fleet suggests that there is an urgent need to replace the Australian fleet with modern efficient vessels if we are to avoid a loss of shipping to other modes and to take advantage of technological developments in shipping design which would further extend the greenhouse advantages of shipping.

- Fleet replacement provides Australia with an opportunity to adopt the latest in ship and engine design to ensure the Australian fleet is at the cutting edge of ship efficiency, which will further enhance the greenhouse credentials of shipping.

I mentioned earlier the Meyrick and Associates report of May 2007 commissioned by the Australian Maritime Group. The MUA considers this report to be the most significant report on coastal shipping to emerge from Government in the past 15 years. In relation to domestic freight shipping the report concludes that:

- The expected increase in Australia's freight task is seen as providing an opportunity for shipping to play a greater role in the movement of domestic freight around Australia's coastline.

- The mainland inter-State container shipping trade across Australia has been the fastest growing trade in Australia over the last 5 years (gaining market share on road and rail).

- The market for non-bulk freight on the East-West and North-South trade lanes is mature and the expected growth is limited in the "business-as-usual" situation. However, the report says that:

  > *"What is exciting is that movement by sea is currently a small part of the overall freight task and even small changes in mode share could lead to significant additional volumes of mainland containers, i.e. there is a large contestable market when critical success factors are achieved for coastal shipping."*

- Coastal shipping can become a viable option for significant annual freight volumes if the critical mode choice factors of price, reliability, availability and transit times are pushed towards market competitive levels relative to rail (which is similar to the conclusion reached in the NIEIR report). Success factors include:

  - Reliability upwards of 90% is likely to be required in order for shipping to be trusted as a market alternative.

  - Shipping can realise substantial mode share through minor streamlining in transit times of 10 to 20%. This means in practice deploying vessels with service speeds of around 21 to 25 knots instead of the more standard 18 to 20 knots.

  - Another critical success factor is the securing of a fixed, ideal day berthing window at the main Australian international ports. Without this, it is very hard to provide the level of quality of service required and be attractive for shippers.

  - Availability of shipping services is a mode choice factor directly within the control of operators for immediate returns in mode share. The number of services between cities needs to be increased to compensate for the lag in transit times.

  - Coastal operators need to complement improvements in critical mode choice factors with significant investment in integrated infrastructure and domestic-sized equipment (i.e. port suitability and container availability)

- There exists the opportunity for an innovative operator (entrepreneur) to create a market by combining domestic containerised flows (using domestic-sized equipment) with bulk and/or breakbulk cargoes using "combi/multi-purpose" vessels.

These reports, individually and collectively, provide a powerful public policy case for a revival of Australian coastal shipping. Should they be ignored, we are likely to witness a further decline in the capacity and capability of domestic shipping in the national freight transport task. If this decline were to continue, the critical mass of Australian shipping may well be lost, with catastrophic consequences for the Australian economy, for the maritime sector, and for the nation's defences and security.

I want to turn away from the economic and industry policy case for Australian shipping and examine domestic merchant shipping in a national defence and national security context.

The merchant navy and merchant seafarers have played a crucial role in a number of the armed conflicts that Australia has been involved in over the past century, including both World Wars, in the Korean conflict and during the Malaysian Emergency.

Of the seventy six merchant ships lost in Australian waters to mines, torpedoes, shelling and bombing, during the Second World War, twenty nine were Australian. Australian seafarers killed on these merchant vessels in Australian waters numbered 1 in 8 in total. The number of Australian merchant seafarers lost on all the oceans whilst serving the nation will probably never be known.

Despite the contribution which the merchant navy has made to growing and defending the nation, the recognition does not currently seem to exist in mainstream popular consciousness in Australia. In fact serving merchant seafarers at times of war battled for many years for full medical and other benefit recognition for their contribution, and in terms of appreciation and recognition through current shipping policy they still battle for that recognition today.

This contrasts markedly for example with the situation in the US – our key defence ally and a key trading partner - where there is a bipartisan support for, and high level acknowledgement of, the important role of the US merchant marine. The US Government celebrates Merchant Navy Day with much the same regard with which Anzac day is celebrated in Australia.

It was pleasing that Labor formally acknowledged in the lead up to the 2007 election campaign that it would recognise a Merchant Navy Day in Australia should it win Government. We will now see that position become a reality.

In 2005 I had the honour, along with Dave Heindel, Secretary of the US Seafarers International Union, to stand beside Vice Admiral David Brewer, Commander of the US Military Sealift Command (MSC) at a US Merchant Navy Day function in Washington, in an act of solidarity between the civilian navy and the US Navy and as a demonstration of the integration of both, in the USA. The Australian embassy sent an apology for non attendance, clearly not one of its priorities. The service was addressed by Elizabeth Chow, the Secretary for Labour in the Bush administration - again highlighting the juxtaposition of views on merchant shipping of the two members of the coalition of the willing.

To demonstrate that integration in the US, it is worth reflecting on the mission of the MSC. It is to provide ocean transportation of equipment, fuel, supplies and ammunition to sustain US forces worldwide during peacetime and in war. It operates more than 100 ships and numerous shore facilities around the world and employs more than 8,000 people worldwide.

It is not lost on the US that the Australian merchant marine seems to be more highly regarded in the US than at home. I would have been more comfortable if the Australian people were represented at this important function of remembrance alongside me.

The US *Merchant Marine Act 1920*, commonly referred to as the Jones Act, gives practical, program and resourcing support to the US merchant navy in recognition of its important role in national security and national defence. In essence, the Jones Act requires that a strong merchant marine be sustained in the US national interest.

The Jones Act not only ensures US coastal cargos are carried on US flagged vessels, but it requires those vessels to be built in the US to US standards and which must be crewed by US seafarers.

The Jones Act principles were applied again in 2006 when President Bush signed into law the *Coast Guard and Maritime Transportation Act of 2006* which provides for developers of Liquefied Natural Gas (LNG) import facilities to get a higher priority in planning and operating approvals if they agree to be supplied by US flagged LNG tankers.

The implication is that high value, high significance cargoes should be transported in secure ships, with secure crews, in the national interest.

The US subsidiary of Australia's major LNG producer, Woodside Energy, has committed to support the US LNG cabotage policy in relation to its proposed Californian offshore LNG supply project.

The US position was again articulated in a Presidential media release of 18 May 2007 to proclaim National Maritime Day 2007. In that media release President Bush said this:

> *"America has a proud maritime history, and the United States Merchant Marine has played a vital role in helping meet our country's economic and national security needs. On National Maritime Day, we honor merchant mariners for their dedication to promoting commerce and protecting our freedom.*
>
> *During times of peace, the US Merchant Marine helps ensure our economic security by keeping the oceans open to trade. Ships operated by merchant mariners transport goods across our Nation's waterways and on the high seas around the world to connect American businesses and consumers with valuable foreign markets and commodities. The skill and expertise of merchant mariners facilitates trade and helps to strengthen our economy.*
>
> *In times of war, the Merchant Marine is the lifeline of our troops overseas. By carrying critical supplies, equipment, and personnel, merchant mariners provide essential support to our Armed Forces and help advance the cause of freedom".*

We have had some recognition in Australia, at least from the Defence Force itself, if not from past Governments. The MUA was acknowledged by former Defence Chief Major General Peter Cosgrove, for the contribution its members made in supporting the post independence mission to Timor Leste.

In a letter to John Coombs, former National Secretary of the MUA, of 15 October 1999, Major General Cosgrove noted that in supporting the INTERFET Force deployment in Timor Leste:

> *"Many civilian ships have carried valuable people, equipment and supplies to the deployed forces, without which our logistic build up would have been severely hampered".*

What this demonstrates is that Australian merchant seafarers and the merchant marine still remains a prominent resource alongside the armed services in humanitarian missions, such as in Timor Leste, as well as in armed conflicts.

I pose this question. Does anybody in this audience believe that regional security in political matters have now been stabilised to the extent that this important support will no longer be required in the future?

Taking again the line through national shipping policy as defined over the last 11 or so years, it would seem that the main policy makers in Canberra, have, at least until recently, believed those threats have been removed. We are more in the hands of foreign and regional shipping interest than at any other point of Australia's history.

Following some robust exchanges between the MUA and the former Minister for Defence in 2006, we have now been invited to participate in the Australian Maritime Defence Council (AMDC) as an observer. The AMDC provides a forum where contemporary merchant marine and Navy interface issues can be discussed. I hope we can achieve full membership of AMDC in the period ahead.

For example, one of the issues currently being considered within AMDC is the issue of the protocols for the boarding of merchant ships by military personnel. It is important that merchant seafarers, as well as Defence personnel are familiar with each others requirements and operating procedures.

The AMDC is an ideal forum to settle these issues and we welcome the invitation to participate in AMDC and the support we are receiving from senior Defence officers.

I now turn to maritime security.

The attack on the World Trade Centre in 2001 provided the impetus for the International Maritime Organisation (IMO) to seriously focus on maritime and port security. The result was an amendment to the Safety of Life at Sea (SOLAS) Convention establishing the International Ship and Port Facility Security Code (ISPS Code).

The MUA has been constructive participant in implementation of the ISPS Code in Australia, through the passage of the *Maritime Transport Security Act* 2003 (later amended to become the *Maritime Transport and Offshore Facilities Security Act 2003)* and regulations.

The MUA maintains an important position on the Maritime Industry Security Consultative Forum (MISCF) and sits on the Maritime Transport Security Working Group along with the Federal Government and industry.

Another important initiative in which we have been a willing participant is the establishment of Port Security Committees which have been set up in virtually all Australian ports. These Committees comprise employers, the MUA, the Federal Office of Transport Security and State Government representatives. They play an important role in monitoring and advocating security at each port.

All these measures have dramatically improved port security, at least on the landward side, so much so that seafarers in port returning to their ship tell me they now have trouble getting a six pack on board, let alone some company for the night.

However, port security is really only as good as the security of the vessels that enter those ports or the security of the cargo on board those vessels.

This is why we have been arguing for some years that there remain weaknesses in Australia's security at the ports. I will briefly document the weaknesses we have identified.

Firstly, lack of Australian ratification of ILO Convention 185. We remain hopeful that we can convince the Government to ratify International Labour Organisation (ILO) Convention 185 - Seafarers' Identity Documents Convention (Revised), 2003.

It is the view of the MUA that Australia's lack of commitment to this ILO Convention limits its moral authority in the international maritime industry to require high level background checking of foreign crews.

If Australia is not prepared to establish a seafarers database and provide biometric recording of its own seafarers (as required by the Convention), it cannot expect to impose more rigorous standards on foreign seafarers wishing to visit Australia on international shipping.

Second, the introduction of the Maritime Crew Visa (MCV). The MUA has held concerns over the introduction of the MCV which will cost Australian tax payers $100 million dollars and has the potential to dilute the effects of the MSIC card and national security when used to replace Australian background checked workers on coastal voyages.

The MUA is of the view that the implementation of the MCV does not adequately plug the current security weakness that allows foreign seafarers to enter Australian waters and ports with security and background checks which do not match the standards applied to Australian seafarers and port workers.

Third, the state of the Australian shipping industry. The decline of the Australian domestic shipping fleet has created a significant weakness in maritime security. As I have discussed, it is widely, though by no means universally, recognised that domestic shipping fleets provide nations with an important security shield.

Australian ships and crews also provide an important surveillance function – well documented in relation to the reporting of illegal fishing vessels in our northern waters.

Australian manned and flagged ships are unquestionably the safest mode of transporting cargos around our coastal zone and must not be allowed to decline any further if we are to maintain high standards of national security.

I now return to the economic arguments. Modelling undertaken by the Apelbaum Consulting Group for the Australian Shipowners Association, reproduced in the report of the *Independent Review of Australian Shipping* (IRAS), shows that if Government shipping policy is not reversed, and Australian ocean-going shipping declines to about 30 vessels, then the industry's export activity will further contract and imports of shipping substantially increase, with a resulting negative impact on the balance of payments.

The National Institute of Economic and Industry Research report, *Australian Coastal Shipping: its Future Role,* models the effect on the national economy of a proposal to increase investment in Australian shipping, using government funds essentially as a transfer of funds which would otherwise be invested in land transport infrastructure.

The report found that, for selected opportunities, investment in shipping rather than land transport has potential to yield a significant dividend in terms of enhanced gross domestic product.

The modelling indicates that investment in shipping passes the cost-benefit test with flying colours. By reducing freight rates it benefits shippers with lower costs, and by reducing capital costs it generates funds which governments can spend to better effect than by amplifying land transport capacity. An investment of $62 million in ships is expected to yield a discounted GDP return, over 30 years, of $2,800 million, provided the infrastructure investment savings are wisely spent.

The overarching conclusion is that selective investment in coastal shipping is likely to yield considerable national interest benefits.

The MUA, as the principal union representing seafarers in the Australian maritime industry, along with the other maritime unions, has participated in a number of national and international initiatives aimed at improving the competitiveness and productivity of the Australian shipping industry.

I will outline the most important of those initiatives to provide a perspective on the positive and constructive approach which we play in supporting our industry, and to demonstrate our legitimacy, our professionalism and our national interest credentials.

Firstly, our rock solid commitment to collective enterprise bargaining and our strong opposition to individual agreements, particularly to Australian Workplace Agreements under the WorkChoices legislation, has delivered to shipping employers wage stability and Consumer Price Index (CPI) tracked wage movements over the last decade.

In other words, our commitment to collective agreements, and the discipline we have exercised among the workforce (our membership), in a period of labour shortages where we could have extracted so called market rates, has been a significant factor in maintaining competitiveness and productivity in the Australian shipping industry. This is despite the fact that Australian collective agreement rates are higher than many of the agreements covering seafarers from the major global labour supply countries like the Philippines, and higher than the International Transport Workers Federation's international seafarer agreements.

Why have we maintained this discipline and not been tempted by the same unethical principles that have characterised the executive pay spiral and some of the AWA segments of the labour market? It's because we firmly believe in the importance of sustainability of the labour market.

We recognise that forcing up the price of labour on a short term basis can be illusory in terms of income security and job security, and ultimately, in terms of net gains for workers. We firmly believe that life-long approaches to wealth generation by wage and salary earners is better served by stability and sustainability, in both employment and income, over the entire employment life cycle.

We therefore have no interest in jeopardising long term sustainable development of the maritime industry by short term exploitation of a particular labour market.

However, such restraint on the part of workers and unions must be accompanied by visible and concrete actions on the part of employers to investment in training and workforce development, genuine career pathing and to cooperative labour relations.

It is also predicated on supportive Government policy that respects internationally agreed rights, such as the right to organise and be represented by a trade union, the right to decent work (as defined by the International Labour Organisation), and on Government policy that addresses social justice issues such as housing affordability, protecting the environment, provision of reliable and affordable public transport, child care and education.

These are surely the foundations of a healthy social democracy, particularly one endowed with such a richness of resources and agricultural land as ours.

We remain committed to sustainable approaches to addressing labour market issues – supply, quality and price – subject to the provisos I outline.

One only has to examine the trends in the wages-profits ratio of GDP to understand why we believe we have a legitimate and sustainable case for a greater share of the wealth generation being distributed to the workforce.

Figure 3 shows that the wages share of total factor income remained relatively stable during the 1990s, at levels similar to those during the 1960s. The highest recorded value of the wages share of total factor income was 62.4% in 1974-75. However, in more recent times, the wages share has been trending down to be 53.7% in 2006-07.

Figure 3: Wages Share of Total Factor Income

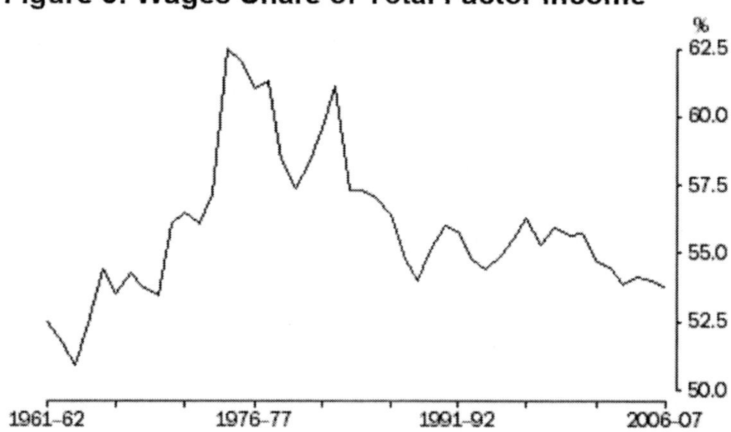

Source: ABS, Australian System of National Accounts, 2006-07 (5204.0)

Similarly, Figure 4 shows that the profits share of total factor income reached 27.2% in 2006-07, representing the highest share recorded since 1959-60. The profits shares recorded since the early 1990s are at a distinctly higher level than those at any time since 1959-60.

Figure 4: Profits Share of Total Factor Income

Source: ABS, Australian System of National Accounts, 2006-07 (5204.0)

The second MUA initiative I want to consider is that within the collective agreements to which we are a party, we have negotiated a raft of productivity benefits and ship performance improvements. Some examples of the unions' partnership approach have been:

- Reductions in crew complement, achieving world's best practice.
- Acceptance of the need for foreign riding gangs to undertake routine shipboard maintenance aimed at achieving better labour productivity and enhanced ship performance.
- Greater flexibility in the use of available labour, such as multi-skilling of the crew and improved rostering (noting the Australian Integrated rating is already the most multi-skilled of all the worlds qualified ratings).
- A strong commitment to training, qualification levels and overall workforce development.

Typical shipping collective agreements we negotiate commit the parties to:

- Operate the Australian crewed fleet to a high standard of efficiency, in a viable and competitive manner, ensuring a high level of customer service.
- Address cost issues and improve the viability of the operation.
- Continually review the agreement, and all workplace practices in order to develop and adopt a culture of continuous improvement.
- Consult regarding the setting of productivity targets to be achieved throughout the life of the agreement, based on competitive benchmarks, to ensure the viability of the business.
- Maintain a supply of skilled labour by ongoing recruitment of Trainee Integrated Ratings.
- Adopt an individual crew member performance appraisal system.

- Adopt a team-based approach to work, with each employee working to the level of their classification, training, competence, certification and applicable legislation in a co-operative effort, to ensure the safe and efficient operation of the vessel.

- Better programming of work through the shipboard operations committee.

- Participation in regular productivity reviews so that appropriate measures aimed at improving and recording the efficiency of the fleet collectively, and individual vessels in a fleet will be developed and implemented. These measures will include, but not be limited to:

  - introduction of vessel performance targets;
  - projects to enhance teamwork; and
  - the achievement of compensation and medical targets.

Third, we continue to work closely with the Australian Maritime Safety Authority (AMSA) in relation to the issuing of safe manning certificates, often agreeing to conditional clauses that allow ships to sail for limited periods with skill sets that are only suitable for a short term. This allows the operator to meet commercial requirements in cases of emergency or of unexpected crewing deficiencies.

Fourth, we have in place an agreement with the Maritime Union of New Zealand, and are negotiating with our counterpart in the UK, for a program of seafarer labour exchange to assist the industry meet peaks in labour demand.

Fifth, we are working with the industry on development of drug and alcohol policy aimed at eliminating the potential for a worker to take their place in the workforce when they are not fit for work and could place themselves and others in danger whilst at the same time recognises the legality of a range of drugs in society.

Sixth, we have established close working relationships with not only the other domestic maritime unions but all the domestic transport unions, through establishment of the Transport Unions Federation (TUF). This alliance promotes a cooperative relationship with employers and enables us to more coherently address policy and industry wise issues. Employers welcome such collective approaches on key issues.

Finally, we have established a full time research, communications and lobbying presence in Canberra, to enable us to better interact with the political process and the Federal bureaucracy.

Internationally, we actively implement the ITF Flag of Convenience (FOC) campaign, aimed at eliminating sub-standard ships and sub-standard labour relations on ships. In a truly global industry operating within a global labour market, all steps that can be taken to raise the floor at a global level are steps towards ensuring there is fair competition in that market, particularly important in evening out the imbalances between the developed and developing nations.

Just to illustrate the extent of labour exploitation in international shipping, ITF FOC Inspectors have recovered US$70.7M in underpayment of wages for international seafarers over the last 4 years alone – and that just from the ones where a complaint has been lodged and follow up action has been successful. What does that tell you about the morality of large segments of international shipping? Probably the same it tells you about the morality of the former Commonwealth Department of Employment and Workplace Relations, the alleged custodian of workplace standards under the Howard regime, which in October 2007 was fined $30,000 for a deliberate breach of the Workplace Relations Act in relation to illegal actions it took against its own employees.

In addition, the MUA has been an active participant in the International Bargaining Forum (IBF) negotiations, which in September 2007 concluded the third international collective seafarers agreement. The IBF is a negotiating forum between the International Transport Workers' Federation and peak global shipping associations represented by the Joint Negotiating Group (JNG – comprising Japanese, Korean and European shipowners). The Agreement covers more than 3,500 vessels employing more than 100,000 seafarers worldwide.

In addition to an 8% wage increase over two years, the 2007 Agreement provides for a $10 per month levy per seafarer on vessels operated by members companies represented on the JNG. The money will be paid into a special fund which will be used to subsidise work for seafarers from traditional maritime nations (developed economy ratings) like Australia, whose seafarers have been costed out of the global market in recent decades. The result is that Australian seafarers will have greater opportunity to work in Australia's international shipping trade.

As a parallel development we will also be lobbying the Government to reform the income taxation arrangements for Australian seafarers working in international trade, so that income earnt while on the high seas is treated as foreign earned income. Together with the new IBF agreement, this reform will add considerably to the opportunities for Australian seafarers.

We recognise the IBF Agreement as a major breakthrough in job security for Australian seafarers, with shipowners recognising the importance of seafarers from Europe, Japan, Australia and America being included in trade between nations. It means that as new exports markets open up, such as LNG sales to Japan and China, Australian seafarers can play a greater part in that trade.

Importantly, the agreement also adopts many of the provisions of the ILO Maritime Labour Convention, which means those provisions will become operational years before they would normally have come into force.

The International Maritime Employers Committee, in a media release of 22 October 2007 stated that that the IBF system had managed once again to deliver an outcome that represented best practice in the industry and a series of measures that would benefit both seafarers and maritime employers.

Again, the significance of these developments is that they level the competitive playing field internationally, while at the same time minimising labour relations risk in international shipping.

It is significant that the Meyrick report (*International and Domestic Shipping and Ports Study*) highlighted a report prepared exclusively for the AMG by Thomson Clarke – Coastal Shipping Case Studies (June 2002) - which shows that the cost penalty of an Australian crew versus a foreign (ITF) crew is quite small, estimated as follows:

- for small coastal vessels (5,500-8,700 DWT) on the studied Australian coastal routes is an additional 10-15% of the total voyage cost; and
- for a handy size bulker (28,000 DWT), an additional 5-6%.

When you contrast that small cost differential, which is becoming relatively less significant as other transport costs rise (eg fuel, port charges, emissions trading schemes, security etc) with the actual freight cost, the conventional wisdom about the alleged competitive disadvantage of Australian shipping simply does not add up. Just to illustrate further, the Maritime International Secretariat Services Limited (MARISEC – a body formed by the International Chamber of Shipping (ICS) and the International Shipping Federation (ISF)) has published a table of typical freight costs for highly traded products.

The MARISEC data shows that the typical cost of transporting a tonne of iron ore from Australia to Europe is about US$12. Some typical costs for other products, on a shipment from say Asia to Australia, are: a TV worth $700, shipping cost $10; a kilo of coffee worth $15, shipping cost $0.15; and a can of beer worth $1, shipping cost is $0.01.

Given these facts, employer business strategies that focus on finding competitive advantage through wages and conditions competition will simply not work.

I want to conclude by focussing on some of the key policy responses that we believe are required if Australian coastal shipping is to be revived and if it is to once again become a serious competitor in the national freight task. The proposals I outline are practical measures that could quite readily be taken up by the Federal Government, working in partnership with the States, employers and the trade unions. They are fiscally conservative, economically responsible, and include.

- A mandate that requires High Consequence Dangerous Goods (HCDGs), and in particular Security Sensitive Ammonium Nitrate (SSAN), to only be transported into and out of Australia, and between Australian ports, on Australian registered vessels which are crewed by Australian seafarers with the highest standards of security clearance and background checking:
    - We think this policy should be supported by a change in the transport regulations applying to SSAN whereby ships' crew responsible for the handling, stowage and transportation of SSAN have the same mandated security checks and authorisations as all other transport workers in the transport chain.
- A change in the provisions for licensing of ships which ensures such ships confirm with Australian industrial standards.
- A revision of the *Ministerial Guidelines for Granting Licences and Permits to Engage in Australia's Domestic Shipping* to eliminate the well documented abuses of the coastal permit system, and that there be a significant improvement in transparency in relation to information and statistics on the granting of coastal trade permits.

- The announcement of a target date, approximately 6 months into the future to allow a period of industry adjustment, after which the *Ministerial Guidelines for Granting Licences and Permits to Engage in Australia's Domestic Shipping* will provide new ship suitability and ship availability tests. The adoption of new tests would significantly reduce the circumstances under which shippers would be permitted to use foreign vessels for the transport of domestic sea freight. We believe the coasting trade provisions in the Canadian *Coasting Trade Act 1992* provide a model for consideration in Australia.

- That the Taxation and Customs duty disadvantages on bunker fuel imposed on Australian vessels operating in the coasting trade relative to international ships granted permits to operate in the coasting trade be eliminated, to provide fair competition to Australian vessel operators as part of a process of encouraging a return to Australian shipping for our coastal freight.

- That section 23AG of the *Income Tax Assessment Act 1936* be amended, and the meaning of 'foreign service' for income tax purposes be clarified so that Australian seafarers are not disadvantaged in their earnings capacity relative to seafarers of other nations when working on foreign-flagged vessels on the high seas.

- That the most appropriate policy options among the menu provided in the May 2007 Meyrick and Associates/Booz Allen Hamilton report to the Australian Maritime Group entitled *International and Domestic Shipping and Ports Study*, the June 2007 National Institute of Economic and Industry Research report to the MUA entitled *Australian Coastal Shipping: Its Future Role* and in the Morris and Sharp report to the Australian Shipowners Association entitled the *Independent Review of Australian Shipping* (IRAS) of December 2001, be considered and implemented, with the aim of providing a positive environment to stimulate investment in Australian shipping to replace foreign ships carrying domestic cargo. These policy options would include, but not be limited to:

  - Pricing of government-owned infrastructure to fully reflect externalities (environmental pollution, congestion, accidents etc.) which would tip the level playing field towards coastal shipping.

  - Efficiency measures that could benefit coastal shipping such as:

    - Reduced manning levels under controlled circumstances involving increased (innovative) vessel automation, specific training, and special operational situations (as occurs in Europe with a number of flag-states). This proposal is premised on the renewal of the Australian coastal fleet with modern sophisticated high technology vessels.

    - Opening-up access of port and connecting intermodal infrastructure to new domestic shipping operators such that dedicated terminals and associated facilities do not have to be constructed where capacity is already available at existing international terminals.

  - Financial measures, which can take a number of forms, either focused on stimulating services or alleviating penalties/barriers, such as:

  - Construction or operational subsidies to help alleviate a cost penalty imposed by a particular aspect of government policy - this occurs in the US for Jones Act trades and in Japan

  - Start-up grants for coastal shipping services on designated routes – a measure which is actively employed by the European Commission to kick start coastal/short sea shipping

  - Waiving or reduction of port access charges for coastal shipping (i.e. reduced port dues for high frequency of calls and exemption of pilotage, as occurs in a number of European government-owned ports)

  - Financial support of coastal/short sea promotional bureaus (often in partnership with industry)

  - Adoption of a tonnage tax to replace existing corporate tax as is now best practice in most nations with a heavy reliance on shipping, provided the taxation advantages to shipowners are accompanied by commitments around seafarer training and employment

- Consideration of the concept of integrated highways, which include so-called 'maritime highways', as a specific initiative (possibly regionally driven) to focus efficiency measures, access policies, and infrastructure investment on designated freight paths or routings. This is being successfully applied in Europe by the European Commission for coastal/shortsea shipping and also in the ASEAN region. A similar model could be applied to the Australian situation taking account of local specifics.

- Integration of shipping into manufacturing supply chains, particularly as pressure grows for more value added downstream processing of our resource products.

- Adoption of strategies which encourage commodity exporters to place a high value on the shipping component of the overall commodity supply chain, from mine to export destination as a means of improving supply chain productivity and efficiency. One aspect of the strategy will require exporters to reconsider the shipping terms of their export contracts, to reduce dependency on Free On Board (FOB) shipping terms whereby the exporter loses all control over shipping – its scheduling, its cost, its standards. The alternative are policies which assist exporters to arrange shipping on a Delivered Ex Ship (DES) basis:

  - Increased usage of DES shipping terms, which provides the exporter with greater influence over the scheduling of the ships, allowing an optimisation of the productivity of the ships will also have the positive side effect of minimising the use of substandard ships, invariably from Flag of Convenience (FOC) registries, operating at the lowest international standards and with crews that are paid at the lowest international seafarer rates of pay, and with the worst conditions of employment and poorest training standards.

- Negotiation of bilateral agreements, particularly in strategic resource sectors such as LNG, uranium, HCDGs to encourage the expansion of Australian participation in Australia's international shipping task (consistent with the principles applying in many of the Industry Action Agendas adopted under the Howard Government).

- The setting of targets for the transition of freight from road to rail, consistent with the approach taken by State Governments to shift freight from road to rail, aimed at shifting a proportion of freight to sea transport, to be adopted as a key measure in reducing greenhouse gas emissions from freight transport.

- That consideration be given to port pricing schemes aimed at increasing the efficiency the sea leg in the supply chain and which also increases the overall competitiveness of a supply chain that includes a sea leg, aimed at improving opportunities for coastal shipping in the movement of Australian domestic freight. In this respect we propose consideration of Ramsey Pricing by the stevedores, noting that such a strategy will probably require a change in the performance criteria in the Port Authority-Stevedoring operator contracts, to include a specific gaol of ensuring coastal sea freight becomes a part of the transport mode mix, if road, and possibly rail, congestion is to be eased.

  Note: *Ramsey pricing is an alternative to cost plus pricing which can be applied where the revenue stream can be differentiated, allowing different prices to be charged against those different streams e.g. international cargo and domestic cargo, or peak and off peak times*

- That the Auslink program and the Corridor Strategy be reformed so that they take account of sea ports, inland ports (intermodal facilities) and importantly shipping (or the sea highways on which shipping operates) in terms of strategic and funding priorities.

- Adoption of new mechanisms at the national level to plan, prioritise and harness investment in the national freight transport plan and in the orderly development of our infrastructure assets. One option is an infrastructure planning model along the lines of Labor's proposed *Infrastructure Australia* body and the national infrastructure audit that will accompany establishment of *Infrastructure Australia.* Such a mechanism will ensure a better national strategic approach to Australia's port planning in the context of improved national freight transport planning.

- Adoption of improved approaches to balance the supply and demand for labour. One method that might be appropriate for consideration in some ports is labour pooling.

- A strong commitment and actions by Government and industry to improve occupational health and safety on Australian ships and at Australia's ports, and in particular, early implementation of a new national code of practice for stevedoring.

- MUA membership (rather than observership) of the Australian Maritime Defence Council (AMDC) to assist in ensuring a high degree of co-operation between the Australian Defence Forces and the Australian merchant fleet.

- Immediate ratification of the ILO Consolidated Maritime Labour Convention and a review of resource allocation to AMSA to ensure it can effectively perform the new elements of its Port State Control function as part of Australia's compliance under the Convention.

- Initiate a review aimed at achieving a single, cooperative national maritime jurisdiction through integration of the various State, Territory and Federal maritime authorities, and reform of the *Navigation Act 1912* (and complementary State/NT legislation).

- Maintenance of an appropriately resourced Australian Maritime Safety Authority (AMSA), Seafarers Safety, Rehabilitation and Compensation Authority (Seacare Authority), National Offshore Petroleum Safety Authority (NOPSA) and Australian Transport Safety Bureau (ATSB), and equivalent State organisations, to ensure they operate efficiently and in a coordinated way to maintain a strict maritime safety regime

- Immediate recognition of the vital role of the International Transport Workers Federation (ITF) in the protection of international seafarers visiting Australia's ports, including acknowledgement of the need to establish a genuine link between all ships and their flag state as a means to combating abuses under the Flag of Convenience (FOC) system

- A strong commitment by Government and industry to skill formation that equips the workforce to deal with the new security environment, new port and ship technologies and the challenges that all stakeholders face in lifting productivity and making Australia's ships and ports more efficient. In order to meet the nation's growing domestic freight task and to support a rapidly expanding offshore oil and gas industry a highly skilled and highly qualified seafaring labour force will be necessary. Accordingly, skills formation policy must be aimed at:

  - ensuring the availability of suitable maritime training and skills development arrangements to meet the skills needs of the maritime industry;

  - ensuring there is a coordinated approach to resourcing and delivery of maritime training across all maritime training institutions;

  - ensuring that any bottlenecks in the distribution of resources to maritime training be resolved, aimed at ensuring training expenditure is allocated to regions and to projects having the greatest demand for seafaring skills;

  - assisting in expediting industry research on reconciling data on maritime industry skills supply and demand with the capacity of training institutions to deliver relevant training;

  - reviewing the charter, management and focus of the Australian Maritime College (AMC) to ensure it is responsive to the needs of the Australian and regional maritime industry;

  - developing a nationally coordinated scheme to ensure that both private and public shipping assets are harnessed in an effort to maximise opportunities for maritime trainees to gain sea time as a necessary part of acquiring a maritime qualification; and

- consulting with regional Governments on the most appropriate ways to involve regional nations in the training and skill formation strategies for the offshore oil and gas industry, to assist in meeting Australia's skills needs and to facilitate skills transfer as a part of Australia's commitment to regional development.

* We want an Industry Innovation Council established for the shipping industry.

The central message I want to leave you with is that the tide is turning in relation to the long period of neglect of Australia's coastal shipping industry.

There is now an overwhelming body of well credentialed policy research that points to the imperative of considering shipping in the national freight transportation task. As well, there is now a clear political commitment for this policy position.

For a country with one of the largest coastlines in the world and in a country situated in a region known for its political volatility, there is increasing recognition of the vital role of domestic shipping in national defence and in national security.

To top that off, shipping is now emerging as a key part of the solution in reducing transport greenhouse gas emissions.

There needs to be an urgent return to sustainable bipartisan political policy securing intelligently our coastal merchant shipping requirements for the future and building a strong foundation for the only future open to Australia as a shipper and a shipping nation.

Thank you

Incat Builds for Japan

Craig Clifford
Incat Australia
incat@incat.com.au

NATCHAN RERA – A NEW WAVE IN SEA TRANSPORT FOR JAPAN

Introduction

Delivered to Japanese operator Higashi Nihon Ferry on August 1, 2007, the first Incat *Evolution One12* 112 metre vessel operates at speeds of approximately 40 knots with capacity for up to 800 persons and 355 cars, or 450 lane metres of trucks and 193 cars. Delivery of a second sister vessel to Japan will follow, ahead of schedule, in 2008.

These vessels are set to revolutionise the Japanese ferry industry. Never before has such a technologically advanced ship operated in the Japanese market. Never before has Incat built such a large ship and for a new customer in an area with such enormous potential for Incat High Speed Craft.

The order is a major part of the Libera Corporation's on-going business development plan for the expansion of regular passenger services between the two ports of Aomori, on the island of Honshu, and Hakodate, on the island of Hokkaido, operated by its subsidiary, Higashi Nihon Ferry.

These new fast ferries are expected to approximately halve the time currently taken for voyages between Hakodate and Aomori by the existing ships and enhance the convenience of regular passenger services.

The 112 metre vessel signals a new era for Incat, particularly as the concentration of our high speed ships to date has primarily been in Europe and the Americas.

The Vessel

Several years ago Incat concluded that to optimise high speed ferry transportation its designs must be larger and more robust in order to provide greater efficiencies and ride comfort. Concentrating its efforts on the development of ever larger capacity high speed vessels, which deliver greater economies and flexibility to operate year round on worldwide routes, Incat pushed the envelope and the 112 metre Evolution One12 Wave Piercing Catamaran was born.

The Incat 112 metre draws on the principles, technology and highly successful in- service experience of the Incat 98 metre vessel and those generations that went before it. The vessel's hull geometry has been optimised for speed and efficiency with power supplied by four V20 MAN 28/33D diesel engines, each rated at a minimum 9000kW at 100% MCR.

The new vessel is certainly a ship of firsts. In addition to being the first of this new large design and the first high speed ferry to be custom-built in Tasmania for Japanese service it is also the first Incat vessel to offer a full length car deck capable of accommodating up to 165 cars above the main vehicle deck, thereby ensuring a full load of freight and buses can be carried in addition to some 193 cars.

Not surprisingly, great emphasis has been placed on interior design and to this end Incat has partnered with MEC Design International, a company with one of the finest records of achievement in Japan. Together, MEC and Incat have worked to offer an interior of ambience, sophistication and individual character which is at the same time functional and easy to maintain.

Japanese High Speed Endeavours

In January 1997 the high speed ferry *Unicorn* was launched at the Shimonoseki Shipyard of Mitsubishi Heavy Industries Higashi Nihon Ferry's Aomori – Hakodate service. The vessel was not a success, its monohull hullform contributing to uneasy passages across the often turbulent Tsugaru Strait.

A much more ambitious program came in 2003 when Japan's Mitsui Engineering & Shipbuilding signed a contract for a Super High-Speed Craft termed as Techno Superliner (TSL) with its owner Techno-Seaways.

The TSL was heralded as a state-of-the-art craft to bring forth a means of sea transport to meet the requirement of next generation. To run at high speeds, reaching almost 40 knots with air lift, the TSL was to be completed in the spring of 2005 for service from Tokyo to Ogasawara in 16-17 hours, a significant time reduction when compared with the 26 hours taken by conventional ship.

However, by July 2005 Japan was abandoning plans for the vessel due to predicted operating deficits, the Construction and Transport Ministry and the Tokyo metropolitan government concluding that they could not subsidise the business plan which was likely to lose money. To save the contract, the ministry and the metropolitan government planned to cover the remaining losses of US$8.9 million a year with an additional subsidy, but the Finance Ministry opposed their plan. Japan's high speed craft endeavours were on hold, for now.

Japan looks to Australia

That a leading Japanese ferry operator should look to Australia for its vessel requirements was a little surprising. In terms of vessel numbers Japan is a world leader in international shipbuilding. As of the end of December 2006, Japan's shipbuilding export order book stood at an all-time high of 55,390,349grt, 10% higher than the previous record of 50,267,000grt set in March 1973.

The vast volume of work on hand means that Japanese yards are currently fully booked through to 2010 - a remarkable milestone achieved in spite of intense competition from Korean and Chinese yards and testament to the continued strength of Japan's shipbuilding industry.

So, why Australia? In fact, why Tasmania and Incat?

Australia's tonnage output pales into insignificance when compared with Japan. But the competitive edge of Australian-built ships is speed (due to the use of lightweight aluminium and composite construction material), innovative design and use of advanced construction technology.

Quite simply, no other shipyard in Japan, or indeed elsewhere, has the experience or track record necessary to see that nation's fast ship aspirations come to pass.

On a global scale Incat accounts for almost 40% of total vehicle-passenger high speed ferries over 50 metres in length. In deadweight terms Incat has built 60% of the world's high speed ferries with capacity over 750 tonnes. Operators employ Incat vessels in conditions ranging from operations in the Arctic Circle in winter, to the tropical climates of the Caribbean, to the hot Spanish summers, to military operations in summer in the Persian Gulf to extreme performance testing in the mid Atlantic (off the coast of Greenland) in winter.

No other yard has achieved this level of success!

When in August 2005 Libera entered the ferry business through the acquisition of Higashi Nihon Ferry the company once again explored high speed ferries as a solution to its requirements. Following a search of world wide high speed craft builders by the trading house Sojitz Marine and Engineering Corporation (SOMEC), Incat was selected and a proposal made to Libera. At SOMEC, the import of the ferry was handled by Mr Katsumi Hattori and Mr Shunsuke Takahashi who quickly got to grips with this first for Japanese imports; an Australian-built ship.

"We were up against a lot of 'firsts,' and it was tough, tough, tough. Once or twice, we thought we'd never make it," they commented.

One of the most difficult things to deal with was introducing this leading edge technology to Japan. According to SOMEC, Japanese experts say that Incat's world-leading aluminium shipbuilding technology is at least 20 years ahead of Japan's. The craft was built to the International Marine Organisation's (IMO) code for high-speed ships, but to import the ferry, both sides had to work and work and work with the concerned Japanese Government (JG) agencies.

Commercial Debut

Despite the challenges, upon which Incat seems to thrive, in September the 112 metre *Natchan Rera* commenced regular service for Higashi Nihon Ferry on the 61-nautical mile run across northern Japan's Tsugaru Strait between the islands of Honshu and Hokkaido. The crossing takes 1 hour and 45 minutes, approximately half the time taken by existing conventional ships.

The Tsugaru Strait connects the Sea of Japan with the Pacific Ocean, and the Seikan tunnel passes under it at its narrowest point (19.5km) between Tappi Misaki on the Tsugaru Peninsula in Aomori, Honshu and Shirakami Misaki on the Matsumae Peninsula in Hokkaido.

From Aomori, heading north to Ushitaki, the sea route is well protected from predominant westerly weather patterns. The maximum fetch to the east across the Mutsu-wan is 50km, while the maximum wave height expected from the east, during 30 knot winds, is a wind driven sea of 2 metres.

The inability of the previous high speed craft, the *Unicorn*, to cope with conditions on Tsugaru Strait saw an early withdrawal and accordingly Higashi Nihon Ferry placed great emphasis on the ability of the *Natchan Rera* to perform well in adverse weather conditions.

Before making their decision to place an order with the builder, the customer visited several successful high speed operations in Europe through September to October 2005. After this visit, they analysed the whole picture, the economics of operation and adaptability of such a service to the Shimokita peninsula. In the end the Wave Piercing Catamaran design was selected as being the optimum vessel for the task. .

Onboard

The passenger spaces on board *Natchan Rera* are located on two decks, Tiers 4 and 3. The latter is the main passenger deck, boasting three distinctive lounges, and is accessed from the vehicle decks, Tiers 2 and 1, by stairways forward and aft and, in a unique first for high speed catamarans, also by Kone escalators port and starboard. Additionally, a TBV marine elevator is fitted to provide barrier free access between Tiers 2, 3 and 4, ensuring passengers requiring assistance enjoy all the craft has to offer.

Passengers entering the accommodation via the escalators, or via shoreside gangways, arrive into the heart of the craft, the Entrance Lounge and here, located around the centrepiece staircase and atrium, is café style seating. Adding to the cosmopolitan air of this space on the port side is a circular café and bar with curved glass bulkheads. Separated by a galley, the café is mirrored on the starboard side by a shop which again features large expanses of curved glass. Aft of the Entrance Lounge are two Business Class lounges, divided by a buffet corner and toilets. Each lounge seats 56 passengers in comfortable aft-facing reclining seats, complete with foot rests. Large windows face onto the aft observation deck from where passengers have a spectacular view of the spray from water jets disappearing astern of the craft.

Immediately forward of the Entrance Lounge is the Children's Lounge, finished with a mural of the cartoon characters featured in the vessel's eye-catching livery. Higashi Nihon Ferry has placed great emphasis upon *Natchan Rera* being a truly friendly family ferry and this lounge is a delight to their younger passengers!

The most striking feature of the Economy Lounge is the promenade walkway, outboard of the main seating area and leading forward along the port and starboard side, wrapping around the front of the accommodation. Seats are also positioned beside the windows and facing out to sea while forward, the promenade opens out with seats facing over the bows. Here passengers are afforded impressive panoramic 300° views of the sea heading towards them at up to 40 knots through sweeping tinted windows!

Executive Style

Above on Tier 4 is the Executive Lounge, the main access to which is via the stunning atrium and staircase in the Entrance Lounge. The eye-catching entrance offers just a glimpse of what lies beyond the full height curved glass bulkheads with its automatic glass doors activated by sensors. Here, in the highest passenger vantage point on board, passengers can sit back and relax in custom designed lover's seats, fitted with a central rotating table complete with integrated LCD TV screen shared between each pair. For those who might wish to take a nap on the short crossing then one of the Ocean Executive reclining seats, complete with calf rest, LCD TV and reading light, is sure to be inviting. The lounge has a dedicated bar, finished with wood panelling and glass counter top.

Like the Economy Lounge below this area is divided by two amenities blocks, containing male and female toilets and showers, and the split wheelhouse with electronics housed in the lower level.

Vehicle Decks and Livery

One cannot fail to be impressed by the sheer size of the main vehicle deck on Tier 1 which, unlike earlier vessels, is completely devoid of mezzanine decks. Incat's first 'double deck' vehicle carrier, an internal ramp leads above to a dedicated full length car deck on Tier 2, leaving the main deck free for high sided vehicles with a load of 12 tonnes per axle.

There is space for 450 metres of freight plus 193 cars. If the vessel was to be filled entirely with cars then up to 355 vehicles may be shipped.

Externally, the craft is beautifully adorned with a livery adapted from paintings made by Natsumi Kawashima, a seven year-old Japanese girl whose grandfather Mr Mutsuo Kawashima is a Japanese-style painter who served as a councillor of the Japan Fine Arts Exhibition and whose father, Mr Wataru Kawashima, is also a Japanese-style painter and is a lecturer at Kyoto City University of Arts. The illustrations depict Natsumi and her friends travelling happily with marine creatures while imagining "The bonding between sea, nature and people".

In Japan, the name Natsumi is often shortened to Natchan, while Rera is from Hokkaido's Ainu language and means wind. Together with the colourful livery it is hoped the friendly name will overturn traditional images of conventional ferries with thoughts of comfortable and delightful sea journeys.

Technical Review

The environmentally progressive *Natchan Rera* is powered by four MAN 20V 28/33D diesel engines, each rated 9000 kW at 1000 rpm and delivering a low weight when compared to other engines in its class. The advantages of engine durability, efficiency, low noise and low maintenance costs make it the engine of choice for Incat vessels, not least of all impressive fuel consumption, burning less kg per cargo tonne per hour than any other high speed catamaran.

The latest combustion chamber technology with electronically controlled pump pipe injection system and optimised cylinder pressures, reduces specific fuel consumption to <190g/kWh. At almost a 7% reduction per kWh over the previous generation of medium speed diesel engines this represents a massive saving on annual fuel bills. The vessel is also environmentally compliant with low NOx of <10g/kWh.

The ferry is fitted with four of the largest transmissions from ZF Marine GmbH, the ZF 60000 NR2H, each with maximum rated power of 12387kW. Special technical solutions were required to satisfy the engine room layout and the demand for the highest reliability. The transmission had to be matched to the interface provided and the ZF solution was an excellent compromise with low weight and small dimensions.

Engine power is converted to propulsion thrust through a new design of waterjet from Wärtsilä. The new LJX waterjet offers, in comparison with other waterjets available today, a 25% reduction in mounting flange diameter, a 10% overall weight reduction and a 35% increase in cavitation margin. It comes with a 1500 mm diameter impeller and an inboard layout for steering and reversing hydraulics.

A substantial weight saving has been achieved through the use of Centa carbon-fibre shafts. Some 70% of weight has been saved compared to that in the use of conventional steel shafts.

Securing reliable and economical power generation on board are four MAN D2876 LE301/HCM 534C 360kW alternators.

Speed, Seakeeping and Safety

Sea trials for *Natchan Rera* were completed in an impressive three days with all systems performing perfectly and meeting best expectations. During the course of trials while running at 100% MCR (maximum continuous rating) speeds of up to 45.4 knots were averaged with 250 tonnes deadweight on board. Loaded with 600 tonnes deadweight an average speed of 41.9 knots was achieved at 100% MCR, almost 3 knots above contract requirement.

In terms of seakeeping the 112 metre offers significant improvements over the already excellent 98 metre class thanks to increased hull separation, increased vessel length and mass together with an increase in tunnel height and refined arch form.

Enhancing the 112 metre Wave Piercing Catamaran's already excellent sea keeping qualities is the Maritime Dynamics/Incat Ride Control System, consisting of transom mounted trim tabs and a retractable T-foil. These, combined with Incat's unique hull form featuring long outer bows, fuller midship sections, stern skeglets and centre bow clearances, means Motion Sickness Incidence has been reduced in higher sea states by reducing pitch, roll and heave, the major contributors to passenger discomfort.

As always, safety is paramount and *Natchan Rera* is equipped with four Liferaft Systems Australia Marine Evacuation Stations, two port and two starboard. The two forward MES serves a total 200 persons each. The two aft MES serve up to a total of 200 persons each. A total of nine 100-person rafts are fitted. Two SOLAS Narwhal inflatable rescue boats, each with 25hp motors, are provided for launch and recovery operations.

With *Natchan Rera* Higashi Nihon has seized a great opportunity. The company can now regenerate from one that simply provides transportation services to one that provides customers with a high-class comfortable hospitality-focused journey.

With the delivery of a sister in April, Incat joins the customer in the hope that the public will travel on both vessels not only for transportation, but also for the enjoyment of being onboard the largest and most fuel efficient diesel powered high speed catamaran in the world.

Ends.

EFFECT OF TUNED MASS ABSORBERS ON LARGE STRUCTURES

John Marco and Ian MacGillivray
DSTO, Department of Defence, PO Box 4331, Melbourne, Australia

john.marco@dsto.defence.gov.au

Abstract

A structure such as a building, ship or vehicle component can produce unwanted vibration in operation when excited at a resonant frequency of the structure. Tuned mass absorbers are simple spring, damper and mass components that can be applied to the structure, targeting the resonant frequency and attenuating the motion of the structure at the problem frequency. These absorbers can be applied to structures in air or water.

This paper will demonstrate, through an example of a hydrofoil component of a ship, how a tuned mass absorber can attenuate its vibrational response at a structural resonance.

Finite element (FE) and boundary element (BE) modal analysis techniques will be employed to show the frequency response of the vibration characteristic before and after application of the tuned mass absorbers. The effect on the overall system response of damping in the structure and tuned mass absorbers will also be considered for some cases.

1. Introduction

A structure operating in its service environment can experience loads that need to be dissipated without causing any damage to the structure. Some applied loads can significantly increase the vibration when a frequency component of the load matches a modal frequency of the structure.

Modification through redesign of the structure can alleviate the resonance phenomenon, but this may be costly or impractical, depending upon the specific structure in question. Another alternative is to add one or more tuned mass absorbers (TMAs) that target the problem frequency [1-4]. A TMA is a combined spring, damper and mass system with a resonance which when applied to the structure can attenuate the motion at that frequency.

When a structure vibrates in a fluid, the resonance frequency to be targeted is generally different to that in air [5]. However, as will be shown later, the addition of a TMA will introduce other modal frequencies into the structure, so care must be exercised to avoid simply shifting the problem to a new frequency. Also, when a TMA is added to a structure it should be attached to a point of maximum response at the resonant frequency, as well as the main framework of the structure, in order for it to perform well.

Although the effect of TMAs is generally well known, the purpose of this paper is to demonstrate numerical modelling of TMAs applied to a relatively large complex structure. We will show how a TMA for a hydrofoil can be tuned to a specific frequency, in order to attenuate the structural motion when excited by an external load. As an introduction, the application of TMAs to a simple plate will first be discussed. Modelling of the effect of a TMA on a water-loaded structure is more difficult than in a vacuum (or in air, when the mass loading of the air can be neglected) but both the unloaded and water-loaded cases will be considered and analysed.

This paper will also show the affects of the structure and TMA damping on the motion of the hydrofoil, in air only, for an applied excitation force. The authors cannot currently model the effect of structural or TMA damping for a water-loaded hydrofoil but the effect of undamped TMAs will be discussed in this case.

A number of numerical techniques are possible [6] for modelling the effect of a TMA. The *ANSYS* [7] numerical FE code, in combination with the vibro-acoustic modelling software *SYSNOISE* [8] for water loading, were used for this work.

2. Tuned Mass Absorbers – Flat Cantilever Plate

A simple flat steel cantilever plate with a mass of 19.5 kg, shown in Figure 1, will be used to demonstrate the use of TMAs to target a specific structural frequency in a vacuum. For this example, an idealised TMA with zero damping has been used. The damping in the plate has also been neglected. The TMA basically only acts on motion normal to the plane of the plate.

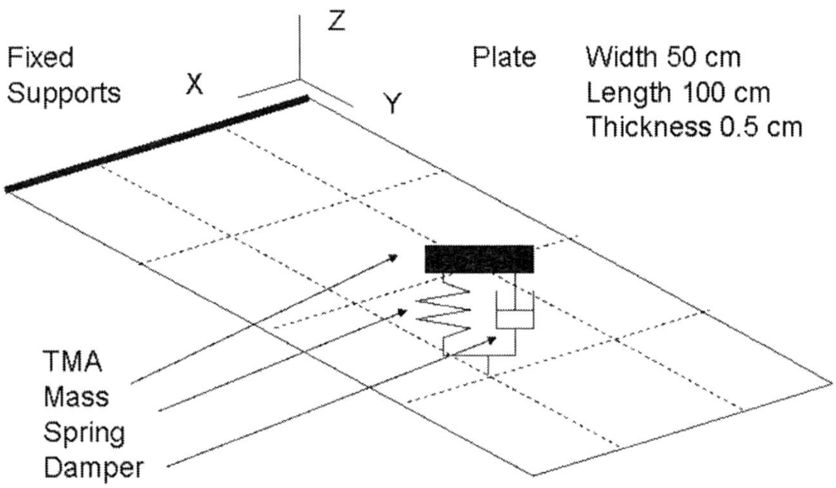

Figure 1. Simple flat cantilever plate with a Tuned Mass Absorber.

The plate without TMAs has a number of structural frequencies, as shown by the vibrational response of the red line in Figure 2. The first bending mode at 4.2 Hz will be arbitrarily targeted by a TMA for vibration response attenuation. Note that the large amplitudes at resonance would not be reached as we have ignored damping in computing the responses of Figure 2.

A TMA with mass m and spring rate k (the 'stiffness') has a resonance frequency f given by the single degree of freedom mass-spring model equation

$$f = (1/2\pi)\sqrt{k/m}.$$

Hence, for an arbitrary mass of 1 kg at a targeted frequency of 4.2 Hz the required spring rate is 696.4 N/m. Figure 2 shows the effect (green line),on the structural modal frequencies and the plate amplitudes due to one TMA, at $x = 30$ cm, $y = 90$ cm, when excited by a load of 1 N at the TMA location. Figure 2 clearly shows that the amplitude of the nodal displacement response at 4.2 Hz has been attenuated by at least four orders of magnitude, but the original resonance frequency has been replaced by two new resonant frequencies at 3.45 Hz and 5.10 Hz, on either side of the targeted frequency. It can also be seen that the amplitudes of the other plate resonance frequencies have not been significantly affected.

An examination of the plate motion shows that at 4.2 Hz the plate remains largely stationary while the TMA mass oscillates perpendicular to the surface of the plate.

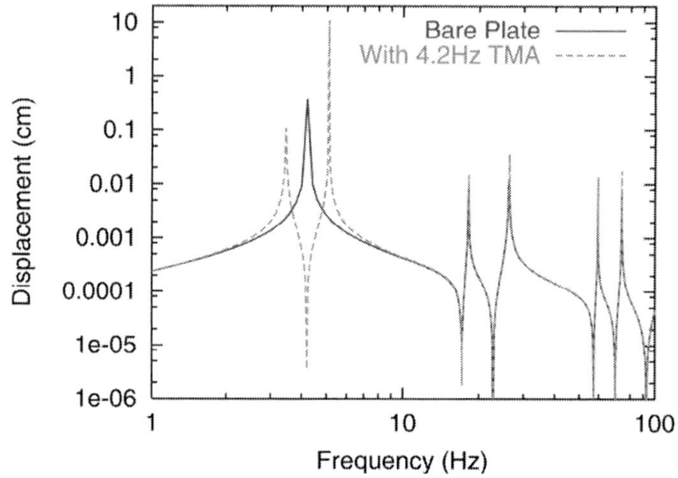

Figure 2. Plate response with and without a 1 kg TMA.

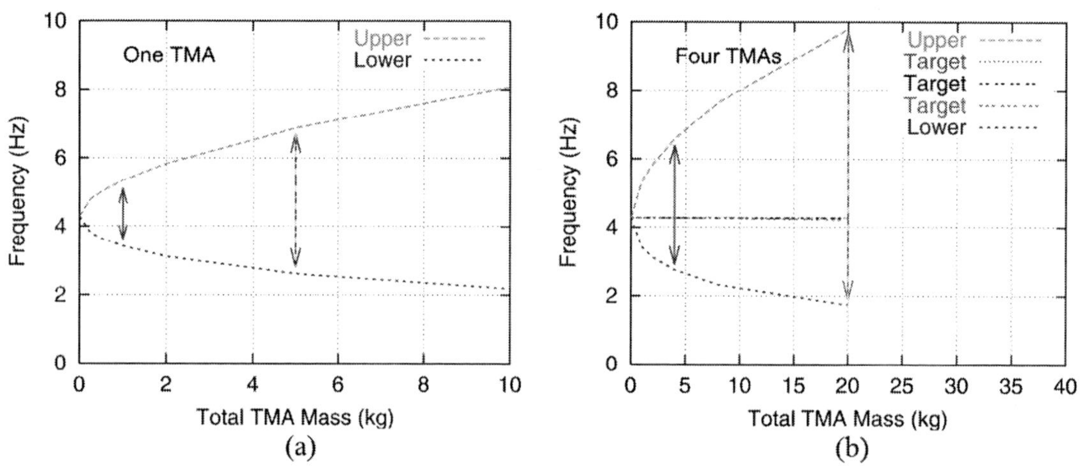

Figure 3. Effect of increasing total TMA mass with (a) one and (b) four TMAs.

Figure 3 shows the effect of increasing the mass or the number of TMAs on the predicted modal frequencies of the plate. Figure 3(a) is a plot for one TMA, showing how the original modal frequency splits into two frequencies of increasing separation as the TMA mass increases. Figure 3(b) shows the effect of four equal-mass TMAs placed at $x = 25 \pm 5$ cm, $y = 85 \pm 5$ cm on the plate. The original mode frequency in this case splits into five, three of which are close to the target frequency, corresponding to three near-horizontal overlapping lines at 4.2 Hz in Figure 3(b). The spread between the lowest and highest new frequencies increases with total TMA mass. For example, the displacement response shown in Figure 2 for the plate with one 1 kg TMA is largely the same if four TMAs of 0.25 kg each are used. Table 1 presents a summary, corresponding to the splitting shown by the vertical arrows in Figure 3.

Table 1. Effect of TMA mass and quantity on the frequency splitting of the modes.

TMA Mass (kg)	TMA Number	Minimum (Hz)	Maximum (Hz)
1	1	3.4	5.3
1	4	2.8	6.6
5	1	2.6	6.9
5	4	1.7	9.8

Therefore some flexibility does exist in designing a TMA system for a structure by manipulating the quantity and mass of the TMAs. Through careful design the new resonance frequencies should not coincide with any of the harmonic frequencies of the applied load.

3. Tuned Mass Absorbers – A Hydrofoil

An aluminium-framed, foam-filled hydrofoil will be considered. Figure 4 shows a structural mode at 222 Hz (in a vacuum) for the hydrofoil. Arrows indicate the direction of movement of the structure at a particular phase of the mode cycle. The mode will be arbitrarily targeted with eight TMAs equally distributed along the rear edge of the structure. For the purpose of this paper a simple TMA configuration of a metallic block attached to a rubber block will be assumed.

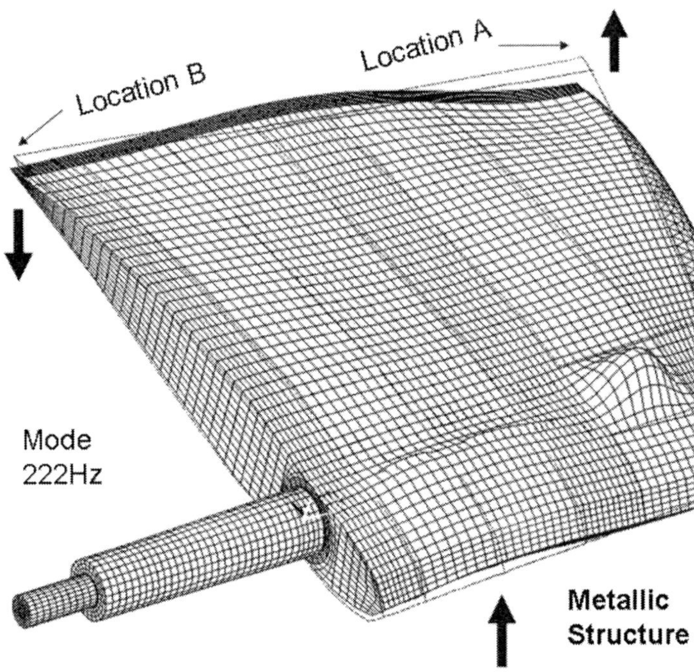

Figure 4. The hydrofoil structure (Finite Element Model representation).

3.1 Hydrofoil in a Vacuum

A number of simulation cases were examined in a vacuum, involving the different structural configurations and response locations as shown in Table 2. Cases in water will be considered in section 3.2. The TMA total mass of 80 kg in configuration 6 represents a few percent of the total mass of the structure.

The numerical approach adopted with the *ANSYS* FE code was to use the eigenvalue extraction method to determine the mode shapes and frequencies of the structure in a vacuum, and to then use the modal superposition method to determine the vibration of the structure to an applied harmonic 1 N load, at location A on the structure, over the frequency range 0 to 350 Hz, in 0.25 Hz increments. Nodal displacements at both locations A and B of Figure 4 were determined.

Table 2. Configuration cases considered in a vacuum.

Configuration Case ID	Structural Damping	222 Hz TMA	TMA Damping	Response Location
1 - baseline	No	No	—	A
2	Yes	No	—	A
3	No	Yes (8 by 1kg)	No	A
4	No	Yes (8 by 1kg)	Yes	A
5	Yes	Yes (8 by 1kg)	Yes	A
6	Yes	Yes (8 by 10kg)	Yes	A
7	Yes	No	—	B
8	Yes	Yes (8 by 1kg)	Yes	B

Estimates were made for the magnitude of damping based upon generic data available in the literature for each component material in the hydrofoil. Material damping for the TMA was based on the damping of a typical rubber material operating in shear.

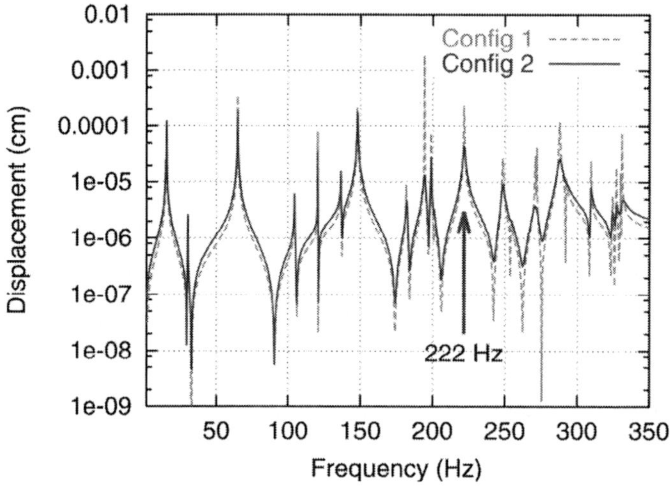

Figure 5. Response of the structure for configurations 1 (baseline) & 2 (with damping).

Figure 5 shows the nodal displacement at location A on the hydrofoil for configurations 1 and 2. The effect of material damping in the structure has, in general, attenuated the peak magnitudes of the response, more so at higher frequencies as expected.

Figure 6 shows the nodal displacement at location A on the hydrofoil for configurations 1 and 3 over the frequency range of 150 to 300 Hz. The effect of the TMAs without damping is a reduction of the nodal displacement at the target frequency of 222 Hz by two orders of magnitude. Eight extra modes are produced adjacent to the original, although these cannot be seen in Figure 6.

Figure 7 shows the nodal displacement at location A on the hydrofoil for configurations 1 and 4. Comparing with Figure 6, the effect of damping in the TMAs is a reduction of the peak nodal displacement around the targeted frequency of 222 Hz.

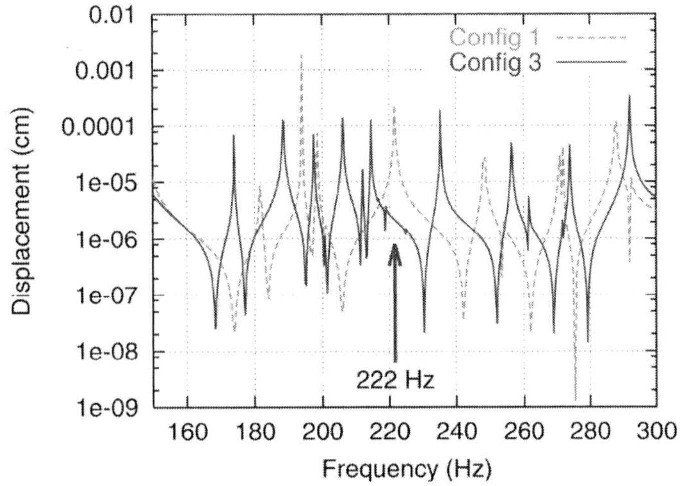

Figure 6. Response of the structure for configurations 1 and 3 (with undamped TMAs).

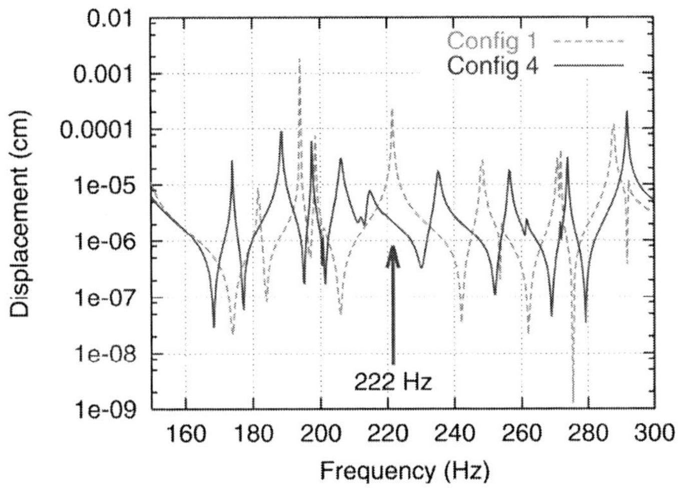

Figure 7. Response of the structure for configurations 1 and 4 (damping only in TMAs).

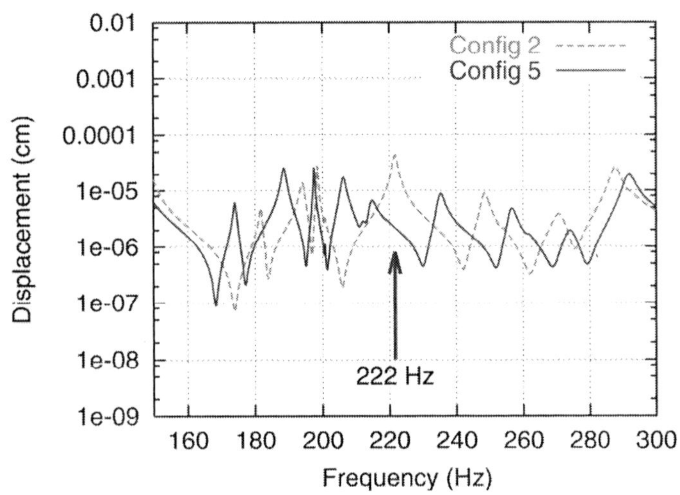

Figure 8. Response of the structure for configurations 2 and 5 (full damping).

Figure 8 displays the response at location A for configurations 2 and 5, showing the effect of TMAs on the 222 Hz mode when damping is included in all components of the structure and TMAs. Since damping has not been neglected, this plot is the most realistic estimate of the effect of eight 1 kg TMAs, and shows that it is possible to lower the response over a broad region near the target frequency

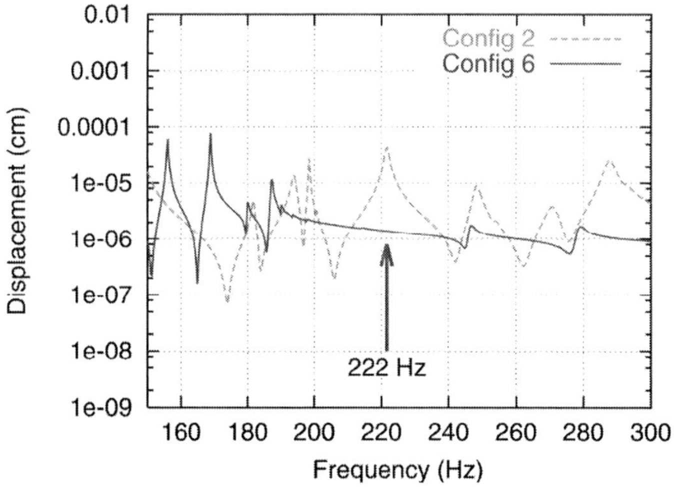

Figure 9. Response of the structure for configurations 2 and 6 (using 10 kg TMAs).

Figure 9 shows the response at location A on the hydrofoil for configurations 2 and 6, showing the effect of eight 10 kg TMAs instead of the eight 1 kg TMAs of Figure 8. The larger masses cause more energy to be dissipated by damping and the effect is a further reduction in response over a broad frequency range near and above the targeted frequency of 222 Hz, when compared with Figure 8.

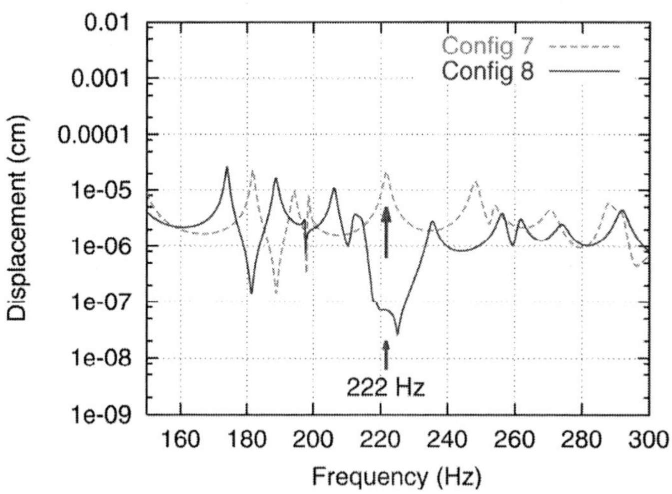

Figure 10. Response of the structure for configurations 7 and 8 (location B).

All previous figures have shown the response of the structure at the load point. Figure 10 shows the response change at the opposite end of the trailing edge of the hydrofoil, at location B, when the eight 1 kg TMAs are introduced. The force is still applied at location A. The figure should be compared with the response at location A, shown in Figure 8. Although the response at location B is quite different the effect of the TMAs in reducing the peak motion responses around the targeted frequency is still apparent.

3.2 Hydrofoil in Water

When the hydrofoil is immersed in water the fluid loading leads to a scaling in the modal frequencies. Particular frequencies are reduced by differing factors depending on the relative effect of the water loading (the added mass) on the particular mode shape. In the case of the in-vacuo 222 Hz resonance chosen as an example in this paper, the effect of the water loading is to reduce the frequency to 143 Hz, which is 64% of the original frequency.

Although *ANSYS* can in principle be used to model the effect of the water loading it has been found by the authors [5] to be easier to couple the *ANSYS* modeling with the vibro-acoustic modelling software *SYSNOISE* [5]. In order to do this, *ANSYS* (using the eigenvalue extraction method) was used to compute the modal response of the structure in a vacuum. The *ANSYS* database of modes produced was then imported into *SYSNOISE* to determine both the modal and vibrational response of the structure (using the modal superposition method) in a fluid. Another advantage of the *SYSNOISE* modelling is that it allows computation of pressure radiated from the structure. A disadvantage of this approach is that *SYSNOISE* can only import real modes and therefore cannot deal directly with the complex modes produced when damping is include in the structure or the TMAs. Damping factors can be applied directly within *SYSNOISE,* but it is difficult to derive these from the ANSYS modelling.

However, it is still possible to partially assess the effect of TMAs targeted at the water-loaded frequency even if damping must be ignored. Table 3 shows the simulation cases that have been examined in water. The baseline configuration 1 (which has no damping) is the same *ANSYS* case as shown in Table 2 when immersed in water. For water loading, the vacuum modal results are imported into *SYSNOISE*.

Table 3. Configuration cases considered in water.

Configuration Case ID	Structural Damping	143 Hz TMA	TMA Damping	Response Location
1 - baseline	No	No	No	A
9	No	Yes (8 by 1kg)	No	A
10	No	Yes (8 by 10kg)	No	A

Figure 11 shows the displacement response of the baseline configuration compared with eight 1 kg TMAs. The eight 1 kg TMAs reduce the response at the target 143 Hz. A more significant effect can be produced with TMAs of larger mass. Figure 12 shows the response comparison with 10 kg TMAs instead of 1 kg TMAs.

The particular mode targeted on this structure radiates sound quite efficiently. Figure 13 shows the sound pressure that would be produced in water in the free field, measured at a 100 m distance normal to the surface of the mid-plane of the bare hydrofoil. In Figure 13, a pressure of 0.001 Pa corresponds to 100 dB re 1 μPa for a 1 N force. In both the case of displacement and the case of the pressure there is a considerable reduction of the response at the target frequency. Although new modal peaks are produced, and the effect of TMA damping is likely to be less for the water-loaded structure than for the structure in a vacuum, a comparison with the figures of section 3.1 suggests that the extra modes would be significantly attenuated when compared with the original targeted mode. Similar results (not shown) are obtained for responses at location B.

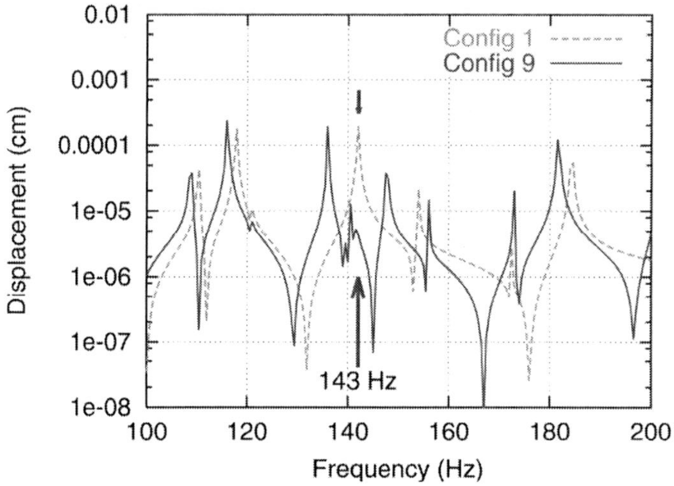

Figure 11. Response of the structure for configurations 1 and 9 in water.

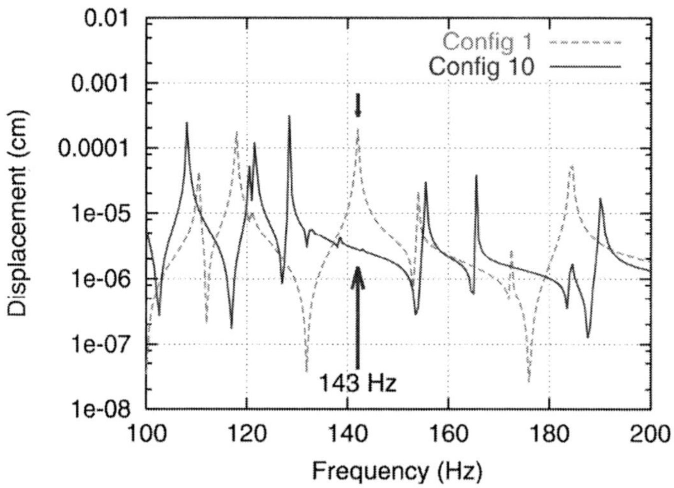

Figure 12. Response of the structure for configurations 1 and 10 in water (10 kg TMAs).

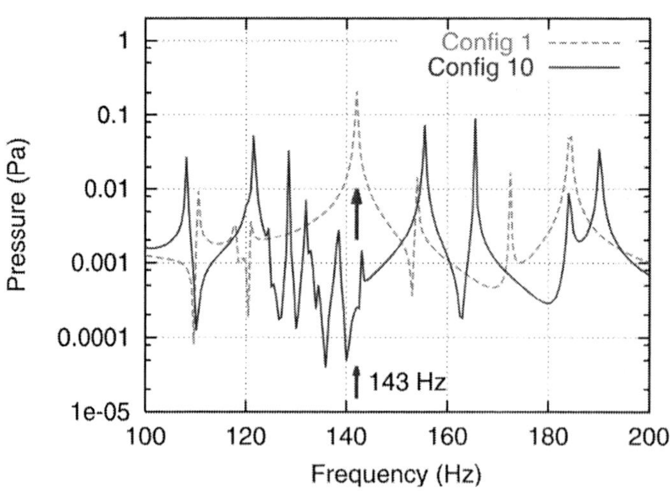

Figure 13. Pressure response of the structure at 100 m for configurations 1 and 10 in water.

4. Discussion

In most structures intrinsic damping plays some part in limiting the peak motion at resonance. In the case of metal structures the damping is often small at the frequencies of interest. Figure 5 shows that the material damping of the base structure only produces moderate attenuation of the modal peaks at the frequencies of interest. Most of this attenuation in the structure is due to the damping by the internal foam, with damping losses around two orders of magnitude greater than the metal.

TMAs can also provide reduction in motion at resonance, as shown in Figure 6, where no material damping has been included at all, although new modal frequency peaks are produced near to the targeted frequency. Material damping (as in the intrinsic damping of a rubber) can be used to reduce the magnitude of the new modal peaks. Figures 7 and 8 show how the inclusion of damping in the eight 1 kg TMAs leads to better general reduction of these peaks near the target frequency of 222 Hz in vacuum.

Although it is well reported in the literature that damping reduces the potential attenuation that can be achieved at the target frequency, note that the attenuation achieved in Figures 7 and 8, at the target frequency, with the inclusion of damping, is not greatly dissimilar to that achieved without damping, probably because the structure is quite complex and has many modes of vibration. In contrast, had damping been included in the flat cantilever plate of section 2, the peak attenuation achieved in Figure 2 at 4.2 Hz would not have been as large (the null would have been less deep) although the introduced peaks would have been lowered.

Changing the masses of the TMAs to 10 kg each, as shown in Figure 9, can achieve a further reduction of around half an order of magnitude in the peaks, in this case essentially fully damping the resonances above the target frequency. This is due to more vibrational energy being available in the larger TMA masses, with a corresponding increase in damping losses per cycle of vibration. The reduction in peak motion is generally observed in all response locations across the hydrofoil, as shown by the red curve in Figure 8 for location A and in Figure 10 for location B.

Figures 11 and 12 show the effect of the TMAs is also similar in water, keeping in mind that the modelling in these cases neglects all material damping.

5. Conclusion

A Tuned Mass Absorber can provide significant help with a resonance problem in a structure by attenuating the peak motion at the problem frequency. This can be achieved by designing the natural frequency of the TMA to match that of the resonance frequency of the structure. This paper has outlined a particular use of TMAs to reduce a resonance in an underwater structure such as a hydrofoil. Material damping can also play a role in helping to reduce the vibrational motion response. Where large attenuations are required at targeted frequencies, as in this example, material damping in the structure alone is not sufficient, and TMAs can be used to achieve the reduction. However, the material damping in the base structure can still be significant in reducing the peak motion and should therefore be considered in the modelling. For the example hydrofoil chosen in this work, a clear reduction in vibrational response at the resonance frequency was achieved using realistic values for TMA masses and the material damping associated with all components.

References

1. C. Maas, B. Mahnken and M. Borgerson, "Application of NVH Countermeasures for Cabin Boom Isolation using Hydraulic Bushing and Silicon Tuned Mass Absorber", SAE 2006 World Congress, April 2006.
2. M.P. Singh, S. Singh and L.M. Moreschi, "Tuned Mass Dampers for Response Control of Torsional Buildings", Earthquake Engineering & Structural Dynamics, Vol. 31, Issue 4, 749-769, Jan 2002.
3. J. Fortgang and W. Singhose, "Design of Vibration Absorbers for Step Motions and Step Disturbances", Woodruff School of Mechanical Engineering, Georgia Institute of Technology, Atlanta, U.S.A.
4. J.K. Vethecan and A.J. Subic, "Vibration Attenuation of Tennis Racquets using Tuned Vibration Absorbers", BAE Systems, Abbotsford, Victoria, Australia.
5. J. Marco and I.R. MacGillivray, "Modal Analysis of Large Structures Submerged in a Fluid", Proceedings of ICS&V14, 9-12 July 2007
6. I.R. MacGillivray and J. Marco, "Review of Numerical Techniques for Modal Analysis in Fluid-Structure Interactions", *INTERNAL DOCUMENT*, DSTO, Australia, 2006.
7. *ANSYS*, Version 11, ANSYS Inc, Southpointe, 275 Technology Drive, Canonsburg, PA 15317 USA, 2007.
8. *SYSNOISE*, Version 5.6, LMS Engineering Innovations, 3001 Leuven, Belgium, 2003.

Progress in Simulation-Based Design

Karsten Fach, Germanischer Lloyd, Hamburg/Germany, karsten.fach@GL-group.com
Volker Bertram, ENSIETA, Brest/France, Volker.bertram@ensieta.fr

Abstract

Simulation-based design increasingly replaces traditional experience-based design. This article gives an overview of techniques now used in advanced industry practice. The article focuses on basics of the techniques, illustrating approaches and state of the art with applications taken from the experience of Germanischer Lloyd.

1. Introduction

The design process is often described as a design spiral following the initial model of *Evans (1959)*. This model has been "refined" by proposing spiraling cones, *Andrews (1981)*, or interconnected spirals, *Mistree et al. (1990)*, but in essence it was the spiral in disguise, Fig.1. This model for ship design has been rejected frequently, as the sequence of analyses and decisions is not as sequential as the spiral models suggest.

Ship design seems to be instead a rather chaotic process of combining various analyses with 'intelligent' or 'creative' generation of solutions, sometimes in parallel, which are then analyzed, evaluated, possibly rejected or modified. The sequence of these elements in the design process changes even for absolutely the same design task from designer to designer. There is no absolute criterion what is the best design process sequence. While different designers may well apply the same tools (or at least tools of comparable performance power), some designers manage to get better results. *Bertram (2004)* sees four basic activities in modern ship design:

1. Generation (creative, intelligent or systematic creation of solutions)
2. Analysis (evaluation, simulation, computation)
3. Visualization (representation, communication)
4. Decision (knowledge, identification, learning)

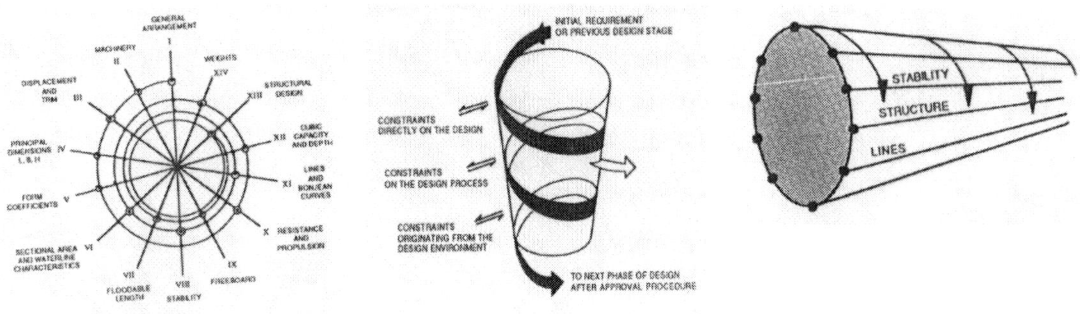

Evans (1959) *Andrews (1981)* *Mistree et al. (1990)*

Fig.1: Process models for ship design

We focus here on the second element, namely analyses as used to evaluate created designs and guide decisions in ship design and operation. We extend in this paper previous work, *Fach and Bertram (2006)*, *Bertram and Couser (2007)*, focussing more on recent developments in hydrodynamic applications due to progress in Computational Fluid Dynamics (CFD).

2. CFD – External flows

2.1 Seakeeping

For many seakeeping issues, linear analyses (assuming small wave height or small wave steepness) are appropriate and frequently applied due to their efficiency. The advantage of this approach is that it is very fast and allows thus the investigation of many parameters (frequency, wave direction, ship speed, metacentric height, etc.). Non-linear computations employing time-domain approaches are usually necessary for the treatment of extreme motions. These simulations require massive computer resources and allow only the simulation of relative short periods (seconds to minutes).

Combining intelligently linear frequency-domain methods with nonlinear time-domain simulations allows exploiting the respective strengths of each approach, *El Moctar (2005)*. The approach starts with a linear analysis to identify the most critical parameter combination for a ship response. Then a non-linear CFD (Computational Fluid Dynamics) analyses determines motions, loads and free surface (green water on deck). We employ the commercial RANSE solver Comet for our purposes, e.g. Fig.2.

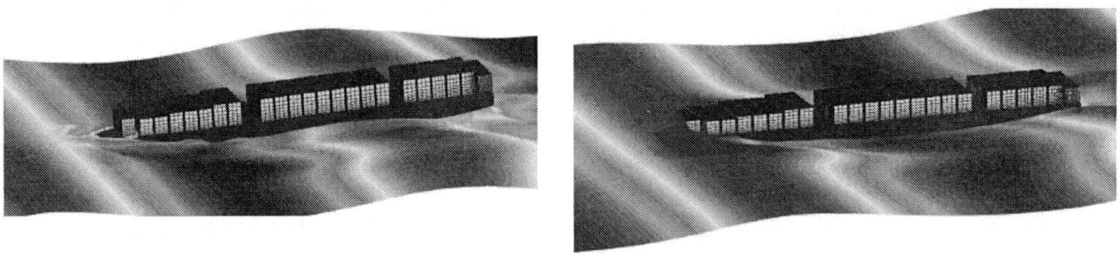

Fig.2: CFD simulation of large containership in extreme waves

1.2 Rudder flows

CFD is by now the most appropriate tool to support practical rudder design, Fig.3. The propeller is typically modelled in a simplified way using axial and tangential body forces. These are external forces distributed over the cells which cover the location where the propeller would be in reality. The sum of all axial body forces is the thrust. The body forces are assumed to vary in radial direction of the propeller only. This procedure is much faster than geometrical modelling of the propeller (by two orders of magnitude) at a negligible penalty in accuracy (about 1%). The procedure has been extensively validated for rudder flows both with and with-out propeller modelling. The same approach for propeller and rudder interaction can be applied for podded drives, *Junglewitz and El Moctar (2004)*. Comet allows also the treatment of cavitating flows, Fig.4. The extensive experience gathered in the last 5 years has resulted in a GL guideline for rudder design procedures, *GL (2005), El Moctar (2007)*.

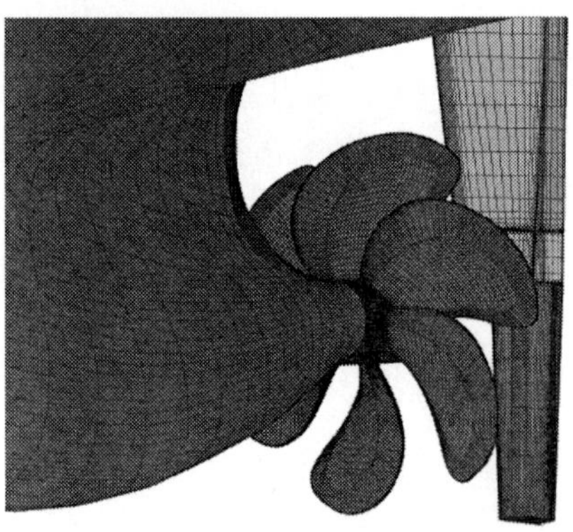

Fig.3: CFD model for hull-propeller-rudder interaction

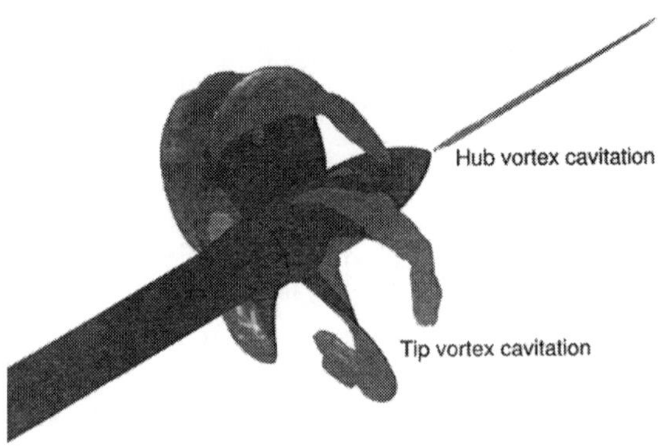

Fig.4: Cavitation predicted at propeller

1.3 HVAC simulations

Aerodynamic flows around ship superstructures can be computed by CFD, Fig.5, although wind tunnel tests still are popular and widely used. CFD offers the advantage of overcoming scale effects which can be significant if thermodynamic processes are involved, *El Moctar and Bertram (2002)*. HVAC (heat, ventilation, air condition) simulations involve the simultaneous solution of fluid mechanics equations and thermodynamic balances, often involving concentrations of different gases. The increasing use of refrigerated containers on ships has motivated advanced CFD simulations using the commercial CFD solver CFX, *Brehm and El Moctar (2004)*, replacing to some extent simple, analytical methods and model tests, Fig.6. Local effects of heat transport cannot be resolved with the typically employed rather coarse grids, but global effects as the difference between pressure and suction ventilation and the influence of natural thermal buoyancy can be reproduced.

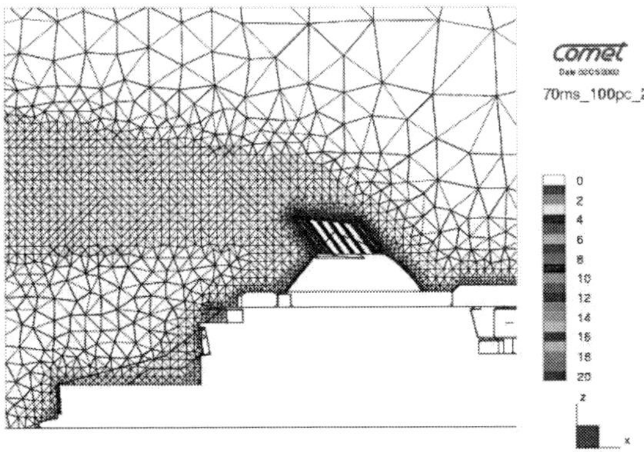

Fig.5: CFD grid for external aerodynamics

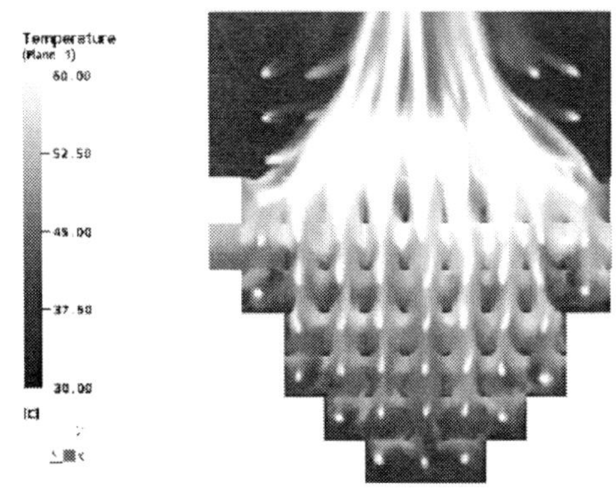

Fig.6: Temperature distribution in cargo hold

1.4 Fire and smoke simulations

At present, zone models and CFD tools are considered for fire simulations in ships. Zone models are suitable for examining more complex, time-dependent scenarios involving multiple compartments and levels, but numerical stability can be a problem for multi-level scenarios, scenarios with Heating, Ventilation and Air Conditioning (HVAC) systems and for post-flashover conditions. CFD models can yield detailed information about temperatures, heat fluxes, and species concentrations, Fig.7. However, the time penalty of this approach currently makes CFD unfeasible for long periods of real time or for large computational domains.

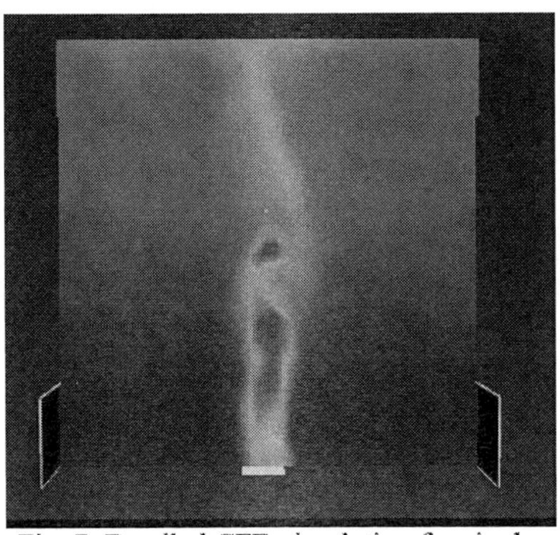

Fig. 7: Detailed CFD simulation for single-room, Single-fire case

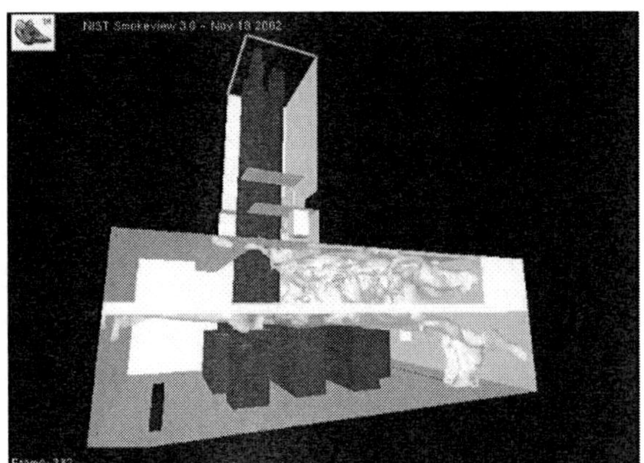

Fig. 8: Fire and smoke simulation in engine room

Applications have graduated from preliminary validation studies to more complex applications for typical ship rooms (accommodation, atrium, engine room), Fig.8, *Bertram et al. (2004)*. The simulations are able to reproduce qualitatively all major fire characteristics, but present combinations of software and hardware do not yet yield reliable quantitative predictions, particularly not for larger and complex geometries. However, a lot more progress can be envisioned in the next decade and the fire simulations appear already suitable to give some general support both for fire containment strategies and for design alternatives.

3. CFD – Coupled Analyses

3.1. Sloshing

Sloshing is a strongly non-linear phenomenon featuring often spray formation and plunging breakers, Fig.9. Only surface-capturing methods can reproduce these features. Extensive experience gathered over the last 5 years allows today the numerical prediction of sloshing loads in ships with great confidence. International validation cases for sloshing are de facto two-dimensional and simple in geometry, *Bertram et al. (2003)*. Measured and computed pressures agree generally well for these validation studies, indicating the high quality of the simulations. The computed sloshing motion agreed also well with videos of the experiments. Complex three-dimensional tanks as found e.g. in the bow have been modelled and pose no principal problems for the simulation tools, Fig.10. More recently, coupled simulations of ship motions in waves with sloshing inside tanks have been performed, Fig.11, *Peric et al. (2007)*. In the simulation, forces and moments exerted by the sea cause the vessel to move, exciting sloshing in the tanks. The sloshing in turn influences the ship motion.

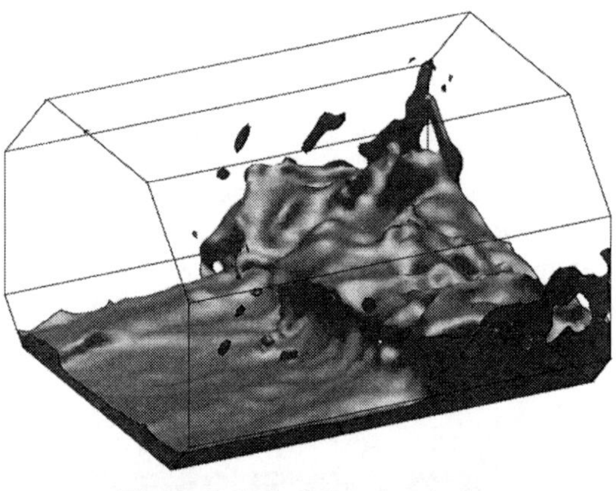

Fig.9: 3-d sloshing simulation

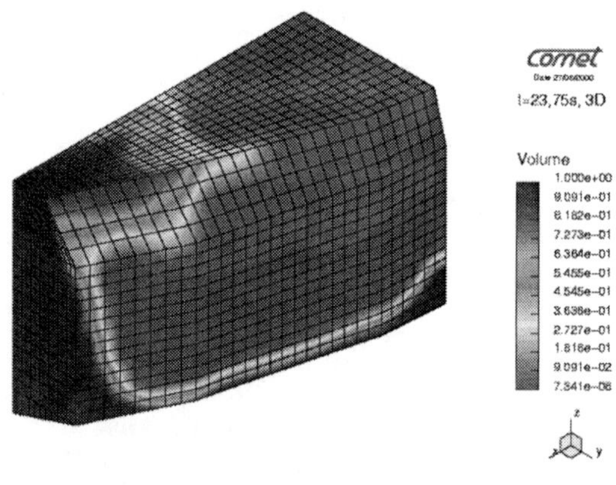

Fig.10: Tank structures near bow

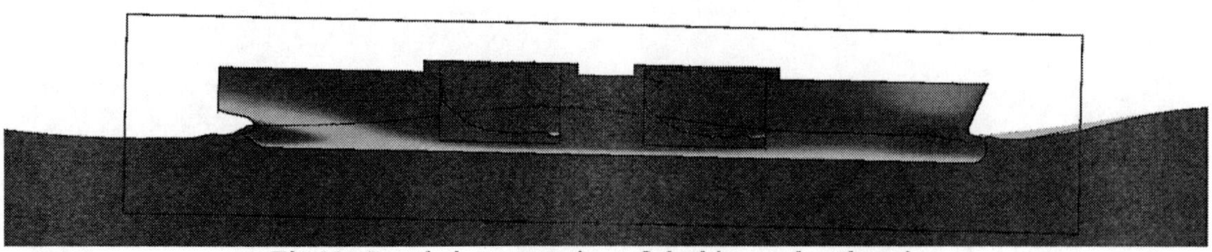

Fig.11: Coupled computation of sloshing and seakeeping

3.2. Evacuation simulation

Evacuation assessment became a major topic at the International Maritime Organization (IMO) after the loss of the 'Estonia', resulting in new requirements for evacuation analyses in an early stage of the design process, *IMO (2002)*. Germanischer Lloyd and TraffGo have developed the software AENEAS for this purpose. Evacuation analyses focus on safety, but the tool can be used also for the optimization of boarding and de-boarding processes, *Petersen et al. (2003)*, or space requirements for promenades on cruise ships and large RoPax ferries. These simulations are very fast, allowing typically 500 simulations within one hour, to gain a broad basis for statistical evaluation. The ship is represented by a simplified grid of different cell types (accessible floor, doors, stairs, obstacles/walls), Fig.12. Passengers and crew are represented by intelligent agents.

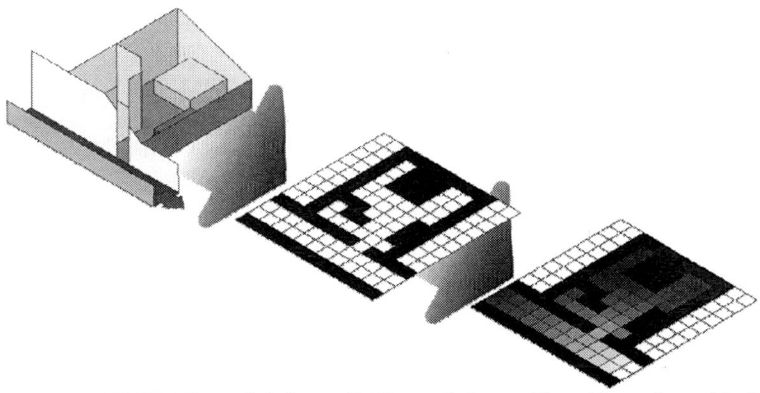

Fig.12: Steps to AENEAS model from CAD model to cells with assigned information

The hazard motivating evacuation will change in time, usually affecting the behaviour of passengers and crew. Then we should simulate both the development of the hazard and its effects on the evacuation. Examples are progressive flooding leading to heel angles and progressive fire and smoke development. Results of fire modelling can be considered in the evacuation simulation. The numerical fire simulation provides time lines for the development of critical conditions (e.g. smoke, temperature) at predefined escape points, such as escape doors at the ends of corridors. Because of the extensive engineering effort necessary for fire simulation, the interaction with AENEAS is arranged in a simplified manner, in which escape routes are blocked sequentially due to smoke or fire. For the process of analyzing fire designs, Germanischer Lloyd has developed an integrated methodology called NESTOR, *Petersen and Voelker (2003),* combining fire simulations with the Multi Room Fire Code, evacuation simulation with AENEAS and a Event Tree Analysis for risk assessment.

The most frequently asked question concerns the influence of the ship motion on the movement of the pedestrians onboard. *Meyer-König et al. (2005)* coupled seakeeping simulations and evacuation simulations in a semi-empirical approach to find the influence of ship motions on evacuation times. Since trim and pitch angles are usually relatively small, their effect is mostly negligible. Roll motions were found to be less critical than static heel for evacuation time.

3.3. Fluid-structure interaction

Finite-element analyses (FEA) require load specifications which for ships involve frequently external hydrostatic and hydrodynamic loads. GL.ShipLoad, *Cabos et al. (2006),* supports efficient load generation for global FEA of ship structures. Hydrostatic and hydrodynamic computations are integrated into the program. GL.ShipLoad supports the generation of loads from first principles (realistic inertia and wave loads for user supplied wave parameters), but the program also aids in the selection of relevant wave situations for the global strength assessment based on bending moments and shear forces according to Germanischer Lloyd's rules. The result is a small number of balanced load cases that are sufficient for the dimensioning of the hull structure.

Advances in computer methods have made 3-d FEA today the standard choice for ship vibration analyses, *Asmussen and Mumm (2001).* The computations require longitudinal mass and stiffness distribution as input. The mass distribution considers the ship, the cargo and the hydrodynamic 'added' mass, Fig.13. The added mass reflects the effect of the surrounding water and depends on the frequency. One can either use estimates based on experience or employ sophisticated hydrodynamic simulations. For local vibrations analyses, added mass needs to be considered if the structures border on tanks or the outer hull plating. Because of the high natural frequencies of local structures, FEA models must be detailed including also the bending stiffness of structural elements.

For very high frequencies (structure-borne noise), the standard FEA approach to vibration analyses is impossible due to excessive computational requirements. For a typical passenger vessel for a frequency of 1000 Hz, a FEA vibration model would lead to several million degrees of freedom. However, the very fact that information is required only averaged over a frequency band allows an alternative, far more efficient approach based on statistical energy analysis. The Noise Finite Element Method (NoiseFEM) of Germanischer Lloyd, *Cabos and Jokat (1998), Cabos et al. (2001),* is based on a related approach. NoiseFEM predicts the propagation of noise by analyzing the exchange of energy between weakly coupled subsystems, Fig.14. Validation with full-scale measurements shows that the accuracy of NoiseFEM is sufficient for typical structure-borne sound predictions for the frequency range between 80 Hz and 2000 Hz, *Wilken et al. (2004).*

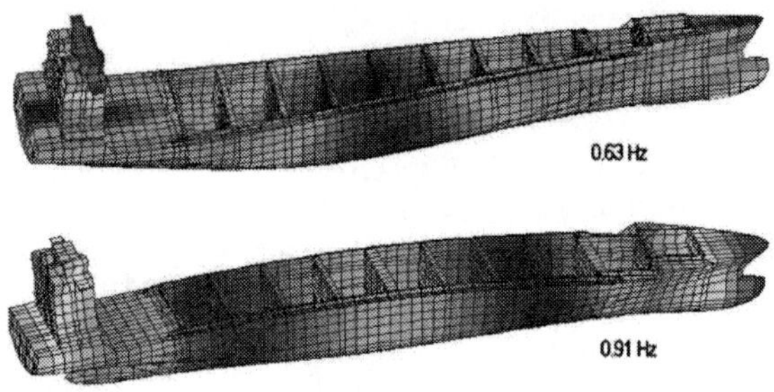

Fig.13: Global FEA of vibrations of containership

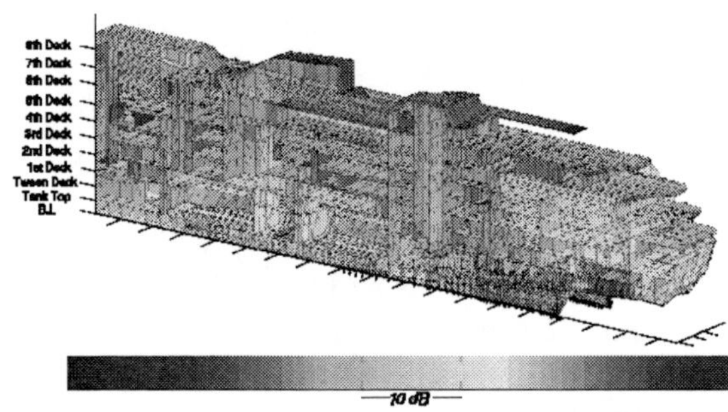

Fig.14: Prediction of structure-borne noise in Blohm&Voss cruise vessel

4. Conclusion

The technological progress is rapid, both for hardware and software. Simulations for numerous applications now often aid decisions, sometimes 'just' for qualitative ranking of solutions, sometimes for quantitative 'optimization' of advanced engineering solutions. Continued validation feedback serves to improve simulation tools as well as it serves to build confidence.

However, advanced simulation software alone is not enough. Engineering is more than ever the art of modelling, finding the right balance between level of detail and resources (time, man-power). This modelling often requires intelligence and considerable (collective) experience. The true value offered by advanced engineering service providers lies thus not in software licenses or hardware, but in the symbiosis of highly skilled staff and these resources.

Acknowledgements

Many colleagues at Germanischer Lloyd have supported this paper with their special expertise, supplying text and/or figures, namely (in alphabetical order) Christian Cabos, Bettar El Moctar, Holger Mumm, Stefan Nusser, Ulf Petersen, Helge Rathje, Pierre Sames, Tobias Zorn.

References

ANDREWS, D. (1981), *Creative ship design*, Trans. RINA 123, pp.447-471

ASMUSSEN, I.; MUMM, H. (2001), *Ship vibration*, GL technology, Germanischer Lloyd, Hamburg, http://www.gl-group.com/brochurepdf/0E094.pdf

BERTRAM, V. (2004), *Towards intelligent simulation-based ship design,* 3rd Conf. Computer and IT Applications in the Maritime Industries (COMPIT), Siguenza, pp.5-16

BERTRAM, V.; COUSER, P. (2007), *CFD possibilities and practice,* The Naval Architect, September 2007, pp.137-147

BERTRAM, V.; EL MOCTAR, O.M.; JUNALIK, B.; NUSSER, S. (2004), *Fire and ventilation simulations for ship compartments,* 4th Int. Conf. High-Performance Marine Vehicles (HIPER), Rome, pp.5-17

BERTRAM, V.; CAPONNETTO, M.; EL MOCTAR, O.M. (2003), *RANSE simulations for unsteady marine two-phase flows*, RINA CFD Conf., London

BREHM, A.; EL MOCTAR, O.M. (2004), *Application of a RANSE method to predict temperature distribution and gas concentration in air ventilated cargo holds*, 7th Num. Towing Tank Symp. (NuTTS), Hamburg

CABOS, C.; EISEN, H.; KRÖMER, M. (2006), *GL.ShipLoad: An Integrated Load Generation Tool for FE Analysis,* 5th Int. Conf. Computer and IT Applications to the Maritime Industries, Leiden

CABOS, C.; JOKAT, J. (1998), *Computation of structure-borne noise propagation in ship structures using noise-FEM,* 7th Int. Symp. Practical Design of Ships and Mobile Units (PRADS), The Hague, pp.927-934

CABOS, C.; WORMS, C.; JOKAT, J. (2001), *Application of an energy finite element method to the prediction of structure borne sound propagation in ships*, Int. Congr. Noise Control Engineering, The Hague

EL MOCTAR, O.M. (2005), *Computation of slamming and global loads for structural design using RANSE*, 8th Num. Towing Tank Symp. (NuTTS), Varna

EL MOCTAR, O.M. (2007), *How to avoid or minimize rudder cavitation,* 10th Num. Towing Tank Symp. (NuTTS), Hamburg

EL MOCTAR, O.M.; BERTRAM, V. (2002), *Computation of viscous flow around fast ship superstructures*, 24th Symp. Naval Hydrodyn., Fukuoka

EVANS, J.H. (1959), *Basic design concepts*, ASNE Journal

FACH, K.; BERTRAM, V. (2006), *High-performance simulations for high-performance ships*, 5th Int. Conf. High-Performance Marine Vehicles (HIPER), Launceston, 2006, pp.455-465

GL (2005), *Recommendations for preventive measures to avoid or minimize rudder cavitation,* Germanischer Lloyd, Hamburg

IMO (2002), *Interim guidelines for evacuation analyses for new and existing passenger craft,* MSC/Circ.1033, International Maritime Organization

JUNGLEWITZ, A.; EL MOCTAR, O.M. (2004), *Numerical analysis of the steering capability of a podded drive,* Ship Technology Research 51/3, pp.134-145

MEYER-KÖNIG, T.; VALANTO, P.; POVEL, D. (2005), *Implementing ship motion in AENEAS - Model development and first* results, 3rd Int. Conf. Pedestrian and Evacuation Dynamics, Vienna

MISTREE, F.; SMITH, W.F.; BRAS, B.; ALLEN, J.K.; MUSTER, D. (1990), *Decision-based design: A contemporary paradigm for ship design*, Trans. SNAME 98, pp.565-595

PERIC, M.; ZORN, T.; EL MOCTAR, O.; SCHELLIN, T.; KIM, Y.S. (2007), *Simulation of sloshing in LNG-tanks*, 26th Conf. Offshore Mechanics and Arctic Engineering (OMAE), San Diego

PETERSEN, U.; MEYER-KÖNIG, T.; POVEL, D. (2003), *Optimising boarding and de-boarding processes with AENEAS*, 7th Int. Conf. Fast Sea Transportation FAST, Ischia, pp.9-16

PETERSEN, U.; VOELKER, J. (2003), *Deviating from the rules – ways to demonstrate an equivalent level of safety*, World Maritime Technology Conf., San Francisco

WILKEN, M.; CABOS, C.; SEMRAU, S.; WORMS, C.; JOKAT, J. (2004), *Prediction and measurement of structure-borne sound propagation in a full scale deckhouse-mock-up,* 9th Int. Symp. Practical Design of Ships and Mobile Units (PRADS), Lübeck-Travemünde, pp.653-659

Preliminary Frigate Design using a Multi-objective Evolutionary Algorithm

Warren Smith and Tapabrata Ray
School of Aerospace, Civil and Mechanical Engineering, University of New South Wales at Australian Defence Force Academy, Canberra. Email {w.smith, t.ray}@adfa.edu.au

ABSTRACT

Using optimisation methods, preliminary ship design is a multidisciplinary task that aims to identify basic ship dimensions and corresponding performance measures that maximize a set of design objectives. The process involves a search through a large variable space in the presence of highly nonlinear constraints arising out of user requirements, statutory norms and physical laws. In this paper, a multi-objective evolutionary algorithm that is capable of effectively and efficiently solving single/multi-objective, unconstrained/constrained optimization problems with mixed variables is utilised. The underlying algorithm relies on a stochastic, population based, elitist, zero order model which explicitly maintains diversity of solutions in both the objective and the variable space. This is unlike most which consider diversity of solutions solely in the objective space. The performance of the algorithm is illustrated using a frigate design problem with an aim to minimize the building cost. The frigate is equipped with an anti-submarine helicopter and appropriate handling system, a medium calibre gun, surveillance radar, a point-defence weapon system and is intended for long-distance patrol duties. Our results are better than previously published results based on a goal programming approach and they highlight the ease of modelling objectives and constraints for such classes of problems.

MOTIVATION AND BACKGROUND

Recent developments (led by Ray) of a multi-objective evolutionary algorithm that is capable of effectively and efficiently solving single/ multi-objective, unconstrained/ constrained optimization problems with mixed variables spawned the notion of improving the performance of a "legacy" ship design model (constructed in the 1990's by Smith). The underlying evolutionary algorithm relies on a stochastic, population based, elitist, zero order model which explicitly maintains diversity of solutions in both the objective and the variable space. This is unlike most other published algorithms which consider diversity of solutions solely in the objective space.

In the 1990s, different aspects of the frigate design model were reported in (Mistree, Smith et al. 1990; Mistree, Smith et al. 1991; Smith 1992; Smith and Mistree 1994; Peacock, Smith et al. 1997). This model was built applying a "decision based design" (DBD) paradigm utilizing the constructs of the Decision Support Problem Technique (DSPT). The technique offers a holistic systems approach that is broadly flexible and adaptable. The DSPT was implemented in the software environment, DSIDES (Decision Support In the Design of Engineering Systems) and the DSPT and DSIDES are well documented in a range of publications including (Mistree, Hughes et al. 1992; Reddy, Smith et al. 1992).
As described in (Xie, Peacock et al. 2002), the relationship of DBD, the DSPT and DSIDES in designing through modelling the real world is portrayed in Figure 1. Once a model is appropriately formulated using the DSPT constructs (typically, a non-linear goal programming styled formulation), DSIDES, with its operations research tools (traditionally an adaptive sequential linear goal programming algorithm delivering vertex solutions), is used to

deduce "model conclusions". These may be seen to be optimal solutions to the closed form mathematical formulation presented. The designer is then at liberty to interpret these conclusions for the real world in the light of the formulated assumptions and approximations.

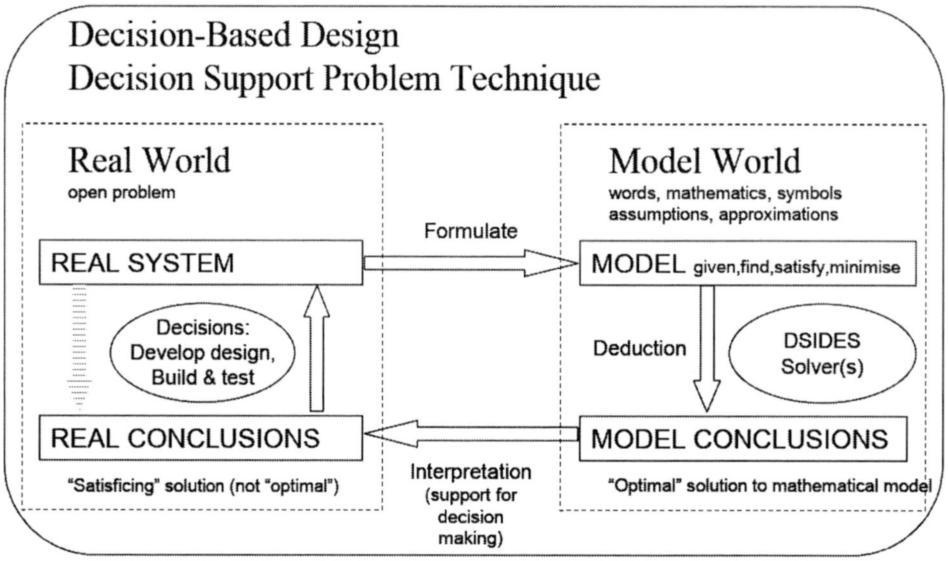

Figure 1 Modelling the Real World (Xie, Peacock et al. 2002)

The focus then of this paper is the application of a "new" evolutionary solver to an existing model with the outcome being better results.

THE FRIGATE MODEL

As proposed in (Smith 1992), assume that:

> *A new frigate is required for undertaking general, long-distance patrol duties off the east coast of Australia. This frigate must have high-speed pursuit capability together with an extensive range and endurance. Weapons and sensors include a large anti-submarine helicopter with an appropriate handling system, a medium-caliber gun, a surveillance radar, a point-defence weapon system, a hull-mounted sonar system and a close-in weapon system. Specifically, the frigate is to be fully operable, including helicopter operations, in seastate 5, have a maximum sustained speed of no less than 30 knots and have an endurance speed of at least 15 knots. The range is to be at least 7000 nautical miles at the endurance speed.*

Consider also that in the pursuit of a "preliminary design" some additional technical design decisions have already been made. These are that the ship will have a combined diesel or gas (CODOG) machinery system located amidships, two propeller shafts, a steel hull and steel superstructure and a 2.55 meters target height between decks. Some further assumptions are:

- the model consists of two strongly coupled Decision Support Problems (DSPs), dealing with hull and propeller dimensioning respectively,
- the basic concept for the hull subsystem is a conventional monohull, and
- the propeller subsystem is designed in accordance with the Wageningen B-Screw Series.

One output of preliminary design is traditionally a general arrangement. This aspect of the design is modeled using a "standard profile" as in Figure 2. The standard profile is a "basis" profile which is scaled to match the principle dimensions. Ship geometry and in particular the area and location of decks, significantly affects the calculation of internal volume, total deck area and system separation and to a lesser extent all other areas of the design. Weapon and system location affects vertical, longitudinal and transverse centers of gravity but also influences weight, internal volume and various system interactions. A Standard Frigate Profile defines the ship geometry in terms of the principal dimensions and is based on both the policy and design practices of a designer. In general, it establishes ship features such as the relative length of decks, location and height of masts and funnels, location and dimensions of the flight deck and location of machinery spaces.

For the light-patrol frigate template, specific assumptions which primarily impact the resistance and powering estimates have also been made as follows:

- standard appendages (e.g., rudders, shaft brackets, skeg, strut bossings, shafts, stabilizer fins and sonar dome) to be fitted,
- normal sections assumed in the afterbody,
- transom not immersed at zero speed,
- no bulbous bow,
- no bow thruster and
- zero design trim.

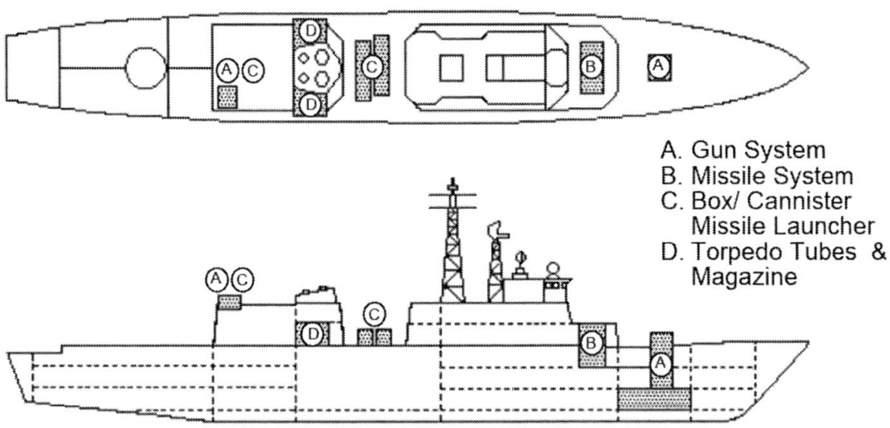

Figure 2 A standard frigate profile

Presented in Tables 1 and 2 respectively are the compromise subsystem word formulations of the frigate hull and propeller models. They are described using the appropriate DSP Technique keywords, namely, Given, Find, Satisfy and Minimize. These keywords and their associated descriptors,

- Given the available Information,
- Find the System Variables,
- Satisfy the System Constraints, the System Goals, the Bounds and
- Minimize the Deviation function,

effectively map the design problem into a form amenable to solution. They provide the uniformity necessary for the application of DSPs to different domains and, more importantly, for taking into account interactions between problems representing subsystems.

Table 1 The Hull Model

Given

Naval Operational Requirements

- Ship Type (frigate)
- Speed goals - maximum sustained speed and endurance speed in knots
- Range - at endurance speed in nautical miles
- Endurance - days at sea
- Seakeeping - maximum seastate for normal ship and helicopter operations
- Payload - weapons, aircraft, command, control and surveillance equipment

Desirable Characteristics

- Machinery Type (CODOG or CODAD or COGAG or Steam Turbine)
- Location of machinery space (aft or amidships) and Number of propeller shafts
- Hull and superstructure material (steel and/or aluminium)
- Deck Height target

Find

The values of ***system variables*** : $\mathbf{x}^* = \{x_i^* : i = 1,\ldots,M\}$ as defined by

The principal ship dimensions

Length (LBP), Beam (B), Draft (T), Depth (D)

The form coefficients

Block Coefficient (CB), Waterplane Coefficient (CW), Prismatic Coefficient (CP)

The hull geometric quantities

LCB, LCF, SDKHT (standard height between decks in meters)

The achievable speeds

VK (maximum sustained speed in knots), VKCR (endurance speed in knots)

and the ***deviation variables*** $\mathbf{d}^-$ and $\mathbf{d}^+$

Satisfy

The following system constraints:

Space

- Displacement is equal to or greater than the estimated weight
- Internal volume is equal to or greater than the required internal volume
- Total deck area is equal to or greater than the required deck area
- Length must exceed aerial, weapon and ship system separation requirements
- The double bottom height is between zero and 2.0 meters

Stability

- The intact stability in the minimum operating condition exceeds the minimum requirement determined in accordance with RAN Stability Criteria

Seakeeping

- The seakeeping rank is greater than that required for the specified sea state for normal ship operations and for helicopter operations
- The natural period of roll is greater than the period of encounter
- The natural period of heave is greater than 120% of the period of encounter relevant to heave, i.e. ship operates in supercritical region
- The natural period of pitch is greater than 120% of the period of encounter relevant to pitch, i.e. ship operates in supercritical region
- Freeboard is greater than the required freeboard at amidships

Form

- The prismatic coefficient is within limits defined by the speed-length ratio
- The LCF is at least a given percentage aft of the LCB
- Bounds for L/B, L/D, B/D, T/D, L/T, B/T, CB/CW, CW/CP and CB/CP

Speed

- The achieved maximum sustained speed exceeds the endurance speed

The following system bounds ($x_j^{min} \le x_j \le x_j^{max}$)

- All variables, except LCB and LCF, must be positive

The following system goals:

- The maximum sustained speed is equal to or greater than the target value
- The endurance speed is equal to or greater than the target value
- The capital cost is equal to or smaller than the target value
- The seakeeping quality is equal to or greater than the target value
- The endurance speed powering is equal to or smaller than the target value
- The maximum sustained speed powering is equal to or smaller than the target value
- The displacement is equal to or smaller than the target value
- The height between decks is equal to the target value

Minimize

- The deviation function, $Z(\mathbf{d}^-,\mathbf{d}^+)=\{f_1(\mathbf{d}^-,\mathbf{d}^+),\ldots,f_k(\mathbf{d}^-,\mathbf{d}^+)\}$

Table 2 The Propeller Model

Given

Ship characteristics

- Draft aft
- Required thrust
- Speed of advance (function of ship speed)
- Thrust coefficient, K_T, for the hull

Machinery Characteristics

- Number of propeller shafts

Propeller Characteristics

- Propeller series data (e.g., Wageningen B-screw series)
- Number of propeller blades and Propeller blade area ratio
- Propeller centreline submergence

Find

The values of ***system variables:*** $\mathbf{y}^* = \{y_i^* : i = 1,\ldots,M\}$ as defined by

PDRAT (propeller pitch diameter ratio), DP (diameter of the propeller), RPM

and the ***deviation variables*** $\mathbf{e}^-$ and $\mathbf{e}^+$

Satisfy

The following system bounds ($y_j^{min} \le y_j \le y_j^{max}$):

- All variables must be positive
- The calculated propeller K_T is within five percent of the hull K_T value
- The diameter of the propeller is less than or equal to some fraction of the draft
- The propeller slip lies between 0.05 and 0.6

The following system goals:

- The difference between the uncorrected propeller shaft RPM and the cavitation corrected RPM is equal to zero (match RPM's)
- The propeller and ship thrust coefficients are matched (match K_T's)

Minimize

- The deviation function, $Z(\mathbf{e}^-,\mathbf{e}^+)=\{f_1(\mathbf{e}^-,\mathbf{e}^+),\ldots,f_k(\mathbf{e}^-,\mathbf{e}^+)\}$

In translating the HULL / PROPELLER model to run under the evolutionary algorithm (EA) as opposed to the adaptive linear programming (ALP) code of DSIDES, a simple wrapper on the legacy analysis FORTRAN code was written and all constraints were modelled identically.

THE EVOLUTIONARY ALGORITHM

The algorithm used in this study is a variant of NSGA-II (Deb 2000; Deb 2002) which has a major difference in the process of population reduction from a size of 2M to M. Our method explicitly maintains solutions that have extremal values of the variables. The process of population reduction is more computationally expensive as compared to NSGA-II, but is useful for problems where maintaining the diversity in the variable space is important. Before introducing the algorithm, it is necessary to introduce the notion of Pareto optimal design and nondominated design. A design $\mathbf{x}^* \in F$ is termed Pareto optimal if there does not exist another $\mathbf{x} \in F$, such that $f_i(\mathbf{x}) \leq f_i(\mathbf{x}^*)$ for all $i = 1,...,k$ objectives and $f_j(\mathbf{x}) < f_j(\mathbf{x}^*)$ for at least one j. Here, F denotes the feasible space (i.e., regions where the constraints are satisfied) and $f_j(\mathbf{x})$ denotes the jth objective corresponding to the design $\mathbf{x}$. If the design space is limited to M solutions instead of the entire F, the set of solutions are termed as nondominated solutions. Since in practice, all the solutions in F cannot be evaluated exhaustively, the goal of multiobjective optimization is to arrive at the set of nondominated solutions with a hope that it is sufficiently close to the set of Pareto solutions. Diversity among this set of solutions is also a desirable feature as it means making a selection from a wider set of design alternatives.

The pseudo-code of the algorithm is as follows.

1. $t \leftarrow 0$
2. Generate M individuals representing a population: $Pop(t) = \{I_1, I_2, \ldots, I_M\}$ uniformly in the parametric space.
3. Evaluate each individual: Compute their objectives and constraints i.e., $f_k(I_i)$ *and* $c_j(I_i)$; for $i = 1,2,\ldots,M$ individuals, $k = 1,\ldots,O$ objectives and $j = 1,\ldots,S$ constraints.
4. **Select** two parents I_1 and I_2 from the population.
5. **Create** two children C_1 and C_2 via crossover and mutation of I_1 and I_2
6. Repeat steps (4) and (5) until M children are created.
7. Evaluate M children.
8. Merge M parents and M Children to form a population of size 2M.
9. **Retain** better performing M solutions from the above 2M solutions.
10. $t \leftarrow t+1$
11. If $t < T_{max}$ then repeat steps (4) through (10) Else Stop. *where* T_{max} denotes the maximum number of generations.

Select: The procedure for selecting parents I_1 and I_2.

a) Select two individuals (I_1 and I_2) from the population of M solutions using a uniform random selection.
b) If I_1 is feasible and I_2 is infeasible: I_1 is selected as the parent and vice versa.
c) If both I_1 and I_2 are infeasible: The one which has the minimum value of the maximum violated constraint is selected as the parent. Compute $max(g_i(x); i = 1, S)$ for I_1 and I_2 and choose the one which has the minimum value. ($g_i(x)$ denotes the constraints).

d) If both I_1 and I_2 are feasible and I_1 dominates I_2: I_1 is selected as parent and vice versa.
e) If both I_1 and I_2 are feasible and neither dominates each other: The parent is a random choice between I_1 and I_2.

Create: We have used simulated binary crossover (SBX) and the polynomial mutation for the real variables as adapted in NSGA-II to create two children from a pair of parents (Deb 1995).

Retain: The procedure to retain M solutions from a set of 2M solutions is:

a) Rank the set of 2M solutions. (The procedure for ranking is described below).
b) If the number of rank=1 solutions (i.e. non-dominated solutions) is less than M, select top M solutions based on their rank and copy them to the new population.
c) If the number of rank=1 solutions is more than M, follow the following steps:
    i. Select the solutions which have a minimum value (assuming minimization in both objectives) in any of the objectives and copy them to the new population.
    ii. For every variable, copy two solutions to the new population which has its minimum and the maximum value if they have not been copied yet (including step i). If the number of solutions copied is more than M/2, than M/2 randomly selected solutions are allowed to stay in the population to maintain adequate space for diversity of solutions in the objective space to exist.
    iii. For the remaining rank=1 solutions, the sequence of who goes in first to the new population is decided as follows:
        - Compute the score by inserting the solution into the new population one at a time. The score is the minimum Euclidean distance computed between the solution attempting to enter with all other existing solutions in the new population based on the objective function space. Scaled values are used for the score computation i.e. the objective space is scaled using the maximum and minimum values in each dimension based on the set of rank=1 solutions.
        - The solution with the highest score is allowed to go into the new population unless it has been copied earlier in which case the solution with the next score goes in.
        - The above listed steps in bullet points are repeated until the new population has a size of M.

Ranking: The procedure for rank computation is as follows:

a) Separate the set of 2M solutions to a set of feasible and a set of infeasible solutions.
b) Perform a non-dominated sorting to assign ranks to the solutions in the feasible set.
c) Rank the solutions in the infeasible set based on their maximum value of the violated constraint.
d) Update the ranks of the solutions in the infeasible set by adding the rank of the worst feasible solution to each.

The assumptions behind the procedures in the algorithm are:

1. A feasible solution is always preferred over an infeasible solution in line with commonly adopted practice.
2. The procedure ensures that the endpoints in the objective space are inserted into the new population and the extent of the non-dominated front is preserved.
3. The procedure explicitly attempts to maintain variable diversity.

Every run of the optimization algorithm requires the inputs for population size, number of generations, probability of crossover, probability of mutation, distribution index of crossover, distribution index of the mutation and a random seed.

THE FRIGATE MODEL EXERCISED

Representations of the tradeoffs arising from a multi-objective cost / power / seakeeping rank formulation are depicted in Figure 3. While the three relational plots in Figure 3 are two dimensional, they are the projection of the multi-objective Pareto surface in a three dimensional cost / power / seakeeping rank space. (The tradeoffs in performance can be seen better with a three dimensional image but such an image is not effectively replicated here.) The Pareto fronts themselves could be further refined through the conduct of further design evaluations in the EA but the observed trends are not expected to change dramatically. Using a population size of 100 evolved over 900 generations resulting in 90000 design evaluations has achieved in separate design instances a cost of 0.4242 (non-dimensional), power of 21180kW and seakeeping rank of 10.59.

The corresponding "best" individual result using ALP from (Smith 1992) for a sprint speed of 32 knots led to an indicative "cost" of 0.4403. This was achieved using a "good" starting point and while trying to balance a range of objectives. The design is highlighted in column D of Table 3. The performance of the ALP algorithm proved very dependent on the quality of the starting point.

In comparison, the single objective designs found in another study using the evolutionary algorithm (EA), seeking individually lowest cost, lowest power and highest rank are shown in columns E-G of Table 3. Significant improvements in performance features have been achieved. For example, when compared with the ALP result, cost can be reduced by 6.4% while also achieving a decrease in installed power of 18%. It is reasonable that the process as it reduces cost also works to reduce powering requirements. As shown in Column F, a marked improvement in seakeeping rank can be achieved (Rank > 12), however, this extreme would come with a 60% increase in cost. Clearly, the compromises that lie between these design solutions would be worthy of exploration. Noting the random and statistical nature of the GA process, the results presented for the single objective work were achieved based on 32 independent runs with population sizes of 100 and 200, 300 and 500 generations, probability of crossover set as 0.8 and 0.9, probability of mutation set as 0.1 and 0.05 and distribution index of crossover 10 and distribution index of mutation 20.

CONCLUSIONS

In comparing the ALP and EA solver results, it is apparent that in the highly complex ship design domain, the EA approach offers advantages related to being able to effectively describe the Pareto boundaries within the design space. The notion of better being made here is on the basis of qualitative feel and decision support from the designer's perspective, as opposed to absolute efficiency of computation. Much has been written in other places discussing the virtues of many operations research / optimisation tools. However, for a user, being able to see the Pareto front and visually interpret the trade-offs in performance is incredibly useful and valuable.

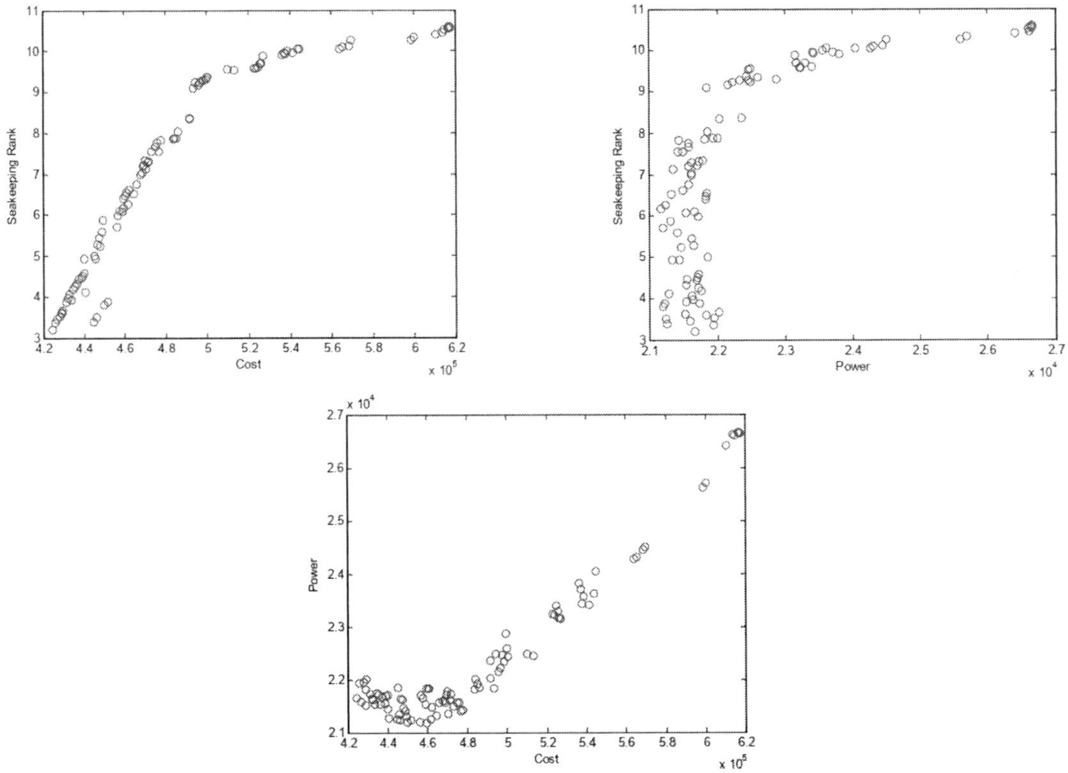

Figure 3 2D representations of the Pareto Surfaces in a 3 dimensional performance space (cost / power / seakeeping)

Table 3 Numerical Results

A	B	C	D	E	F	G
	Lower Bound	**Upper Bound**	**DSIDESOriginal Best 32 Knot**	**EA Lowest Power**	**EA Highest Rank r**	**EA Lowest Cost**
LBP	115	135	124.3	125.637	134.993	119.453
BEAM	10	18	12.29	12.303	14.8732	12.0445
DRAFT	3	5	4.145	4.23795	4.99999	4.14099
DEPTH	7.35	10.25	7.846	7.82185	9.66042	7.62586
CB	0.4	0.6	0.4618	0.44807	0.4538	0.45979
CW	0.6	0.8	0.7384	0.73532	0.76282	0.74341
CP	0.5	0.7	0.5645	0.57244	0.577	0.59343
LCB	50	54	52.88	51.8008	50.2491	50
LCF	52	58	54.49	55.5921	52.0471	52.0029
SDKHT	2.45	2.75	2.516	2.47462	2.74444	2.46812
PDRAT	0.7	1.3	1.296	1.27891	1.29131	1.27967
DP	3.0	5.0	3.582	3.6793	4.14567	3.77475
RPM	200	250	250	240	222	238
VK(Fixed)	20	35	**32**	**32**	**32**	**32**
VKCR(Fixed)	12	18	**15**	**15**	**15**	**15**
Cost			**0.4403**	**0.4440**	**0.6937**	**0.4120**
Installed (kW)			**26084**	**20703**	**30447**	**21484**
Seakeeping Rank			**3.22**	**3.26**	**12.22**	**3.31**

What is also implied is that the work done previously can easily be leveraged into the EA environment. In a lot of ways, the significant modelling effort is encapsulated in the domain performance models. In this way, the work done in the 1990's is still as useful today as it was when first coded. Thankfully, Newton's Laws have not changed.

In summary, using optimisation methods, preliminary ship design is a multidisciplinary task that aims to identify basic ship dimensions and corresponding performance measures that maximize a set of design objectives. The process involves a search through a large variable space in the presence of highly nonlinear constraints arising out of user requirements, statutory norms and physical laws. Using a legacy frigate model and an evolutionary algorithm approach, our recent results are better than those previously published and they highlight the ease of modelling objectives and constraints for such classes of problems.

REFERENCES

Deb, K., Agrawal, R. B. (1995). "Simulated Binary Crossover for Continuous Search Space." Complex Systems.(9): 115-148.

Deb, K., Pratap, A., Agarwal, S., Meyarivan, T. (2002). "A fast and elitist multi-objective genetic algorithm: NSGAII." IEEE Transaction on Evolutionary Computation(6): 181-197.

Deb, K., Pratap, A., Agarwal, S., Meyarivan, T. (2000). A fast elitist non-dominated sorting genetic algorithm for multi-objective optimization: NSGAII. Parallel Problem Solving from Nature VI Conference, Paris, France.

Mistree, F., O. F. Hughes, et al. (1992). Compromise Decision Support Problem and the Adaptive Linear Programming Algorithm. Structural Optimisation: Status and Promise. M. P. Kamat. Washington, DC, AIAA. **150:** 251 - 290.

Mistree, F., W. F. Smith, et al. (1990). "Decision-Based Design: A Contemporary Paradigm for Ship Design." Transactions SNAME **98**: 565-597.

Mistree, F., W. F. Smith, et al. (1991). Designing Decisions: Axioms, Models and Marine Applications. Fourth International Marine Systems Design Conference, IMSDC, Kobe, Japan, Society of Naval Architects of Japan.

Peacock, D., W. F. Smith, et al. (1997). Minimal Ship Motion Hull-Form Design for High Speed using Multi-Criteria Optimisation Techniques. 4th International Conference on Fast Sea Transportation, FAST97, Sydney, Australia, Baird Publications.

Reddy, R., W. F. Smith, et al. (1992). DSIDES User Manual, Systems Design Laboratory, Department of Mechanical Engineering, University of Houston.

Smith, W. F. (1992). The Modeling and Exploration of Ship Systems in the Early Stages of Decision-Based Design. Interdisciplinary Operations Research Program. Houston, Texas, University of Houston.

Smith, W. F. and F. Mistree (1994). The Development of Top-Level Ship Specifications: A Decision-Based Approach. 5th International Marine Design Conference, IMDC and Summer Meeting of the German Society of Naval Architects, Delft, The Netherlands.

Xie, S., D. Peacock, et al. (2002). Modelling and Exploring Hierarchical Systems Engineering Models with a Preliminary Ship Design Focus. 9th AIAA/ISSMO Symposium and Exhibit on Multidisciplinary Analysis and Optimization, Atlanta, GA, USA, American Institute of Aeronautics and Astronautics, AIAA.

Australian's Future Maritime Air Warfare Destroyer Capability – Balancing National Interests

Warren King
Defence Materiel Organisation, Australia

Following the selection of the design for Australia's next generation Air Warfare Destroyers, the AWD Alliance has now commenced the detailed work required to deliver this quantum leap in naval capability to the Royal Australian Navy. The HOBART class of AWDs will provide air defence for accompanying ships, land forces and infrastructure in proximate coastal areas, and self-protection against attacking missiles and aircraft. The Aegis Combat System, incorporating the state-of-the-art phased array radar, AN/SPY 1D(V), in combination with the SM-2 missile, will provide an advanced air defence system capable of engaging enemy aircraft and missiles at ranges in excess of 150km. Each ship will require around three million man hours to complete, with over 1,000 local companies involved.

As it moves into the Build Phase of the project, the Alliance is gaining a clearer picture of the impressive capabilities of Australian industry and the broad range of challenges which lie ahead for Australia's shipbuilding industry. These challenges will require a unified response from government and industry if Australia is to continue undertaking advanced naval construction on a national interest basis effectively and efficiently.

Technically complex naval projects are carried out within tight budget and schedule constraints. If we can demonstrate that these programs can successfully meet all of the schedule, budget and quality criteria then the Australian public is more likely to support similar future programs such as the next generation of submarines and surface combatants to replace the Collins submarines and the ANZAC ships. If we can generate this support and provide this certainty we should be able to maintain the skills and resources that have hitherto been threatened by the stop-start nature of Australian naval shipbuilding.

Long term infrastructure development and risk reduction mechanisms have long been accepted as cornerstones of a forward-looking naval shipbuilding plan. However, there is now an increasingly obvious requirement to focus attention on the development of skilled engineers and tradespersons and experienced project managers to support an industry that is currently growing faster than its supporting mechanisms.

Continuity of employment and continuous development of Australia's skills base will be of vital importance if we are to leverage off the growth in industry capabilities that will result from the AWD Program and link it to future projects. The distribution of work, outsourcing and sustainment of industry will also need to be considered carefully if the strategic naval shipbuilding and repair sector is to remain viable and sustainable into the future. This in turn will encourage the continued development of wide-ranging professional and trade skills to meet the needs of those industries.

One of the issues for us is to optimise Australian industry involvement to ensure that the necessary through-life support skills and resources are in place for the ships when they are delivered. The class is expected to be in service until the mid 2040s and as the client we are looking for solutions which will reduce our through life costs while maximising platform availability and capability.

From the customer's perspective the management of future industry resource levels must promote competitive industrial environments that are conducive to the exercise of best practice management and demonstrate value for money. There will also need to be careful consideration of the level of investment in Australian technology development to ensure it is consistent with the need for efficient long-term sustainment while at the same time avoiding the pitfalls of over-capitalisation.

The bottom line is that we must balance our desire to achieve the best outcome for Australian industry with the need to provide the war-fighters with the best possible chance of winning. That inevitably means that we will continue to source much of our high end capability offshore, while we seek to also foster our industrial aims through a focus on particular areas of the capability production process, and on through life support.

And we need to do this in an environment which will inspire tomorrow's generation of shipbuilding professionals and tradespersons to be a part of similar challenges in the future.

The Pacific 2008 International Maritime Conference

Port State Control

Sydney – February 2008

Australia's Port State Control Inspection Program

Presented by Lindsay Emmett
Principal Marine Surveyor, Ship Inspection group
Australian Maritime Safety Authority

Synopsis

This paper provides an overview of the port State control activities undertaken by the Australian Maritime Safety Authority, (AMSA), along with Australia's relationship with the Asia-Pacific and Indian Ocean Memorandums of Understandings.

The paper then goes on to examine Australia's Port State Control (PSC) program in detail, and the results of the PSC inspections from the 1st of January to the 31st of December 2006.

Australian Maritime Safety Authority

The Australian Maritime Safety Authority is the Australian Governments self-funded statutory authority responsible for a wide range of maritime services. Apart from PSC inspections, AMSA implements the development and oversight of maritime regulatory requirements, the provision of navigational infrastructure, maritime and aviation search and rescue arrangements for an area of approximately one ninth of the world's surface, and for the management of Australia's national plan to combat oil pollution of the sea.

AMSA supports various strategies in an effort to combat the current problems with ship safety standards. These activities take into account AMSAs responsibilities associated with Australia's role as both a flag state and a port state.

In those areas where it is possible to actively influence events, AMSA maintains a belief in the ability of the IMO to foster a more effective maritime safety regime.

Port State Control Overview

Port State control is the regime under which nations inspect foreign-flagged vessels that visit their ports. Nations carry out these inspections to ensure that foreign vessels trading to their ports meet internationally accepted standards, and that qualified crews operate the vessels. Port State control also plays an important role in identifying and ultimately eliminating sub-standard shipping, and their operators.

The United Nations Convention of the Law of the Sea (UNCLOS) provides all nations with many rights in relation to ship registration, freedom of navigation on the high seas and innocent passage through waters of coastal states. However, every right requires a corresponding responsibility. It should be clear that primary responsibility for ensuring compliance with international conventions rests with flag States. Even the 1929 edition of SOLAS incorporated control provisions to allow officers duly authorised by a Government to verify that the certificates issued to a ship under that SOLAS convention were valid.

But experience has clearly demonstrated that not all ships fully comply with these conventions, as many flag States are unable or unwilling to maintain full and continuous control over their ships.

While some nations continue to ignore their responsibilities in relation to vessels carrying their flag, then coastal and port States will implement strategies aimed at protecting their resources, ports, and environment from the impact of unsafe ships.

For a nation to commit itself to a port State control program, without initially ensuring that it fulfils its own role as a flag State not only brings its objectivity into question, but also undermines the effectiveness of port State control. There would be no need for port State control if all nations fulfilled their flag State responsibilities.

The primary responsibility for safe ships lies with the operator and flag State. However, whilst some owners and administrations remain unable, or in some cases unwilling to meet their obligations, port State control will remain an important enforcement approach in the eradication unsafe substandard shipping from the trade routes of the world?

Port State control is the inspection by officers duly authorised by a nation's maritime authority to inspect foreign ships visiting the countries ports to ensure that the vessel complies with the international maritime safety and pollution prevention conventions. In general terms, a vessel is substandard if the safety of the ship is in question, or if it is believed that the ship is a threat to marine environment, or if the welfare of the crew is compromised.

Today, the establishment of regional cooperation in the control of ships has enhanced port State control around the world. Such cooperative programs include among others, the Tokyo and Indian Ocean Memorandums of Understanding (MOU), of which Australia is a member. Membership of these organisations is beneficial in the harmonisation of port State control inspections.

The benefits to Australia and shipowners of Australia's participation in the MOUs is that when a foreign vessel visiting an Australian port is subject to a PSC inspection, AMSA will carry out the inspection in a manner consistent to those carried out by other members of the MOU.

Additionally, members of an MOU have direct access to inter-regional PSC information held on the Asia-Pacific Computerised Information System (APCIS), along with the Paris MOUs computerised information system (SIRENAC). Access to these information systems assists in targeting vessels and avoids duplication of PSC inspections. Additionally, the Indian Ocean database, IOCIS is now in the final stages of development and should be available shortly.

The effectiveness of information exchange is the core to any MOU on port State control. Regional cooperation on PSC has the most effect when the information that one member (or country) collects on a ship is promptly made available to other countries in the region

Australia is fully committed to the aims of both the Tokyo and Indian Ocean MOUs, and the benefits it can bring to ensuring safer shipping in the respective regions.

Port State Control in Australia

Objectives

Australia has for many years carried out safety inspections on foreign vessels visiting Australian ports. However, it was not until the formation of the Paris MOU, and subsequent establishment of the Tokyo MOU that these safety or PSC inspections took on a more formal structure.

Australia's PSC program complies with both the spirit and intent of the control provisions contained in the relevant international conventions. In addition, Australian domestic legislation, i.e. the Navigation Act, provides the authority for AMSA surveyors to board vessels at any time to investigate issues that have the potential to jeopardise safety, or the marine environment.

In addition to complying with the Australian Governments safety objectives, AMSA's PSC Program focuses on the aims of the Tokyo and Indian Ocean MOUs through the operation of uniform and consistent PSC programs and inspections.

AMSA conducts PSC inspections in accordance with international guidelines and within the limitations of its authority under modern administrative law. In conducting PSC inspections, Australian port State control officers follow a set of instructions to surveyors along with a PSC manual. The basis of these instructions is a number of resolutions promulgated by both the IMO and ILO.

Consistency, uniformity, and objectivity are the keys to a successful and credible PSC program. To achieve these ends, AMSA conducts a comprehensive training and auditing program for AMSAs PSC officers in order to achieve consistency in PSC inspection decision-making. AMSA continually strives to enhance performance in these areas to ensure Australia's PSC Program continues to gain credibility from both Australian and foreign interests.

AMSAs objective is to inspect at least 50% of foreign vessels visiting Australian ports. The basis of this percentage is the number of eligible ships visiting our ports during a given year. For this purpose, an eligible ship means one that AMSA has not inspected during the previous six months (three months and tankers over fifteen years old, and passenger ships of any age) immediately preceding the date of arrival at a port

Resources

Forty-two surveyors located in fourteen ports around Australia conduct PSC inspections.

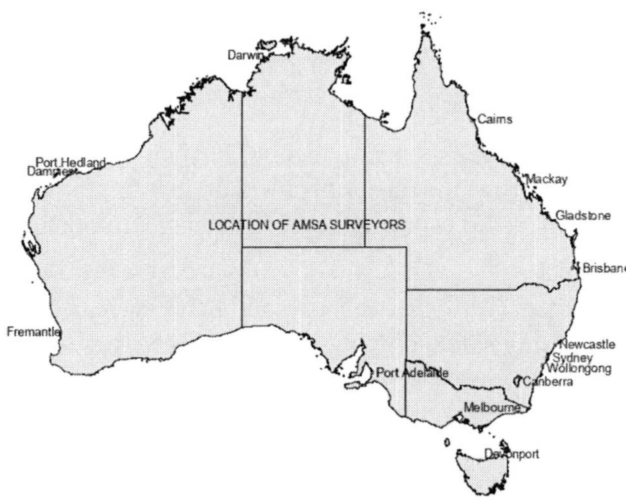

Figure 2

During the year 2006, 3688 foreign flagships made 20793 calls to Australian ports. 3424 of these were eligible for inspection, and 2538 of these were inspected one or more times during the year, i.e. AMSA inspected 74% of the eligible ships at least once during the year.

Bulk carriers constitute the majority of ships inspected and accounted for some 58% of inspections; tankers constituted 6% of the vessels inspected.

AMSA's Risk Management Program

There are many factors that have an influence on ship condition, including ship age, type, Flag, Class, etc. The issue is to identify which of these factors has the most influence on ship condition.

Some targeting approaches in use overseas focus on Flag, for example, with further weightings for combinations of ship type and age. Detailed analysis should indicate whether such approaches have merit or whether other factors should also be used.

The results to date of AMSA's analysis suggest that a ship's age is the most important objective guide to its condition.

Ship type also has a bearing, as there are considerable differences in detention rates between different ship types.

These variations in detention rates between ship types can be partly, but not fully, explained by variations in the average age of each ship type. Some ships types lead a harder life than others, reflected in detention rates that are higher than what would be suggested by the average age of that type (e.g., bulk carriers). The converse also applies, with some ship types having lower-than-expected detention rates (e.g., passenger ships).

During 2000 AMSA commenced a detailed strategic analysis of the AMSA PSC records from the previous 6 years to identify the factors that influence the seaworthiness of ships in order that risk management can be applied to the PSC inspection program by identifying higher risk ships. The analysis found a strong correlation between age and detention rate.

This profile when linked to the age profile of ships being inspected by AMSA lead to the refinement of target categories for PSC inspection

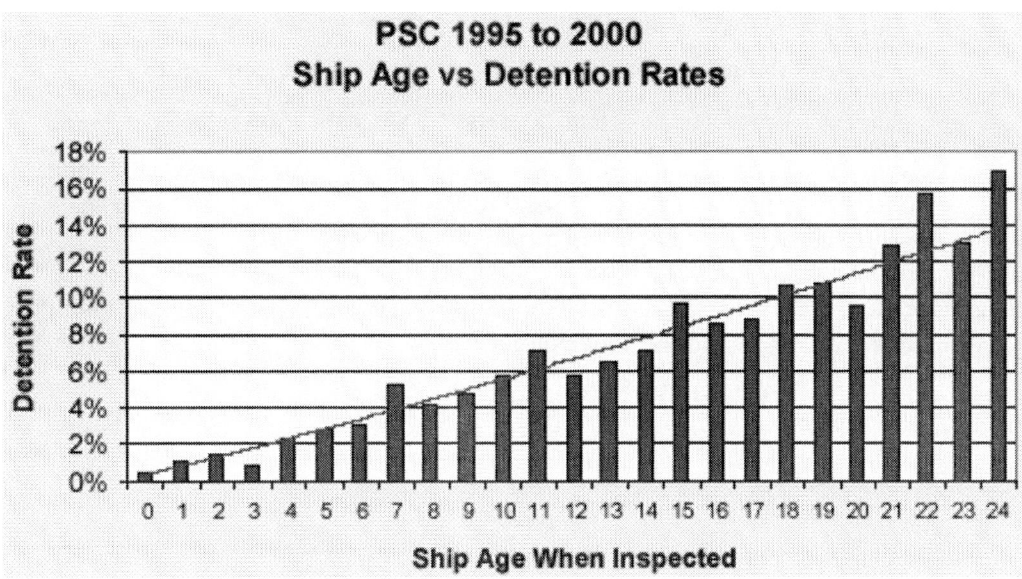

There are also variations in detention rates between Flag states, Class Societies and, once again, these variations can be partly, but not fully, explained by the age profile of the ships with each Flag or Class society.

Changes to Inspection rate Targets

AMSA recently adopted a new basis for setting inspection rate targets for the Port State Control (PSC) ship inspection program. These new inspection rates will directly reflect AMSA's objective, statistically based, risk ranking of ships eligible for inspection.

Background

For many years, Australia has operated its PSC program in a manner consistent with international practice by aiming to inspect at least 50% of all eligible ships. The basis for this target is essentially historical practice, however, and is largely arbitrary, lacking any currently applicable objective rationale. Under this approach, most ships are considered eligible for inspection every six months, although that is reduced to three months for passenger vessels and oil tankers over 15 years of age.

One refinement by AMSA to this broad target inspection rate was the adoption in 2001 of objective statistical methods to apply risk rankings to ships – an approach that has been improved since then following a detailed statistical analysis of past PSC inspection data.

Two key features of this enhanced risk management system are:

- The allocation of a risk ranking (based on several ship characteristics, and not just its age) to all arriving ships, that indicates the probability of each vessel being found to be unseaworthy if inspected; and
- Individual inspection rate targets for ships depending on their broad (age-related only) risk grouping
    - A – High Risk (ships over 15 years of age) - 80% inspection rate
    - B – Medium High Risk (10 to 14 years) - 60% inspection rate
    - C – Medium Low Risk (5 to 9 years) - 40% inspection rate
    - D – Low Risk (less than 5 years) - 25% inspection rate

  Given the age profile of foreign flagships visiting Australia, these individual target inspection rates for each risk group were still aimed to achieve an overall inspection rate of about 50% of eligible ships.

The inspection rates of ships according to their broad age-related risk have, when viewed nationally, exceeded the above targets.

Port State Control – The Process

To assist in targeting foreign vessels visiting Australia, surveyors have access to a database called 'ShipSys2000'

ShipSYS2000

- **Australia chose to keep it's own database**
- **Contains not only port State control, but flag State activities, incident reporting, port arrivals and database query tools.**
- **Uploads to APCIS via batch protocol.**

I will not spend too much time on how ShipSys2000 operates. But a surveyor need only look at those arrivals in his port to target 'eligible ships', but arrivals in all ports are also available. ShipSys2000 has an inbuilt risk factor calculator; the 'risk factor' equals the percentage likelihood of a ships detention in Australia. AMSA also uses ShipSys2000 to produce statistical reports of inspections.

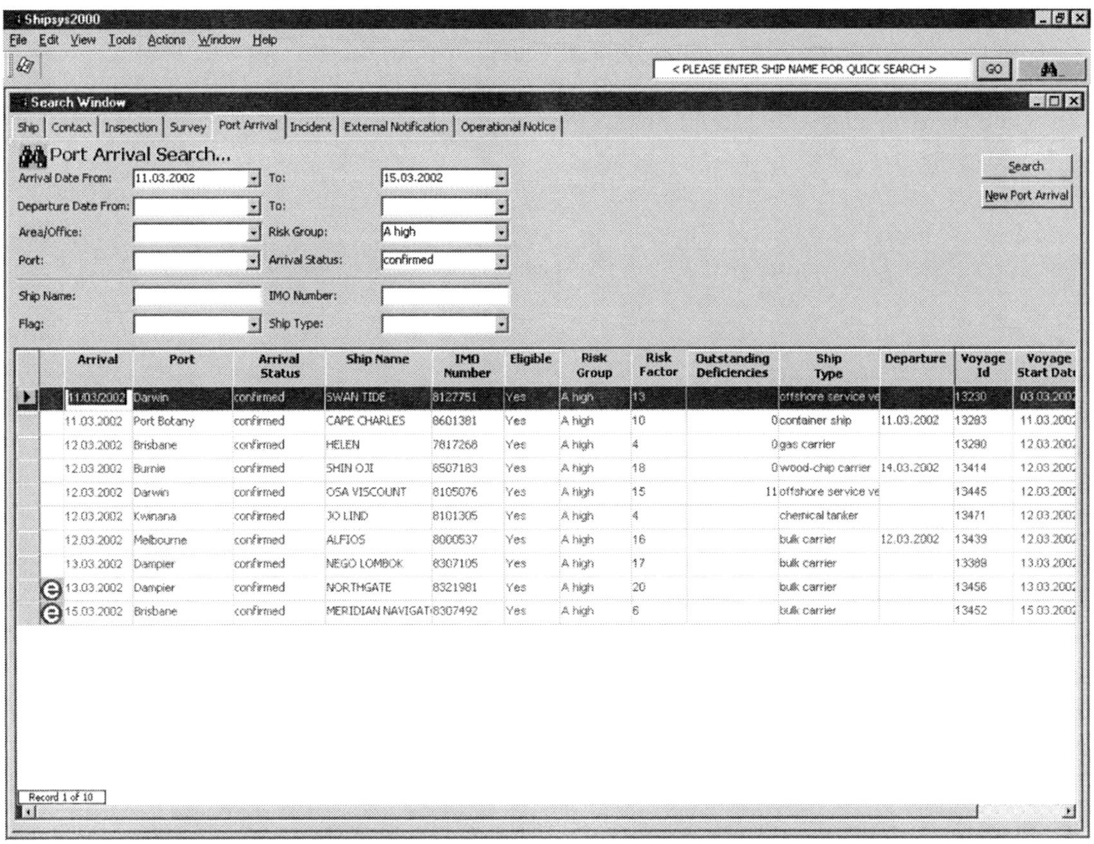

As previously mentioned, an eligible in this case means a ship that had not been inspected in the last 6 months (3 months in the case of tankers over 15 years old and passenger ships of any age), and AMSA sets a minimum level of 50% inspections.

Port State Control
Eligible Ship

- **Not inspected by AMSA during the last 6 months – 3 months for passenger ships irrespective of age, and 3 months for oil and chemical tankers 15 years and over.**
- **2006 – 2538 PSC inspections = 74%**
- **Overall target rate = 50%**

Once a surveyor targets a vessel, he will carry out a port State control inspection. Time does not permit me to discuss the details of the conduct of a port State control inspection in great depth.

But briefly, Australia carries out port State control inspections in much the same manner as an inspection carried out by any other member of the Tokyo and Indian Ocean MOU's.

A surveyor has a Ship Inspection Record (SIR) book that includes most of the documentation and guidance needed to carry out the inspection. As an example, this is a small section of the standard initial checklist from the current SIR Book … it requires the surveyor to visit a minimum number of specific areas on the ship in order to ensure an adequate overall impression is gained of it's condition.

SECTION A - STANDARD INITIAL INSPECTION CHECKLIST

Note: On completion, all areas MUST be marked with either a ✔, a Code Letter and/or a Comment.
The area box at the beginning of each section may be ticked to indicate the surveyor has considered all items.

		Additional items inspected or other comments
Area 1 - External Hull & Pre-Boarding	☐	....
1.1 Hull Condition	☐	....
1.2 Evidence of Tank Leakage	☐	....
1.3 Hull Markings	☐	....
1.4 Accommodation Ladder & Side Netting	☐	....
Area 2 - Certificates	☐	....
2.1 Ships Certificates	☐	....
2.2 ISM Code	☐	....
2.2.1 ISM Code - Designated Person Defined	☐	....
2.2.2 ISM Code - Manuals, Audit Reports & Document Control	☐	....
2.2.3 ISM Code - Document Emergency Procedures	☐	....
2.3 Crew Qualifications	☐	....
Area 3 - Wheelhouse	☐	....
3.1 Voyage Charts	☐	....
3.2 ARPA Radar	☐	....
3.3 Appropriate Publications	☐	....
3.4 Compass Deviation Records	☐	....
3.5 Echo Sounder	☐	....
3.6 Fire Detection Panel	☐	....
3.7 Navigation Light Panel	☐	....
Area 4 - Radio	☐	....
4.1 Correct MSI Reception	☐	....
4.2 On Air Radio Check Radiotelephony	☐	....

In addition, a surveyor may focus on particular items, such as closing appliances on vessels leaving the North West of Australia in the cyclone season.

From time to time, AMSA will conduct Focussed Inspection Campaigns targeting specific items such as radio and lifesaving equipment etc.

Also, AMSA undertakes Concentrated Inspection Campaigns (CIC) in conjunction with members of the Tokyo and Indian Ocean MOUs. Last week Australia completes a CIC on lifesaving equipment initiated by the Indian Ocean MOU, and coordinated by Iran.

PSC inspection outcomes

Detentions

AMSA detains a ship under the Australian Navigation Act when the AMSA surveyor considers the deficiencies observed during an inspection render the ship unseaworthy or substandard at the time of the inspection.

Serious deterioration of the hull, overloading, or defective equipment such as lifesaving, radio, and fire fighting appliances are causes to render a ship unseaworthy. A surveyor may detain a vessel which seriously breaches the provisions of ILO147 if considered to pose a safety or health hazard. AMSA's marine surveyors use their professional judgement to determine whether or not to detain a ship.

When intervention action is taken to detain a ship, AMSA follows the international convention requirements of informing the Consul or the nearest diplomatic representative of the vessel's flag State and the appropriate classification society. The IMO will also receive details of the intervention. AMSA also publishes the details of the intervention on the AMSA Internet web site.

Detention Rates by Ship Type

Ship type	Inspected	Detained	Detention Rate
bulk carrier	1788	96	5.4%
chemical tanker	92	3	3.3%
combination carrier	11	0	0.0%
container ship	314	11	3.5%
gas carrier	63	0	0.0%
general cargo/multi-purpose ship	210	11	5.2%
heavy load carrier	16	0	0.0%
high speed cargo craft	1	0	0.0%
livestock carrier	39	2	5.1%
MODU & FPSO	2	0	0.0%
offshore service vessel	24	0	0.0%
oil tanker	194	4	2.1%
other types of ship	13	1	7.7%
passenger ship	27	0	0.0%
refrigerated cargo vessel	11	0	0.0%
ro-ro cargo ship	12	0	0.0%
ro-ro passenger ship	2	0	0.0%
special purpose ship	9	1	11.1%
tanker, not otherwise specified	4	0	0.0%
tugboat	23	0	0.0%
vehicle carrier	144	6	4.2%
wood-chip carrier	81	3	3.7%
Totals	**3080**	**138**	**4.5%**

Summary of Detentions and Inspections between 2002 and October 2007

	2002	2003	2004	2005	2006	2007 to October
PSC Inspections	**2842**	**2827**	**3201**	**3508**	**3512**	**2907**
Detentions	**158**	**188**	**173**	**154**	**138**	**135**
Detention Rate	**5.6%**	**6.7%**	**5.4%**	**4.4%**	**3.9%**	**4.6%**

When comparing the detention figures since 2002 there appears to be a general reduction in the number of detentions. There could be a number of factors in this trend, and AMSA believes that the continual refining of the targeting system has a major influence, and to some extent the influence of the ISM Code.

Deficiencies

A deficiency is recorded when the condition of the ship's hull or its equipment does not conform to the requirements of the relevant IMO safety or pollution prevention conventions, or where hazards to the health or safety of the crew exist which are considered to be in breach of ILO conventions.

AMSA surveyors find deficiencies in many ships. These may not pose an immediate hazard to the safety of the ship, its crew or passengers, however they need to be rectified. During 2006, AMSA surveyors observed some 8972 deficiencies on various ships. Since 1996, there has been a steady decrease in the average number of deficiencies found per inspection. This is positive sign that the quality of ships inspected is improving.

Detainable Deficiencies by Category

Deficiency Category	No. of detainable deficiencies	Detention rate as a %
FIRE SAFETY MEASURES	77	27.5
LOAD LINES	49	17.5
STABILITY, STRUCTURE AND RELATED EQUIPMENT	44	15.7
LIFE-SAVING APPLIANCES	29	10.4
ISM-RELATED DEFICIENCIES	26	9.3
MARPOL - ANNEX I	18	6.4
RADIO COMMUNICATIONS	17	6.1
SOLAS-RELATED OPERATIONAL DEFICIENCIES	10	3.6
CARRIAGE OF CARGO AND DANGEROUS GOODS	4	1.4
CERTIFICATION AND WATCHKEEPING FOR SEAFARERS	2	0.7
PROPULSION AND AUXILIARY MACHINERY	2	0.7
SAFETY OF NAVIGATION	2	0.7
Total	**280**	

Fire fighting equipment is still major item of concern where surveyors find most deficiencies. It is still amazing that ships arrive at Australian ports with defective engine room fire flaps and oily water separators.

Additionally, the relatively high incidence of false DSC alerts tends to compromise search and rescue services. This clearly points towards lack of officer capability in operating critical safety equipment. AMSA will continue to target this and other areas of crew competence in critical areas of operation.

2000 to 2005 - Average Number of Deficiencies per Inspection

Year	%
2002	**2.62**
2003	**2.38**
2004	**2.35**
2005	**2.50**
2006	**2.91**

Training

The maintenance of uniform port State control standards within Australia is always an ongoing problem for the Australia Maritime Safety Authority.

Operationally, AMSA has four regions, namely, North, South, East, and West. And the function of the Ship Inspection Group located in Canberra is to co-ordinate the consistency, uniformity, and objectivity of the four regions. Due to the variety of vessels inspected by AMSA, uniform PSC standards are essential in all the regions.

To achieve this, AMSA has in place a comprehensive training and auditing program.

Major Objectives and Strategies

- **Consistency, uniformity, and objectivity**
- **Training and audit**
- **Up to date PSC instructions and manuals**
- **Risk management approach to targeting**
- **Continuous improvement**

Training

- **AMSA recruits new surveyors with a minimum Master Class 1, Chief Engineer Class 1 or Naval Architect qualification and relevant experience.**
- **Initial training**
  –One-on-one training by a PSC trainer
  –Based upon Ship Inspection Manual and Tokyo MOU Manual
  –Minimum number of PSC as observer
  –Minimum number of PSC under observation

–**Specialist short courses**

Training

- **New short courses as required in order to meet new developments in industry**
- **Periodic workshops**
- **5 year rolling program of various courses**Audit
- **Regional Manager's review surveyor performance regularly**
- **Four Senior Surveyors are PSC Auditors and audit surveyors outside their region.**
- **Audit reports provided to Manager Ship Inspections (who is also able to audit surveyor performance when considered necessary)**
- **Audit**
    - **Adverse audit findings may result in further training and monitoring**
    - **Audit program forms a part of the internal audit required under ISO 9001:2000**
    - **PSC program also subject to external audit as a part of the accreditation process**

Challenges

We should remember that port State control is an imperfect system operating in an imperfect world. It is neither a cure all nor does it absolve others from their responsibilities. The active participation and recognition of their responsibilities by ship operators, owners, charterers, underwriters, classification societies, and flag States is essential in the eradication of substandard shipping.

The sheer size of Australia, with ships calling at numerous ports, many in remote and difficult to access locations presents AMSA with considerable challenges. AMSA will endeavour to ensure that no Australian port is so remote that the presence of a high risk ship would go undetected. And AMSA will continue to refine its Ship Inspection Decision Support System.

It is apparent that port State control has proven to be a valuable tool in addressing many of the current maritime safety problems. However, we must not become complacent; we must remember that port State control is only a single strategy in a total program aimed at raising the level of maritime safety. Australia's PSC program will continually refocus, and refine our risk management systems. Australia is increasingly identifying human factors as a root cause of ship safety problems. Young well constructed and equipped ships are becoming substandard on account of human factors. PSC risk management systems will need to identify the human factors affecting ship safety as well as the hardware issues that we have become very astute at detecting.

AMSA with the full support of the Australian Government will continue to conduct our port State control program in a firm, fair, independent, and fearless manner. We will look for ways to reward those that operate their ships in safe, competent, and environmentally sensitive manner. We will be tireless in seeking out those that try to operate outside the boundaries of the international conventions.

AMSA will continually work with our International and Regional partners to strengthen the regional arrangements to ensure a robust port State control system that will make it increasingly difficult for unsafe ships to freely ply the trade routes of the world.

For more information on AMSA's PSC program, I would invite you to visit AMSA's web site at www.amsa.gov.au.

******

DETERMINATION OF SHIP GROUNDING AND ACCIDENT SCENARIOS OF A GEOGRAPHICAL AREA USING RISK ANALYSIS

[1]A.R. Mohd Sidek, [1]B. Gangadhara Prusty, [2]Tapabrata Ray

[1]School of Mechanical and Manufacturing
University of New South Wales
Tel: +61 2 9385 5939 (Office)
Fax: +61 2 9663 1222
E-mail: z3206578@student.unsw.edu.au ; g.prusty@unsw.edu.au

[2]School of Aerospace, Civil & Mechanical Engineering
ADFA, University of New South Wales
Tel: +61 2 6268 8248 (Office)
Fax: +61 2 6268 8276
E-mail: t.ray@adfa.edu.au

ABSTRACT

Collision and grounding are among the most common causes for ship disasters; and will continue to happen despite continual efforts in improving design, construction and maintenance of vessels. Ship collision and grounding accidents may lead to major disasters involving loss of human life or property damage, as well as environmental hazards and pollutions. As a consequence, there is a growing interest in investigating safety assessment in the shipbuilding and shipping environment. Minimising the damage and impact of an accident to the environment involves the process of predicting the accidents and assessing their consequences, which include damage extent, oil outflow and hull girder strength.

This paper reviews the various methods of estimating the collision and grounding probabilities of ships with reference to the collision and grounding studies of vessels operating in a specific geographical area, involving description of traffic composition along the navigational routes as well as descriptions of grounds within the geographical areas. Causation probability peculiar to the specific geographical area needs to be carefully considered to represent an accurate local traffic situation which could lead to collision and grounding. A general risk model can then be developed in order to evaluate the frequency of grounding and collision accidents and the results will form part of the future study of estimating the damage extent of such composite high speed vessels. The loading scenarios involving composite vessels in accident and groundings can then be realistically addresses and formulated.

Keywords: ship grounding and accident, risk analysis, composite marine vessel

1.0 INTRODUCTION

Ship accidents and groundings are low probability occurrence events with high and damaging consequences to life and property as well as the environment. Breakdown of total losses by ship types of conventional design for the year 1995-2000 from Lloyds's world casualty statistics illustrates the general pattern of casualties of world fleet as depicted in Figure 1.

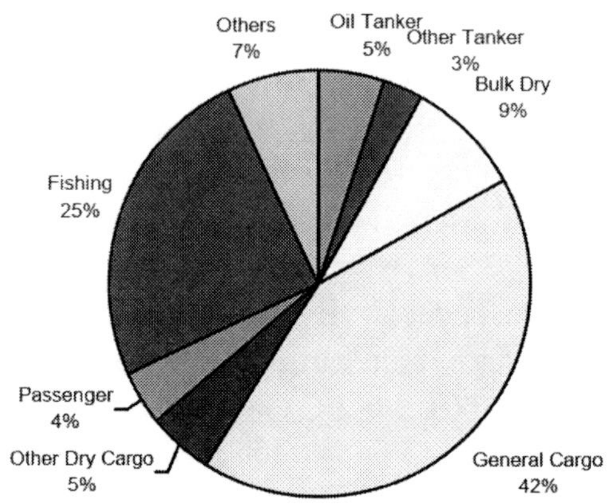

Figure 1: Breakdown of Total Loses of Ship Types, 1995-2000 [1]

In the case of high speed vessel inclusive of mono-hull, catamaran, surface effect ship, hydrofoil and hovercraft, the number of various incident categories statistics is shown in Table 1. These sources of data are from internationally reported accidents, various national authorities and various company log incidents where they are generally considered to be confidential in nature. These data indicated that collision and grounding of such vessels contribute to about 36% of the various significant incidents.

Various effort and studies appears to have been directed towards the understanding of the mechanics of collision and groundings of conventional vessels but has not been fully extended to high speed craft. Due to the different nature of high speed crafts, encompassing major design aspects such as speed, new hull forms, types of materials used in its constructions and tentatively new passive safety measures employed, the risk for these crafts are generally different from the risk of accident and grounding of conventional vessels.

Currently, the high-speed craft rules require risk analysis to be performed and new non-standard tanker designs require detailed oil out-flow analysis. The current International Maritime Organisation (IMO) regulations trend is generally towards probabilistic performance based standard which requires predicting environmental performance and safety of a specific ship design. Various probabilistic standards in damage stability regulations for passenger ships and cargo ships together with interim guidelines for oil tankers under Regulations 13F(5) of Annex I of MARPOL 73/78 has been introduced by IMO.

Summary of HSC Incident Category Statistics								
Accident Category	**Number of Incidents**	**Percent of total incidents**	**Fatalities**	**Major Injuries**	**Minor Injuries**	**Equiv. Fatalities**	**Equiv. Fatalities per incident**	**Fatalities per incident**
Accident to personnel	10	1.47	0	4	7	0.47	0.05	0.00
Collision	162	23.80	41	177	631	65.01	0.40	0.25
Contact	143	21.06	10	90	221	21.21	0.15	0.07
Explosion	6	0.88	0	21	14	2.24	0.37	0.00
Fire	46	6.77	14	6	51	15.11	0.33	0.30
Flooding or Leakage	8	1.18	0	0	0	0.00	0.00	0.00
Grounding or Stranding	84	12.87	18	186	364	40.24	0.48	0.21
Loss of Hull Integrity (LOHI)	20	2.95	0	32	27	3.47	0.17	0.00
Machinery or Elec. Failure	117	17.23	0	0	0	0.00	0.00	0.00
Other/Unknown	83	12.22	16	42	29	20.49	0.25	0.19
Total	**679**	**100**	**99**	**558**	**1344**	**168.24**		

Table 1: High Speed Craft Incident Category Statistics [2]

2.0 HIGH SPEED CRAFT

For the past forty years, there has been a steady rise in the marine industry's use of reinforced composite materials especially for high speed craft for its lighter weight requirement and advantages. The wider exploitation has been driven by the flexibility of fibre reinforced composites for the construction of hulls of complex shape plus the attractive properties and environmental stability of the material. Significant trends in the marine vessels production have shown the use of composites moving away from the simplest manufacturing process namely contact moulding to the more sophisticated closed mould processes. This is followed by the progressive increase in the size of structures that can be manufactured. The development of new marine vehicle types other than displacement types, potentially to attain speeds, require weight reductions as one of its performance driven parameters in design philosophy. As these high performance vehicles require design from first principles, they offer a huge opportunity for advanced composite materials to be used in the pursuit of the lightest marine vehicles. Following this, various developments of advanced composite materials has been investigated, ranging from hybrid to reinforcement configurations as well as new structural concepts form 'monocoque' and foam supported shell concept.

The majority of contemporary composite marine structures, such as glass reinforced plastics, are very conservatively designed on the basis of available classification society rules and regulations [3,4,5]. In cases of high performance large vessels and special purpose naval ships, the intended application of these rules might lead to unacceptably large overall weight. For high performance marine vehicles, a more rational performance based design approach towards safety due to accidents and groundings needs to be followed as opposed to design by rules. In performance based design towards safety due to accidents and groundings, state of the art research has been reviewed by Wang et.al. [6] for vessels made of isotropic materials, mainly steel. The main motivation is due to the introduction of Interim Guidelines [7] in response to US Oil Pollution Act of 90 (OPA 90). A broad spectrum of issues has also been investigated relating to vessel collisions and groundings. It ranges from the survey of historical accident data [8,9,10], various types of ship grounding [11,12], stranding [13,14,15] and raking [16,17,18], side collisions [19,20,21] as well as head-on collisions [22,23,24].

As a consequence, in the summer of 1996, IMO published "Draft Guidelines for Application of Formal Safety Assessment (FSA) to the IMO rule-making process". Thus, high-speed craft rules would require risk analysis to be performed while new non-standard designs of tankers require detailed oil outflow analysis in case of accident and grounding.

3.0 FORMAL SAFETY ASSESSMENT GUIDELINES

A formal safety assessment is normally formulated as guidelines to safety issues common to a particular ship type (high speed vessel) or to a specific type of hazard (fire or explosion on board of a ship). Various incentives could be derived from such an assessment ranging from improved performance of the current fleet of vessels, ability to measure the changes in performance as well as a basis of mechanism from which to control and predict the most likely scenarios that could result in any forms of accident and grounding. The acceptance of the final risk which is influenced by various factors ranging from political, social, economic and type of activities will finally form the final decision making process. Framework for decision of reliability of risk is categorised under three regions: intolerable, ALARP (As low as reasonably practicable) and broadly acceptable.

The Formal Safety Assessment guidelines by IMO comprised of at least the following steps:

1. Identification of hazards
2. Risk analysis
3. Risk control options
4. Cost benefit analysis
5. Recommendations for decision making

Formal ship safety assessment guidelines have been approved by IMO. Various submissions of applications of formal safety assessment demonstrating the successful case studies have been made to the governing body. Amongst them are the notable submissions by MCA, UK on the studies on high-speed craft [25] and to the trial application of bulk carrier [26]. The formal safety assessment has also been applied in supporting the various rules making process [27]. The identification of hazards such as collision and grounding as performed in the study of high-speed craft, involves various techniques ranging from initially using various historical data and followed by expert elicitation using the Delphi techniques. Risk contribution tree were then structured and quantifications of hazard risk were made using the historical data.

The study is limited to twin hulled high speed passenger carrying ferries but not limited to internationally operating craft, and inclusive of significant national routes while excluding the river ferries. Other scopes defined include no geographical boundaries, vessel systems employed and relevant regulations. As such, limitations defined in the study are of utmost important towards the quantification of the associated risks.

4.0 RISK ANALYSIS IN A SPECIFIC GEOGRAPHICAL BOUNDARY

In the effort to quantify the collision and grounding risks in a given traffic of a limited geographical boundary such as rivers, various research towards developing the initial design tools to estimate probabilities of collisions and groundings which could lead to consequential damage and pollution could be employed in the intended study. Overview of steps in a comprehensive collision risk analysis is as shown in Figure 2 below:

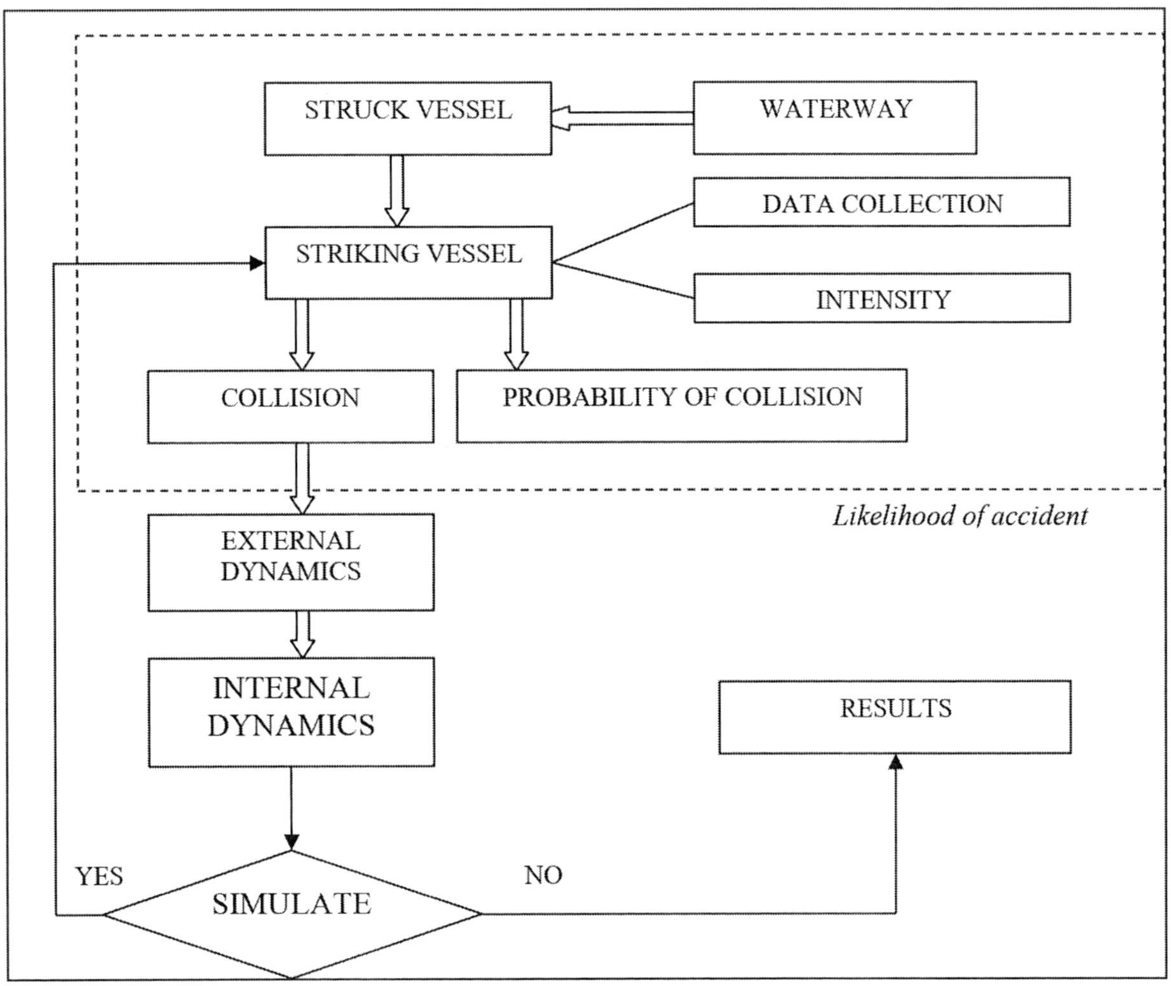

Figure 2: Comprehensive Collision Risk Analysis

The first aspect of the comprehensive risk analysis is the determination or estimation of the likelihood of incidents, defined as events that could result in an unintended collision or grounding in a specific waterway. Various possible approaches of estimating number of accidents per year involving specific types of vessels operating within the waterway are:

1. **Statistics of incidents from various database:**
   Database currently available includes MAIB (Marine Accident Investigation Bureau), United Kingdom Ministry of Transport, Australian TSB (Australian Transportation Safety Board), The Transportation Safety Board of Canada, USCG (United States Coast Guard) and MAIA (Japan's Maritime Accident Inquiry Agency). Various studies have been conducted using past statistical data in the determination of likelihood of accidents e.g. HARDER model [28] and by Lutzen and Simonsen [29]. Theses databases encompassed various categories of incidents involving various types of vessels and large ships. As such, number of vessels especially high speed craft within the particular waterway of interest might be sparse and limited which can lead to the calculated statistical probabilities of accidents and groundings to be highly inaccurate.

2. **Expert opinion**
   This approach has long been used when limited by data albeit it could be subjective in nature.

3. **Predictive calculations**
   This approach enables the prediction on unfavourable conditions i.e. the occurrence probabilities although limited only to known scenarios of a specific waterway. Systematic and risk based ship accident and grounding analysis and approaches were initially done in 1974 by Fujii et al. [30] and MacDuff [31]. Various accident and grounding studies were conducted by MacDuff at Dover Straits, UK as well as calculating the theoretical probabilities of ship accident and grounding in that particular area. Their approach in estimating the potential number or probabilities of grounding or ship-ship collision at a particular area starts with the assumption that no aversive manoeuvres are initially made on a specific traffic volume of a specified geometrical waterway. The potential probabilities of accident or grounding are then multiplied by causation probabilities derived from past accident records, in order to find the actual number of accident and grounding.

   Gluver and Olsen [32] named such approach as scenario approach representing a situation whereby the analysis of probability of accident or grounding per year at a particular site is based on accident scenarios from accident records collected or observed and imbedded in the statistical analysis. Gluver and Olsen also proposed an alternative approach known as the synthesis approach. The probability of ship collision and grounding are calculated based upon various conditions of scenarios where specific error occurs prior to collision and grounding. This will require identifying and analysing all significant accident scenarios and putting a condition by means of causation probabilities on the accident scenarios considered. The determination of causation factor of a ship-ship collision on a specific accident scenario can be determined by means of Bayesian Network as shown by Friis-Hansen & Pederson [33,34].

In general, the first step in the predictive calculations is to determine the potential number of collision or groundings if no aversive method is taken.

The second step is determination of the causation factor that models the effect of crew and equipment related actions to avoid accidents and groundings. Factors depend strongly on navigational aids on the shipping lanes, weather condition visibility and equipment on the bridge. Methods of finding the causation probabilities may include, for example fault trees or Bayesian networks.

The probability of collision is finally determined as shown by Figure 3 below:

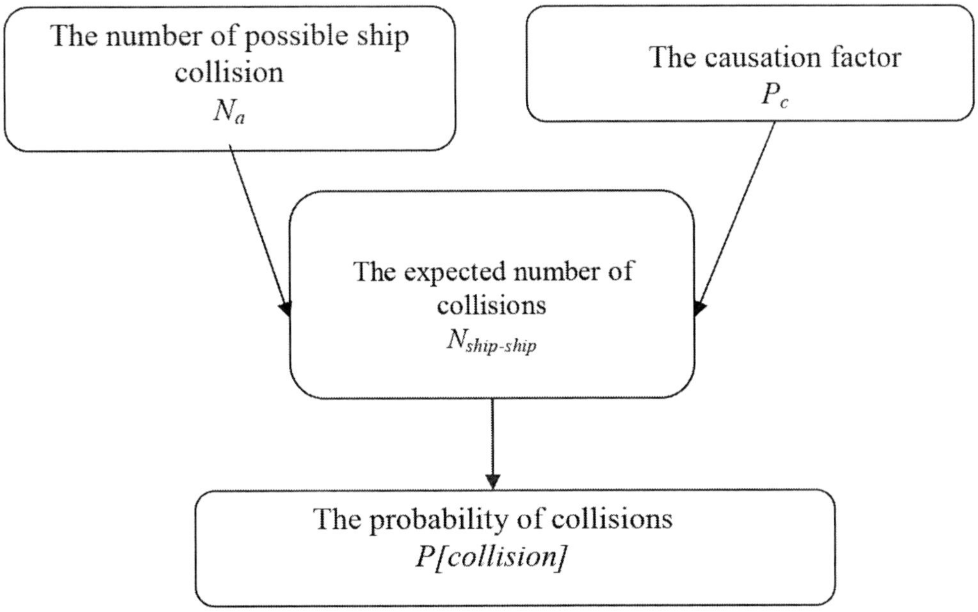

Figure 3: Probability of Collisions

where,

$$P[collision] = 1 - e^{-N_{ship\text{-}ship}} \quad (1)$$

$$N_{ship\text{-}ship} = P_c \, N_a \quad (2)$$

$$N_a = \sum_i \sum_j \int \int_{\Omega(z_i z_j)} \frac{Q_i^{(1)} Q_j^{(2)}}{V_i^{(1)} V_j^{(2)}} f_i^{(2)}(z_i) f_j^{(2)}(z_i) V_{ij} D_{ij} \, dA \Delta t \quad (3)$$

and,

$Q_j^{(2)}$ is the traffic flow i.e. the number of ships per unit of time of class "j' in waterway 2

V_j^2 is the associated speed.

$f_j^{(2)}$ is the lateral distribution of the ship traffic of class "j" in waterway 2

D_{ij} is the geometrical diameter

V_{ij} is the relative velocity

4. Comprehensive risk analysis

This approach can be considered as rational with consequence estimation of the ship accidents and groundings. Various real accident data are embedded for benchmarking purposes. The complete formulations of such approaches are implemented in software **GRACAT** (Grounding and Accident Analysis Toolbox) developed by Technical University of Denmark [35] and also "**IALA** Waterway Risk Assessment Programme" (IWRAP) developed by the Canadian Coast Guard. Variations differences exist especially in the various methods adopted in the analysis of external and internal dynamics of accidents and groundings.

In the case of collisions, the various influencing factors can be separated into three main groups: the waterway and environment system, the involved vessels and human factors as depicted by the Figure 4.

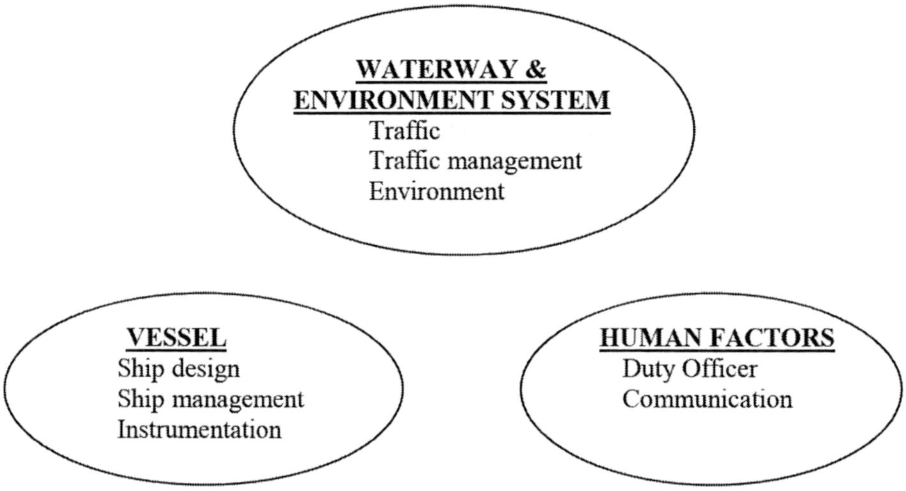

Figure 4: Factors Influencing Collisions

The details of these various factors are pertinent to the waterway of interest, for example the various ship types and design and the composition of such vessels with various other types of vessel. Similarly, the grounding risk analysis and the likelihood of occurrence and the various influencing factors need to be properly addressed, namely the locations of soft beds within the waterway as a typical example.

5.0 CONCLUSION

This study could assist in highlighting the risk assessment of a selected vessel made of composite material with anisotropy property in a specific scenario and hazards involving grounding and accidents. The risk analysis can be performed on a specific waterway where known numbers of accident are recorded and generally considered high in terms of occurrences. In view of the localised conditions of the selected geographical area and taking into account the potential uniqueness of the vessel design, using the existing causation probabilities derived from statistical values could inherently lead to inaccuracies. Thus these unique causation probabilities in the variable of vessel shapes, speed and other governing factors are to be derived using the Bayesian Network analysis. The final largest calculated probabilities of collisions could form the worst loading scenarios that can then be applied in further studies of external and internal dynamics of such collisions. The results of the study would later enable us to address the issue of the different nature of high speed craft encompassing major design aspects such as speed, hull forms unique to its specific design, types of materials used in its constructions and tentatively new passive safety measures employed. The risk for these crafts are generally different from the risk of accident and grounding of conventional vessels and thus the study will highlight the exact scenarios that would be further studied in order to understand the real dynamics of collision and grounding of such high speed vessel.

REFERENCES:

1. MSC 77/25/4, Comparison of general cargo ship losses and fatalities, Submitted by RINA to IMO/MSC, 77th session, March 2003.
2. MCA Research Project 504, 2004, HSC Incident Database Analysis Summary Report, Southampton, UK.
3. Rules and Regulations for The Classification of Yatchs and Small Craft., (1981), Lloyds Register, London.
4. Rules for Classification of High Speed Craft., (1985), det Norske Veritas, Norway.
5. Rules for Building and Classing Reinforced Plastic Vessels., (1978), American Bureau of Shipping, New York.
6. Wang Ge, Spencer J., Chen Y., Assessment of a ship's performance in accidents. Marine Structures 2002;15:313-333.
7. "Interim Guidelines for Approval of Alternative Methods of Design and Construction of Oil Tankers under Regulation 13F(5) of Annex I of MARPOL 73/78", Resolution MEPC.66(37), Adopted September 14.1995.
8. Minorsky VU., An analysis of ship collisions with reference to nuclear power plants. J Ship Research 1959;3(2):1-4.
9. IMO. Interim guidelines for approval of alternative methods of design, construction of oil tankers under Regulation 13F(5) of Annex I of MARPOL 73/78, 1995.
10. Kite-Powell HL, Jin D, Jebsen J, Papakonstantinou V, Patrikalakis N. Investigation of Potential risk factors for grounding of commercial vessels in US ports. Int J Offshore Polar Eng 1999;9:16-21.
11. Pederson PT. Ship grounding and hull girder strength. Marine Structures 1994;7:1-29.
12. Sterndorf MJ, Pedersen PT. Analysis Grounding experiments on soft bottom. J Mar Sci Technol 1996;1(13):174-81.
13. Simonsen BC. Mechanics of ship grounding. Ph.D. Thesis, Technical University of Denmark, 1997.
14. Wang G, Arita H, Liu D. Behaviour of a double hull in variety of stranding or collision scenarios. Mar Strut 2000;13:147-87.
15. Amdahl J, Kavlie D, Johansen A. Tanker grounding resistance. Proceedings of The Sixth International Symposium on Practical Design of Ships and Mobil Units(PRADS'95), vol. 2, Seoul, 1995. p. 1072-83.

16. Simonsen BC. Ship grounding on rock: I & II. Mar Struct 1997;10: 519-84.
17. Zhang S. The Mechanics of ship collision. Dr. Thesis, Technical University of Denmark, 1999.
18. Ohtsubo H, Wang G. An upper-bound solution to the problem of plate tearing. J Mar Sci Technol 1995;1:46-51.
19. McDermott JF, Kline JF, Jones RG, Maniar EL, Chiang WP. Tanker structural analysis for minor collision. Trans SNAME 1974;82: 382-407.
20. Akita Y, Ando N, Fujita Y, Kitamura K. Studies on collision-protective structures in nuclear powered ships. Nucl Eng Design 1972;19: 365-401.
21. Zhu L. Dynamic inelastic behaviour of ship plates in collision. Ph.D Thesis, University of Glasgow, 1990.
22. Pedersen PT, Valsgard S, Olsen D, Spangenberg S. Ship impacts: bow collisions. Int J Impact Eng 1993;13:163-87.
23. Paik JK, Pedersen PT. Modelling of internal mechanics in ship collision. Ocean Eng 1996;23:107-42.
24. Ohtsubo H, Suzuki K. The crushing mechanics of bow structure and its optimal design against head on collision. In: Kim H, Lee JW, editors. Proceedings of the Sixth International Symposium on Practical Design of Ships and Mobile Units, vol. 2, 1995. p. 1060-71.
25. IMO. Formal safety assessment trial application to high speed passenger vessels. MSC 68/14/2, submitted by the U.K. to the FSA Working Group, London, U.K., 27 May 1997.
26. IMO. Bulk carrier safety, proposal for a formal safety assessment of bulk carriers. MSC 70/4/Addl, submitted to the IMO by the U.K. MCA, September 1998.
27. IMO. Interim guidelines for the application of formal safety assessment to the IMO rule-making process. IMO/MSC Circular 829, IMO, London, U.K., 17 November 1997.
28. Lutzen, M. (2001). Ship Collision Damage, Ph.D. Thesis. Technical University of Denmark, Lygby, Denmark.
29. Lutzen, M. and Simonsen, B.C. (2003). Grounding Damage to conventional vessels, World Maritime Technology Conference, San Fransisco, CA, October 2004.
30. Fujii Y., Yamanouchi H, Mizuki N., Some Factors affecting the Frequency of Accidents in Marine Traffic. II: The Probability of Stranding, III: The Effect of Darkness on the Probability of Stranding. J. Navigation 1974; 27.
31. MacDuff T., The Probability of Vessel Collisions. Ocean Industry, September 1974;144-8.
32. Gluver H., Olsen D., Current Practice in Risk Analysis of Ship Collision to Bridges. In: Gluver, Olsen, editors. Proceedings of the International Symposium on Advances in Ship Collision Analysis. Copenhagen, Denmark, 10-13 May, 1996, p. 85-96.
33. Friis-Hansen P., Pederson P.T., Risk Analysis of Conventional and Solo Watch Keeping. International Maritime Organisation (IMO) Maritime Safety Committee by Denmark, 69th Session, 1988.
34. Friis-Hansen A., Bayesian Networks as a Decision Support Tool in Marine Applications. Ph.D Thesis, University of Denmark, 2000.
35. Friis-Hansen P., Simonsen, B.C., Software for Grounding and Collision Risk Analysis. Marine Structures 2002;15:383-401.

The study of the effective education technique of ship management system and the present conditions of the Japanese fleet management

H. Kido
Maritime Technical College Japan, Hyogo Japan, kido@mail.mtc.ac.jp
K.Sakaguchi
Veritas Japan, Hyogo Japan

ABSTRACT

When considering about seafarer's education system, it is needed accord with the fact about the review of the education scheme, develop the new qualification systems, the foreign education from correspondence to the changes of the manning system for the mix crew system and the acquirement of the skill for the ship management.

As a concrete idea, there are the improvement of the ship management ability, education and training to improve individual seamen's quality, the maintenance of the driving management standard for every ship type.

If the inspection system / a technology level / information communalization / maintenance skill and factor such as the spot training education do not cooperate on a system mutually, the ship management cannot get the synergy. It seems that merely ship is old as for the main factor of the serious sea disaster not only these functions are not affected enough. If it is devised a diagnosis and the prolongation of ship's life plan of the existing ship early and do not carry it out at the same time, it is caused a delay in navigation by many ships is expected easily. However, it is considerably difficult the range of ship management educates effectively widely, and to train the crewmembers and needs a lot of time and document.

Here, as a methods of contributing to upbringing seafarers and keep qualification ship, we try to developing effective ship management education technique by used the research report of present conditions of the Japanese fleet management and visual expression by computer software.

1. THE PRESENT CONDITIONS OF THE JAPANESE FLEET MANAGEMENT

Japan is managing by import a lot of foods and sources from foreign countries, export some kind of product to other countries caused by a condition such as lack domestic resources and country surrounded by sea.

Therefore, there's no question as to the importance of that the overseas shipping business which is loaded import/export cargo by seaborne is based on the prop up Japanese economy.

However, Japanese shipping companies have no other choice to go to the way of multi nationalization crew ship. Because of intensifying international competition in overseas shipping business is very hard. Ocean going Japanese flag ships has about 100 vessels at present, which is decreasing to 1/10 during 20 years.

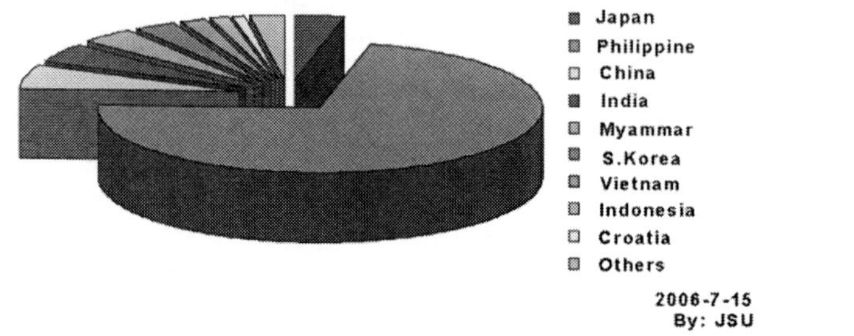

Unit: No.of crew members

Japan	Philippine	China	India	Myammar	S.Korea	Vietnam	Indonesia	Croatia	Others	Total
1,504	29,171	2,392	1,910	1,596	1,211	849	536	366	969	40,504
3.71%	72.02%	5.91%	4.72%	3.94%	2.99%	2.10%	1.32%	0.90%	2.39%	100.00%

Fig.1.1 Distribution of Crew Members in Japan Fleet

The situations for such conditions, the approval system of international marine officer was established in 1998 that was foreign marine officer can onboard a Japanese flag ship without Japanese marine license due to revised the Law of Ships' Officers and Boats' operators in Japan under the Amended STCW Convention. At present, the crew of Japanese flag ship tends to change from two Japanese as Capt. and C/E to foreign officers. Actual ocean going ships are modernization-multi nationalization crew ship and flag of convenience vessel mainly. Fig.1.1 shows "Distribution of Crew Members in Japan Fleet ". According to these background, as for a qualification of Japanese marine officer are requested an education and a managing technique. Land, Infrastructure and Transportation Ministry Designated Statistics No.29 '05. They are accelerated progress

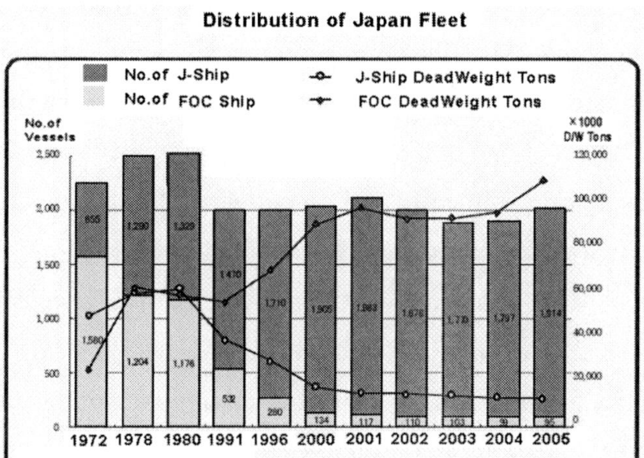

Fig.1.2 Annual variation of J Flag Vessel

with multi nationalization crew ship. Fig.1.2 shows Annual variation of J Flag Vessel from 1972 to 2005. Also, ship management system is changed from distributed control to total control by ship management company, so, as for a qualification of marine engineer officer is requested them not only managing the engine room equipment but also managing the hull maintenance and covered network communication system. As a method of compensate the seaman-ship (marine tech), training and educational simulator are needed that is requested as possible as realistic performances.

2. THE FUTURE SITUATIONS OF THE OCEAN GOING VESSEL AND CREW IN JAPAN [EX: LNG CARRIER]

Fig. 2.1 shows the order number of LNG carrier by 2008. According to this estimation, engineering officers who have some knowledge of steam turbine plant and LNG systems are needed round 1200 persons by 2008 in the world. Though, these qualified engineers cannot find and deliver to the LNG market because the lecturers and training facilities are not sufficient at this present. Mainly causes are comes up by the oil shock in 1980's. Similar situations are appeared in other ocean going vessels. Japan's major shipping companies are going to construct more than 500 vessels within 2 years as shown in Table-2.1, and plus 200 vessels will be more delivered until 2010.

This tendency not only LNGC world, bust also container, ore-bulk, crude oil tanker and other kind of vessels are becoming short crew members. Fig.2.2 shows forecast of "Distribution of seafarer's Supply & Demand condition on 2005, 2010 and 2010.

Therefore the education and training organization should be prepared turn out as soon as possible.

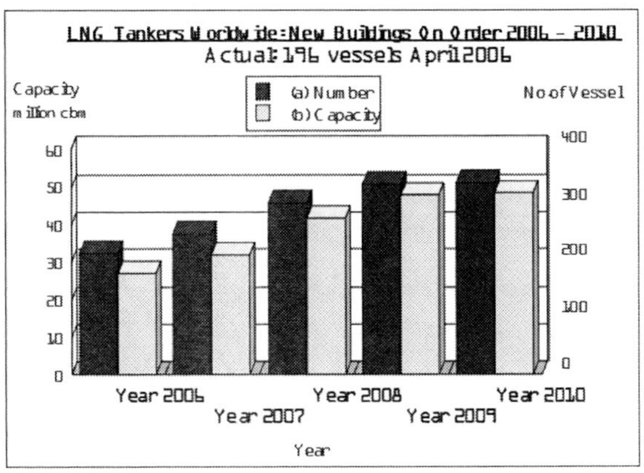

Fig.2.1 Order Number of LNGC by 2008

Table-2.1

Condition of New Ship Construction on 3 major Companies	NYL	MOL	KL
	2003 -2007	2003 -2009	2003 -2007
Container	15	50	10
PCC	16		15
ORE BULK	72		
VLCC	17	193	47
PRODUCT	14		
LNG	18		
Others	8		
Total = 503	160	243	100

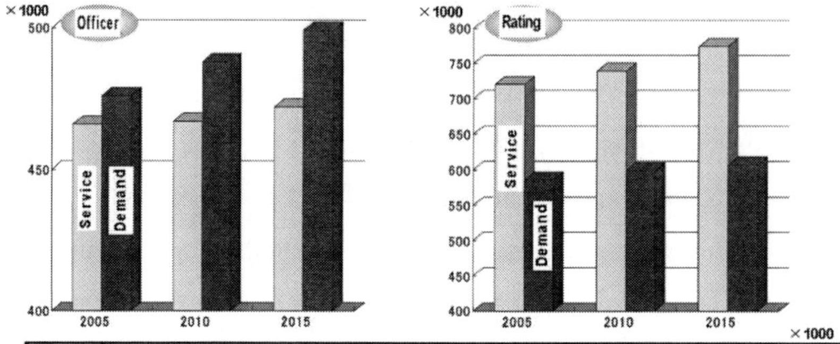

	2005			2010			2015		
	Service	Demand	excess or deficiency	Service	Demand	excess or deficiency	Service	Demand	excess or deficiency
Officer	466	476	−10	467	488	−21	472	499	−27
Rating	721	586	+135	740	598	+142	774	607	+167

In case of assumed that world fleet increase 1 % per year each
By BIMCO[ISF 2005 Research of seafarer's supply & demand] Data

Fig. 2.2 Distribution of seafa

Fig.5.2 computer using illustrations

Fig.5.2 are consisted two part of functions. The first one is included all related ISM concrete roles. And the other one have the contents of ISM manual and check list. If ISM item is selected, open selected item windows such as Fig.5.3 or Fig.5.4. All figures are consisted "Proper Management" part and against, "Wrong Management" part. According to compare with both situations, we can try work out what is the ISM step by step. Especially, these education and training methods are found to be useful for maritime developing countries. Because of, students training reports of our MTC, MLIT-Japan Scheme are collected during the training period and shipping company according to some exam about needed qualification of the marine engineer on actual ocean going vessels which is noted on the STCW95 Code, there are some admits of

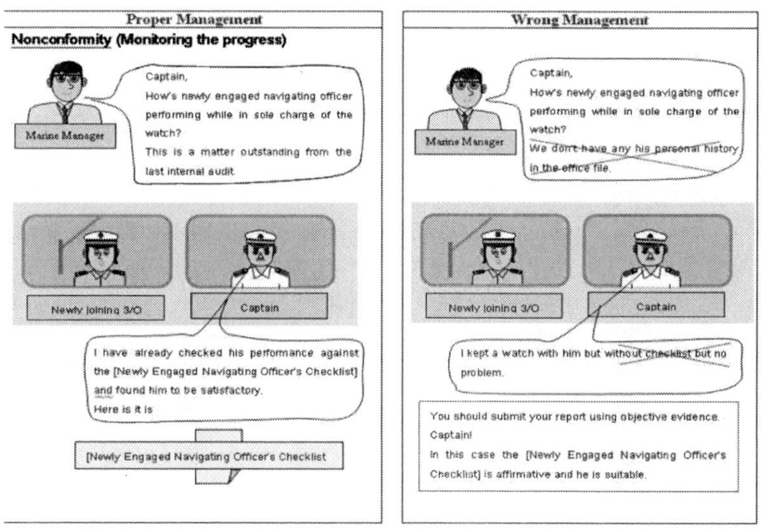

Fig.5.3 Proper Management" part and "Wrong Management

improvement such as maintain and treatment of the equipment which is included control system.

Especially, basic knowledge admits of improvement like a control theories, electric and electronics system. In additions, interest level of knowledge and skill for the outside field of marine engineering is need more efforts, too.

For examples; seaworthiness of the ship and monitor compliance with legislative requirement like a Marine Pollution Prevention Law.

For these lack factors, it is supposed that there are knowledge of the basic mathematics and physics, lack of technical vocabulary, outlook on ethic or a difference of the consciousness of the generation.

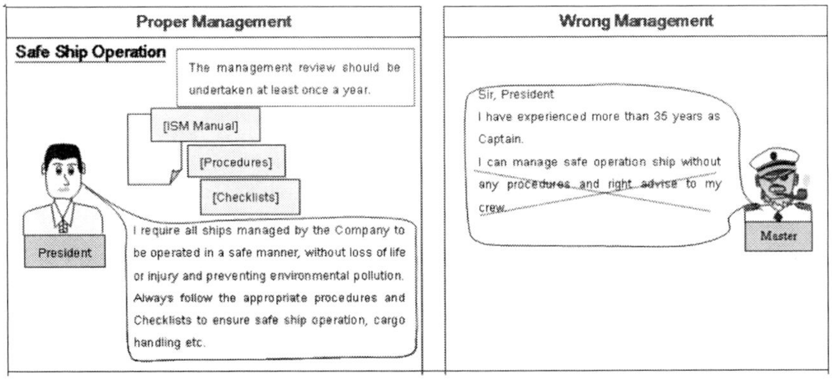

A comics-like technique as a method that it makes up for these gaps is thought to be effective in the intuition.

Fig.5.4 Comparison SMS situations

5.2 Learning support method by the Check Card method

We try to make four pattern of check card which are Check List of Ship Survey, Check List of PSC, Check List of ISPS and Check List of ISM as shown in Fig 5.5(a~f).

These cards are consisted Check Title, Check Item, and Check detail, Conformed Regulation or Protocol per each check item. On the check list of PSC Inspection, they are consisted Check list for most common deficiencies, check item, common deficiency, related questionnaires and check (yes/no) per each check item.

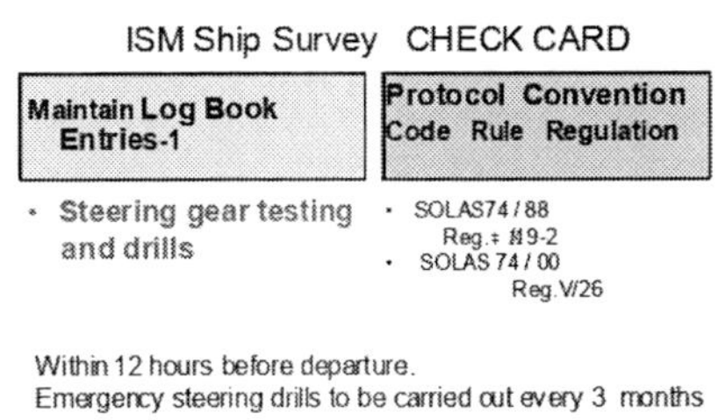

Fig.5.5 (a~b) Check List of Ship Survey,

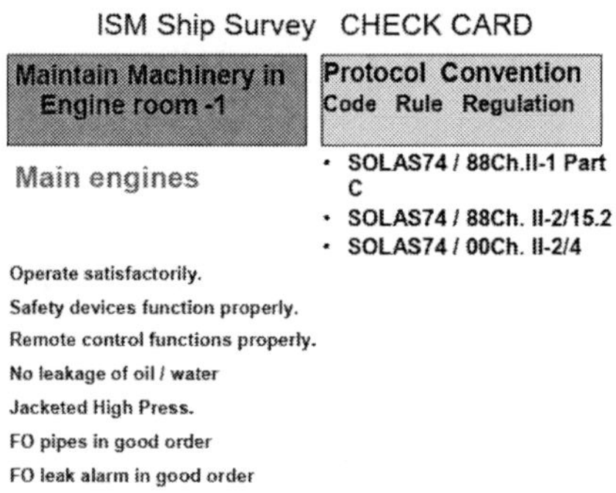

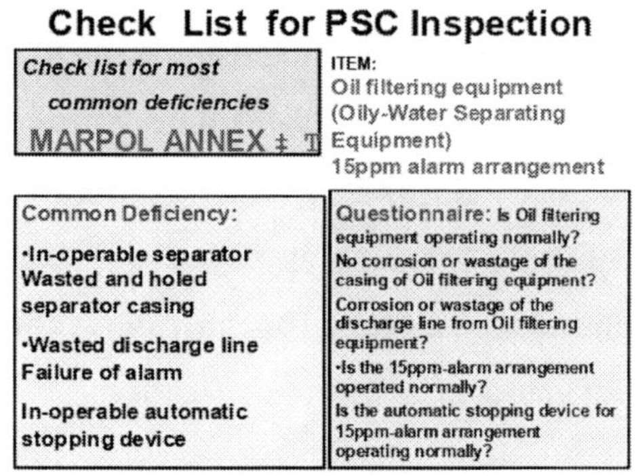

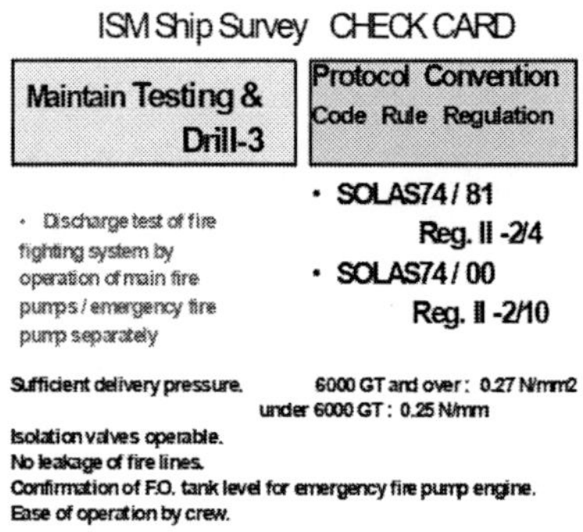

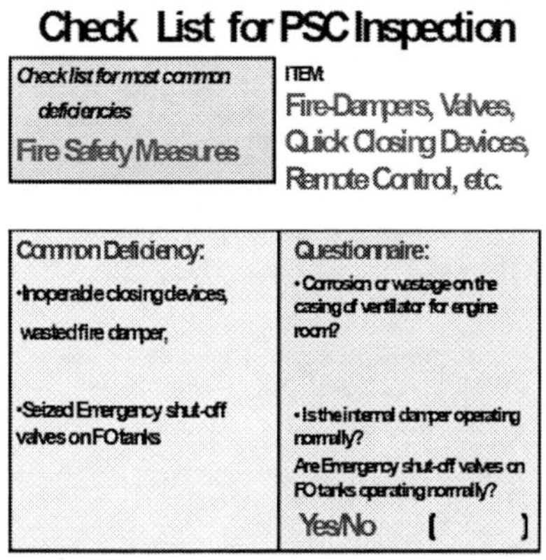

Fig.5.5(c~f) Check List of Ship Survey, Check List of PSC, Check List of ISPS and Check List of ISM

□□ACKNOWLEDGEMENT

The Japanese and foreign seafarers educational system need face actualities maritime education such as the revisions of scheme for education on view point of handling the modifications for crew managing system and duty assignment system.

This paper studied contents the necessary qualification for seaman-ship of Japanese and foreign marine engineer. And it shown educational items which was through author's experiences that is training to some foreign student in Marine Technical College during 18 years as a focus on the situation of handling the modifications for multi nationalization crew ship.

The objectives are ensuring safety, prevent human injury or loss of life, and avoid damage to the environment, in particular, the marine environment, and to property. The method by the computer using illustrations combination with the method using the Check Card are easy mastered for develop countries seafarers we hope.

Robust Ship Designs - Balancing Uncertainty - Impacts of Future Trends

Tapabrata Ray and Warren Smith
School of Aerospace, Civil and Mechanical Engineering, University of New South Wales at Australian Defence Force Academy, Canberra.
Email { t.ray,w.smith}@adfa.edu.au

Abstract
Ship design traditionally has been based on a sequential and iterative approach. With the availability of efficient and effective optimization methods in recent years, several researchers have attempted to solve ship design problems using optimization methods that allowed them to deliver designs while considering multidisciplinary interactions within a short time span. However, most of these approaches using optimization fail to identify designs that are robust i.e. those less sensitive to variations caused due to fluctuations in freight rates, cargo discharge rates, fuel oil prices etc. Outlined in this paper is a robust design optimization framework for preliminary ship design. The underlying optimization algorithm is a stochastic, elitist, zero-order, population based method that maximizes one or more design objectives and simultaneously minimizes their variance. A bulk carrier design example is discussed in this paper and differences between an "optimum" design and a "robust" design are highlighted. The importance of constraints in a robust design optimization problem is also highlighted using the example. The model can be used to predict the impact of possible future design trends and provide quantitative answers to a series of questions such as "what is the likely speed of bulk carriers if the fuel oil costs go up by 5%?" or "should a ship owner be more concerned with an increase of 1% in fuel oil price or a 1% increase in the port charges?" The modular nature of the framework allows a designer to include his preferred estimation methods for power, weight etc. and use it for a different class of vessel.

BACKGROUND

Ship design is a multidisciplinary and multifaceted activity which converts the information that characterizes the needs and requirements of the ship into knowledge about the ship. Optimization methods have been an integral part of ship design, largely because effective design synthesis can be performed on the basis of simultaneous consideration of all the design requirements.

Although there are numerous reports on the use of optimization methods for preliminary ship design (Keane et al.(1991), Ray et al(1995), Ganeshan (1999), Neu at al (2000) none seem to search for a robust design. Traditionally, such problems have been formulated as a single objective optimization problem such as maximization of Net Present Value, minimization of Required Freight Rate etc. subject to a set of constraints. The performance of the optimal solution is then studied under expected parametric variations (fuel oil price, port charges, load factor etc.) to identify its sensitivity or robustness.

A formulation that attempts to locate an optimal solution to a single objective performance maximization problem cannot guarantee to locate robust optimal solutions. The term "robust" in the present context refers to a solution that has low sensitivity to parametric variations and the aim in a design exercise is to locate a solution that has maximum performance and minimal performance variance. This paper introduces an evolutionary algorithm that attempts to solve the multi-objective, constrained optimization encountered in robust design optimization problems. The algorithm is fundamentally a population based, zero order, elitist stochastic model with similarities to Non-dominated Sorting Genetic Algorithm (NSGA-II). The algorithm unlike NSGA-II explicitly maintains the diversity in the variable space which is useful for problems in design where multiple parametrically varying designs might have same or similar performance measures.

The notion of robust design was originally introduced by Taguchi (1986) (1987) where robustness was defined using two types of inputs to any design system: (a) control factors i.e. inputs that can be easily controlled or manipulated by a designer e.g. design variables and (b) noise factors i.e. inputs that are difficult or beyond the control of a designer and are typically decided by the operating environment. The methods for robust design can be broadly classified into two major groups: (a) methods relying on gradient based optimization that attempts to work with a single solution (b) stochastic, population based methods which work with a set of solutions instead. Both use iteration based approaches to improve solutions. There have also been attempts to identify robust solutions using compromise decision support systems and hierarchical systems approach. Excellent reviews and detailed discussion on various aspects of robust design appear in Messac and Ismail-Yahaya (2003), Jin and Sendhoff(2003), Chen et al.(1996) and Du and Chen(2000). The details of the algorithm are presented in Section 2 while the bulk carrier design problem is discussed in Section 3. Major conclusions are summarized in Section 4.

MATHEMATICAL FORMULATION

A constrained robust optimal design problem is presented as follows:

Minimize

$$f = [f_1(\mathbf{x}) \quad f_2(\mathbf{x})]$$

Subject to:

$$g_i(\mathbf{x}) \geq a_i \quad, i = 1,2,\ldots,q$$
$$h_j(\mathbf{x}) = b_j \ , j = 1,2,\ldots,r$$

where $f_1(\mathbf{x})$ is the objective function evaluated at a point P, $f_2(\mathbf{x})$ is the variance of the objective function evaluated at K neighbors of P, $\mathbf{x} = [x_1 \quad x_2 \quad \ldots \quad x_n]$ is the vector of n design variables, q is the number of inequality constraints, and r is the number of equality constraints.

The equality constraints are transformed to a set of inequalities (with a tolerance δ) in order to use a unified formulation for all the constraints:

$$c_i = \begin{cases} 0 & if\ i^{th}\ const.vio.,\ i = 1,2,...,s \\ a_i - g_i(\mathbf{x}) & if\ i^{th}\ const.vio.,\ i = 1,2,...,q \\ b_i - \delta - h_i(\mathbf{x}) & if\ i^{th}\ const.vio.,\ i = q+1, q+2,...,q+r \\ -b_i - \delta + h_i(\mathbf{x}) & if\ i^{th}\ const.vio.,\ i = q+r+1, q+r+2,...,s \end{cases}$$

Thus r equality constraints will give rise to $2r$ inequalities, and the total number of inequalities for the problem is denoted by s, where $S=q+2r$. For each candidate solution, the constraint satisfaction vector is given by $\mathbf{c} = [c_1 \quad c_2 \quad \dots \quad c_s]$ where $c_i = 0$ indicates the ith constraint is satisfied, whereas $c_i > 0$ indicates the violation of the constraint. The constraint matrix for a population of M individuals assumes the form

$$CONSTRAINT = \begin{bmatrix} c_{11} & c_{12} & \cdots & c_{1s} \\ c_{21} & c_{22} & \cdots & c_{2s} \\ \vdots & \vdots & \ddots & \vdots \\ c_{M1} & c_{M2} & & c_{Ms} \end{bmatrix}$$

In order to asses the neighborhood feasibility of the ith individual, $c_{i\,s+1} \quad c_{i\,s+2} \quad \cdots \quad c_{i\,2s}$ is computed where $c_{i\,s+1}$ denotes the number of violations of the first constraint among K neighbors. Thus the modified constraint matrix for the constrained robust optimal design problem assumes the form

$$\begin{bmatrix} c_{11} & c_{12} & \cdots & c_{1s} & c_{1s+1} & c_{1s+2} & \cdots & c_{12s} \\ c_{21} & c_{22} & \cdots & c_{2s} & c_{2s+1} & c_{2s+2} & \cdots & c_{22s} \\ \vdots & \vdots & \ddots & \vdots & \vdots & \vdots & \ddots & \vdots \\ c_{M1} & c_{M2} & \cdots & c_{Ms} & c_{Ms+1} & c_{Ms+2} & \cdots & c_{M2s} \end{bmatrix}$$

The notion of using the above constraint matrix based nondominance was first introduced by Ray et al (2001). Before introducing the details of the optimization algorithm, it is necessary to introduce the notion of Pareto optimal design and nondominated design. A design $\mathbf{x}^* \in F$ is termed Pareto optimal if there does not exist another $\mathbf{x} \in F$, such that $f_i(\mathbf{x}) \le f_i(\mathbf{x}^*)$ for all $i = 1,...,O$ objectives and $f_j(\mathbf{x}) < f_j(\mathbf{x}^*)$ for at least one j. Here, F denotes the feasible space (i.e., regions where the constraints are satisfied) and $f_j(\mathbf{x})$ denotes the jth objective corresponding to the design $\mathbf{x}$. If the design space is limited to M solutions instead of the entire F, the set of solutions are termed as nondominated solutions. Since in practice, all the solutions in F cannot be evaluated exhaustively, the goal of multiobjective optimization is to arrive at the set nondominated solutions with a hope that it is sufficiently close to the set of Pareto solutions. The pseudo-code of the algorithm is provided below.

(a) $t \leftarrow 0$
(b) Generate M individuals representing a population: $Pop(t) = \{I_1, I_2, \dots, I_M\}$ uniformly in the parametric space.
(c) Evaluate each individual: Compute their objectives and constraints i.e., $f_k(I_i)$ and $c_j(I_i)$; for $i = 1,2,\dots,M$ individuals, $k = 1,\dots,O$ objectives and $j = 1,\dots,S$ constraints.

(d) ***Select*** two parents P1 and P2. (The procedure for selection is described below).
(e) ***Create*** two children C1 and C2 via crossover and mutation of P1 and P2.
(f) Repeat steps (d) and (e) until M children are created.
(g) Evaluate M children.
(h) Merge M parents and M Children to form a population of size 2M.
(i) ***Retain*** better performing M solutions from the above 2M solutions.
(j) $t \leftarrow t+1$
(k) If $t < T_{max}$ then repeat steps (d) through (j) Else Stop. *where*T_{max} denotes the maximum number of generations.

Select

(a) Select two individuals (S1 and S2) from the population of M solutions using a uniform random selection.
(b) If S1 is feasible and S2 is infeasible: S1 is selected as the parent and vice versa.
(c) If both S1 and S2 are infeasible: One which has the minimum value of the maximum violated constraint is selected as the parent. Compute $max(g_i(\mathbf{x}); i = 1, S)$ for S1 and S2 and choose the one which has the minimum value. ($g_i(\mathbf{x})$ denotes the constraints).
(d) If both S1 and S2 are feasible and S1 dominates S2: S1 is selected as parent and vice versa.
(e) If both S1 and S2 are feasible and none dominates each other: The parent is a random choice between S1 and S2.

We have used simulated binary crossover (SBX) and the polynomial mutation for the real variables as adapted in NSGA-II to create two children from a pair of parents (Deb and Agrawal, 1995).

Retain

The procedure to retain M solutions from a set of 2M solutions is presented below:

(a) Rank the set of 2M solutions.
(b) If the number of rank=1 solutions (i.e. non-dominated solutions) is less than M, select top M solutions based on their rank and copy them to the new population.
(c) If the number of rank=1 solutions is more than M, follow the following steps:
    i. Select the solutions which have a minimum value (assuming minimization in both objectives) in any of the objectives and copy them to the new population.
    ii. For every variable, copy two solutions to the new population which has its minimum and the maximum value if they have not been copied yet (including step i). If the number of solutions copied is more than M/2, than M/2 randomly selected solutions are allowed to stay in the population.
    iii. For the remaining rank=1 solutions, the sequence of who goes in first to the new population is decided as follows:
        1. Compute the score by inserting the solution into the new population one at a time. The score is the minimum Euclidean distance computed between the solution attempting to enter with all other existing solutions in the new population based on the objective function space. Scaled values are used for the score computation i.e. the objective

space is scaled using the maximum and minimum values in each dimension based on the set of rank=1 solutions.

2. The solution with the highest score is allowed to go into the new population unless it has been copied earlier in which case the solution with the next score goes in.
3. The steps (1) and (2) are repeated until the new population has a size of M.

Rank

The procedure for rank computation is as follows:

(a) Separate the set of 2M solutions to a set of feasible and a set of infeasible solutions.
(b) Perform a non-dominated sorting to assign ranks to the solutions in the feasible set.
(c) Rank the solutions in the infeasible set based on their maximum value of the violated constraint.
(d) Update the ranks of the solutions in the infeasible set by adding the rank of the worst feasible solution to each.

The assumptions behind the procedures in the algorithm are:

(a) A feasible solution is always preferred over an infeasible solution
(b) Step (i) ensures that the endpoints in the objective space are inserted into the new population and the extent of the non-dominated front is preserved.
(c) Step (ii) is a means to maintain variable diversity i.e. to include a possibility of retaining variable values which might be useful.

Every run of the optimization algorithm requires the inputs for population size, number of generations, probability of crossover, probability of mutation, distribution index of crossover, distribution index of the mutation and a random seed.

BULK CARRIER DESIGN EXAMPLE

Sen and Yang (1998) originally introduced this bulk carrier design example. They considered the minimization of transportation cost, minimization of lightship mass and the maximization of annual cargo transport capacity as three objectives for the design. The formulation presented in this paper considers the minimization of transportation cost per tonne with constraints on the annual cargo transport capacity to lie within 1500000 and 1600000 tonnes. There are 6 continuous variables (length of the ship L, draft of the ship T, depth of the ship D, Block coefficient C_B, breadth of the ship B and the speed of the vessel V) in this problem. The complete formulation is presented in Appendix. I. The bounds for the variables are $0 \leq L \leq 500$, $0 \leq T \leq 50$, $0 \leq D \leq 50$, $0.63 \leq C_B \leq 0.75$, $0 \leq B \leq 100$ and $14 \leq V \leq 18$.

Case A: Minimization of Annual Transportation Cost per Tonne

The single objective optimization problem was run using the our algorithm with population sizes of 100 and 200, number of generations of 100 and 200, probability of crossover of 0.7 and 0.9, probability of mutation of 0.05 and 0.1, distribution index of crossover of 10 and 20 and the distribution index of mutation of 10 and 20. From the set of 128 independent runs, the best design had an annual transportation cost of

8.37442 pounds per tonne while the worst design had it as 8.55894 pounds per tonne. The mean, median and standard deviation are 8.46668, 8.46668 and0.092625.

The variables corresponding to the best design and its performance measures are presented in Table 1.

Table 1: Details of the Optimal Design and Robust Optimal Design

Variables	Case A	Case B
LBP (m)	303.487	310.562
B (m)	37.2548	38.8094
T (m)	16.8432	16.6755
D (m)	24.6896	28.5736
CB	0.742441	0.747827
V (knots)	14.8079	14.7278
Performance		
Steel Weight (Tonnes)	22043.6	25100
Outfit Weight (Tonnes)	2153.98	2350.99
Machinery Weight (Tonnes)	2114.64	2108.52
Displacement (Tonnes)	144922	154060
Deadweight (Tonnes)	118610	124501
Power (HP)	35462	35348.1
Daily Fuel Consumption (Tonnes)	161.907	161.387
Fuel Cost per Round Trip (Pounds)	239177	239707
Sea days per Round Trip	14.069	14.1456
Port days per Round Trip	12.4834	13.0705
Round Trips per Annum	13.1815	12.86
Building Cost (Pounds)	3.62008e+007	3.85543e+007
Annual Cargo Carried (Tonnes)	1.51368e+006	1.55227e+006
Annual Cost (Pounds)	1.26762e+007	1.31098e+007
Transportation Cost per Tonne	8.37441	8.44556

A simulation was then used to assess the robustness of the design (Table 1: Case A) under simultaneous variations in cargo handling rate, fuel oil price and port charges. The distribution of transportation cost per tonne is presented in Figure 1 as a result of simultaneous variations of cargo handling rate, fuel price and port charge factor using a 100 point simulation. The distribution of annual transportation cost per tonne is presented in Figure 1 under the assumed probabilistic variations.

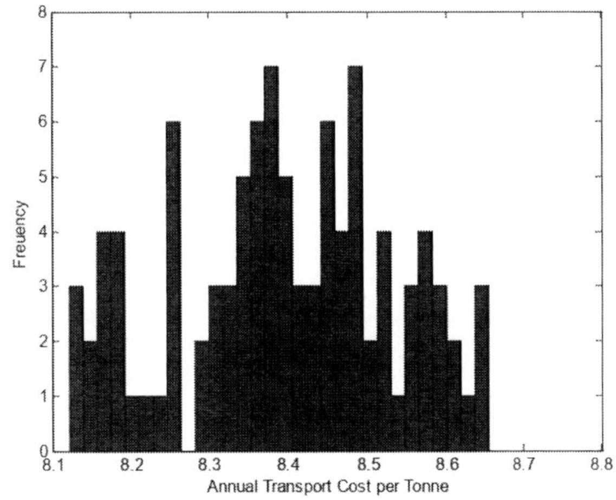

Fig. 1. Distribution of Annual Transportation Cost per Tonne under assumed Variations for the Optimal Design

The mean annual transportation cost per tonne was 8.374411 and the standard deviation was 0.846290. It is worth noting that 24 instances out of 100, violated the annual cargo transport capacity constraint.

Case B: Minimization of Annual Transportation Cost per Tonne and Minimization of its Standard Deviation

The results of the multiobjective, constrained, robust design formulation is presented in Table 1(Case B). The mean annual transportation cost per tonne was 8.445557and the standard deviation was 0.852654. It is worth noting that all instances satisfied the annual cargo transport capacity constraint. As can be observed from the above example that the robust design is marginally inferior in mean performance and also variance in performance with respect to the optimal solution, but it respects the constraints under assumed variations in operating conditions. For different instances of the problem, a robust design might have low variance as compared to the optimal solution which would indicate that the robust solution lies on a flat region as opposed to an optimal solution lying on a peak.

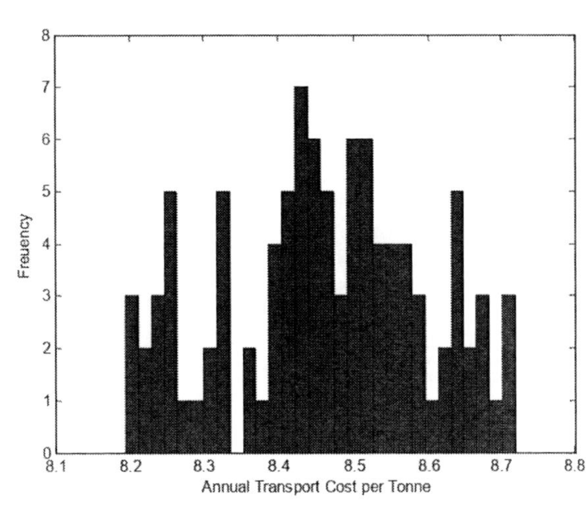

Fig. 2. Distribution of Annual Transportation Cost per Tonne under assumed Variations for the Robust Design

Once the robust design is obtained using the proposed method, the relative influence of the parameters can be computed using a simulation. The ratio of variance of the performance measure(e.g. annual transportation cost per tonne) and the variance of the input parameter is an indication of the influence of the parameter, where a larger value would indicate a large influence.

Case C: Maximization of Annual Transport Capacity and Minimization Annual Transportation Cost per Tonne

The framework seamlessly allows the designer to solve single and multiobjective problems. The result of the multiobjective case is presented in Figure 3. It is worth noting that the annual transportation cost per tonne has a large range of variation and it exhibits a large change in slope around 10 pounds per tonne. Such information is critical as it allows identification of designs that could carry close to 1600000 tonnes of cargo annually at an operating cost of around 10 pounds per tonne. The results of a single objective optimization is not capable of providing such an insight.

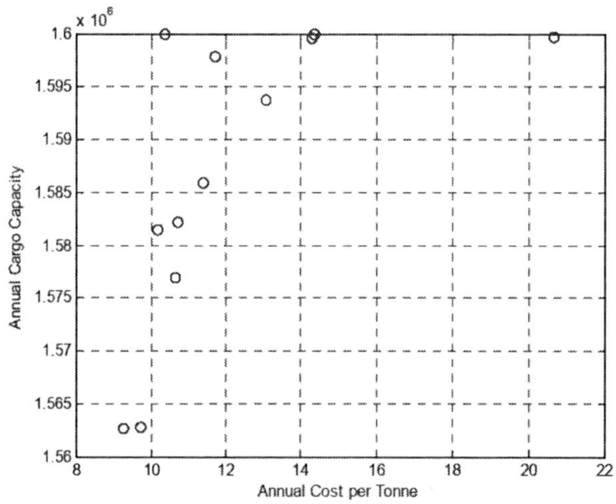

Fig. 3. Solutions to the Multiobjective Formulation

SUMMARY AND CONCLUSIONS

The paper highlights the differences between an optimal design and a robust optimal design. Subsequently, an approach to deal with constrained robust optimal design optimization problem is presented. The approach is based on formulating the problem of robust design optimization as a multiobjective optimization problem with an aim to simultaneously maximize performance and minimize its variance. A bulk carrier design example is introduced to highlight the differences between an optimal design and a constrained robust optimal design. The methodology of quantifying the influence of the varying parameters is suggested and the versatility of the algorithm is highlighted using single objective and multiobjective bulk carrier design optimization problems. The modular nature of the framework allows seamless integration with detailed analysis modules and probabilistic models for expected parametric variations.

REFERENCES

1. Chen, W., Allen, J.K., Tsui, K-L, and Mistree, F., A Procedure for Robust Design, *ASME Journal of Mechanical Design*, Vol. 118, No. 4, pp. 478-485, 1996.
2. Deb, K., Agrawal, R. B. Simulated binary crossover for continuous search space. Complex Systems. 9: 115-148, 1995.
3. Deb. K., Pratap, A., Agarwal, S., Meyarivan, T. A fast and elitist multi-objective genetic algorithm: NSGAII. IEEE Transaction on Evolutionary Computatio., *6*: 181-197, 2002.
4. Deb. K., Pratap, A., Agarwal, S., Meyarivan, T. A fast elitist non-dominated sorting genetic algorithm for multi-objective optimization: NSGAII. Proceedings of the Parallel Problem Solving from Nature VI Conference. Paris, France, pp. 849-858, 2000.
5. Du, X. and Chen, W., Towards a Better Understanding of Modeling Feasibility Robustness in Engineering Design, *ASME Journal of Mechanical Design*, Vol. 122, No. 4, pp. 385-394, 2000.
6. Ganesan, V., A Model for Multidisciplinary Design Optimization of Containerships, Department of Aerospace and Ocean Engineering, Virginia Polytechnic Institute and State University, Blacksburg, July 1999.
7. K. Deb, An efficient constraint handling method for genetic algorithms, *Computer Methods in Applied Mechanics and Engineering*, 186, pp. 311-338, 2000.
8. Keane, A.J., Price, W.G., and Schachter, R.D., Optimization Techniques in Ship Concept Design, *Transactions of the Royal Institute of Naval Architects*, Vol. 133, Part A, pp. 123-139, 1991.
9. Messac, A. and Ismail-Yahaya, A., Multiobjective Robust Design Using Physical Programming, *Structural and Multidisciplinary Optimization*, Vol. 23, No. 5, pp. 357-371, 2000.
10. Neu, W.L., Hughes, O., Mason, W.H., Ni, S., Chen, Y., Ganesan, V., Lin, Z., and Tumma, S., A Prototype Tool for Multidisciplinary Design Optimization of Ships, Ninth Congress of the International Maritime Association of the Mediterranean, Naples, Italy, April 2000.
11. Ray, T., Tai, K. and Seow, K.C., Multiobjective Design Optimization by an Evolutionary Algorithm, *Engineering Optimization*, 33(4), pp.399-424, 2001.
12. Ray, T., Gokarn, R.P., and Sha, O.P., A Global Optimization Model for Ship Design, *Computers in Industry*, Vol. 26, pp.175-192, 1995.
13. Sen, P., and Yang, J.B., Multiple criteria decision support in engineering design, Springer-Verlag, London, 1998.
14. Taguchi, G. Introduction to Quality Engineering, Krauss International Publications, 1986.
15. Taguchi, G. System of Experimental Design, Vol. 1 and Vol. 2, Krauss International Publications, 1987.
16. Yang, J.B., Minimax reference point approach and its application for multiobjective optimization, European Journal of Operational Research 126 541-556, 2000.

APPENDIX

Transportation Cost $= \dfrac{A\_COST}{A\_CARGO}$
$\text{A\_COST} = \text{CAP\_COST} + \text{RUN\_COST} + \text{VOY\_COST} \times RTPA$
$\text{CAP\_COST} = 0.2 \times \text{SHIP\_COST}$
$SHIP\_COST = 1.3(2000 \times (STEEL\_MASS)^{0.85} + 3500 \times (OUT\_MASS) + 2400 \times P^{0.8})$
$\text{STEEL\_MASS} = 0.034 \times \text{L}^{1.7} \times B^{0.7} \times D^{0.4} \times C_B^{\ 0.5}$
$\text{OUT\_MASS} = \text{L}^{0.8} \times B^{0.6} \times D^{0.3} \times C_B^{\ 0.1}$
$\text{MAC\_MASS} = 0.17 \times \text{P}^{0.9}$
$P = DISP^{2/3} \times V^3 \times \dfrac{1}{\dfrac{C2 \times V}{(gL)^{0.5}} + C1}$
$\text{C2} = \text{-}10847.2 \times \text{CB}^2 + 12817 \times \text{CB} - 6960.32$
$\text{C1} = 4977.06 \times \text{CB}^2 - 8105.61 \times \text{CB} + 4465.51$
$DISP = L \times B \times T \times C_B \times 1.025$
$\text{LSM} = \text{STEEL\_MASS} + \text{OUT\_MASS} + \text{MAC\_MASS}$
$\text{DWT} = \text{DISP} - \text{LSM}$
$\text{RUN\_COST} = 40000 \times DWT^{0.3}$
$\text{VOY\_COST} = \text{F\_COST} + \text{P\_COST}$
$\text{F\_COST} = 1.05 \times \text{DC} \times \text{SD} \times FP$
$\text{DC} = \text{P} \times 0.19 \times 24 / 1000 + 0.2$
$SD = \dfrac{RTM}{24 \times V}$
$RTPA = \dfrac{350}{SD + PD}$
$PD = 2 \times \dfrac{C\_DWT}{C\_RATE} + 0.5$
$C\_DWT = DWT - FL - CSW\,; FL = DC \times (SD + 5)$
$CSW = 2 \times DWT^{0.5}\,; P\_COST = RATE \times DWT^{0.8}$
$GM = 0.53 \times T + \dfrac{(0.085 \times C_B - 0.002)B^2}{T \times C_B} - 1.0 - 0.52 \times D$
$L/B \geq 6\,; L/D \leq 15\,; L/T \leq 19\,; 3000 \leq DWT \leq 500000$
$\dfrac{V}{(gL)^{0.5}} \leq 0.32$
$T \leq 0.45 \times DWT^{0.31}; T \leq 0.7 \times D + 0.7$

RTM = 5000 nautical miles, FP = 100 pounds / tonne, C_RATE = 20000 tonnes / day, RATE as 6.3 and g = 9.8065 m/s^2

Geometric Non-Linear Analysis of Stiffened Structures: A Review

RE Ojeda[1], BG Prusty[2], N Lawrence[1]

[1] Australian Maritime College, TAS, Australia

r.ojeda@amc.edu.au, n.lawrence@amc.edu.au

[2] University of New South Wales, NSW, Australia

g.prusty@unsw.edu.au

ABSTRACT

This paper presents a review of the state-of-the-art in analytical and numerical tools for the geometric non-linear analyses of stiffened structures. The different approaches and idealizations used for modelling stiffened structures, such as the "smeared approach" and plate-and-beam idealization, are reviewed and commented upon. Commonly used solution methods are examined and categorized according to their main application (i.e. static analysis, stability analysis, dynamic analysis, etc). Shortcomings in the existing literature are discussed, and the requirements for further work in this field are identified.

INTRODUCTION

A lightweight and efficient structure is an essential element in the design and successful operation of a high-performance craft. Structural weight reduction can be achieved by using materials with a high stiffness to weight ratio and an optimal structural configuration. The most common configurations of high performance craft structures are either thin flat or curved panels strengthened by rows of stiffeners or a composite sandwich shell stiffened by ring frames and bulkheads.

The progress in the analysis of stiffened structures has been reported in the past by several authors [1-5]. However, these reviews were focused mainly on the linear response of such structures. Moreover, a vast amount of literature has been since then accumulated on the diverse aspects of the study of stiffened structures. Therefore, a critical review of the pertinent technical literature is required to highlight the advances and the state-of-the-art in the field of the non-linear analysis of stiffened plates and shells.

The present review was conducted as part of a project to develop a finite element program for the non-linear analysis of arbitrary orientated stiffened panels as encountered in marine, aerospace and civil engineering structures. The cited references were selected to illustrate the variety of analytical tools being used in the broad area of stiffened structural analysis and are not necessarily the only significant contribution on the subject. The discussion is kept on a descriptive level, and for all the mathematical details, the reader is advice to refer to the cited literature. This review will serve as a useful guide and reference for researchers, academics and structural analysts. Undoubtedly, a survey of this type will not do justice to all the contributions for which the authors apologizes.

STRUCTURAL ANALSYIS OF STIFFENED PANELS

A variety of methods and computer programmes are available for the analysis of stiffened panels, ranging from simple closed form solutions to complicated three dimensional discretised solutions. The analysis cost typically increase with the level of detail modelled and the fidelity of the analysis procedure used[6]. The most common analysis models or methods are as follows:

1. Analytical and semi-analytical methods
2. Finite strip methods
3. Finite element methods

1. Analytical and semi-analytical methods

The simplest approach to analyse stiffened structures is to use an analytical model in which the structure is converted into equivalent plate/shell with constant thickness by smearing out the stiffeners. The model is suitable for obtaining load paths, stiffness constraints and overall or general buckling load estimates. However, these models cannot be used for stress calculations or to capture local and stiffener buckling failures[6]. Moreover further difficulties appear when using this approach if the stiffeners are not identical in both directions since the resulting thickness becomes non-uniform.

Sheinman and Simitses [7] presented a methodology for the buckling analysis of imperfect, thin, cylindrical shells uniform axial compression. This work was later expanded by Simitses et al. [8] for the non-linear stability analysis of pressure-loaded imperfect stiffened cylinders. The methodology is based on the smeared technique and the von Karman-Donell nonlinear theory. The field equations were reduced to a system of ordinary differential equations using the Galerkin procedure. These equations where then solved using a finite difference scheme.

Other examples of this analysis approach have been presented by Ji and Yeh [9] for the general non-linear buckling solution of axis-symmetric ring and stringer stiffened cylindrical shells, and by Savoia and Reddy [10] to track the post-buckling response of stiffened, cross ply laminated, cylindrical shells. Srinivasan and Ramachadran [11] further developed this method to analyse plates where the stiffeners are not provided orthogonally. The author proposed that, for this case, the equivalent "smeared" plate should be treated as an anisotropic one.

The "smeared stiffener" approach was also used by Shen [12] for the post-buckling analysis of imperfect stiffened laminated cylindrical shells under combined external pressure and thermal loading. The author used a boundary layer theory of shell buckling which includes the effects of non-linear pre-buckling deformations, non-linear large deflections in the post-buckling range and initial geometrical imperfections of the shell.

Nonetheless the "smeared approach" fails to capture the local effects of the stiffeners, thereby making it unsuitable to study the response of stiffened shells where the stiffeners are spaced unevenly or are of different cross section; hence the applicability of the "smeared" approach is very much restricted.

In order to overcome these difficulties Steen et al. [13-17] have proposed three semi-analytical models to study the post-buckling stiffness and ultimate strength of stiffened plate panels: an unstiffened model, an orthogonally stiffened model and an arbitrary stiffened model. These models can be classified as semi-analytical in the sense that they are based on the recognized plate theory of Maguerre in combination with numerical techniques for solution of governing equations. Both the unstiffened and the orthogonally stiffened model are capable of assessing post-buckling strength whereas the arbitrary orientated stiffened model is limited only to linearized analyses.

2. Finite strip method

The finite strip method[18] was devised for structural analysis in the late 1960's. This method treats the stiffened panels as an assemblage of plates or shells. For each component in the assemblage, the field equations are solved and boundary conditions at the interfaces are matched between adjoining members. The field equations for each component can be solved exactly or approximately using assumed displacements.

Conventional finite strip method allows different segments to have different properties, but cannot permit variations of thickness or properties in each segment. This restricts the models to have lengthwise or breadth wise uniform properties. Local details (e.g. cutouts, localized loads) are difficult to model [6]. For example Khalil et al. [19] used the finite strip method to analyse non-linear dynamic behaviour of stiffened plates. The authors included the effects of large deflections and strain rate sensitive yielding in their analysis. Although this works obtained good correlation with experimental results it is limited to the analysis of plates reinforced with uni-directional stiffeners.

The finite strip method is also suitable for the analysis of stiffened panels manufactured from layered composite materials. Loughlan [20] and Loughlan and Delaunoy [21] used the finite strip method to study the effects of fibre orientation on the buckling of composite stiffened plates. A multi-term finite strip approach was required to predict the complex buckling behaviour of composite stiffened panels under shear loads. The method was later extended by Loughlan [22] for the buckling analysis of composite box sections. In both cases only structures reinforced with stiffener running in one direction where considered.

The spline finite strip method was developed to overcome some of the limitations of the conventional finite strip method. Cubic B-spline function is adopted in one direction which eliminates most of the drawbacks of conventional finite strip formulations. This approach has been proven suitable for the vibration[23], buckling [23, 24], and transient[24] analyses of prismatically stiffened composite plates as well as for the vibration and buckling analysis of stiffened sandwich plates [25]. A recent example of the application of this spline finite strip method to the study of stiffened plates is the work of Sheikh and Mukhopadhyay [26, 27]. The authors extended the analysis capabilities their previously developed linear formulation [28, 29] to perform large deflection and non-linear transient vibration analysis of stiffened plates. Von Kármán's large deflection plate theory was used in the formulation with a total Lagrangian description of motion. The non-linear equations were solved using the Newton-Raphson iteration technique where Newmark's method was used for the time integration. The main advantage of the formulation lies in the treatment of the stiffeners which does not need to be placed along the nodal lines the plate strip. However, the main limitation of this approach is that the stiffener ends had to pass through two opposite edges of the strips.

3. Finite element method

The finite element method [30] is one of the most accurate and versatile techniques, but it can take large amount of computer storage and processing time [26]. One method of accounting for the presence of stiffening beams, attached to plates and shells in a finite element model, is to approximate these members by the same element types as used for the shell. For efficiency, a finite element model should give sufficiently accurate results with minimum number of degrees of freedom [31].

Spagnoly and Chryssanthopoulos [32] studied the elastic buckling and post-buckling behaviour of widely-stiffened conical shells under axial compression. Nine-node Lagrangian doubly curved shell elements, in accordance with the Mindlin-Reissner kinematic assumptions were used to model both the shell and stringer walls. Symmetry conditions were used to keep a low number of degrees of freedom in the analysis. However, in the absence of repeated symmetry conditions, the finite element analysis of such models becomes very resource intensive due to the large number of elements and the geometric non-linearities inherent in the analysis.

The inefficiency of idealising stiffened structures with only plate, shell or solid finite elements has led researchers to search for accurate and computationally efficient beam-plate and beam-shell models.

However, an admissible beam-shell model demands compatible displacements fields for both the beam and the shell elements. Therefore, the interpolation functions approximating the displacement fields in beams must be the same as the ones approximating the corresponding displacement field in the shell elements. This approach, known as the discrete stiffener approach was used by Liao & Reddy [33] to present a degenerated continuum-based, laminated, isoparametric, stiffened shell element in conjunction with a degenerated, isoparametric beam element to investigate the static, geometric, non-linear response of composite stiffened shells. The degenerated shell and beam elements were obtained by imposing appropriate kinematic constraints on the three-dimensional isoparametric solid element as presented by Bathe & Bolourchi [34]. Similarly, Qun et al. [35] used the discrete stiffener approach to develop a finite element program for the static analysis of concentrically stiffened isotropic plates and shells with large deflection and rotations. A generalized conforming triangular flat shell element with drilling degrees of freedom and the three-dimensional beam element with Hermite interpolation was used to represent the stiffened structure.

The discrete stiffener approach has not been limited to the global analysis of stiffened shells. The local behaviour of discretely stiffened laminated composite cylindrical shells has been investigated by Kassegne & Reddy [36]. The layer-wise theory [37] was used to model the composite laminate. The layer-wise model is essentially a 3D model in a 2D format, where the 3D displacement field is expanded as a function of a surface wise 2D displacement field and a one-dimensional interpolation through the thickness. Hence it is capable of representing the local layer behaviour. The authors acknowledged that the discrete layer-wise model might be too expensive computationally for shells with a large number of stiffeners.

A significant disadvantage of the approaches discussed so far is that the orientation of the stiffeners is restricted by the mesh, because the beam and shell elements are connected to each other at their nodal points. Therefore, a beam element can be placed only along the boundaries of the plate or shell element. Hence the distribution of stiffeners modelled by beam elements has a very restrictive influence on the meshing of the skin, which may be very inefficient, especially when repeated analyses are to be performed to estimate the optimum spacing of stiffeners.

Koko & Olson [38] used a super-element approach for the large deflection elasto-plastic analysis of orthogonally stiffened plates. The super-elements are designed to contain all the basic modes of deformation so that only one plate element per bay and one beam element per span are needed to analyse a stiffened structure, therefore reducing the storage requirement and solution times. The authors further extended the proposed method for the non-linear modal and transient analyses of stiffened plates [39, 40]. Later, and inspired by the satisfactory performance of the super-elements for the non-linear analysis stiffened plates, Jiang & Olson [41] further developed the super-element approach to study the non-linear behaviour of stiffened shells. The proposed method was limited to the study of isotropic rectangular plates and shells, stiffened in mutually perpendicular directions. Hence, and similar to the discrete stiffener approach, the super-element approach is restricted by the condition that the stiffeners must pass through the edges of the super-element.

Another interesting approach was reported by Günay [42]. The author presented a geometrically non-linear finite element analysis of laminated stiffened cylindrical shallow shells using composite shallow shell element with stringer-type stiffeners. In his analysis, two side meshing system was generated to represent cylindrical shell with stiffeners in a three-dimensional co-ordinate system. No attempts were made to model stiffeners of various shapes and/or having arbitrary orientation in the plate/shell panel.

An optimal beam-plate model should involve, as explained by Barut [43], the placement of the beam element within the plate or shell element at an arbitrary orientation. This requires the definition of the displacement fields at any point within a slave element (beam element) in terms of the field variables of a master element (plate or shell element). Since the beam kinematic field is described by the shell kinematic field, only the master plates or shells are needed to discretize the entire stiffened shell structure. Thus eliminating the modelling difficulties attributed to the conventional beam-plate and beam-shell models. This approach is referred to as the arbitrary-orientated stiffener approach and has been widely applied for more than two decades to the linear analysis of stiffened plates [44-51] and stiffened shells [52-60]. However reported work on the application of this approach to geometrically non-linear analysis of stiffened plates and shells is scarce.

The first application of this approach for the geometrically non-linear analysis of stiffened plates was reported by Rao et al. [61]. The authors presented the finite element static analysis of the large deflection response of isotropic stiffened plates using an isoparametric quadratic stiffened plate bending element. The stiffened element was a development of the linear formulation presented by Mukhopadhyay et al. [62, 63]. The authors excluded the contribution of the stiffener non-linearities in the formulation. This was later introduced by Chattopadhyay et al. [64]. In this later work, the large deflections of laminated composite stiffened plates were analysed using an eight noded isoparametric element. The element formulation was based on Reissner-Mindlin's hypothesis with a total Lagrangian description of motion. In both works the arbitrary orientated stiffener approach was applied so that the stiffener may lay within the element, with the limitation that the stiffener ends had to pass through two opposite edges of the element. The presented formulation was not capable of modelling stiffeners of arbitrary cross-section.

The arbitrary-orientated stiffener approach was applied to the geometrically non-linear analysis of stiffened shells by Goswami and Mukhopadhyay [65]. Geometrically non-linear finite element static, modal and transient analyses of laminated composite stiffened shells were carried out using a nine noded Lagrangian curved shear flexible isoparametric stiffened element with five degrees of freedom per node. Again the formulation accounts for arbitrarily orientated stiffeners with the limitation that stiffeners have to pass through two opposite edges of the element. Hence it is not possible to model stiffeners that are required to pass through two adjacent edges of the shell element. Additionally, the formulation is restricted to the analysis of shallow shells.

In view of this restriction, Samanta and Mukhopadhyay [66] presented a formulation for the large deflections of deep isotropic stiffened shells. The analysis was conducted using a combination of Allman's plane stress triangle and the discrete Kirchhoff triangle plate bending element which was previously used by the authors to solve linear examples of stiffened shells [52, 67]. The element initially had 24 degrees of freedom which were later reduced to 18 using appropriated constraint equations. The proposed formulation was limited to thin, isotropic plates and shells reinforced with stiffeners of rectangular cross-section only. The element, being faceted, differs slightly from the actual shell and hence the coupling of stretching and bending is excluded from the element due to the absence of element curvature.

Ray and Satsangi [68] presented an arbitrary orientated finite element formulation for the geometrically non-linear analysis of laminated hat stiffened plates. The stiffened panel was discretized using eight node plate and three node beam isoparametric elements. Although the authors suggested that the stiffener can be incorporated at any arbitrary orientation within the plate element, no examples were presented to support this claim.

Barut et al. [43] recently presented a stiffened shell element formulation for the geometric non-linear analysis of composite laminated stiffened shells. It is a four node, triangular, C^0 anisoparametric element with five degrees of freedom at the corner nodes and two degrees of freedom at the centre node. The formulation accounts for transverse shear deformation and material anisotropy in both the shell and stiffener. The laminated stiffeners may be arbitrary in orientation and cross section. Whilst the element is capable of taking large deflections and rotations into account, the presented formulation has not been proven capable of solving deep shell examples. Although the study discusses the incorporation of arbitrarily shaped and arbitrarily orientated stiffeners, only one example for laminated cross-ply stiffened shell with rectangular stiffeners was covered.

CONCLUSIONS

The literature review on the non-linear analysis of composite and sandwich stiffened structures therefore reveals that:

1. Analytical or semi-analytical solutions are the simplest way to analyse the non-linear response of stiffened structures. However complex non-linear analyses such as the ones of involving plates reinforced with multiple stiffeners at arbitrary orientations is very difficult, if not impossible.

2. Design level accuracy can be obtained by using the finite strip method for the non-linear analyses of stiffened structures for a fraction of the cost of full finite element analyses but with limitations regarding the orientation of the stiffener within the strips.

3. The arbitrary-orientated stiffener approach appears to be the most flexible and efficient finite element technique to model stiffened structures in the preliminary design stage.

4. Although the arbitrary-orientated stiffener approach is reported to have been widely applied to the linear analysis of stiffened structures, its application for the geometrically non-linear analysis of such structures is scarce. Moreover, current arbitrary-orientated non-linear formulations have yet to be proven capable of modelling stiffeners passing through two adjacent edges of the plate element.

5. Little information is available on the geometrically non-linear analysis of FRP stiffened structures using the arbitrary-orientated stiffener approach, and, to the authors' knowledge, no work has been carried out to analyse sandwich stiffened structures using the arbitrary-orientated stiffener approach.

REFERENCES

1. Satsangi S.K. and Mukhopadhyay M., (1989), "A Review of Static Analysis of Stiffened Plates", Journal of Structural Engineering, 15, 117-126.
2. Mukhopadhyay M. and Mukherjee A., (1989), "Recent Advances on the Dynamic Behavior of Stiffened Plates", The Shock and Vibration Digest, 21, 6-9.
3. Sinha G. and Mukhopadhyay M., (1995), "Static and Dynamic Analysis of Stiffenend Shells - a Review", Proceedings of Indian National Science Academy, 61, 195-219.
4. Mukhopadhyay M. and Sinha G., (1992), "A Review of Dynamic Behavior of Stiffened Shells", The Shock and Vibration Digest, 24, 3-13.
5. Norwood C., (1995), "The Free Vibration of Ring Stiffened Cylinders - a Critical Review of the Unclassified Literature", Defence Science and Technology Organization, DTSO-TR-0200.
6. Lamberti L., Venkataraman S., Haftka R.T., and Johnson T.F., (2003), "Preliminary Design Optimization of Stiffened Panels Using Approximate Analysis Models", International Journal for Numerical Methods in Engineering, 57, 1351-1380.
7. Sheinman I. and Simitses G.J., (1977), "Buckling Analysis of Geometrically Imperfect Stiffened Cylinders under Axial Compression", AIAA Journal, 15, 374-382.
8. Simitses G.J., Sheinman I., and Girl J., (1979), "Nonlinear Stability Analysis of Pressure-Loaded Imperfect Stiffened Cylinders", Journal of Ship Research, 23, 123-126.
9. Ji Z.-Y. and Yeh K.-Y., (1990), "General Solution for Nonlinear Buckling of Nonhomogeneous Axial Symmetric Ring- and Stringer-Stiffened Cylindrical Shells", Computers & Structures, 34, 585-591.
10. Savoia M. and Reddy J.N., (1994), "Post-Buckling Behavior of Stiffened Cross-Ply Cylindrical Shells", Transactions of the ASME, 61, 998-1000.
11. Srinivasan R.S. and Ramachandran S.V., (1977), "Linear and Nonlinear Analysis of Stiffened Plates", International Journal of Solids and Structures, 13, 897-912.
12. Shen H.-S., (1998), "Postbuckling Analysis of Imperfect Stiffened Laminated Cylindrical Shells under Combined External Pressure and Thermal Loading", International Journal of Mechanical Sciences, 40, 339-355.
13. Steen E. and Byklum E., (2005), "Ultimate Strength and Postbuckling Stiffness of Plate Panels Subjected to Combined Loads Using Semi-Analytical Models", International Conference on Marine Research and Transportation, Ischia Island, Naples, Italy.
14. Steen E., Byklum E., Vilming K.G., and Ostvold T.K., (2004), "Computerized Buckling Models for Ultimate Strength Assessments of Stiffened Ship Hull Panels", PRADS, Lubeck/Travemunde.
15. Steen E., Ostvold T.K., and Valsgard S., (2001), "A New Design Model for Ultimate and Buckling Strength Assessment of Stiffened Plates", PRADS,

16. Byklum E., Steen E., and Amdahl J., (2004), "A Semi-Analytical Model for Global Buckling and Postbuckling Analysis of Stiffened Panels", Thin-Walled Structures, 42, 701-717.
17. Steen E., Byklum E., and Vilming K.G., (2004), "Computer Efficient Non-Linear Buckling Models for Capacity Assessments of Stiffened Panels Subjected to Combined Loads", International Conference on Thin-Walled Structures, Loughborough, UK.
18. Cheung Y.K. and Tham L.G., (1998), "Finite Strip Method (New Directions in Civil Engineering)".
19. Khalil M.R., Olson M.D., and Anderson D.L., (1988), "Nonlinear Dynamic Analysis of Stiffened Plates", Computers & Structures, 29, 929-941.
20. Loughlan J., (1994), "The Buckling Performance of Composite Stiffened Panel Structures Subjected to Combined in-Plane Compression and Shear Loading", Composite Structures, 29, 197-212.
21. Loughlan J. and Delaunoy J.-M., (1993), "The Buckling of Composite Stiffened Plates with Some Emphasis on the Effects of Fibre Orientation and on Loading Configuration", Composite Structures, 25, 485-494.
22. Loughlan J., (1996), "The Buckling of Composite Stiffened Box Sections Subjected to Compression and Bending", Composite Structures, 35, 101-116.
23. Wang S. and Dawe D.J., (1997), "Spline Finite Strip Analysis of the Buckling and Vibration of Composite Prismatic Plate Structures", International Journal of Mechanical Sciences, 39, 1161-1180.
24. Dawe D.J., (2002), "Use of the Finite Strip Method in Predicting the Behaviour of Composite Laminated Structures", Composite Structures, 57, 11-36.
25. Yuan W.X. and Dawe D.J., (2004), "Free Vibration and Stability Analysis of Stiffened Sandwich Plates", Composite Structures, 63, 123-137.
26. Sheikh A.H. and Mukhopadhyay M., (2002), "Linear and Nonlinear Transient Vibration Analysis of Stiffened Plate Structures", Finite Elements in Analysis and Design, 38, 477-502.
27. Sheikh A.H. and Mukhopadhyay M., (2000), "Geometric Non-Linear Analysis of Stiffened Plates by the Finite Strip Method", Computers & Structures, 76, 765-785.
28. Sheikh A.H. and Mukhopadhyay M., (1993), "Free Vibration Analysis of Stiffened Plates with Arbitrary Planform by the General Finite Strip Method", Journal of Sound and Vibration, 162, 147-164.
29. Sheikh A.H. and Mukhopadhyay M., (1992), "Analysis of Stiffened Plate with with Arbitrary Planform by the General Spline Finite Strip Method", Computers & Structures, 42, 53-67.
30. Zienkiewicz O.C. and Taylor R.L., (2000), "The Finite Element Method". 5th ed,
31. Houlston R. and Slater J.E., (1991), "Global and Local Modelling of Naval Panels Subjected to Shock Loads", Computers & Structures, 40, 353-364.
32. Spagnoli A. and Chryssanthopoulos M.K., (1999), "Elastic Buckling and Postbuckling Behavior of Widely-Stiffened Conical Shells under Axial Compression", Engineering Structures, 21, 845-855.
33. Liao C.-L. and Reddy J.N., (1990), "Analysis of Anisotropic, Stiffened Composite Laminates Using a Continuum-Based Shell Element", Computers & Structures, 34, 805-815.
34. Bathe K.-J. and Bolourchi S., (1980), "A Geometric and Material Nonlinear Plate and Shell Element", Computers & Structures, 11, 23-48.
35. Qun Z., Mu. L., and Wenqui K., (1999), "Application If the Generalized Conforming Flat Shell Element to Geometrical Non-Linear Analysis for

Composite Stiffened Shell Structures", Communications in Numerical Methods in Engineering, 15, 399-412.

36. Kassegne S.K. and Reddy J.N., (1998), "Local Behaviour of Discretely Stiffened Composite Plates and Cylindrical Shells", Composite Structures, 41, 13-26.

37. Reddy J.N., (1987), "A Generalization of Two-Dimensional Theories of Laminated Composite Laminates", Communications in Applied Numerical Methods, 3, 173-180.

38. Koko T.S. and Olson M.D., (1991), "Non-Linear Analysis of Stiffened Plates Using Super Elements", International Journal for Numerical Methods in Engineering, 31, 319-343.

39. Koko T.S. and Olson M.D., (1991), "Nonlinear Transient Response of Stiffened Plates to Air Blast Loading by a Superelement Approach", Computer Methods in Applied Mechanics and Engineering, 90, 737-760.

40. Koko T.S. and Olson M.D., (1992), "Vibration Analysis of Stiffened Plates by Super Elements", Journal of Sound and Vibration, 158, 149-167.

41. Jiang J. and Olson M.D., (1994), "Nonlinear Analysis of Orthogonally Stiffened Cylindrical Shells by a Super Element Approach", Finite Elements in Analysis and Design, 18, 99-110.

42. Günay E., (1999), "Finite-Element Analysis of Laminated Stiffened Cylindrical Shallow Shell", Applied Composite Materials, 6, 381-395.

43. Barut A., Madenci E., Tessler A., and Starnes J.H.J., (2000), "A New Stiffened Shell Element for Geometrically Nonlinear Analysis of Composite Laminates", Computers & Structures, 77, 11-40.

44. Ray C. and Satsangi S.K., (1996), "Finite Element Analysis of Laminated Hat-Stiffened Plates", Journal of Reinforced Plastics and Composites, 15, 1174-1193.

45. Ray C. and Satsangi S.K., (1999), "Laminated Stiffened Plate- a First Ply Failure Analysis", Journal of Reinforced Plastics and Composites, 18, 1061-1076.

46. Kumar Y.V.S. and Mukhopadhyay M., (2002), "A New Modelling Tool for 3-D Finite Element Analysis of Frp Ships", International Shipbuilding Progress, 49, 5-35.

47. Kumar Y.V.S., Mukhopadhyay M., and Sarkar T., (2002), "Development of a New Efficient Stiffened Plate Element for Dynamic Analysis of a Three-Dimensional Ship Structure", Journal of Engineering for the Maritime Enviroment, 216, 45-56.

48. Kumar Y.V.S. and Srivastava A., (2003), "First Ply Failure Analysis of Laminated Stiffened Plates", Composite Structures, 60, 307-315.

49. Satsangi S.K. and Ray C., (1998), "Structural Analysis of Ships' Stiffened Plate Panels in Bending", International Shipbuilding Progress, 45, 181-195.

50. Chattopadhyay B., Sinha P.K., and Mukhopadhyay M., (1993), "Finite Element Analysis of Blade-Stiffened Composite Plates under Transverse Loads", Journal of Reinforced Plastics and Composites, 12, 76-100.

51. Barik M. and Mukhopadhyay M., (2002), "A New Stiffened Plate Element for the Analysis of Arbitrary Plates", Thin-Walled Structures, 40, 625-639.

52. Samanta A. and Mukhopadhyay M., (1998), "Finite Element Static Analysis of Stiffened Shells", International Journal of Applied Mechanics and Engineering, 3, 55-87.

53. Prusty B.G., (2001), "Static, Dynamic, Buckling and Failure Analyses of Composite Stiffened Shell Structures: A Finite Element Approach", Indian Institute of Technology, Kharagpur.

54. Nayak A.N. and Bandyopadhyay J.N., (2002), "On the Free Vibration of Stiffened Shallow Shells", Journal of Sound and Vibration, 255, 357-382.
55. Prusty B.G., (2001), "Analysis of Stiffened Shell for Ships and Ocean Structures by Finite Element Method", Ocean Engineering, 28, 621-638.
56. Prusty B.G. and Satsangi S.K., (2001), "Finite Element Transient Dynamic Analysis of Laminated Stiffened Shells", Journal of Sound and Vibration, 248, 215-233.
57. Prusty B.G. and Satsangi S.K., (2001), "Finite Element Buckling Analysis of Laminated Composite Stiffened Shells", International Journal of Crashworthiness, 6, 471-484.
58. Prusty B.G., (2003), "Linear Static Analysis of Composite Hat-Stiffened Laminated Shells Using Finite Elements", Finite Elements in Analysis and Design, 39, 1125-1138.
59. Prusty B.G., Ray C., and Satsangi S.K., (2001), "First Ply Failure Analysis of Stiffened Panels - a Finite Element Approach", Composite Structures, 51, 73-81.
60. Prusty B.G., Satsangi S.K., and Ray C., (2001), "First Ply Failure Analysis of Laminated Panels under Transverse Loading", Journal of Reinforced Plastics and Composites, 20, 671-684.
61. Rao D.V., Sheikh A.H., and Mukhopadhyay M., (1993), "A Finite Element Large Displacement Analysis of Stiffened Plates", Computers & Structures, 47, 987-993.
62. Mukhopadhyay M. and Satsangi S.K., (1984), "Isoparametric Stiffened Plate Bending Element for the Analysis of Ships' Structures", Transactions of the Royal Institution of Naval Architects, 126, 144-151.
63. Mukhopadhyay M., Satsangi S.K., and Mukherjee A., (1990), "A New Isoparametric Plate Element for the Analysis of Ship Structures", International Shipbuilding Progress, 37, 79-117.
64. Chattopadhyay B., Sinha P.K., and Mukhopadhyay M., (1995), "Geometrically Nonlinear Analysis of Composite Stiffened Plates Using Finite Elements", Composite Structures, 31, 107-118.
65. Goswami S. and Mukhopadhyay M., (1995), "Geometrically Non-Linear Transient Dynamic Response of Laminated Composite Stiffened Shells", Journal of Reinforced Plastics and Composites, 14, 618-640.
66. Samanta A. and Mukhopadhyay M., (1999), "Finite Element Large Deflection Analysis of Shallow and Deep Stiffened Shells", Finite Elements in Analysis and Design, 33, 187-208.
67. Samanta A. and Mukhopadhyay M., (2004), "Free Vibration Analysis of Stiffened Shells by the Finite Element Technique", European Journal of Mechanics A/Solids, 23, 159-179.
68. Ray C. and Satsangi S.K., (1998), "Geometric Nonlinear Static Analysis of Laminated Hat Stiffened Plates", Journal of Reinforced Plastics and Composites, 17, 1264-1280.

A Guide to the Assessment of Vessel Wash within Sheltered Waterways

Gregor J. Macfarlane[1] and Greg Cox[2]
[1] Australian Maritime College, Launceston, TAS, Australia
[2] Kamira Holdings, Newcastle, NSW, Australia

ABSTRACT

Sheltered waterways are being subjected to increasing pressure from recreational activities, including recreational boating, as well as public transport and marine tourism. The characteristics of the waves generated by such craft can be very complex and their effect on shoreline environments is poorly understood. To date the development of vessel operating criteria for mitigating foreshore impacts has been largely vessel and site-specific, making transposition of operating criteria between different sites very difficult.

In order to develop a rational means of assessing and regulating vessel wash, the relevant scientific, environmental and social issues should be understood and embraced. The basics of vessel wash generation are summarised to introduce the potential difficulties when attempting to measure and parameterise wash components. The likely impacts on the natural and man-made environment are discussed, as well as a range of wash criteria that are currently in use. Lastly, an overview is provided of commonly measured wash parameters and how they may be incorporated into operating requirements.

1.0 INTRODUCTION

Sheltered waterways have been used successfully for transportation and trade for thousands of years, however the development of high-speed recreational and commercial vessels over the past fifty years has introduced erosive mechanisms that can create waterway damage faster than it can be studied and regulated. The response of research and regulatory bodies has been highly reactive and characterised by pockets of site-specific research with little attempt at standardisation.

Many existing wave wake criteria are based on over-simplified concepts and may provide only limited protection against the most common environmental wash problem – foreshore erosion. Further study to extract the erosive components of a vessel's wash and how they may influence erosion mechanisms is required.

2.0 VESSEL WASH IN SHELTERED WATERS

Prior to the introduction of high-speed vessels, most waterways were either not subjected to increased wave energy levels from vessel wash or speeds were regulated to maintain environmental and operational requirements. The wash energy of a vessel travelling at slow speed is directly related to the wave drag of the hull, which is a function of the 6th power of vessel speed. Small changes in speed can therefore substantially reduce or increase wash energy, making regulation relatively easy.

High vessel speeds generate waves that are of different form to slow speeds. Waves with long periods are generated, introducing a wave climate to sheltered waterways against which they have little resistance. Vessel waves and their environmental effects are discussed briefly in this section, summarised from Macfarlane and Cox (2007).

2.1 Vessel waves

In deep water, defined as a water depth greater than half the wavelength, vessels generate two wave trains; transverse waves (often termed stern waves) that propagate along the sailing line and divergent waves (often termed bow waves) that propagate at an angle to the sailing line. In the region close to the vessel, these wave trains interact. Beyond a lateral distance from the vessel of about three boat lengths the wave pattern is dominated by the divergent waves, which decay in height at a slower rate than transverse waves for a given lateral distance from the sailing line.

The maximum wave heights and corresponding periods generated increase with increasing speed, peaking at a condition when a vessel's length Froude number is approximately 0.5. Once into the high-speed range, maximum wave heights decrease gradually and the corresponding periods remain approximately constant. However, the high-speed condition generates long-period leading waves ahead of the highest wave that can cause significant environmental disturbance.

In shallow water, depth Froude number governs the wave patterns generated. When the depth Froude number equals unity, the wave energy of a vessel (whether slow speed or high speed) is concentrated into one wave (or occasionally more than one). This wave is highly damaging and this vessel-speed-to-water-depth ratio is to be avoided.

At high-speed in shallow water, termed supercritical when the depth Froude number is greater than unity, a vessel's transverse wave train dies away and the divergent waves propagate at an angle to the sailing line dependent on the water depth. Long-period leading waves are also generated.

2.2 Vessel Speed Regimes

The wave wake generated is directly related to a vessel's wavemaking resistance. There are three distinct speed ranges where a vessel's wash changes in magnitude and interactions between these speed conditions and the three depth-related wash regimes (sub-, trans- and super-critical) can create an unacceptable wash. Figure 1 shows graphically these speed regimes in terms of a non-dimensional speed-length ratio, length Froude number.

In practical terms, the first regime (displacement speed) is best observed in slow, heavy vessels. These vessels experience a practical upper limit of their speed, termed hull speed, which can only be exceeded with a substantial increase in engine power. High-speed craft, which have a power-to-weight ratio such that they can travel faster than their hull speed, first experience a resistance hump just above hull speed before settling into the high-speed regime. For a planing hull, this will be the onset of planing.

Displacement Speed

All vessels have a displacement speed range where the length of the transverse waves generated is less than the waterline length. For deep water operation the upper limit of this speed range is when the length Froude number is approximately 0.4, which reduces to:

$$\text{Speed, v (knots)} = 2.43\sqrt{L} \qquad \text{Equation (1)}$$

This maximum displacement speed, or hull speed, represents the condition where the longest wave generated equals the waterline length of the vessel. To travel faster than this, the vessel must begin to climb its own bow wave (a common analogy).

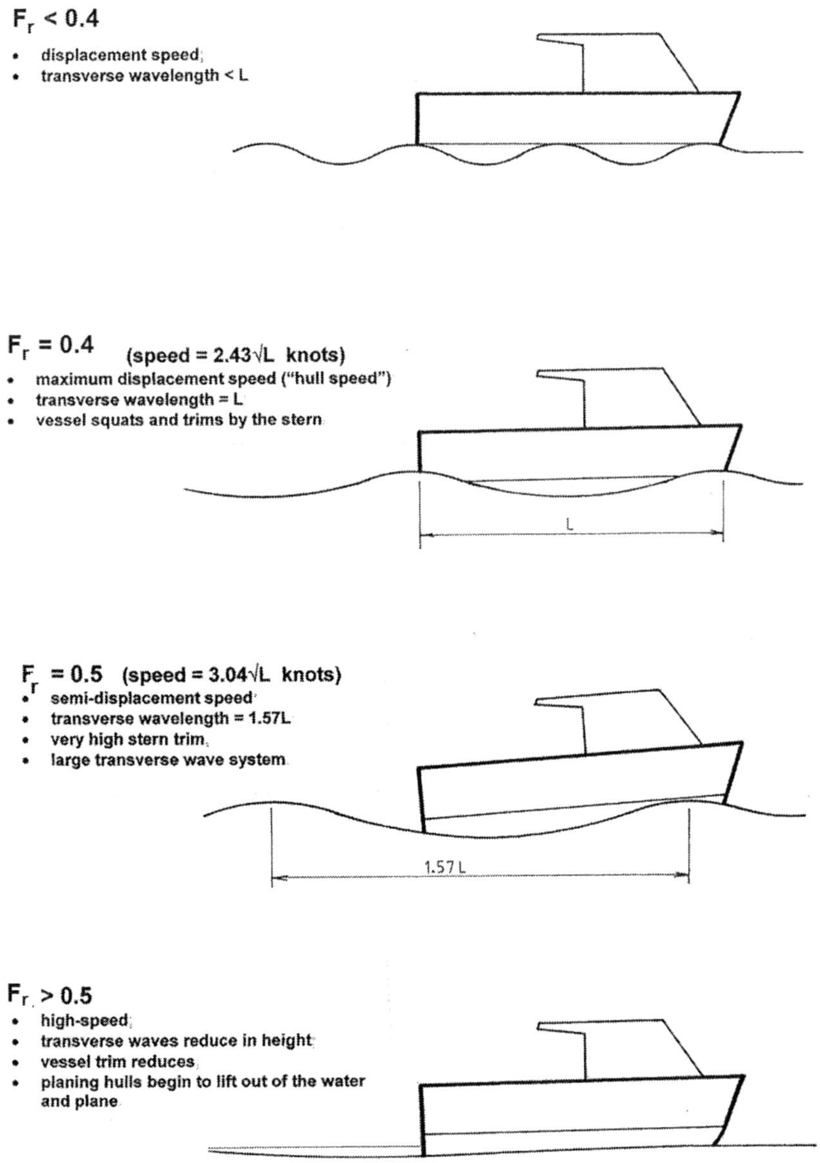

Figure 1: Vessel speed regimes

Semi-Displacement Speed

At increased vessel speeds within the displacement speed region, the running trim also increases as the transverse waves move aft of the transom. Vessel wavemaking resistance is high, with the non-dimensionalised wavemaking coefficient peaking at a length Froude number of approximately 0.5.

Wave wake height increases to its maximum and divergent wave periods increase steadily. A particular operating condition to be avoided is the combination of a length Froude number of 0.5 and depth Froude number of 1.0, when maximum specific wavemaking resistance and depth effects coincide. This condition occurs when the water depth is approximately 0.25L and the vessel speed is approximately $3.04\sqrt{L}$ (knots).

In general, semi-displacement speeds, often referred to as hump speeds in planing craft terms (when the vessel appears to climb over the hump before planing), create damaging wash.

High Speed

As the length Froude number increases above 0.5, the specific wavemaking resistance slowly reduces. The maximum wave height reduces and the corresponding wave period levels to a relatively constant value. For a planing hull form, the vessel will be approaching its fully-planing regime. Round bilge multihull forms are assumed to behave as high-speed displacement forms.

It is often said that high-speed vessel wake is preferable to that of the semi-displacement speed, where waves are high and steep. This may appear to be the case, but the high-speed condition produces the longest wave periods, which may have as much or greater effect on shorelines and shoreline structures as wave height. It is important to remember that in the deep water condition, wave height attenuates with lateral distance from the sailing line but period remains approximately constant.

2.3 Environmental Effects

Sheltered waterways exist in a state of dynamic equilibrium, with the three greatest natural dynamic influences being wind waves, tides and currents. Occasionally, major events such as flooding will create a new state from which a new dynamic equilibrium develops, U.S. Army Coastal Engineering Research Center (1984).

When an external source of wave energy is introduced, such as vessel wash, a waterway may either reach a new equilibrium or its natural erosion/deposition processes may become unstable, resulting in gradual erosion.

Waves with different characteristics have different effects on the environment. Wind waves are characterised by low to moderate wave heights, and short periods. As an example, the wind wave height and period for varying wind speeds and fetch lengths are shown in Table 1. These figures are generated by hindcasting - applying a standard set of equations to generate wave data given the wind speed and fetch (the distance the wind has been blowing over water). The formulas used for predicting the spectral wave height and peak spectral wave period are provided in U.S. Army Coastal Engineering Research Center, Chapter 3 (1984).

Fetch (m)	5m/s	10m/s	20m/s
100	26mm / 0.5s	62 / 0.7	144 / 0.9
500	59 / 0.8	137 / 1.1	321 / 1.5
1000	83 / 1.0	194 / 1.4	452 / 1.9
10000	250 / 2.0	586 / 2.8	1304 / 3.8

Table 1: Hindcast wind waves (wave height in mm / wave period in seconds)

It is clear that the height of wind waves grows by several hundred percent but the period increases at a much slower rate with increasing wind speed. This has three particular consequences.

Firstly, wind waves cause discomfort to occupants of small craft. The waves are close together and relatively high, so they are considered to be steep. The wave period is often similar to the roll period of small vessels, causing them to roll synchronously in beam seas when travelling at zero forward speed.

Secondly, the growth in energy in wind waves tends to be more height dependent than period dependent. From linear potential flow theory, the mean energy contained per wave, per unit width of wave front for a simple single-frequency sinusoidal wave is given by Equation 2, Dean and Dalrymple (1991):

$$E = \frac{\rho g^2 H^2 T^2}{16\pi} \qquad \text{Equation (2)}$$

As can be seen in Equation 2, wave energy is equally a function of both wave height and wave period. A shoreline naturally subjected to wind waves may occasionally experience waves with a large height, but the corresponding period will remain relatively low. Similarly, such a shoreline may be able to withstand vessel wash, provided the wash is characterised by moderate wave height but low corresponding wave period.

Thirdly, shoaling waves increase in height before breaking. However, for waves with periods less than about 3 seconds, the shoaling in most practical situations is considered to be negligible (less than 10% height increase), and for waves with periods less than 2 seconds the waves will be close to breaking before any shoaling occurs, U.S. Army Coastal Engineering Research Center (1984). Wind waves, particularly in sheltered and semi-enclosed waters, tend to maintain their deep-water height before breaking. This shoaling difference between short and long period waves can be seen at a surf beach, where the long-period swells stand up before breaking but the wind-driven chop simply breaks on the shore.

The last point is of particular interest, as it is commonly believed by many people that vessel wash height is the primary determinant of erosion potential – wave height being a more visual indicator of wake waves. However, wave period possibly has a greater effect in sheltered and semi-enclosed waters. Such shorelines have natural, dynamic mechanisms to withstand wind waves, but the sudden introduction of vessel wash containing much longer-period waves may upset the balance, Cox (2000).

In summary, the introduction of wave climates generated by modern, high-speed vessel designs into waterways that would normally experience only wind waves has been a primary cause of environmental degradation.

3.0 WAVE WAKE CRITERIA

3.1 Introduction

The development of wave wake criteria has gradually built with the growing interest in vessel wash issues. Nowhere in the world does there exist a comprehensive methodology for assessing wash-related impacts and introducing vessel operating regimes, without resorting to lengthy and expensive field studies.

Marine regulatory authorities have been overwhelmed by the complexity of the issues, which extend well beyond their traditional maritime roles. The knowledge base from which to draw is both limited in size and depth. This is a relatively new science and one that is a long way from developing conclusive answers.

3.2 Criteria requirements

In an effort to quantify vessel-generated waves, Macfarlane and Renilson (1999) suggested that any proposed measures meet the requirements listed in Table 2.

1	Independent of the length of the data sample
2	Able to be used to compare one vessel against another vessel
3	Easy to understand
4	Representative of wave wake problems
5	Easy to measure
6	Independent of the exact distance from vessel sailing line

Table 2: Requirements of proposed wave wake measures

In addition to the requirements listed in Table 2, the proposed wave wake criteria must have the following attributes:

- The number of variables must be limited if the criteria are to be applied without resorting to individual vessel testing;
- The criteria and their variable inputs should not be over-simplified to the point where their effectiveness is diluted;
- The degree of subjectivity in deriving the value of the variables must be reduced, otherwise the criteria will become open to manipulation and abuse; and,
- Any empirical equations derived must have a proper mathematical basis and obey the laws of similitude (i.e., must have certain properties that retain their mathematical integrity with varying input data so as not to be specific only to the data from which they were derived).

It is considered highly unlikely that any Government agency or commercial entity would be willing to commit funds towards a full wash-testing program on every waterway with existing or potential wash problems. The aim of wave wake research should be to define wave wake characteristics that are independent of any given site and are formulated in a manner such that they can be transferred between sites with confidence.

The ultimate aim of the regulatory authorities would be to have a set of criteria that can be applied over a variety of waterway types without the need to resort to the testing of individual vessels on specific routes, though this will always remain the best option for sensitive areas where a high degree of certainty is required.

3.3 The need for wave wake criteria

The imposition of operating regimes to reduce the consequences of vessel wave wakes is a relatively new phenomenon worldwide. Some long-standing vessel operating limits can be found, but are almost always based on blanket speed restrictions as opposed to other more scientific measures.

Macfarlane and Cox (2007) stated that there are several reasons for introducing operating restrictions, including:

- Limiting the effects of wash on moored vessels;
- Limiting the effects on maritime structures such as marinas (and their moored vessels) and seawalls;
- Ensuring that other waterway users are not adversely affected by vessel waves;
- Reducing the impact of shoaling waves on foreshore users;
- Reducing foreshore erosion; and,
- Using speed restrictions as an indirect means of addressing other issues, such as safety of navigation, noise and even unsociable behaviour.

Almost all of the most recent wash-limiting criteria are targeted at either large, high-speed ferries on coastal routes or have been formulated to specific vessels on specific routes. Few are aimed at a broad range of vessel types, where a blanket speed limit is often viewed as the simplest method to define, implement and police.

3.4 Existing wave wake criteria

Most existing wave wake criteria have been brought about because of the operation of high-speed craft, often in areas of limited water depth. Some locations where wave wake criteria have been implemented are discussed by Macfarlane and Cox (2007), including:

- Gordon River, Tasmania, Australia
- Sydney Harbour, New South Wales, Australia
- Parramatta River, New South Wales, Australia
- Swan River, Western Australia, Australia
- Coastal ferry services, Denmark
- Thames River, England, United Kingdom
- Rich Passage, Puget Sound, United States of America
- Marlborough Sound, New Zealand
- Noosa River, Queensland, Australia

Of all the examples of existing criteria above, only those implemented in Denmark and New Zealand meet all six of the suggested requirements listed in Table 2. The criteria adopted in Rich Passage and the Gordon and Noosa Rivers meet the first five requirements but relate to a specific lateral distance from the vessel's sailing line. It is considered appropriate to define a specific lateral distance in such cases where the width of the waterway is very limited and defined, as is the case in these examples.

3.5 Types of commonly applied criteria

Macfarlane and Cox (2007) discussed the above examples of wave wake criteria, the majority of which apply more to the operation of high-speed craft in areas of limited water depth. Glamore *et al* (2005) also discussed the benefits and limitations of some of these existing criteria. From these examples and others known to be in existence, the types of commonly applied criteria can be grouped into the following six categories:

- Blanket speed limit
- Wave height
- Wave height and period
- Wind wave approach
- Wave power
- Wave energy

Each of these commonly applied criteria are briefly discussed below.

Blanket speed limit

This is a universally applied criterion that is easy to both implement and police. There are two main drawbacks of the approach. Firstly, it can unfairly penalise certain vessels that may not produce as much wash as others, which is of concern for commercial vessel operations and does not encourage operators/designers to optimise their designs for minimal wave wake.

The second point stems from the first, in that it single-handedly ignores the fundamental principles of ship hydrodynamics. The point at which a vessel begins to produce a significant wave wake is dependent not only on its speed, but also its waterline length. Wave generation is a length Froude number dependent relationship and blanket speed limits ignore this. It is quite possibly the case that a proposed speed limit may have the correct intent for some vessels, but may allow others to operate at speeds greater than they should.

Wave height
This overly simplistic concept was considered appropriate a decade or more ago and unfortunately still contaminates popular thinking on wave wake matters. Lesleighter (1964) makes pertinent comments on the anecdotal estimates of wave heights from water ski vessels on the Hawkesbury River and how human nature exaggerates the relevance of obvious indicators like wave height.

Maximum wave height is used as an operating guideline for the Thames and Parramatta Rivers. Between 1998 and 2005 this measure was also adopted to assess commercial vessels operating on the lower Gordon River. The Parramatta River and Sydney Harbour ferries have been required to meet a wave height criterion of around 300mm. The lateral separation of the measurement point was often not specified and the earlier complementary wave period measure was dropped from the design requirements. It has become evident that this height measure alone is inadequate when assessing the environmental viability of high-speed vessels in sheltered waterways.

A review of published data in this field was undertaken by Macfarlane (2002) with the results highlighting that there are many varied forms and definitions for wave height, all of which can have considerable influence on the value derived. The magnitude of the variations make it very difficult to directly and/or fairly compare vessels and could make the difference between a vessel meeting or exceeding a specific criteria.

It is quite unlikely that criteria based on wave height alone will continue to be implemented by maritime authorities having prior experience in wash matters.

Wave height and period
The generalised Sydney Harbour limit of 400mm height and 3 seconds period would appear to be based on the local wind wave climate in the lower harbour and so has some historical as well as practical basis. The application of such limits is site-specific.

Wind wave approach
This was the investigative approach used on the Swan River, where the annualised energy of vessel wash was compared to the annualised energy of wind waves, Pattiaratchi and Hegge (1990). Provided the introduced wash is a small fraction of the background energy, changes should be minimal.

This is not unreasonable, provided the components of the vessel wash energy (i.e., height and period) are within the bounds of the wind wave components. Shorelines may withstand an increase in energy from waves of the same kind, but not from waves of a different kind. Simply comparing summated energy levels may not produce a fair comparison.

Wave power
The peak wave power approach is used in the Danish criteria Kofoed-Hansen and Mikkelson (1997). The Danish Hydraulics Institute is a world-renowned coastal engineering organisation and thus approaches its wash problems as such.

An equation was developed which contained two constants that relate to wave height and period respectively. These constants were derived from analysing the wash from conventional ferries operating in the region, where there were no negative effects reported. These constants are therefore site-specific.

The principal aim of this power criterion is to limit wave shoaling and not foreshore erosion. It should therefore be clear from a coastal engineering perspective that peak wave power will best produce the correct shoaling relationship, as wave shoaling is a function of the inverse square root of the wave group velocity and therefore the inverse square root of wave period. This power-based equation and its constants are therefore only useful in specific applications.

Since 2005, all new commercial vessels operating on the lower Gordon River have been assessed against a similar peak wave power equation. In this case, the two site-specific constants in the equation were defined following significant analysis of full scale on-site vessel wave wake and bank erosion data, Bradbury (2005).

Wave energy

Of the more advanced measures, wave energy has gained the greatest acceptance in the profession. It is certainly the most widely quoted potential wash indicator. Wave energy is defined in Equation 2.

Both wave power and wave energy have been shown to be useful indicators of erosion potential. Power is a measure of the wave energy expended over a given time whereas energy is a simple measure of the energy content of a particular wave. Wave power is a parameter commonly used in coastal engineering where coastal processes occur over long periods and therefore may be better characterised by a time-based parameter such as power. However, wave energy is often used when assessing the wash from a passing vessel as this can define a discrete event that has a definite start and finish, compared to naturally occurring waves such as wind waves that are often better analysed over time. Similarly, tidal events are often used as an argument against energy-based criteria, as tides are both long in period and can be high in range, yet may cause little erosion relative to their "energy". It is reasonable to argue that tidal shorelines already have features to resist tidal impacts (tides are cyclical and not incidental), and tidal flows do not have a wave-like structure.

There is enough evidence from published and commercial-in-confidence reports to suggest that a wave energy approach is the most appropriate basis for wave wake criteria at this point in time. Over recent years the Washington State Ferries criteria has become a de facto standard that is regularly applied for projects in the United States and is rapidly being adopted as the industry standard. As previously noted, care must be taken when applying this particular criteria as its development was very site specific.

3.6 Multiple criteria approach

The Washington State Ferries criteria combining wave height and wave energy is an example of the application of multiple criteria. However, the combination of maximum wave energy and maximum wave height may not be an ideal combination.

As a vessel's speed increases, the maximum wave height increases, peaking at a length Froude number of approximately 0.5. Beyond this, the vessel is considered to be operating in the high-speed range and maximum wave height decreases. Wave period, taken as the period of the highest wave (which may not necessarily be the maximum wave period), exhibits similar behaviour, except that period remains approximately constant in the high-speed range. Maximum wave energy, being a function of the square of both height and period, is therefore governed by the gradual decrease in high-speed wave height, so exhibits an even faster rate of decline at high-speed.

Multiple criteria relying on energy and height may be compromised by the decrease in both parameters in the high-speed range, which suggests that it may be possible to design a vessel with a speed/length relationship sufficiently high such that its maximum wave height and energy fall below the criteria limits.

In contrast, the multiple criteria proposed by Macfarlane and Cox (2004) utilises maximum wave energy and period of the highest wave, with period expressed as vessel waterline length. It can be shown that the period of the highest wave of a vessel travelling at high speeds can be reduced to a function of vessel waterline length. Introducing this period criterion ensures that long vessels, which are capable of producing low wave height and energy, are not allowed to operate in sensitive areas where their long-period waves could cause environmental problems.

It is quite probable that one set of criteria may not adequately cover all possible waterway types.

4.0 CONCLUSIONS

In order to develop wave wake criteria, certain simplified parameters that characterise a vessel's wash potential must be used; otherwise the total range of variables may prove too large to be of practical use.

Of the available measures, the height and corresponding period of the highest wave in a vessel's wash, measured at a specific lateral distance from the sailing line, are the most appealing of parameters as they consistently reflect wave wake intensity and can be statistically represented. Traditional single parameters such as wave height alone appear to be inconsistent and misleading as to their potential to cause waterway damage. Multiple parameters that correctly account for the relative environmental influences of wave height and wave period are increasingly gaining popularity.

From a review of wave wake criteria in use worldwide, older methods that relied on wave height alone are being superseded by measures involving wave power or energy. This reflects the growing understanding that wave period and not just height is a major determinant of wave wake severity.

It is also quite possible that wave power or energy alone may be an incomplete measure and combinations of energy/power, wave height and wave period criteria may be necessary to completely cover all possible wave wake impacts.

REFERENCES

Bradbury, J., (2005). *Revised wave wake criteria for vessel operation on the lower Gordon River*, Unpublished Nature Conservation Branch Report, Tasmanian Department of Primary Industry, Water & Environment.

Cox, G., (2000). *Sex, lies and wave wake*, Proceedings of the International Conference on the Hydrodynamics of High Speed Craft – Wake Wash and Motions Control, Royal Institution of Naval Architects, London, United Kingdom.

Dean R.G. and Dalrymple, R.A., (1991). *Water wave mechanics for engineers and scientists*, Advanced Series on Ocean Engineering - Volume 2, World Scientific.

Glamore, W.C., Hudson, R. and Cox, R.J., (2005). *Measurement and analysis of boat wake waves: Management implications*, Proceedings of the 17th Australasian Coastal and Ocean Engineering Conference, Adelaide.

Kofoed-Hansen, H. and Mikkelson, A.C., (1997). *Wake wash from fast ferries in Denmark*, Proceedings of the Fourth International Conference on Fast Sea Transportation, Sydney, Australia, pp. 471-478.

Lesleighter, E.J., (1964). *Hawkesbury River - The effect of speedboat activities on bank erosion*. NSW Department of Public Works, Report No. 106, Sydney, Australia.

Macfarlane, G.J., (2002). *The measurement and assessment of sub-critical vessel generated waves*, Master of Philosophy Thesis, Australian Maritime College.

Macfarlane, G.J. and Cox, G., (2004). *The development of vessel wave wake criteria for the Noosa and Brisbane Rivers in Southeast Queensland*, Fifth International Conference on Coastal Environment – Environmental Problems in Coastal Regions, Alicante, Spain.

Macfarlane, G.J. and Cox, G., (2007). *An introduction to the development of rational criteria for assessing vessel wash within sheltered waterways*, IMarEST Journal of Marine Design and Operations, Part BII, 2007.

Macfarlane, G.J. and Renilson, M.R., (1999). *Wave wake – A rational method for assessment*. Proceedings of the International Conference on Coastal Ships and Inland Waterways, Royal Institution of Naval Architects, London, United Kingdom.

Pattiaratchi, C. and Hegge, B., (1990). *Impact of ferry and large vessel traffic on Swan River foreshore erosion*, Centre for Water Research Report WP 452CP, University of Western Australia.

U.S. Army Coastal Engineering Research Center, (1984). *Shore protection manual*, Volumes I and II, U.S. Government Printing Office, Washington, DC.

NOMENCLATURE

E	Energy per metre of crest length [Jm-1]
Fr	Length Froude number [v(gL)-1/2]
Fr_h	Depth Froude number [v(gh)-1/2]
h	Water depth [m]
H	Wave height [m]
L	Waterline length [m]
T	Wave period [s]
v	Vessel speed [ms-1]
g	Acceleration due to gravity [9.81ms-2]
ρ	Density of water [kgm-3]

Hydrodynamic Design of High Speed Multi-Hulls and Motion-Control Devices Using a Panel Method

Dmitry Sadovnikov
Austal, Australia

ABSTRACT

Minimization of hull resistance and ship motions are important problems for the design of modern vessels. Resistance calculations by CFD, based on RANS equations, are time consuming and need much attention to grid generation, turbulence model selection, calculation of the free surface and dynamic trim. That is why simpler panel methods are still in use because they can give a quick result on the wave-making resistance and pressure distribution.

A panel code has been used at Austal for optimization of multi-hulls for several years. For this investigation, panels were distributed on the hull and free surface. The final wave pattern, running trim and sinkage of the hull were determined in an iterative procedure. Solutions for the nonlinear boundary condition were found in most cases. The linear boundary condition was used instead where the diverging solution took place usually due to a pronounced stern breaking wave. A developed method for correlation between the calculated wave-making resistance and the residual resistance from tank testing or trials was used to improve the accuracy of the predicted resistance.

The panel code was also used for hydrodynamic optimization of motion control foils and fins. The calculated pressure distribution gave an indication of the cavitation inception for compared foils with different profiles and plan forms. The study showed which standard families of profiles provided highest cavitation-free lift versus anticipated angles of attack for foils with flaps and rotatable foils.

1. INTRODUCTION

Accurate prediction of the hull resistance, estimation of the achievable speed and optimisation of the hull form for minimising the resistance are crucial points in the design of every new vessel because structural design of the hull cannot be started before the final hull lines are developed. Hydrodynamic design of the ship hull usually needs to be completed within a short period of time, thus speeding up the vessel delivery and allowing more time for the detailed design and construction of the vessel.

Calculations of the resistance using contemporary CFD methods solving RANS equations may take a long time and not give final certainty of the speed estimate if they were not correlated to existing data for similar hull shapes tested in a towing tank or full-scale. RANS-based computations become more complicated for high-speed vessels because the dynamic trim and sinkage need to be found.

Simple panel methods utilising the potential theory of inviscid fluid can be successfully used for the hydrodynamic design and optimisation of slender hull shapes including catamarans and trimarans. These methods are intended for calculations of the wave-making resistance.

The residual resistance of the vessel can be found from computed wave-making resistance applying correlation factors derived from appropriate data on tank testing of scaled models or sea trials. Commercially available software Shipflow® has been used at Austal for calculation of the wave-making resistance and optimization of multi-hulls for several years.

Three recent Austal fast ferries with numerically optimized hulls shapes are presented in Figure 1 through Figure 3. The length of these catamarans varied from 47 to 88 meters and the operational speed ranged from 40 to 50 knots. The largest Austal vessel with numerically optimized hull shape was a 127m trimaran described in paper [1].

Figure 1. Austal 88-metre 38-knot passenger-vehicle ferry

Figure 2. Austal 47-metre 42-knot passenger ferry

Figure 3. Austal 65-metre 50-knot passenger-vehicle ferry

Fast ferries cannot be operated at high speed in waves unless efficient motion-control systems are fitted. Austal have developed a series of standard T-foils and roll fins which have been used on most of Austal vessels over the last ten years. An example of a T-foil is shown in Figure 4. Rotatable foils/fins or fixed foils/fins with deflectable flaps can be used to control the generated lift force which may be limited by cavitation at high speeds. In the present study, the panel code used for the wave-making resistance calculations was also applied to hydrofoils for predicting the cavitation onset and finding optimum configurations at various inflow conditions.

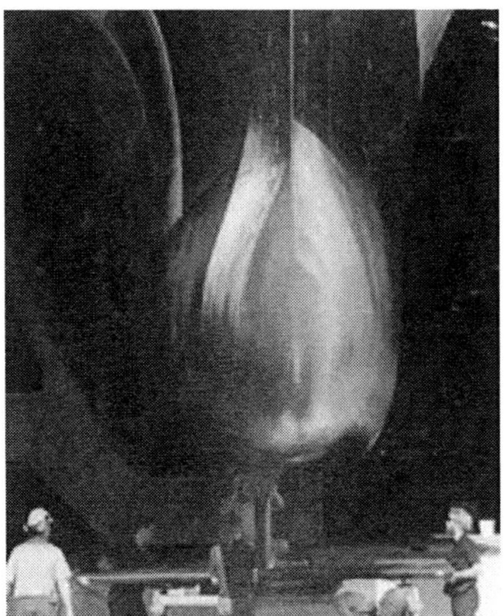

Figure 4. Austal T-foil for motion control

2. PREDICTION OF RESISTANCE AND OPTIMIZATION OF HULL FORM FOR MULTI-HULL VESSELS

2.1 General Procedure

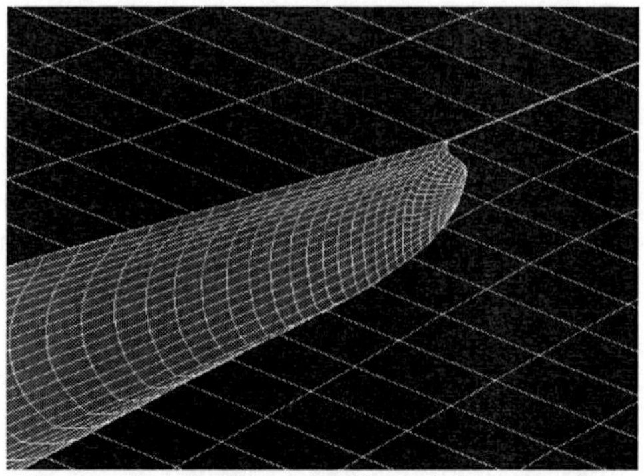

Figure 5. Mesh on hull and free surface

Prediction of resistance started with the calculation of the wave-making resistance by the panel code. Meshes on the ship hull and the free surface depended on the hull geometry and Froude number. A finer mesh was required in areas of a rapid geometry change, such as bows, bilges with high curvature, stern wedges, etc. An example of the mesh on the bulbous bow of a catamaran is presented in Figure 5. More panels on the free surface were used at lower Froude numbers when the wave length was shorter compared to the hull length, while the number of free surface panels was reduced at higher Froude numbers. Panels were distributed over one half of the ship hull and the free surface due to the symmetry of the flow about the vessel centre-plane. Shipflow® includes a procedure which simulates separation of flow at the transom by introducing a group of panes extending aft from the transom. In this way the transom is effectively dry (at atmospheric pressure) and there is no requirement for an additional hydrostatic pressure correction.

Computed wave-making resistance is usually lower than the actual residual resistance of the vessel. Based on the classic formulation of the residual resistance as the difference between the total bare hull resistance and its viscous resistance, some other resistance components shall be added to the wave-making resistance. These components are typical for high-speed vessels and include resistance of sprays, spray rails (where fitted) and the induced drag due to the dynamic hull lift associated with partial planing at high Froude numbers.

Potential flow computations by panel methods do not take into account some additional effects, such as the boundary layer on the ship hull and wave breaking. Moreover, the static wetted surface is usually used in the resistance calculations and therefore some inaccuracy in the viscous resistance prediction may occur due to the difference between the actual dynamic and the static wetted surface areas and also inaccuracy of the friction formula. The corresponding difference between the actual and predicted viscous resistance was included in the residual resistance in the present method. The boundary layer on the ship hull obviously increases the virtual displacement of the vessel and therefore its wave-making resistance. Hence, the boundary layer on the ship hull increases the residual resistance in all cases. On the other hand, the additional components of the residual resistance due to wave breaking and the difference in the wetted surface areas may be either positive or negative because the panel code predicts a large-amplitude non-breaking wave instead of the breaking wave and the dynamic wetted surface area can exceed or be below the static one. Components of the residual resistance adopted in the current resistance prediction method are presented in Figure 6.

The sequence of calculations of the vessel total resistance and speed is shown in Figure 7. The residual resistance was calculated by multiplying the computed wave-making resistance by correlation factors which are considered in the following paragraph. The total resistance was found as the sum of the residual resistance, viscous resistance, air drag and appendage drag. The viscous resistance was calculated as follows:

$$R_v = 0.5\rho v^2 A_{st}\left(C_f(1+k)+C_a\right) \tag{1}$$

where: ρ – water density; v – vessel speed; A_{st} – static wetted surface area; $(1+k)$ – form factor; C_f – friction coefficient calculated by the ITTC formula; C_a – correlation allowance.

Finally, the vessel speed was estimated as the intersection of the total resistance curve and the propulsor thrust curve. The latter included the hull-propulsor interaction in the form of the wake and thrust deduction factors.

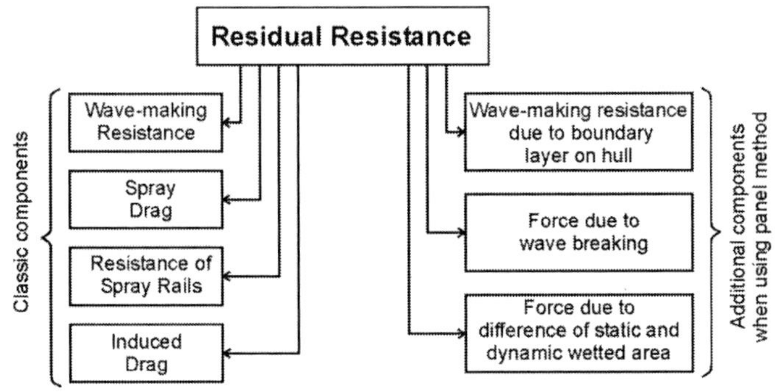

Figure 6. Components of residual resistance

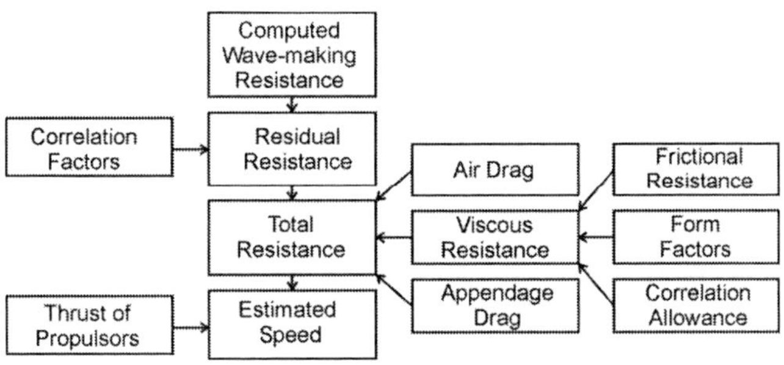

Figure 7. Scheme of resistance calculation and speed prediction

2.2 Correlation of Computed Wave-making Resistance

As was shown in the previous paragraph, the computed wave-making resistance R_w does not completely represent the residual resistance R_r. That is why correlation factors k_c were introduced to calculate the residual resistance in the following form:

$$R_r = R_w \cdot k_c \tag{2}$$

The correlation factors k_c indirectly took into account the residual resistance components depicted in Figure 6 except the wave-making resistance. Moreover, k_c factors allowed for any inaccuracies in the wave-making resistance computed by the panel method.

The correlation factors can be found by comparing the computed wave-making resistance with the residual resistance from tank testing or full-scale trials of a similar vessel (prototype hull). The prototype hull and the hull in question ideally have similar Fn, $L/V^{1/3}$, L/B, L/T, L/T_{tr}, L/S, where Fn – Froude number; L – waterline length; V – submerged volume; B – waterline beam; T – hull draft; T_{tr} – transom immersion, S – separation between hulls of multi-hull vessels.

The correlation factor is therefore a function of the above dimensionless parameters and may also depend on the running trim angle of the vessel. Values of the correlation factors are typically between 1 and 2, increasing at higher Fn and $L/V^{1/3}$.

2.3 Wave Breaking and Convergence of Computations

The free surface around high-speed multi-hull vessels often contains areas of wave breaking which is in most cases caused by the interaction between diverging waves from adjacent hulls. An example of a typical computed wave pattern around a fast catamaran is presented in Figure 8. Wave breaking may occur at the flow centre-line between the demi-hulls and behind the vessel. The aft central breaking wave crest moves forward and two new areas of wave breaking right behind demi-hulls may appear at lower Froude numbers as shown in Figure 9. Wave breaking was more pronounced for deeper transom immersion, higher running trim angles and the resistance hump speeds.

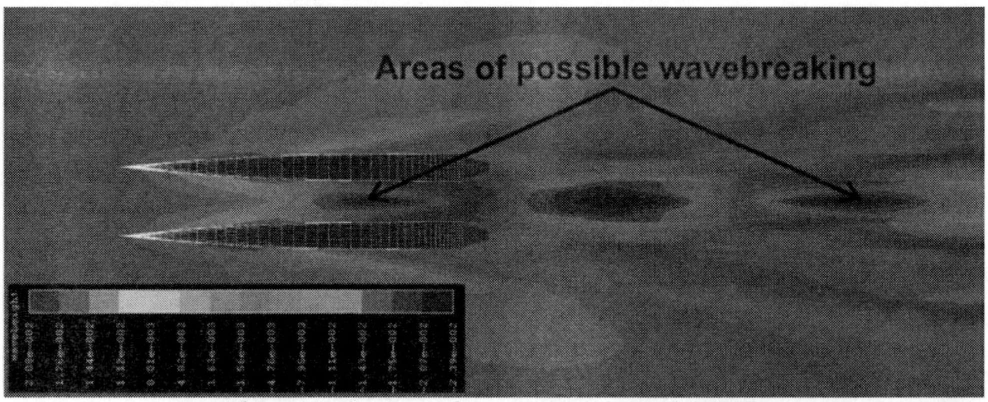

Figure 8. Computed wave pattern at Fn=1.03

The above wave breaking areas contain the spilling-type breakers where the wave energy dissipates, resulting in reduced heights of the breaking wave itself and non-breaking waves downstream. Despite some attempts to apply simplified physical models of the spilling breaker and the potential flow methods for computing 2D free surfaces with wave breaking were undertaken by the author in [2,3], the simulation of breaking waves in the framework of practical 3D potential panel methods seems to be complicated or unachievable.

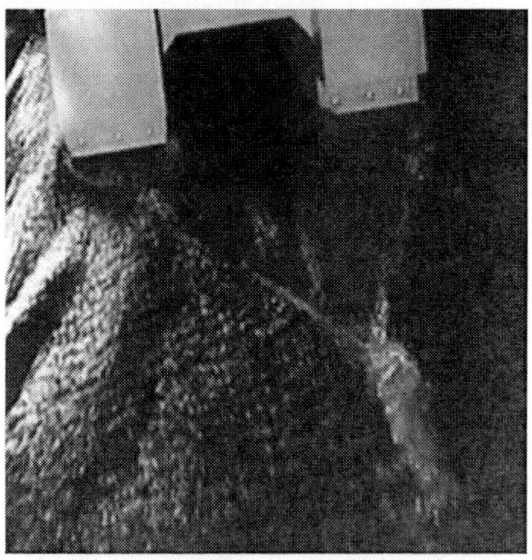
Figure 9. Wave breaking in towing tank at Fn=0.55

Wave breaking was not simulated in the present study, resulting in some cases in high-amplitude non-breaking waves causing difficulties with convergence of nonlinear computations. Coarser free-surface meshes and higher relaxation factors can be used for improving convergence of nonlinear computations. However, the nonlinear solution with reasonable sizes of free surface panels was not found in some cases and the linear boundary condition on the free surface was used instead, resulting in convergence of computations in all cases. However, correlation factors k_c were found to be higher for the linear computations. As a rule, convergence of nonlinear computations was better for optimum hull forms with minimised wave-making resistance.

3. CAVITATIONAL CHARACTERISTICS OF MOTION-CONTROL UNITS

3.1 Computation of Pressure Distribution on Hydrofoils

Pressure distribution on the motion control foils was calculated using Shipflow® panel code applied to lifting surfaces with the linear model of the trailing vortex sheet. Two lift groups of panels were input for the foil, representing its upper and lower surfaces. These groups had the same number of stations and points, i.e. the number of panels in the longitudinal and transverse directions on the upper and lower surfaces were identical. The trailing vortex sheet started at the Kutta-points near the trailing edges and formed a flat plane parallel to the incoming flow. The method can be called ½ -nonlinear because lifting panels were located on the actual surface of the foil (not on the chord surface) and the simplified linear model of the trailing vortex sheet was used.

Lift and induced drag were calculated using two approaches. The first one implied integration of pressures on the foil surfaces, while the second approach used parameters of the trailing vortex sheet in the Trefftz-plane located far behind the foil. The difference in lift and induced drag for the two approaches may reach several percent because some lift can appear on the trailing vortex sheet which is linear and does not follow the actual streamlines. In the present study, the preference was given to the first approach because pressure distributions were of particular interest.

It was found that the higher order panels which allowed for the actual curvature of the foil surface may cause unwanted peaks ("saw") in the pressure distribution especially for foils with deflected flaps. Therefore the first order panels, which are flat and with constant distribution of hydrodynamic singularities, were used in the calculations. Unfortunately, the utilised panel method was unable to take into account symmetry of lifting surfaces and therefore both port and starboard sides of symmetric foils were calculated, thus increasing the required computer memory by 4 times. It was found that the optimum number of lifting panels on a foil was between 7,000 and 9,000. Lower number of panels may be insufficient for finding peaks in pressure distributions.

An example of a mesh on a foil without flaps is presented in Figure 10. The vortex strip extended by several chord lengths downstream, the number of panels across half of the foil span was 31 and 120 panels were located on its

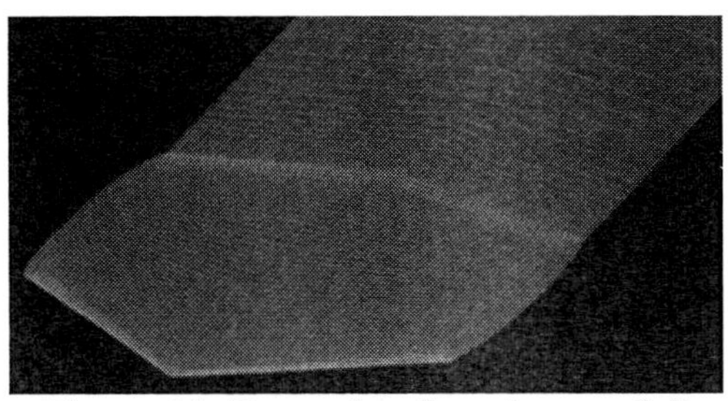

Figure 10. Example of meshes on foil and trailing vortex plane

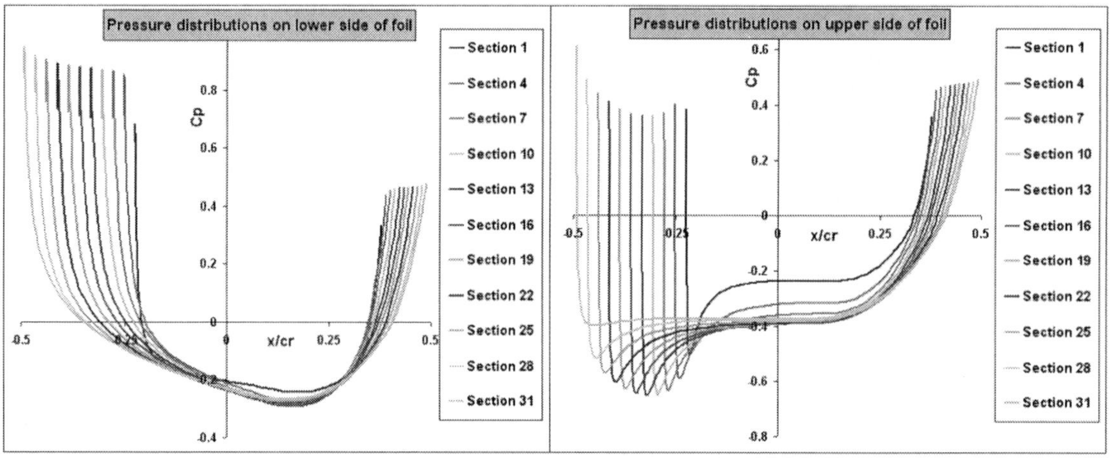

Figure 11. Pressure distribution on 3D foil

each transverse section (profile). Up to 140 panels were used on every transverse section of foils with deflected flaps. Computed pressure distribution at some transverse sections of the foil with NACA-16015 profiles and flaps in the neutral position at the angle of attack α=4° are presented in Figure 11, while the cavitation buckets for the foil with various flap deflections were plotted in Figure 12.

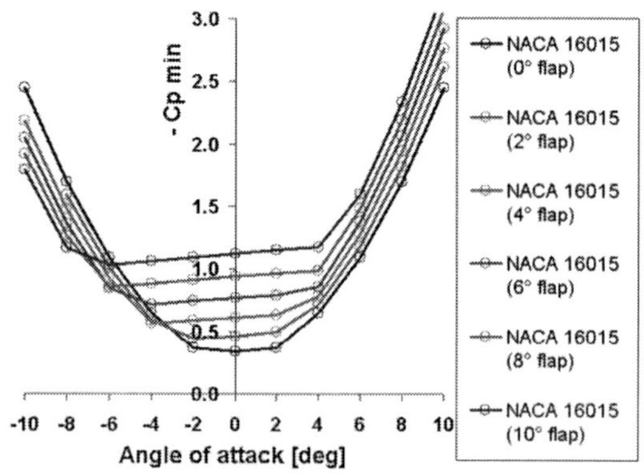

Figure 12. Cavitation buckets for 3D foil

Sections 1 and 31 were the closest to the tip and root of the foil accordingly. The pressure coefficient C_p was determined as follows:

$$C_p = \frac{p - p_\infty}{0.5 \rho v_\infty^2} = 1 - \left(\frac{v}{v_\infty} \right)^2 \tag{3}$$

where: p – local pressure on the foil; p_∞ – flow pressure far upstream; ρ – fluid density; v – local velocity on the foil; v – flow speed far upstream.

The value of C_{pmin}, which is the minimum value of C_p on the foil surface, characterised the cavitation inception on the foil. Cavitation appeared when $-C_{pmin}$ equalled to the flow cavitation number σ formulated as follows:

$$\sigma = \frac{p_\infty - p_v}{0.5 \rho v_\infty^2} \tag{4}$$

where: p_v – saturated vapour pressure.

Lift and cavitation of the foil may be affected by the ship hull if they are located close to each other. In this case, the ship hull above the foil can be simulated as shown in Figure 13.

Computed C_{pmin} values versus flap angles were plotted in Figure 14 for the isolated foil and the foil located at the distance h=0.3c_r under the hull with U-shaped sections, where c_r is the maximum chord length of the foil. There was an insignificant hull effect on minimum pressures on the foil when flaps were deflected up, while the hull presence resulted in lower minimum pressures when the flaps were deflected down. The hull influence became much less when the hull-to-foil clearance was increased by extra 0.5c_r.

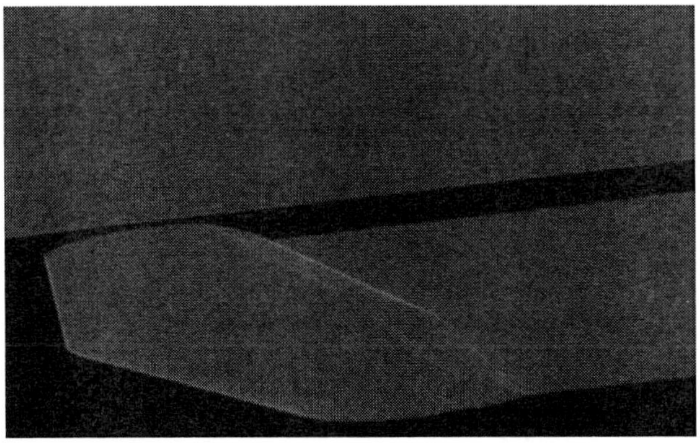

Figure 13. Foil under ship hull

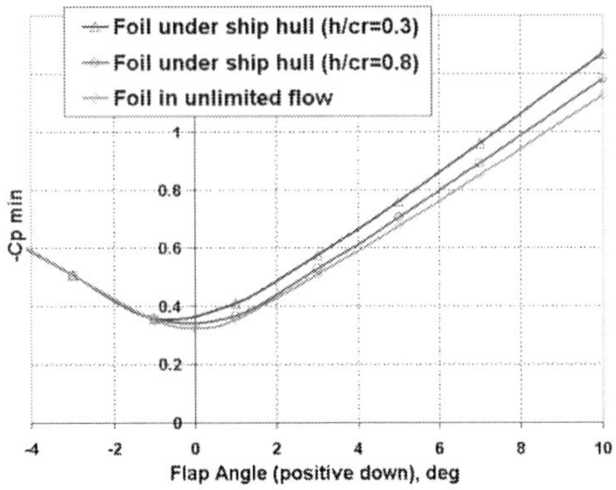

Figure 14. Effect of ship hull on hydrofoil cavitation inception

3.2 Effect of Foil Plan Form on Cavitation Inception

Foil plan form influences the pressure distribution and cavitational characteristics. Foils with higher sweep usually have superior cavitational characteristics. Pressure distributions for four plan forms of rotatable foils, presented in Figure 15, were calculated and compared. The foils had NACA-16015 profiles and the same projected area and span. Calculated cavitation characteristics of four foils at the angle of attack α=8° are presented in Figure 16. A significant difference in C_{pmin} values took place when the lift coefficients was above 0.2 where high-sweep foils had superior cavitational characteristics in comparison with the rectangular plan form. Foil 3 with extreme sweep had the best cavitational performance, while foils 1 and 2 shared similar $C_{p\ min}$ values and were much better in respect of cavitation inception than the rectangular foil 4. Minimum pressures took place in the root section on the rectangular foil 4 and shifted towards tips for sweep foils 1, 2 and 3. The dependence of the lift coefficient versus the angle of attack was found to be very similar for the four compared foils, thus demonstrating a minor effect of the plan form on the lift force of foils with the same aspect ratio prior to cavitation inception.

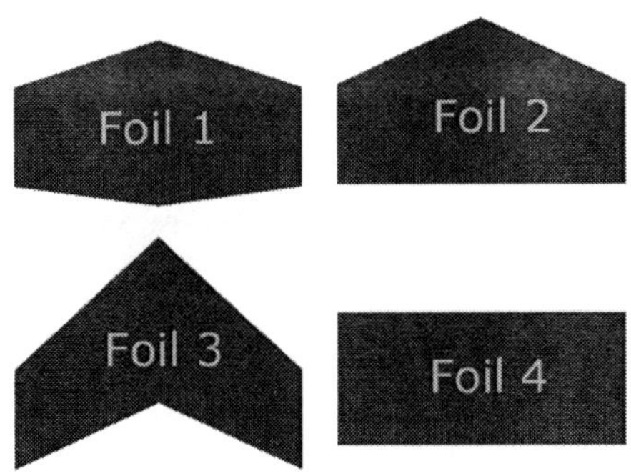

Figure15. Compared plan forms

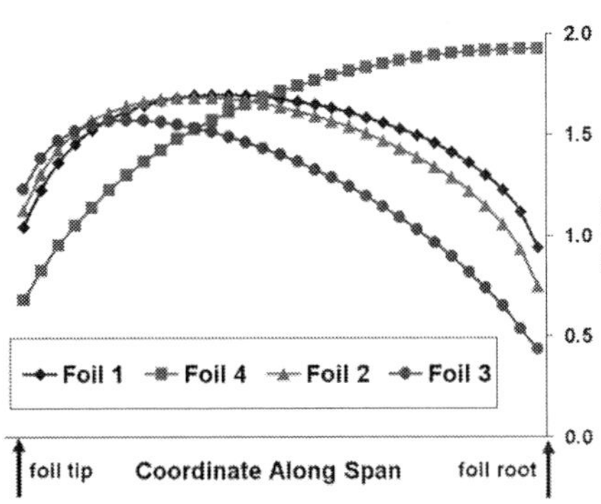

Figure 16. Cavitational characteristics of foils with different plan forms

3.3 Inflow Parameters of Motion-control Foils

Inflow angles of motion-control units may vary in a wide range due to pitch, heave and roll of the vessel and variable flow velocity in waves. The realistic estimation of operational inflow angles is important for determination of cavitational characteristics and design of ride-control units. A method was developed at Austal for calculation of the inflow angles and cavitation numbers for ride-control units in regular waves. Software for the ship motion prediction Veres® was used to calculate response-amplitude operators and phases of pitch, heave and heave velocity of the vessel with the operational ride-control system. Flow velocities in waves were calculated using the linear wave theory. Finally, foil immersion under the free surface, inflow angle and cavitation number were found for assumed height and period of regular waves.

An example of calculations for a catamaran with a couple of T-foils in regular head seas is presented in Figure 17. The vessel speed, wave height and wave period were 30 knots, 2.5m and 8 seconds respectively. All parameters corresponded to the foil location and the phase angles were measured between the wave crests and the foils. In this example, the foil inflow angles varied from -6° to about +7° and the cavitation number σ was about 0.9 at peak values of the inflow angle. The vessel's bow moved downwards when the maximum inflow angle of +7° occurred, while the minimum inflow angle of -6° took place when the bow moved upwards.

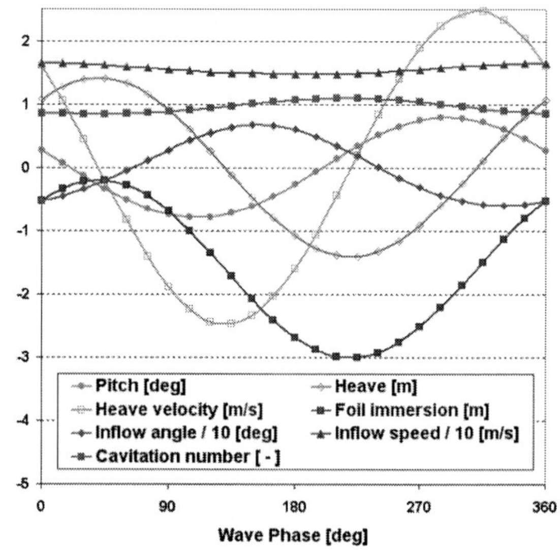

Figure 17. Ship motions and T-foil inflow parameters

3.4 Comparison of Cavitation Inception on Foils with Flaps and Rotatable Foils

Computed cavitational characteristics of the foil without strut, depicted in Figure 10, with flaps and the rotatable option of this foil were used for the comparison between the two options. Two families of symmetrical profiles (NACA-00xx and NACA-160xx) with the thickness-to-chord ratios of 9, 12 and 15% were used in the computations. These two families of profiles were selected because they represented very different profile shapes, NACA-00xx and NACA-160xx having the maximum thickness closer to the leading edge and the middle of the chord respectively. Geometry comparison between the two profiles at 15% relative thickness is presented in Figure 18.

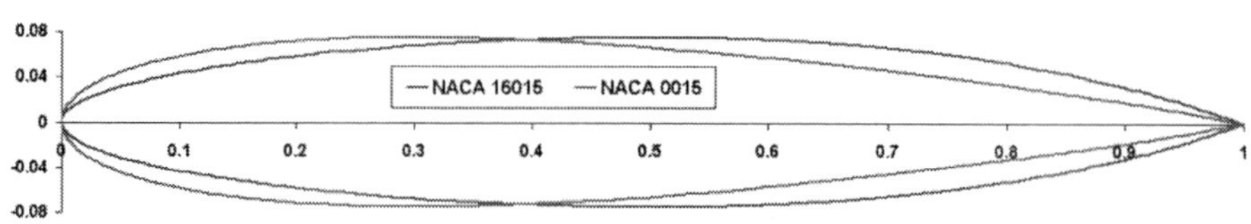

Figure 18. Comparison of NACA-0015 and NACA-16015 profiles

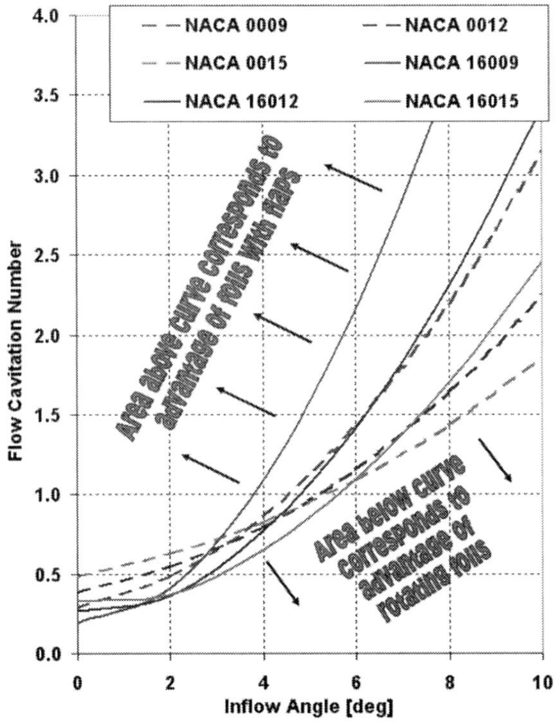

Figure 19. Comparison of cavitation inception on foils with flaps and rotatable foils

Curves plotted in Figure 19 corresponded to equal C_{pmin} values for the rotatable option and the foil with flaps at the same angles of attack. In fact, every curve in Figure 19 represented cavitation inception on the rotatable foil if the angle of attack was equal to the inflow angle. If a peak value of the inflow angle exceeds the maximum cavitation-free angle of attack of the rotatable foil, the foil can be rotated to keep the actual angle of attack within the cavitation-free limit. In the same case, the fixed foil with flaps will experience cavitation of the leading edge because the actual angle of attack of this foil is always equal to the inflow angle. On the other hand, the fixed foil with flaps has advantages when the peak inflow angles are below a critical value corresponding to cavitation inception at the neutral position of the flaps because a higher increase of the cavitation-free lift can be achieved by deflection of the flaps (i.e. by making the foil cambered) than by rotating the non-cambered foil.

4. CONCLUSIONS

Panel methods correlated to tank tests of scaled models and sea trials can be successfully used for predicting resistance and speed of fast multi-hull vessels. However, using the computed wave-making resistance without correlation may lead to significant underestimation of resistance. Panel codes applied to lifting surfaces are helpful for comparison of lift and cavitation inception of different options of motion-control units and prediction of the hull effect on closely located foils and roll fins. The rotatable foil option has an advantage over foils with flaps at large inflow angles and low flow cavitation numbers.

5. REFERENCES

1. Armstrong T. and Holden K., (2003), "A New Generation of Large Fast Ferry – from Concept to Contract Reality", Proceedings of the 7th International Conference on Fast Sea Transportation FAST'2003, Vol. 1, Ischia.

2. Sadovnikov D. and Trincas G., (1998), "Nonlinear Simulation of Breaking Waves with Spilling Breakers by a Boundary Integral Method", Proceedings of the 22nd Symposium on Naval Hydrodynamics, Washington D.C.

3. Sadovnikov D. and Trincas G., (1998), "Numerical Simulation of Breaking Waves Past a Vortex", Proceedings of the 2nd International Shipbuilding Conference ISC'98, Vol. B, St. Petersburg

Naval Ship Assurance – A New Approach

James Buckley
BMT Defence Services Ltd, UK
James.Buckley@nsass.org.uk
Supporting Authors:
Paul James / Ben Cuckson
Lloyds Register EMEA, UK
Ben Cuckson
Lloyds Register EMEA, UK
Steve Marshall
Ministry of Defence, UK

UK APPROACH TO NAVAL SHIP ASSURANCE

Introduction

It is widely recognized that Navies around the world are subject to increasing pressure from their Governments to reduce the cost and establish a more consistent approach to the design, operation and assurance of naval fleets.

The UK MoD (MoD), in conjunction with Class and industrial stakeholders has been investigating, and increasingly adopting, a more commercially based approach for the assurance of naval ship acquisition and operation in an effort to reduce costs, whilst maintaining safety, environmental, and military capability. The approach being taken in the UK is based on a Naval Ship Assurance (NSA) framework. Significant work has been undertaken towards development of the NSA framework.

This paper will begin by describing the UK MoD's approach to NSA and how it is working to define a framework for its implementation. The paper concludes by outlining some potential adaptations in order to suit other Navies' ship assurance requirements or other applications.

What is Naval Assurance?[1]

Assurance can be defined as an overall process that provides confidence from an independent party that compliance with key requirements or standards is achieved for a platform through life. It will generally comprise of a series of assurance events or activities, which demonstrates compliance with legislative requirements or key risks for safety, (e.g. lifesaving), ship function (e.g. electrical generation), environmental, capability (e.g. specific weapon systems), commercial or programme risk. Assurance should be established from concept through to disposal through design approval, equipment approval, build survey, acceptance into service and through life survey.

Acceptance can be defined as a specific process that provides confidence contractual requirements have been achieved for a specific activity. Historically is has been associated with delivery of the ship to obtain the owners agreement that the contract requirements have been achieved. More recently it has been extended to cover key project milestones such as concept, design and manufacture.

Development of the UK MoD Approach to Naval Ship Assurance[1]

The MoD has traditionally developed and maintained its own comprehensive set of naval standards which were implemented by contractors and assured by naval overseers. These standards were maintained through a combination of feedback from operation and from research and development.

With reduced resources for the maintenance of these naval standards, the MoD has made increasing use of civilian standards on naval ships However the selective use of merchant standards, if applied without due consideration, can result in duplication and gaps (e.g. landing craft davits on HMS OCEAN, National Audit Office Report 864/1993).

In 1998 the MoD recognised the need to re-align its arrangements for ship safety leading to the recommendation that MoD practice be harmonised with civilian practice; a key aspect of which was to form Naval Authorities that would equate in purpose and approach to Statutory Authorities for merchant shipping.

Recognising that the MoD was no longer resourced to rely solely on developing and maintain its own standards, a project was initiated in 1995 with Lloyd's Register to develop rules for naval ships. The rules were first published in 1999 and have continued to develop with feedback of experience gained from their application.

NAVAL SHIP ASSURANCE FRAMEWORK

The benefits of a Naval Ship Assurance Framework[2]

The benefits of the proposed framework are:

Performance: benchmarked against best merchant practice, improved accuracy of application, naval requirements clearly communicated;

Cost: elimination of overlaps and gaps between purely merchant practice and traditional naval practice drives out risk and supports early acceptance, and avoids costly modification to ships once in-service; and

Time: a much simplified framework that is dependent on far fewer key publications reduces the time required for MoD to select options, industry to apply in design and construction, and assurance organisations to review and confirm compliance; leading directly to further cost savings.

The Framework

The framework brings together the elements of Naval Assurance into a coherent structure to enable the user to identify how to apply the requirements, codes and standards and is pertinent to both submarines and surface ships.

This Naval Assurance Framework is for the overall assurance of Through Life Capability Management, encompassing: Safety, Environment, Operability, Supportability, Affordability and Programme.

Tailoring General Requirements to Ship Specifications

With reference to Figure 1 below, the framework consists of a central cone that makes the link between the capability requested by the customer and the capability delivered in the platform. At the top of the central cone is the customer requirement input - i.e. requirements from Director Equipment Capability (DEC). Working down the cone, this is translated, through the procurement process of preparing general naval requirements and general system specifications, to produce detailed system specifications that can be contracted against.

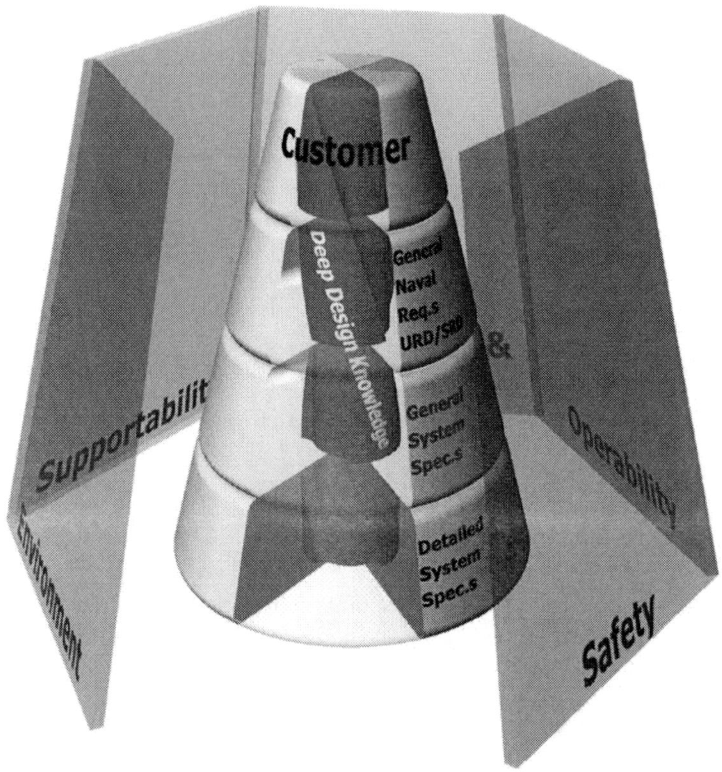

Figure 1 - Naval Ship Assurance Framework

The process for developing the General Naval Requirements into a set of General System Specifications and Detailed System Specifications is aided by the General Naval Specification document (GNS).

A key benefit of using the General Naval Specification as a template is the incorporation of learning from experience and so avoiding costly mistakes that are a repeat of history.

The General Naval Specification supports the layers of the specification process, so providing the ability to exit at different levels appropriate to the procurement process and the shipbuilding contractual arrangement.

With reference to the central cone, for the 1st layer down from the customer the General Naval Specification document provides a template for a project team to develop their own tailored and specific User Requirement and System Requirement documents. Use of this template aids the project team to ensure that the output is comprehensive and coherent. If allied across a Navy it will provide consistency and provides a place to store lessons learned from experience.

In the 2nd and 3rd layers, the General Naval Specification document provides a template for a project team to take the User and System Requirements to develop system specifications, which are then developed into detailed system specifications that could form the basis of contracting. Implicit in these levels are the development of acceptance criteria against the requirements.

At the core of the central cone is the backbone of Deep Design Knowledge. This is a knowledge base that aids the generation of the capability requirements, through development of the User Requirements and System Requirements and down to final acceptance of the ship / submarine into service.

A key part of this design knowledge base is the Maritime Platform Characteristics (MPCs). These enable the 'sponsor' to define the 'attributes' that they expect the solution to the requirements set to exhibit. It is not the function of MPCs to tell the project team delivering a new capability how to derive its solution; however, they provide a check list against which any solution could be assured. In addition, MPCs contain a wealth of information that can be accessed by the project team to help ensure they are on the right track when working their way towards the specification tailored for that platform.

Another example of the content of this design knowledge base is ship specific design knowledge which aids the project team in developing the specification; for example the Landing Craft Design Guide, Guide to the Naval Ship Code, Stability for Surface Ship Acquisition publication MAP01-024. Other content might cover systems, functions or stages of the procurement process.

Tailoring Lenses of the Framework

Figure 2 - Applying the Safety Lens

Around the central cone (in Figure 1) are lenses defining the codes and standards used throughout the design process. The lenses can be viewed from either direction.

The project team views them from the inside and sees the "regulatory" framework that it has to apply and which impacts on the solution. The regulator views them from the outside and sees the "regulatory" framework reflected in the requirements documents and specifications that have been tailored to provide the necessary assurance for that ship.

The lenses providing the framework are: Safety, Environment, Operability, Supportability, and Affordability & Programme, as shown in Figure 1. These provide a route map for guiding the development of the specification through to the end certification and acceptance, which records attainment of the required standard and so contributes to the overall assurance. Figure 2 shows the safety lens expanded showing Policy, Codes and Standards for surface ships.

The link between the development of customer requirements into detailed system specifications with the "regulatory" framework are key for the defence project team and industry to ensure all bases have been covered and to provide the required documentary evidence of assurance to the top level military customer.

The choice of lenses is coherent with the categories found in both the General Naval Specification template and the design knowledge contained in the Maritime Platform Characteristics. This does not preclude the use of other lenses but provides the initial framework for managing the assurance.

These diagrams depict the concept of the assurance framework. The detail will adapt to the platform under consideration, surface ship or submarine, and the specifics depending on choices made by the project team as they navigate the framework. The framework is comprised of several key document sets, which are at different levels of maturity. A key aim is to raise the maturity level of all documents as we ensure that the framework is comprehensive, coherent and clear.

DEVELOPING AND IMPLEMENTING THE FRAMEWORK

What does it actually mean?

The framework currently provides a high level structure to guide project teams through the assurance process. This structure needs to be developed at the detailed level. It needs to be credible and to recognise the environment in which the project teams are operating and the process for making use of the framework needs to clearly provide benefit balanced with any additional effort required to make use of a structured approach.

What does it look like?

At the most basic level the framework provides a template for defining the assurance requirements for a project by listing subjects and key standards. This template can then be used and populated for a specific project, identifying more detailed requirements. The example below shows how a service requirement is translated into a fatigue requirement which then identifies key standards and assurance activities.

Naval Assurance Framework Template				Project Specific Requirements	
Framework document	**Level**	**Title**	**Frame work standard**	**Requirements**	**Criteria**
General Naval Specification	User Requirement Document	Operational Area	Not Applicable	4000 Days North Atlantic	
	System Requirement Document	Platform	Not Applicable	Fatigue assessment	
	Ship Specification	Structure	Naval Ship Rules	FDA Notation	Classification certificate
	System Specification	Structure Fatigue	FDA Procedures	FDA2 Spectral fatigue analysis FDA3 at 8 detailed locations	Design appraisal Enhanced survey during build CM notation
Through Life Management Plan		Structure Fatigue	Naval Survey Guidance	Survey of CM points	Classification Survey

Table 1 - Implementing the Framework - An Example

Not redefining ship design

The framework is not a redefinition of the ship design process and it is not intending to create new processes but to provide a structure for processes that should already be conducted; for guiding project teams as they collate the evidence for assurance and consider the evidence holistically.

Using the framework

The framework provides an opportunity to map out key requirements for a project and develop a baseline requirements/standards set upon which cost capability discussions can take place. If the framework is used consistently within a Navy from project to project, these decisions can be benchmarked against legacy projects or at the time of a mid life upgrade, against the original design intent.

Some areas of the framework will have mandatory requirements that relate to safety or environmental compliance. It is important that these are clearly identified. They may well differ between navies and will certainly evolve over time in the same way that commercial legislation changes. Tailoring of the mandatory requirements can occur but, generally, equivalent levels of performance will have to be demonstrated.

For other areas, mandatory performance or capability requirements can be specified but a Navy will need to exercise extreme care when doing this, as mandatory capability can drive project cost. It may be better to specify minimum and desired capability.

Many requirements will be optional and need to be negotiated during the development of a project; the framework offers a consistent and familiar basis on which to do so.

For example, the framework has a place holder for a steering system. The framework will identify mandatory requirements for safety e.g. Redundancy and rate of turn (NSR Vol II

Pt6 Ch1) but also performance maintenance and operability requirements (GNS Ref Part 1 - S2329/2330 Manoeuvring and Part 2 - 508 Steering Systems and Stabilisation Systems). A decision has to be made about the type and specification of the steering gear on this particular project and this can be recorded in the framework. Ultimately this will form the specification for the ship builder but it can also be used to identify assurance activities (class equipment certificate), acceptance events (test and trials) and record the decision process.

Ten years later when there is a plan to upgrade the capability of the ship the framework can be a reference for the original design intent. The upgrade can then be assessed as compliant or demonstrated as an improvement on that original intent. In a parallel project the framework can be used to clarify requirements which could be copied or used as a bench mark for a new solution.

This example demonstrates use of the framework at a low level (specification level), but similar principles can be applied at a higher level in the framework (user requirement level)

Applicability to in service

In implementing this approach to assist managing naval assurance it is important that it is of use not only to the new build project team but also to the in-service project team, particularly for those making significant mid life upgrades but also in support of general operation. In essence it needs to provide support to project teams understanding how they navigate their way to collate the necessary assurance as part of their through life capability management in whatever form that management might take place. When in-service experience demonstrates particular maintenance problems the lessons learned can be stored in a guidance document for use within the framework, to benefit future projects. Where there is a failure, due to an unforeseen mode of operation, design rules within the framework can be updated.

NAVAL SHIP ASSURANCE SUPPORT SERVICE - NSASS

Background

BMT and LR have formed an alliance with the MoD to support them in developing, managing and enhancing the framework for naval ship specifications, standards and assurance; the NSA framework.

In addition NSASS will undertake tasks to populate the framework, either through the development of specific standards or contributing to existing standards through established maintenance processes.

The identified need arose from the MoD Sea Systems Group, focused predominantly on safety and the key hazard certificates. The Naval Ship Assurance Support Service has been borne out of this requirement.

Whilst Naval Assurance is significantly wider than the framework for key hazard safety assurance "Naval Assurance is about providing the overall assurance underpinning Through Life Capability Management", the alliance through NSASS provides a route to implement the framework's safety aspects on behalf of the MoD.

The broad remit for NSASS is captured in the following objectives:

Key Objective 1 - Develop a framework for ship safety assurance that is coherent, comprehensive and clear.

Key Objective 2 - Continuously monitor and improve the framework.

Key Objective 3 - Make the framework accessible to MoD regulatory bodies, MoD Integrated Project Teams and Industry and provide guidance for effective navigation.

Within this is the role of influencing the key document sets that populate the framework. This includes the General Naval Specification (GNS), Naval Ship Code (NSC), LR Naval Ship Rules (LR NSR) and Naval Standards.

There are already well defined mechanisms for the maintenance and development of these document sets and the role of NSASS is not to duplicate or contravene these but to provide input through established processes based on lessons learned from application of the framework and standards in a project. NSASS has a role to understand the place and boundaries of new standards within the framework plus take an holistic view of naval assurance and review the standards set for completeness, coherency and clarity.

Benefits Brought by NSASS

The key benefits in supporting the NSA framework through the NSASS arrangement are:

The NSASS team directly involves Commercial and Defence personnel, with experience of specification and standard development and implementation, working on the generation and consolidation of the framework;

The NSASS supports alignment of non-MoD frameworks (i.e. Commercial models) with bespoke MoD processes, to reduce the long term risk of divergence;

Through NSASS, key industry standards and specialist design development knowledge and capability is provided as a managed input to the NSA framework;

Through NSASS, experienced manpower, is focused and made available to provide framework development, implementation and specification / standards development work which continuously improves and assists the implementation of an effective NSA framework.

BMT and LR have a strong and long record of achievement and pedigree in the provision of the document sets that underpin the NSA framework. This collaboration of our two organisations delivers an in depth understanding of the challenges presented by the assurance of naval surface ships.

ADAPTING THE NAVAL SHIP ASSURANCE FRAMEWORK FOR OTHER APPLICATIONS AND NATIONS

Submarines

Whilst the NSA framework has been developed in the context of surface ships, it has also deliberately been constructed to ensure that the overall concept and high level framework is as applicable to submarines. At the detailed level the specific documents and guidance for project teams will be different but the overall issues and routes to providing assurance are similar.

Work is currently ongoing to explore the development of a submarine General Naval Specification. There are also class naval rules for submarines.

As yet there is not a submarine version of Naval Ship Code however this does not prevent application of the framework as the lens can be developed to interface with existing Naval Authority policy and standards prior to such a document being generated.

Other Nations

The framework has also been developed to make it generic enough that at the high level framework it is not specific to an individual country's situation. So the framework concept has been depicted in a way that recognizes there is a process for developing the customer requirements into a specification and a physical product and that this is within the context of a regulatory system.

NSASS can work with any navy to take the high level requirement and develop the lower levels of the framework to incorporate a navy's specific policies and standards.

The terminology is purposefully generic that the country specific terminology can be inserted, together with the country specific composition of policy and standards.

Generic Placeholders in the Framework

In essence the framework is designed to provide a concept and mechanism for depicting the assurance situation and almost a 'plug and play' approach to generating a country specific solution. The detail has to be developed to ensure that the adapted framework is comprehensive, coherent and clear.

CONCLUSION

Comprehensiveness, Coherency and Clarity in Supporting Assurance

This Naval Assurance Framework is for the overall assurance of Through Life Capability Management, encompassing: Safety, Environment, Operability, Supportability, Affordability and Programme.

The framework brings together the elements of naval assurance into a coherent structure to enable the user to identify how to apply the requirements, codes and standards and is pertinent to both submarines and surface ships.

The key drivers for an effective framework are that it is comprehensive, coherent and provides clarity to the regulators and project teams.

The alliance through NSASS provides a route to implement the framework's safety aspects on behalf of the MoD. It is in its early days of developing this framework and adapting it in light of experience of its use in implementation on new procurement and in-service ships, and is beginning exploration of its transfer into submarine assurance.

The MoD has travelled a long way in adopting a more commercially based approach for naval assurance. The NSASS alliance is the next stage in that journey. There remains significant ground to be covered as the Naval Ship Assurance Framework is implemented. The NSASS team would be interested in receiving feedback that the reader, may have in relation to naval assurance in your situation, as we tackle an issue that affects many navies.

Please make contact with the authors through the NSASS team:

Email: enquiries@nsass.org.uk Website: www.nsass.org.uk Tel: +44 (0)1225 473572

REFERENCES

1 Rudgley R (2006) Naval Ship Specifications, Standards and Assurance: A new framework - UK MoD Sea Systems Group Poster - Issue 04.

2 Rudgley R (2006) Naval Ship Specifications, Standards and Assurance. UK MoD Sea Systems Group Reference SSG/181/4/1/1/1 - 23 September 2006.

SUBMARINE POWER AND PROPULSION - TRENDS AND OPPORTUNITIES

John Buckingham CEng MIMechE
Christopher Hodge CEng FIMarEST
Timothy Hardy CEng MIMarEST
BMT Defence Services Ltd, Bath, United Kingdom
johnb@bmtdsl.co.uk

1 INTRODUCTION

1.1 The aspiration to improve submarine Power and Propulsion System (P&PS) performance without hazarding the platform's safety and increasing whole life cost continues to challenge the designers of the world's SSKs.

1.2 In-service costs can be reduced through judicious manning reductions achieved through improved automation and improved Availability, Reliability and Maintainability (ARM). The technologies under review will seek to improve these aspects in the context of the use of more COTS type equipment.

1.3 Recent technological developments and improvements have created the potential to improve overall power and propulsion performance and therefore overall submarine capability with a reduced risk to these hazards.

1.4 To bring to maturity, prove and ultimately integrate such technology into a submarine design requires a firm understanding of the actual technology in terms of the benefit it offers and its limitations, complemented by knowledge of integrating such technology within the host submarine.

1.5 This paper will begin by defining the future challenge for SSK designers and then consider how this could be translated into P&PS requirements. The technologies that may allow such a requirements set to be met and the anticipated performance that may be achieved are then explored. The paper will emphasis energy storage, power generation and propulsion trends and opportunities.

1.6 The paper will conclude by proposing a vision for a future putative submarine that exploits appropriate technologies that would enter service in the near (e.g. 2020) term.

2 THE CHALLENGE

2.1 The current challenges facing today's SSK designers can be considered to be:

- To achieve an SSK design which allows for a much greater submerged role for prolonged periods and with greater reach and endurance.
- To achieve this with no loss of military performance or increased threat of safety to the crew.
- To identify the opportunities to achieve this with currently developing technology;

- To identify the programme for the implementation of such a design.

2.1.1 Such a set of challenges can only continue to be met with the adoption of new technologies: chiefly, energy storage, power generation and electric propulsion. The first two topics are firmly part of the Air Independent Power (AIP) technology set and since one of the keystone AIP presentations by Donaldson [Ref. i] and before him Thornton [Ref. ii], this technology has matured into viable and in-service SSK designs.

3 BACKGROUND/CONTEXT

3.1 SSK Population

3.1.1 The world's population of SSK has been growing through the increased number of user nations and through the acquisition of larger and more capable platforms. New submarine are likely to have a longer range (submerged and surfaced) than their counterparts of twenty years ago. These trends are a consequence of the need for the SSK to perform across wider spans of ocean and to be ever more independent from external support. Whilst the speed and range may not be co-incident as with an SSN, long ranges at slow speeds and high speeds at low ranges are possible with an SSK such that an operational deployment can still have the same duration as an SSN, being limited only by stores capacity and human endurance.

3.2 Type 214

3.2.1 The German HDW Type 214 export-design is a good example of the current state-of-the-art of SSK design in terms of P&PS. The design has evolved from the excellent Type 212A to become an export variant of notable success.

3.2.2 The use of two Siemens BZM 120 (120kW) Polymer Electrolyte Membrane (PEM) fuel cells units with onboard stores of oxygen and hydrogen gas has allowed a considerable performance to be achieved.

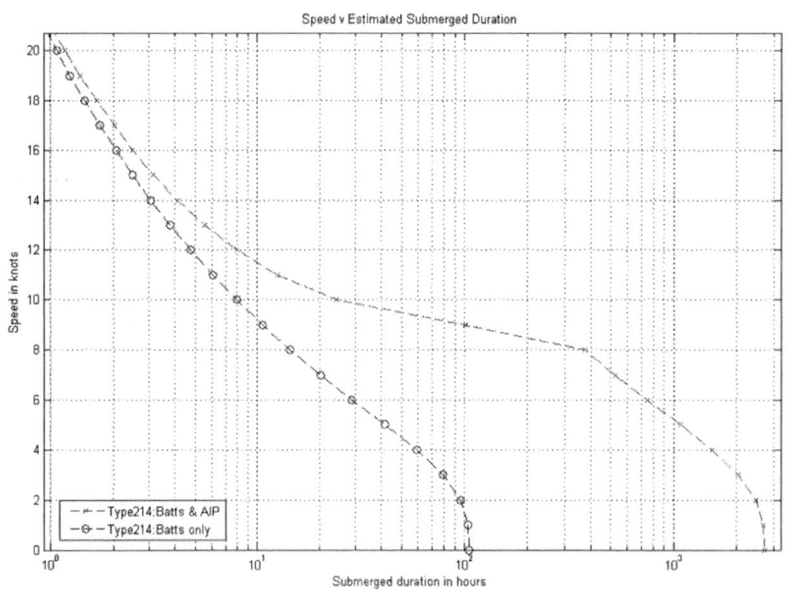

Figure 1 – Type 214: Estimated Submerged Propulsion Performance

3.2.3 Figure 1 shows the estimated performance of the HDW Type 214 submarine with battery only operation and with batteries and fuel cells operating together. The simple assessment shows clearly (on the x-axis in log format), how the AIP provides a significant increase in reach for a wide range of speeds below the top speed. For speeds below 4 knots the duration would be dominated by the ships onboard power demands which exceed those of the propulsion.

4 REQUIREMENTS

4.1 General

4.1.1 The key to achieving a successful SSK design is to fully capture the whole panoply of P&PS requirements at the outset so that the requirements can be turned to a virtue through technology synergies. The P&PS requirements will stem from operational analysis for which the following aspects are key:

- Top speed and time at top speed;
- Submerged range at the most economic speed of advance;
- Poise: Submerged endurance when at rest in elapsed time;
- Ice operations;
- Littoral Operations;
- Manning;
- Platform size.

4.1.2 These requirements will now be explored in a little more detail.

4.2 Speed

4.2.1 The top speed of many of the modern SSK designs is around 20 knots. This speed is partly selected due to the self-noise created above this speed which makes it difficult to detect other vessels. Due to the era of modern high-speed torpedoes and other munitions, the top speed of an SSK is not set to outrun a torpedo as the power penalty would be immense.

4.2.2 Technical opportunities may permit higher powers to be put into the water with the same footprint, this is not likely to lead to higher sustained top speeds. Higher rated motors would also allow for more rapid acceleration and better thrust for manoeuvring purposes.

4.3 Range

4.3.1 The key requirement is range, not top speed: a good range will permit operations far from base-port, as well as extended duration in theatre once the submarine has arrived. The transit to the theatre may allow snorting and be on batteries to avoid depleting the AIP fuels and oxidant. Self-noise on such occasions is still critical but to arrive in a realistic timescale, the speed in transits needs to be greater than the submerged speed of economic advance where the propulsive power equals the ships base cruise electrical power demand. Transit speeds of 8 to 10 knots are therefore to be expected with snorting between once and twice a day. A reasonable future transit requirement would be 12,000nm in 50 days at 10 knots.

4.3.2 When on patrol in theatre, the AIP would be used and the duration at speeds between 4 and 5 knots can now be over 3 weeks. A reasonable future requirement would be 30 days at such speeds giving a patrol period of up to 80 days.

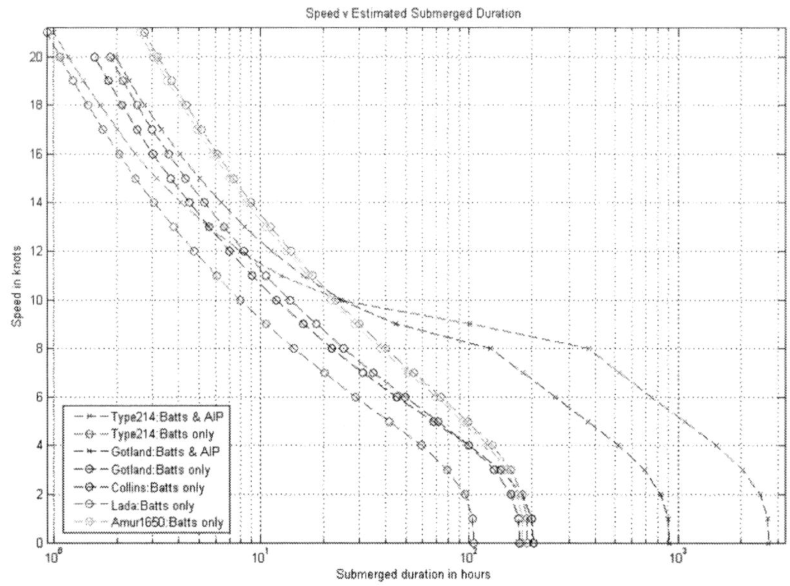

Figure 2 - Estimated Submerged Speed v Endurance for a Range of SSKs

4.3.3 The actual performance of the current set of modern SSK designs is difficult to discern but some basic analysis has allowed their overall likely performance to be estimated. Figure 2 shows how the smaller Type 214 battery set is compensated by the AIP capability. In general the analyses indicated that an AIP feature will allow an extra 6% top speed as the batteries are assisted by the AIP but once exhausted the AIP cannot sustain top speed.

4.3.4 Figure 2 also shows that the future AIP-based Type 214 will be able to stay immersed for over 1 month. This compares with the Type 212A record of 2 weeks submerged without snorting with fuel-cells between Germany and Spain in April 2006 [Ref. iii].

4.4 Poise

4.4.1 The ability to stay on the sea bed for extended periods of time for deployment of UUV and their recovery and simply for the reconnoitring of passing shipping with a possible view to trailing them is also a key part of the requirements set for an SSK.

4.4.2 The SSK would be expected to stay put for one to two weeks and thus the AIP would have to cope with the steady ships services electrical demand.

4.4.3 Future operations will probably have a higher component of UUV deployment and recovery [Ref. iv] which will require the boat to poise, loiter or stay grounded to avoid propulsion noise and to avoid navigation errors and any contact with other maritime traffic in the littoral environment.

4.4.4 The time on station will have to be extended to allow such reconnoitre operations to be concluded and such operations may take-place far from a home or friendly port. One or both of such events would tend to lengthen the patrol duration and a 50 day submerged patrol is a valid design target. The putative design should therefore allow 3 weeks submerged without snorting [Ref. v].

4.5 Under Ice Operations

4.5.1 The opening up of the North-West Passage in North America and the claims being made over territories in that region has raised the likelihood of a greater military presence in the arctic in the near future. Those countries which wish to demonstrate their ability to police such areas would be likely to need a submarine which could handle operations under 60cm first year ice and this would seem to be a suitable "first-pass" requirement.

4.5.2 The provision AIP for such under ice operations would offer a much greater degree of safety margin than conventional batteries allow.

4.6 Littoral Operations

4.6.1 The usual role of an SSK in peacetime is the need to monitor events near shore. This leads to a real need for stealth and so low indiscretion and noise reduction and elimination are key objectives.

4.6.2 With the improved passive sonar performance of many MOTS products and a greater number of potential adversarial SSK's at sea, it is evermore necessary to sustain stealth. These operating conditions have changed the need for power and the storage of energy.

4.6.3 Therefore the use of diesel engines to recharge batteries is increasingly seen as a limiting feature to achieving the necessary stealth. The AIP would provide the direct power to the ships services in this mode. The excess power from the AIP would also be used to replenish batteries so that top speeds could continue to be reached if so demanded.

4.6.4 In the past, power delivery and energy storage has been predicated around sprint conditions for evasion after a strike. Whilst this continues to be a key design requirement, noise from hull turbulence is difficult to avoid at speeds above 20 knots and so although such speeds may be required for sprint, they are not necessary for transits.

4.6.5 Consequently the challenge is less power-driven but centred more around energy storage and submerged air-independent methods to recharge the energy storage.

4.6.6 A suitable requirement would be to consider how best to recharge the batteries without the noise of diesel engine operations. This might comprise an alternative passive surface-breathing power-source (SBPS) such as a fuel cell which is much quieter. Its reduced noise would allow for longer snorting without the risk of indiscretion. Therefore the SBPS could be rated at a lower power than the current diesel engines.

4.7 Manning

4.7.1 The number of crew is predicated by a range of factors. Those required to sail the vessel and man its combat system are determined by factors outside the scope of this paper. The number onboard has a significant impact on ships service and hotel power demand. However there is certainly scope for the manning of the Marine Engineering Department to be carefully managed through suitable training, the selective use of automation and the careful choice of equipment.

4.7.2 The former Royal Navy Upholder class vessel had 11 members in the Marine Engineering Department. A realistic target would be nearer to 7 and this might be achieved by replacing the diesel engine and the lead acid batteries with more modern alternatives. A reduction of 4 staff would lead to significant relaxation of accommodation, hotel services and provisioning demands.

4.8 Size

4.8.1 The construction and outfit of larger submarines do cost more money, especially if the available space is all consumed with equipment. A good design will comprise a boat length which incorporates the necessary weapon fit and crew accommodation without the need for an unduly large proportion being allocated to P&PS equipment. However experience of a range of craft does indicate that such equipment usually consumes 30% of weight and 50% of volume [Ref. vi].

4.9 Summary

4.9.1 This section has identified the main features which influence the requirements set for an SSK and has offered some indicative future requirements which will be used to assess the scope for benefits from modern day and emerging technologies.

4.9.2 The key aspect that emerges is that the design of the P&PS of a modern class SSK is determined by energy considerations whereas the ability to put power into the water may have been a leading challenge of past designs.

5 TECHNICAL OPPORTUNITIES

5.1 General

5.1.1 The main topics of interest are: Energy Consumption; Power generation; Energy storage; Propulsion Motors.

5.1.2 Many developing energy storage and power generation technologies have now been trialled in automotive applications where high cycle rates and concerns for public safety have made great inroads into their reliability and ability to graceful degradation.

5.2 Energy Consumption

5.2.1 The solution to the current issue lies behind the combined use of the range of ideas and approaches identified above. The key lies with the use of technologies which allow:

- More efficient and compact electronics;
- Whole submarine to operate in a manner which is less consuming of energy;

- Equipment to match its operating load (variable speed pumps and fans);

5.2.2 An average consumption of 1kW per person could be achieved as a cruise average but with additional peak power demands of up to 100kW due to increased operational tempo.

5.3 Power Generation

General

5.3.1 The selection of power generation now appears to be reduced to the following: Closed-cycle diesel engines, Diesel engines; Fuel cells; MESMA; Stirling engines.

Closed-Cycle Diesel Engines

5.3.2 The scope for the use of closed-cycle diesel engines has declined with the significant reduction of the Dutch submarine building capability who developed the Spectre design [Ref. vii]. Although they continue to be considered for specialist commercial applications [Ref. viii] they do not have a growing industrial base to draw upon for technological development and they also have a low overall thermal efficiency of around 30%.

5.3.3 However they do offer an AIP capability at a very affordable cost. They can use the same fuel as the main diesel engines and they would require little new training for the crew.

Diesel Engines

5.3.4 The use of diesel engines for snorting at Periscope Depth (PD) to recharge the batteries presents a noisy solution to a necessary function. Although such engines could be acoustically enclosed the confines of a submarine make this difficult to achieve and the future pressing stealth targets mean that a diesel has to be seen as a prime target for replacement.

5.3.5 To allow an alternative to fit in the same footprint as a 1580kW 16V396 MTU genset it would have to have a volume which is less than 10 m^3 (or less than 150kW/ m^3).

PEM Fuel Cells

5.3.6 The use of PEM Fuel Cells in SSK has been championed by the German firm of HDW [Ref. ix] in their Type 212A and Type 214 designs[Ref . x]. In the Type 214A designs as used by Greece, the two Siemens BZM120 fuel cells are rated at 120kW each and weigh 900kg with a volume of 500 litres each.

5.3.7 The design of the Fuel Cell permits a building-block approach and the use of a greater number of smaller rated units will offset concerns about single point of failure.

5.3.8 Considering that the fuel cell is a static device with few moving parts it may be a reasonable decision to employ between two and four units to service the likely demand. However these units are often subject to high inlet fuel and oxygen pressure to increase the concentrations and performance. Therefore compressor reliability performance is also of the utmost consideration.

5.3.9 Part load efficiencies also need to be considered and a multiple unit approach would allow the right number to be on-line to match demand. Start-up times are in the order of minutes and as there will always be battery back-up this is an acceptable aspect.

5.3.10 The fuel needs to be free of sulphur and this leads to use of volume-intensive high-quality methane or hydrogen. The extended life of units is not yet fully proven and this may have a cost implication in years to come.

SOFC Fuel Cells

5.3.11 The Solid-Oxide Fuel Cell (SOFC) is a popular concept for a future power generation application as its high operating temperature (> 600°C) makes its catalysts largely resistant to sulphur pollution in the fuel. Such technologies are being developed by Rolls Royce, Siemens, Alstom and Wartsila [Ref. xi].

5.3.12 The SOFC has a theoretical electrical efficiency (or yield) of 50% [Ref. xii] which is better than the PEMFC. Its higher operating temperature also gives it scope for operations with a turbo-charger (or gas turbine-type device) which serves to increase the overall efficiency.

5.3.13 The device offers a role as the surface-breathing power-source (SBPS) to replace the diesel genset. The SOFC would be much quieter for operability and stealth purposes and its lack of reciprocating parts would make the upkeep burden lower. However current designs are for land-based applications and are in their field trials stages. There are still issues with design for acceptable degradation and this is subject to ongoing studies. [Ref. xiii].

5.3.14 Due to their early stage of development the current designs are not small and the equivalent SOFC for a 2MW diesel gensets would be five times more bulky: clearly not a proposition yet. Much of the bulk is due to the Balance of Plant (BOP) with a single Rolls-Royce 250kW generator module being about 2m tall and 1.5m diameter [Ref. xii].

5.3.15 The fuel for many SOFC applications is still natural gas but for marine applications the use of methanol is being research by Wartsila as part of the METHAPU project [Ref. xiv]. Methanol is flammable with a flash point of 11°C which is considered to be far too low for safe storage inboard in on a submarine. For this and other safety reasons, methanol could be stored outboard as hydrogen is currently stored on the HDW SSK designs. Methanol has a calorific value of 23MJ/kg compared to 42MJ/kg for F76 and so almost twice as much fuel by volume would have to be carried but this would be countered by the greater efficiency of the SOFC.

5.3.16 An SOFC design solution for snorting power generation is therefore likely to require outboard storage of methanol in addition to the hydrogen for the PEMFC. The removal of the inboard diesel fuel storage would allow for the extra balance of plant associated with the methanol-driven SOFC. The time for the SOFC to achieve operating temperature can be up to 30 minutes. However this is acceptable as the submarine can run on batteries and the AIP system until the SOFC is at operating temperature for snorting at PD.

MESMA

5.3.17 MESMA (Module d'Energie Sous-marine Autonome) comprises a steam turbine system whereby the heat from burnt ethanol creates steam. The design is based on a Rankine cycle which has a typically efficiency of 25% which is poor compared to a fuel cell.

5.3.18 The design has been operationally proven at sea in an Agosta class of the Pakistan Navy (PN). The system will be fitted to PN's Scorpene design by DCN of France. A simple assessment of the limited data available suggests the Scorpene performance will be similar to that shown in Figure 3.

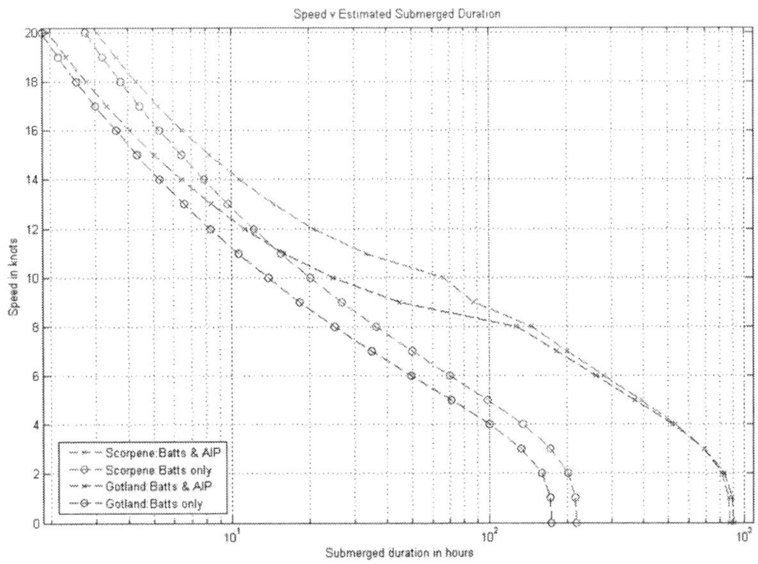

Figure 3 - Estimated Scorpene & Gotland Speed v Endurance

5.3.19 Figure 3 indicates that the MESMA allows a tripling of the range at patrol speed (4 knots). However with MESMA waste combustion products are discharged overboard at pressure and so there may be a noise penalty associated with this method of gas management.

Stirling Engines

5.3.20 Although well established as a feasible AIP device since their introduction into the Gotland class in 1992 [Ref. xv], they do not appear to have been further developed for SSK use, not withstanding the innovative efforts being conducted by Whisper Tech of NZ. (Where multiple cylinder arrangements are being progressed to provide 1kW sets but at a low power density of 5W/litre [Ref. xvi].)

5.3.21 Figure 3 shows the estimated performance of the Swedish Gotland class when operating with and without the currently installed battery set and AIP. The plot shows how the 150kW Stirling engine AIP starts to give the SSK endurance below 8 knots thus allowing extended poise in the Baltic Sea operating area where transit times to the patrol areas would be small.

5.3.22 It is likely that the trend away from Stirling engines is due to their poor industrial infrastructure and their upkeep and noise which would be greater than a passive fuel cell system notwithstanding enclosures and modern rafting methods.

5.4 Energy Storage

5.4.1 Energy storage can considered to be: Batteries; Fuel Storage; Oxygen storage.

Batteries

5.4.2 The technology of batteries is undergoing a sea-change with the emergence of the Zebra product from Rolls-Royce and the development of lithium ion designs [Ref. xvii]. This comes on top of the use of sodium sulphide batteries in the HDW Type 212A class.

5.4.3 The Zebra battery [Ref xviii].] is a sodium nickel chloride battery which has been developed for marine use by Rolls-Royce. Zebras and the sodium sulphide batteries allow for distributed energy storage throughout the boat and may enable dc systems with less installed copper. They can produce twice the energy of a lead acid battery at the one hour rate as can be seen in the cell-based data in Table.1.

Description	Power Density	Specific Power	Energy Density	Specific Weight
	kW/litre	kW/kg	Wh/litre	Wh/kg
Lead acid	0.12	0.08	90	44
Zebra	0.24	0.16	167	114
Sodium Sulphide (NaS)	0.021	0.17	170	117
Lithium ion	0.22	0.11	270	120
Siemens PEMFC BZM 120	0.257	0.13		

Table.1 - Battery & Fuel Cell Technologies

5.4.4 Lithium ion designs have successfully been developed for automotive applications [Ref. xix]. Their energy density is over twice that of lead acids batteries and it is less than half the weight for the same energy at the 5 hour discharge rate. A unit which is 50cm by 40cm by 40cm has energy of 21kWh and can develop 100kW continuously (i.e. five hour discharge) or 200kW for short periods of time. In energy density and specific weight terms it promises to be better than the sodium sulphur batteries as used in the German 212A class.

5.4.5 The lithium ion batteries would be used both centralised and for distributed energy storage. Lithium ion batteries would be:

- rated to a higher current than other battery types;
- durable to experience more full-charge cycles than NaS or Zebra;
- capable of sudden changes in demand;
- have a documented developed record of in-service reliability.

5.4.6 The collective increase in stored energy will make for more operational flexibility and allow for oceanic operations such as those undertaken by the Collins class. The use of lithium ion batteries for the centralised main battery banks would more than double the stored energy.

Distributed Energy Storage

5.4.7 The use of Zebra batteries in forward spaces near to the consumers will allow for a greater installed energy storage with a small impact on the total volume. Such a distributed energy storage arrangement if adequately sized could allow for fewer pumps with no need for alternative and emergency supplies to each essential equipment.

5.4.8 Such arrangements should mean that boat capability could be increased without leading to their enlargement which would itself become a penalty in the littoral.

5.4.9 The greater the energy storage, the lower rated can be the power source as it need only address the time-averaged power demand.

Fuel Storage

5.4.10 Kerosene could be stored inboard or even outboard. Ever since the adoption of the Walter cycle by the Russians and then the exploratory UK submarines the Explorer and the Excalibur, (Ref. xx),. there have been records of kerosene or other fuels being stored outside the pressure hull in collapsible bags. However inboard storage allows for a reliable water-free fuel stability.

5.4.11 Methanol could be used for the SOFC but consideration could be given to the use of liquefied natural gas which would be fully contained and isolated from atmospheric air unlike a methanol-based solution. However energy storage density favours the methanol approach.

5.4.12 It is envisaged that hydrogen will continue to be stored in metal hydrides for the foreseeable future. However for larger energy storage and vessels there may be issues with scaling up the current arrangements.

Oxygen storage

5.4.13 Whatever the power generation design there will be always be a need for stored oxygen in one form or another.

5.4.14 For modern day SSK design the emphasis is on energy and hence large amounts of oxygen needs to be embarked to allow for extend range. The HDW approach is to have one single bottle of liquid oxygen (LOX) located inboard. In this way the bottle can be monitored and contained in the event of a mishap. The criticality of the supply of oxygen suitably heated and fed to the AIP is such that the pipework is kept to a short length for hygiene reasons. Oxygen pipework needs to be kept clean of hydrocarbons as experience with the HMS CHALLENGER demonstrated [Ref. xxi].

5.4.15 The use of LOX at temperatures near 100K would in theory allow for synergy with the use of a High-Temperature Superconducting (HTS) motor aft. However for independent reliability and also safety concerns, it is likely that such a synergy in terms of reduced cooling plant would be difficult to achieve although there would doubtless be a cryo-plant cross-connection for emergency use.

5.4.16 There has been much research in the field of compact energy dense oxygen storage. The kind of medium can vary from oxygen candles through to liquid oxygen. In 2006, Davies et al [Ref. xxii] showed the wide range of over 10 different storage solutions and showed LOX to be amongst the best practical storage solutions.

5.4.17 For simplicity and convenience two large LOX tanks would be located inboard each one dedicated to a PEMFC unit. The amount of stored oxygen is the limitation for submerged operations on AIP. Consideration should also be given to using bottle air as a back-up should the LOX be depleted. This could be replenished by the HPAC when snorting.

5.5 Propulsion Motors

5.5.1 The developments of more power-dense / compact electric propulsion motors and their matching converters should allow the rated shaft power to be increased at no penalty to the weight or volume of the installation. Companies such as Converteam Ltd [Ref. xxiii] are known to be developing designs where the converters are located inside the rotor thus saving a great deal of space.

5.5.2 But how is this mean of placing extra power into the water to be used? If the current SSK are effectively blind at speeds above 20 knots due to the flow-noise over the sonar, then what use is a faster vessel? The increasing use of network-centric battle-space technologies means that future SSK will be able to gather better navigation data and speeds above 20 knots would be acceptably safe for short distances. If this does come to pass, then the extra power capability could be use to good effect for a rapid transit or very short sprint bursts (30mins at higher speeds possibly to allow a target to be attacked.

5.5.3 The current surface speed of 10 knots for many SSK is slow for keeping part of a convoy with merchant ships. Notwithstanding the issue of extra power from the surface-breathing power-source, a motor of double the power and 58% more weight for the same design would allow 12.5 knots to be achieved as well as extra thrust for manoeuvring and acceleration

6 PUTATIVE SUBMARINE

6.1 A future putative submarine is offered which is based on the AIP Type 214. It would comprise the following:

- A larger displacement (3500 tonnes) to offer ocean-going performance;
- A brushless DC propulsion motor rated at 6MW (Twice the rating of the Type 214).
- PEMFC for submerged endurance[1];
- Liquid oxygen for submerged fuel combustion 1;
- Hydrogen located outboard 1;
- A SOFC for snort power generation;
- Methanol for fuelling the SOFC;
- Zonal power supply units with lithium ion other batteries;

[1] As is currently the case.

- Four fewer staff due to removal of reciprocating machinery (i.e. diesels).

6.1.1 Figure 4 shows how the putative submarine might perform for submerged endurance v speed.

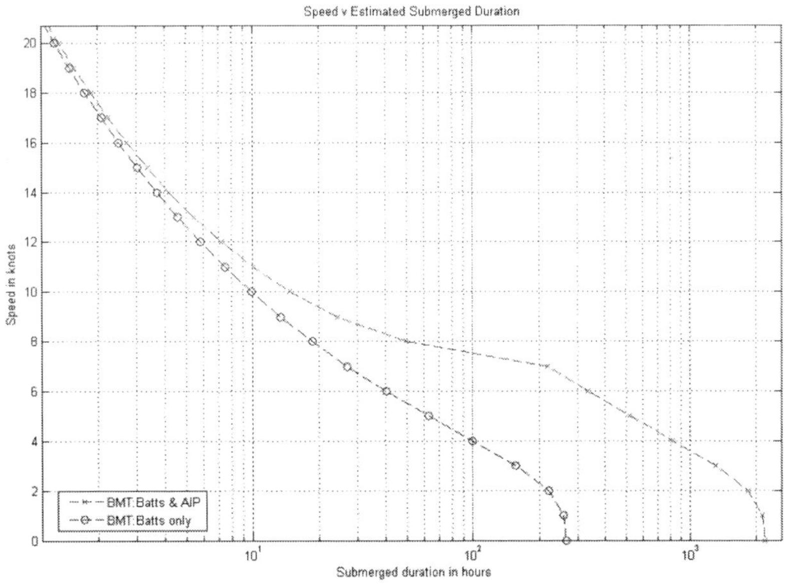

Figure 4 - BMT design performance

7 REFERENCES

7.1 The opinions expressed in this paper are those of the authors and not those of BMT Defence Services Limited.

---

i Donaldson A J. "Air-Independent Propulsion Systems". Rolls-Royce. 5-Nov-98. UCL.

ii Thornton. Lt Cdr G B. "A design tool for the evaluation of AIP in submarines". Thesis for BS Marine Engineering by USN. 1979.

iii http://www.fuelcellsworks.com/Supppage5052.html

iv Hardy T et al, "Effective, integrated UUV launch and recovery from conventional submarines" BMT Defence Services, UDT Europe 2007.

v http://En.Wikipedia.ord/wiki/Type_214_submarine

vi Wrobel P G "Design of the Type 2400 Patrol Class Submarine". RINA. 1984.

vii http://www.dutchsubmarines.com/rd/r&d_ccd.htm

viii Kokes Marine Technologies Inc.: Closed cycle diesel engines ix Hauschildt P. et al "AIP Systems for Submarines: The Benefits of PEM Fuel Cells today". . HDW.

x Hammerschmidt. Dr A "Fuel Cell Propulsion of Submarines".. Advanced Naval Propulsion Symposium 2006, October 30- 31, 2006, Arlington, VA, USA

xi Wartsila press release on 250kW SOFC units. 4 Dec 2006.

xii "Pressurization of IP-SOFC Technology for Second Generation Hybrid Application".

PIP-SOFC NNE5-2001-00791.

xiii Koch S, "Solid Oxide Fuel Cell Performance Under Severe Operating Conditions" Risø National Laboratory, Denmark. Published in the proceedings of the 6th European Solid Oxide Fuel Cell Forum 28 June - 2 July 2004, page 299 Lucerne / Switzerland

http://www.ecn.nl/docs/library/report/2005/rx05083.pdf

xiv http://www.methapu.com/

xv Jane's Fighting Ships 2006-2007.

xvi Whisper Gen Limited of New Zealand.

xvii Green Dr K et al. "High-performance rechargeable batteries as power sources for submarine propulsion" JNE 37(2). 1997.

xviii Rolls-Royce Zebra battery.

xix PML Flightlink Ltd. Mini QED battery. May 2007.

xx Walter cycle reference

xxi UK MoD Journal of Naval Engineering, 32(3), 1990

xxii Davies K et al. "UUV FCEPS Technology Assessment & Design Process". University of Hawaii. Oct 2006.

xxiii Converteam Ltd.

Minor Warship Roles - How technology is leading to a new vessel type

Andy Kimber, Will Giles
BMT Defence Services Limited, Bath UK
Info@bmtdsl.co.uk

ABSTRACT

Minor warships are a category of naval vessel covering a range of roles, including Mine Counter Measures (MCM), offshore patrol and survey. These vessels may be deployed in a war role, within threat environments, but are often small and of a specialized nature. To date, the roles have generally been achieved through the design of specialized vessels for each role with little commonality. This is particularly true of the MCM vessel, a highly optimized platform due to the demanding and specific nature of its role.

Previous attempts to provide a common platform to meet these varying roles have met with mixed results. The emergence of new technologies, particularly the expected wide spread use of off-board, unmanned systems offers new potential for a single common platform to be reconfigurable across these roles; the role optimization becomes focused on the off-board systems and not the mother platform.

This paper discusses the work conducted into a concept for a multi-role small surface combatant focused on the roles of MCM, maritime security, hydrographic and environmental assessment and patrol. The objective of this work was to identify how much commonality can be achieved in the platform, the art of the possible in terms of a small, globally deployable warship and the advantages and disadvantages of navies adopting the multi-role approach. Key aspects investigated include: the type of hullform adopted, comparing monohull and multi-hull performances to achieve global deployment requirements, issues of modularity, arrangement of multi-missions spaces, and deployment of unmanned vehicles to perform some missions.

BACKGROUND

Changing MCM Technology

To day many navies operate the classic Glass Reinforced Plastic (GRP) hulled mine hunter vessel. Designed to operate within or close to the minefield, these vessels are built of single skin GRP, have noise and vibration isolated main machinery and other fixtures, fittings and equipment constructed of non-metallic materials, all optimised so as not to compromise the signature of the vessel.

Mine hunting is conducted using the combination of hull mounted and towed mine sensors such as the RN Sonar 2153 and Remotely Operated Vehicles for identification, inspection and disposal. The latter are often deployed using cranes and require the vessel to operate within the immediate suspected minefield area.

The future of MCM is being rapidly changed by the introduction of Unmanned Vehicles, both Unmanned Underwater Vehicles (UUV) such as the HUGIN (Figure 1) and REMUS ranges and Unmanned Surface Vehicles (USV) such as the UK's Flexible Agile Sweeping Technology (FAST) demonstrator. Already on trial successfully in a number of navies, the introduction of unmanned vehicles is seen as a key enabler to "remove the man from the minefield". A result of this strategy will be the reduced need to manage the mothership's signatures and hence negate the use of GRP or non-magnetic steels and noise / vibration reduction measures. What this will mean in practise is that MCM systems may be deployed from a variety of navy platforms rather than needing dedicated MCM vessel's with very low signature characteristics, as in the case of the US Littoral Combatant Ship. In theory this will allow the MCM capability to be fielded with the combat force and without additional vessels.

However, there is a drawback to exclusively deploying from surface combatants; whilst conducting MCM operations the combatant is likely to be restricted in its ability to conduct other operations and the required volume for MCM systems will be at a premium alongside other mission equipments. Therefore, another concept that may be introduced alongside the deployment of MCM systems from surface combatants is the development of the "Auxiliary Surface Combatant (ASC)".

Simplistically, these vessels may be considered the replacement for the legacy MCM vessels, but the introduction of the new technologies offers the chance to make these vessels more flexible in mission profile, whilst still retaining the concept of a single role vessel aimed at low cost. The concept therefore is a very simple platform, using off-board and modular systems to conduct MCM, route survey or surveillance activities.

Figure 1 Hugin UUV in containerised Launch and Recovery System (*left*) and on board a MCM Vessel (*right*) **(images courtesy of and © Kongsberg)**

Wider Capabilities

The MCM vessel has often been utilised as a patrol ship to supplement its utility; examples include RN Hunt Class vessels which have been used for fishery protection and some were modified as dedicated patrol platforms, whilst the Canadian MCDV class already seeks to address utility through modular systems.

The roles for the ASC are those that represent a poor utilisation for a costly surface combatant such as a frigate, or would detract from them conducting their primary mission. In this sense, the ASC becomes a "second rate" combatant, capable of supporting the fleet world-wide but conducting either single role or low threat tasks to reduce the burden on the surface combatant fleet. For the RN, they would represent a return to a type of vessel prevalent before the NATO navies became optimised for the cold war.

Typically, these roles would include:

- Mine Counter Measures;
- Mine Counter Measures Tasking and Support;
- Survey and Rapid Environmental Assessment (REA);
- Coastal and Offshore Patrol;
- Maritime Security Operations (MSO) including Counter Interdiction, persistent surveillance and presence;
- Training.

In comparison to existing MCM and OPV types, in particular the Canadian MCDV which seems to bear a resemblance to the proposed concept, there are some key enablers required for the ASC not provided in these existing vessels:

- To provide the global reach required, the ASC needs both range and seaworthiness;
- Speed has to be sufficient to deploy with a task group and to conduct operations ahead of that group;
- Armament has to be sufficient for self-defence and to offer limited and modular offensive capability.

Mission Context

The ASC will be required to transit globally both as a dispersed unit and as part of a task group. As such, it will require an appropriate level of seaworthiness, cruise speed and range. Specifically, it will be required to conduct ocean transit with minimal risk of damage or diversion, be capable of deploying at equivalent speed to the task group without unreasonable demands on replenishment and to be able to transit at moderate speed over long distances (e.g. UK to Caribbean) without support.

However, as the ASC roles are principally littoral in nature, it would not be required to conduct missions in an open ocean environment. Hence, mission based performance requirements (sprint speed, deployment of off-board systems and helicopters, use of weapons) would be limited to offshore / littoral environments, e.g. Sea State 4 to 5.

ASC ENABLERS

Modularity

As an affordable and therefore relatively small combatant type, the ASC would rely on modularity to offer design utility. Each ASC would operate as a single role unit; the range of roles would be achieved by tailored equipments fits. Advantages to the use of modular systems versus batching are:

- The ability to flex the mix of role vessels;
- Simplification of technology up grading and potentially spiral development of equipments fits;
- The porting of equipment between platforms during non-operational periods to reduce the required number of expensive systems.

There are already a number of modular systems available; however, it is important to define what "type" of modularity is required. One possible description is four levels of modularity:

- Construction modularity
- Configuration modularity (e.g. MEKO®-class ships)
- Mission modularity (e.g. StanFlex series of vessels)
- Battle (network) modularity

A rapid re-role requirement is not envisioned for the ASC (refer to Mission Crews below), and hence modularity will primarily be concerned with bounding mission systems and interfaces, configuration control of both mission packages and platforms but as separate entities and in providing suitable routes for exchange but within a controlled environment, e.g. an up keep period in a home base or similar facility. Hence, it relates to "Configuration" modularity, perhaps with a touch of "Mission" modularity for some roles.

A key factor for the ASC is likely to be the deployment of off-board sensors, most probably using UAV's / UUV's / USV's. Both the mine warfare and survey roles are likely to be performed using one or a combination of types of vehicle and the ability to deploy either modular vehicles or re-role by changing vehicles is an interesting prospect. To achieve this, the ASC needs to achieve flexibility in:

- Deployment and recovery of off-board systems;
- System control;
- System maintenance;
- Possible system reconfiguration if the vehicle is modular itself.

Mission Crews

The concept of a mission crew, individuals who embark with specific mission equipment in addition to a basic "float / move" complement, has been proposed for future modular concepts. However, there are significant drawbacks to such a concept:

- Time for mission crew to work up on new vessel;
- Familiarity with specific equipment on ship (can often be subtly different to that on which training is undertaken);
- How to maintain skills in mission crews not required at sea ("mission skills fade").

This is a key reason why ASC vessels may preferably be configured to specific roles for long periods, rather than re-role on a regular basis. Hence, whilst there would be a common core complement that is the same for all variants, the role specific personnel can be integrated within the complement given that the vessel is not expected to re-role outside of normal crew rotations. However, it is noted that some roles will involve non-complement staff (e.g. fleet protection troops, marines, MCM support etc.) and some missions will require surge accommodation.

Aviation

Provision for a helicopter is a key feature for the MSO role but is not essential for other roles. Whilst provision should be made for limited embarkation of a suitable helicopter (e.g. Lynx), this space may also be more flexible. Hence, it could be used for other equipments when operating in a role that does not require an organic helicopter. It is noted that facilities to land a helicopter may be advantageous for MCM and survey although as a minimum a VERTREP capability is required. Additionally, facilities should also be provided for surveillance UAV's.

Survivability and Threat Environment

Depending on the role and mission, the ASC could be conducting operations within:

- A benign environment, e.g. training, civil hydrographics;
- A low threat environment against lightly armed opponents (small arms, small calibre weapons, small missiles or rockets), e.g. maritime security, surveillance and patrol operations;
- A high threat environment, e.g. when conducting MCM or REA in theatre for a medium or large scale campaign.

Against the threats, the ASC survivability should reflect the environment and its relative value, e.g. in low threat conditions it is likely to be a dispersed unit and therefore should be capable of continued operation with degraded or limited degradation of capability. In high threat environments it may be accepted that the vessel is either not likely to be a primary target or can be accepted as having no residual capability after a significant event.

Hydrodynamics

The required hydrodynamic performance is driven by the need to offer a very high degree of seaworthiness to achieve global deployability, balanced against resistance optimisation in order to achieve the maximum speed and a manageable fuel load for the required range.

Suggested requirements include:

- Maximum speed 25kts (track and stop suspect vessels);
- Cruise speed 18kts (for Task Group operations);
- Range 5000nm@18kts (Task Group operations), 7000nm@12kts (dispersed unit);
- Sustain transit in Sea State 6, operational in Sea State 4 / 5.

Other factors will include consideration of draught given the focus towards littoral missions and achievement of a satisfactory stability standard. The hull design should seek to achieve the required speed, range, draught and stability requirements whilst maximising the seaworthiness.

BASELINE SOLUTION - "VENATOR" CONCEPT

Having developed a basic set of requirements, BMT Defences Services together with BMT SeaTech and BMT Nigel Gee, began the process of developing a concept design with the aims of:

- Assessing the practicalities of a small, reconfigurable minor warship;
- Assessing the size required for the various payloads;
- Assessing the seaworthiness issues and potential size for global reach.

The principal characteristics and arrangement of design are indicated in Figure 2; a key driver behind the initial choice of size was a change in the damage stability rules of the UK MOD standard (Reference 1) that occurs at a length of 92metres, changing from a 2 compartment standard to a damage length 15% of the waterline. The arrangement was developed around the mission working area aft which features the following:

- A stern ramp for launch and recovery of RIB / USV's of up to 11 metres length;
- Facilities for a containerised UUV launch and recovery system for MCM reconnaissance or survey UUV's;
- A large garage opening on to the quarter deck for the stowage of mission equipment;
- A flight deck suitable for a small helicopter (e.g. Lynx) over the garage top;
- A telescopic Hangar.

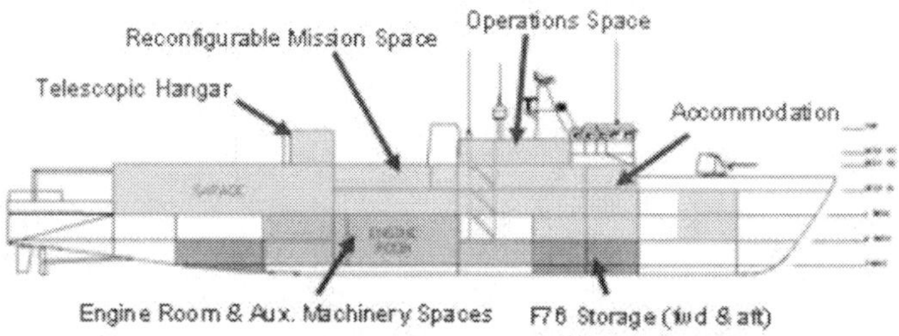

Overall Length, metres	93.25
Waterline Length, metres	90
Moulded Breath, metres	15.12
Design Moulded Draught, metres	3.95
Design Displacement, tonnes	2680
Scantling Displacement, tonnes	3039

Figure 2 Baseline Design Arrangement and Characteristics

Various configurations of Hangar, vehicle launching and recovery and flight deck were considered, however the selected length quickly suggests that the selected arrangement offers the optimum solution.

The adoption of a garage for stowage of mission equipment offers a more protected environment when the vessel is in transit although the required height to both stow and move items results in a garage of approximately 6 metres or 2 deck heights. Hence, the flight deck is placed higher than would conventionally be adopted and attention is required into how the garage / flight deck is incorporated into the primary structure of the vessel.

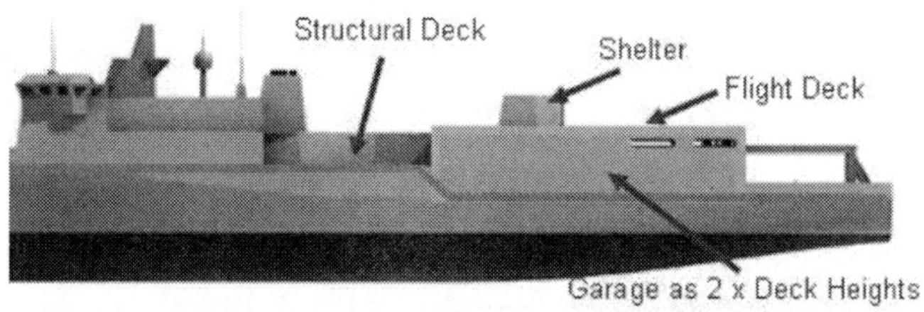

Figure 3 Garage and Flight Deck details

MODULAR SYSTEMS

A variety of modular payload and mission system standards have evolved, mostly bespoke to specific navies. However, increasingly there is a trend for the military to use existing packing standards as this eases integration with the wider transport infrastructure. The most widely used standard is the 20 foot container. It is also not necessary to force all items into the same dimensioned packages and an early decision was that the large unmanned vehicles should have bespoke stowage (e.g. large RIB based USV's and the large reconnaissance UUV's).

ISO Containers

Most modularity concepts discussed to date focus on the use of ISO containers. There are clear benefits to these containers:

- Established in the wider transport infrastructure;
- Already widely utilised by the military;
- Durable containers offering excellent protection against weather and damage;
- High volume;
- They can be used as enclosed offices and workshops.

However, when considered in the context of the ASC, there are also some disadvantages to containers:

- Voluminous and can lead to poor packing densities;
- Difficult to move without significant material handling equipment;
- Unversatile- fixed, large sizes could mean even small containers such as quadcons are too large for this application;
- Very difficult to manoeuvre in confined spaces;
- Excessive height is needed to lift one container over the other.

The above disadvantages are particularly applicable to the ASC because:

- The garage is a confined space with a small footprint;
- Significant height restrictions due to the flight deck and associated structure above the garage;
- It would be beneficial to be able to move equipment when underway;
- Versatility is a critical requirement in storage solution;

Sub-Intermodal Packing

When the wider range of commercial and military transport packaging is considered, beyond the 20ft, then "sub intermodal" packaging appears a very attractive solution. Here, units smaller than the standard container are considered but they retain compatibility to be stowed within 20ft containers for transport when required. The solution discussed here is typical and represents an in-service example.

The Mobile Shelter Systems Storage and Transport Frames (STF) are essentially caged storage racks built to be compatible with all commonly used means of transport. They are built in several sizes and are designed to permit quick loading and unloading of cargo from standard ISO containers, quadcons, tricons and other pallets.

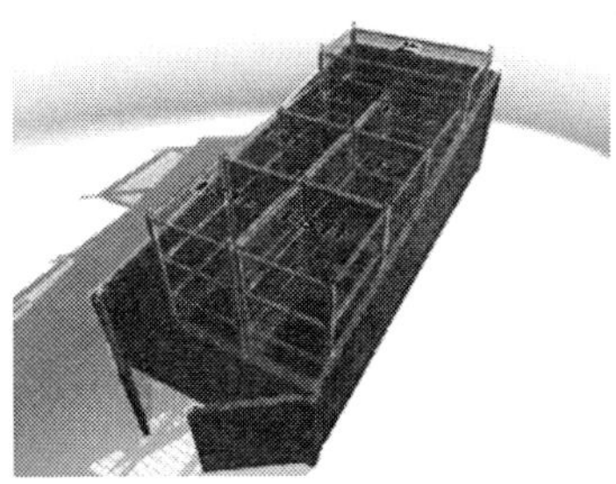

Figure 4 Storage and Transport Frame developed by MSS (images courtesy of and © Mobile shelter Systems)

These racks provide all the advantages of the ISO container and dispel several of the disadvantages:

- Equipment can still be transported using ISO containers;
- Containers can be unloaded at dockside and the racks loaded onto the ship;

- Racks are much smaller and lighter and therefore more versatile both in mobility (forklifts can lift and move racks) and in storage into a small space (such as a garage).

The size selected for the VENATOR concept is a 2 tonne unit, which offers the potential for integration within the military supply change (RAS, VERTREP) and it is stackable within the two deck high garage, offering an even greater stowage capacity on the vessel.

Controlled Environment Solution

It is also necessary to provide reconfigurable "internal" space for MCM support spaces, unmanned vehicle control and planning / briefing / training. Whilst containers may be used within the garage space, limited internal area, water and weather-tightness, soundproofing, electricity, lighting and climate control are just some of the factors that would need to be addressed.

A preferable solution could be to incorporate a controlled, internal space that may be easily adapted. This is incorporated in the design within the superstructure. By placing the space immediately forward of the garage, direct access from the garage may be arranged for ease of re-role.

Re-Role of Equipment

Significant loading and unloading is required during re-role and some of the key assumptions developed regarding re-role are:

- Rapid re-role is not perceived as a requirement;
- Re-role will take place in a facility geared towards the task (i.e. mooring alongside with cranage and trained workforce available);
- Modular concept to facilitate an easy integration of equipments.

As has been discussed, a garage opening aft to a stern ramp and a flight deck above is a natural configuration for this size and type of vessel. However, if the flight deck is entirely above the garage a crane cannot lift equipment or stores directly into the garage.

Potential solutions to embarkation / disembarkation of equipment from the garage include:

- A "landing area" allowing cranage to "land" equipment onto the working deck and to utilise a secondary lifting system to move the items in their prescribed location within the garage;
- Introduce a lift between the flight deck and garage. Hence a crane lowers onto the lift, the lift can retract and a secondary lifting system locates items to their prescribed location within the garage;
- A retracting or opening flight deck which will allow access by a crane directly into the garage.

The second and third suggestions are costly, voluminous and introduce complexity to the vessel. The first solution is inexpensive and simple provided an effective secondary lifting system can be utilised within the garage. Hence, the baseline design includes an open quarterdeck area aft of the garage.

Onboard Material Handling

The onboard material handling is required for internal movement during re-role, movement of the unmanned vehicles and RIB's to the points of deployement / recovery and general logisitics re-supply. Various solutions have been considered as described in the following paragraphs.

Forklift

A forklift truck could be used during re-role to locate items inside the garage and offers a robust and practical solution. A forklift could be lowered onto the open transom area by crane and work in tandem with the crane to load and unload equipment. Once the task is complete the crane could unload the forklift.

Sled on Track

A track system could be laid into the deck with longitudinal and transverse freedom of movement. A sled can then move on the track through winches and pulleys. This system would require the stern ramp to accommodate the cradle and the RIB's / USV's would be deployed from the cradle on to the ramp. The sled system has the advantage of safer operation in higher sea-states. However, the sled system introduces complexity in use and a lack of versatility as sleds may have to be bespoke for a wide range of equipments. Arranging access throughout the garage could also pose a problem.

Gantry system

A gantry provides superior ease of use and versatility. However, a weight suspended on a gantry would swing as the ship responds to wave action and this is an area that requires further analysis and consideration against the user requirements and the actual conditions under which vehicles and equipment will be moved.

It has been discussed how equipment would be lifted on to the open quarterdeck for embarkation during re-role and the stern ramp is also external, hence material handling is required external to the garage. Two potential solutions are available:

- Run the rails of the gantry outside the garage and aft to the transom, e.g. so the gantry can lower RIB's onto the ramp;
- Install a second lifting method such as a davit to bridge the gap between the gantry and the ramp.

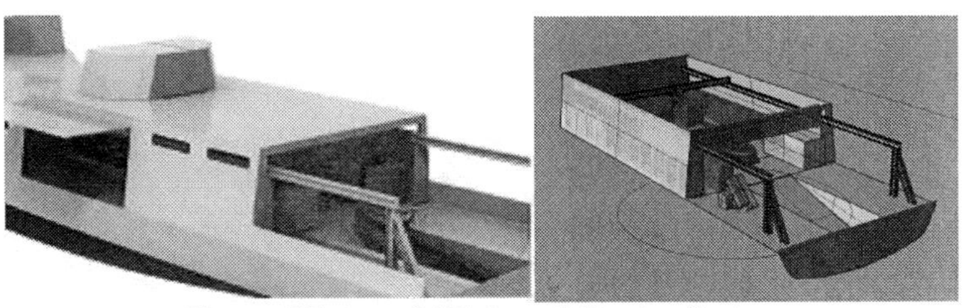

Figure 5 Internal / External Gantry System

The second option is convenient because the davit allows a second deployment method for vehicles. However, when utilising the ramp deployment becomes a more complex and timely 3-stage process (gantry + davit + ramp). If the gantry rails can be run externally to the transom this would introduce simplicity and good versatility. The gantry could also aid in re-role by moving equipment from the external landing area to inside the garage.

GLOBAL DEPLOYMENT

The impact of global deployability has been considered through a parametric investigation of the baseline monohull and comparison to a SWATH solution. Nine parametric forms have been assessed for seakeeping, powering and stability performance based on variations in length (90m, 100m and 110m) and length / beam proportions (variously L/B from 6 to 8). As would be expected, the results show that the choice of hull characteristics allows the designer to optimise towards specific performance targets, but no one hull offers the best performance at all aspects. Hence, the selection of a hull is directly linked to a full understanding of the user's requirements and the correct balance between of the relative performance of the hull in order to maximise its operating envelope.

Overall, it is concluded from this initial analysis that a 90m will not offer true global deployability, particularly if transiting with larger surface combatants (utilising Reference 2). A vessel approaching 110m has been shown to be more preferable and not necessarily significantly more expensive, due to potentially reduced powering requirements compared to a 90metre vessel (provided the mission payload is unchanged of course).

A short investigation into a SWATH alternative suggests that whilst this would offer far superior seaworthiness, speeds in excess of 20knots would make this form unattractive due to high resistance. Hence, from a hydrodynamic stand point, a monohull best meets the requirements set.

SUMMARY

With Project VENATOR BMT has sought to understand the operational context, design drivers and potential solution space for a globally deployable, single role but reconfigurable vessel to replace a range of minor warship types. The presented concept offers a glimpse of the types of vessel and has also identified the key design drivers to be considered by future operators:

- What is the required speed?
- What is the true definition of "globally deployable" within the concept of operations?
- What are the principal drivers in the choice of a modular systems, speed of reconfiguration, ease of configuration or minimal ship impact?

Above all, the study has highlighted that seaworthiness and complexity of incorporating such a large and diverse payload make the platform design and its requirement specification the challenge for this concept.

REFERENCES

1 Defence Standard 02-109 Stability Standards for Surface Vessels
2 STANAG 4154 Common Procedures for Seakeeping in the Ship Design Process

Energy Scavenging Unmanned Ocean Vehicles Affordable Persistent Ocean Surveillance

Payne Kilbourn
President, Unmanned Ocean Vehicles, Inc.
PKilbourn@msn.com

STATEMENT OF NEED

Military, government and commercial communities are increasingly understanding the importance of gaining knowledge of the world's oceans, for many reasons. But the overwhelming expanse of the oceans is makes gaining sufficient data for extensive understanding exceptionally difficult. Space-based systems can cover large areas, but are expensive, are affected by climatic conditions, and can't collect sufficient surface and subsurface data. Manned ships are expensive to build, man and operate. Buoys require ships to deploy them and provide limited coverage. They are either moored, or if drifting, they cannot be directed to a desired area.

Naval Needs. Naval forces require affordable persistent sensor systems to support Intelligence, Surveillance; and Reconnaissance (ISR) and Battlespace Awareness in both the open ocean and littoral domains. The US Navy has identified seven high priority USV missions, in order of priority:

Mine Countermeasures
Anti-Submarine Warfare
Maritime Security
Surface Warfare
Special Operations Forces Support
Electronic Warfare
Maritime Interdiction Operations.

The Naval Studies Board reported in 2005, "Although current USV systems are used in a remotely controlled mode, opportunities for these systems, when used in an autonomous and adaptive control mode, are significant."

Governmental Needs. Governments the world over are expanding their interest in the ocean. Unwanted immigration and terrorist threats require coastal security and monitoring of shipping. Climate and weather prediction requires detailed data collection for ocean science. Marine sanctuaries and protection of aquatic species and marine resources require ocean monitoring and protection. For example, in the U.S., the President's Ocean Policy requires integrated ocean observing system sensors. Unmanned surface vehicles have the potential to meet these many needs and requirements.

Commercial Needs. The oceans offer commercial enterprises many opportunities. Offshore drilling for oil and minerals is the predominant one today. But there are other developing needs. These include off-shore fish and shellfish farming, ocean archaeology, laying and supporting communication cables, tourism, fresh water farming, and renewable energy harvesting. Deep ocean tasks almost require robotics and unmanned vehicles. But surface and near-surface unmanned vehicles can provide a cost-effective alternative to manned ships.

ENERGY SCAVENGING UNMANNED OCEAN VEHICLE

General. Persistent ocean surveillance demands vehicles with exceptionally long endurance. Manned vehicles have to be big to store lots of fuel, or rely on underway replenishment. Nuclear power provides near-unlimited endurance but is problematic for unmanned vehicles. Small unmanned vehicles have limited fuel and therefore do not have the necessary endurance. Scavenging energy from the ocean environment has the potential to give small unmanned vehicles long range and unlimited endurance. By collecting wind, solar, motion and wave energy, an unmanned surface vehicle can be self-propelled, self-deployable, and capable of conducting long range surveillance. The vehicle can use collected energy to provide electrical power for sensors and communications.

The Technology. Unmanned Ocean Vehicles, or UOVs, integrate wind, solar, and motion energy technologies to create a step-change in long-endurance autonomous marine vehicles. Solar energy and wind energy can both be captured and transformed into electrical energy, which in turn is available for propulsion. For maximum efficiency, wind energy is used directly for propulsion using rigid sails equipped with solar panels.

The UOV approach includes:

1. Rigid sails covered with solar panels that maximize the conversion from wind energy to propulsive energy. No other technology can provide the propulsive force provided by average winds at sea.
2. Solar panels on all horizontal surfaces to maximize the availability of electrical energy for hotel/sensor/communications load, and for propulsion via electric motor.
3. Battery storage in the hull bottom to enable operations at night and in low wind. Batteries replace ballast required by wind-powered vessels.
4. Electrical energy generation from propeller when under wind power. A propeller, driven by a high efficiency DC motor, will be used for propulsion in low-wind conditions. When wind power is used directly for propulsion, the propeller will rotate freely, and the DC motor will generate electrical energy.

Although the underlying technologies associated with the use of wind and solar energy are not new, optimizing the system of systems is critical to an autonomous vehicle that can utilize these technologies for primary propulsion and to satisfy an overall power budget. The base technology has already been utilized in manned vessels and in an autonomous vessel. A manned vessel, shown below, the hybrid wind and solar powered "Solar Sailor" is being operated commercially as a ferry in Sydney Harbor.

The UOV Company is developing an unmanned vehicle in cooperation with the Australian company that is operating the Solar Sailor ferry pictured above. The company constructed a full scale Engineering Development model that is the basis for a full prototype and subsequent production models.. A photograph of the 20 foot vehicle is below. All elements of a production vehicle were included. It demonstrated autonomous operation, wireless communications capability, electrical power generation, and sensor data collection. It remains a test bed for systems integration, and experimentation, including alternative wing and sail configurations and control mechanisms.

One Example.

Forecasting the intensity of tropical cyclones is crucial for modern society. However, getting the data is problematic. To improve intensity models, more and better real-time data is needed on the structure and intensity of tropical cyclones, and on their open ocean environment. Environmental surveillance sensor sets must operate in tropical cyclones, should be steered into their path, and be capable of collecting extensive surface and subsurface environmental data. The sensors require a rugged, reliable platform that functions in extreme ocean conditions.

A UOV capable of extended operation in the open ocean is outfitted with a full sensor suite. Capable of self-deploying long distances, it is pre-positioned along historical tropical cyclone paths. One or more UOVs are then vectored into the path of developing or developed storms, and penetrate selected storm quadrants, or into the center. It remains in or near the storm, or passes through it to collect data on conditions in the storm's wake.

The vehicle is tailored to operate in hurricane conditions, and survive extreme hurricane anomalies. The vehicle has multiple-month endurance, large payload capacity (volume and weight), and virtually inexhaustible electrical power for sensors and communications. The vehicles can be outfitted with any number of ocean data collection systems already in service on buoys and manned ships, including tethered underwater sensors.

The vehicle can be launched dock-side, and voyage to distant ocean areas. A preliminary design is about 8.5 meters (28 feet) long and weighs about 3200 kg (7000 pounds). It should be shippable in a standard container. One view is shown in Figure 1. The design uses a proven sea-worthy hull. A small rigid-wing tapered sail is sized for tropical storm winds. A motor generator is connected to a propeller. It can propel the vehicle for tactical maneuvering or during low wind conditions. Most of the time, it is a generator, harvesting energy from the propeller when the vehicle is propelled by wind.

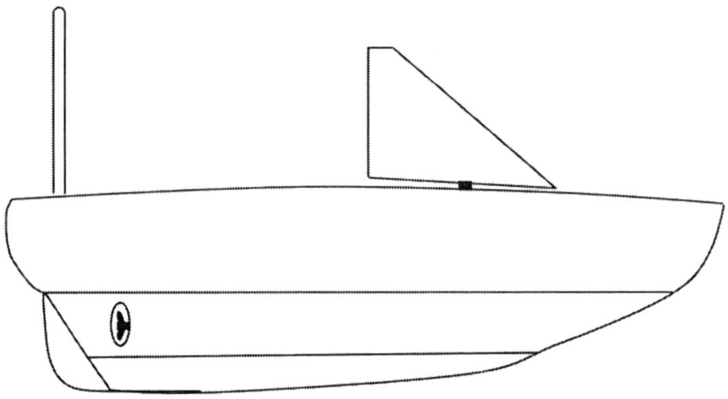

Figure 1

The vehicle power generation system is centered on multiple batteries that provide continuity of power. A stable hull requires a significant amount of ballast. For most boats, poured lead or steel is used. However, batteries can provide ballast instead. As an example, the preliminary design of the vehicle uses 1500 kg (3300 pounds) of batteries, providing 44 kilowatt hours of reserve electrical power. Primary electrical power generation is from solar energy. Secondary electrical power comes from the motor acting as a generator when the vehicle is propelled by the wind. A schematic of the power generation and distribution system is shown in Figure 2. The arrangement of the solar panels is illustrated in Figure 3.

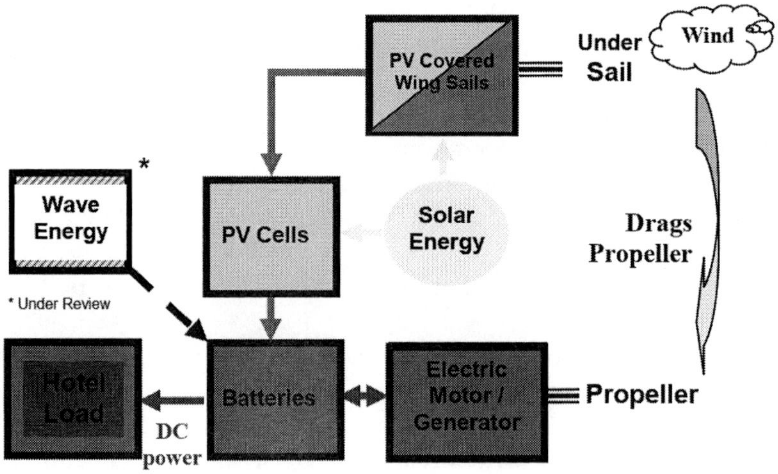

Figure 2: Power diagram.

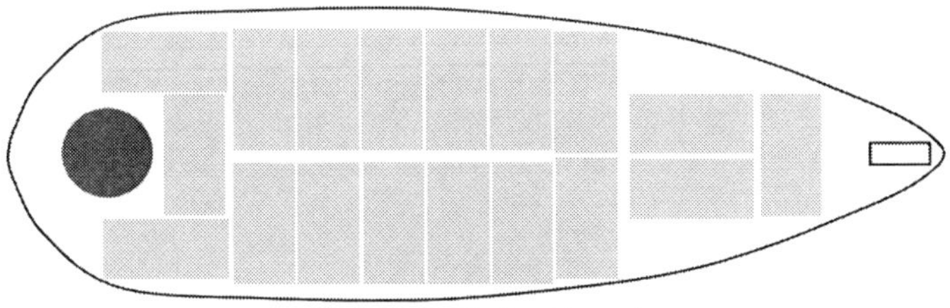

Figure 3: Solar Panels

The following tables show the expected performance of a 8.5m (28 foot) vehicle. The calculations are based on conservative assumptions. The estimates use the following.

(1) Nominal sun is 4 hours sunlight per day at 30 degree sun elevation.
(2) Nominal wind is 20 knots continuous wind (trade winds)
(3) Photovoltaic efficiency from sun to storage battery is 15%.
(4) Wind force (power) is based on driving force at a 60 degree sailing angle.
(5) "Motion Power" is from dragging propeller during sailing.

Table 1
Vehicle Characteristics

Length Overall	8.5 m (28 feet)
Waterline Length	7.6m (25 feet)
Beam	2.3m (7'6")
Weight	3200 kg (7000 pounds)
Battery Capacity	44 KWH
Battery Weight	1500 kg (3300 pounds)
PV Cell Area (not including sails)	10.9 M^2
Sail Area	1.5M^2

Table 2
Nominal Power Generation

Total KWH/Day	7.7
From PV Cells	7.2
From Motion	0.5

The electrical power generated from nominal sun and wind does not include insolation from water reflection, and does not include insolation on the wing sail.

Table 3
Vehicle Hull Propulsion

Speed (knots)	Thrust (kg) [pounds]	Watts (50% efficiency)
2	2.5 [5.5]	40
3	6.3 [13.8]	130
4	12.2 [26.8]	320
5	20.8 [45.7]	700

A displacement hull with a 7.6m (25 foot) waterline length requires little power at low speeds. Drive power from the battery assumes the drive motor and propeller are 50% efficient.

Table 4
Wind Power

Wind Speed (knots)	Driving Force (kg) [pounds]	Sailing Speed (knots)	Motion Power (watts)
10	2.7 [5.9]	1.4	0
20	10.5 [23.0]	3.6	20
30	24.0 [53.0]	5.2	30
40	43.2 [95.0]	6.6	100
80	173 [380]	7.5	100

At wind speeds below 5 knots power generation through fluid motion is minimal. At 20 knots and above, power generation is expected to be significant. It can also be variable, wherein the vehicle controller could opt to put more drag on the vehicle's forward motion to generate additional power, or conversely, less drag for less power. The vehicle could supplement wind power with electric drive in some situations. In hurricane winds, the vehicle should achieve hull speed. Some luffing of the sail may be required.

Table 5
Electrical Hotel Loads

Navigation and Control	20 watts/hour
Communications Equipment	15 watts/hour
Sensors	15 watts/hour
Total	50 watts/hour

The power requirements for on board control and electronics can cover a wide range. For this analysis, the values represent average consumption for a full passive sensor suite and a high data rate communications system. Table 6 demonstrates vehicle endurance using the characteristics tabulated in the first five tables. The propulsion column indicates draw from the battery for a vehicle speed of 2.0 knots in no/light winds.

Table 6
Endurance

Condition Wind	Sun	Propulsion (watts)	Watt-Hours Balance (KWH/day)	Endurance (Days)
Nominal	Nominal	0	+ 5.3	Unlimited
Nominal	None	0	0.0	Unlimited
None	Nominal	75	+ 2.4	Unlimited
Half	Half	20	+ 1.2	Unlimited
Quarter	Quarter	30	- 0.6	50
None	None	75	- 3.6	12

(2) Prototype Vehicle Control.

A power management system will be required to control and distribute electrical energy that is generated by the system. Figure 8 shows a schematic of the system.

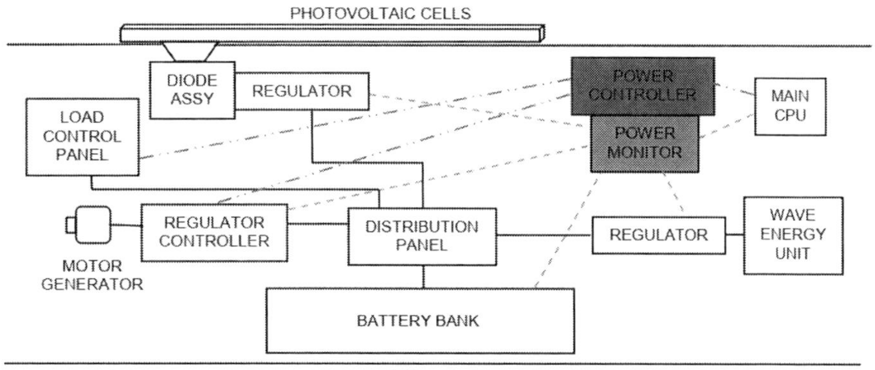

Figure 8

The system is centered on a power monitor that is linked to a power controller. The system will need to assess incident power generation, and then control power distribution as a function of the vehicle mission assignment, the state of the batteries, and the availability of harvestable energy. This effort will be controlled by a central vehicle control system. Identification of the wave energy unit is under review. The system will be controlled by a complete electronic control system, schematically represented in Figure 9.

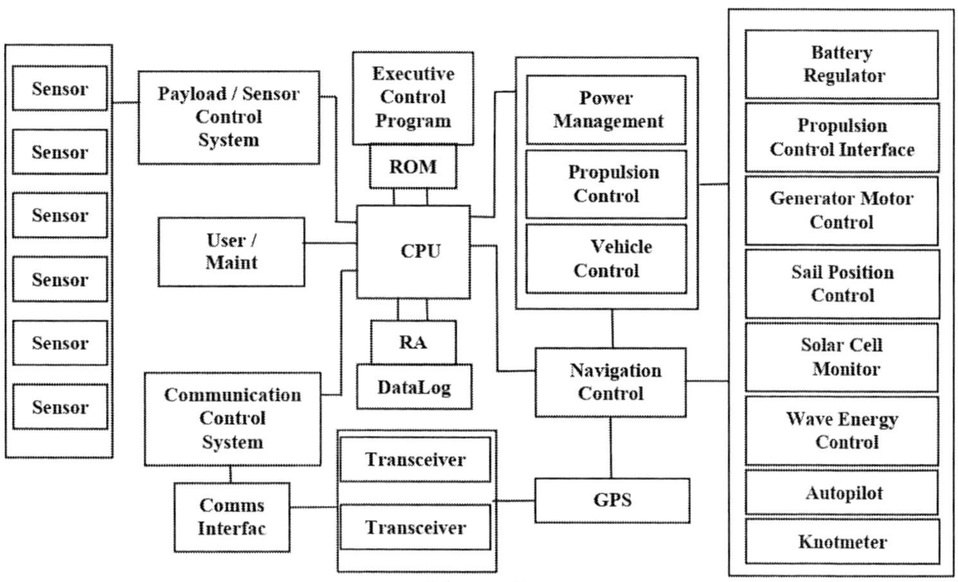

Figure 9.

A final piece of the vehicle design that is critical to its serviceability and survivability is the physical arrangements and control of the wing sail. Figure 10 shows a sketch of the system. It relies on the minimum number of moving parts while still providing positive control.

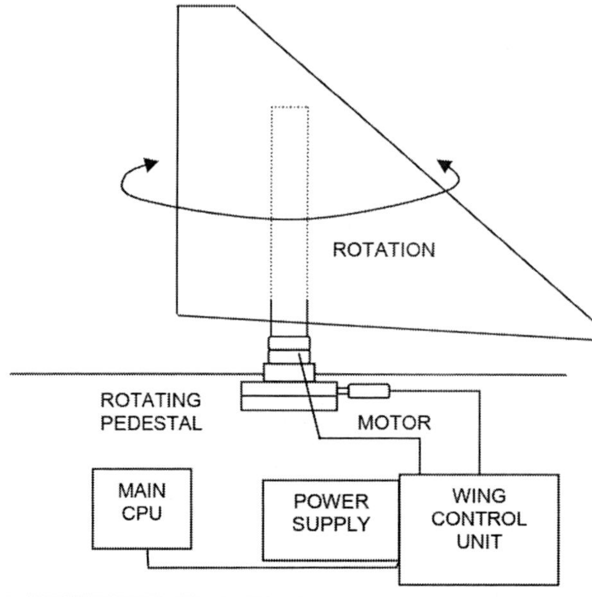

Figure 10.

(3) Vehicle Control Algorithms.

The vehicle control architecture will utilize a layered set of algorithms with the portioning indicated in Table 7.

Table 7: Major classes of control algorithms.

Mission management	Selects mission mode based on sea-state, sunlight, wind, etc; manages waypoints and generates heading/speed commands.
Cruise control	Maintains course and speed with closed-loop control of sails, rudder, and motor.
Power control	Manages power storage, generation, and use. Provides generation and consumption data to mission management subsystem
Data management	Processes incoming data from all sensors, reformats (as needed) and time-stamps all data, assesses data quality and discards bad data, logs all data for post-mission analysis.
Exception management	Monitors all fault conditions, initiates a set of preprogrammed actions, including system shut-down.

The control architecture will be based on Labview, a real-time control software development environment from National Instruments, or other commercial control system. It will adapt the control system developed for the company's Engineering Development Model.

Collision is unlikely due to the very low density of ocean traffic except in shipping lanes. Small boats for centuries have navigated vast ocean tracts at night and in poor visibility without collision. Nonetheless, the vehicles will need to maintain situational awareness and be capable of maneuvering to avoid collision. This will be done using simple topside passive sensors and collision avoidance decision aids. By over-rating the DC motor for short periods, the vehicle possesses sprint capability to allow rapid maneuverability for collision avoidance.

The vehicles will be subject to piracy, theft and vandalism. Operational vehicles will require tamper-proof systems to protect vehicle technology and to transmit assured-delivery notices to owners if the vehicle is breached. Other security features will need to be included.

"Biofouling" will need to be addressed in the design and materials of the vehicle and onboard sensors. Slow speed long term operation at sea can permit marine life to accumulate on underwater surfaces. Other types of fouling of the hull and on topside instruments and surfaces could occur (e.g., debris, sea birds). This potential must be deliberately addressed in the vehicle design and construction. The vehicle will also have a "Fouling" mode of operation. If the system senses underwater or topside fouling, the vehicle will follow a pre-programmed series of maneuvers to extricate itself from any debris or other cause of fouling.

UNMANNED OCEAN VEHICLES - THE FUTURE.

Missions.

UOVs are the beneficiary of four significant recent technological developments. These include:

1. Solar Panel construction and efficiency improvements.
2. Advance scientific sensors with low power requirements.
3. Global communications.
4. Advance computers, networks and data storage, at low power.

These four advances make an Unmanned Ocean Vehicle configurable for many different missions. The technology is scalable to different platform sizes, making the system cost effective depending on customer needs.

UOVs can be equipped to conduct most missions that currently require manned vessels, and their low production and operating cost allows fielding them in large numbers to cover large ocean areas.

Advantages.

Each vehicle has multiple-month endurance, large payload capacity (volume and weight), and virtually inexhaustible electrical power for sensors and communications. The models offer a range of performance and cost. Custom models could be designed for specific applications. Commercial satellite communications link the vehicles to shore for command and control, and for receiving collected sensor data. The vehicles can be outfitted with any number of ocean data collection systems already in service on buoys and manned ships.

The vehicles can conduct many of the missions currently done by manned ships, but at far less cost and without risk to personnel. They can be launched dock-side, and voyage to distant ocean areas. They can be arrayed in a barrier in shipping lanes or a few hundred miles off the coasts and harbor entrances. Because they remain at sea for very long periods, UOVs can collect data on a time-share or multi-task basis. For example, continuously collect water and climate data while also performing security surveillance; or bottom mapping while mine hunting. UOV is the affordable system for gathering critically needed ocean scientific data around the world, today.

E. Payne Kilbourn
President, Unmanned Ocean Vehicles, Inc.
Virginia, USA

The President of Unmanned Ocean Vehicles, Inc. is E. Payne Kilbourn. He brings broad and in-depth management, technical and ocean experience to the company. A graduate of the US Naval Academy with a degree in Naval Architecture, he concentrated on small vessel design and performance. He has extensive ocean experience on wind-driven vessels as a competitive ocean racing sailor, including severe weather experience.

He served on five attack submarines in the Atlantic and Pacific. He was the sonar and communications officer on USS Sturgeon, and conducted three deployments and two special operations in the Atlantic and Mediterranean. He was engineer officer on USS Narwhal, a one design nuclear submarine with a unique propulsion plant, and was the senior watch officer for three deployments and two special operations, including arctic operations, in the Atlantic and Mediterranean. He was executive officer of USS Birmingham, and conducted an extended North and South Pacific deployment, and two special operations, including weathering two typhoons. He was commanding officer of USS Omaha (SSN 692) in the Pacific from 1993-1995, and conducted a Pacific deployment, two special operations, and an inter-fleet transfer from the Pacific to the Atlantic.

Other experience includes operations analysis of conventional naval warfighting on the Joint Staff, direct support to the Vice Chief of Naval Operations, and serving as the USN liaison to NATO headquarters for technical and operational standardization. He was the Navy Federal Executive Fellow to the Brookings Institution. His last tour in the Navy was as OPNAV N772, setting requirements for attack submarine systems, including sensors, sonars, weapons and payloads. He fully understands the challenges of operationalizing technology solutions in the sea services.

After retiring as a Captain in 2003 with 27 years of active service, he co-authored an international patent covering new technology unmanned ocean vehicles, and founded the company to develop energy-harvesting marine vehicles. He directly managed the design of several Unmanned Surface Vehicles, and the fabrication and testing of the first engineering development model, including integration of all systems, and writing the algorithms and software for vehicle control.

The use of warship vulnerability analysis tools as a measurement of blast and ballistic protection requirements.

Michael Buckland
Maritime Platforms Division
Defence Science and Technology Organisation
Melbourne, Australia
Michael.Buckland@dsto.defence.gov.au

ABSTRACT

The recent trend towards performance based requirements by naval regulators has forced a rethink on how vulnerability design features are specified and the performance criteria, against which they can be measured. The development of requirements for the Royal Australian Navy (RAN) to protect its naval vessels against the effects of conventional weapon events i.e. requirements for blast protection, ballistic protection, shock protection, and zoning, separation, and redundancy poses challenges in assigning the performance parameters against the level of protection for a particular asset, and what capability is essential to be retained after an explosive event.

This paper will provide an overview of the use of the Defence Science and Technology Organisation (DSTO) developed Vulnerability Assessment Method CVAM as a tool to measure the required minimum retained capability post a conventional weapon event in order to meet the requirement set for naval ships. The tool can be used to identify both systems and structure which may be modified to maximise the performance of a design to both external and internal conventional weapon detonations.

INTRODUCTION

The survivability of a warship is the combination of its vulnerability, susceptibility, and recoverability. The survivability problem is usually seen as a trade-off between these factors, in that the overall survivability may be increased by reducing either or both the vulnerability or susceptibility of the warship. Susceptibility refers to the likelihood that the warship may be detected and engaged by the weapons systems of the enemy. Susceptibility reduction measures include signature management, control of electromagnetic emissions and the use of soft-kill techniques such as decoys and jammers. Vulnerability refers to the likelihood that the warship may be damaged or destroyed following the impact of enemy weapons and this post-hit military capability is an important military characteristic for the warships operational effectiveness.

There is a natural tendency to assume that the tactics and susceptibility reduction measures of one's own warships will always be sufficient to prevent the impact of enemy weapons. But past experience has shown that combat damage will occur, the USS COLE and the USS STARK being reminders of this. It is thus imperative that minimum levels of protection are built into the design to prevent the loss of essential capability from even a small incident or what is referred to as a "cheap kill".

The ability of the warship to recover from such damage and continue to fight effectively is determined by the combination of the following design features and human attributes;

- Water-tight compartmentation
- Blast-resistant structures
- Separation of systems and components
- Redundancy of systems and components
- Shock hardening levels
- Ballistic protection
- Fire management and suppression
- Smoke removal
- Flood mitigation and management
- Crew numbers, motivation and training

VULNERABILITY REQUIREMENTS

From the viewpoint of achieving warship designs with minimum vulnerability, the optimum approach is that of the United States Navy (USN) which is able to specify minimum quantitative vulnerability requirements to its ship builders. In this way, the ship is designed from first principles with this requirement in mind.

On the other hand, for a country, like Australia, which is reliant on overseas designed ships, it needs to have a well developed set of requirements which are able to ensure the required minimum level of both structural detail and systems vulnerability is acquired for the operational environment of its warships. It also needs to have the ability to realistically measure the performance of the design that is being acquired.

The two basic requirements to produce a warship design with low vulnerability are to;

a. Ensure that the critical components/systems layout meets the minimum zoning, separation and redundancy requirements to survive the expected damage zone from a weapon event, or

b. Limit the damage zone produced by a weapon event by means of hardening the components to the blast, shock and fragment effects of the warhead, or prevent the spread of damage by adding blast and ballistic structural protective measures.

Vulnerability requirements must be based on the weapons or weapons types the ship is expected to encounter in performance of its mission which is usually stated within the Capabilities documentation (e.g. Operational Concept Description (OCD)). The requirements must state that ships be designed to retain some stated level of mission capability after a hit by one or more of these weapons. As with all designs, risk/capability trade-offs will occur. However, an essential requirement for a warship is its ability to resist and/or control the effects of a weapon while retaining a pre-determined level of operational capability. It is essential that a minimum level be stated within the requirements.

It is setting this minimum level of capability that is sometimes difficult to conceptualise, e.g. is the minimum requirement for a warship to survive fully from the effects of a typical Anti-Ship Missile in a crew space, or from a rocket propelled grenade in the operations room. In knowing that all systems are not fully redundant on a ship, are we willing to lose capability from one small but localised event or put in design features such as providing redundancy or protective boundaries around that system?

It needs to be recognised that there are threats and resulting conceded vulnerabilities that a ship would not be expected to survive (depending on the ship size and mission) e.g. a large missile strike in the magazine. However, structural design features and system layouts can minimise the loss of capability of the ship.

The nature of the threats faced by naval vessels have changed. Although designers need to consider "blue water" threats such as submarine launched torpedoes and anti-ship missiles, it is far more likely that in the near future naval vessels will be threatened by low technology weapons such as simple mines, easily available missiles and high-speed boat attacks. Operations in the littoral will make platforms more vulnerable to a "cheap kill" and allowances must be made in the design for protective measures to limit damage from these threats.

Given that conflicts are likely to occur in littoral waters, it is unrealistic to expect that warship damage can be completely avoided. During the Falkland's conflict, for example, approximately two-thirds of Royal Navy (RN) warships in the theatre suffered some form of battle damage despite the availability of organic air cover [1].

Therefore, the vulnerability requirements of warships will need to evolve as both the threat set and vulnerability technologies develop. In the future, it is expected that a greater emphasis will be placed on the protection of the crew from both internal and external detonations, which will lead to a higher level of structural protection such as ballistic protection surrounding critical crewing spaces. The recent Army experience has been to heavily armour its land vehicles against Improvised Explosive Devices (IED's) to provide protection to the crew, rather then to provide protection of the vehicle systems, which is secondary effect consideration.

The functional performance specifications need to be readily measurable and assessable. This requires realistic failure criteria for a range of components and structure to the explosive events. For shock damage, this is well established in that failure occurs due to the vertical acceleration of the components and individual components can be tested and protected by use of shock absorbing mounts or the mounting arrangement within the structure. Full-scale testing of the completed system is still needed, due to the difficulty in simulating the inter-connectivity between the components making up a system. For blast and fragmentation damage, full-scale testing is not a practical option and simulation of the damage is the only cost effective measure.

The operational environment of a warship means that the vulnerability requirements for a warship go beyond that of a conventional ship design. The vulnerability of a warship can be greatly reduced by making relatively simple and inexpensive changes to its structural details and configuration. For example, the resistance of bulkheads to blast overpressure can be more than doubled by changing the edge attachment details [2]. Depending on the type of weapon, this could result in the structure surviving the effects of a weapon detonation at a stand-off factor of less than half the distance (greatly reducing the total damage radius), or from a weapon perspective, will provide equivalent protection from a weapon of more than twice its size for the equivalent stand-off (structure can now survive the effects of a larger warhead). By confining the damage to a smaller zone, the ship designer is able to have greater flexibility in the layout of systems within the ship to maintain the required capability post event.

In developing performance measures it is difficult but essential to capture lessons learnt from previous combat damage events and ensure the same mistakes are not repeated. The USS STARK survived the immediate effects of an ASM, however the resulting fire highlighted the poor performance of aluminium alloys. Enquiries into the USS COLE concluded that certain additional design improvements to the ship class could improve crew survivability in the future [3]. For example, the internal communication system would be less vulnerable if a wireless system was employed to replace the current hard wire system. The failure mode logic and increased capacity of the pumps used for removing water from flooded spaces should be considered. Flooring material tends to shatter into glass-like shards that cause injury. Autopsy and injury data indicated that the major cause of multiple blunt forces was flying metallic and non-metallic fragments.

VULNERABILITY ASSESSMENT

The ability to be able to assess if a particular design meets these minimum requirements relies on realistic simulations of the warhead/ship scenarios, as it is unrealistic to provide a large number of live fire scenarios which could possibly occur. The use of modern simulation techniques to undertake vulnerability assessments of new and existing designs is well established.

The usual method of stating the vulnerability performance is a probability of kill. However it needs to be clearly stated what is actually meant by this value, i.e. a probability of survival for the ship of 0.9 may indicate that 90% of detonation points located randomly across the design will result in the essential capability being retained. Alternatively it could indicate that on average for a warhead detonation of a set type and size, 90% of the capability is on average functional. The interpretation of this value will lead to different consequences when protective measures to improve the performance are to be implemented.

It is also important to set out the process and failure criteria in obtaining the probability of kill Pk value as both the damage contours and the damage algorithms used to simulate the weapons effects will vary with different assessment methodologies. For example, a requirement that the mobility capability has a Pk of 0.8 after a detonation of a warhead of a certain size may pass one assessment methodology but fail an alternative methodology. Recently, Navy Systems Command implemented essential capabilities that the Air Warfare Destroyer (AWD) is to retain after being hit by warheads of varying sizes. To achieve this, each of the primary capability areas were broken down into smaller sub-functions which were required to have a Pk of 0 after an event. This made it clear if the design failed to meet the requirements and areas of concern which may need to be modified.

Generally, vulnerability assessment techniques are a balancing act between providing physically exact models and the need to rapidly assess thousands of different scenario combinations in the shortest time-frame. The simplest and least time consuming is the use of standard blast damage contours overlaid on the design as shown in Figure 1. This method which is able to quickly identify zoning, separation and redundancy issues.

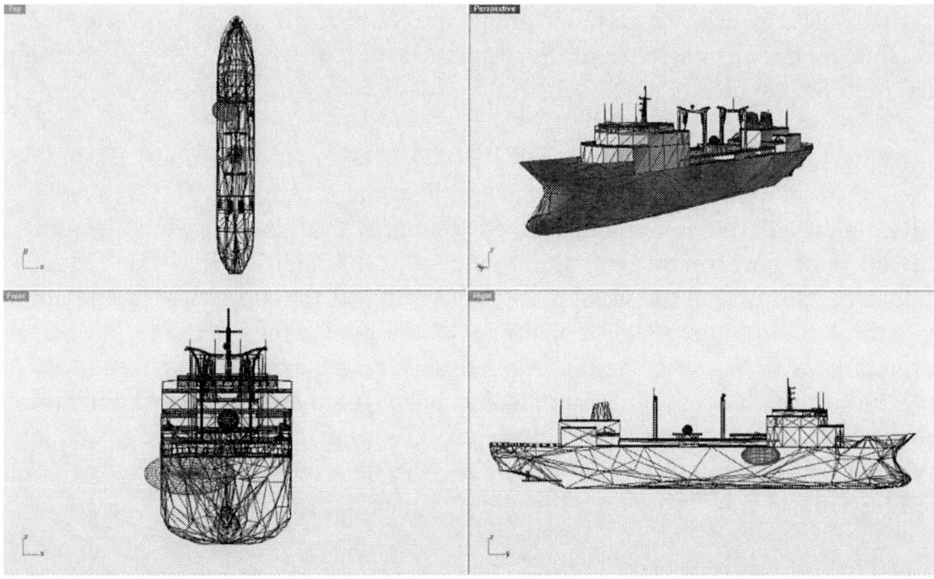

Figure 1 Damage contour overlay. Damage contour represents 100% damage.

At the preliminary stage of the design process, the use of blast damage contours is a very efficient method in comparing the relative performance of a number of designs. As the design progresses and becomes more evolved, a more detailed and physically correct methodology can identify critical areas of the design which may require attention. For example the initial damage contour assessment will treat all structural detail, bulkheads, hatches and doors equally; a more detailed analysis as shown in Figure 2 is able to determine the progress of damage through the structure and can therefore highlight the need of, for example, blast resistant doors to mitigate the spread of blast overpressure along passageways.

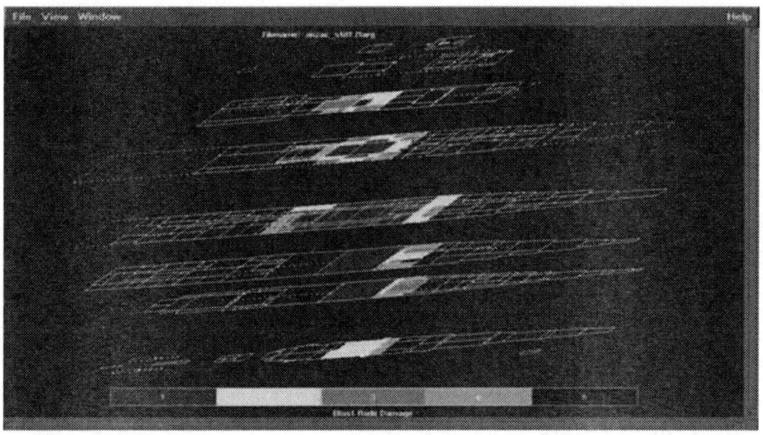

Figure 2 Vulnerability analysis output identifying bulkheads

For more detailed simulations, more precise representations of the structure are required. This is being addressed with the CVAM Vulnerability Assessment tool.

CVAM SHIP VULNERABILITY CODE

The CVAM ship vulnerability assessment suite of computer programs developed by Maritime Platforms Division of DSTO provides a methodology for modelling ship structures and systems using three-dimensional compartment models and detailed fault-trees to describe the functionality of the major systems. Damage to the structure and systems of the ship is simulated using a variety of weapons effects algorithms which have been developed at DSTO over the past fifteen years. Algorithms for explosive blast and fragmentation effects are based on the results of field experiments, finite-element code predictions and data derived from the Ship Survivability Enhancement Program (SSEP) conducted aboard the former HMAS Derwent during 1994 [4]. This data set is also enhanced by data received by DSTO through international defence forums.

One of the major drawbacks in being able to assess ship designs at the early evaluation stage is the time required in developing a ship model with enough detail to produce a realistic assessment. The simulation requires an accurate representation of the design, showing the general arrangement of the compartments, information on the structural elements, and the location and interconnectivity of components. CVAM has addressed this by making use of the modelling and display features inherent in the CATIA product modelling package. CVAM has been developed as an add-on module to the CATIA product modelling package; which has allowed the vulnerability assessor to make use of the detailed information stored in electronic product models of the ship supplied by the ship designer.

The CVAM code requires the use of ship electronic product models being supplied or converted to the STEP (STandard Exchange Protocol) format. That allows the transfer of product data from the supplier's native product modelling software package without being tied to the CATIA software. The data formats and detailed data requirements required by the electronic model for use in CVAM have been formulated into the Navy Systems Command requirements for electronic product models. These documents are in the Data Identification Description documents for the Electronic Product Model [5] and the Vulnerability Report [6].

A major advantage of the CVAM code is that it allows an accurate representation of both the structure and components within the design. This allows for more realistic simulations of the spread of blast and fragment damage throughout the warship. CVAM currently provides methods to assess both above water, internal and external detonation effects, and below water effects.

The CVAM assessment methodology has been designed around a number of individual assessment modules. These modules are initially accessed via a start-up menu as shown in Figure 3. This allows for the logical progression through the vulnerability assessment.

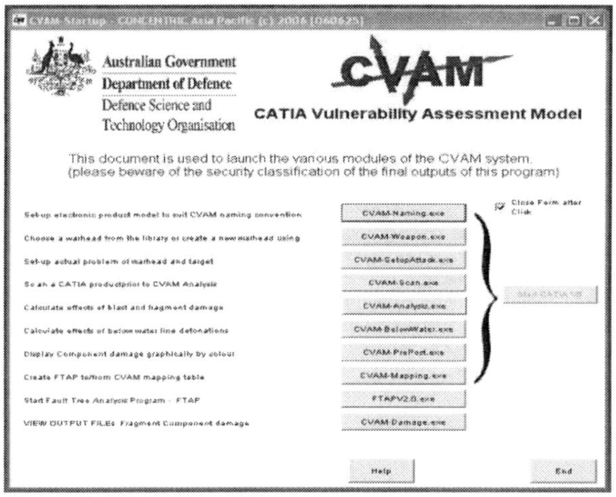

Figure 3 CVAM Start Menu.

To successfully model the effects of the warhead/ship interaction, both the warhead and ship need to be modelled realistically. The CVAM weapon creation module shown in Figure 4 allows the user to access a database of predefined weapons previously created or allows users to input their own variables.

The warhead creation module requires the warhead characteristics of the weapon, such as weight and type of explosive, and fragmentation pattern characteristics of fragment mass and velocity to be input into the simulation. Fragment zonal data is usually determined from arena testing of the warhead or else from warhead predictive software based on the shape and casing material of the warhead.

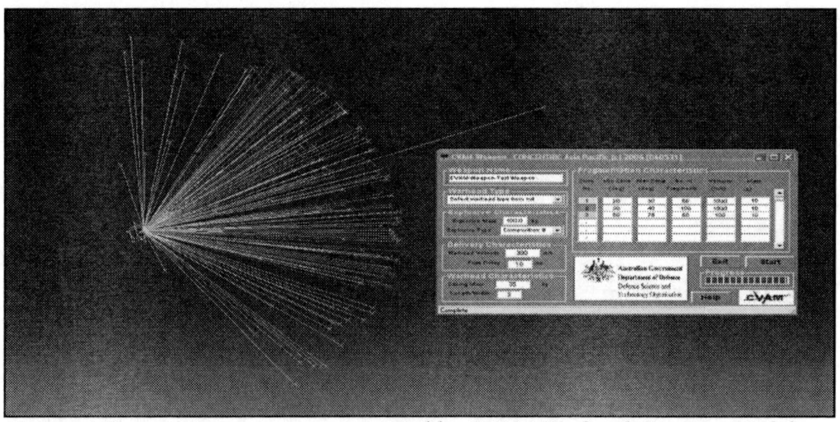

Figure 4 Fragmentation pattern generated by CVAM Warhead Creation Module.

The electronic product model of the ship may be built using either solid or surface elements. Although CVAM processes the information slightly differently depending on the type of elements being used, CVAM is able to access information by using a unique data structure and naming convention. The user is able to rapidly check if the model supplied has been appropriately named using the module shown in Figure 5. This allows for the interrogation and checking of the essential elements required by the vulnerability assessment to correspond to the vulnerability data identification description documents. It also facilitates the modification of any models received into the appropriate structure in a timely manner.

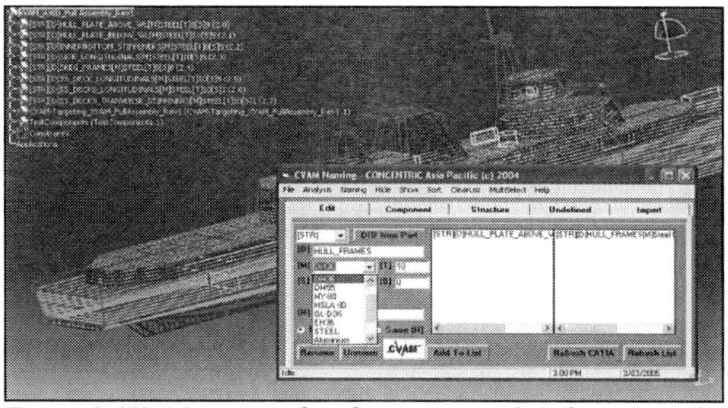

Figure 5 CVAM structural and component identification module.

Once both the warhead and the electronic ship models have been developed, the CVAM users are able to create the scenario best suited to the problem they are trying to simulate. Options based on the warhead attack direction and warhead fusing options are able to be programmed into the module as shown in Figure 6. Also single detonation points can either be saved or programmed into the database to allow for rapid processing of scenarios.

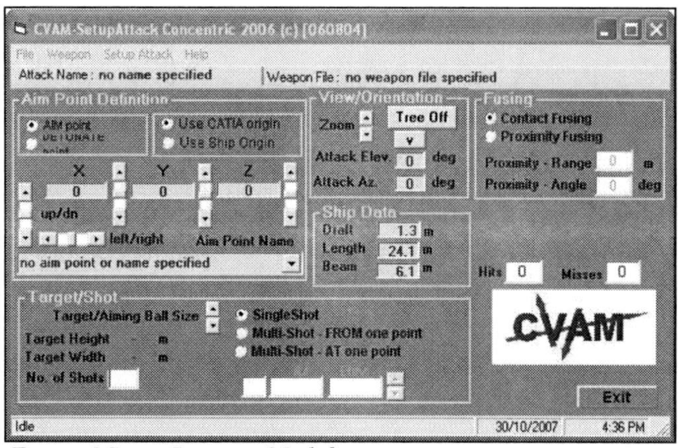

Figure 6 Setup Scenario Module.

Depending on the scenario being analysed CVAM, will provide probability of kill values for both structural and component damage. Damage from both above and below waterline can be analysed. Figure 7 shows the fragmentation paths through the ship structure, details of each path and its intersection with both structural elements and components can be extracted from a database.

Figure 7 CVAM output showing fragmentation paths

CVAM has been developed to allow the separation of the weapon effects algorithms which may be of a classified nature. This file can be stored separately to the rest of the code, allowing the user to run a scenario, and the results database interrogated at a later date.

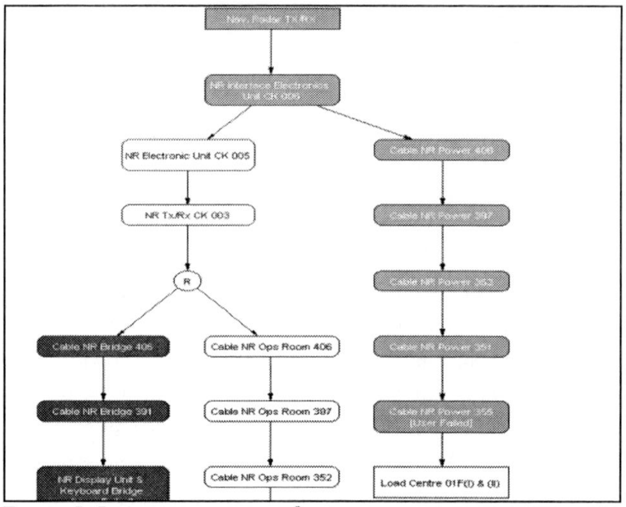

Figure 8 Inter-connectivity of components.

To facilitate the rapid generation of fault trees which describe the functionality of the major fighting systems, a windows based graphical fault tree editor known as FTAP V2 has been directly linked to the CVAM component kill output file. These components can be identified in a fault-tree as shown in Figure 8, allowing critical elements to be identified (failure of left-hand branch in Figure 8 is not critical due to redundancy, failure of right-hand branch has resulted in a critical failure).

Rather than assessing the vulnerability of individual ship systems, it is conventional to link the systems together into major capability areas such as the Anti-Air Weapon (AAW) system, the Anti-Submarine Weapon (ASW) system, the Anti-Surface Warfare (ASuW) system and the Mobility system (encompassing navigation, steering and propulsion). Results presented in this way are more directly usable for tactical assessments and operational planning. In the assessment of new designs, the major capabilities are broken down to sub-systems to refine the assessment to more manageable systems identifying essential capabilities.

CONCLUSION

The use of vulnerability assessment methods at an early stage of the design process is essential to ensure that the vulnerability is kept to a minimum and that the vulnerability requirement performance measures are met. Retrofitting of blast and ballistic structural elements can be costly and impractical. It is imperative that the location of these elements is identified at the earliest stage of the design. The lack of detailed information on the critical systems at the early evaluation stage can make the identification of areas to be protected difficult. As the design matures, it is important to have the design reassessed regularly to identify any weakness in the system integration methods.

CVAM has been developed by DSTO to provide a tool for the assessment of conventional weapons effects on a ship. Instead of developing unique vulnerability models, CVAM makes use of existing electronic product models which can be used by a range of simulation tools.

REFERENCES

1. Walsh B.E., *Vulnerability Assessment Manual for Conventional Weapons*, DSTO-GD-0349, 2002.
2. Walsh B.E., *Blast-Resistance of Bulkheads for current and future RAN Warships*, DSTO-TR-0910, 1999.
3. *Live Fire Test and Evaluation Overview Annual Report FY2001*, http://www.globalsecurity.org/military/library/budget/fy2001/dot-e/accessories/01lft.htm, 2007.
4. Walsh B.E., Burman N.M., and Buckland M.E., *Ship survivability Enhancement Program (SSEP) – Trial of Conventional Weapons Data*, DSTO-GD-0105, 1996.
5. DID-REP-VULNERA-V1.1- Vulnerability Report.
6. DID-DWG-EPM-V1 - Electronic Product Model.

Reduction of Roll Motion of a Surfaced Submarine in Beam Seas

A Davies[1,2], D Ranmuthugala[2], J El-Atm[2], Y Tso[3]
[1] Royal Australian Navy (RAN), Launceston, Tasmania, Australia
[2] Australian Maritime College (AMC), Launceston, Tasmania, Australia
[3] Defence, Science, and Technology Organisation (DSTO), Melbourne, VIC Australia.

1. INTRODUCTION

In 2004, the Canadian submarine *HMCS Chicoutimi* was immobilised on the surface in sea state eight due to an onboard fire. This prompted the Defence, Science, and Technology Organisation (DSTO) to instigate a joint project with the Australian Maritime Collage (AMC) to investigate the parameters influencing the roll of intact and damaged surfaced submarines in beam seas, and to identify methods to reduce the roll motion.

The parameters investigated included: the effect of entrained water in the submarine top casing, (including the size of its draining ports); orientation of the stern control surfaces between the "x" and "+" configurations; addition of "wing tips" to the control surfaces; addition of bilge keels; and the effect of adverse trim due to damaged conditions. The experimental investigation consisted of testing a scaled model submarine in AMC's towing tank and test basin. The model was tested in fixed beam seas and in a free-to-drift condition, while subjected to a variety of wave velocities. The damaged condition, when required, was simulated by ballasting the forward compartment. The experiments determined the best options for the top casing (including port opening sizes and the effect of entrained water), and quantified the effects of various appendages for different encounter frequencies (ω_e) and vessel headings. The effects of adverse trim on the above results were also analysed.

A newly developed vessel hydrodynamic software package by Defence Research and Development Canada-Atlantic (DRDC Atlantic) in collaboration with DSTO was also used to theoretically model the submarine and analyse its motions. This allowed comparison with the experimental results and validation of the package.

2. EXPERIMENTAL SET-UP

A generic scaled submarine model was developed jointly by AMC and DSTO to study the roll behaviour of surfaced submarines. The model has a ballast tray within it (Fig 1), enabling adjustment to the model's vertical, longitudinal, and transverse centres of gravity to ensure appropriate static position. In addition, the ballast tray was orientated such that the model's *KG* and moment of inertia (I_{xx}) were consistent throughout the project. For the damaged condition, a mass was added to the forward section to simulate a flooded section between 10-12% of its length (see Fig 11).

In order to investigate the submarine's roll response due to the orientation of the vessel's stern control surfaces, the stern cone could be rotated. This resulted in either a "+"or "x" configuration, while a blank ("o") stern cone provided a reference setting (Fig 9). The two different tail cone pieces were ballasted to have an identical mass, which ensured that interchanging the tail would result in minimum variations to the parameters.

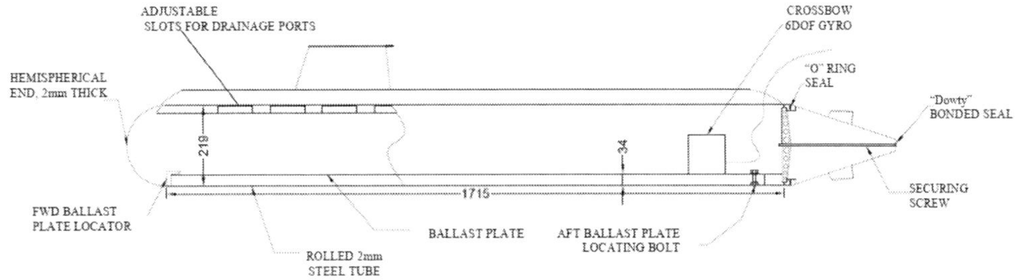

Fig 1 - AMC 05-06 Generic Submarine Model

Since the study included investigating the affects that entrained (free draining) water within the casing has on the roll of a surfaced submarine, drainage ports (Fig 1) in the ballast tanks were setup and tested at three different settings, i.e. 100% open, 50% open and watertight. This was also used to investigate the flow in the casing, thus enabling the free surface in the casing to act as an anti-roll tank.

The investigation into the affect of added appendages on roll required the independent addition of wingtips (winglets) to the stern control surfaces and bilge keels to the parallel body. The bilge keels were sized according to Ref 3, with the alignment optimised for flow in a submerged condition resulting in the bilge keels running straight along the hull. The bilge keels are 1/3 the model length, with a 45 degree angle of entry and exit, and a depth of 20 mm. The model had removable wingtips of an aerofoil shape fitted on to the control fins and optimised for forward motion.

Testing was carried out in three phases, i.e. investigating the roll motion characteristics of a surfaced submarine in beam seas (Ref 5), the same for a damaged submarine (Ref 2), and the affects of added appendages on roll damping (Ref 1).

Test Facilities

The experiments were carried out in AMC's hydrodynamic test facilities, i.e. the Towing Tank and the Model Test Basin. The former is 100 m long, 3.55 m wide, with variable water depth up to 1.6 m. The facility has a powered carriage to tow models up to a maximum speed of approximately 4.6 m/s. Waves are produced by a hydraulically driven wave-maker positioned at one end of the tank. The computer software that controls the wave-maker paddle can produce a wide variety of waveforms, including: regular waves, most commonly available wave spectra, and user defined wave spectra or sequences. A large wave energy dissipation device (beach) is fitted to the opposite end of the tank.

The model test basin is 35 m long, 12 m wide, with a variable water depth up to 1 m. The bottom of the basin is flat, providing the ability to conduct experiments in very shallow water depths, if required. The combination of the increased width, solid flat bottom, variable water depth, model towing winch, and wave generation capability makes this facility extremely versatile. Waves are produced by an electrically driven wave-maker positioned on one side of the basin. The computer software that controls the 16 individual wave-maker paddles can produce a wide variety of wave forms, including: regular waves, irregular waves, and user defined wave spectra or sequences.

Larger wave heights were tested in the towing tank due to the limitations and capabilities of the test basin. Sea states four to nine were tested in the towing tank, while sea states four to six were tested in the test basin. Analysis into wall effects is also of importance, therefore comparing data between the tow tank and the test basin under the same test condition allowed the detection of errors. All free-to-drift tests were conducted in the test basin, since its larger breadth allowed a wider degree of un-obstructed motion. Although both facilities are capable of producing a large number of wave height and wave frequency combinations, for model testing purposes it was imperative that the combination resulted in regular wave trains, i.e. did not break or spill.

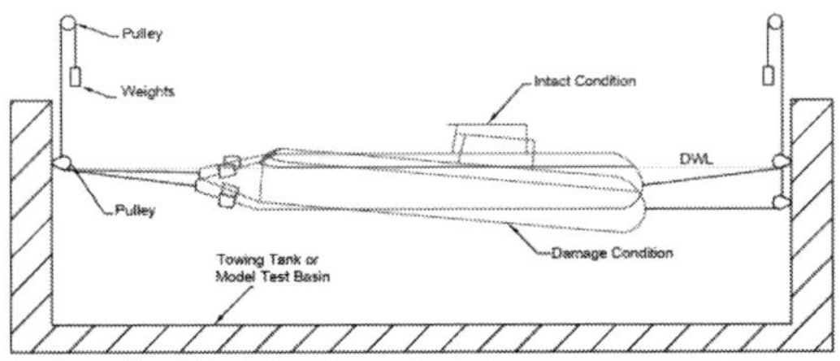

Fig 2 - Model Set-Up

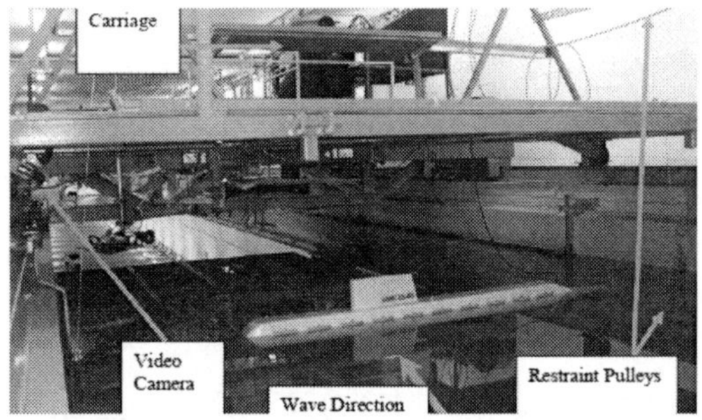

Fig 3 – Intact Model in Tow Tank

The model was restrained with a cable from the bow and the stern that were connected at the centre of roll of the model, (Figs 2 and 3). These cables were then threaded through pullies at each side of the tank and restrained by weights, which depended on the wave amplitudes being tested. Data was gathered using a six-DOF gyro located within the model, measuring vessel motions in terms of velocity. A static wave probe in the tank provided the wave amplitude and wave frequency. Integrating the raw data response curve for each cycle allowed the calculation of the roll amplitudes. In order to accurately measure roll amplitudes based on the response curve, a program was developed in *LabVIEW*.

3. EXPERIMENTAL RESULTS

The results shown are a selection from those obtained during the testing (given in Ref 1,2 and 5), with comparisons made between the intact and damaged conditions.

Effect of Flooding Rates of the Casing in both Intact and Damaged Conditions

The tests for the three different casing slot heights in the damaged and intact cases, revelled some distinctive characteristics in the roll response graphs for both conditions (Figs 4 and 5). General characteristics include:

- Response shifts to lower natural frequencies for intact and damaged conditions as the wave amplitudes increase (Fig 5).
- Roll RAO's decrease with increasing wave amplitudes (Fig 4).
- Lower natural frequencies in the damaged condition for casing at 50% and 100% open, and higher natural frequencies for 0% open (Fig 5).

At constant sea state and variable casing openings, (Fig 4):

- For intact and damaged conditions, with casing open at 50%, roll response is greater when compared against other casing open conditions.

At constant casing opening and variable sea states, (Fig 5):

- At lower sea states, the damaged condition exhibits smaller roll response compared to the intact condition.
- Higher sea states cause the opposite effect, with the damaged condition having a greater roll response compared to the intact condition.

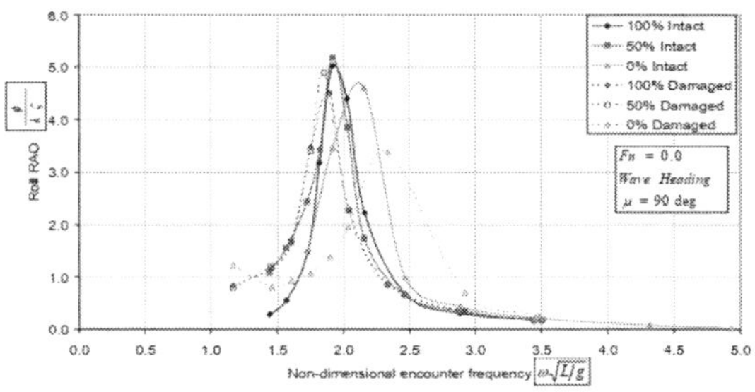

Fig 4 - Casing Openings, Intact vs. Damaged (Sea State: 4, Height: 54 mm)

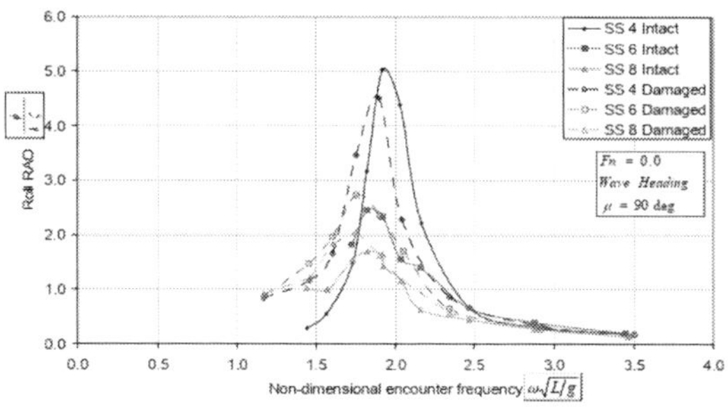

Fig 5 - Casing Opening, Intact vs. Damaged (Constant Casing Position)

The shift in natural frequency to a lower value is attributed to the increase in added mass of the vessel during rolling motions, (equations [1] and [2]). Since added mass of a vessel is directly related to the accelerations of the underwater hull, and caused by the energy or excitation force imposed by the wave, the magnitude of the accelerations change significantly. If Δ is vessel displacement, k_{xx} is the roll radius of gyration, and GM_T is the metacentric height, then its virtual mass moment of inertia ($a = I'_{xx}$) and the roll natural frequency (ω_ϕ) are:

$$a = I'_{xx} = (\Delta + added\ \ mass)k_{xx}^2 / g \qquad \ldots\ldots [1]$$

$$\omega_\phi = \sqrt{\Delta GM_T / a} \qquad \ldots\ldots [2]$$

Since the metacentric height of a submarine does not change significantly during rolling motions, it can be confidently assumed that the natural frequency is related to wave height and roll radius of gyration. Higher wave amplitudes will have a greater exciting force impacting on the vessel and therefore cause greater velocities and accelerations. In the damaged condition, the vessel has a higher roll radius of gyration due to the coupling effect of yaw. In addition, in a bow down plane, it will have greater added mass than for the intact condition, due to the asymmetrical wetted hull body. These two characteristics, linked with accelerations due to waves, cause a shift in the natural frequency to a lower value. Further, the magnitude of overall shift in the resonant peak is greater in the damaged condition than in the intact condition, due to a higher roll radius of gyration and asymmetric underwater body.

At a casing opening of 0%, the natural frequency in the damaged condition is higher than for the intact case. This is due to an increase in displacement and limitations in vertical mass distribution, i.e. the centre of gravity is moved lower in the 0% casing position. In turn, the metacentric height (GM_T) is increased, thus changing the natural frequency. The roll response is greater for the casing slot positioned at 50% open for both intact and damaged cases. Although it is not immediately clear why this phenomenon occurs, it maybe a function of the internal sloshing frequency coinciding with the vessel's natural frequency. The roll amplitude (ϕ_a) is given by,

$$\phi_a = \frac{\left(\alpha'_m + F_{Slosh}\right)}{\sqrt{\left(1-\Lambda^2\right)^2 + 4k^2\Lambda^2}} \qquad \text{...... [3]}$$

Where α'_m is the maximum wave slope with respect to heading and time, F_{Slosh} is the sloshing force, Λ is the turning factor (i.e. ω_e/ω_ϕ), and k is the non-dimensional damping factor. If the sloshing frequency is in phase with the vessel's natural frequency, the magnitude of F_{Slosh} will be large. In the damaged condition, the magnitude of F_{Slosh} causes a 7% increase in roll response compared with a 4% increase in the intact case. This may indicate that the sloshing frequency depends on the position of casing slots and the amount of fluid present in the casing itself, which also agrees with the findings in Ref 4. Note: a loss of buoyancy is experienced when the casing is 50% and 100% open, and the model is rolled in excess of 5^0, due to the ingress of water into the casing.

At low wave frequencies, with casing slots at 50% and 100% open (both having the ability to passively drain), the intact and damaged conditions exhibited small signs of wave force excitation due to internal sloshing in the top casing as shown in Fig 6. A similar response was seen for the condition of casing 0% open (Fig 7), however the secondary peak is more pronounced. Larger roll amplitudes are also noted for the 0% open, damaged case.

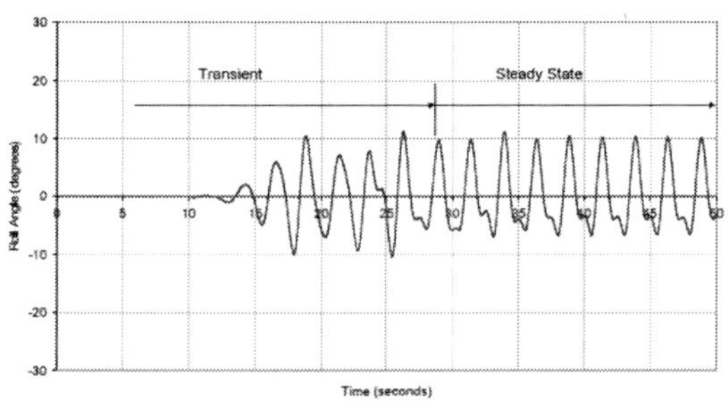

Fig 6 - Damaged Condition Time History, Casing 50% Open, H_w = 214 mm

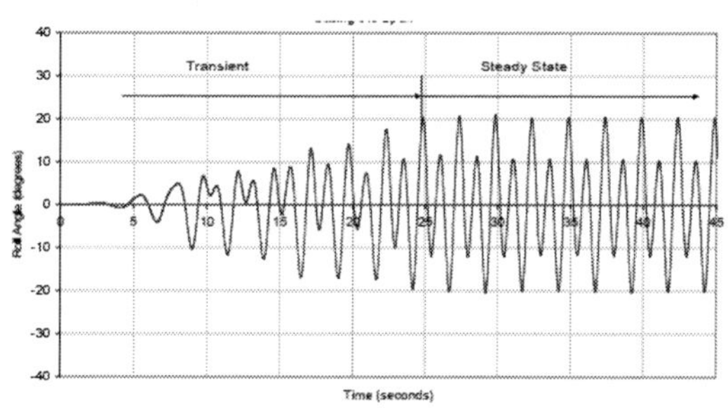

Fig 7 - Damaged Condition Time History, Casing 0% open, H_w = 214 mm

While the sloshing force has a slight impact on roll motions, it is believed that the effect of entrained water has a greater influence on the overall mass moment of inertia (a). Reference 5 shows that for sea states eight and nine (unlike that for low sea states), the roll amplitude at resonance for the 50% and 100% open conditions are not significantly greater than that for the 0% case. This could be due to the effect of the entrained water having a greater affect on the mass moment of inertia (a) rather than the restoring coefficient (c) on roll amplitude (ϕ_a) in equation [3]. (Note: $k = b/2a\omega_\phi$, where b is the dampening coefficient and $\omega_\phi = \sqrt{[c/a]}$). However, the wave force imposed on the submarines sail will cause a greater effect on the overall roll motions.

Effect of Vessel Heading

In response to a request from the RAN, the effect of heading angle on the roll response was investigated by testing the model at four headings from beam seas through to 50°. As expected and shown in roll theory, the maximum roll occurs when the vessel is in beam seas and drops linearly to heading angles of 70°, 60° and 50° as shown in Fig 8.

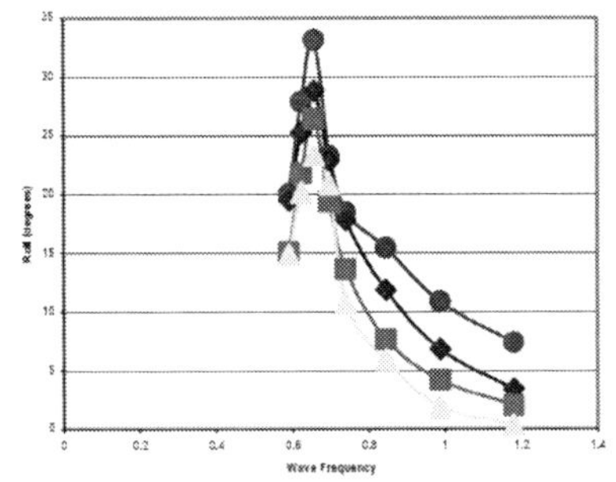

Fig 8 - Effect of Heading in Sea State 6

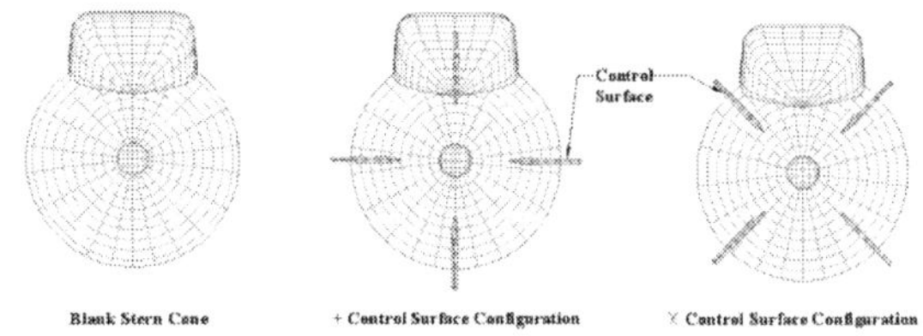

Fig 9 - Stern Cone Configurations

Figs 8 & 9 - Source: Ref 5

Effect of Control Surface Orientation

Another design feature of a submarine that affects the roll is the orientation of the control surfaces on the stern cone of the vessel. To evaluate the effect that these control surfaces have on the roll, three conditions were tested, i.e. the "+" and "x" stern cone configurations, plus a blank ("o") stern arrangement as a control test to compare against the other two forms (Fig 9).

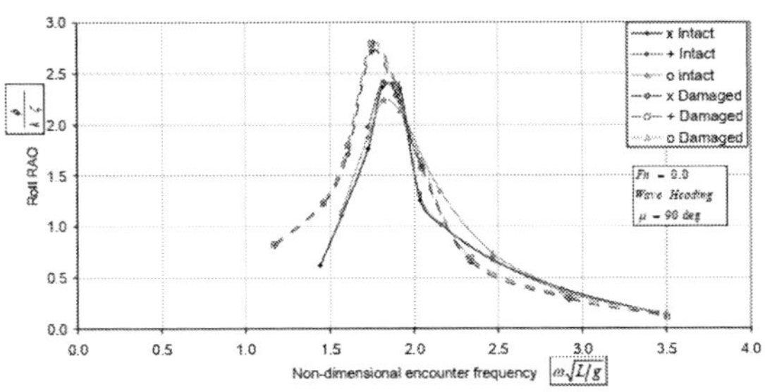

Fig 10 - Tail Cone, Intact vs. Damaged, SS 6, Hw = 143 mm, Casing 100%

Fig 11 - Damaged Condition in Tow Tank

Upon comparing the damaged and intact test results (Fig 10), there are some distinctive characteristics noticed in the roll response graphs for both conditions. General characteristics of the roll response graph include:

- Roll response of the damaged condition is greater than that of the intact condition.
- Natural frequency is lower for the damaged condition.
- Effect of tail configuration has greater influence on the magnitude of roll motions in the intact condition.
- The "x" tail is most effective in reducing roll motions in damaged condition.

The roll response in the damaged condition is larger than in the intact condition and is attributed to the immersion of the control surfaces. At a level trim, the control surfaces are almost completely submerged, however trimming the vessel causes the stern to rise (Fig 11) and reduce the wetted area of the control surfaces. The magnitude of roll damping is attributed to the wetted surface area of appendages (A_{app}), the distance to (the base of) the appendage from the centre of gravity (d_1), and the vessel's form of bilge (i.e. hard chine or round bilge). The more roll damping present, the less the vessel will roll. This is also recognised in Ref 2, where damping is given by,

$$damping = \frac{0.55\left[A_{app}\sqrt{W} + 0.0024\,L\,B\sqrt{d_1}\right]d_1^{\,2.5}}{\Delta B^2} \quad \text{...... [4]}$$

Where W is the width or height of appendage, L is the length of waterline, and B is the breadth of waterline. In the damaged condition, the "x' stern configuration provides greatest roll damping as there are always two control surfaces submerged, while the "+" stern has its horizontal appendages continually entering and re-entering the water when the vessel is rolling. Since the size of control surfaces is small with respect to the vessel's length and displacement, the difference in roll damping is also small. It is concluded that the effect of the tail orientation has only a minimal effect on the roll damping and hence the roll motions in damaged condition for a surfaced submarine.

Effect of Drift

The model was allowed to drift freely in the model test basin, with the only restraint being the data cable that was suspended from a pole above the model, thus allowing the model to move in six-DOF. These tests indicate how the surfaced submarine would behaves if propulsive power is lost in beam seas. From Fig 12 it is noted that the roll amplitudes are larger at resonance for the damaged condition and the effect of tail orientation on results is very similar for both conditions. From Fig 13, the general characteristics for the yaw response show:

- The rate of yaw and angular displacement is greater in the damaged condition.
- Tail orientation has significant impact on yaw response for the intact condition.
- At higher wave frequencies in the intact condition, the vessel yaws away from the oncoming waves, i.e. encountering following seas after 30 seconds.
- In the damaged condition, the opposite effect occurs with the vessel yawing into the oncoming waves, i.e. encountering head seas after 30 seconds.
- Yaw rate is nearly identical for the damaged condition at encounter frequencies of 0.7 Hz and greater.

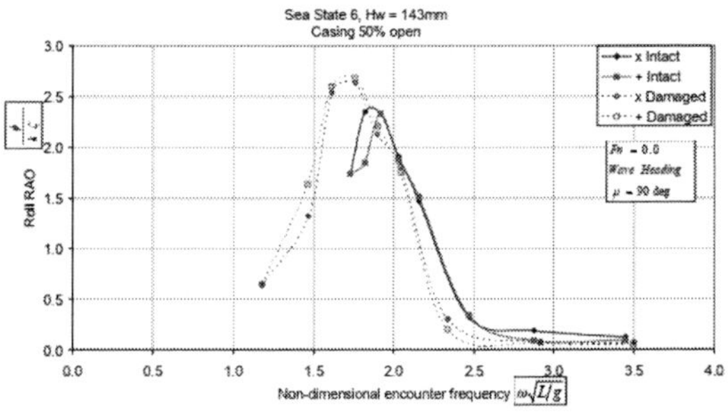

Fig 12 - Free to Drift, Roll after 30 sec, Hw = 143 mm, Casing 50%

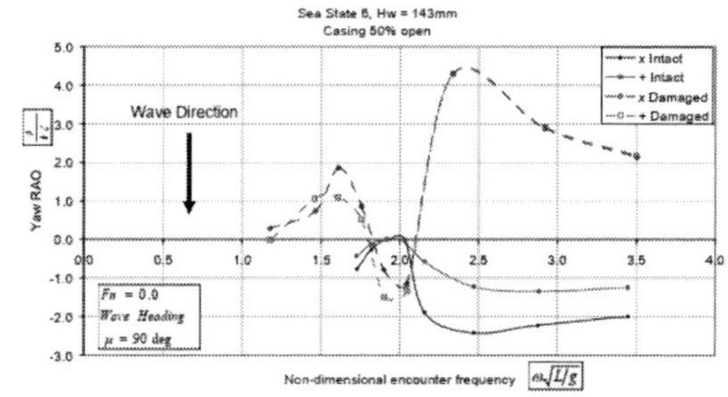

Fig 13 - Free to Drift, Yaw after 30 sec, Hw = 143 mm, Casing 50%

Maximum roll response is apparent when the submarine is in beam seas. At an encounter frequency close to the natural frequency, the vessel yaw rate remains relatively small and larger sway or drift velocities are present. Through video analysis, it is seen that coupling exists between roll and yaw motions. It was seen that at low frequencies for the damaged condition, the vessel will yaw into the oncoming waves. However as the encounter frequency increases, the vessel begins to yaw away and slowly experiences "following seas". Fig 13 clearly shows this response in terms of yaw RAO and follows a similar trend to the intact case, although the peaks in the lower frequency region are more exaggerated.

The most interesting characteristic for the damaged tests was at an encounter frequency greater than 0.8 Hz, where the submarine no longer yaws away from the wave direction rather it yaws to face the oncoming waves. In turn, the vessel is now encountering head seas. Directional stability is maintained through the top fin and immersed control surfaces. Also, since the forward body is trimmed bow down, additional drag is created forward; therefore as waves excite the vessel, the stern pivots around the bow. As wave frequencies increase, the surfaced submarine yaws at a higher rate into the waves.

4. WING TIPS AND BILGE KEELS

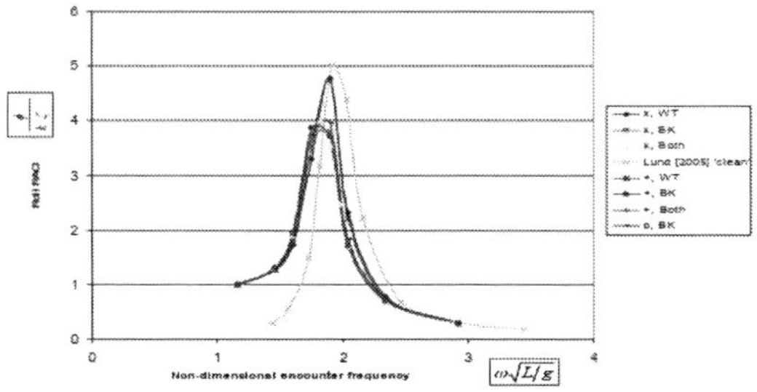

Fig 14 – Sea State 4, Roll RAO

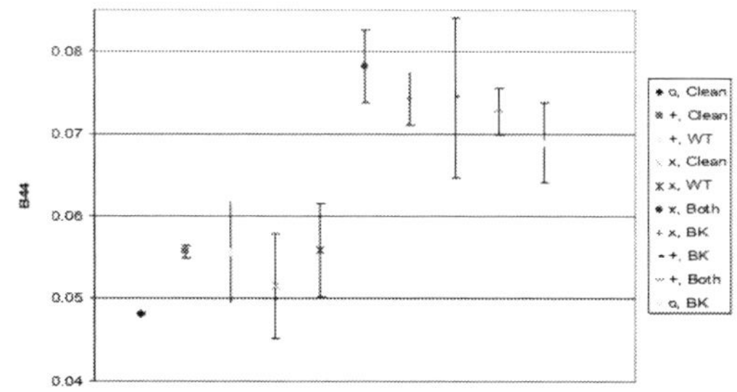

Fig 15 - Roll Damping Coefficients for Tail and Appendage Configurations

The last phase of the tests was to evaluate the affects of adding wing tips and bilge keels to the model in beam seas, through a series of roll decay tests. These have just been completed and are currently being analysed, however some results showing the effects of the appendages are presented below. At low sea states (Fig 14), it is seen that the wing tips in the "x" stern configuration have a minor effect in reducing the peak of the RAO, while the bilge keels have a substantial effect in reducing this peak. From Fig 15, comparing the roll damping coefficients shows the improvement due to the installation of wing tips, and the larger improvement with the fitting of bilge keels. This is expected in this lower sea state, since the addition of any roll damping device to a submarine hull will be an improvement, as the submarine hull offers very little in the way of roll damping.

At the largest sea state tested, the bilge keels appear to have an adverse affect on the roll motions, increasing the peak height of the RAO (Ref 1). The reasons for this increase in roll motion is attributed to the lack of reserve buoyancy and the lack of roll damping of the submarine shape, thus the bilge keels increase the excitation effects of the wave motions. The wing tips in this case provide better damping than the bilge keels in lower frequency waves, thus suggesting that to minimise roll motions in larger sea states the installation of wing tips is beneficial. The added benefit of the wingtips is that they also offer directional stability.

5. SHIPMO3D

ShipMo3D is a new object-oriented library for simulation of ship motions in waves developed by DRDC Atlantic (Ref 6). It is a panel method for solution of hydrodynamic forces giving robust and accurate results for both slender and non-slender hull forms. Forces from rudders, bilge keels, and other appendages are included to ensure accurate roll predictions. Nonlinear buoyancy and incident wave forces are used to maintain reliability in higher sea states, while still maintaining faster than real time computational speed.

The program was provided by DRDC Atlantic to simulate the model submarine in order to validate the code (the first time it was used in anger) and to assist in the investigation. In this simulation, ShipMo3D predictions were carried out using a linear 3D frequency domain method, since the motion amplitudes are fairly small. Thus, the nonlinearities in the hydrodynamic coefficients were not predicted with real confidence. The comparison was made for roll response by analysing the data using transfer functions or RAOs. The numerical hull was modelled with its casing position at 0% open. A ShipMo3D model is shown in Fig 16.

A complete set of results for changing wave heights of numerical prediction against experimental data is given in Ref 2. Fig 17 shows the comparison for the effect of the tail configuration under damaged conditions, where the resonant peaks of the numerical prediction are close to experimental values, and there is little difference between the roll responses of various tail configurations.

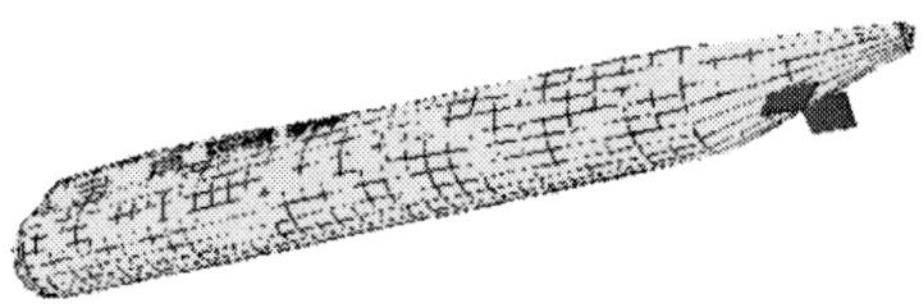

Fig 16 – 'x' tail orientation in Damaged condition

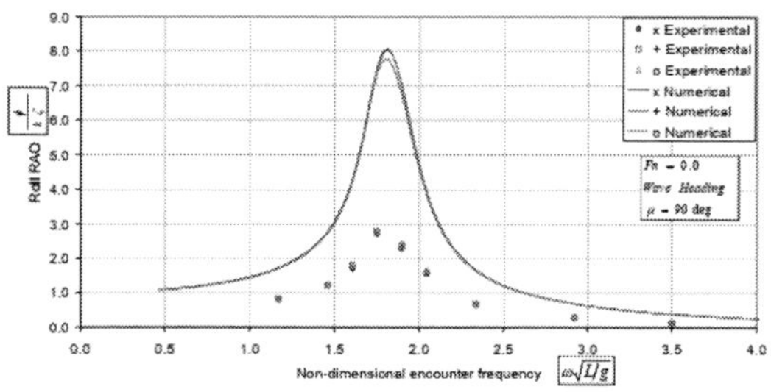

Fig 17 - Numerical and Experimental Comparison, Effect of Tail Configuration in Damaged Condition

However in all cases, ShipMo3D over predicts the peaks for roll response, as seems to be the case with many other theoretical programs of a similar nature, (e.g. Formation Design Systems, Seakeeper®). This may be due to the complexity in solving the non-linear behaviour for vessels in roll motions. The shift in natural frequency is not evident in the numerical prediction as the wave height increases; however the roll response reduces similar to the experimental data. In addition, the numerical method estimates the intact condition to have greater roll response, which does not agree with experimental data

Therefore, a numerical package such as ShipMo3D or Seakeeper® are not ideal to predict absolute vessel roll motions, however they are useful to find broad characteristics of the vessel's behaviour. In addition, it was seen that ShipMo3D is very sensitive to the vessel's centre of gravity and roll radius of gyration, where a small change in either, results in large variations in the overall response.

The numerical results indicate that the roll response is significantly less in the damaged condition. This suggests roll damping due to the hull body is much greater than the contribution form appendages and may explain the discrepancy for the effect of tail configuration in the damaged condition experimental findings.

6. CONCLUSION

At smaller wave heights, the damaged condition exhibits less roll amplitudes than the intact condition, however this changes as the wave height increases. For both conditions with casing at 50%, roll motions are greatest for all wave heights. The natural frequency reduces with increasing sea state. However there is an increase in the damage condition when the casing is closed. This is attributed to an increased displacement to keep a bow down trim of five degrees.

The effect of tail orientation for both the 'x' and '+' tail configurations had little effect on roll damping, while the 'o' tail configuration had approximately 30% less roll damping. This indicates viscous damping contributes a large portion of overall roll damping. The 'x' tail is marginally better for reducing roll motions when damaged. The damping due to the immersed hull body is deemed a greater contributor to damping than the control surfaces.

The ShipMo3D model over predicted roll response peaks and did not show the shift in natural frequency with change in wave height. In addition, it gave a lower roll RAO for the damage condition than the intact one, contradicting experimental data. The tail configuration tests showed better results with the natural frequency aligned with experimental data. The roll RAO prediction agrees with the findings of the roll decay experimental data. ShipMo3D is not ideal to predict absolute vessel roll motions, but is useful to find broad characteristics of the vessel's behaviour.

The addition of bilge keels helped reduce roll amplitudes at low sea states, but has an adverse effect at large sea states. Although the wing tips do improve damping in all cases, the improvement is marginal.

In free to drift tests, it was found that the vessel remained beam on to the oncoming waves around its natural frequency for both conditions. A threshold frequency range was found for which the surfaced submarine in damage condition would yaw into the wave train, thus encountering head seas. Because of the drag created by the bow when trimmed, the stern will yaw around the bow and inherently face the vessel in head seas.

More work is currently being carried out to develop a comprehensive database to understand the behaviour of surfaced submarines, especially in damaged conditions.

7. ACKNOWLEDGEMENTS AND REFERENCES

The authors would like to thank Mr Warren Lund, Dr Kevin McTaggart, Mr Gregor MacFarlane and the staff of the Hydrodynamics Centre at AMC for their contributions and assistance throughout the project.

1. Davies, A., *The Effect of Appendages on Roll Damping on a Surfaced Submarine*, BE Thesis, AMC, 2007
2. El-Atm, J., *Roll Motion of a Damaged Surfaced Submarine in Beam Seas*, BE Thesis, AMC, 2006
3. Ikeda Y. and Umeda U., *Application of Slender Body Theory to Rolling Motion of a Moving Ship with Bilge Keels*, SNAME, Japan, May 1988
4. Lee, B. and Vassalos, D., *An Investigation into the Stabilisation Effects of Anti-Roll Tanks with Flow Obstructions*, Int. Shipbuilding Progress, Vol. 43, No. 433, 1996
5. Lund, W., *Parameters that Effect the Roll of a Surfaced Submarine in Beam Seas*, BE Thesis, AMC, 2005
6. McTaggart K., *ShipMo3D User Manual for Simulating Motions of a Freely Maneuvering Ship in a Seaway*, Defence R&D Canada – Atlantic, Canada, 2006

Advanced CFD Simulation Of Submarine Dynamics

Kyong-Huhn Lee & Craig Pregnalato
BMT WBM, Melbourne Australia
cpregnalato@wbmpl.com.au

ABSTRACT

This paper presents a novel approach to submarine design through the use of computational fluid dynamics. The dynamics of a submarine are investigated when fully submerged and also when in the vicinity of the water surface. A coupled solution method is demonstrated that combines the unsteady flow solver with a six degree-of-freedom solver that updates the motion of the submarine taking into account the hydrodynamics forces acting on it. Overlapping meshes are used to accurately predict the motion of the submarine whilst allowing relatively long integration times. By performing fully coupled simulations of the submarine, it is possible to predict and analyse in detail the vessel's maneuvering performance as well as to optimize its design through a very cost-effective process.

INTRODUCTION

The maneuvering of submarines and other marine vessels using computational modelling is a task that is receiving widespread attention in the maritime industry. Predicting the maneuvering performance of submarines using numerical simulations is an appealing option to designers who have traditionally relied upon tow-tanks and other experimental facilities to validate their designs. Thus, a tool that can accurately compute the hydrodynamic forces acting on the submarine hull, propeller and control surfaces, as well as use these forces to predict the motion of the vessel, is of great interest to the naval community.

Computational fluid dynamics (CFD) has become an attractive instrument in modern times for predicting hydrodynamic forces and investigating the near wake flow of submarines. CFD simulations have been used in a wide range of investigations including maneuvering, propulsion, seakeeping and resistance. Recent advances have demonstrated the accuracy that can be obtained in capturing the near wake flow of surface vessels using unsteady Reynolds-averaged Navier-Stokes (URANS) simulations to large eddy simulations (Yang et al., 2007).

It has been common practice in the past to neglect the influence of the pressure field around the hull of a vessel with a view to efficiently computing the flow around the body in a given fixed position or in uniform motion. However, this approach is not suitable from a design point of view since precise predictions of the hydrodynamic forces are unattainable due to the close union of the fluid forces and the vessel's motion. This obstacle has recently been resolved with the employment of 6-DOF solvers that couple the hydrodynamics with the rigid body dynamics to yield solutions that accurately reflect the fluid-structure interaction.

Early simulations using a coupled 6-DOF solver were performed by Azcueta (2002) for sailing yachts in steady flows as well as incident regular waves from different directions. In this work, a commercial CFD solver was used to demonstrate the capability of the coupled 6-DOF method when investigating yacht responses to oblique waves. Similar techniques were employed by Peric et al. (2007) to simulate sloshing in LNG tanks for floating tankers in head waves. These results demonstrate the possibility of computing coupled simulations of the flow around floating vessels and in its internal tanks, along with the vessel's motion.

Although coupled 6-DOF CFD simulations have a long history in the aerospace industry, e.g. for predicting the trajectories of stores separating from aircraft (Murman et al., 2003), these methods have relatively recently been employed in the maritime sector to fully submerged bodies. Venkatesan & Clark (2007) present novel simulations of submarine maneuvers for both a constant depth and heading as well as a horizontal overshoot maneuver. In this study, the propulsion of the submarine was included by physically modelling the propeller and its rotation using sliding mesh blocks. Good qualitative agreement was found between the simulations and published experimental data, however neglecting the effect of sternplane deflections appeared to affect the accuracy of the results. Pankajakshan et al. (2002) showed that quite good experimental agreement can be obtained when including all control surface deflections. Nevertheless, the method shows promise for the numerical modelling of submarine and similar underwater vehicles, and reinforces the suitability of CFD simulation tools to aid in the design and analysis of submarines.

The objective of the present study is to assess the potential of an existing fluid flow and 6-DOF solver when applied to submarine motion. Specifically, the dynamics of a submarine when surfacing are investigated to predict the rigid body response when excited by head waves. Particular emphasis is placed on evaluating different mesh strategies in order to yield a computational model that is most fitting from a design and analysis perspective. Although no experimental data currently exist for comparison, the results demonstrate the simulation capability and promote further validation work.

COMPUTATIONAL MODEL

The simulations presented in this paper were performed by solving the unsteady Reynolds-averaged Navier-Stokes equations with the standard two-equation k-ε eddy viscosity model. The solver is second-order accurate in both space and time. Submarine dynamic simulations represent a two-step process that require the solution of the URANS equations for the fluid motion as well as a 6-DOF equation for the rigid-body motion, and the method presented herein is well suited to this application.

The equations governing the fluid motion are solved first using a URANS solver as detailed in Demirdzic & Muzaferija (1995). These equations define the conservation of mass, momentum and other scalar quantities (such as turbulence) and are respectively:

$$\frac{\mathrm{d}}{\mathrm{d}t}\int_V \rho \,\mathrm{d}V + \int_S \rho(\mathbf{v}-\mathbf{v}_b)\cdot\mathbf{n}\,\mathrm{d}S = 0$$

$$\frac{\mathrm{d}}{\mathrm{d}t}\int_V \rho\mathbf{v}\,\mathrm{d}V + \int_S \rho\mathbf{v}(\mathbf{v}-\mathbf{v}_b)\cdot\mathbf{n}\,\mathrm{d}S = \int_S (\mathbf{T}-p\mathbf{I})\cdot\mathbf{n}\,\mathrm{d}S + \int_V \rho\mathbf{b}\,\mathrm{d}V$$

$$\frac{\mathrm{d}}{\mathrm{d}t}\int_V \rho\phi\,\mathrm{d}V + \int_S \rho\phi(\mathbf{v}-\mathbf{v}_b)\cdot\mathbf{n}\,\mathrm{d}S = \int_S \Gamma\nabla\phi\cdot\mathbf{n}\,\mathrm{d}S + \int_V \rho b_\varphi\,\mathrm{d}V$$

where $\mathbf{v}_b$ is the vector velocity of the control volume surface, $\mathbf{T}$ is the effective stress tensor, $\mathbf{b}$ is the body force vector, φ is a scalar quantity, Γ is the diffusivity coefficient and b_φ represents sources or sinks of φ. Since submarine maneuvering simulations inherently require the modelling of moving boundaries, a space-conservation law must also be satisfied and is given by:

$$\frac{\mathrm{d}}{\mathrm{d}t}\int_V \mathrm{d}V - \int_S \mathbf{v}_b\cdot\mathbf{n}\,\mathrm{d}S = 0$$

In this paper we focus on submarine dynamics near a free surface, and in order to allow for deformation of the free surface a VOF-type interface-capturing scheme is used to solve for the volume fraction c of the gas phase governed by the equation:

$$\frac{d}{dt}\int_V c\,dV + \int_S c(\mathbf{v}-\mathbf{v}_b)\cdot\mathbf{n}\,dS = 0$$

The sharpness of the interface between the water and the air is obtained using a high-resolution interface capturing (HRIC) discretization scheme. For further details of the numerical method for the flow solver, see Muzaferija & Peric (1999) and Ferziger & Peric (2003).

The second step in the solution process is to update the motion of the submarine. This is accomplished by solving a 6-DOF set of equations that govern the rigid body motion of the vessel and are composed of the linear (translation) and angular (rotation) dynamics, represented respectively by the following equations:

$$\frac{d(m\mathbf{V})}{dt} = \mathbf{F}$$

$$\frac{d(\mathbf{I}\cdot\boldsymbol{\omega})}{dt} = \mathbf{M}$$

In these equations m is the mass of the submarine, **I** is its moment of inertia, **V** is the velocity of the submarine's center of mass, $\boldsymbol{\omega}$ is its angular velocity, **F** is the total force acting on the submarine and **M** is the net moment. The total hydrodynamic force **F** is comprised of both pressure and viscous (shear stress) contributions.

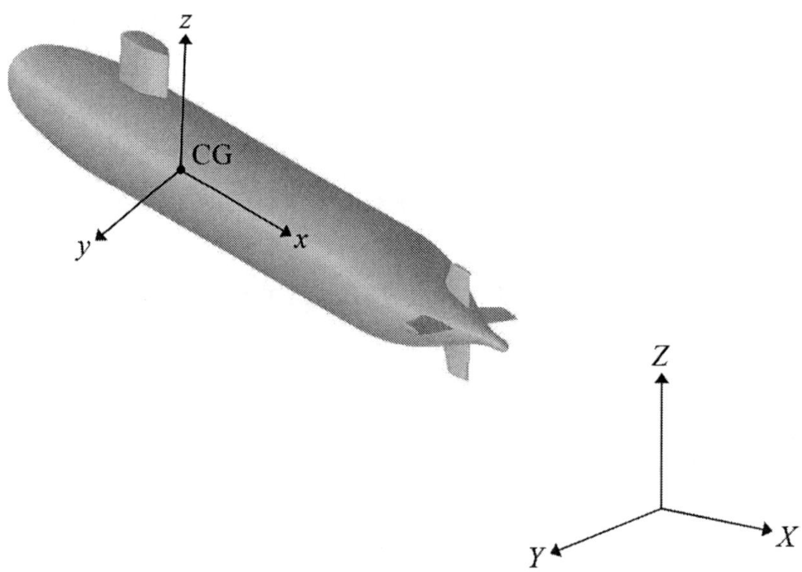

Figure 1. Inertial (X,Y,Z) and body-fitted non-inertial (x,y,z) coordinate systems used for the submarine simulations.

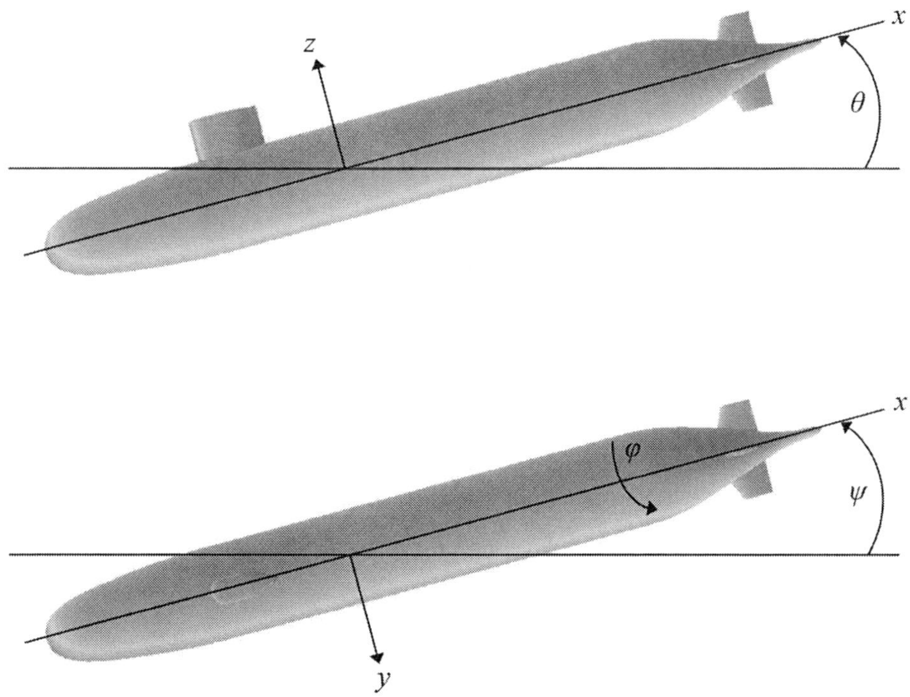

Figure 2. Definitions of the pitch, roll and yaw angles, denoted by θ, φ and ψ respectively.

Figure 1 and Figure 2 show the coordinate systems and Euler angles respectively that define the vessel motion. A simplified submarine of length 14m and displacement 18.3 tonnes was modelled, and in order to reduce the size of the computational mesh whilst allowing relatively long integration times, a half model of the submarine was simulated with the plane of symmetry being the *x-z* plane. Thus, although the method allows full 6-DOF motions, in the present analysis the roll and sway motions were not taken into account.

Two different mesh strategies are employed for the present simulations. The first approach involves moving the computational mesh with the rigid body as it translates and rotates in response to the wave motion. In the second approach, an overlapping (or oversetting) grid is used for comparison to investigate any improvements that this method may offer. Similar Chimera meshing techniques have recently been used, for example, to investigate tank sloshing on floating vessels (Peric et al., 2007). In the current overlapping method, a Cartesian background grid was created relative to the fixed inertial coordinate system (X,Y,Z). A finer body-fitted mesh governed by the non-inertial coordinate system (x,y,z) attached to the submarine was also created in order to better predict the hydrodynamic forces acting on the vessel. Figure 3 shows the grids used for the simulations.

The total number of elements for the moving mesh model is approximately 310000, and 393000 for the overlapping mesh model. The upper, lower and side surfaces are modelled as slip walls, and the flow outlet is modelled with a zero-gradient pressure outlet boundary condition. A uniform velocity of 5m/s was specified at the inlet boundary. The head waves at the inlet are superimposed on to the mean flow velocity and are computed from linear wave theory by specifying the wave amplitude A, wavelength λ and the wave direction α relative to the heading of the submarine. For the present simulations, the parameters used are A = 0.3m, λ = 10m and α = 0°.

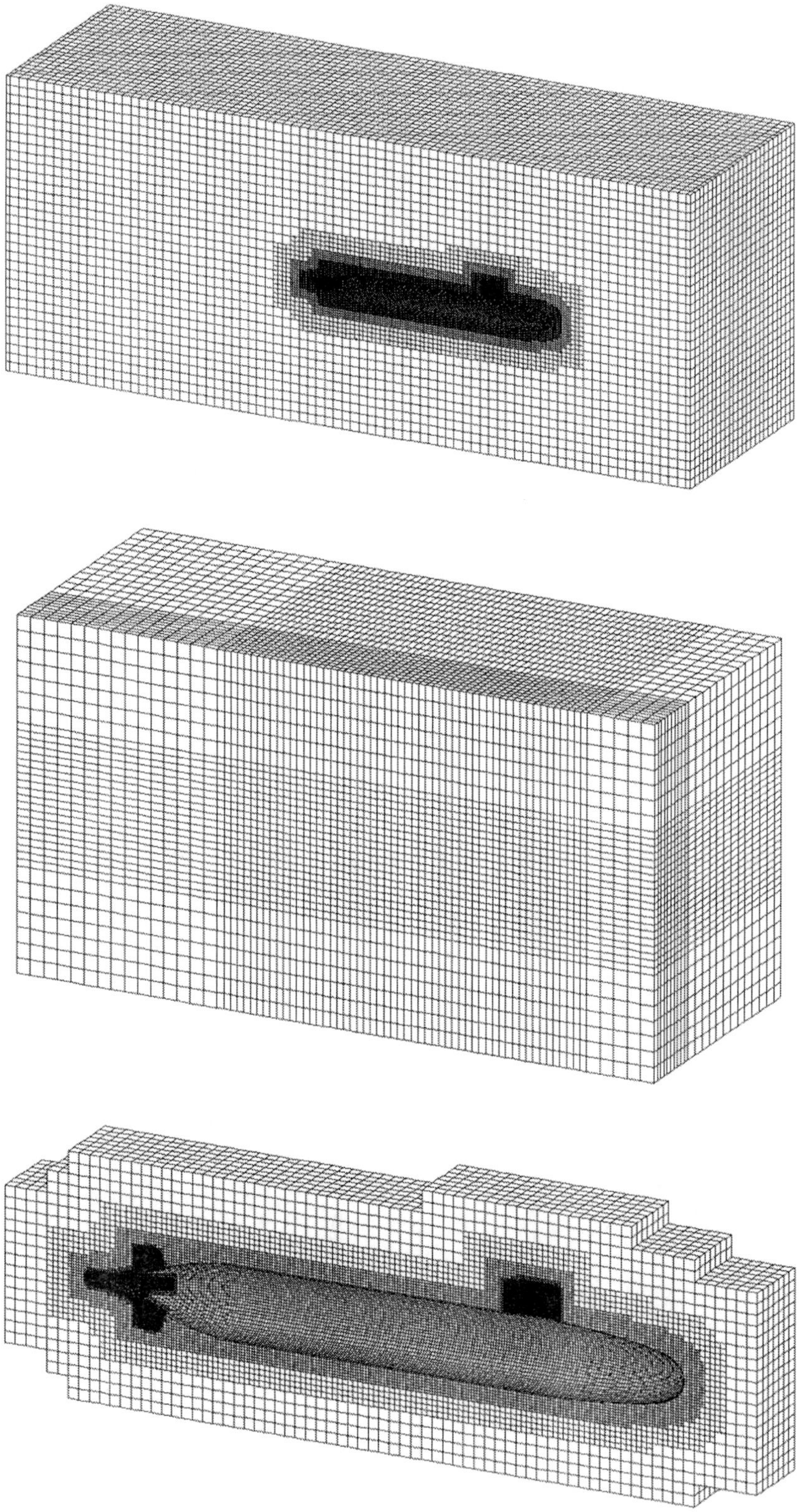

Figure 3. Grids used for the CFD comparisons: single, non-overlapping mesh (top); background Cartesian overlapping mesh (middle); body-fitted overlapping mesh (bottom).

RESULTS

Ass stated in the previous section, couples 6-DOF simulations were performed in this investigation for both a single moving mesh and an overlapping mesh. However, only the overlapping mesh results are communicated here in order to demonstrate the method and to reduce duplication. At the start of the simulation, the submarine was initially positioned below the water surface so as to introduce hydrodynamic non-equilibrium. The subsequent motion of the vessel was then computed in response to the incident waves. Figure 4 shows the computed z-displacement of the submarine's center of gravity (CG) as well as the pitch angle with time. The initial large displacements and rotations are evident as the vessel surfaces and readjusts to the hydrodynamic forces.

Figure 5 highlights the computed resistance and heave forces on the submarine. For the half model analysed here, the weight of the submarine is roughly 90kN. This weight is correspondingly balanced by the mean heave force, which likewise is approximately 90kN from examination of Figure 5. In the same way, the average thrust of the submarine may be calculated to be approximately 6.2kN (based on its velocity of 5m/s), and this driving force is balanced by the computed resistance of roughly 6.2kN as illustrated in Figure 5.

In order to compute the full 10s of flowtime shown in these figures, approximately 80 hours of CPU time was required for the overlapping mesh model on a standard single processor workstation. For the single moving mesh model, around 60 hours of CPU time was required on the same platform. This increase in computing time for the overlapping mesh model is primarily due to the larger mesh used for this simulation. However, it should be noted that this is more than offset by the significant reduction in model set-up time that was observed for the overlapping mesh. The increase in complexity in creating a suitable high-quality grid for the single mesh model becomes prohibitive especially when considering multiple iterations in the design process.

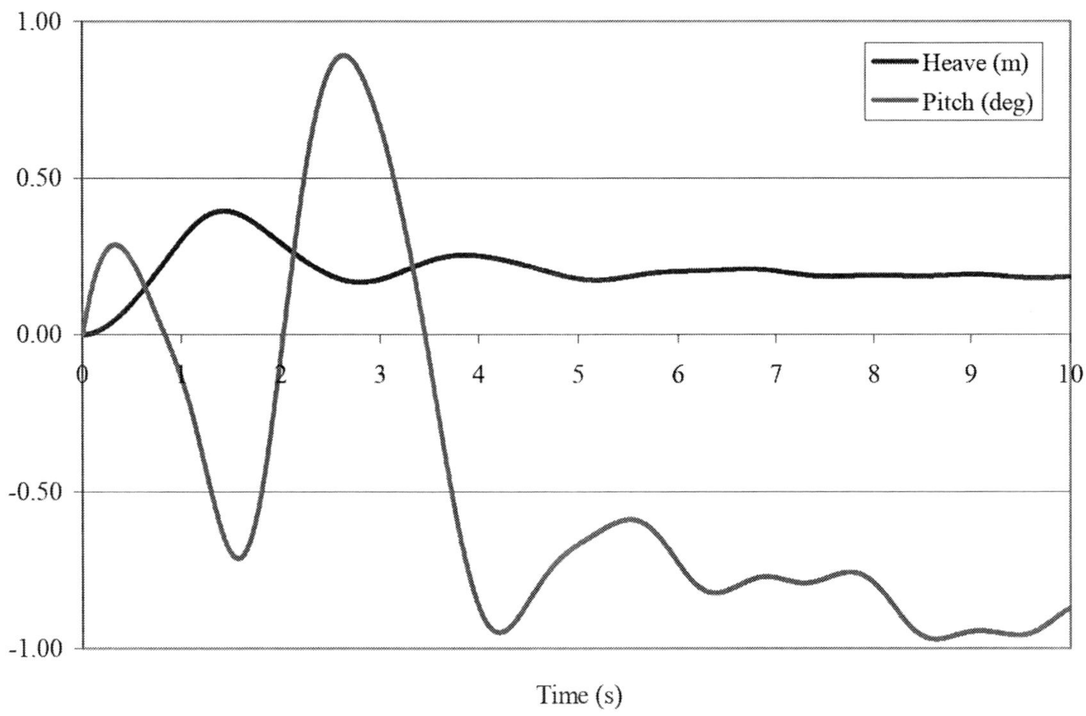

Figure 4. Computed heave and pitch motion of the submarine.

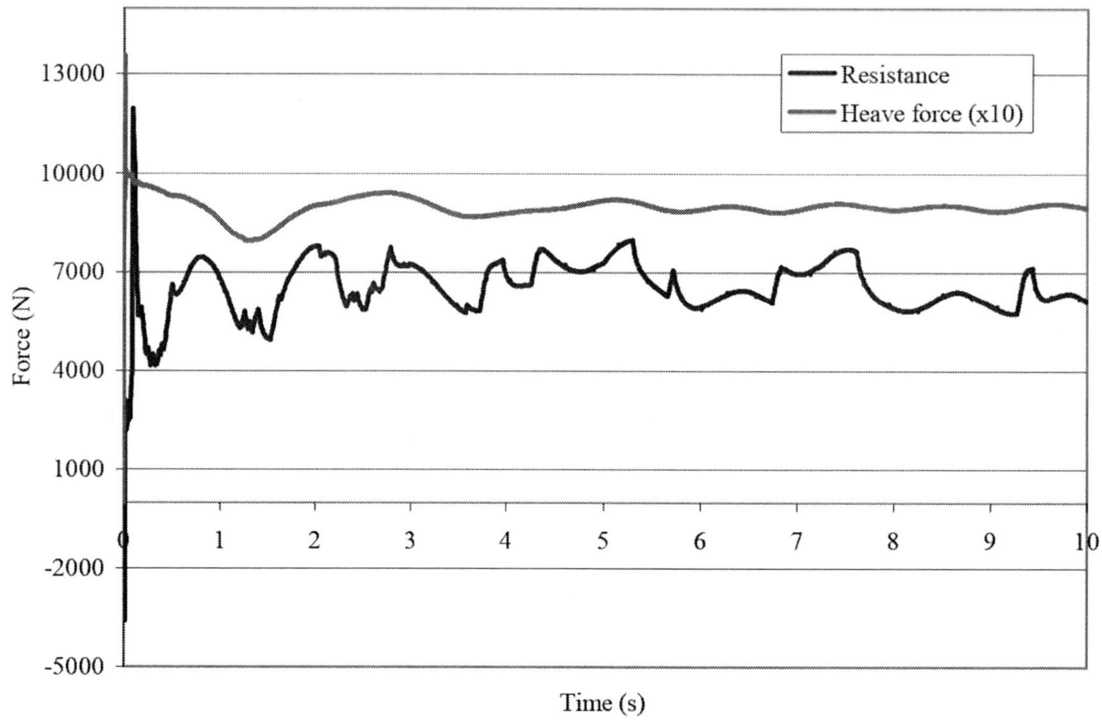

Figure 5. Computed forces on the submarine.

Figure 6 shows three different time instances of the simulation at t = 0, 1.45 and 2.75s respectively. In this plot, the contour colours indicate the water elevation. With respect to Figure 4, these snapshots correspond to the initial position of the submarine, the highest heave displacement of the submarine, and the lowest heave displacement after start-up. The effect of the submarine's displacement on the wave height is clearly seen as the waves impinge and then roll off the surface of the vessel.

Even with the relatively coarse grids used in the present study, the robustness of the solver in resolving the free surface is duly noted. This is true of both meshes used in these simulations. In addition, although the resolution of the background Cartesian grid (used for the overlapping mesh simulations) as well as the single moving mesh was not fine enough to accurately reproduce the waves due to the presence of the vessel, the simulations nevertheless are considered appropriate for the purposes detailed herein.

Although only small transverse and angular displacements were identified (due to the small-amplitude incident waves), the meshes were deemed adequate for these simulations. However, in the case of large angular displacements, the single mesh approach will become increasingly unsuitable due to fact that the grid rotates with the body and the requirement that a sufficiently fine mesh is created that is capable of resolving the water-air interface throughout the simulation history. This weakness does not apply to the overlapping mesh approach, since the background grid is fixed and may be created so as to accurately resolve the free surface in response to the wave heights that are being modelled. The rotation of the body-fitted grid in this approach is handled separately and independently from the background mesh.

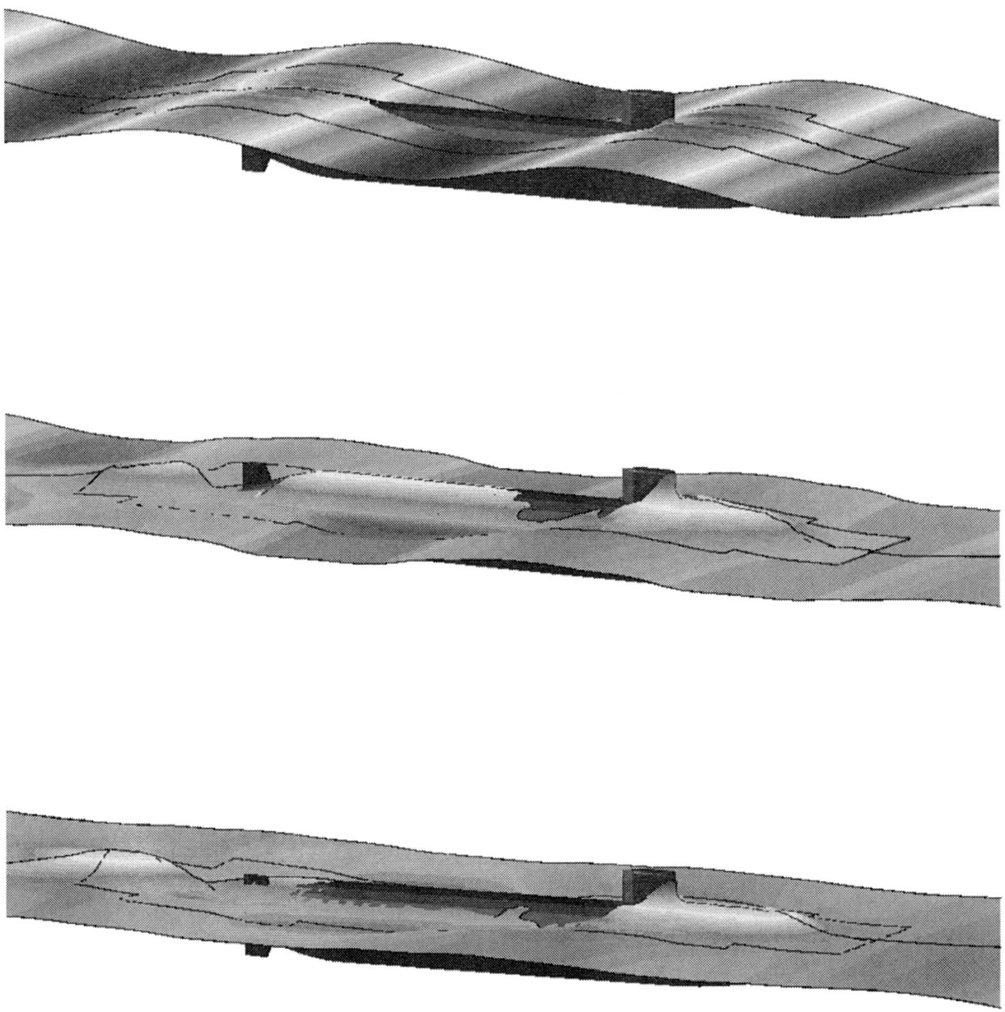

Figure 6. Contours of the free surface level coloured by wave height during the start-up submarine motion: initial position of vessel, t = 0s (top); largest heave displacement, t = 1.45s (middle); smallest heave displacement, t = 2.75s (bottom).

CONCLUSION

This study has demonstrated the use of a coupled fluid flow/6-DOF solver in analyzing the dynamics of a submarine in the early stages of surfacing. Two different mesh strategies were employed, a single moving mesh and an overlapping mesh. It was found that a major problem with the single moving mesh approach is that for large angular displacements of the vessel, a relatively fine mesh will be required in order to accurately resolve the interface between the water and air, since with this approach the interface is not known *a priori* and numerical diffusion becomes significant. Furthermore, where the motions between relative bodies are required, this approach is unsuitable. On the other hand, it was demonstrated that the overlapping mesh approach can be used quite effectively and is preferable despite a slight increase in computation time. Future work will involve a detailed grid refinement study and the incorporation of propulsion and control surface motion effects.

REFERENCES

Azcueta, R. (2002) "RANSE simulations for sailing yachts including dynamic sinkage and trim and unsteady motions in waves" Proc. High Perf. Yacht Design Conf., Auckland.

Demirdzic, I. and Muzaferija, S. (1995) "Numerical method for coupled fluid flow, heat transfer and stress analysis using unstructured moving meshes with cells of arbitrary topology" Comp. Meth. Applied Mech. Eng., vol. 125, pp. 235 – 255.

Ferziger, J. & Peric, M. (2003) "Computational Methods for Fluid Dynamics" 3rd ed, Springer Berlin.

Murman, S.M., Aftosmis, M.J. and Berger, M.J. (2003) "Simulations of 6-DOF motion with a Cartesian method" Proc. 41st AIAA Aerospace Sciences Meet., AIAA 2003-1246.

Muzaferija, S. and Peric, M. (1999) "Computation of free surface flows using interface-tracking and interface-capturing methods" Chapter 2 in Nonlinear Water Wave Interaction (eds. O. Mahrenholtz & M. Markiewicz), WIT Press Southampton.

Pankajakshan, R., Remotigue, M.G., Taylor, L.K., Jiang, M., Briley, W.R. and Whitfield, D.L. (2002) "Validation of control-surface induced submarine maneuvering simulations using UNCLE" Proc. 24th Symp. Naval Hydrodyn., Fukuoka.

Peric, M., Zorn, T., el Moctar, O., Schellin, T.E. and Kim, Y. (2007) "Simulation of sloshing in LNG-tanks" Proc. 26th Int. Conf. Offshore Mech. Arctic Eng., OMAE2007-29555.

Venkatesan, G. and Clark, W. (2007) "Submarine maneuvering simulations of ONR Body 1" Proc. 26th Int. Conf. Offshore Mech. Arctic Eng., OMAE2007-29516.

Yang, J., Sakamoto, N., Wang, Z., Carrica, P. and Stern, F. (2007) "Two phase level-set/immersed-boundary Cartesian grid method for ship hydrodynamics" Proc. 9th Int. Conf. Num. Ship Hydro., Ann Arbor.

Seakeeping Operator Guidance

J.L. Colwell and D. Stredulinsky, Defence R&D Canada - Atlantic, Dartmouth NS, Canada

ABSTRACT

In many cases, experienced naval officers have little difficulty in determining the best ship speed and heading to take for performing motion-sensitive tasks; however, this is not necessarily the case in conditions with a complex seaway that has multi-directional and/or multi-modal frequency characteristics, nor when conflicting performance requirements must be satisfied. "Seakeeping operator guidance" is a methodology for obtaining and presenting the naval operator with information on ship motions in two distinctly different ways: the first is tactical operator guidance, which provides information for planning the best ship tactics to adopt before performing motion-sensitive activities; and, the second is real time operator guidance, which monitors ship motions while the activity is actually being performed, in order to warn the operator when safe operating conditions are being exceeded, or are about to be exceeded. An accurate definition of the wave height, frequency and directional parameters is critical for providing reliable tactical operator guidance; however, this is not easy to accomplish with existing shipboard sensors. A new methodology called wave data fusion is being considered to provide this capability, and if successful, will enable the development of an integrated and reliable seakeeping operator guidance system.

INTRODUCTION

Seakeeping operator guidance provides information for input to decision making when planning and performing shipboard motion-sensitive activities. Information used for mission planning, and especially for determining ship tactics (i.e. speed and heading) in advance of performing motion-sensitive activities, is called "tactical operator guidance". Information on current conditions with respect to safe operating limits while actually performing motion-sensitive activities is called "real time operator guidance". The reliability of the tactical operator guidance methodology is directly related to the accuracy of the seaway characteristics that are used as input for calculating anticipated ship motions; however, it is not easy to provide accurate definition of the seaway using only shipboard sensors.

This paper describes tactical and real time operator guidance concepts, and presents a number of examples from current DRDC Atlantic research and development activities for providing this capability to the Canadian Forces. Major limitations on the reliability of tactical and real time operator guidance systems are discussed, and the importance of accurate seaway characterization for tactical operator guidance is highlighted. A new methodology called "wave data fusion" is described as a potential way ahead for obtaining accurate seaway definition, by combining information from ship-based wave radar with other shipboard information sources, including measured and calculated ship motions. If successful, wave data fusion will enable the development of an integrated and reliable seakeeping operator guidance system. This approach may also offer interesting possibilities for future real time operator guidance systems that "look ahead" in space and/or time, but these advanced concepts are beyond the scope of the current paper.

Ship Classes used for Operator Guidance Examples

Seakeeping operator guidance examples are presented for research projects involving two Canadian Forces ship classes: the HALIFAX Class Canadian Patrol Frigate (CPF) and the KINGSTON Class Maritime Coastal Defence Vessel (MCDV). Also, important experimental work has been carried out on HMCS Nipigon (recently decommissioned) and on Canadian Forces Auxiliary Vessel (CFAV) Quest. All four ship types have steel, monohull construction. The general particulars of these ships are as follows: HALIFAX Class, full load displacement $\Delta = 4700$ tonnes (t), overall length $L = 134.1$ m, and midships beam $B = 16.4$ m; KINGSTON Class, $\Delta = 970$ t, $L = 55.3$ m and $B = 11.3$ metres; HMCS Nipigon, $\Delta = 3000$ t, $L = 113$ m, and $B = 9.2$ m; and, CFAV Quest $\Delta = 2200$ t, $L = 76$ m and $B = 12.6$ m.

TACTICAL OPERATOR GUIDANCE

Tactical operator guidance provides information for planning and decision making, in advance of performing motion-sensitive activities. This is especially useful in complex multi-directional seaways, where experience does not always provide reliable guidance, and in situations where conflicting requirements must be reconciled. The basis for tactical operator guidance is to provide accurate estimates of anticipated ship motions, for the current or forecast future ocean wave and wind environment. The examples presented here use the DRDC Atlantic frequency domain, strip theory, ship motion computer program SHIPMO7 [1] to calculate ship motions for a variety of ship speeds and headings. The calculated ship motions are then compared with known ship and system performance limits, or ship motion criteria, to determine operational capability.

Two tactical operator guidance examples are presented. The first example on "slam avoidance" describes a case where current sea conditions are combined with calculated ship responses and known slamming characteristics of the KINGSTON Class, to identify combinations of ship heading and ship speed with high slamming probabilities that should be avoided. A second example considers a "Transit and Patrol" scenario [2], where many different seakeeping criteria are evaluated simultaneously.

Tactical Operator Guidance Example 1: KINGSTON Class Slam Avoidance

The KINGSTON class was originally intended to operate primarily in coastal waters; however, in practice these ships often operate in open ocean conditions in which severe slamming can occur. These ships are relatively susceptible to slamming damage in way of the flat bottom hull sections shown as an "area of concern" in Figure 1. Sea trial measurements and finite element analysis were conducted to determine slamming pressures likely to cause structural damage. To reduce the incidence of slamming and associated structural loads, operational guidance in the form of polar plots was developed to indicate which combinations of ship speed and relative sea direction can lead to high hourly exceedence probabilities for slamming pressures, as shown in Figure 2, for slamming pressures calculated by SHIPMO7 [1]. Areas of red color on the plots indicate combinations of ship speed and relative wave heading to be avoided. A technical bulletin was supplied to the operators with a series of paper plots for different wave heights and periods. The plots were found to be very sensitive to wave steepness. With input of existing wave conditions and the current ship loading condition, such plots can be generated by an operator guidance system and displayed to the operator showing the current ship 'location' on the plot. The operator could also input hypothetical wave conditions to asses how a changing environment may limit ship operations.

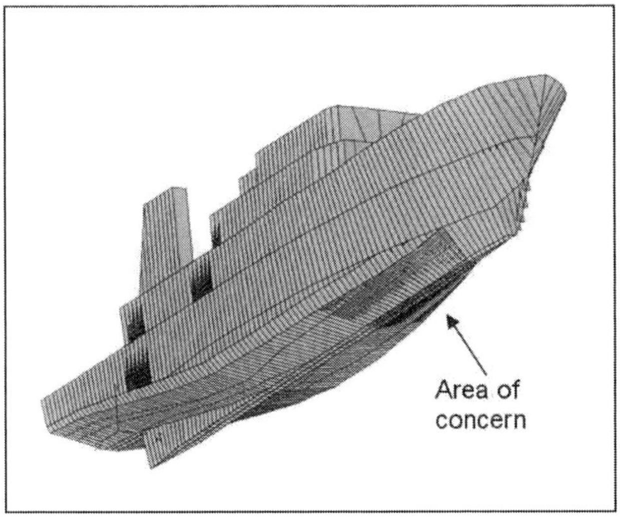

Figure 1: KINGSTON Class Paneled Hull

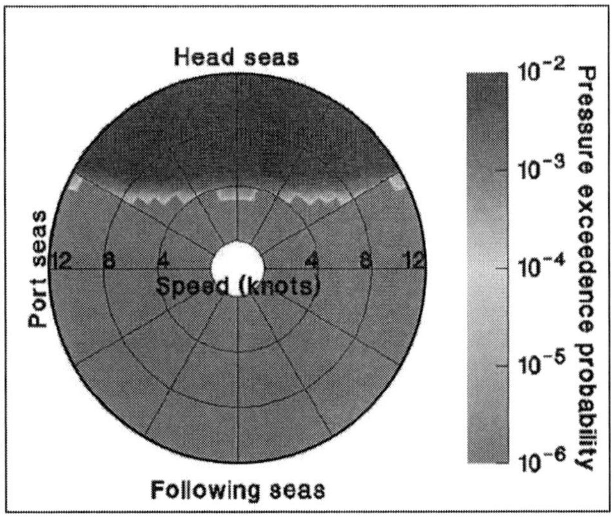

Figure 2: KINGSTON Class Slam Pressure Exceedence Plot

Tactical Operator Guidance Example 2: HALIFAX Class Transit and Patrol

Figure 3 shows an example "operability plot" for a HALIFAX Class frigate, based on a "Transit and Patrol" mission concept [2]. The display format shown in Figure 3 has a geographic orientation, with a heading of North (000° True) at the top, and concentric rings to denote ship speed, here varying from zero at the centre to thirty knots at the outer ring. The criteria exceedence colour scale on the right of Figure 3 shows the ratio of expected motion divided by motion limit value, so that a value of 1.2 indicates the criterion has been exceeded by 20%. This operability plot was produced *in situ* during a seakeeping operator guidance demonstration trial on a HALIFAX Class frigate in November 2006. The expected motions were calculated using SHIPMO7 [1] for the seaway being experienced at the time. Visual observations, combined with weather forecast information, identified that this was a multi-directional seaway, with rapidly developing wind-generated seas coming from the West with significant wave height of approximately 5.0 metres and peak period of 11.9 seconds, and with decaying swells coming from the South-East (course 135° True), with significant wave height of approximately 3.0 metres, and peak period of 15.0 seconds.

The operability plot shown in Figure 3 contains motion calculations for ship speeds from 0 kt to 30 kt, in 2 kt increments, and ship headings from 0° to 360°, in 10° increments, resulting in 16 x 37 = 592 discrete "cells" in the operability plot. For each cell, the following seven motion parameters are evaluated: roll angle (max. 10°); pitch angle (max. 4°); flight deck vertical acceleration (max. 0.2 g); flight deck lateral acceleration (max 0.1 g); Motion Induced Interruptions (MII's)[1] at the flight deck (max. 1 per minute); bow slamming (max. 1 per minute); and deck wetness (max. 1 per minute). The ratio of calculated motion amplitude divided by that parameter's limit value is called the operability ratio, OR, and the colour of each cell is defined by the largest OR value amongst the seven motion parameters. For example, if the calculated flight deck vertical acceleration at 20 knots for a heading of 270° is 0.23 g, then the operability ratio is OR = 0.23/0.20 = 1.15. In this case (for 20 knots and 270°), all other motion parameters have lower OR values, and so the colour for this cell is orange-red, as defined by the Criteria Exceedence colour scale at the right side of Figure 3, and the "limiting parameter" is flight deck vertical acceleration.

[1] The loss of a person's postural stability due to tipping or sliding is a Motion Induced Interruption.

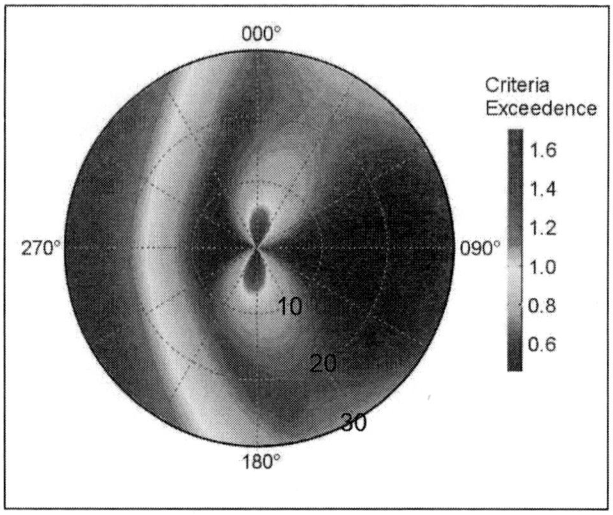

Figure 3: HALIFAX Class Transit and Patrol Operability Plot

The large red zone on the left hand side of Figure 3 for ship speeds above about 20 knots and headings from approximately North-West through South-West, corresponds to the ship heading more or less directly into the 5 metre significant wave height seaway coming from the West. Flight deck vertical acceleration is the main limiting parameter throughout this zone (although OR values for slamming are also well into the red for most of this zone). The red "figure of eight" zone in the central part of Figure 3 is limited by the MII criterion, and corresponds to the cases with the ship heading either North or South at slow speed, and with the large seaway from the West coming from the beam, resulting in relatively large lateral motions, and so personnel balance and mobility is an issue.

Tactical Operator Guidance Limitations

The two main limitations on the accuracy, and hence reliability, of tactical operator guidance are: (i) the accuracy of the ship motion calculations which produce the information for decision-making; and, (ii) the accuracy of the seaway definition used as input for the motion calculations. For any given real ship, such as the HALIFAX and KINGSTON classes, any inaccuracies in ship motion calculations can be minimized by supplementing theoretical calculations with measured ship motion data from model tests and sea trials. On the other hand, it is currently not possible to obtain accurate wave measurements from shipboard sensors alone, and so a new approach called "wave data fusion" is being investigated to provide increased reliability of shipboard wave measurement, as discussed later in this paper.

REAL TIME OPERATOR GUIDANCE

Real time operator guidance provides information for safety in real time, by continuously evaluating current ship motions while actually performing a motion-sensitive activity. This is especially useful when existing physical cues available to the operator are either absent or unreliable. An important characteristic of real time operator guidance is that each different system is very dependent on the actual tasks being performed, for both the ship motion criteria that define safe conditions, and for the types of cues that should be provided to the operator.

Two examples of real time operator guidance are presented. The first example considers a methodology to calculate and display the severity of hull slamming for the KINGSTON Class, and the second example considers the identification and display of "quiescent periods" for helicopter landings on the HALIFAX Class in high sea states. The real time operator guidance systems considered here do not require an accurate seaway definition, as they rely solely on real time ship motion measurements. More advanced (and as yet unproven) concepts for looking ahead in time and/or space to identify approaching quiescent periods or anticipated severe motions are not considered in this paper, but are interesting candidates for future projects.

Real Time Operator Guidance Example 1: KINGSTON Class Slam Severity

On most ships, including the HALIFAX Class and CFAV Quest, the transition from mild, through moderate, and to severe slamming is apparent to the operator through a variety of sensory cues. For example, mild slams are associated with a low frequency "shudder", moderate slams often add a distinctive booming sound to the list of cues, and severe slams are often associated with high-frequency hull whipping, and a sudden and unexpected appearance of the Commanding Officer on the bridge. On the KINGSTON class, there are no obvious signs to differentiate between mild, moderate and severe slams [3], as the apparent hull response, in terms of motion, sound and vibration, is the same. Thus, it is difficult to assess when the ship is in a high-risk condition. Note that, the development of real time operator guidance on slam severity for the KINGSTON Class recognizes that while the tactical operator guidance system described earlier may identify that particular conditions should be avoided, it is not always possible to do so. Real time operator guidance on slam severity will enable the Commanding Officer to evaluate the risks associated with severe slamming vs. the urgency of the mission.

An important goal for this project was to find a 'global' parameter that can be easily measured in real time, and which is a reliable indicator of slam severity. This is an alternative approach to the significantly more expensive option of incorporating and processing data from a grid of strain sensors to monitor the ship's structural response to slamming. From sea trial measurements for the KINGSTON class, it was found that there was a strong correlation between the centerline vertical bow acceleration and the probability of the occurrence of slamming, as shown in Figure 4. Also, the work related to developing the tactical operator guidance polar plots discussed earlier found a close relationship between the hourly slamming pressure exceedence probability and RMS vertical bow acceleration at the ship centerline. This showed little variation with significant wave height, peak wave period, ship speed, or relative sea direction and thus it is expected that the vertical bow acceleration can provide a good measure of the slam severity. A thermometer display is proposed such as that shown in Figure 5. Two pointers are shown, one related to the RMS acceleration over a chosen period (shown here for 10 minutes) and a 'Current' indicator, updated using the vertical bow acceleration amplitude for each wave cycle. Work is underway using structural response measurements from sea trials to validate this approach. The sea trial structural response measurements and analysis are also being used to determine if one or more representative locations for strain gauges can be found to provide direct structural response feedback to the operator.

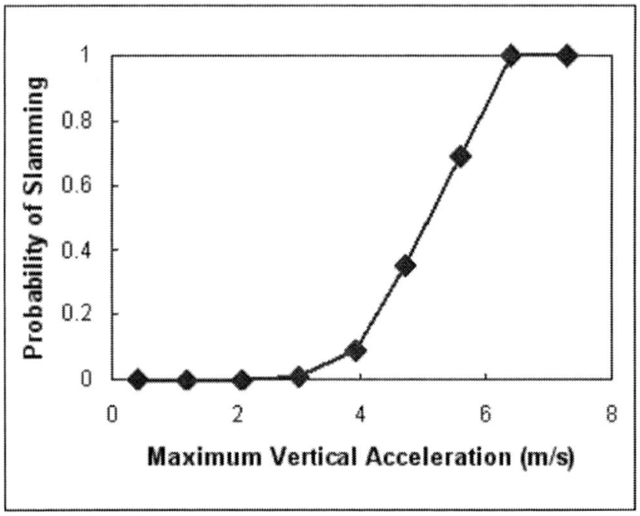

Figure 4: KINGSTON Class Slamming, measured probability of slamming vs. measured maximum vertical acceleration.

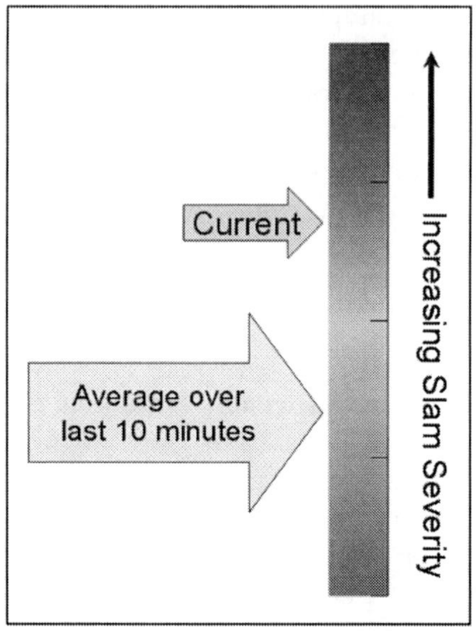

Figure 5: KINGSTON Class Slam Severity Indicator

Real Time Operator Guidance Example 2: HALIFAX Class Helicopter Operations

The Flight Deck Motion System concept was developed at DRDC Atlantic as a real time operator guidance system for landing helicopters in high sea states. Reference [4] describes the Flight Deck Motion System concept in detail, including the Quiescent Period Indicator (QPI) that is presented here as an example of a multi-parameter, real time operator guidance display. Figure 6 shows the changing of quiescent states for two ship motion parameters, roll angle and flight deck vertical acceleration (FDVA). The top panel of Figure 6 shows two minutes of measured roll angle data, with the roll angle limit values shown by the horizontal red lines. The colour bar located immediately below the roll angle data shows the "state of quiescence" for roll angle changing from red (out of limits), to green (within limits) and in some cases yellow (within 20% of the limit). The quiescent period indicator (QPI) itself is simply a display that shows the colour of the current quiescent state. The second panel in Figure 6 shows similar time series data and quiescent period status for flight deck vertical acceleration (FDVA), and the bottom section of Figure 6 shows how the roll angle and FDVA quiescent period information are combined to produce the two-parameter QPI. This same approach can be followed to incorporate more motion parameters (at least those for which reliable limit values are known), and to model other activities where different motion parameters might be evaluated, such as on-deck handling and weapon loading.

Real Time Operator Guidance Limitations

The major limitation for developing reliable real time operator guidance is the determination of appropriate motion criteria, or similar limits for other parameters such as slam severity and deck wetness. In most cases, the current state of the art for modeling and simulation is not sufficiently advanced to provide absolute performance criteria (although relative performance can often be reliably assessed), and so criteria for real time operator guidance systems must be determined by observing and measuring real system performance, at full scale.

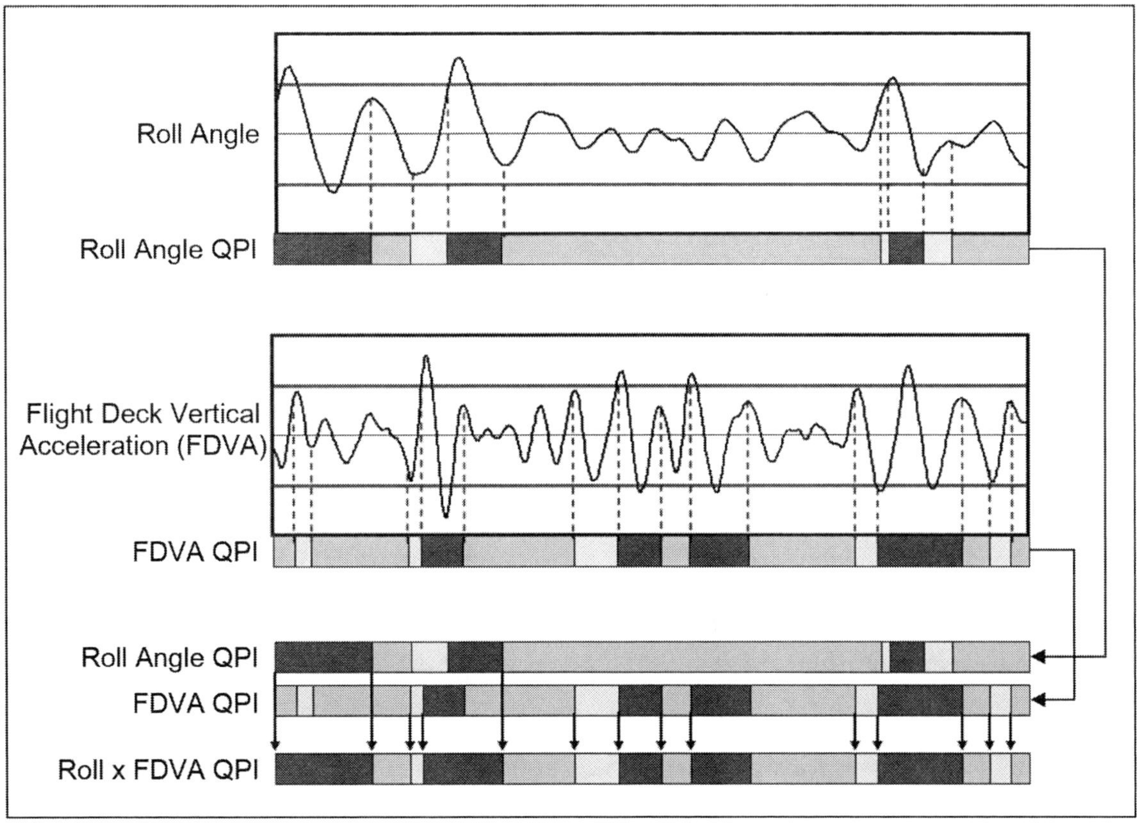

Figure 6: Quiescent Period Indicator (QPI) for roll angle, for flight deck vertical acceleration (FDVA), and for combined roll angle and FDVA (Roll x FDVA)

In order for a real time operator guidance system to be of practical benefit, there must be a reasonable opportunity to respond to a warning that unsafe conditions are imminent or at hand. In some cases, it will be possible to safely secure an activity and wait for better conditions before resuming the task; however, in other cases, it may only be possible to stand back and try to keep out of harm's way. If there is no safe way to either control the activity or to avoid the consequences of performing the activity while the motions are "out of limits", then a real time operator guidance system is of limited use. In this case, the equipment and/or procedures used to perform the activity should be modified to increase safety.

WAVE DATA FUSION

Complex wave conditions are typically encountered when decaying, low frequency swells from one direction are combined with growing or "fully developed" wind-generated waves from a different direction. Average annual statistics from a multi-year wave climatology study for the East coast of Canada [5] shows that approximately 44% of the time the seaway had substantial multi-directional characteristics, with wave energy coming from two or more distinctly different directions. The same study also shows that approximately 59% of the time, the seaways were multi-modal, with two or more distinct peaks in wave frequency. Reliable tactical operator guidance under these conditions requires accurate knowledge of the seaway conditions. Directional wave spectra provide the most comprehensive input data for operator guidance systems. Wave buoys provide the most accurate and reliable method of measuring directional wave spectra, however, it is not practical to deploy wave buoys during normal ship operations. Thus the goal of this work has been to obtain seaway information from shipboard measurement systems and observations, perhaps augmented with wave forecast and measured data that can be transmitted to the ship, for example, through satellite links over the internet. As discussed in the following subsection it may be necessary to use the fusion of data from more than one shipboard system to obtain reliable directional wave spectra.

Shipboard Wave Measurements

While trained observers can provide estimates of the wave height, period, and direction, these estimates are often not accurate enough for operator guidance purposes. Figure 7a compares visual estimates of wave height (Hv) by a trained observer, as a function of measured significant wave height (Hs) during a sea trial on HMCS Nipigon, off the East coast of Canada in 1997 [6]. The visual observations are normalized by dividing the visual wave height by the significant wave height measured with a Triaxys directional wave buoy in hourly intervals. While the mean of the visual observations was only three percent higher than measured, the standard deviation was 20 percent with some observations in error by as much as 40 percent. Past comparisons of observed vs. measured wave data [7] suggest that visual estimates of "average" wave height made by trained observers are generally within about 6 percent of measured significant wave height, but visual observations of wave period should not be regarded as reliable.

Commercial wave radar systems are available to indirectly measure the directional wave spectrum by processing images of the back scatter from the waves with standard X-band navigational radars; however, DRDC trials of several commercial systems have shown that while the systems generally provide reasonable wave frequency and direction measurements, often the wave heights have been inaccurate. The wave radar system 'calibration' is sensitive to the type of radar scanner and ship-scanner geometry. It is DRDC's experience that even when a sea trial covering a range of sea conditions is used to calibrate parameters, that the height measurements are often inaccurate. Figure 7b shows the significant wave heights from wave radar (normalized by the Triaxys wave buoy height) for a trial on the CFAV Quest in 2006 in which the wave radar calibration parameters were adjusted after the trial to get the best overall fit to the sea trial data. Each point on the figure represents a 30-minute period. Even with the 'best-fit' calibration of heights, there is large scatter in the normalized wave height data (mean of 1.1 and standard deviation of 0.3), and some points are close to twice the wave buoy height measurement. The normalized mean wave direction results from the Quest 2006 sea trial plotted in shown Figure 7c show reasonably good agreement between the wave radar and the Triaxys directional wave buoy (mean of 0.98 and standard deviation of 0.05). The normalized peak wave period is shown in Figure 7d and exhibits fair agreement between the wave radar and buoy peak period (mean of 1.1 and a standard deviation of 0.17). The outlying peak period points tended to occur in cases of multiple wave components of similar energy where the wave radar and wave buoy showed different components as having the highest energy. A subsequent trial on CFAV Quest in March 2007, using these 'best-fit' calibration parameters for the 2006 trial, produced wave radar measurements of significant wave heights ranging from less than half that of the buoy during some periods, to over double the buoy significant wave heights during other periods.

Scaling of directional wave radar spectra

In-house work is currently underway to consider the fusion of wave radar data with other shipboard data sources, to try to improve the height measurements obtained from a wave radar system. Hydrodynamic codes such as SHIPMO7 [1] have progressed to the stage that they can provide reliable ship motion predictions in a directional seaway. Under the assumption that the wave radar can provide a sufficiently accurate directional wave spectrum 'shape', which provides reliable frequency and directional information, this directional spectrum could be used as input to a hydrodynamics code to predict various motion response parameters. A comparison of the predicted and measured responses could then be used to scale the energy of the wave radar spectrum to 'match' the predicted and measured ship motion responses. The absolute ship motion response is easily measured using ship motion sensors at lower wave frequencies but as wave lengths become smaller relative to the ship length then the absolute motion response becomes small. On the other hand, the relative motion between the ship and waves at higher frequencies can be measured using "look-down", over-the-bow microwave radar systems [8]. Thus, a combination of absolute ship motions and ship-wave relative motions may provide the best approach for scaling the wave radar directional wave spectrum, and so indirectly determine the significant wave height.

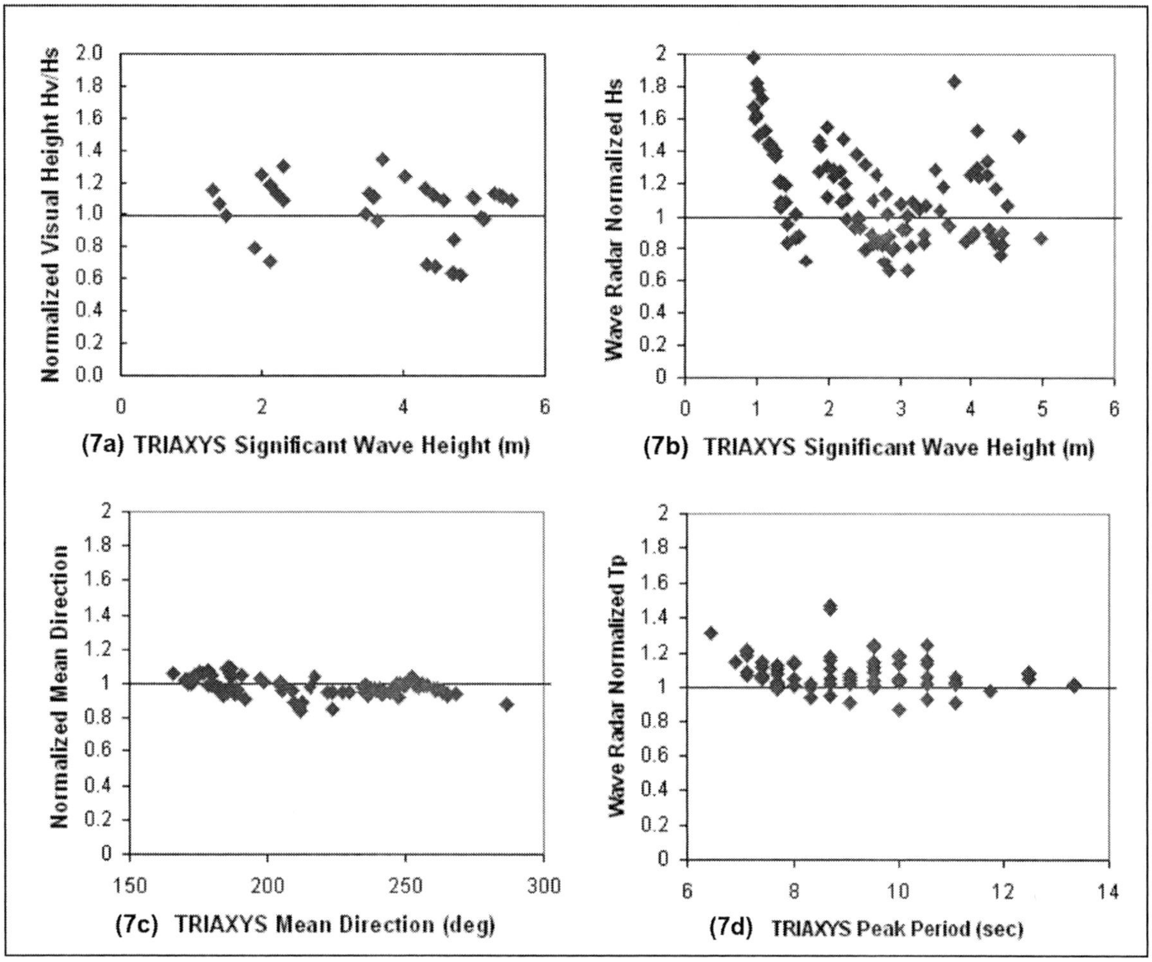

Figure 7: Comparison of TRIAXYS directional wave buoy measurements with visual observations and wave radar measurements.

CONCLUDING REMARKS

The main goal of current research and development on seakeeping operator guidance at DRDC Atlantic is to develop and validate an integrated system comprised of ship motion sensors, wave radar and theoretical ship motion calculation methods to provide reliable tactical and real time operator guidance. The generic nature of tactical operator guidance is well-suited for inclusion in an integrated system. As discussed earlier, the implementation of real time operator guidance is very dependent on the actual activities being performed; however, it should be possible to develop a common methodology for motion measurement, analysis and display that is appropriate for most purposes.

Various options for wave data fusion are currently being explored through analysis of existing data from past DRDC trials on several ships with wave radar. Future sea trials are currently being planned to validate the emerging wave data fusion methodology, and DRDC Atlantic is involved in a number of initiatives with Canadian Forces sea trials that will include opportunities to demonstrate and validate the seakeeping operator guidance technologies described here.

ACKNOWLEGEMENTS

The authors thank Dr. Kevin McTaggart and Dr. Eric Thornhill of DRDC Atlantic for their contributions on real time criteria for slam severity, and also thank Mr. Mike Dervin of National Defence Headquarters, Canada, for his important leadership role in this work.

REFERENCES

1. McTaggart, K.A. (1997). SHIPMO7: An Updated Strip Theory Program for Predicting Ship Motions and Sea Loads in Waves. DREA TM 96/243.

2. NATO Military Agency for Standardisation. 1997. Common Procedures for Assessing Seakeeping in the Ship Design Process. NATO STANAG 4154 (Edition 3),.

3. Colwell, J.L. 2002. MCDV Seaworthiness Study - EVAL A01/25: Seakeeping Trial Procedures, Environmental Data and Ship Motions. DRDC Atlantic TM 2002-172.

4. Colwell, J.L. 2004. Flight Deck Motion System (FDMS): Operating Concepts and System Description. DRDC Atlantic TM 2004-003.

5. Juszko Scientific Services. 1994. Statistical Description of the East Coast Wave Climate Using 10-Parameter Spectra. DREA CR 94/410.

6. D.C. Stredulinsky, N.G. Pegg, Motion and Load Predictions and Measurements on HMCS Nipigon, RINA 2000 Transactions, Part D, January 2001.

7. Lloyd, A.R.J.M. 1989. Seakeeping: Ship Behaviour in Rough Weather. Ellis Horwood Limited. Chichester UK.

8. Stredulinsky D. and Thornhill E. 2007. Experiences with Shipboard Measurement of Seaway Wave Characteristics. Canadian Marine Hydromechanics and Structures Conference 2007. St. John's, Newfoundland.

The Changing Face of International Marine Casualty Investigation, New Ways for Modern Times

Author: Peter Foley, Deputy Director Surface Safety Investigations
Australian Transport Safety Bureau

peter.foley@atsb.gov.au

Introduction

This paper outlines the genesis of modern marine casualty investigation and recent work by the Australian Transport Safety Bureau in consultation with other Flag State administrations, under the auspices of the International Maritime Organisation's Flag State Implementation Sub-Committee, to amend and annex the 'Code for the Investigation of Marine Casualties and Incidents' (the Code), to the Safety of Life at Sea Convention.

Safe transport, whether by sea, road, rail or by air underpins our modern society and we depend upon it for our standard of living. The public demand a high level of safety and, in the event of an accident, that it should be investigated. Transport accidents may be investigated by various bodies for various purposes. Chief amongst these are:

- To understand what caused the accident so as to prevent similar accidents in the future – the 'safety investigation'.
- To prove beyond reasonable doubt that there has been a breach of the law (criminal).
- To establish and apportion liability (civil).
- To assess the competence of individuals or organizations and to take appropriate disciplinary action (administrative).

Safety investigations are recognised as providing the best means of understanding the failures in a system and, thereby, how the system can best be adjusted or changed to remove the points of failure.

Increasingly, the concept of an independent, objective technical investigation is being accepted internationally as the best way to achieve safety improvements in the aftermath of a transport accident. These are termed 'no blame' investigations as their object is to identify and address safety issues rather than to sanction either administratively or criminally and nor should they be used to apportion liability in civil matters following an accident. Such investigations do not enjoy the *gravitas* of legal/judicial procedures and evidence is not heard in open court. But safety investigations are transparent and less expensive and usually quicker to come to their findings.

This is not to say that the judicial process has not been effective in the marine field. Formal investigations, *British Crown* and *Herald of Free Enterprise* and many others, have been effective in looking beyond blaming the individuals to the underlying factors. Some civil cases, most notably *Lady Gwendolen*[i] and *Marion*[ii], have made their mark in looking to organisational factors and issues of the limitation of liability, but not human factors.

The main problem with formal judicial inquiries is their cost. The immediate attraction of a technical investigation by a specialist investigation authority is the saving of money for the same outcome. It is not unusual for the cost of one judicial inquiry to run into many millions of dollars, often well in excess of a specialised safety investigation authority's annual budget. In addition, technical investigation findings are always open to review and comment. Certainly as public documents they are open to rigorous peer review. Investigations are also open to scrutiny, (in Australia) by the Senate Select Committee system, the coronial process in the event of fatalities, and the Royal Commission process should the Parliament so desire.

The Australian Transport Safety Bureau (ATSB) is Australia's operationally and legislatively independent transport safety investigation agency. The ATSB conducts transport safety investigations into air, marine and rail accidents, under the provisions of the *Transport Safety Investigation Act 2003* (TSI Act). The Australian Government and Parliament recognised, in the provisions of the TSI Act, the necessity and priority of expert technical safety investigations. Unless the incident can be shown to have been caused by extreme criminality, in other words not an accident at all, the ATSB's investigation will prevail. This priority does not exclude other forms of investigation, but ensures that the ATSB has access to all evidence.

The TSI Act was developed from pre-existing aviation and marine investigation legislation, with particular emphasis on the International Civil Aviation Organization provisions in Annex 13. The provisions also have embraced best practice from other like-minded countries, USA, Sweden, Canada, New Zealand in adopting provisions that provide, under accident situations, primacy for the ATSB.

Marine safety investigations – recent history

In the last thirty years there has been a gradual but significant move away from regulator-based investigations and judicial inquiries. In jurisdictions, such as the US and Canada, the investigator is created as a separate legal body from the regulator, which remains within the broader Transport Portfolio. In 1967 the US government set up the National Transportation Safety Board. Similar boards were created as independent multi-modal transport safety investigation bodies in Sweden (1978 – initially aviation, but multi-modal since 1990), Finland (1986), Canada and New Zealand in 1990 and the Netherlands in 1999. In Japan, Korea and France independent, mode-specific marine investigation organisations have been created separate from the government-based regulator.

In Australia, following the inquiry into the grounding of *TNT Alltrans* on Lady Musgrave Reef in March 1985, a tripartite committee was set up to examine Preliminary Inquiries and Courts of Marine Inquiry in Australia. The Committee reported in 1989 and, based on its recommendations, the Marine Incident Investigation Unit (MIIU) was created on 1 January 1991 with the statutory position of the Inspector of Marine Accidents. The MIIU operated under the provisions of

Navigation (Marine Casualty) Regulations formed under the *Commonwealth Navigation Act 1912* and was independent of the marine regulator; the Australian Maritime Safety Authority.

The Australian Transport Safety Bureau was formed on 1 January 1999 with the amalgamation of the MIIU with the then, Bureau of Air Safety Investigations and some of the non-regulatory components of the Federal Office of Road Safety. Marine incidents continued to be investigated under the *Navigation (Marine Casualty) Regulations*. The MIIU had grown to a staff of six with four investigators, including the Inspector, and had access to the additional resources of 'human factors' and 'technical analysis' that were available to the aviation investigators.

On 1 July 2003, the *Navigation (Marine Casualty) Regulations* were repealed. Since then, marine investigations have been conducted under the *Transport Safety Investigation Act (2003) and Regulations*. The TSI Act is multi modal, contains provisions which provide significantly more clarity than the previous legislation on issues such as the purpose and independence of transport safety investigations, better protection for information provided by witnesses under investigation and various new and/or enhanced powers in relation to ATSB investigations. In addition, the TSI Act and regulations contain provisions, and a jurisdiction for the first time, which allows the ATSB to investigate interstate rail accidents.

Systemic investigation

Contemporary safety investigations focus on root causes and organizational and system factors. This shift from a focus on individual fault to organizational analysis has been due, not to the legal profession or the technical expert, but to the disciplines of sociology and psychology. Turner (1978), looked at disasters and accidents and their effect on society, on the premise that such events are measured in the disruption experienced by an ordered society[iii]. Barry Turner reviewed 84 disasters and accidents and developed a model of seven stages of disaster: notionally normal; incubation; precipitating event; onset; salvage and rescue; and full cultural readjustment.

> *'it becomes clear to anyone concerned to understand more about the origins of disaster will need to study the forces which affect the manner in which organizations handle communications about potentially dangerous energy sources of all kinds - . . .'*[iv]

Following Turner's work James Reason's seminal work *Human Error*[v], mirrored the concept of an *incubation* period with the metaphor of *resident pathogens* and *latent factors*. Reason focuses on the way errors are made in the context of an organisation allowing an accumulation of erroneous decision making to by-pass defensive strategies. Other leading writers in the field include Jens Rasmussen and Patrick Hudson together with Charles Perrow (1984) (complex and linear systems and tight and loose coupling) [vi], Wagenaar and Groeneweg (1987)[vii], Richard Allinson[viii], who has a background in philosophy, rather than psychology and Sidney Dekker[ix].

Safety investigations may be seen as predicated on reversing safety management principles to form safety investigation principles. Reason (1997) identifies three approaches to safety management – *The Person Model; The Engineering Model;* and *The Organizational Model*[x].

A person model approach to safety investigation focuses on individual behaviour and, in particular, the acts or omissions and decisions of individuals measured against a 'reasonable peer'. It tends to be associated with blame.

The engineering model approach tends to focus on human/machine incompatibility and has obvious applications in complex activities involving the high skill levels of a small operational workforce. It may identify failures at the machine/human interface.

The organisational safety investigation model analyses how the system became vulnerable to the hazards that are present in any activity. It allows a proper focus on the latent factors that may lie dormant within a system for a long time until the right combination of factors occur releasing energy at the wrong place, at the wrong time and at the wrong rate.

Reason (1997) acknowledges that criticism of the person model should not be taken as indicating that the organisational approach is not without problems[xi].

In looking at the causal factors in terms of collective/organisational terms, individual responsibility is not given the emphasis it may deserve and there is a risk that individual corrective action may be overlooked. However, in terms of safety outcomes, while not perfect, such investigations give the best results.

Accident investigation, both in the marine and aviation fields, has developed recommended investigation procedures based on work by Edwards and Hawkins[xii] (the SHEL model), which explores the inter-relationship between individuals and the work environment, and organisational issues based on James Reason's 'Swiss Cheese' model[xiii]. Both the International Maritime Organization (IMO) and the International Civil Aviation Organization (ICAO) have adopted the Reason Model as the basis for systemic investigations.

In terms of safety investigation, there has been a realisation that major accidents are not caused by a failure in isolation of design, procedures, equipment, humans, or environmental effects alone. The causes of an accident rest with the complexity of the system. Even quite simple systems may require an initiating action to by-pass defences or locate flaws in the system to increase the probability of an adverse outcome.

The IMO Code for the investigation of marine casualtie and incidents

In June 1992, marine accident investigators from 15 countries met in Canada and agreed that the maritime industry must have a uniform approach to marine casualty investigation. If not, the lessons to be learned from accidents would be potential lost and lives and property put at risk because the maritime industry was unable to move with the times. From this meeting a proposal that the International Maritime Organization should include the item in its work program.

The result was, that on 1 December 1997, the IMO Assembly accepted Resolution A.849 (20), the Code for the Investigation of Marine Casualties and Incidents. The Code espoused a consistent casualty investigation methodology relying on the principles of 'cooperation', 'independence' and 'no-blame' for marine accident. It also contained some practical check lists for marine casualty investigators.

Work continued on guidelines for the investigation of human factors issues and a systemic or organisational approach to casualty investigation and was adopted as an annex to the Code in February 2000 (A.884 (21)). The annex contains investigation guidance material based on the SHEL model and Reason's 'Swiss Cheese' model and generic error modelling system (GEMS).

The logic behind the organizational approach to serious marine accidents is compelling. While Flag State authorities were increasingly adopting the principles contained in the Code, it was not mandatory, and there was serious consideration at the Marine Accident Investigators International Forum (the peak body international body of marine casualty investigators) as to how an amended/improved set of provisions could be submitted to IMO for possible adoption as an integral part of the Organization's Safety of Life at Sea Convention (SOLAS) Convention. It was evident that a common international approach to marine casualties would provide better and more consistent data to allow the development of better safety measures through the IMO.

In May 2006, Australia had submitted a paper to the 14th meeting of the International Maritime Organization's Flag State Implementation (FSI) Sub-Committee proposing that the provisions of the Code be reviewed, modified and ultimately that the Code be Annexed to SOLAS as a set of mandatory and recommended provisions for casualty investigation. The intention was to provide a clearer mandate (and a set of guiding principles) to Flag and Substantially Interested States, when investigating marine casualties, in particular very serious casualties, than had existed before in the IMO's various resolutions and the United Nations Convention on the Law of the Sea. This accorded with the long standing practice in international civil aviation and the provisions of Annex 13 of ICAO's Chicago Convention.

Further work on the Code was performed at the FSI meeting by the Casualty Analysis Working Group, and it was agreed that an inter-sessional correspondence group, with terms of reference agreed by the Sub-Committee, would be formed to continue the work and be coordinated by Australia. The terms of reference for the correspondence group included, in particular, to review the Code taking into consideration the recent IMO/ILO resolution LEG.3(91), Guidelines for the Fair Treatment of Seafarers in the Event of an Accident (the Guidelines).

The Guidelines, developed by a Joint IMO/ILO Ad Hoc Expert Working Group, are recommended practice to be observed in all instances where seafarers may be detained by public authorities in the event of a maritime accident. The guidelines state that seafarers are recognized as a special category of worker. Given the global nature of the shipping industry and the different jurisdictions with which they may be brought into contact, they need special protection, especially in relation to contact with public authorities. The stated objective of the Guidelines is to ensure that seafarers are treated fairly following a maritime accident and during any investigation and detention by public authorities and that detention is for no longer than necessary.

The Guidelines also give advice on steps to be taken by all those who may be involved following an incident: the Port or Coastal State, Flag State, the seafarer's State, the shipowner and seafarers themselves. The emphasis is on co-operation and communication between those involved and in ensuring that no discriminatory or retaliatory measures are taken against seafarers because of their participation in investigations. The Guidelines say that all necessary measures should be taken to ensure the fair treatment of seafarers.

The correspondence group consisted of representatives from 13 countries and two observer organisations. Several iterations of the Code were circulated based on the input of the various correspondents and in addition a new regulation was drafted for inclusion in the SOLAS convention to give effect to the provisions of the amended Code. A report was then prepared and submitted for the consideration of the Sub-Committee at its 15th meeting.

The 15th meeting of the FSI Sub-Committee meeting occurred in June 2007, where the Casualty Analysis Working Group was again then convened to consider the provisions of the amended Code in detail. Some 60 delegates formed the working group which was chaired by the US. One of the central issues, again, was addressing the provisions of the Guidelines for Fair Treatment of Seafarers in the Event of an Accident and ensuring the consistency of the provisions in the Code.

By the end of the meeting the text of the provisions of draft Casualty Investigation Code (short title) and covering IMO Maritime Safety Committee (MSC) resolution was agreed. The Sub-Committee proposed annexing the Code to Chapter XI-1 of SOLAS. This means its application will only be mandatory in relation to ships engaged on an international voyage. But not those ships that are:

(a) ships of war and troop ships;
(b) cargo ships of less than 500 gross tonnage;
(c) ships not propelled by mechanical means;
(d) wooden ships of primitive build;
(e) pleasure yachts not engaged in trade; or
(f) fishing vessels.

Further, the Casualty Investigation Code contains Mandatory Standards and Recommended Practices.

The Mandatory Standards include:

(a) a requirement for the Flag State to investigate all Very Serious Marine Casualties: being a Death, total loss of the Ship or Severe Damage to the Environment;
(b) provision for the Flag State to consult with other Substantially Interested (particularly the Coastal State) for a Substantially Interested State to lead the investigation into a Marine Casualty;
(c) provision for cooperation between States;

(d) a requirement for the submission of a report the IMO

The Recommended Practices, which would not be obligatory, include:

(a) the establishment of principles for a Marine Safety Investigation covering the independence of the investigation, its safety focussed nature, cooperation, priority and scope;
(b) a recommendation to investigate Marine Casualties and Marine Incidents which do not constitute a Very Serious Marine Casualty;
(c) recommendations for the collection of evidence;
(d) recommendations for protecting the confidentiality of evidence.

Is was also agreed that the draft Casualty Investigation Code be submitted to the MSC's next meeting for approval and the following meeting of the MSC for adoption. The subsequent meeting of the MSC approved the draft Code so that it now set to come into force by the end of 2008.

This is a watershed in global marine casualty investigation because for the first time there will be a mandatory requirement to conduct a marine safety investigation into very serious marine casualties. The Casualty Investigation Code will have effect in Australia for investigations conducted by the ATSB under the *Transport Safety Investigation Act 2003*. The Mandatory Standards and Recommended Practices in the draft Casualty Investigation Code are consistent with the provisions of the TSI Act.

Conclusion

Societal expectations and attitudes dictate that safety investigations must be seen as being dynamic and as keeping abreast with contemporary expectations and evolving safety issues. The past decade, in particular, has seen significant changes in the way in which marine casualties are investigated world wide but still there are challenges for safety investigation authorities. While shifts in public attitudes are relevant, investigators must enjoy a charter or constitution of independence to resist short term or sectional interests that may seek to pervert or compromise safety investigations.

The adoption of the Casualty Investigation Code as an instrument of SOLAS, will mean that for the first time, there will be a requirement to investigate all significant marine casualties using systemic safety investigation principles. For international marine regulation, which has been largely driven by casualties and their investigation in the past, this means that contemporary and emerging safety issues may be more readily identified and addressed. Ultimately, this will lead to a safer, cleaner international maritime transport industry.

---

[i] Sheen, J. (1965), *The Lady Gwendolen*, Lloyd's List Law Reports, 1965. Vol.1.
[ii] Brandon, LJ. (1982), The *Marion*. Lloyd's Law Report, 1982. Vol 2
[iii] Turner, B. (1978) *Man-Made Disasters*, Wykeham Publications.
[iv] Turner, B. & Pidgeon N., (1997) *Man-Made Disasters*, 2nd Edition, Butterworth-Heinemann, 166.
[v] Reason, J. (1990) *Human Error*, Cambridge University Press , Cambridge.
[vi] Perrow, C., *Normal Accidents – Living with High-Risk Technologies*, Basic Books, Inc. New York

[vii] Wagenaar W., and Groeneweg, J., 1987, Accidents at Sea: Multiple causes and impossible consequences, *International Journal of Man-Machine Studies*, 27, 587-598.
[viii] Allinson, R.E., *Global Disasters*, Inquiries into Management Ethics,Prentice Hall (1993)
[ix] Dekker, S., (2002) *The Field Guide to Human Error Investigations*, Ashgate, Aldershot.
[x] Reason, J. (1997). *Managing the Risks of Organizational Accidents,* Ashgate,
[xi] Ibid 223.
[xii] International Maritime Organization (2000) Annex to Resolution A.884 (21)
[xiii]Ibid

Novel antifouling surfaces mimicked from nature

Andrew Scardino

Maritime Platforms Division, Defence Science & Technology Organisation, Melbourne, VIC, Australia

Andrew.Scardino@dsto.defence.gov.au

Biofouling is a universal problem affecting all structures in the marine environment. The effects of biofouling include increased drag and fuel consumption, delayed transports times, higher cleaning costs, more frequent dry docking cycles and the translocation of invasive marine species. The effective control of fouling is therefore crucial to minimise the costs to the shipping industry and to mitigate the spread of invasive species. Previously the control of fouling was predominately based on heavy metals and pesticides. There is now a wide recognition that non-toxic alternatives are required to minimise the impact to non-target marine species. In this study natural antifouling systems are investigated. The surface of the blue mussel, *Mytilus galloprovincialis*, is used as a model to determine what characterises a fouling resistant surface. Microtextured ripples (1-2 μm) were discovered on the surface of the blue mussel, which correlated to low fouling cover. The surface features of 36 other species of shellfish were characterised and a variety of microtopographies were discovered. The surface roughness properties of these shell surfaces correlated to fouling resistance and removal. When the microtopographies of selected shell surfaces were mimicked, those surfaces that provided fewer attachment points reduced the settlement of a wide range of fouling organisms. Extracts derived from the surface of the blue mussel have also been successful in deterring the attachment of certain microfoulers. Other novel non-toxic solutions to control biofouling are presented including extreme wettabilities and foul-release surfaces. The implications of bio-inspired design for future antifouling technologies are discussed.

INTRODUCTION

Biofouling is the colonisation processes of a surface in the aqueous environment and occurs most commonly in the marine environment. Within minutes of immersion a surface will be conditioned by a film of macromolecules such as glycoproteins and polysaccharides[1]. After several hours the surface becomes colonised by bacteria and after a few days by diatoms, protozoa and other unicellular eukaryotes. The attachment by diatoms through extracellular polymers creates a biofilm matrix on the surface, known as microfouling or slime[2]. A more complex community develops from the microbial biofilm to include multicellular primary producers. The final stage of biofouling involves the settlement of large marine invertebrates and macro-algae. This macro-fouling stage can take from weeks to years and is often dynamic and complex with many species settling to form a complex three dimensional community. The most common macro-fouling organisms are barnacles, bryozoans, algae, tubeworms and molluscs. The biofouling sequence is described in detail in[3] and the stages of fouling outlined by[4] and more recently reviewed by[5,6].

Biofouling poses a major problem to all marine industries. The economic cost of fouling is substantial and includes expenses such as dry-docking, hull maintenance, increased fuel consumption (up to 40%[7]), delayed transport times and increased cleaning[5]. The economic consequences of fouling have lead to numerous solutions for biofouling control, which began many centuries ago. More sophisticated antifouling paints were developed in the 19th century generally containing a mixture of copper and rosin and in the 1950's the incorporation of organotins greatly improved the antifouling performance of ships (particularly the tri-butyl-tin (TBT) moiety). In 1974 TBT self-polishing copolymers (TBT-SPC) were patented[8]. This breakthrough was seen as the definitive antifouling solution as the polishing rate could be controlled for various vessel operating speeds[9], and the polishing produced low hull roughness which reduced fuel consumption and emission[10]. Furthermore, the paints are not corrosive to aluminium and steel[11], have short drying times, are durable[12] and dry docking periods are extended to 5 years[13].

Unfortunately TBT-SPC antifouling systems have proved to be far from perfect. These paints have very serious effects on non-target organisms in the marine environment including reduced growth, imposex and malformations in molluscs[14-16] and accumulation in mammals and fishes[17]. These adverse effects have forced the International Maritime Organisation to ban the use of TBT-SPC[7]. At present the application of TBT based antifouling paints is banned and its presence on all vessels is to be removed, or sealed in, by 2008[18]. These actions have lead to urgency in the search for a non-toxic environmentally friendly antifouling alternative.

Biomimicry is the copying of natural strategies to solve mankind's problems. For antifouling purposes there is a plethora of natural fouling resistant marine organisms that can be used to inspire novel surface design. Copying surface structures alleviates concern of release of toxins into the marine environment and potential harm to non-target species. The biomimetic approach to antifouling is the focus of this paper, with an emphasis on physical fouling defence mechanisms in marine molluscs.

THE BLUE MUSSEL

Although fouling organisms quickly settle on most submerged surfaces, some marine organisms are able to keep their surfaces relatively fouling free. One such species is the blue mussel, *Mytilus galloprovincialis*. To investigate which properties of the mussel surface was responsible for fouling deterrence the surface of the blue mussel was examined under Scanning Electron Microscopy (SEM) and Atomic Force Microscopy (AFM). It was found that the surface of intact blue mussel shells have a homogeneous microtextured ripple (Figure 1), which had a mean height of 1.9 μm and a mean height of 123 nm[19].

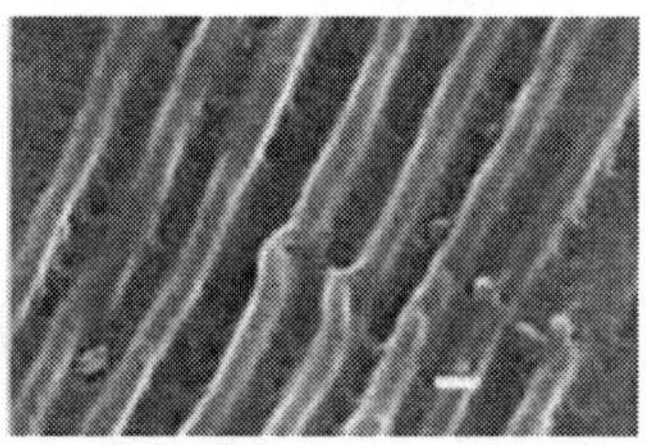

Figure 1: The microtextured surface of the blue mussel, *Mytilus galloprovincialis*. Scale 1 μm

The surface of the blue mussel was further examined by sectioning the outermost layer of the shell, known as the periostracum. Shells were cleaned, fixed in epoxy resin, sectioned with a diamond saw and examined under light microscopy (400x). The periostracum was found to be persistent across most of the sectioned shells, with a mean thickness of 10 μm. To assess the extent of the periostracum cover on the shell surface, shells were stained with the dye Coomasi Blue which binds to the protein in the periostracum. The blue mussel shells had a periostracum cover in excess of 90%. Another shellfish species, the pearl oyster *Pinctata imbricata*, was compared to the blue mussel to determine periostracum cover, thickness and topography of shell. The pearl oyster had a thinner periostracum (1-2 μm), less persistent covering (60%), no surface microtopography and a shell surface with a higher mean roughness (300 nm compared to 95 nm)[19].

Although the blue mussel was shown to have a homogeneous microtextured surface, and a thicker and more persistent periostracum than the pearl oyster, it is unknown if these properties are responsible for fouling resistance. To compare the ability of the two species shells to resist fouling a field trial was conducted for 3 months in Port Stephens, NSW. Shells were assessed for the rate of fouling, it was found that the pearl oysters were fouled more than 5 times the rate of the blue mussel shells by the conclusion of the field trial. Over 80% of the pearl oyster shells were fouled in comparison to just 10% of the blue mussel shells[19]. This demonstrates a clear correlation between fouling resistance and the presence of a microtextured periostracum. To determine if a microtextured surface is a common antifouling property a range of marine mollusc species were examined.

MARINE MOLLUSCS FROM THE GREAT BARRIER REEF (GBR)

The natural surface microtextures from 36 species of marine molluscs from the GBR that generally had low levels of fouling *in-situ* were examined. The surfaces of the shellfish were characterised by generating surface roughness parameters using Laser Scanning Confocal Microscopy (LSCM). The six surface roughness parameters examined were: 1. average roughness (Ra), 2. average waviness (Wa), 3. skewness of the surface roughness profile (Rsk), 4. skewness of the surface waviness profile (Wsk), 5. texture aspect ratio (Str), and 6. fractal dimension (Fd) and surface hydrophobicity.

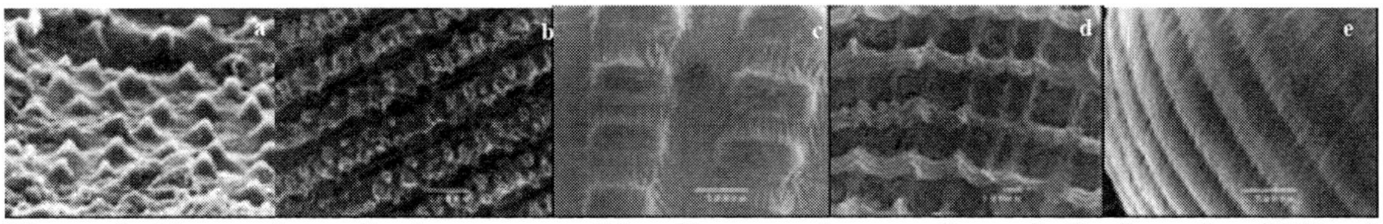

Figure 2: The variety of microtopographies discovered on mollusc shells from the GBR, a) *Tellina plicata*, scale 5 μm, b) *Diadema savignyi* scale 100 μm, c) *Gafrarium tumidum* scale 500 μm, d) *Donax faba* scale 100 μm, e) *Nerita plicata* scale 500 μm,

A variety of microtextured shapes and scales were discovered on the marine molluscs (Figure 2). To quantify the surface roughness of the mollusc shells six roughness parameters were derived. The software package Matlab 6.0 was used to calculate Ra, Wa, Rsk and Wsk, while the software package Optimas 6.1 along with code developed by Dr Zhongxiao Peng[20] was used to quantify Str and Fd of the imaged surfaces. The Ra and Rsk for roughness and waviness are amplitude parameters that describe the amplitude and height distribution of the surface which was calculated after separating the original surface profile into three profiles being form, waviness and roughness as illustrated in Figure 3. The fractal dimension measures the irregularity or complexity of the surface texture while texture aspect ratio characterises the periodic features or the uniformity of the surface texture patterns. These six parameters allow for analysis of quantified relationships between surface topography characteristics. The equations used to generate these parameters and the implications of high and low values for a surface are listed elsewhere[21].

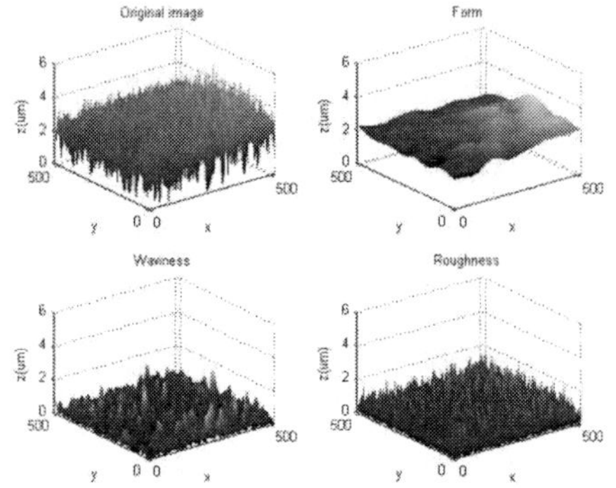

Figure 3: Original surface profiles of *Mytilus galloprovincialis* separated into its three profiles being form, waviness and roughness. The 3-D scale is measure in microns. The x and y axis are 0-500 μm and the z axis (height) is 0-6 μm.

The mollusc surfaces examined displayed small scales of roughness and waviness, predominately in the nanometer range. Most surfaces were highly complex with Fd values close to the maximum. Surfaces had either periodic texture or a greater distribution of peaks, but not valleys. All surfaces were highly isotropic and therefore had less directionality and more randomness. The surfaces had a moderate hydrophobicity. The mollusc surfaces characterised differed at the class, family and species level. Bivalves had significantly higher waviness profiles and were more positively skewed in the roughness and waviness profiles whilst gastropods had significantly greater fractal dimensions. The Ra, Wa, Rsk and Fd parameters differed significantly between families, whilst at the species level all parameters differed significantly except hydrophobicity. Several of the surfaces parameters were correlated together.

To determine which roughness parameters are most important to fouling resistance a field experiment was conducted to determine which mollusc shells exhibited the highest degree of fouling resistance. The critical roughness features of the best and worst performing species could then be identified and used for recommendations for future biomimics and as a basis to determine the underlying mechanisms that determine fouling resistance.

The fouling resistance experiments were conducted at Ross Creek, Townsville. Five replicates of each of the 36 species were submerged for 3 months and photographed weekly to assess the level of fouling. At the completion of the field trial the shells were removed and exposed to a water jet to measure the ease of removal of fouling organisms. The water jet (Gerni – Crown compact) was locked into place on a stage vertically above the test surfaces. The pressure from the water jet acting on the shell surface was calculated by $Pa=N/m^2$, where Pa = the surface pressure in Pascals, N = force in Newtons. The water jet was delivered at a flow rate of 6L/min from a height of 75cm. The weight of the water jet was measured from a strain gauge (Scale Components P/L; Model: STC - 250 kg; calibration # - mV/V: 2.9991; serial # - W13397) and was constant (1.23 kg) for varying distances from the surface. The pressure acting on the shell surfaces was 76kPa.

The fouling cover on the 36 mollusc species was found to vary significantly after 3 months exposure. The worst performing species had over 80% fouling cover, in contrast the most fouling resistant species had less than 30% fouling cover (Figure 4). The level of fouling resistance also varied between Classes and Families of molluscs.

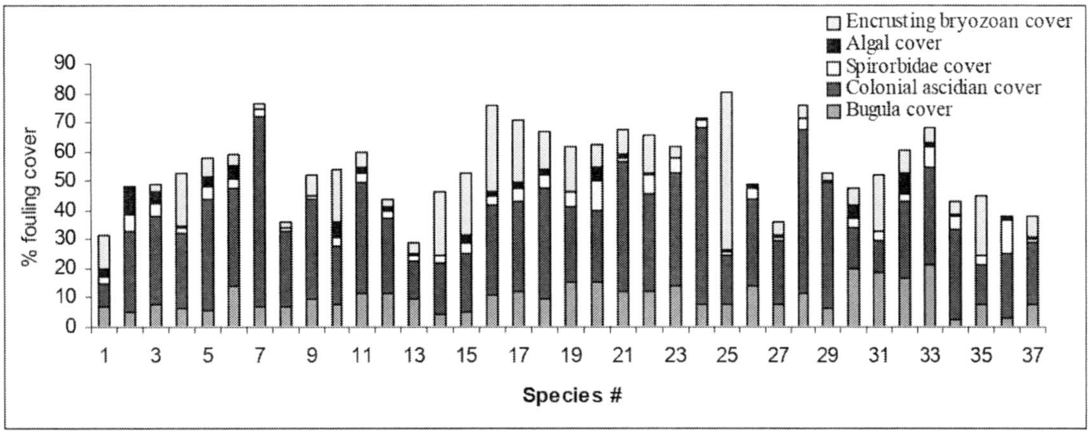

Figure 4: Mean species specific fouling cover on the 36 species and PVC after 12 weeks exposure

The fouling resistance and fouling removal of the 36 species was then correlated to the surface roughness parameters generated. Several significant correlations were found (Table 1) which indicates that some roughness parameters are more important for fouling resistance and others are responsible for maximising fouling release. These findings help to guide the development of future biomimetic antifouling surfaces.

Table 1: Summary of significant differences at the Class, Family and Species level between total fouling cover, species specific fouling cover, and total percent fouling removal. Significant correlations between roughness parameters and fouling cover and removal are also listed. Y= Significant difference, N= No significant difference

	Total cover	***Bugula neritina***	**Colonial ascidian**	**Spirorbid**	**Algae**	**Encrusting bryozoan**	**% removal**
Class	Y	N	N	N	N	Y	N
Family	Y	Y	Y	N	N	N	N
Species correlation	Rsk , Wsk Fd	-	-	Fd	Rsk, Wsk, wettability	Rsk , Wsk	Wa Algal cover.

There were a number of significant correlations between total fouling cover, species specific fouling cover, and the individual roughness parameters generated for the 36 mollusc species. Total fouling cover was positively correlated to the fractal dimension of a surface ($r = 0.376$, $P = 0.022$) and therefore the more fractal or complex a surface is the more likely it is to have a higher overall fouling cover. In contrast, total fouling cover was negatively correlated to Rsk ($r = -0.357$, $P = 0.030$) and Wsk ($r = -0.371$, $P = 0.024$). This means that surfaces with a greater distribution of peaks about the mean line had lower overall fouling cover. Algal cover was negatively correlated to hydrophobicity ($r = -0.395$, $P = 0.021$) and therefore surfaces which were most hydrophobic had lower algal cover. In accordance with total fouling cover, spirorbid tubeworm cover was positively correlated to fractal dimension ($r = 0.342$, $P = 0.039$). Encrusting bryozoan cover was negatively correlated to both Rsk ($r = -0.329$, $P = 0.046$) and Wsk ($r = -0.367$, $P = 0.025$). This means those surfaces with few peaks about the mean line had the highest encrusting bryozoan cover. In contrast macro-algal cover was positively correlated to both Rsk ($r = 0.469$, $P = 0.003$) and Wsk ($r = 0.423$, $P = 0.009$). Of the roughness parameters determined for the 36 species only mean waviness was significantly correlated to total percent fouling removal ($P = 0.008$,). This means that species with higher mean waviness profiles had the highest removal of total fouling after exposure to a water jet.

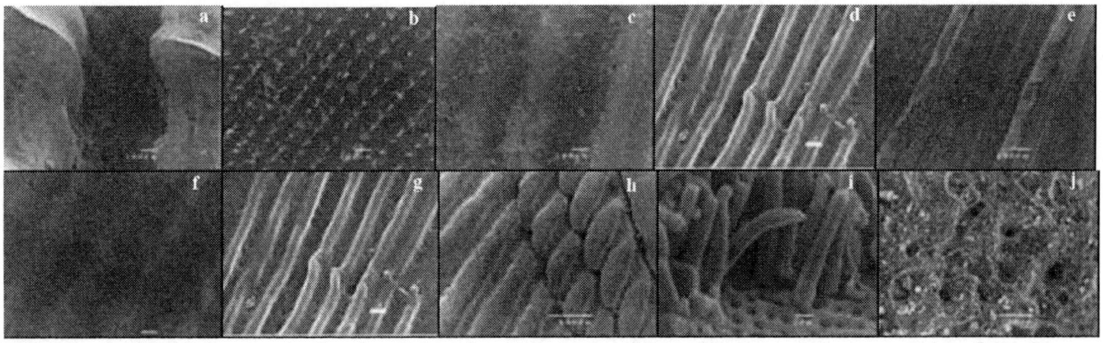

Figure 5: SEM images of selected microtopographies discovered on marine bivalve and gastropod shells. Figures a-e are the top ranked species in terms of fouling resistance; figures f-j are the top ranked species in terms of fouling release. The scale of each picture varies based on the topographical dimensions of the shell a, *Dosinia juvenilis* ribs 400-500 µm wide (scale 100 µm) **b**, *Tellina inflata* spines 50 µm long, 10 µm wide and spaced 50 µm apart (scale 100 µm). **c**, *Mactra olorina* untextured surface (scale 100 µm) **d**, *Mytilus galloprovincialis* ripples 1-2 µm wide (scale 1 µm) **e,** *Acrosterigma reeveanum.* ridges, 500 µm wide, 200 µm deep (scale 100 µm). **f**, *Septifer bilocularis ridges* 200 µm wide ribs 200 µm wide, 100 µm apart (scale 500 µm) **g**, *Mytilus galloprovincialis* (scale 1 µm). **h**, *Hemidonax* sp. (scale 500 µm).**i**, *Gafrarium tumidum* spines 5 µm long, 1 µm wide (scale 1 µm) **j**, *Divaricella ornate* perforations, 5 µm diameter (scale 10 µm).

MICROTEXTURED BIOMIMICS & ATTACHMENT POINTS

The broad ranges of fouling resistances observed from the surfaces of the 36 molluscs tested leads to questions as to whether resistance to fouling is due to physical and/or chemical forces associated with the shell surface. To address this two bivalves with different surface structures and fouling resistant properties are chosen for biomimicry to test whether the copied surfaces (thereby eliminating potential surface chemistry) will perform as well as the natural surfaces. Microtextured biomimics were created based on the surface ripples found on the blue mussel *Mytilus galloprovincialis* (Figure 1) and the pyramid topography of the bivalve *Tellina plicata* (Figure 2a). Test surfaces were produced using methodologies adapted from[22]. The microtextured surface patterns were laser etched onto Polyimide film (Kapton HN®, thickness 125 µm, test surface 20 mm x 20 mm) by MiniFAB (Ltd) using a KrF excimer laser beam (248 nm, fluence 1 J/cm^2. impulse, 20 ns per impulse, X10 mask magnification) coupled with an optical device which projects the laser beam through a metal photomask producing 1 µm wide bars with a 2 µm pitch. The surfaces were 2 µm ripples modeled from *M. galloprovincialis* (2 µm wavelength, depth 0.4 µm), 4 µm ripples slightly larger than *M. galloprovincialis* microtexture (4 µm wavelength, depth 1.2µm), and 4 µm projections modeled from *T. plicata* (4 µm wavelength, depth 0.5 µm). Rippled surfaces were made using a sine wave microtexture. Projections were made using a 3D sine wave by crossing two sine wave ripples at right angles to each other. Smooth polyimide film was used for controls. The accuracy of the surfaces was confirmed using SEM.

The microtextured biomimics were tested against four diatom species of differing cell sizes. The mean number of cells attached to the biomimics varied significantly between the smooth polyimide and micro-textured polyimide treatments. Furthermore, the response to the surfaces was species specific. Cell sizes for the four diatom species and the number of adhesion points on the microtextured polyimide treatments are shown in Table 2. Theoretical attachment points are designated as multiple where diatoms fit within surface microtexture, or for smooth controls where the shape of pinnate diatoms allows for multiple attachment sites (Figure 6). It is assumed that the highest number of attachment points is possible where the diatom fits within the most 'attractive' microtexture[23]. Points of attachment were determined based on the width of the diatom cell and the wavelength of the surface microtexture when longitudinal to microtexture as cells consistently align with the surface microtexture. For microtextured peaks the length of the cell was also used to determine theoretical attachment points as the longer the cell the more peaks are available for attachment.

Table 2: Cell sizes and the number of attachment points to polyimide microtextures.

Species	Cell length	Cell width	# adhesion points on 2 μm ripples	# adhesion points on 4 μm ripples	# adhesion points on 4 μm peaks
Fallacia carpentariae	3 μm	1 μm	multiple	multiple	multiple
Nitzschia paleacea	7.2 μm	2 μm	multiple	multiple	multiple
Navicula jeffreyi	12 μm	4 μm	3	2	6
Amphora sp.	14.7 μm	7.3 μm	4	2	6

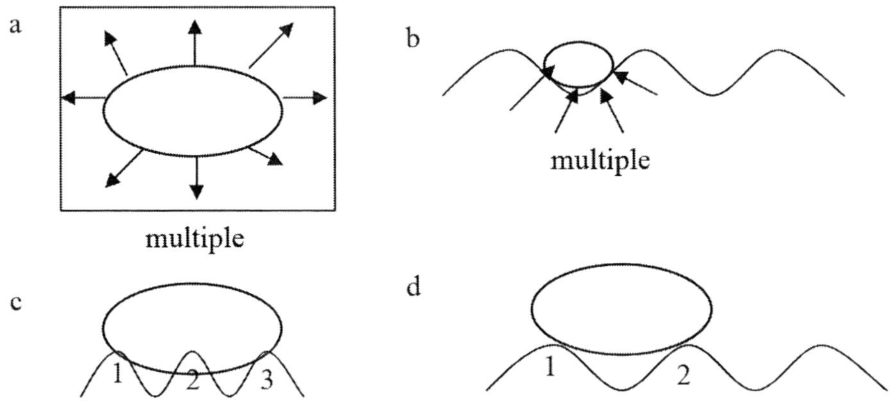

Figure 6: A schematic illustrations of theoretical attachment points for settlement by a) all diatoms on a smooth surface – multiple attachment points b) *Fallacia carpentariae* on 2 μm ripples – multiple attachment points c) *Navicula jeffreyi* settling on 2 μm ripples – 3 attachment points d) *Amphora* sp. settling on 4 μm ripples – 2 attachment points

Attachment to the microtextured biomimics was consistently highest when there were multiple numbers of attachments available. Cells that could fit inside the microtextures settled in higher numbers than cells which could only sit on top of the textures. Attachment was reduced on the microtextured biomimics compared to smooth surfaces. For the smallest diatom, *Fallacia carpentariae*, there was no difference in mean number of cells attached between any treatments. This species has a cell size less than the wavelength of all microtextured treatments and as such the cells of this species have a multiple attachment points on all test surfaces. *Nitzchia paleacea* has a cell width slightly less than 2 μm, the perfect fit for 2 μm microtextured ripples. The cells of this species were observed to line up longitudinally inside the 2 μm microtextured ripples and incidentally had twice the number of cells adhering to this treatment than any other. With increasing cell size the number of attachment points was reduced particularly on the 4 μm ripple surfaces where for both *Amphora* sp. and *Navicula jeffreyi* there were only two theoretical attachment points. Both species had significantly lower attachment to the 4 μm ripple surfaces than the smooth control. The response of the tested fouling diatoms to microtextured polyimide supports the application of attachment point theory to diatoms, whereby reduced numbers of attachment points results in lower rates of settlement[24].

MICROTEXTURED MATRIX

To further test the ability of microtextured biomimics to reduce fouling settlement a second set of experiments were conducted using several different sized fouling organisms on a microtextured matrix. The major fouling classes used comprised a wide range of cell and propagule sizes (7-321 μm), which are tested for attachment preferences across a broad range of microtexture wavelengths (4-512 μm). The aim of this study was to determine if attachment point theory holds for broad classes of fouling larvae and algal propagules in static choice bioassays. The six fouling organisms used are the raphid diatom *Amphora* sp., the green algae *Ulva rigida*, the red algae *Polysiphonia sphaerocarpa* and *Centroceras clavulatum,* the tubeworn *Hydroides elegans* and the bryozoan *Bugula neritina*.

A matrix of microtextured patterns was manufactured by MiniFAB Pty Ltd in 1.5 mm polycarbonate by laser micromachining using an excimer laser system (8000 series, Exitech, UK) equipped with a Lambda Physik LPX210i laser source operating at a wavelength of 248 nm with krypton fluoride (KrF) as the excitation gas. The master template for the polycarbonate pieces was 100 mm x 100 mm and consisted of nine 30 mm x 30 mm squares each with a sinusoidal shaped pattern which varied in both the period (distance from peak to peak) and depth (distance from peak to trough) of the sinusoidal shape. Full details of the machining process used to produce the microtextured matrix are detailed elsewhere[25]. In brief it involved delivering a laser beam onto a chrome-on-quartz mask. The mask was controlled by an X-Y stage allowing precise lateral and vertical movement.

256 μm	**16** μm	**64** μm
***512** μm	**smooth**	**32** μm
8 μm	**128** μm	**4** μm

Figure 7: The arrangement of microtextured squares on the Polycarbonate matrix. The number in each square represents the depth and period of microtexture with the exception (*) of the 512 μm microtextured square which has a period of 512 μm and a depth of 256 μm due to machining constraints.

The fouling organisms chosen in this study cover the major groups of fouling classes and larval/spore sizes, and demonstrate the influence of attachment points and surface microtexture on attachment. There is a broad range of attachment preferences by the different fouling organisms to varying scales of microtexture (in this case referring to attachment of diatoms, settlement and germination of algae, and settlement and metamorphosis of larva).

The most important finding across all the organisms, except *C. clavulatum*, is that attachment is lower on textures slightly smaller than the size of the fouling larvae/cells compared to other microtextures. Furthermore, on the texture immediately greater than the size of the larvae attachment was in general strongly increased. For the diatom *Amphora* sp., (cell size 7 μm), attachment is lower on the 4 μm texture and higher on the 8 μm texture and the green algae, *Ulva rigida*, (gamete size 7 μm) attached in lower numbers on the 4 μm texture. However, these differences were not significant. The red algae *Polysiphonia sphaerocarpa*, (spores size 51 μm) attached in significantly lower numbers on the 32 μm texture while for the bryozoan, *Bugula neritina* (larval size 321 μm), attachment was significantly lower on the 256 μm texture and significantly higher on the 512 μm texture. These findings provide species-specific support for attachment point theory.

Table 3: Summary of attachment to microtextured polycarbonate for each species tested. Lower = lower proportion of cells attached compared to other microtextures, Higher = a greater proportion of cells attached compared to other microtextures, Equal = a similar proportion of cells attached compared to other microtextures, n/s = no significant difference in attached cells between microtextures, * = a significant difference in attached cells between microtextures at $\alpha = 0.05$.

Species	**Size (μm)**	**Attachment to first texture smaller than cell size**	**Attachment to first texture larger than cell size**
Amphora sp.	7	(4 μm) Lower	(8 μm) Higher
Ulva rigida	7	(4 μm) Lower	(8 μm) Equal
Centroceras clavulatum	37	(32 μm) Equal	(64 μm) Equal
Hydroides elegans	129	(256 μm) Lower*	(512 μm) Higher*
Bugula neritina	321	(128 μm) Lower*	(256 μm) Higher*

With many antifouling surfaces moving towards foul-release, rather than non-stick coatings, it is important that surface roughness/microtexture does not increase adhesion strength and the forces required for fouling removal. A surface with fewer attachment points, and one which does not provide microrefuges, may reduce the shear forces required for fouling removal, and therefore would be an excellent foul-release candidate. Finally, *in situ*, fouling larvae and propagules come in all sizes, and as such a successful surface will need a combination of microtextures with fractal dimensions to cover micro and macro fouling organisms[26]. The applicability of attachment point theory and the influence of surface microtexture against multispecies fouling consortia in laboratory trials, and in field trials, is now the next step in assessing the longer-term efficiency of microfabricated biomimics.

FUTURE DIRECTIONS

This paper has demonstrated the promising alternatives biomimetic antifouling technologies provide. With the imminent removal of TBT from all vessels and the longer-term phasing out of all toxic coatings the urgency for a non-toxic solution is increasing and biomimetics will contribute to these answers. Whilst this paper has used marine molluscs as a model for biomimetic technologies, there are many other marine organisms which have evolved antifouling strategies which are waiting to inspire new ideas and designs for fouling control. It is widely recognised that multiple defence strategies are necessary to combat the myriad of fouling organisms; therefore studying several natural models of fouling resistance can only contribute to the knowledge of biomimetic technologies.

Based on the promising results of biomimicked marine mollusc surfaces I have identified four key areas for future research into biomimetic antifouling technologies. Firstly, the key roughness parameters identified from mollusc surfaces that influence fouling need to be tested on biomimicked surfaces. Designing a surface with a high degree of isotropy, high skewness values, with a greater distribution of peaks and large waviness profiles, should facilitate fouling resistance and the ease of fouling detachment. The incorporation of these features into a silicone based foul release surface requires testing in the field and in flow chambers. Secondly, attachment points have been demonstrated to be important to fouling attachment for both micro and macrofouling propagules. Designing fractal textures to account for multiple organism sizes will be a requirement for successful fouling resistance in the field. The foul-release ability of reduced attachment points is also unexplored and quantified analysis in flow chambers is necessary. Thirdly, it is clear that other factors combine with surface microtexture to produce extended fouling resistance on natural bivalve shells. For the mussel *Mytilus galloprovincialis* it appears that the periostracum plays a major role in fouling deterrence. Determining the nature of the periostracal properties that contribute to fouling deterrence is a vital link to determine the mechanisms of action. A non-destructive method to denature the Mytilid periostracum will determine the extent to which the periostracum contributes to fouling resistance. This will lead to proteins and other compounds being tethered to foul-release surfaces to combine the chemical and physical surface properties which provide multiple defence strategies. Lastly, Nano-coatings are a new concept and the promising results from static assays[21] will require further research into the influence of flow and the longevity of resistance in the field. The mechanism of action of fouling deterrence on nano-scale surfaces is yet to be demonstrated and the influence of nano-bubbles on the surface of these coatings will be the focus of future investigations.

ACKNOWLEDGEMENTS

I would like to thank the input and support of Professor Rock de Nys and the School of Marine & Tropical Biology, James Cook University. Dr Zhongxiao Peng, Des Hudleston, Dr Nick Paul were important collaborators in determining mollusc shell surface characteristics and correlations to fouling resistance and removal. The Centre for Marine Biofouling & Bioinnovation, UNSW kindly allowed bio-assays to be conducted to test biomimics. Dr Hua Zhang and Professor Lamb, UNSW developed the nano-scale Superhydrophobic coatings. Dr Rick Wetherbee and Paul Molino provided resources to conduct diatom assays and Jana Guenther, JCU conducted the tubeworm assays. Dr Simon Ovenden has led the analysis of the blue mussel periostracum.

REFERENCES

1. Bakker, D. P., Klijnstra, J. W., Busscher, H. J. & van der Mei, H. C. The effect of dissolved organic carbon on bacterial adhesion to conditioning films adsorbed on glass from natural seawater collected during different seasons. *Biofouling* **19**, 391-397 (2003).
2. Callow, J. A. & Callow, M. E. in *Antifouling Compounds. Progress in Milecular and sub-cellular Biology: Marine Molecular Biotechnology.* (eds. Fusetani, N. & Clare, A. S.) pp. 141-170. (Springer-Verlag Press., 2006).
3. Wahl, M. Marine epibiosis. 1. Fouling and antifouling: some basic aspects. *Marine Ecological Progress Series* **58**, 175-189 (1989).

4. Abarzua, S. & Jakubowski, S. Biotechnological investigation for the prevention of biofouling. 1. Biological and biochemical principles for the prevention of biofouling. *Mar. Ecol. Pro. Ser.* **123**, 301-312 (1995).
5. Yebra, D. M., Kiil, S. & Dam-Johansen, K. Antifouling technology - past, present and future steps towards efficient and environmentally friendly antifouling coatings. *Progress in Organic Coatings* **50**, 75-104 (2004).
6. Krug, P. J. in *Progress in Molecular and Subcellular Biology - Subseries Marine Molecular Biotechnology.* (eds. Fusetani, N. & Clare, A. S.) 1-54 (Springer-Verlag Press., 2006).
7. Champ, M. A. A review of organotin regulatory strategies, pending actions, related costs and benefits. *Science of the Total Environment* **258**, 21-71 (2000).
8. Milne, A. & Hails, G. Patent #1457590 International Paint Plc, Great Britain. (1974).
9. Anderson, C. D. in *Proceedings of protecting the ship whie safeguarding the environment. 5-6 April* (London, 1995).
10. Swain, G. in *International symposium on seawater drag reduction. 22-24 July* (Newport, 1998).
11. Pal, S. & Misra, S. K. Organotin copolymers for self-polishing antifouling paints (SPC's). *Paintindia* **43**, 19-24 (1993).
12. Hunter, J. E. & Cain, P. in *Inst of Marine Engineers (IMarE), Pt I, IMAS 96, 10th Intn Maritime & Shipping Conf, Session V, Paper 16, 108(5) 22-24 October* 61-70 (London, 1996).
13. Anderson, C. D. in *IBC UK Conferences Limited* 1-12 (London, 1998).
14. Alzieu, C. in *Proceedings of Oceans 86, Organotin Symposium* 1130-1134 (Marine Technology Society, Washington D.C, 1986).
15. Evans, S. M., Leksono, T. & McKinnell, P. D. Tributyltin Pollution - a Diminishing Problem Following Legislation Limiting the Use of Tbt-Based Anti-Fouling Paints. *Marine Pollution Bulletin* **30**, 14-21 (1995).
16. Gibbs, P. E. & Bryan, G. W. Reproductive failure in populations of the Dog-Whelk, *Nucella-Lapillus*, caused by imposex induced by Tributyltin from antifouling paints. *Journal of the Marine Biological Association of the United Kingdom* **66**, 767-777 (1986).
17. Kannan, K. & Tanabe, S. Elevated accumulation of tributyltin and its breakdown products in bottlenose dolphins (*Tursiops truncatus*) found stranded along the US Atlantic and Gulf coasts - Response. *Environmental Science & Technology* **31**, 3035-3036 (1997).
18. Champ, M. A. New IMO convention to control anti-fouling systems on ships. *Sea Technology* **42**, 48-50 (2001).
19. Scardino, A., de Nys, R., Ison, O., O'Connor, W. & Steinberg, P. Microtopography and antifouling properties of the shell surface of the bivalve molluscs *Mytilis galloprovincialis* and *Pinctada imbricata*. *Biofouling* **19**, 221-230 (2003).
20. Yuan, C. & Peng, Z. Surface Characterisation Using Wavelet Theory and Confocal Laser Scanning Microscopy. *Journal of Tribology (in press)* (2004).
21. Scardino, A. J. Biomimetic Fouling Control. *PhD thesis: School of Marine Biology & Aquaculture, James Cook University* (2006).
22. Duncan, A. C., Weisbuch, F., Rouais, F., Lazare, S. & Baquey, C. Laser microfabricated model surfaces for controlled cell growth. *Biosensors & Bioelectronics* **17**, 413-426 (2002).
23. Callow, M. E. et al. Microtopographic cues for settlement of zoospores of the green fouling alga *Enteromorpha*. *Biofouling* **18**, 237-245 (2002).
24. Scardino, A. J., Harvey, E. & De Nys, R. Testing attachment point theory: diatom attachment on microtextured polyimide biomimics. *Biofouling* **22**, 55-60 (2006).
25. Harvey, E. C. & Rumsby, P. T. in *Proceedings from SPIE-International Society for Optical Engineering, 3223* 26-33 (1997).
26. Schumacher, J. F. et al. Species-specific engineered antifouling topographies: correlations between the settlement of algal zoospores and barnacle cyprids. *Biofouling* **23**, 307-317 (2007).

Explorer AUV missions in coastal Newfoundland

R. Lewis[1], S. Adams[1], N. Bose[1,2], J. Anderson[3]

1. Memorial University of Newfoundland, St. John's, NL, Canada
2. Australian Maritime Hydrodynamics Research Centre, AMC, Launceston, Tasmania
3. Fisheries and Oceans Canada, St. John's, NL, Canada

ABSTRACT

The Marine Environmental Lab for Intelligent Vehicles (MERLIN Lab) at Memorial University of Newfoundland (MUN) operates a survey class autonomous underwater vehicle (AUV) available for scientific research within a wide range of disciplines. The *MUN Explorer* is 4.5 meters in length, 0.69 meters in diameter, weighs 650 kg and displaces 660 kg. This International Submarine Engineering Ltd. (ISE) Explorer class vehicle has the capacity to handle 200 kg of scientific payload. With a range of up to 100 km at speeds of up to 2.5 m/s and depths to 3000 metres, the vehicle is ideal for environmental surveys where bathymetric and mapping sonar, physical and chemicals sensors, cameras and acoustic devices are carried. The paper focuses on the capabilities of the *MUN Explorer AUV* and plans to build collaborative projects. Data will be presented on vehicle performance during habitat mapping in coastal Newfoundland.

Habitat mapping was conducted with Fisheries and Oceans Canada by using a single frequency acoustic system adapted to the AUV. The aim is to evaluate the potential to cost effectively classify seabed habitats as well as to monitor fish and zooplankton abundance and distributions in association with these habitats. The work detailed in this paper concerns the preliminary evaluation of the AUV as a research tool in this context.

I. INTRODUCTION

The MERLIN Lab has been operating the *MUN Explorer* AUV for just over a year. It was originally purchased from International Submarine Engineering Ltd. in Port Coquitlam B.C., Canada, as a survey class underwater vehicle. The *MUN Explorer* is 4.5 meters in length, 0.69 meters in diameter, weighs 650 kg and displaces 660 kg. It has the capacity to handle 200 kg of scientific payload. The vehicle has a potential range of up to 100 km at speeds of up to 2.5 m/s and depths to 3000 metres which makes it an ideal platform for environmental surveys where bathymetric and mapping sonar, physical and chemical sensors, cameras and acoustic devices are carried. Memorial University's previous deployments have included using an Applied Microsystems Conductivity-Temperature-Depth Sensor coupled with a Cyclops Fluorometer, Rhodamine WT, Chlorophyll *a* or Turbidity sensors in order to track simulated and actual effluent waste plumes (Niu et al., 2007).

Operations with the MUN Explorer in September 2007 focused on an acoustic payload as opposed to a chemical sensor payload. The BioSonics DTX Digital Scientific Echosounder (provided by Fisheries and Oceans Canada) was integrated for the purpose of fisheries habitat mapping for an area where a pre-existing multibeam echosounder data set of the bathymetry existed. These data had been gathered from a ship. The motivation for the latest deployments using the acoustic sensor was two fold. The primary goal of the deployments was to evaluate the dynamical stability and accuracy of the AUV when navigating in an area known for strong geostrophic currents (B. deYoung, pers. comm.). The secondary goal was to assess the

Figure 1 MUN Explorer AUV

quality of the collected acoustic data from the AUV compared to that collected from a ship or towfish system.

The identified area of interest was Haystack Bank located in northern Placentia Bay, Newfoundland and Labrador, Canada (47°38.43 N, 54°0.0 W) (Figure 2). The bank is an area measuring approximately 7 km by 4 km with a variation of water depths from 250m to 18m.

The second section of the paper highlights the onboard systems used to facilitate the automated sampling of the echosounder. The third section describes the methodology and evaluation of the dynamical performance and navigational accuracy of the vehicle. The fourth section includes some of the digital acoustical data. A summary and conclusion are contained in the fifth section.

II. SYSTEMS

Typical BioSonics DTX operation includes the Surface Unit, an embedded PC with LINUX OS, which is the signal generator and controls the operation of the transducer(s); a user provided PC running Windows XP to facilitate configuration of the Surface Unit and data storage; a Global Positioning System (GPS) feed; and a digital transducer. To facilitate the integration of the DTX unit with the AUV, it was necessary to modify the existing *MUN Explorer* system. Solutions for the lack of GPS availability underwater and the inability to remotely operate an onboard PC in real time needed to be identified, designed and implemented.

The National Research Council, Institute of Ocean Technology (NRC-IOT), provided an Ixsea Photonic Inertial Navigation Systems (INS) or PHINS unit. The unit has the ability to process output data in a GPS protocol using a NMEA 0183 format as required by the BioSonics DTX. The PHINS also provided enhanced navigation and positional accuracy for the *MUN Explorer* compared with the onboard Watson Attitude and Heading Reference System (AHRS-E304). This is an added benefit of using the PHINS unit which is a preferred INS for the Explorer. Both units remained on board for the duration of the tests.

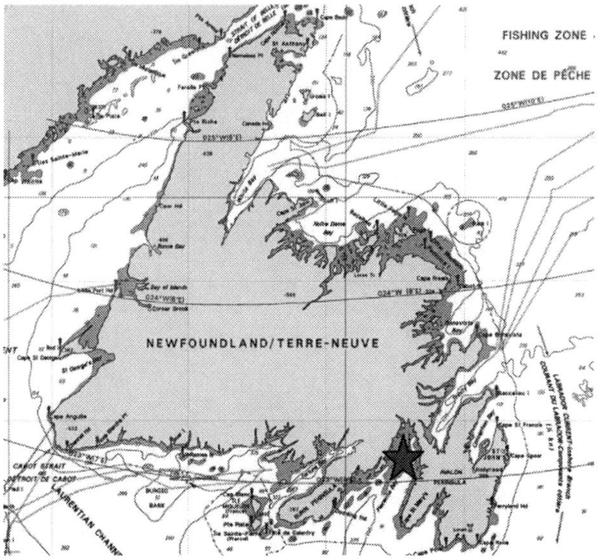

Figure 2 Map of the island of Newfoundland and the study area (marked by the star)

In order to facilitate the remote initialization, operation, and shutdown of the BioSonics Surface Unit, an embedded Compact Peripheral Component Interconnect (CPCI) computer system was integrated to run parallel with the existing vehicle control computer onboard the AUV. The CPCI system employed a Windows XP OS and used software based on IFREMER's (Institut français de recherche pour l'exploitation de la mer) Contrôle de Charge Utile or CCU software. The function of this software is to remotely and automatically control onboard sensor/payload computers and to manage data collection and storage.

III. AUV DYNAMICS

The ISE Explorer class vehicles are capable of subsurface navigation with respect to the seafloor in either a depth or in an altitude mode. In depth mode, the vehicle will maintain a user specified depth (distance below the surface) for a given transect or leg of its mission. When in altitude mode, the vehicle will navigate at a constant user specified range above from the sea floor.

The dynamical behavior of the MUN Explorer was assessed by configuring a mission composed of a series of North to South, South to North, East to West and West to East transects. These entailed vehicle operation parallel and orthogonal to the geostropic currents and over varied sea floor topography. Figure 3 shows multibeam echosounder bathymetry data for the Haystack Bank area. The principal goal was to evaluate vehicle stability when the vehicle was prompted to maintain a consistent distance from the sea bottom at a constant speed. The transects were arranged such that the vehicle would swim with, against and orthogonal to the currents. A hardware constraint for these missions was a maximum depth set point of 75 to 80 meters due to the BioSonics DTX transducer depth limitations. As well, for some transects a minimum depth set point of 5 meters was instituted in order to keep the vehicle fully immersed and less subjected to surface and wind effects.

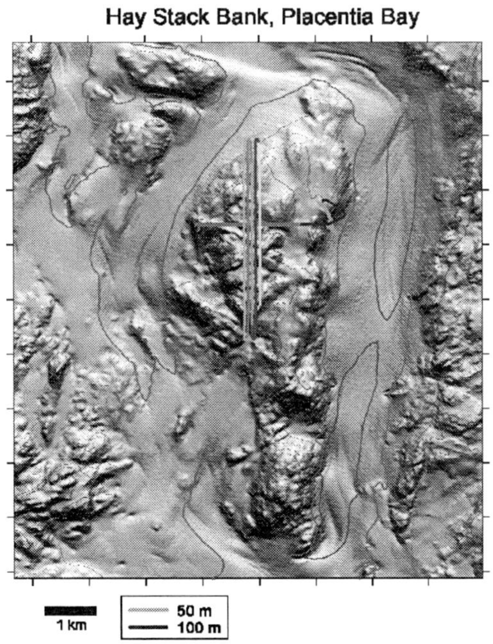

Figure 3 Multibeam data of Haystack Bank (Courtesy of Dr. J. Anderson, DFO)

Figure 4 demonstrates both methods of depth navigation to ensure the vehicle remained at 5 m from the surface while maintaining the specified altitude of the overall mission. The reason for the ceiling requirement was due to the desire to completely eliminate surface effects from the whole dive while traversing over a shallow area (less than 20 m depth).

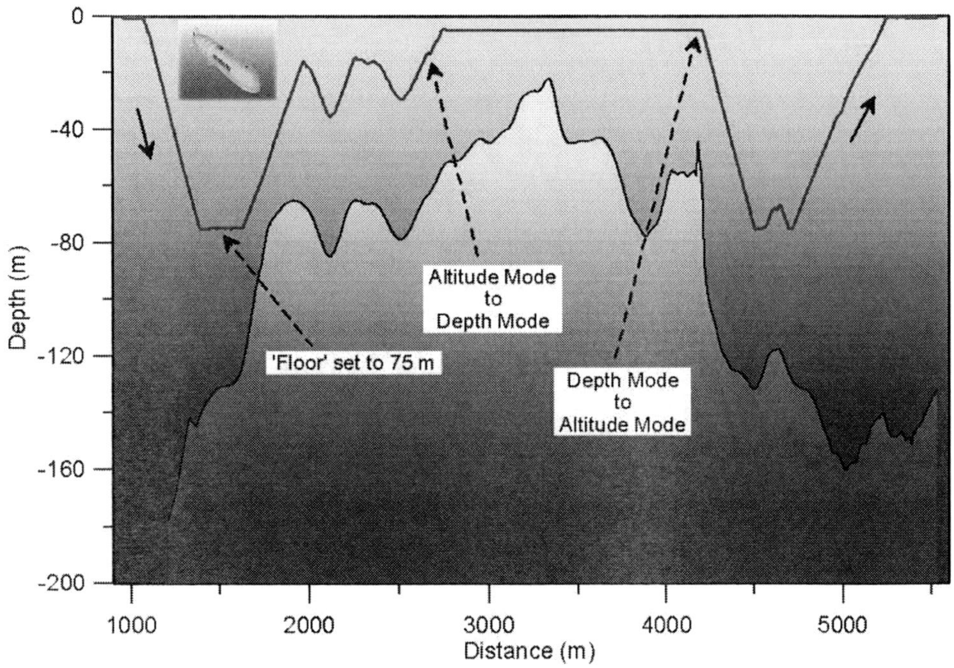

Figure 4 Haystack Bank transect using altitude and depth modes (North-South Line)

The Explorer also has different modes of traversing between depths, i.e. different modes for diving or ascending. In heave mode the vehicle will travel vertically through the water column while maintaining a zero pitch. In pitch mode the vehicle will traverse the water column using a non-zero pitch. This contrast of modes has an important influence upon the angle of incidence for bottom looking acoustic transducers. Figure 5 shows a comparison conducted in each mode, using altitude mode throughout the path with the exception of a small portion over the shallow region of the bank where the vehicle maintained a 5 m depth. The Heave Mode represents the West–East transect that the vehicle flew in a 50 m altitude mode. The Pitch Mode represents the East-West transect conducted along the same track as the previous track, in pitch mode.

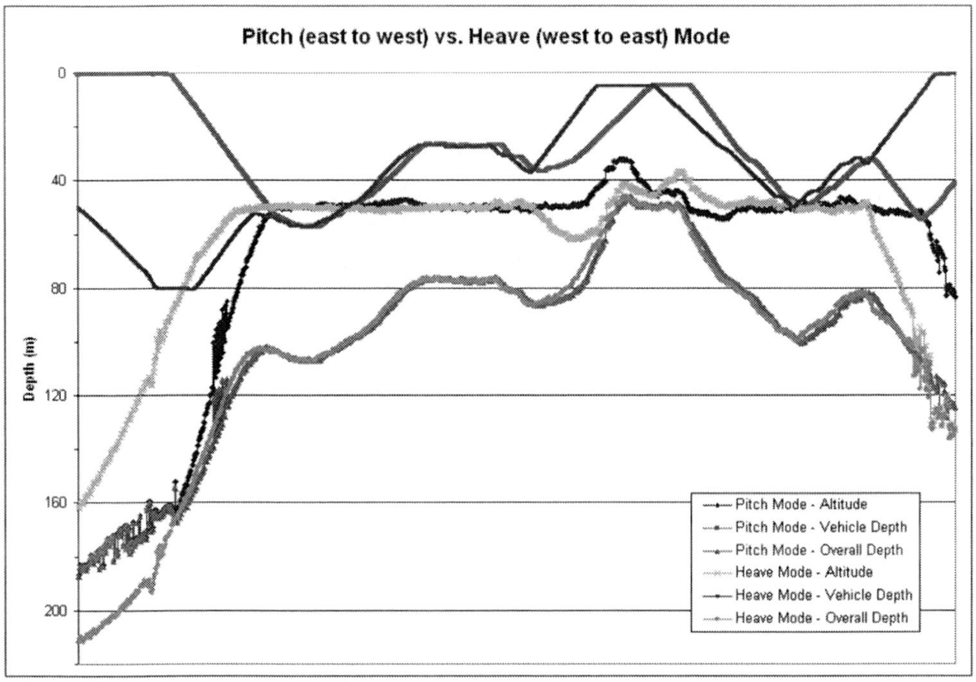

Figure 5 AUV path in pitch and heave modes

From this graphical representation, several conclusions can be made. The overall depth for both transects coincides very well. The altitude perceived by the vehicle (through an RD Instruments Workhorse Doppler Velocity Log) was consistent for regions of the bank that maintained a relatively small gradient (represented by the blue lines for each mode). Deviations from each other occurred at the edges of the bank, or the steeper sections of the transect. It can be noted that the pitch mode dive was able to maintain a higher gradient than the heave mode dive at the west side of the bank. This is also evident in the vehicle depth as shown by the pink/purple lines). The only exception is close to the pinnacle of the bank where the heave mode altitude varies from the pitch mode altitude. This could be an artifact of the directionality of the vehicle, i.e., the AUV traveled from the West to East in the heave mode transect and therefore encountered this feature from a different perspective than the East to West transect.

IV. ECHOSOUNDER

The BioSonics Scientific Digital Echosounder hardware configuration employed for these missions utilized a single split beam transducer with a 120 kHz operating frequency. Frequencies between 10-300 kHz are typically used for the purposes of bottom classification with the ancillary ability to detect fish. The results of these types of studies correspond to the shape and geological nature of the seabed and to the marine organisms present which can include finfish, invertebrates and benthic species (ICES, 2007).

Another goal of the Haystack Bank tests was to verify the effectiveness of the AUV as a platform for fisheries acoustic testing. It is known that AUVs can be inherently quieter than a ship (Fernandes et al., 2000). Further, by taking advantage of the proven vehicle dynamics, one can maintain a steady, constant altitude from the seafloor thereby obtaining a consistent resolution of the data. Finally, any noise from the surface effects of the platform can be eliminated by maintaining a minimum depth during the mission.

The following is a presentation of preliminary results achieved in the Haystack Bank region. Figure 6a shows a small section of the echosounder data captured when the AUV was maintaining a constant altitude of 25 m. At the small hill, the vehicle was forced to surface to maintain its altitude. The surfacing coincides with some noise in the data set. As the vehicle descends, the surface noise is eliminated. In contrast, Figure 6b shows the same track, but with the restriction that the vehicle did not surface during any point during the mission. The vehicle followed the seabed more effectively and the surface noise is not present.

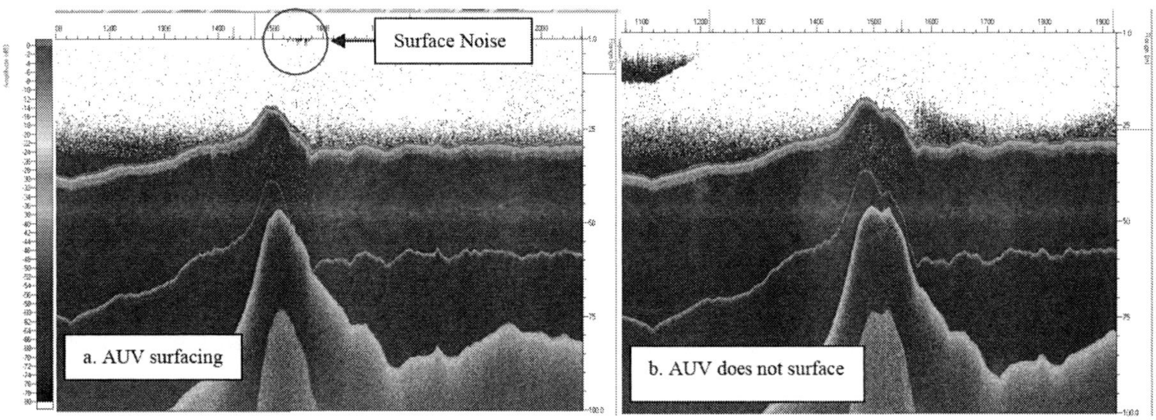

Figure 6 Biosonics DTX data over the same hill, with and without surface effects

V. CONCLUSIONS

The *MUN Explorer* was able to perform the vehicle dynamic tasks while collecting quality fisheries echosounder data efficiently and effectively. The transects performed in the various modes showed the dynamical capability of the AUV given the various mission objectives over complex terrain. Further, the missions proved that the AUV could be used as an alternative platform to a ship for acoustic data collection. Surface noise normally found in data from ship-based missions can be eliminated. A more in depth analysis of the results will be conducted to determine the extent of potential benefits that the AUV can bring to a fisheries survey. However, this initial study has shown that the MUN Explorer is an effective tool in these surveys.

VI. ACKNOWLEDGEMENTS

The authors thank Fisheries and Oceans Canada for the funding of the operations and ship time on the CCGS Shamook whose crew was very helpful in conducting missions. We thank International Submarine Engineering Ltd. and the Department of Underwater Systems at IFREMER (Institut français de recherche pour l'exploitation de la mer) for their continued technical support.

VI. REFERENCES

Fernandes, P. G., Brierley, A. S., Simmonds, E. J., Millard, N. W., McPhail, S. D., Armstrong, F., 2000, Fish do not avoid survey vessels, *Nature*, 404: 35–36.

ICES, 2007, Acoustic seabed classification of marine physical and biological landscapes. ICES Cooperative Research Report No. 286. 183 pp.

Niu, H., Adams, S., Husain, T., Bose, N., Lee, K., 2007, The Application of Autonomous Underwater Vehicles in Offshore Environmental Effects Monitoring, 2007 Canadian International Petroleum Conference, June 12-14, 2007, Calgary, AB, Canada.

Modelling 100,000 years of Tropical Cyclone Waves

Thomas A. Hardy[1], Luciano Mason[1], Jason McConochie[2]
[1]Marine Modelling Unit, Australian Maritime College, Launceston TAS, Australia
[2]Metocean Engineer, Woodside Energy Ltd, Perth WA 6000
t.hardy@amc.edu.au

Abstract: A state-of-the-art modelling system required adaptation and development of a series of models including synthetic tropical cyclone, windfield, and wave models to determine wave information for significant wave heights with return periods from 100 to 10,000 years for the oil and gas producing regions of northern and northwest Australia. A very large ensemble of approximately 450,000 individual synthetic tropical cyclones (100,000 years) was created to represent the tropical cyclone population of the study area. Methods were tested in order to reduce the number of simulated storms to a feasible level to accurately define the return period above the 100 year level.

INTRODUCTION

Woodside Energy Limited (WEL) contracted the Marine Modelling Unit (MMU) to provide estimates of the one in ten thousand year wave conditions at a number of locations throughout the oil and gas exploration and production region offshore of the north-western Australian coastline (Figure 1).

The scope of the project is summarised as follows: Technology that has been used to produce statistics for waves during tropical cyclones in the Great Barrier Reef (GBR) region (Hardy *et al.,* 2003) and storm surge statistics for the east coast of Queensland (Hardy *et al.,* 2004) will be adapted for this project. This will include development of a synthetic tropical cyclone database, based on the methodology for generation of synthetic cyclone tracks, as described in James and Mason, 2005, which will provide inputs to a parametric windfield model. This windfield model will in turn force a third generation wave model, WAMGBR (Hardy *et al.* 2000). A very large number of synthetic tropical cyclones will be simulated over a large area in the oceans immediately to the north-west of Australia, in order to produce statistics on extreme waves.

A 100,000 year simulation was envisioned which would provide, in effect, ten renditions of the desired 10,000 years of record. This redundancy will act to provide more confidence in the return period results near the upper end (10,000 years) of the curves. It is important to emphasise that this study does not attempt to predict tropical cyclone occurrences for the next 100,000 nor even 10,000 years. The large number of synthetic tropical cyclones is generated to represent those that could occur in the study region during present day climatic conditions.

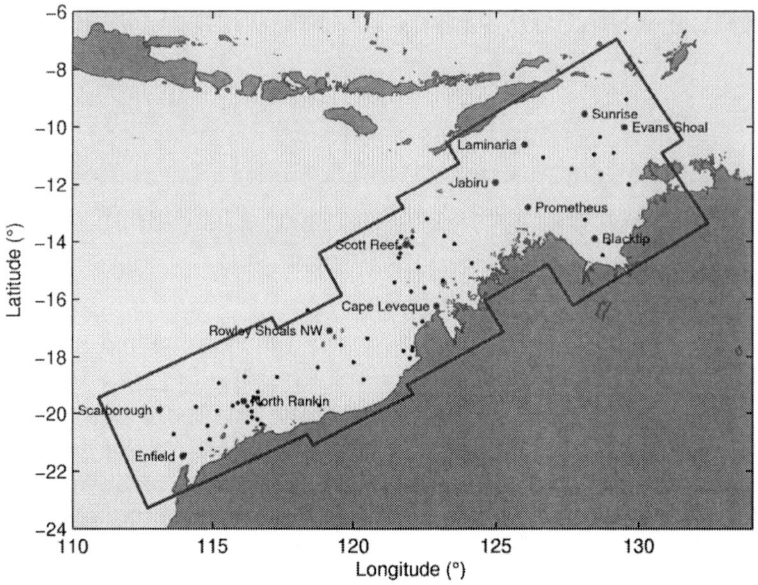

Figure 1: North West coast of Australia showing modelled region outline by red polygon.

MAIN RESEARCH ISSUES

The main research issues are as follows:

1. Adapt and extend the synthetic tropical cyclone track and pressure model, *SynCyc*, (James and Mason, 2005) from the Coral Sea to the North-West Australian region.
2. Improve the specification of values for the determination of size and shape of a tropical cyclone windfield.
3. Identify those storms that contribute to the return periods above the 100 year level at each of the required locations, since it will not be feasible to simulate the whole ensemble of more than 450,000 tropical cyclones across this very large region.
4. Determine a method to limit the physical size (radius to maximum winds) of the synthetic storms given that radius is a very important parameter in wave generation and the limited data record does not define the upper limit.

WAVE MODEL *WAMGBR*

WAMGBR is based on the widely-used wave model, WAM-cycle 4 (Komen *et al.,* 1994) with a number of significant alterations. Hardy *et al.* (2000) describe these alterations, which include: (i) a numerical scheme with implicit capabilities, which allows for longer time steps and therefore more efficient modelling, while maintaining WAM's diffusion characteristics; (ii) an offset in the discrete angle values, which provides more symmetrical diffusion characteristics; (iii) a rotated spherical coordinate grid system, which permits the use of more efficient grids; and (iv) a reef parameterisation scheme that provides a representation of sub-grid scale features and hence dissipation.

The use and accuracy of *WAMGBR* in modelling tropical cyclone waves has been verified in a number of wave modelling studies, both in the GBR region (Hardy *et al.*, 2000, 2003) and on the North-West Shelf (MMU, 2002).

The model was verified (Figure 2) against both T.C. *Tiffany* a NWS tropical cyclones and more importantly against the very severe Gulf of Mexico Hurricanes. These storms were *Ivan* (September 2004), *Katrina* (August 2005), and *Rita* (September 2004); the respective minimum measured central pressures were 910, 902 and 897 hPa.

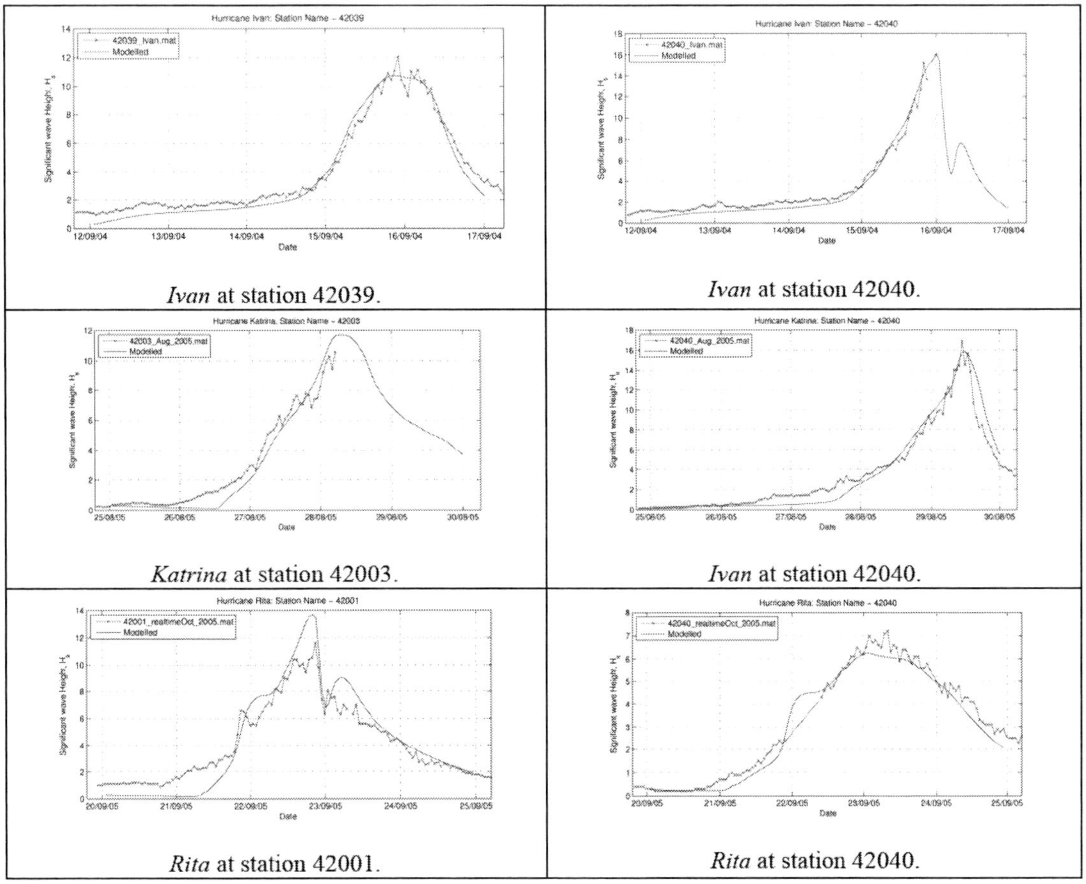

Ivan at station 42039. *Ivan* at station 42040.

Katrina at station 42003. *Ivan* at station 42040.

Rita at station 42001. *Rita* at station 42040.

Figure 2: Comparison time series between measured and modelled significant wave heights for selected severe tropical cyclones and hurricanes.

TRACK AND PRESSURE MODEL

The historical dataset is not adequate in length to define the tropical cyclone population that threatens the study region, especially if very severe storms are the portion of the population that is of interest. This required the development and testing of a model to generate a very large number of synthetic tropical cyclones. The model *SynCyc* (James and Mason, 2005) developed originally for the Coral Sea (James and Mason, 2005; Hardy *et al.,* 2004a,b; Hardy *et al.,* 2003) was adapted for use for this study. *SynCyc* has undergone significant development for this project.

The model contains three equations, one each for latitude and longitude of the storm's centre and one for the central pressure. These equations give the change with respect to time in the values of these three variables and were used to create a time series (three hour time step) during the life of each synthetic storm. The equations contain the addition of several linear components, each of which is the product of a coefficient and the value of latitude, longitude and central pressure, as well as the time rate of change (velocity) and second time rate of change (acceleration) of these three parameters, all evaluated at the previous time step. Additional parameters used include the time and spatial varying zonal and meridional components of the synoptic wind, local water depth and an estimate of the spatial and seasonal varying maximum potential intensity (MPI) of tropical cyclones. In addition there are some nonlinear components resulting from the product of a coefficient and products of the above variables. Not all of these components are included in the final set of equations and of those included not all have a nonzero value of the coefficient in all three equations. The values of the coefficients are determined by a least squares fit to the data.

Unlike the Coral Sea projects for which one set of equations sufficed for generating synthetic tropical cyclones throughout that region, the present project required eight regions each with its unique set of coefficients. Ideally it would have been best to have one model for the whole region and if the process of tropical cyclone genesis, development and decay were completely understood this would be possible. However such knowledge is incomplete and the multiple regions aided in allowing variations that are known to occur, but for which complete understanding is not available.

At this point the model is deterministic, but an important realistic variation was incorporated by randomly selecting with replacement from the set of residuals that result from the single step comparison of the model with the time series of position and pressure of historical tropical cyclones. One could think of this random variable as accounting for the unknown influences on tropical cyclone movement and intensity that are not included in the deterministic portion of the model.

The final model configuration (number of sub-regions and the components to include in each equation) was determined after an exhaustive iterative process comparing model performance to historical data.

The simulation of each synthetic storm begins with initial values of position (latitude and longitude) and central pressure deficit, as well as the change with respect to time of these three parameters. The seventh initial parameter is the time of year. These seven initial values for each synthetic storm are obtained from the random selection within a seven dimensional "sphere" with a multidimensional "radius" determined from standardised (de-meaned) and normalised (by standard deviation) values of nearest-neighbour "distances" in these seven parameters. For the creation of each synthetic storm an historical storm was randomly selected with replacement and its initial values for these seven parameters defined the "centre" of the hyper-sphere. For storms which are crossing the boundary at initialization, the dimension of the hypersphere is reduced to 6, as the latitude and longitude of a boundary point are related.

At each time step after initialisation the values of latitude and longitude of the storm's centre and its central pressure deficit are determined using the model equations. The inclusion of the randomly selected residuals and randomised initialisations helps generate realistic paths as seen in Figure 3, which shows starting points and paths of 150 historical and 150 randomly selected storms.

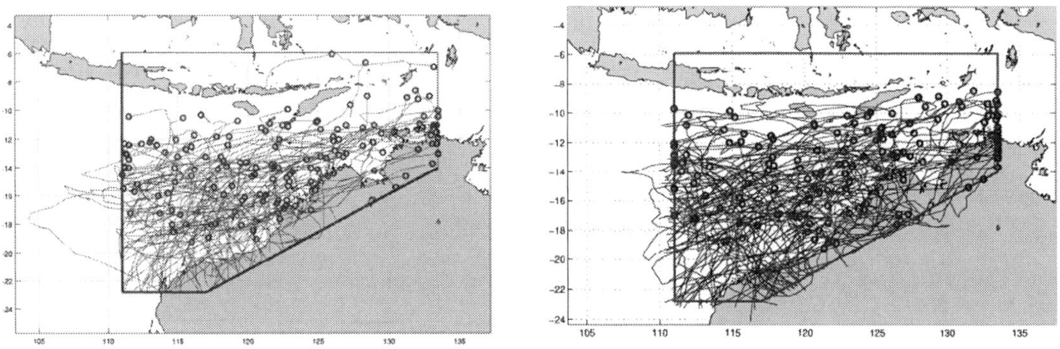

Figure 3: Starting positions (circles) and tracks of the 150 historical (left) and one sample of 150 synthetic (right) tropical cyclones that started inside or crossed into the initialisation zone. Storms that crossed into the zone were started on the boundary.

The time series of over 450,000 synthetic tropical cyclones were generated representing 100,000 years of record in order to be able to create a robust estimate of the 10,000 year values of waves and winds required in the study.

How good are the results of this implementation of *SynCyc* to northwest Australia? The foundation of the modelling is the 33 year WEL data set, and although it has been checked and refined (Harper, 2002, WNI, 2003), it is still a relatively short data record. We cannot do anything about that and comparison with this data is our primary testing mechanism. An exhaustive set of data comparisons was conducted in the development of model as was described above. Some of these comparisons are included in Figure 4, which includes statistical comparisons. In Figure 4 a cumulative distribution function (CDF) for each of maximum intensity, speed, heading, total time and mean annual time is included, as well as a plot of frequency of maximum intensity. This data is in a small section of the study area called a "comparison area" shown in green outline.

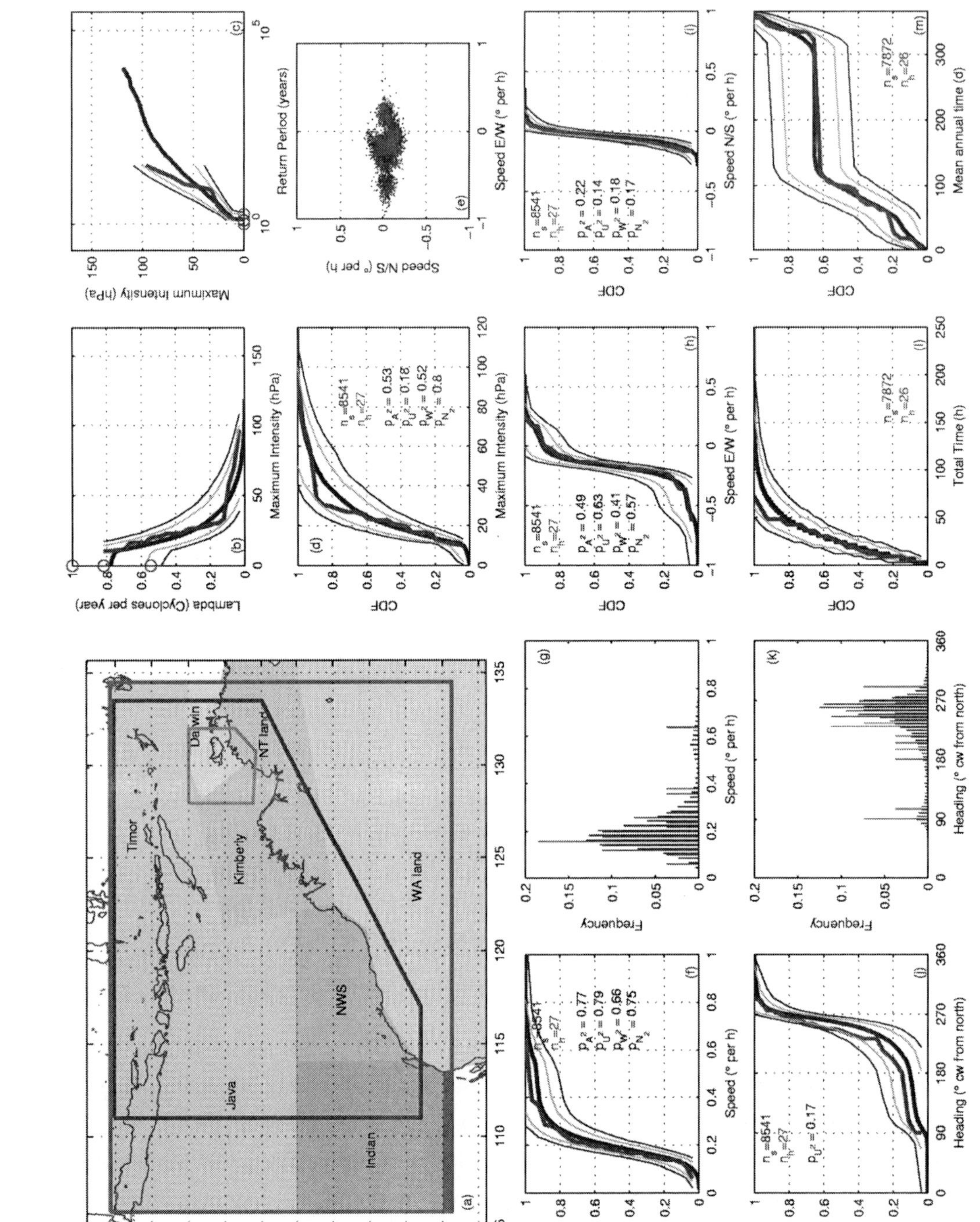

Figure 4: Example comparison between historical data (red) and model results (blue) collected inside the green polygon in (a). Thin blue curves in graphs are 90% confidence bands about individual ordinates. The medium blue curves bound 90% of a set of 5000 curves, each with the same number of points as the historical set, which were obtained from a bootstrap-like analysis of the synthetic results. Goodness-of-fit statistics (*P* value) for W^2 (Cramer-von Mises), A^2 (Anderson-Darling), U^2 (Watson), N_2 (Neyman) are included on the CDF figures.

It is important to note that the process does not slavishly reproduce the historical data, but allows variations that would occur in a longer historical data series of multiple sets of 33 year periods. Thus, multiple renditions of 33 years of synthetic data contain some sets that are very much more severe and some that are very much less severe but most that are close to the values of the historical data. The spread of this variation about the historical data is indicated in the confidence intervals in Figure 4. A global quantitative answer for "goodness of fit" is not possible since the study region is large and the tropical cyclone characteristics are variable across the region. Thus multiple figures, each for a specified target location are needed.

The best sense of the quality of the output of *SynCyc* perhaps can be understood by realising that these statistical analyses in Figure 4 were conducted with data from storms only while they were inside these relatively small comparison zones. These storms had a wide variety of initialisation locations and many would have traversed through several model sub regions, each with its separate set of model coefficients before arriving at the location where the data was collected. The fact that the historical data curves fall within the 90% limits of the multiple samples of synthetic data for these widely varying test locations inspires confidence in the modelling process.

WIND MODELLING

Using a timestep of 10 minutes, timeseries of wind speed, direction and atmospheric pressure were calculated by *CycWind,* for all 454,546 cyclones in the synthetic tropical cyclone database at 95 locations throughout the study region.

Using meteorological data from 24 Bureau of Meteorology stations, the computed return periods of the wind fields at these locations from all synthetic cyclones have been analysed to determine if their statistics are consistent with measured values. An example of these (Figure 5), shows good agreement between the synthetic and measured data. Note that wind speed values from the original measured data sets have been modified by using multiplicative factors recommended by Harper (2005a).

Although the computation of cyclonic wind fields can be thought of as an intermediate step in the production of wave modelling results, the synthetic data obtained is a valuable resource in its own right, and has been shown to give good agreement with statistical measures of actual wind records from the region.

WAVE MODELLING

Due to the computational expense of running the wave model at the required spatial resolution, it was not feasible to simulate all 450,000+ storms on all of the grids (Figure 6). At any single location only a relatively small number of storms will contribute to return periods in the desired range (above 100 years). A single storm could cause significant waves heights in this range at all points in all grids, or at none of the points in any of the grids. The problem is that the region of interest is geographically large, and methods for determining which storms cause these more severe wave heights are not obvious. The storms that cause the largest waves at locations of interest had to be identified, so that the number of simulations at high resolution can be minimised. After selection of the most severe storms approximate 95,000 numerical simulations were conducted.

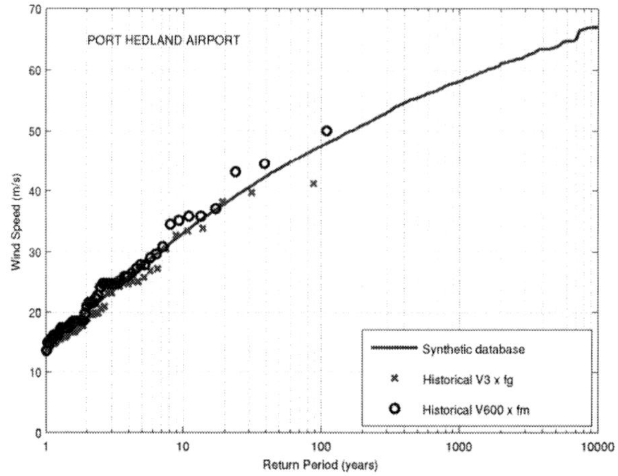

Figure 5: Return period curves for wind speed at Port Hedland airport. The computed return period curve for the synthetic data (blue curve) is compared with modified gust data (red crosses) and mean data (black circles)

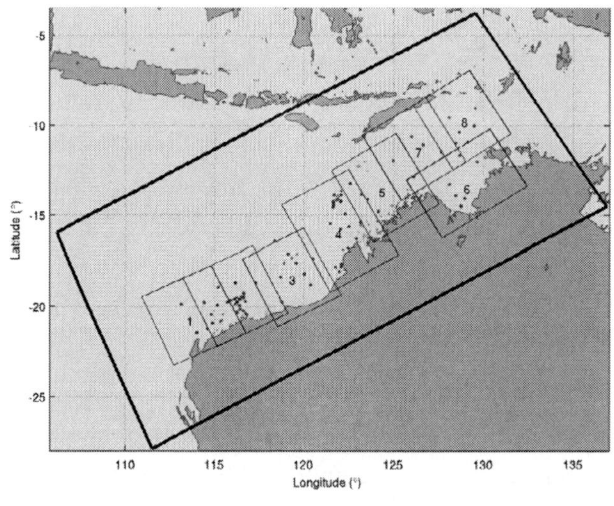

Figure 6: The A grid (20′ resolution) and the eight nested B grids (5′ resolution). Red dots indicate wave modelling output locations.

An example of wave information is for a specific location is given in Figure 7. Included in this figure are: a return period and encounter probability (for lives of 50, 100 and 200 years) for significant wave height; polar plots of magnitude as a function of direction for both significant wave height and peak wave period; and a scatter plot of peak period as a function of significant wave height. Such data sheets were produced for 75 locations.

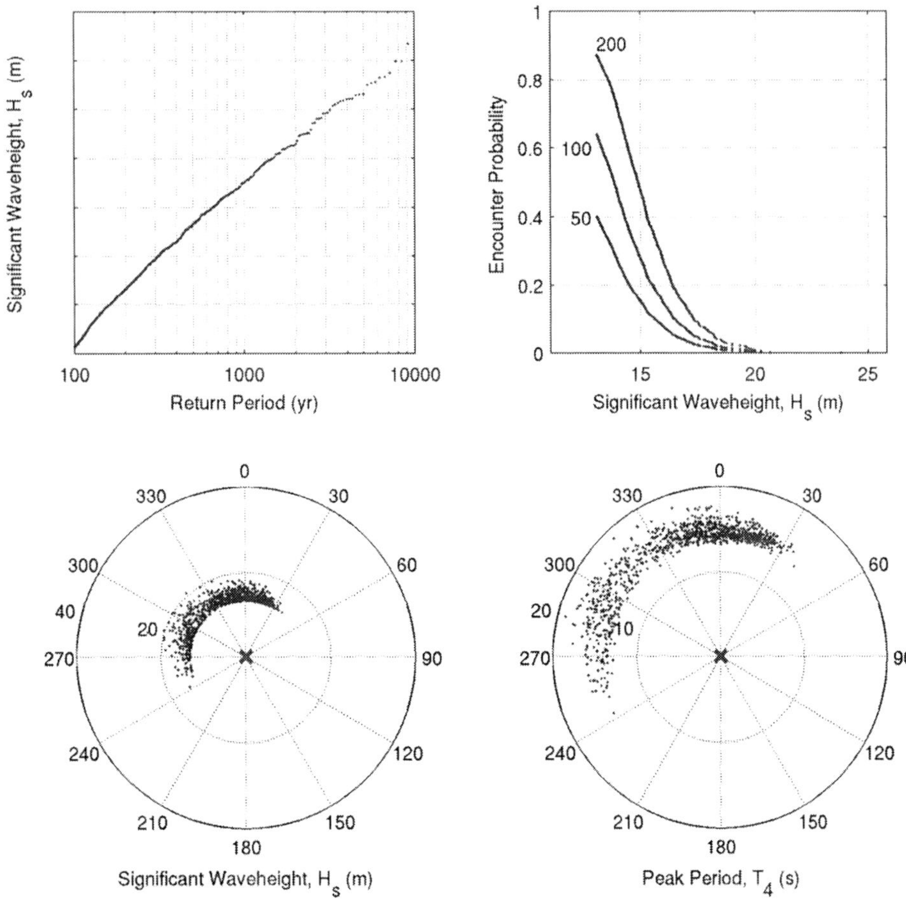

Figure 7: Example of the wave climate information obtained from the *WAMGBR* modelling at a single location.

ACKNOWLEGEMENTS

This study was funded by Woodside Energy Ltd and the support of Stan Stroud of that organisation is gratefully acknowledged.

REFERENCES

Hardy, T.A., Mason, L.B. and McConochie, J.D. (2001) "Great Barrier Reef tropical cyclone wave database." *Coasts and Ports 2001*, I.E.Aust., pp. 151–156.

Hardy, T.A., McConochie, J.D. and Mason, L.B. (2003) "Modeling tropical cyclone wave population of the Great Barrier Reef." *Journal of Waterway, Port, Coastal and Ocean Engineering*, ASCE, **129**, 104–113.

Hardy, T.A., Mason, L.B. and McConochie, J.D. (2000) "A wave model for the Great Barrier Reef." *Ocean Engineering*, **28**, 45–70.

Hardy, T.A., Mason, L.B., Astorquia, A., and Harper. B.A. (2004) "Tropical Cyclone-Induced Water Levels and Waves: Hervey Bay and Sunshine Coast." *Queensland Climate Change and Community Vulnerability to Tropical Cyclones: Ocean Hazards Assessment Stage 2*, Queensland Departments of Emergency Services, Natural Resources and Mines, 115 pp.

Hardy, T.A., Mason, L.B. and Astorquia, A. (2004) "The Frequency of Surge Plus Tide during Tropical Cyclones for Selected Open Coast Locations along the Queensland East Coast." *Queensland Climate Change and Community Vulnerability to Tropical Cyclones: Ocean Hazards Assessment Stage 3*, Marine Modelling Unit report, prepared for Bureau of Meteorology and Environmental Protection Agency, James Cook University.

Harper, B.A. (2003) "Tropical cyclone wind and pressure model for the 10^{-4} waves study". Systems Engineering Australia Pty Ltd, Report No. PR005, prepared for Woodside Energy Ltd, 40 pp.

Harper, B.A. (2005a) "Historical wind speed analyses". Systems Engineering Australia Pty Ltd, Report No. MO002, prepared for Woodside Energy Ltd, 7 pp.

Holland, G.J. (1980) "An analytic model of the wind and pressure profiles in hurricanes". *Monthly Weather Review*, **108**, 1212–1218.

James, M.K. and Mason, L.B. (2005) "Synthetic Tropical Cyclone Database", *Journal of Waterway, Port, Coastal and Ocean Engineering*, ASCE, **131**, 181–192.

Komen, G.J., Cavaleri, L., Donelan, M., Hasselmann, H., Hasselmann, S. and Janssen, P. (1994) "Dynamics and Modelling of Ocean Waves." *Cambridge University Press*, 532 pp.

Marine Modelling Unit (2002) "An Assessment of the Potential State-of-the-Art Accuracy of Spectral Wave Modelling of Tropical Cyclones in North-West Australia." Marine Modelling Unit Report, prepared for Woodside Energy Ltd.

McConochie, J.D., Mason, L.B., and Hardy. T.A. (1999) "A Coral Sea cyclone wind model intended for wave modelling." *Coasts and Ports '99*, I.E.Aust., pp. 413–418.

McConochie, J.D., Hardy, T.A. and Mason, L.B. (2004) "Modelling tropical cyclone over-water wind and pressure fields." *Ocean Engineering*, **31**, 1757–1782.

WNI (2003) "Review of North West Shelf and Timor Sea Tropical Cyclone Database for the 10^{-4} Waves Study", report prepared for Woodside Energy Ltd.

Hydrofoils Applied to Canting-keel Yachts

Richard S. Milne
Thales Australia (Naval) Pty Ltd, Australia
richard.milne@thalesgroup.com.au

Phillip J. Helmore
University of New South Wales, Australia
p.helmore@unsw.edu.au

ABSTRACT
Hydrofoils have been applied to many different types of marine craft, including yachts. However, most of these have been for attempts at speed records or by inventors and, notably, the international Moth class, but so far not to modern racing yachts. They provide a number of advantages, and this project has investigated the feasibility of application to a modern canting-keel maxi yacht.

The investigation began with drawing up a set of lines for a modern canting-keel maxi yacht. A resistance was then predicted, both with and without foils, using the Delft Systematic Yacht Series and a three-dimensional analysis of the foil-borne resistance. A towing rig was designed and constructed, and several models built and tested on open water and in a swimming pool to compare the resistance and performance, both with and without foils. Finally, a radio-controlled sailing model was constructed and tested, both with and without foils.

The results showed a clear speed advantage for the foil-assisted maxi yacht. Just as with canting keels, fitting foils to an ocean-racing maxi will not be without its problems. However, the evidence is compelling, and it is considered only a matter of time before someone sets a precedent with a full-scale yacht, which others will follow.

1. INTRODUCTION

The use of hydrofoils on sail-powered vessels is not new. Hydrofoils are used in the quest for speed, and the hydrofoil helps vessels achieve this with its high lift/drag ratio. The first sailing vessel to use a hydrofoil was *Monitor*, built in 1955 (International Hydrofoil Society 2006) and shown in Figure 1.

Since then, hydrofoils have been applied to many different types of craft, including monohulls, proas, catamarans and trimarans.

Due to the complexity and fragility of previous foil systems, their use has been confined mainly to the realms of sailing speed record attempts and inventors. However, with the use of composites and high-tensile fibres, modern foils are much more robust and simpler in design.

In recent years the introduction of canting keels to racing yachts has brought a marked increase in speed. The concept, like many innovations, attracted criticism and resistance from the establishment but has been embraced by those with more liberal views in the pursuit of speed.

The authors consider that hydrofoils are waiting in the wings for their turn to push monohull sailing speed limits higher. They therefore embarked on a testing program to investigate the feasibility of fitting foils to a modern canting-keel yacht.

2. SAILING HYDROFOILS

2.1 Previous Hydrofoils

Some of the notable sailing hydrofoils have included the following:

- *Monitor*
  Built at the then huge cost of $20 000, *Monitor* was essentially a monohull with a set of ladder-type main foils on outriggers to port and starboard and a small ladder foil aft which acted as a rudder. She reached 25 knots when launched, and 30 knots a year later (IHS 2006).

Figure 1 *Monitor* sailing on Lake Mendota, USA (Photo Edwin Stein)

- *Williwaw*
  This vessel was built in 1968 and made many ocean voyages (Keiper 1996). She was essentially a sloop-rigged 10 m trimaran, with ladder foils on the side hulls and stern, and a large V-foil at the bow of the main hull.The foil take-off speed was 10–12 kn, and she regularly sailed at 15–22 kn.

- *The Ugly Duckling*
  This vessel was built as a proa in 1982 for a serious attempt on the world sailing speed record, then 34 kn (Smit 1982). The vesel was built for a low budget, using a Tornado catamaran rig and plywood hulls. She achieved 24 kn when fully foilborne, but never broke the world record. However, she showed what could be achieved, and subsequent proas have held the world sailing speed record over the years.

- International Moth Class
  The Moth is an international class of dinghy and is a development class with liberal views on design. In the quest for speed, the class has adopted the use of hydrofoils, as shown in Figure 2. The results are dramatic, with speed increases of between thirty and one hundred percent having been achieved over their non-foilborne predecessors (International Moth Class Association 2006).

Figure 2 Hydrofoil Moth (Veal 2006)

- *L'Hydroptere*
  This French-designed trimaran (the name is Greek for water-wing) has convincingly demonstrated just what a hydrofoil vessel can do. She has broken many speed records, including an open-water speed record of 47.2 kn and crossed the English Channel in less time than the first aircraft to do so! (*L'Hydroptere* 2006)

Figure 3 *L'Hydroptere* at speed (www.hydroptere.com)

There have been many others, but the sample of vessels described has shown that increases in speed are achievable with the application of hydrofoils.

In recent years the introduction of the canting keel into the world of racing yachts has brought with it a marked increase in speed for those vessels. A modern maxi-yacht with a canting keel regularly travels at speeds of 15–25 kn (International Sailing Federation 2006). It is in this speed range that a hydrofoil should be beneficial in reducing the total resistance of the yacht. Through the inherent efficiency of a foil, a speed increase can be achieved due to a higher lift/drag ratio in comparison with that of a planing hull.

The authors embarked on a project to investigate the feasibility of fitting hydorfoils to a modern maxi-yacht with a canting keel. The key to a successful foil-borne vessel is the power/weight ratio, which is directly related to the righting moment/weight ratio when using a conventional sail configuration. The righting moment, in turn, is directly related to the span of the foils.

The arrangement for the current concept of a foil-assisted monohull was commenced with research of the foil configurations of previous foil-assisted sailing vessels. *L'Hydroptere* gave much inspiration, with her use of angled outer foils and a T-type foil on the rudder. The angled outer foil gives lateral resistance and provides passive ride-height control due to the affects of dihedral. A T-foil attached to the rudder is an efficient way of stabilising the pitch of the craft.

3. DESIGN

3.1 Modelling

The basis for the design project was chosen to be the 30 m maxi-yachts *Alfa Romeo* and *Wild Oats*, designed by Reichel-Pugh. The basic dimensions of these vessels are available on the web (Reichel-Pugh 2006) but, understandably, no lines plan was available from the designer. However, a hull shape was drawn up, using the Rhino CAD package, having the correct dimensions, displacement and general hull shape. We are confident that this shape, while not being an exact replica, is sufficiently close to be able to test the concept, and to show what happens when the vessel becomes foilborne.

3.2 Scaling

For the manufacture of a model of the 30 m maxi-yacht, a scale ratio of 1:25 was selected, giving a model length of 1.2 m.

Making model to the correct scale displacement proved the most challenging task of all. The racing displacement of the 30 m maxi-yacht is 26 000 kg, which translates to $26\,000/25^3 = 1.664$ kg at 1:25 scale. The challenge lay in keeping the model this light, yet strong enough to handle the sailing forces.

Initial estimates of masses which would be required were as follows:

Item	**Mass (kg)**
Keel	0.915
Fin and rudder	0.130
Sails and rig	0.110
Deck and support structure	0.110
Radio-control unit	0.140
Hull	0.259
Total	**1.664**

With a 1.2 m hull, a mass of only 259 g is a very hard task. Strength scales as the square of size, while mass scales as the cube (the square-cube law), so the material for our 1:25 scale model should be relatively twenty five times stronger than that of the maxi-yacht! If it were not for the non-scaling of materials, then the model would be almost impossible to make.

3.3 Resistance Prediction

Having determined the size of the model, preliminary estimates of the resistance were made using the Delft Systematic Yacht Series for the hull and Design Foil for the foil. The Delft Series II resistance prediction is available in Gerritsma et al. (1991) and was programmed. Although not catering for modern vessels travelling at over 20 kn, the Delft Series predicted resistance up to the speed at which the foil system would take over.

At more than 20 kn the yacht would become foilborne, and the Design Foil program (Dreese 2006) was used to calculate the resistance of the foils. Design Foil is a two dimensional CFD program that is simple to use and proved a useful tool in selecting the foil geometry. A NACA4412 foil was chosen for its characteristics at different angles of attack (other than at optimum). Results of the resistance calculations are shown in Figure 4.

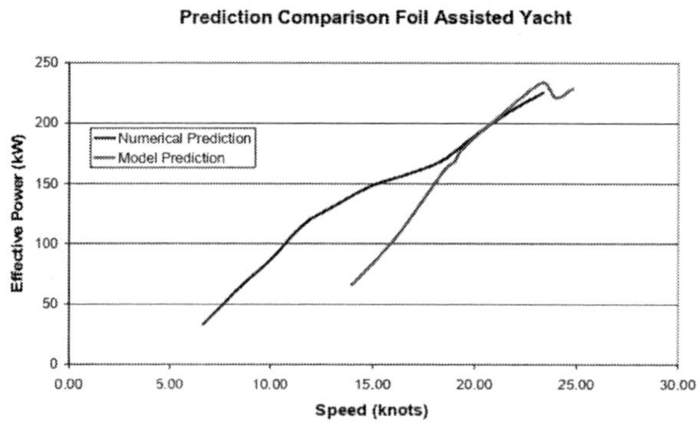

Figure 4 Resistance calculations from Delft Series and Design Foil

It is not possible to include all of the forces which will act on a hydrofoil-assisted yacht. However, the results of this resistance prediction have been useful in the design of this concept. The resistance predictions have added weight to the viability of this concept.

3.4 Arrangement

The arrangement used for the design of this concept is shown in Figure 5. It is envisaged that implementatin of this concept would incorporate active main foils to give a constant ride height; much in the same way as the foil of a modern Moth dinghy works. This results in a foil depth which greatly diminishes the free-surface effect, thus reducing wavemaking resistance. An active foil will also deliver good seakeeping characteristics (a smooth ride), by eliminating wave impact loads.

The incorporation of retractable foils would solve the problem of foil resistance at speeds below foilborne, and solve most berthing problems.

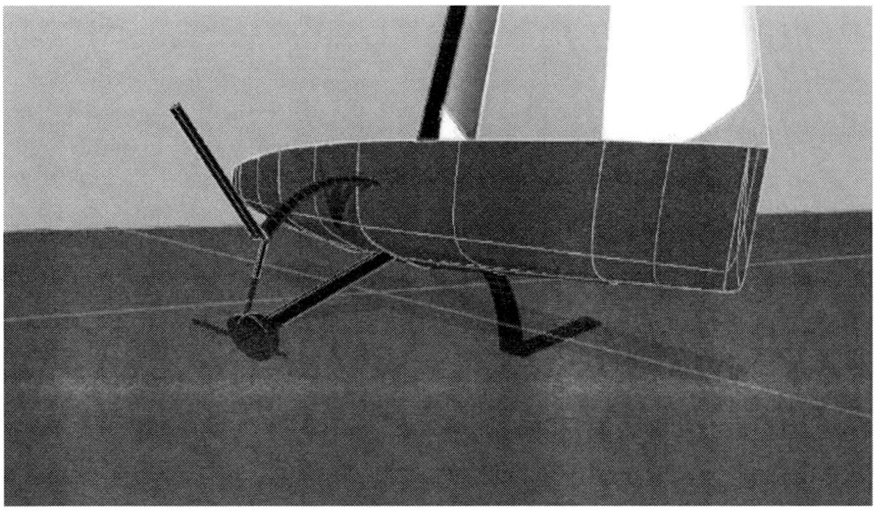

Figure 5 Concept of the hydrofoil-assisted maxi-yacht

4. CONSTRUCTION OF MODELS

The building of the prototype male plug and female mould gave a good insight into the effort required to create a full-size mould for production. Building of a male plug is a task which requires the utmost precision and quality of finish. Any imperfection, mark, or blemish in the male plug will be carried over to the mould and, most likely, to every production hull to come from that mould. Spending the time to get it right from the start saved a large amount of time later. Much of the time was spent checking dimensional precision and fairness of the surface. Any variation required filling and sanding until a level of perfection was reached.

The prototype was built using a foam sandwich structure. Strips of Divinycell foam were attached to a plywood mould with double-sided tape and covered with a layer of fibreglass and resin as shown in Figure 6. The prototype was lifted off the mould and glassed on the inside.

Many weeks were then spent filling and sanding the outside until perfect finish and shape was acquired, as shown in Figure 7.

Figure 6 Plug under construction, still attached to building frame

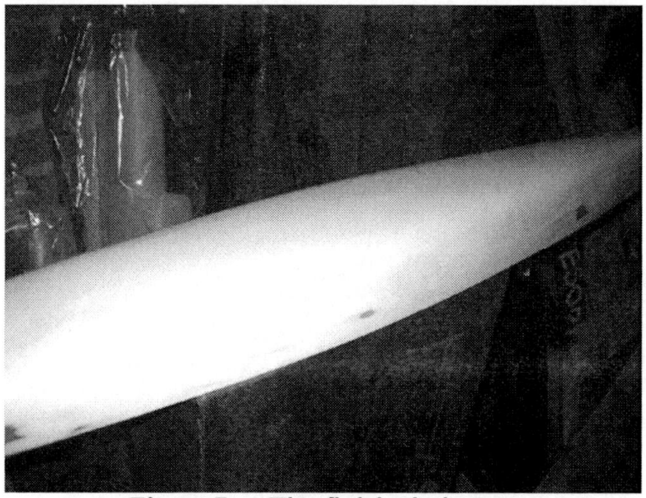

Figure 7 The finished plug

To form the female mould, a thick gelcoat followed by many layers of chopped-strand fibreglass and a set of frames produced a good mould, as shown in Figure 8. From this mould exact copies could be made.

Figure 8 The female mould

As previously mentioned, making the model to the correct scale mass proved the most challenging task of all The challenge lay in keeping the model light yet strong enough to handle the sailing forces. In all, seven hulls were made and only one was light and strong enough to be used as a sailing model; another was slightly heavier and was able to be used as a towing model.

Hull No. 4, shown in Figure 9, ended up being the lightest hull. The first layer was 200 gsm, with a vacuum bag being used to hold the core down after just wetting the contact surface of the core before installing. The final layer of 25 gsm glass was applied with care, trying not to use too much resin as it would fill the cavities of the Nomex. The light weight glass was difficult to work with, as it was easily snagged and pulled out of shape. Even with a lot of care, a little extra resin was used. This hull on release from the mould had a mass of 260 g, but was quite stiff and strong. After fitting out, fin, rudder and radio control added, it came to a total weight of 780 g and was used for the live sailing model. This meant with the rig and keel weight added, the model was 1.770 kg, thus being slightly overweight for the model scale, but close enough to use for sailing.

Figure 9 Hull No. 4 showing Nomex honeycomb core
foils later moved to lower position

In the three subsequent models, fine silk was used as a reinforcement for the inner skin and a lighter 90 gsm fibreglass was sourced for the outer skin. The outer skin and the Nomex laying went very well, but there was some spillage of resin onto the Nomex during the laying of the inner silk skin. Out of the mould, Hull No. 6 had a mass of 340 g, and this hull was fitted with a canting keel to allow the model to run at the desired angle of heel in the towing tests, as shown in Figure 10.

Figure 10 Towing model with canting keel, note foil on centreline

5. TOWING TEST RIG

We had no towing-tank facility available for low cost, and so designed and built a test rig to measure the resistance of the model for a range of speeds while providing the motive force to bring the vessel to the foil-borne condition. The resulting machine turned out to be easy to use, portable and reliable.

The concept was that the model would be pulled through the water by a light nylon fishing line which was wound in by the machine. The machine, in turn, gave a tension reading on a linear spring scale from this nylon line and thus gave the force required to tow the model. Although the nylon line in the water added a small component of resistance, this was disregarded.

The towing apparatus needed various speed settings in order to obtain the different speeds through the water. Speed control was achieved by the use of an electronic controller with negative feedback of the voltage output. This speed controller was built with parts readily available from an electronics store. With a constant voltage, regardless of the load, the model could be towed at a constant velocity.

A simple voltage regulator was incorporated to stabilise the towing speed. Regulated power delivery is very useful when selecting a desired speed especially if the speed is in the early stages of the planing or foilborne regions. As a model approaches foilborne speed, it encounters a resistance hump and the speed controller applies extra power to pass through this hump and into the foilborne mode. Once the model is foilborne, the resistance is decreased, resulting in reduced power requirements. This allowed us to obtain results near the resistance hump.

However, a simple controller would keep on supplying the power required to pass through the hump, even when foilborne, and would cause the model to accelerate to much higher speeds. Also, if the simple controller were to be set at the power requirement of just being foilborne, then the model would never pass through the resistance hump. A simple controller would miss out on some of the most important data needed for this project.

Trial-and-error was used in selecting the correct electrical variable resistor setting which would give a desired power output range needed for the models.

The completed towing apparatus is shown in Figure 11.

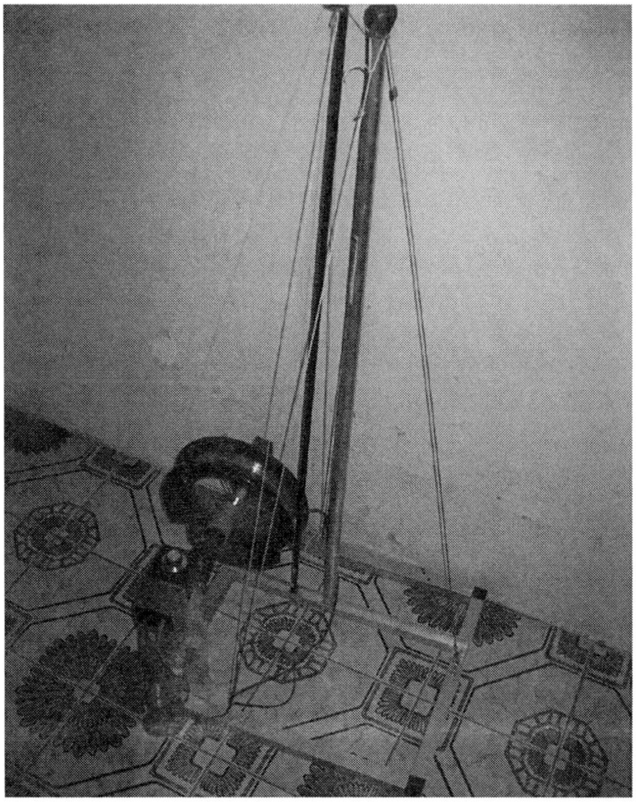

Figure 11 Towing apparatus

The crux of the towing apparatus is its ability to give a reading without the influence of internal frictio. Elimination of internal the friction component is achieved with a vertical mast which is perpendicular to the incoming nylon line. On top of the mast is a lightweight pulley, over which the line travels, and then down parallel to the mast and onto the winding spool. Any friction due to the pulley on the top of the mast is transmitted vertically through the mast and not in the direction of the incoming line, so the friction causes no moment on the mast structure. The mast is able to rotate about its base and move in the direction of the incoming line. A spring gauge is attached to one side of the mast to give readings of varying incoming line tensions.

A radio-controlled rudder on the model enabled it to be steered in a straight line towards the towing machine. A slight deviation off course (five degrees or less) would not significantly influence the results. An advantage of using the nylon line is that there is a small amount of elasticity in it whereby any small waves or any slight deviation of the model does not instantly translate into a change in the tension at the towing apparatus. Thus a constant tension can be measured while the model is still some distance from the towing apparatus.

6. EXPERIMENTS

6.1 Towing Tests

A preliminary set of trials with the towing rig was conducted in the open harbour in order to refine the process and iron out any bugs in the towing and resistance-measuring apparatus. After a period of trial and error, the towing tests proved successful, and it was time to test the model in a more controlled environment.

The Swimming Pool at the University of New South Wales was selected for the trials, one of the advantages being that distance and time could be accurately measured. Many towing runs were conducted without foils to simulate the conventional hull, with speed and resistance being recorded. Runs were then made with the foils fitted and, again, speeds and resistances were measured. An example is shown in Figure 12.

Figure 12 Model foilborne with tips just breaking the surface at 1.74 m/s

Once the vessel was foilborne, any small waves which were present seemed to have very little impact on the way the vessel behaved. Thus small wave conditions might be a more ideal testing scenario for the model in both foilborne and conventional-yacht configurations, as it simulates the true operating environment of an ocean-going yacht.

The model required a speed of around 1.74 m/s (16.9 kn at full scale) to become foilborne. However, once foilborne the speed could be slowed to 1.55 m/s (15.0 kn at full scale) and still remain foilborne.

The resistance results are shown in Figure 13, and bear a marked similarity to the numerical predictions which were made in the early design stage (Figure 4), thus verifying the predictions.

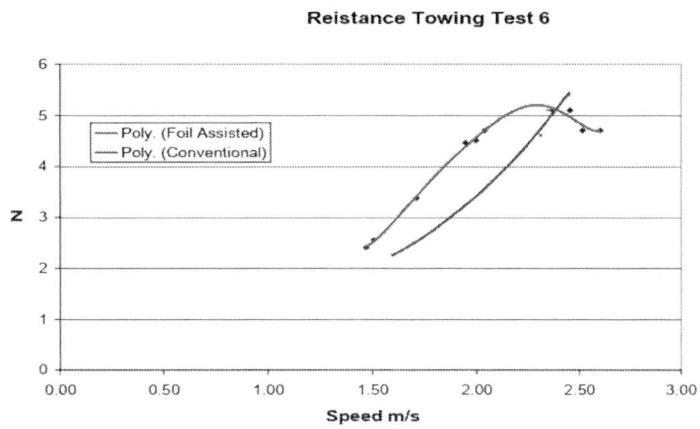

Figure 13 Resistance results for conventional and hydrofoil hulls

6.2 Sailing Tests

A set of sails (main and genoa) and rig was designed, built and fitted to the vessel, as well as radio-control gear. The overall mass ended up approximately 130 g (8%) above the scaled displacement.

Scaling of the wind strength also plays a part in modeling sailing trials. A maxi-yacht sails in winds of up to 40 or 50 kn, so the model could be expected to sail in 10 kn, with average wind speeds of 5–6 kn. This is a light breeze, and there are often periods of associated calm. The mast on the model is 1.5 m tall, and the wind gradient can be extreme at this level.

In all, five sailing trials were conducted, each one an improvement on the previous trials and with many lessons being learned along the way. The first trail was a near-disaster, with the model almost sinking! However, the proof of the pudding is in the eating, and the fifth trial was the most successful. The model became foilborne at about the same speed as the towed model (estimated rather than measured). Figure 14 shows the model up on the foils.

Figure 15 Model reaching on foils

7. CONCLUSIONS

The concept of a hydrofoil racing yacht has been investigated in this project. A 30 m Reichel-Pugh design has been used as a basis, and resistance calculations done to show that speed advantages by way of lower resistance) can be achieved in the range of 15–25 kn. Two models were then built at 1:25 scale, and towing and sailing trials conducted. The results show that this concept works — at least in model form.

Implementation at full scale may not be easy; there will be problems with structure, control and berthing arrangements. Some of the problems can be solved relatively easily, such as by using retractable foils to reduce resistance below foilborne speeds and obviate berthing difficulties; some problems are more difficult to solve. However, canting keels are not without their problems either (and not all have yet been solved), but canting keels are now firmly entrenched for the go-fast enthusiasts. Many factors give a full-sized yacht advantages over the model where hydrofoils are concerned: the ability to trim sails, have an active main foil, and an adjustable rear foil.

It is the author's belief that when one enthusiast takes the plunge and fits hydrofoils to a full-sized yacht, then the speeds achieved will convince more to follow.

8. REFERENCES

Dreese, J. (2006), *Design Foil Software*, URL www.dreesecode.com, accessed 15 May.

Gerritsma, J., Keuning, J.A. and Onnink, R. (1991), The Delft Systematic Yacht Hull (Series II) Experiments, *Proceedings Tenth Chesapeake Sailing Yacht Symposium*.

International Hydrofoil Society (2006), *Sailing Hydrofoils*, URL www.foils.org/gallery/sail.htm, accessed 20 August.

International Moth Class Association (2006), *2004 European Moth Chanpionships*, URL www.moth-sailing.org/racing/2004_european.htm, accessed 12 February.

International Sailing Federation (2006), *Rolex Sydney to Hobart Yacht Race 2005*, URL www.sailing.org, accessed 8 July.

Keiper, D.A. (1996), *Hydrofoil Voyager:* Williwaw, *from Dream to Reality and Towards the Sailing Yacht of the Future*, Hinsdale Press, USA.

L'Hydroptere (2006), News, URL www.hydroptere.com, accessed 12 April.

Reichel-Pugh (2006), *Designs*, URL www.reichel-pugh.com/index_designs.html, accessed 20 March.

Smit, L.W. (1982), Proa *The Ugly Duckling*, *Proceedings International Conference on Sailing Hydrofoils*, Danish Ship Research Laboratory, RINA Small Craft Group Paper No. 3, 12 November.

Veal, R. (2006), *The 20-knot Club*, International Moth Class Association, URL www.moth.iointegration.com/homesp.htm, accessed 2 October.

Update on van Oortmerssen's Resistance Prediction

Phillip J. Helmore
University of New South Wales, Australia
p.helmore@unsw.edu.au

ABSTRACT
van Oortmerssen in 1971 published the results of a regression analysis of the resistance of ninety-three models of tugs and trawlers which had been tank-tested at the Maritime Research Institute Netherlands. This was in the form of an expression for the residuary resistance of the vessel based on parameters available at an early stage of design. His method remains one of the few suited to such hullforms and quickly found favour with those involved in prediction of their resistance. The method is included in many commercial resistance-prediction packages, and is still widely used.

However, there was a number of errors in the original publication. Depending on how these errors are treated, it is possible to come up with different values of the total resistance, and almost every known implementation gives differing results for the resistance. Many of these errors have been resolved by correspondence with the author and with MARIN, and it is the intention here to record corrections to the known issues.

In addition, it has been found that there are some combinations of parameters which give anomalous results. For these combinations, the resistance does not increase monotonically with speed as might be expected for this type of displacement hullform, but shows a distinct "hump" as might be expected for planing hullforms. The anomalous results have been investigated to determine the combinations of parameters for which they are produced, and a method of dealing with the results is proposed.

1. INTRODUCTION

1.1 General

The prediction of resistance is important in most ship designs for the estimation of either the speed which will be achieved with a given power, or the power which will be required to achieve a given speed. It becomes particularly important if there are penalties in the contract for under-achievement of the required speed.

There have been many fine analyses of resistance data for different hullforms, and the range of hullforms is constantly increasing. However, vessels with hullforms having fine entrances and broad, flat sterns, such as tugs, trawlers and offshore supply vessels, have had less attention paid to their resistance characteristics than most. Designers of these vessels therefore have a harder time of it in predicting their resistance with confidence.

Much of the early resistance data was presented in the form of tables or graphs for use in the design office. One such was the publication of resistance data for trawler hullforms by Ridgeley-Nevitt (1956, 1963 and 1967), whose students tank tested a systematic series of high displacement/length ratio trawler hullforms at the Webb Institute. The results were presented as contour plots of residuary resistance coefficient, C_R, vs displacement-length ratio and prismatic coefficient at each speed-length ratio. Helmore and Swain (2006) published the results of a regression analysis of the Ridgeley-Nevitt data.

However, the determination of the required values for a particular new design from tables or graphs is time-consuming and labour-intensive. The availability of digital computers in the fifties and sixties made regression analysis of large data sets possible, and the availability of personal computers in the eighties made calculations in the design office much quicker. These are things which we now take for granted.

The first application of statistical regression analysis to ship resistance data was made by Doust and O'Brien (1959), who analysed the results for some 130 unrelated trawlers which had been tank tested at the Ship Division of the National Physical Laboratory, UK. Since then there have been many regression analyses of ship resistance data carried out.

One of the early ones to follow was van Oortmerssen (1971), who published the results of a regression analysis of the residuary resistance of a set of 93 unrelated models of tugs and trawlers which had been tank tested at the Maritime Research Institute Netherlands (then the Netherlands Ship Model Basin). He obtained a parametric expression for the residuary resistance in terms of hull characteristics which are usually available at an early stage in the design process. The method has become widely used because of its utility.

The other known publication of resistance data for these hullforms is that of Calisal and McGreer (1993), who published the results of a small systematic series for double-chine, low length/beam ratio hullforms based on the purse seiners on the west coast of Canada. The results were published as a parametric expression for the residuary resistance coefficient obtained by regression analysis. However, the bow shape of the parent hullform was modified to reduce the half-angle of waterline entrance and to remain developable. The low angles of waterline entrance tend to give low resistance at high speeds.

1.2 Issues

van Oortmerssen's method was quickly adopted by tug and trawler designers in the eighties because it could be easily programmed and then gave quick predictions of resistance which were found to be reliable in practice. The method has been subsequently included in many commercial resistance-prediction packages, and is still widely used.

However, there was a number of errors in the original paper. Some of these errors are obvious, and have been known almost since publication, but some are not so obvious. It is in the treatment of the not-so-obvious errors that differences can arise and, depending on how these errors are treated, it is possible to come up with different values of the total resistance. As a result, the known implementations of the method all seem to give differing results for the predicted resistance.

The author has corresponded with Dr van Oortmerssen and with MARIN over a number of years, and has resolved most of the issues, and it is the intention here to record corrections to the known issues.

In addition, Holtrop (2001) noted that MARIN had received a comment some years previously that, for some combinations of parameters, the resistance predicted for the vessel by van Oortmerssen's method did not increase monotonically with speed. For displacement vessels such as these, it would be expected that the resistance *would* increase monotonically with speed. The present author also found such a case, and embarked on an analysis of the extent of the problem.

2. VAN OORTMERSSEN'S METHOD

2.1 Summary of the Method

van Oortmerssen gave the equation for the prediction of the residuary resistance of a vessel as follows:

$$R_T / W = c_1 e^{-m/9F_n^2} + c_2 e^{-m/F_n^2} + c_3 e^{-m/F_n^2} \sin(1/F_n^2) + c_4 e^{-m/F_n^2} \cos(1/F_n^2) \qquad (1)$$

where W is the weight of the vessel, F_n is the Froude number, $m = 0.14347\, C_P^{-2.1976}$, C_P is the prismatic coefficient, and the c_i coeffcients were given by:

$$c_i = [d_{i,0} + d_{i,1} lcb + d_{i,2} lcb^2 + d_{i,3} C_P + d_{i,4} C_P^2 + d_{i,5} L_D / B + d_{i,6} (L_D / B)^2 + d_{i,7} C_{WL} + d_{i,8} C_{WL}^2 + d_{i,9} B/T + d_{i,10} (B/T)^2 + d_{i,11} C_M]/1000 \qquad (2)$$

where L_D is van Oortmerssen's "displacement length" = $(L_{WL} + L_{BP})/2$, lcb = LCB as a percentage of L_D from midships, B is the beam, T is the draft, $C_{WL} = i_E L_D/B$, i_E is the half-angle of waterline entrance, C_M is the maximum section coefficient, and the $d_{i,j}$ coefficients were given in van Oortmerssen's Table II.

F_n, lcb and C_P are all based on the displacement length, L_D. The ranges of values of the six parameters of the models were:

L_D/B	3.0–6.2	B/T	1.9–4.0	lcb	(–8%)–(+2.8%)
C_P	0.5–0.725	C_M	0.73–0.97	i_E	10°–46°

2.2 Errors

In the text, the following errors have been known for many years:

1. In van Ortmerssen's Equation (30), shown as Equation (1) above, the right-hand side should be enclosed in brackets and all should be divided by1000 for consistency with his Table II.
2. In Table II, the sign of the coefficient $d_{3,5}$ should be positive instead of negative.
3. In Equation II-1 in Appendix 2, "…$c_4 + f_4$)" should be replaced by "…$c_4 f_4$)"

In the example calculation in Appendix 2 for a tug of displacement length 29 m at a speed of 13 kn:

4. The values of the c_3 and c_4 coefficients should be interchanged.
5. The values of f_3 and f_4 should be 0.005 and 0.066, based on the given figures.
6. The value of c_3 is –79.33, and f_3 is 0.005so the product $c_3 * f_3$ must be negative, not positive as shown.
7. The value given for C_F is for a length L_D of about 21 metres, not for the 29 metre displacement length of this vessel.
8. The values given for C_P and C_M cannot be obtained simultaneously with the values given for ∇, L_D, B and T. Assuming C_P to be correct, then:
   $C_M = C_B/C_P = \nabla / C_P L_D B T = 376/[0.609(29/3.45)(29/3.45/2.85)] = 0.8587$

This compares with the value of 0.817 given in the paper. The calculations of the c_i coefficients depend on both C_P and C_M, so the calculated c_i values will vary with both.

9. It is clear from the values given for c_1 to c_4 that van Oortmerssen used $C_P = 0.609$ and $C_M = 0.817$. However, since C_P and C_M are not consistent with the values given for ∇, L_D, B and T, the values of the c_i will be different for a consistent set of values.

Subsequent correspondence (van Oortmerssen 1986 and 2000, and Holtrop 2001) has confirmed these errors. Data for van Oortmerssen's example tug are no longer available, and so the issue of C_P and C_M is not easily decided.

2.3 Resolution

In order to resolve the difficulties, we start with a consistent set of data for the same example vessel, taking only the stated value of C_P and ignoring the stated value of C_M:

$L_D = 29$ m	$L_D/B = 3.45$	$B/T = 2.85$	$lcb = -2.55\%$ of L_D
$C_P = 0.609$	$i_E = 23°$	$S = 284$ m^2	$\nabla = 376$ m^3
$v = 13$ kn	$\Delta C_F = 0.00051$		

We now calculate the required value of C_M and proceed:

$$C_M = C_B/C_P = \nabla/C_P L_D BT = 376/[0.609(29/3.45)(29/3.45/2.85)] = 0.8587$$
$$m = 0.14347 C_P^{-2.1976} = 0.14347*0.609^{-2.1976} = 0.4267$$
$$C_{WL} = i_E L_D/B = 23*3.45 = 79.35$$

The values of $d_{i,j}$ are taken from van Oortmerssen's Table II, and the c_i are calculated from Equation (2) above, giving:

$$c_1 = 2.096\times10^{-3} \quad c_2 = 248.3\times10^{-3} \quad c_3 = -68.87\times10^{-3} \quad c_4 = 42.33\times10^{-3}$$

At 13 knots, $Fn = v/\sqrt{(gL_D)} = 13\times0.5144/\sqrt{(9.80665*29)} = 0.3966$

$$f_1 = e^{-m/9F_n^2} = 0.7398 \qquad f_2 = e^{-m/F_n^2} = 0.06634$$
$$f_3 = e^{-m/F_n^2}\sin(1/F_n^2) = 0.004992 \qquad f_4 = e^{-m/F_n^2}\cos(1/F_n^2) = 0.06615$$

$$c_1 f_1 = 2.096*10^{-3}\times0.7398 = 1.55*10^{-3}$$
$$c_2 f_2 = 248.3\times10^{-3}\times0.06634 = 16.471\times10^{-3}$$
$$c_3 f_3 = -68.87\times10^{-3}\times0.004992 = -0.344\times10^{-3}$$
$$c_4 f_4 = 42.33\times10^{-3}\times0.06615 = 2.801\times10^{-3}$$

$$R_R/W = c_1 f_1 + c_2 f_2 + c_3 f_3 + c_4 f_4 = 1.55\times10^{-3} + 16.471\times10^{-3} - 0.344\times10^{-3} + 2.801\times10^{-3} = 20.48*10^{-3}$$

$$R_R = (R_R/W)\rho\nabla = 20.48*10^{-3}*1025*376 = 7893 \text{ kgf}$$
$$Rn = vL_D/\nu = (13\times0.5144)\times29/1.18831\times10^6 = 1.632\times10^8$$
$$C_F = 0.075/(\log Rn - 2)^2 = 0.075/(\log 1.632\times10^8 - 2)^2 = 0.001943$$
$$C_F + \Delta C_F = 0.001953 + 0.00051 = 0.002453$$
$$R_F = 0.5\rho S v^2(C_F + \Delta C_F) = 0.5\times(1025/9.80665)\times284\times(13\times0.5144)^2\times0.002453 = 1628 \text{ kgf}$$
$$R_T = R_R + R_F = 7893 + 1628 = 9521 \text{ kgf} = 93.37 \text{ kN}$$

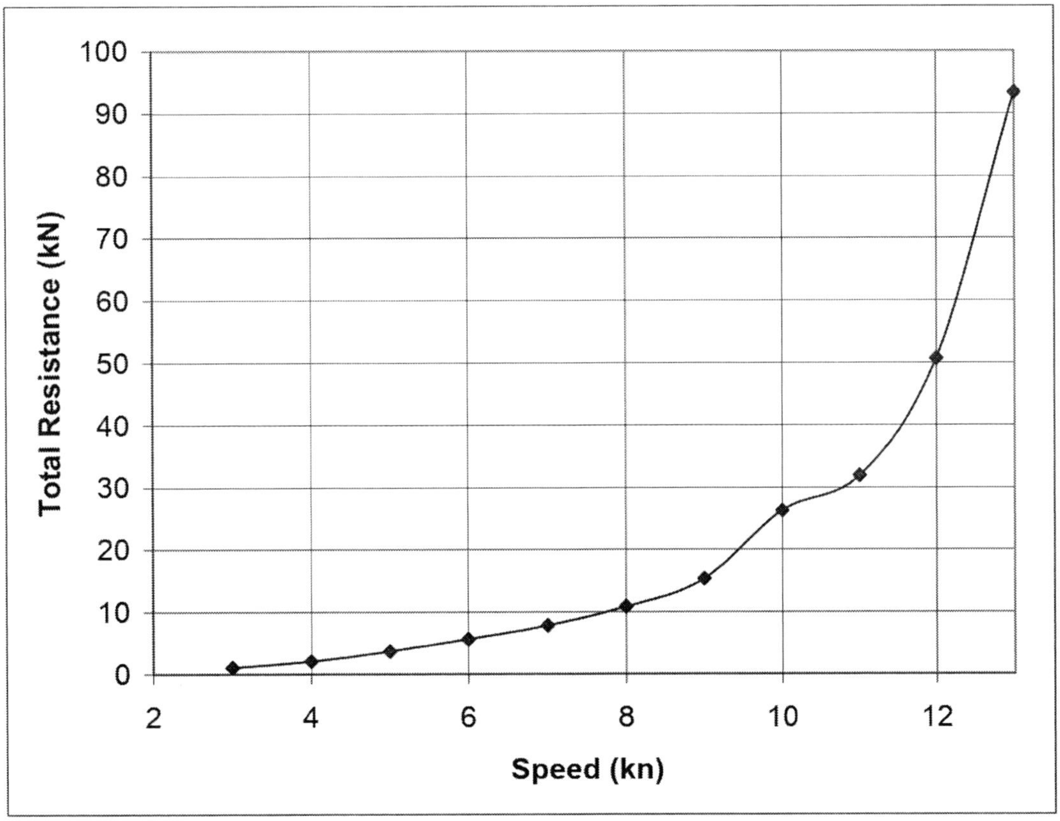

Figure 1 Total resistance vs speed for van Oortmerssen's 29 m example tug

This compares with a total resistance value at 13 kn of 9755 kgf = 95.66 kN in the original calculation. The above calculations are believed to eliminate all of the errors in van Oortmerssen's published example calculation for the 29 m tug. The spreadsheet calculations for a range of speeds up to 13 knots are shown in Appendix 1, and graphed in Figure 1.

3. VARIATIONS OF PARAMETERS

3.1 General

van Oortmerssen did not state a specific range for the variation of Froude number for his method. However, he shows variations of his functions f_3 and f_4 for Froude numbers up to 0.50, and it would seem reasonable to assume that his equation applies up to that limit. The resistance calculation for his example tug was done for a speed of 13 kn, corresponding to $Fn = 0.3966$.

van Oossanen (1979) in his discussion of various resistance prediction methods, says that some extrapolation of van oortmerssen's equation to higher speeds is permissible because of the theoretical nature of the basic expression. van Oortmerssen based his theory on that of Havelock for the waves produced by a travelling two-dimensional pressure disturbance, but simplified so that it could be formulated in terms of the vessel's parameters.

However, MacPherson (1991) sounds a note of caution in his summary of various resistance prediction methods. He says that the basis for the Havelock theory in van Oortmerssen's method is questionable, and a speed-dependent analysis (like Havelock) typically has trouble matching the unusual C_R curve shape at Froude numbers below about 0.3.

The Havelock theory gave rise to the sin and cos terms in van Oortmerssen's equation, and these attempt to account for the humps and hollows in the residuary resistance. Figure 1 shows that, for the example 29 m tug, they do a respectable job.

Inspection of the sin and cos terms shows them to have arguments of $1/F_n^2$. Since F_n is dimensionless, and radians are effectively dimensionless, it is considered that $1/F_n^2$ is already in radians and can be used directly. That this was van Oortmerssen's intention is borne out by taking $1/F_n^2$ in radians for a prismatic coefficient of 0.75 and graphing the functions f_3 and f_4 in Figures 2 and 3, which reproduce his results. However, one known commercial implementation of van Oortmerssen's method considers that $1/F_n^2$ is not dimensionless, and converts $1/F_n^2$ to radians by dividing by 57.296. The effect of so doing on van Oortmnerssen's functions f_3 and f_4 can be seen in Figures 2 and 3.

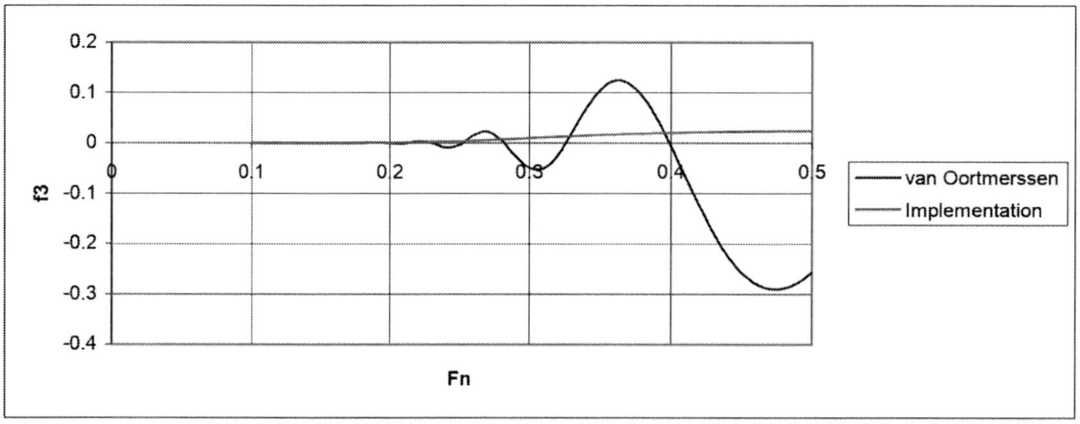

Figure 2 Effect on f_3 of treating the argument $1/F_n^2$ in radians or degrees at $C_P = 0.75$

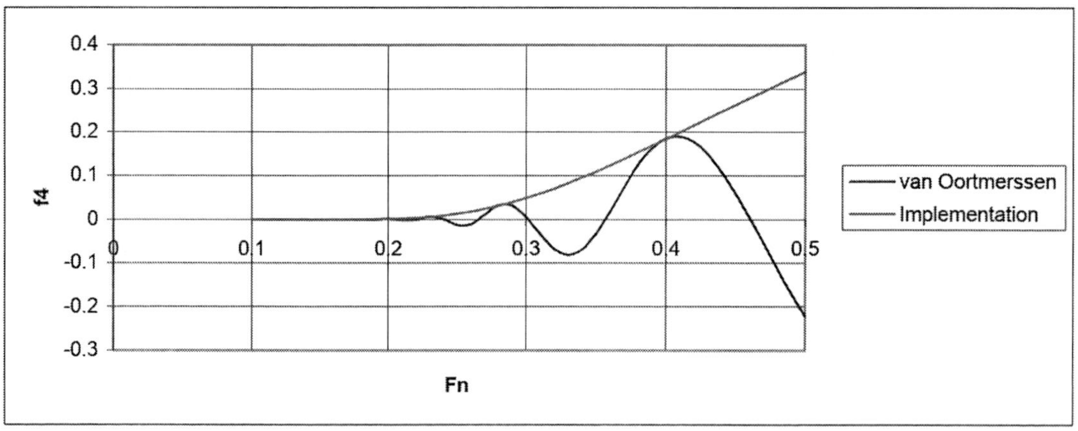

Figure 3 Effect on f_4 of treating the argument $1/F_n^2$ in radians or degrees at $C_P = 0.75$

Treating the argument $1/F_n^2$ in degrees gives monotonically increasing vaules of f_3 and f_4, and does not introduce any humps and hollows into the residuary resistance as intended by van Oortmerssen.

3.2 Combinations of Parameters

Holtrop (2001) reported that MARIN had received a comment some years previously that, for some combinations of parameters, the resistance predicted for a vessel by van Oortmerssen's method did not increase monotonically with speed, i.e. that the "hollow" of the residuary resistance produced values which reduced as speed increased, before increasing again at higher speeds. For displacement vessels such as tugs and trawlers, it would be expected that the resistance *would* increase monotonically with speed.

This author found such a case while setting an assignment for his students to use van Oortmerssen's method to predict the resistance of a 34 m tug. The six parameters of the tug are as follows:

L_D/B	3.032	B/T	2.643	lcb	–5.1768%
C_P	0.7070	C_M	0.8086	i_E	25°

The parameters for this vessel are all within van Oortmerssen's limits, although L_D/B is close to the lower limit of 3. The calculations are presented in Appendix 2, and the results are graphed in Figure 4.

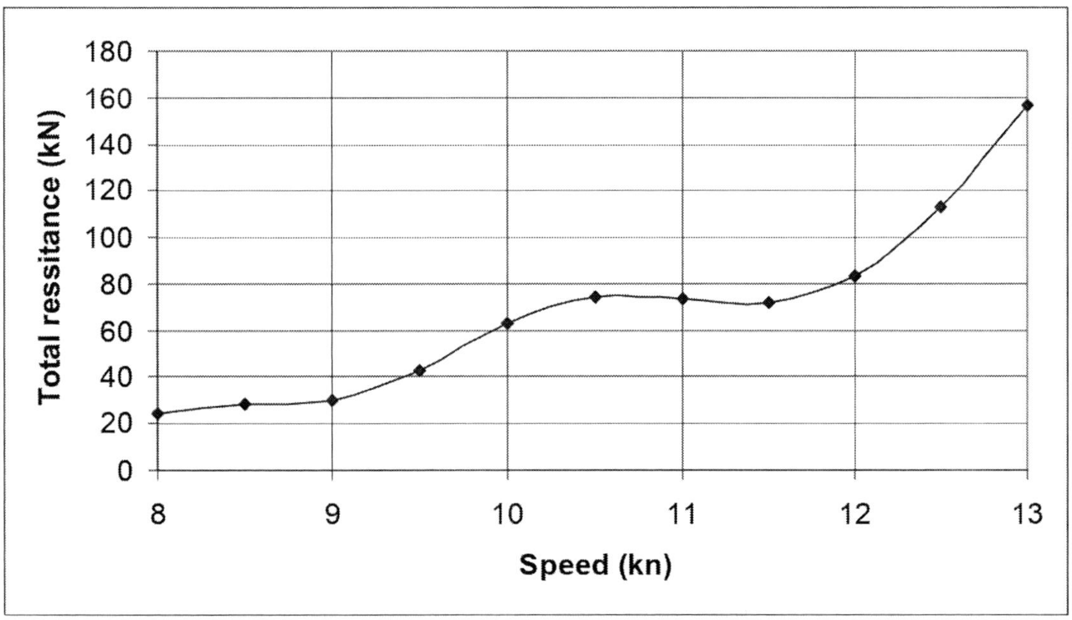

Figure 4 Total resistance vs speed for 34 m tug

As can be seen from Figure 4, the R_T values decrease from about 10.5 kn to 11.5 kn, before increasing at higher speeds. The hollow in the curve can be removed by increasing the beam to give L_D/B = 3.301, resulting in equal values of R_R at 10.5 and 11 kn, but this approach also changes other parameters, (B/T, i_E and, incidentally, the displacement-length ratio), and might not be what is wanted for a particular design. The change in parameters also increases the resistance at the higher speeds.

3.3 The Data Set

In order to investigate the extent of combinations of parameters for which this scenario occurs, it would be useful to examine the original data to see if any particular combinations were not included in the set. However, the original data is no longer available or, at least, not easily accessible (van Oortmerssen 2000, Holtrop 2001).

Inspection of van Oortmerssen's histograms of numbers of vessels with particular parameters shows that some parameters had only one vessel towards the end of the range, while others had several. As an example, the L_D/B ratio had one vessel in the range 3.0–3.4 at the low end, but six vessels in the range 5.8–6.2 at the high end, with at least nine in each range of 0.4 in between. What is not brought out is whether there were particular *combinations* of parameters which were *not* included. For example, it might be expected that vessels with lower values of L_D/B might have higher values of i_E, but the histograms do not show whether there were any vessels with, say, $i_E < 30°$ for $L_D/B < 4$.

Not having access to the data, we have to start somewhere, and so have included all of each range in the set to be examined here.

One parameter of particular interest is the displacement-length ratio $\Delta/(0.01L)^3$ tons/ft^3 or, in its non-dimensional equivalents, the slenderness ratio $L/\nabla^{1/3}$ or the fatness ratio $\nabla/(L/10)^3$. However, since the author is more familiar with displacement-length ratio, we will work with that. Tugs and trawlers are pushing the boundaries of this ever higher (840 is not unknown for a modern tug), and the author was interested to see if there were any limits placed by the ranges of parameters in van Oortmerssen's data set.

It may be shown that $\Delta/(0.01L)^3 = 28567 C_P C_M/[(L_D/B)^2(B/T)]$ tons/ft^3, where the factor 28567 includes a water density of 1.025 t/m^3 and conversion factors from metric SI units to give the result in tons/ft^3.

Substituting maximum values of C_P and C_M, and minimum values of L_D/B and B/T:

$$\Delta/(0.01L)^3_{max} = 28567\times0.725\times0.97/(3^2\times1.9) = 1175 \text{ tons/ft}^3$$

Substituting minimum values of C_P and C_M, and maximum values of L_D/B and B/T:

$$\Delta/(0.01L)^3_{max} = 28567\times0.5\times0.73/(6.2^2\times4) = 68 \text{ tons/ft}^3$$

There are no known tugs or trawlers with such extreme values of $\Delta/(0.01L)^3$, so it is clear that the extreme values of the contributing parameters in the data set did not occur together. However, there appears to be no direct restriction on practical values of $\Delta/(0.01L)^3$ for tugs and trawlers.

3.4 Investigation

The author then set about investigating the data set to find out for which combinations of parameters non-monotonically-increasing values of R_T occurred.

van Oortmerssen's parameter in the equation for residuary resistance, $C_{WL} = i_E L_D/B$, is dependent on i_E and L_D/B, and so i_E was taken as the independent parameter. Combinations of five values of each of the six parameters were tested, as shown in Table 1, giving 5^6 = 15 625 combinations.

Table 1 Combinations of parameters

Value	Lowest	Low	Mean	High	Highest
L_D/B	3.0	3.8	4.6	5.4	6.2
B/T	1.9	2.425	2.95	3.475	4.0
lcb%	–8	–5.3	–2.6	0.1	2.8
C_P	0.5	0.5563	0.6125	0.6688	0.725
C_M	0.73	0.79	0.85	0.91	0.97
i_E	10°	19°	28°	37°	46°

The test for monotonic increase of resistance with speed was applied to the residuary resistance from van Oortmerssen's equation; i.e. the calculated R_R was accepted if it increased or remained the same with increasing speed; it was rejected if it decreased with increasing speed. Thus, with the addition of the frictional resistance (which increases monotonically with speed) the total resistance also increases monotonically with speed and the graph visually does the same.

3.5 Results

As might be expected, problems occur for combinations of high length/beam ratios with high angles of waterline entrance, and for low length/beam ratios with low angles of waterline entrance. These combinations were probably not represented in van Oortmerssen's original data set, and are unlikely to be used.

Other results are not easy to summarise. For combinations of L_D/B, B/T and $lcb\%$, problems are encountered for some combinations of C_P and C_M at some values of i_E and not at others. No pattern is apparent. Two examples are shown, in Table 2 for $L_D/B = 3.8$, $B/T = 2.95$ and $lcb\% = -5.3$, and in Table 3 for $L_D/B = 5.4$, $B/T = 4.0$ and $lcb\% = 0.1$.

Table 2 Monotonic increase in R_T for $L_D/B = 3.8$, $B/T = 2.95$ and lcb% = –5.3

$L_D/B = 3.8$	$B/T = 2.95$	lcb% = -5.3

$i_E = 10°$	Cp				
Cm	0.5	0.5563	0.6125	0.6688	0.725
0.73	✓	✓	x	x	x
0.79	✓	✓	x	x	x
0.85	✓	✓	x	x	x
0.91	✓	✓	x	x	x
0.97	✓	✓	x	x	x

$i_E = 19°$	Cp				
Cm	0.5	0.5563	0.6125	0.6688	0.725
0.73	✓	✓	x	x	✓
0.79	✓	✓	✓	x	✓
0.85	✓	✓	✓	✓	✓
0.91	✓	✓	✓	✓	✓
0.97	✓	✓	✓	✓	✓

$i_E = 28°$	Cp				
Cm	0.5	0.5563	0.6125	0.6688	0.725
0.73	✓	✓	✓	x	✓
0.79	✓	✓	✓	✓	✓
0.85	✓	✓	✓	✓	✓
0.91	✓	✓	✓	✓	✓
0.97	✓	✓	✓	✓	x

$i_E = 37°$	Cp				
Cm	0.5	0.5563	0.6125	0.6688	0.725
0.73	✓	✓	x	x	x
0.79	✓	✓	x	x	✓
0.85	✓	✓	✓	x	✓
0.91	✓	✓	✓	✓	✓
0.97	✓	✓	✓	✓	x

$i_E = 46°$	Cp				
Cm	0.5	0.5563	0.6125	0.6688	0.725
0.73	✓	✓	x	x	x
0.79	✓	✓	x	x	x
0.85	✓	✓	x	x	x
0.91	✓	✓	x	x	x
0.97	✓	✓	x	x	x

✓ = R_R ↑ monotonically with speed

x = R_R not ↑ monotonically with speed

Table 2 Monotonic increase in R_T for $L_D/B = 3.8$, $B/T = 2.95$ and lcb% = –5.3

$L_D/B = 5.4$						$B/T = 4.0$					lcb% = 0.1
$i_E = 10°$						$i_E = 19°$					
	Cp						Cp				
Cm	0.5	0.5563	0.6125	0.6688	0.725	Cm	0.5	0.5563	0.6125	0.6688	0.725
0.73	✓	✓	✓	x	x	0.73	✓	✓	x	x	x
0.79	✓	✓	✓	x	✓	0.79	✓	✓	✓	x	x
0.85	✓	✓	✓	x	✓	0.85	✓	✓	✓	x	x
0.91	✓	✓	✓	✓	✓	0.91	✓	✓	✓	x	x
0.97	✓	✓	✓	✓	✓	0.97	✓	✓	✓	x	x

$i_E = 28°$						$i_E = 37°$					
	Cp						Cp				
Cm	0.5	0.5563	0.6125	0.6688	0.725	Cm	0.5	0.5563	0.6125	0.6688	0.725
0.73	✓	✓	x	x	x	0.73	✓	✓	x	x	x
0.79	✓	✓	x	x	x	0.79	✓	✓	x	x	x
0.85	✓	✓	✓	x	x	0.85	✓	✓	x	x	x
0.91	✓	✓	✓	x	x	0.91	✓	✓	x	x	x
0.97	✓	✓	✓	x	x	0.97	✓	✓	x	x	x

$i_E = 46°$					
	Cp				
Cm	0.5	0.5563	0.6125	0.6688	0.725
0.73	x	x	x	x	x
0.79	x	x	x	x	x
0.85	x	x	x	x	x
0.91	x	x	x	x	x
0.97	x	x	x	x	x

✓ = R_R ↑ monotonically with speed

x = R_R not ↑ monotonically with speed

Some resistance curves display a marked deviation from the expected form. An example is shown in Figure 6 for the following combination of parameters (high L_D/B and low i_E):

L_D/B	5.4	B/T	4.0	lcb	0.1%
C_P	0.6688	C_M	0.73	i_E	28°

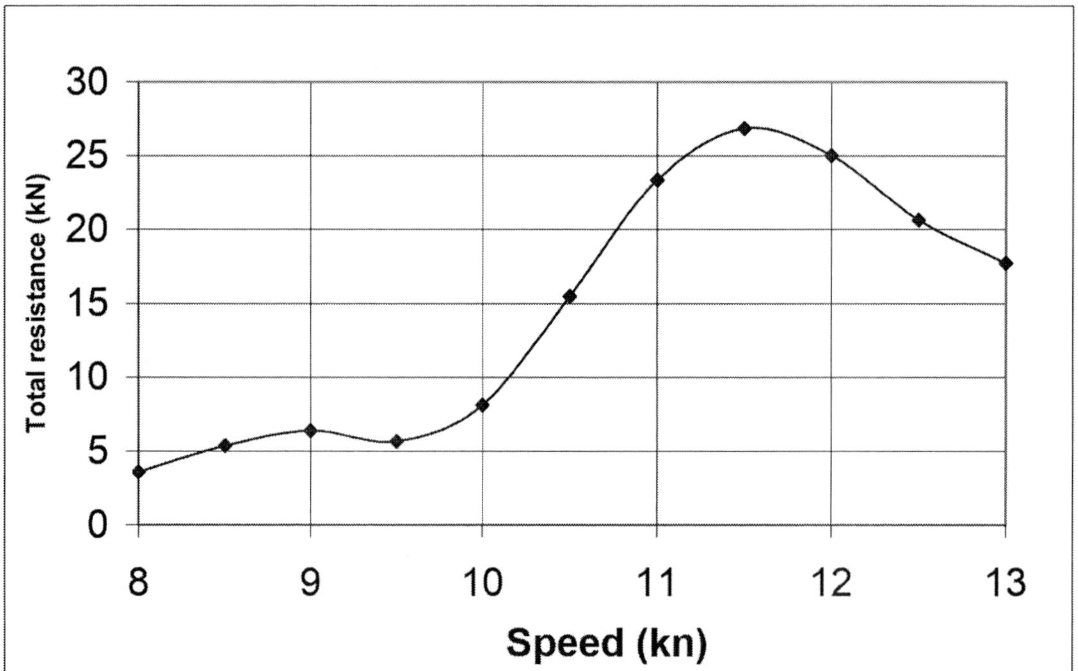

Figure 5 Total resistance vs speed for 34 m tug

The resistance for this vessel is much lower than the previous vessel because of the higher length/beam ratio and, hence, the lower displacement on the same length. The curve is unusual for the location of the resistance hollow at 13 kn ($Fn = 0.373$).

3.6 Solution

With the overall results being difficult to summarise by way of rules for particular combinations of parameters to avoid, the author proposes a pragmatic approach instead:

(a) If the resistance curves exhibit the characteristics of Figure 5, then the results should not be used, as it is likely that the parameters are outside the range of van Oortmerssen's original data set.

(b) If the resistance curves exhibit the characteristics of Figure 4, then data points on either side of the point of inflexion on the resistance curve should be deleted and the curve faired through the remaining points, as shown in Figure 6.

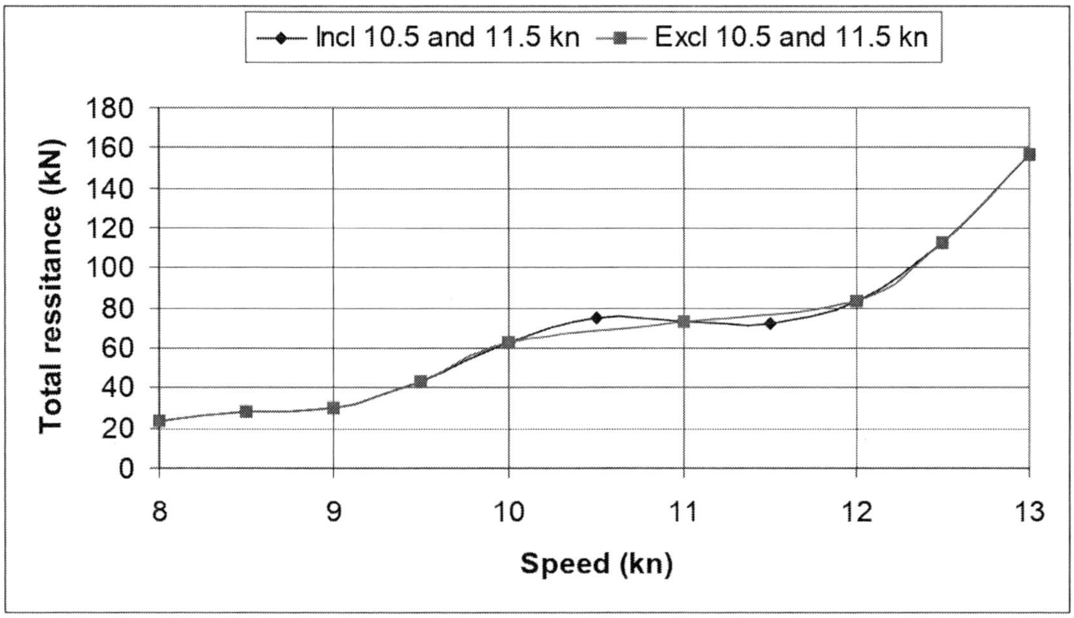

Figure 4 Total resistance vs speed for 34 m tug

4. CONCLUSIONS

Errors in van Oortmerssen's (1971) paper on the prediction of the resistance of tugs and trawlers have been investigated, and a consistent set of results for his example 29 m tug have been generated.

In addition, issues on the combinations of parameters which can be used with the method have been investigated, leading to the conclusion that high length/beam ratios combined with high angles of waterline entrance, or low length/beam ratios combined with low angles of waterline entrance should be avoided. Other combinations which provide non-monotonically-increasing values of resistance are harder to summarise. Rather, a pragmatic approach is proposed, whereby some predicted points are deleted and the curve faired through the remaining points.

In the author's estimation, van Oortmerssen's method is still the best around for these types of vessels, especially for small consultancies which have limited access to high-powered methods of resistance prediction.

5. REFERENCES

Calisal, S.M. and McGreer, D. (1993), A Resistance Study on a Systematic Series of Low *L/B* Vessels, *Marine Technology*, v.30 n.4, SNAME, Jersey City, October.

Doust, D.J. and O'Brien, T.P. (1959), Resistance and Propulsion of Trawlers, *Trans. North-East Coast Institution of Engineers and Shipbuilders*, v.75, NECIES, Newcastle, UK.

Helmore, P.J. and Swain, P.M. (2006), Regression Analysis of the Ridgeley-Nevitt Trawler Series Resistance, *Proceedings Pacific 2006 International Maritime Conference*, Sydney.

Holtrop, J. (2001), MARIN, personal correspondence, March.

MacPherson, D. (1991), Reliable Performance Prediction: Techniques using a Personal Computer, *Marine Technology*, v.30 n.4, October.

Ridgeley-Nevitt, C. (1956), The Resistance of Trawler Hull Forms of 0.65 prismatic Coefficient, *Trans. Society of Naval Architects and Marine Engineers*, v.64, SNAME, New York.

Ridgeley-Nevitt, C. (1963), The Development of Parent Hulls for a High Displacement-Length Ratio Series of Trawler Hull Forms, *Trans. Society of Naval Architects and Marine Engineers*, v.71, SNAME, New York.

Ridgeley-Nevitt, C. (1967), The Resistance of a High Displacement-Length Ratio Trawler Series, *Trans. Society of Naval Architects and Marine Engineers*, v.75, SNAME, New York.

van Oortmerssen, G. (1971), A Power Prediction Method and its Application to Small Ships, *International Shipbuilding Progress*, v.18 n.217.

van Oortmerssen, G. (1986), Netherlands Ship Model Basin, personal correspondence, October and December.

van Oortmerssen, G. (2000), Director Dutch Centre for Mathematics, personal correspondence, March.

van Oossanen, P. (1979), Resistance Prediction if Small High-speed Displacement Vessels: State of the Art, *Proceedings Australian Symposium on Ship Technology*, RINA/ImarE, The Univesity of New South Wales, Sydney.

APPENDIX 1

Calculations for van Oortmerssen's Example Tug

Length WL	30.5 m		
Length BP	27.5 m		
Beam WL	8.406 m		
Draft moulded	2.949 m		
LCB fwd of midships L_{WL}	0.0105 m		
Displacement volume	376 m^3	5-3150	
Wetted surface	284 m^2	15-1400	281.0 if unknown
Prismatic co-efficient on L_{WL}	0.5790		
1/2 angle of WL entrance	23 deg	10.0-46.0	
Water density	1.025 t/m^3		
Water viscosity	1.18831 x 10^{-6} m^2/s		
Correlation allowance	0.00051		

Displacement length $L_D = (L_{WL}+L_{BP})/2$	29 m	
Displacement	385.4 t	
L_D/B	3.450	3.0-6.2
B/T	2.850	1.9-4.0
LCB fwd of midships L_D	-0.7395	
LCB% L_D	-2.550	(-8.0)-2.8
Prismatic co-efficient on L_D	0.6090	0.50-0.725
Midsection co-efficient	0.8587	0.73-0.97
C_{WL} entrance	79.35	
1000C1	2.096	
1000C2	248.3	
1000C3	-68.87	
1000C4	42.33	
m	0.4267	

Speed	**Fn**	**Cf+Ca**	**Rf**	**Rr/W**	**Rr**	**Rt**	**Pe**
kn	on L_D	ITTC57	kN		kN	kN	kW
3	0.0915	0.002922	1.01	0.00001	0.03	1.04	1.6
4	0.1220	0.002818	1.74	0.00009	0.33	2.06	4.2
5	0.1525	0.002741	2.64	0.00027	1.03	3.67	9.4
6	0.1830	0.002681	3.72	0.00051	1.93	5.65	17.4
7	0.2135	0.002633	4.97	0.00076	2.87	7.84	28.2
8	0.2440	0.002592	6.39	0.00117	4.42	10.81	44.5
9	0.2746	0.002557	7.98	0.00194	7.33	15.31	70.9
10	0.3051	0.002526	9.73	0.00437	16.51	26.24	135.0
11	0.3356	0.002499	11.65	0.00537	20.28	31.93	180.7
12	0.3661	0.002475	13.73	0.00979	36.99	50.72	313.1
13	0.3966	0.002453	15.97	0.02048	77.40	93.37	624.4

APPENDIX 2

Calculations for 34 m Tug

Length WL	34 m		
Length BP	31.5 m		
Beam WL	10.8 m		
Draft moulded	4.1 m		
LCB fwd of midships L_{WL}	-1.07 m		
Displacement volume	829 m^3	5-3150	
Wetted surface	463 m^2	15-1400	450.6 if unknown
Prismatic co-efficient on L_{WL}	0.681		
1/2 angle of WL entrance	25 deg	10.0-46.0	
Water density	1.025 t/m^3		
Water viscosity	1.07854×10^{-6} m^2/s		
Correlation allowance	0.0005		

Displacement length $L_D = (L_{WL}+L_{BP})/2$	32.75 m	
Displacement	849.7 t	
L_D/B	3.032	3.0-6.2
B/T	2.634	1.9-4.0
LCB fwd of midships L_D	-1.695	
LCB% L_D	-5.176	(-8.0)-2.8
Prismatic co-efficient on L_D	0.7070	0.50-0.725
Midsection co-efficient	0.8086	0.73-0.97
C_{WL} entrance	75.81	
1000C1	2.213	
1000C2	126.4	
1000C3	-32.29	
1000C4	42.43	
m	0.3074	

Speed kn	Fn on L_D	Cf+Ca ITTC57	Rf kN	Rr/W	Rr kN	Rt kN	Pe kW
8	0.2296	0.002518	10.12	0.001644	13.70	23.81	98.0
8.5	0.2440	0.002500	11.34	0.002022	16.85	28.19	123.3
9	0.2584	0.002484	12.64	0.002059	17.16	29.80	138.0
9.5	0.2727	0.002469	13.99	0.003459	28.82	42.82	209.3
10	0.2871	0.002455	15.42	0.005742	47.85	63.27	325.5
10.5	0.3014	0.002442	16.91	0.006921	57.67	74.58	402.8
11	0.3158	0.002429	18.46	0.006606	55.04	73.50	415.9
11.5	0.3301	0.002417	20.08	0.006225	51.87	71.95	425.7
12	0.3445	0.002406	21.76	0.007379	61.49	83.24	513.9
12.5	0.3588	0.002395	23.50	0.010696	89.13	112.63	724.3
13	0.3732	0.002385	25.31	0.015770	131.41	156.73	1048.2

AMPHIBIOUS SHIPS

Phill Brown

Program Manager, Amphibious Deployment and Sustainment Program

Defence Materiel Organisation

On the 9th of October last year, the Commonwealth signed a contract with Tenix Defence Pty Ltd for the supply to the Royal Australian Navy of two 27,000-tonne Amphibious Landing Ships, or LHDs.

The acquisition will be administered by Joint Project 2048 Phase 4A/B, 'Joint' meaning that it is not just a Navy ship, but a Navy and Army capability. Navy will operate the ships, Army the watercraft and both services the helicopters; Army forces are transported and inserted; and the joint force headquarters aboard is multi-service, including RAAF air traffic control. So the services cooperated in the development of the capability.

Today I'm going to give you an overview of the Ships. I'll talk about the capability they represent – what the ADF can do with them. I'll outline how we involved Australian industry and I'll describe the systems – primarily the electronic and landing systems. I'll give you a project timeline for the requirement, and a rough idea of what we expect industry to be able to do for the ADF.

Finally, I'll mention the companion projects to the ships, because I don't want industry to think of this as a one-off opportunity, but rather part of an ongoing program of available Defence work that can help grow and sustain the Australian companies that we need in order to support our equipment in service.

If there appear to be some gaps in my information I would ask you to be patient – my address will be followed by a comprehensive overview from Tenix of how they are approaching the project, and I am sure that will answer many questions.

On the industry element, I'd like to state at the outset that Defence has no say, and does not seek any say, in the selection of subcontractors for the construction phase. That is rightly a commercial matter and the province of the prime contractor, Tenix.

Defence's current amphibious capability is provided by two Landing Platforms Auxiliary (HMAS *Manoora* and *Kanimbla*) and a Heavy Landing Ship (HMAS *Tobruk*), which are due to be withdrawn from service between 2012 and 2017.

These ships are capable of carrying and landing equipment and personnel, but only in fairly calm water and with relatively few helicopters. This limits their operational flexibility and their ability to meet the demands outlined in the *Defence 2000* White Paper for the defence of Australia and the immediate neighbourhood - such as seizure of entry points for larger follow-on forces; raids or similar operations; transportation of military forces, humanitarian operations, assistance to the civil power and a range of evacuation operations.

JP2048 Phase 4A/B will replace these vessels as their life of type expires and in so doing will greatly increase the amphibious capability of the Australian Defence Force. But the LHDs won't be the sole amphibious element of the force. There are companion projects to complete the capability and I'll address them briefly at the end.

The ADF's amphibious requirements include the need to transport and support a group for the first ten days of operations. This means the ability to deploy approximately 2000 troops, of which 1200 can be landed ashore through a mix of helicopters and organic landing craft. Up to 800 joint personnel will remain onboard the LHD providing command, aviation, medical and logistic support from the ships.

When we examined the question of what sort of ship we needed, we found that its design and size would be determined largely by the nature of the force that needs to be carried, the means by which it would need to be put ashore and the speed with which this could be done (by air, by watercraft or both).

We surveyed the options late in 2003 and found there were only two designs potentially able to meet our requirements: the 26,800-tonne projection ship designed by the Spanish company Navantia and the 22,000-tonne *Mistral* class LHD designed by the French company Armaris.

These were new and unfamiliar ships to us, in a procurement activity with a lot of potential risk. In dealing with that risk we were going to follow the recommendations of Mr Malcolm Kinnaird's Review of Defence Procurement, the recommendations of which had been adopted by Government in 2003. Accordingly, we attacked risk from the start – technical risk, schedule risk and therefore cost risk, which flows from the first two, primarily schedule.

We also involved Australian industry from the beginning. We conducted a comprehensive risk reduction and design program with the designers, Australian shipbuilders and a cost estimator. We established that the designs were suitable, and that an overseas build could be achieved within budget, but that building in Australia would significantly exceed budget. That last was no surprise – a short, cold-start production run on facilities that have to be adapted for the purpose is never likely to be as cheap as an established production line where the customer is just another couple of units.

So we had to devise a strategy to maximise opportunities for local industry while getting value for the taxpayer's dollar. That's not a political issue, by the way – we needed to lay the basis for industry's support of the ships.

First Pass approval by Government was achieved in August 2005. The Government approved a strategy that allowed Australian shipbuilders to bid, did not specify any particular level of Australian Industry Involvement, and sought innovation to reduce the cost of an Australian build.

So we came up with a three-pronged strategy to maximise the opportunities for Australian industry:

a) through-life support,

b) mandating certain Australian products as costed options, and

c) options to substitute Australian materials and minor equipment.

After working with the designers to further develop the designs to Australian requirements, Defence issued a Request for Tender to the Australian builders Tenix and Thales Australia in April 2006. Tenix teamed with the Spanish designer Navantia to propose a solution based on the *Juan Carlos* design, and Thales teamed with the French company DCN-Armaris for the *Mistral*.

Tenders closed in October 2006 and after the results of the evaluation went to Government at Second Pass, the Tenix proposal was selected as preferred.

I want to mention a couple of points for the benefit of industry here: Firstly, 'preferred tenderer' does not mean 'contractor' – there is usually a lot of definition work and negotiation that must take place before the Commonwealth will be prepared to commit. Second, we don't decide on something because it is biggest or cheapest or even best, which is usually the opposite of cheapest. If the contenders meet the requirement and all other things are equal, value for money will usually decide it.

In this case, both tenders met the requirement and were competitive.

Now let's look at the ships.

The *Canberra* Class LHDs are modified versions of an LHD designed by the Spanish shipbuilding company Navantia for the Spanish Navy.

The cutaway of the "Projection Ship" shown in this slide depicts the Australian LHD. It is baselined on the Spanish BPE presently under construction.

The design has been modified to meet essential ADF requirements. The BPE was designed as a multi-purpose air-capable platform to be used for amphibious operations, tactical transport and humanitarian relief missions. Incidentally, it was also designed to be able to be converted to operate as a second aircraft carrier for the Spanish navy, but the equipment necessary to that role is not included in the Australian version.

There is a considerable size increase from the LPAs to the LHDs, with the latter being roughly three times the displacement, longer, far greater in the beam, and with immense interior space compared with what we are used to.

The physical size carries with it a significant increase in capability that the two LHDs will provide the ADF. As an example, one LHD will be able to provide three times the assistance KANIMBLA provided in the wake of the 2004 Boxing Day tsunami. I will give more details during this presentation.

The design has four main decks as follows: - From the bottom:

A dock for four medium size watercraft and four to six rigid hull inflatable boats (known as RHIBs) and a heavy vehicle deck is located near the ship's waterline. The heavy vehicle garage is located forward of the dock.

A main accommodation deck is located above dock/heavy vehicle hangar. This includes cabins, sanitary spaces, medical spaces, galley and pantries, offices, mess rooms and recreation rooms.

A combined aviation hangar (aft) and upper garage deck for light vehicles (forward) is located above the main accommodation deck. This combined space provides flexibility to embark either additional vehicles or helicopters. For example, the allocated hangar space can accommodate 8 medium-sized MRH90 Multi-Role Helicopters or ARH90 Armed Reconnaissance helicopters or, for that matter, MSH90 Maritime Support Helicopters; but as many as 16 medium helicopters or 10 large CH-47 twin-rotor Chinooks can be carried at the expense of 34% of the total vehicle space.

The flight deck is capable of operating all helicopters in the ADF inventory. Six MRH-90 size helicopters or four CH-47 Chinook helicopters can operate from it simultaneously.

This is an all-electric ship with the two diesel engines and one gas turbine producing electricity to run the two propulsor pods and all ship systems.

The pods can rotate through 360 degrees when at slow speed, providing excellent manoeuvrability for a ship of this size.

Commonality is a theme in acquisition across the ADF and we exploit it wherever possible – the fewer types and makes of equipment we have the cheaper and easier it is to maintain or replace, the lower our training load and the better in-service support we'll get. In the generation/propulsion area, the gas turbines are GE LM2500s, as used in the FFGs and *Anzacs*.

The other equipment is updated or new models.

- Diesel Generation sets: two MAN/BMW x 7448 kW
- Gas turbine: one LM2500 x 22.5 kW
- Emergency Generator: one Volvo Penta 1396 kW
- Propulsion: two Siemens azimuthing 11 MW Electric Pods aft and two tunnel mounted 1.5 MW thrusters in the bow

We expect to attain 8,000 nautical miles range at 15 knots and at least 9,000 at 12 knots.

The six-spot flight deck allows the simultaneous insertion of a Company Group supported by Armed Reconnaissance Helicopters. This allows the security force ahead of the main landing to provide protection during the vulnerable stages when the rest of the force approaches the shore. The capacity for a simultaneous Company Group helicopter lift also provides a useful capability to undertake amphibious raids or rapid seizures of important sites such as landing zones.

The Light Vehicle Garage (Deck 1) can withstand loads imposed by vehicles weighing less than 16 tonnes. Access is via a fixed 14° internal ramp, the Hangar, the Lightweight Cargo Elevator capable of carrying 16 tonnes and the forward aircraft elevator, which is capable of carrying 27 tonnes.

The Heavy Vehicle Garage (Deck 4) is forward of the Dock with access being provided via a steel beach fixed ramp with 15° maximum slope, two starboard vehicle ramps, and the internal ramp and Lightweight Cargo Elevator from Deck 1. The Heavyweight Cargo Garage is intended for the stowage of heavy vehicles and sea containers and will support all ADF vehicles including the M1A1 Abrams tank.

The two garages have overhead gantry cranes fitted to assist in handling sea containers and pallet loads

Crew numbers have yet to be finalised but are likely to be in the order of 370. The crew will be a mix of Army and Navy and is also likely to include an RAAF Air Traffic Controller.

Each LHD is equipped with 1403 bunks. Accommodation for the crew is:

- Officers - 2 berth ensuite cabins
- Senior sailors - 4 berth ensuite cabins
- Junior Sailors - 6 berth cabins with communal ablutions

Embarked Forces will be accommodated in a range of 2 and 4 berth ensuite cabins and 9 to 18 berth cabins with communal facilities.

In addition to the dining rooms, there are 18 separate recreation spaces of various sizes for crew and embarked forces spread around the ship, and two gymnasiums.

Operational Capability

The ships can:

- Form the core of an Amphibious Task Force, along with escorts and AO/AOR (tankers and supply ships).
- Provide a command platform for Commander Amphibious Task Force (CATF) and Commander Landing Force (CLF) with the most powerful Joint communications and situational awareness suite in the ADF –
  - Primary and secondary joint operations rooms
  - Primary and secondary planning rooms
  - Voice/data circuits,
  - 5 Local area networks (LANs).
- Provide airspace and air traffic control over the Amphibious Objective Area.
- Coordinate offensive support including fighter ground attack, naval gunfire support, land artillery and electronic attack.
- Project a combat team from over the horizon by helicopter, supported by Tiger Armed Reconnaissance Helicopters.
- Project a second combat team within 30 minutes of the return of the first wave.
- Project two mechanised combat teams ashore using four landing craft per ship, operating from an enclosed dock.

The heavy vehicles and equipment for an amphibious lodgement will be transported ashore in watercraft. Watercraft are relatively slow and their loading is greatly affected by sea conditions. The welldock improves watercraft operations by providing shelter for watercraft while loading.

The Hospital

The ships provide a Level 3 hospital with resuscitation, intensive care and high-dependency facilities equal to the best onshore hospitals in Australia.

The hospital provides:

- a Resuscitation room with 6 beds,
- 2 Operating Theatres,
- an Intensive Care Ward with 8 beds,
- a Hospital Ward with 20 medium dependency beds,
- a Temporary Ward with 28 low dependency beds,
- a Dental Surgery, and
- doctors' consultation rooms.

The LHD Combat System comprises the infrastructure and information processing elements of:

The **Combat System:** responsible for ship situational awareness, planning and execution for self-defence and attack, but in the case of the LHD, also responsible for enabling, controlling and monitoring many of the functional capabilities specific to the characteristics of an amphibious-type ship.

The **Communication System:** responsible for internal and external communications, but in the case of the LHD Mission System, also responsible for enabling a large part of the Embarked Force's C4ISTAR capability and a range of Air Traffic Control (ATC) and Watercraft control capabilities.

The LHD will provide the communications bearers, infrastructure and terminals for strategic, operational and tactical communications to Joint Operations Command, Joint Task Force Commander, the Amphibious Task Force, and forces involved in Shaping and Support Operations under the command of the Naval Component Commander (NCC), Air Component Commander (ACC), and Special Operations Component Commander (SOCC).

The systems also provide interfaces and transmissions capabilities to allow interoperability with US and Coalition forces.

The integrated Combat System provides functionality and interfacing in order to provide a number of Mission Capabilities, including:

- Ship Operations: protection and attack, surveillance and reconnaissance, navigation, communications;
- Amphibious Operations: air assault, surface assault, air traffic control, embarked joint force C2; and
- Interoperability: including the development and distribution of a consolidated, real-time tactical picture that encompasses the entire amphibious capability operation on the ship, at sea, in the air and on land, and can simultaneously interact with these local environments and remote civilian and defence command infrastructure.

The SAAB Combat System solution will provide commonality between the LHD and the existing ANZAC CMS. The 9LV Mk3E CMS to be installed on the LHD will be installed across the remaining ANZAC class ships as part of the ASMD Upgrade.

The LHD will have a fully integrated Combat System with the SAAB Combat Management System at the core. Situational awareness will be provided through the various sensors including, Radar, Infra-Red Search and Track, and tactical data links. Information from these sensors will be processed and displayed by the CMS on individual operator consoles and large screen displays.

Command and control for the ship command and embarked force commanders will be provided through extensive communications and data link channels. Four 25mm Typhoon guns that are fully integrated with the Combat Management System will provide self defence against air and surface asymmetric attack. The Nixie decoy system will be a standalone system to provide defence against torpedo attack.

Scope for growth in the combat system will be provided as Space and Weight for the NULKA decoy and the INMARSAT maritime communications system. Other systems that will be Fitted For But Not With are: Communications and Radar ESM, and Common Data Link (CDL), which can provide for video datalinks from UAVs and other compatible sources.

The Combat System will provide capabilities to deliver a comprehensive Communication Plan to undertake an amphibious operation and there will be a significant amount of capability over numerous satellite communications channels. C2 and situational awareness will be enhanced through tactical data links compatible with ADF and allied units including Land units through BCSS and BMS systems to be introduced through Land 125/75 and JP 2072.

LAN connections will be available through external communications and the MTWAN, compatible with SEA 1442, over all security classifications into the wider ADF network.

Embarked force staff will be able to utilise these communications and C2 capabilities to undertake all activities necessary for command of the amphibious operations.

The LHD will provide capacity to embark and support units such as Intel, EW and JICO that are required to support specific operations.

The hulls will be built and partly fitted out in Spain, then brought to Australia by even larger ships. Here, the superstructures will be built and integrated with the hulls at Williamstown. The combat systems will have baseline commonality with the Anzacs and will be installed in Williamstown, where the frigates were built, but will be developed and tested by SAAB in Adelaide. The Tenix presentation will give you the detail.

Working groups in the Defence Materiel Organisation will be developing the equipment, logistic support and training arrangements over the next few years.

The ships are due to commission between 2013 and 2015. As they do, Navy will pay off first *Tobruk* and then either *Kanimbla* or *Manoora*.

What does all this mean for Australian industry? Defence and Tenix will shortly conduct an industry 'roadshow' to explain what we need. However, and without anticipating Tenix's excellent presentation, let me discuss some industry aspects from the Defence point of view.

The value of Australian Industry Involvement in this project is estimated to be about $500 million, or almost 25% of the contract signature value. Of this, the bulk will be taken up by Tenix as the prime and by the main Australian subcontractors. However, they will have their own subcontracts to be let to Australian industry, Tenix especially.

We understand that products Australian companies may be called on to provide include a range of steel, including plate, shaped products and may fittings and equipment; paint and painting; ballistic glass; plumbing supplies; bridge consoles; bulk materials; and services related to set-to-work and test and trials.

The business opportunities might not seem vast, but with the Air Warfare Destroyer running concurrently, Defence demands on Australian industry at a time of near-full employment will be considerable. We urge small-to-medium firms to seize opportunities.

Why do we want Australian industry participation? In short, self interest: we need to maintain these ships in-country through their service lives, which is expected to be up to 40 years. That's 40 years of steady, sustained maintenance, repair, equipment replacement and upgrade, systems and software development, supplies, fuel, logistics systems and so on. It's steady, it's sustained and it will help to sustain the companies we need to support the ships. We are developing our support requirements now and expect to call for the first tenders in about two years.

And it's not small potatoes: the money spent on maintaining a ship in service over its life is typically several times the cost of acquisition.

The schedule from effective date to final acceptance is seven years and three months. This would normally be a short timeframe to build, fit out, test and deliver two large and complex ships from scratch, but we have the advantage of multiple production lines and concurrent work: while work is commencing on the hulls at the existing production line in Spain, the facilities and project arrangements for the other work can be set up here in Australia.

The effective date – the date on which contract performance commenced – was 23 November 2007. The various system design reviews will take a little over 18 months – 20, to be exact. Again, a short time, but that reflects the intensive design work done as part of the risk management strategy, firstly in the Risk Reduction and Design Study before First Pass, and then in the Design Development Activity we conducted before tenders were called for.

The design reviews will cover ship, platform, mission, combat and communications systems, concluding with a detailed design review of the overall mission system. First steel is to be cut next September.

The first hull is to arrive at Williamstown by ship in July 2012 and the second in February 2014. We expect to take delivery of the ships in December 2013 and July 2015, with final acceptance due in February 2015. Concurrently with this program, the land-Based Test Site (LBTS) will be established and tested and is expected to be ready for acceptance by September 2012, three months before the first hull arrives.

There are two related procurements that I refer to as 'companion projects' each being another phase of Joint Project 2048. They are estimated to be worth a total of about $300m to $400m in round terms and 2006 dollars.

The 'closer' companion, Phase 3, aims to procure watercraft to interface with the welldock of the LHD and thereby be able to fill the role of ship-to-shore connector. There are a few requirements that are non-negotiable, including the ability to carry the M1A1 Abrams tank, and the ability to operate in Sea State 4, roughly winds of up to 20 knots and waves of two to 2.5 metres coming every five or six seconds. The value of this acquisition is estimated at $150m to $200m in the current Defence Capability Plan. We released the Requests for Proposal in November and we are about to commence examining responses. We hope to issue a Request for Tender about January 2009 and to recommend a preferred tender to Government about the end of 2009, with delivery about the same time as the LHDs, that is, in the 2012 – 2014 timeframe.

The other companion project is Phase 4C, which will aim to acquire a strategic sea lift vessel. This will enable the ADF to transport bulk equipment, supplies and forces into a theatre of operations and provide significant ongoing support to deployed forces. Strategic sea lift is not really part of an amphibious capability but is complementary to it. The value of this project is estimated at $150m to $200m. It is still early days in the capability development process, with requirements currently being refined. Early planning is for delivery in the 2016 – 2018 timeframe.

A word of caution on the dollar figures for these two projects. They are for planning purposes and are estimates only, based on the information available at the time. They in no way represent a budgetary provision and may change later.

In conclusion, the delivery of HMA Ships *Canberra* and *Adelaide* in around 2013 and 2015 will provide a significant increase in the ADF's amphibious capability. The challenge for the ADF and for industry will be to work together to realise the potential of that capability. For those who wish to contribute to the construction and support of these ships, I would again commend to you the Defence and Tenix Roadshow to commence on 18 February 2008.

Use of Slocum Glider Technology in Coastal and Marine Applications

V. V. Anokhin (GHD Pty Ltd, MIEAust, vadim.anokhin@ghd.com.au),
J. R. Romero (GHD Pty Ltd, jose.romero@ghd.com.au), and
R. Steedman (GHD Pty Ltd, raymond.steedman@ghd.com.au)

ABSTRACT

The paper describes an Autonomous Underwater Vehicle (Slocum glider) and outlines potential applications in coastal and marine environments. The instrument offers a new and reliable method to collect high quality spatial and temporal data for ocean environmental monitoring. Driven in a sawtooth pattern by variable buoyancy, the glider moves both horizontally and vertically, collecting temperature, conductivity and depth data in addition to measurements from a range of sensors that can be placed in the science bay. The capabilities of the glider makes it ideally suited for long-term marine monitoring and oceanographic investigations. Communications with the glider occur via a global satellite system that provides near real-time data transmissions and if needed vehicle control. The glider surfaces to transmit data and to receive new instructions at regular intervals. It can be deployed for periods of 15 to 30 days that generally yields a 600 to 1500 km range. The glider can be used for a wide range of marine applications, which include water quality monitoring, tracking plumes and ocean monitoring during storms. This paper illustrates some of these applications during deployments in the region of a wastewater outlet in coastal waters and the oceanography of the Western Australian continental shelf.

INTRODUCTION

The development of the Autonomus Underwater Vehicles (AUVs) was motivated by a visionary paper of Strommel (1989), who described a network of remotely controlled gliders that "migrate vertically through the ocean by changing ballast, and they can be steered horizontally by gliding on wings at about 35 degrees angle... During brief moments at the surface, they transmit their accumulated data and receive instructions... Their speed is generally about 0.5 knot." Since then, his visionary idea has been implemented in a number of underwater vehicles. The most common are Seagliders (Eriksen et al., 2001) of the University of Washington, Spray gliders of the Scripps Institute of Oceanography (Sherman et al., 2001), and Slocum gliders of the Webb Research Corporation (Jones et al., 2005). An overview of available glider technology is provided by Rudnick et al. (2006).

The long range and duration capabilities of the gliders make them ideally suited for ocean monitoring at the regional scale. The gliders can be programmed to patrol for weeks at a time, surfacing to transmit their data to shore, while new instructions are uploaded at regular intervals. Driven in a sawtooth vertical profile by variable buoyancy, the glider moves both horizontally and vertically, collecting data from an array of sensors. The glider provides a cost effective method of collecting high quality and resolution oceanographic data continuously 24 hours per day.

Underwater gliders have been successfully applied in a number of investigations. These include ocean climate measurements in coastal and deep ocean environments (Jones et al., 2005), long-term autonomous deployment for water quality monitoring (Bachmayer et al.,

2006), monitoring of wastewater, thermal and saline discharges from ocean outfalls (Oceanica, 2007), internal waves and tidal measurements for design and operation of off-shore structures, monitoring of pollutant spills (hydrocarbons), plume propagation and formation water discharges, tracking of dredge turbidity plumes, tracer release studies, deployments during extreme storms and tropical cyclones and many others (Rudnick et al., 2005)

One of the key benefits of a glider technology is reduced offshore safety risk. The exposure to offshore hazards concerning adverse weather, seamanship and boat operations is minimised by reduction in personnel sea hours. Deployment and retrieval of instruments is limited to several hours in coastal applications by a few personnel with relatively small boats.

In this paper we describe initial applications of the Slocum glider for commercial purposes in Australia. The first deployment assessed the functionality of the Slocum glider for monitoring wastewater outfalls based on manufactures' information. The second deployment was to characterise the oceanography of the continental shelf near Rottnest Island, Western Australia.

SLOCUM GLIDER DESCRIPTION

Slocum glider is an autonomous robotic vehicle propelled by changing buoyancy and moving horizontally on wings. The glider changes its displacement with a hydraulic pump, which enables it to glide up and down along a see-saw path at descent (ascent) angles that can exceed 25 degrees. The Slocum glider described here was developed by Webb Research Inc. (Eriksen et al., 2001).

The glider can be operated to a depth of 4-200 meters; the autonomous deployment period can reach 30 days. It is 150 cm long, 21 cm in diameter, weighs 52 kg and has a scientific payload of 5 kg (Fig. 1). The hydraulic pump, which controls glider's buoyancy, can be optimised for shallow (4-30 meters) or deep (20-200 meters) applications. The shallow water pump controls buoyancy more rapidly, which allow improved control of the vehicle, while the deep water pump is more energy efficient. The operation and technical characteristics of Slocum gliders are summarized in Graver et al. (2003).

Communications with the glider occur through high frequency bandwidth radio (operational radius of 4-5 km), two-way Iridium satellite telemetry or one-way Argos transmitter. Once the glider is connected to the control station via the radio or the Iridium satellite network, it can upload collected data and diagnostics parameters or receive new mission instructions. The system allows real-time data downloads that provide improved monitoring capabilities and that enable tracking the location and operational state of the vehicle. Argos transmitter provides information about vehicle's location to aid in recovery.

The glider carries a science bay, which is an exchangeable 5 L payload (Fig. 1). The payload can be configured for a number of sensors, which are managed by an independent computer for scientific data retrieval. The data acquisition module processes and communicates with the main vehicle computer, which in turn, accesses and transmits the information. In its simplest configuration the glider carries Seabird Conductivity, Temperature and Depth sensors (CTD by Seabird Electronics).

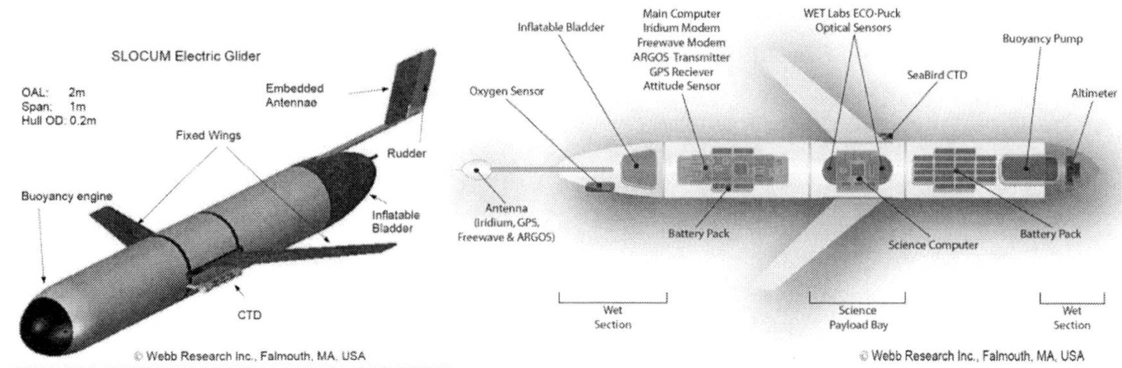

Figure 1 Schematic Diagram of Slocum glider. Copyright Webb Research Inc.

In addition, the following sensors can be included as customized science payloads: NFTU sensor measuring Chlorophyll fluorescence (ex/em 470/695nm) and turbidity, ECO triplet PUCK sensor (fl3slo) measuring Chlorophyll fluorescence, CDOM fluorescence and Rhodamine fluorescence or ECO triplet PUCK sensor (bb3slo) with backscattering measurements at wavelengths 470 (blue), 532 (green) and 660 (red) nm. An Aanderra dissolved oxygen sensor is fitted at the tail end of the glider, measuring oxygen saturations and concentrations. The measurement frequency is approximately 0.5 Hz. Glider sensor specifications are summarised in Table 1.

Table 1 Sensor specifications based on manufacturer information

	Parameter	SI Units	Range	Resolution	Initial accuracy
Seabird Electronics, CDT sensor CBE 49	pressure	m	0 - 300	0.002% of full scale range	0.1% of full scale range
	temperature	^{o}C	-5 to +35	0.0001	0.002
	conductivity	S m-1	0 - 9	0.00005	0.0003
Aanderra oxygen optode 3835	oxygen concentration	µg/L	0 - 16	<1	<8 (or 5%)
	oxygen saturation	%	0 - 120	0.4	<5%
WET Labs, ECO FLNTU Puck (optical)	Chlorophyll-a	µg/L	0.01 - 50	0.01	linearity r^2 ~ 99%
	turbidity	NTU	0.01 - 25	0.01	linearity r^2 ~ 99%
WET Labs, ECO Triplet optical backscatter, wavelength	470nm (blue)	$m^{-1} sr^{-1}$	0.0024 -5	$1.2x10^{-5}$	-
	532nm (green)	$m^{-1} sr^{-1}$	0.0024 –5	$7.7x10^{-6}$	-
	660nm (red)	$m^{-1} sr^{-1}$	0.0024 -5	$3.8x10^{-6}$	-
WET Labs, ECO triplet PUCK optical (fl3slo)	Chlorophyll-a	µg/L	0.01-50	0.01	linearity r^2 ~ 99%
	Phycoerythrin	ppb	0.04-175	0.04	-
	CDOM	ppb	0.18-375	0.18	-

A range of other sensors can be fitted on the glider to suit the specific requirements of the deployment. These include acoustic modems, bathyphotometers, photosynthetically active radiation (PAR) sensors, hyper-spectral spectrometers, radiometers and others.

Depth-averaged currents are estimated by dead reckoning of glider's position from its intended waypoint location and the actual surfacing position. The depth –averaged currents are calculated each time the glider surfaces and obtains a GPS fix. When the glider remains on the surface for an extended period of time, such as during data upload/download phase, it acts as a surface drifter, thus yielding a rough estimate of surface layer velocities.

The glider is operated via the Dockserver developed at Ruttgers University (Jones et. al 2005). Low-density data, vehicle diagnostics and position coordinates of the glider are transmitted via Iridium and Argos satellite telemetry at regular time intervals. This low-density data is processed nearly in real-time and posted on the Internet, which allows tracking of the vehicle flight progress and incoming data quality.

GLIDER APPLICATION

Monitoring wastewater outfall in shallow waters

A deployment to assess the functionality of the Slocum glider was conducted by monitoring the region in the vicinity of wastewater outlets in 8-15 meters depth off of Perth.

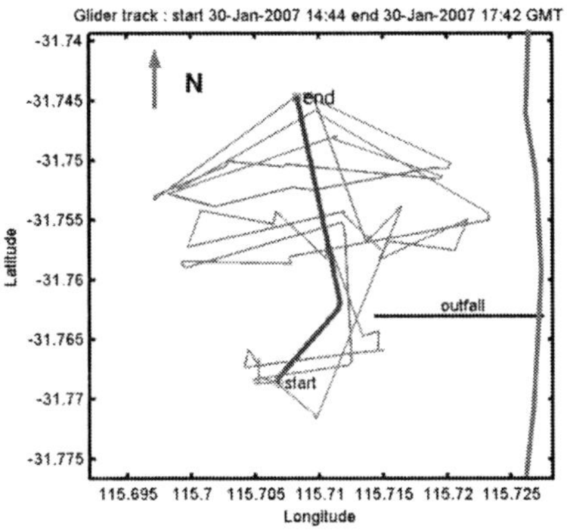

Figure 2 Location of the deployment site and selected glider transect

The dimensions of the study area were approximately 3 km (along-shore) by 2.5 km (cross-shore) (Fig. 2). The outlets diffusers are located approximately 1.5 km from the shoreline. The outlets are located in an approximate depth of 10 m. The average depth of the area is approximately 12 m. The sediment composition consists of fine sands with occasional seagrass meadows. Tidal velocities in this area are small occasionally exceeding 0.1 m s^{-1} though wind-induced velocities can be substantially greater.

The glider was deployed on 30 January at 10:38 and retrieved approximately 2 days later on 1 February at 12:54 Perth local time. Over the deployment period the tidal range was approximately 0.75 m with low tide in the early mornings and high tide in the early evenings. Meteorological conditions were characterised by moderate summer air temperatures (20-27°C). Westerly to southerly winds were relatively low (10-15 km/hr) throughout most of the trial.

During the trial the vehicle was configured with an altimeter, conductivity, temperature, pressure, turbidity, chlorophyll-a (chl-a, via fluorescence) and dissolved oxygen sensors. A total of 500 valid science profiles (downcast only) were collected over the two days of the experiment. The glider flight path was over 30 km in length with regular satellite transmittal of the vehicle's position and low density data to allow operators to track location and monitoring in a real-time manner. During the trial the glider completed two circuits (1 circuit per day) of cross-shore and along-shore transects to map the extent of the plume with salinity and the effects of the plume on algal standing stocks with the fluorometer.

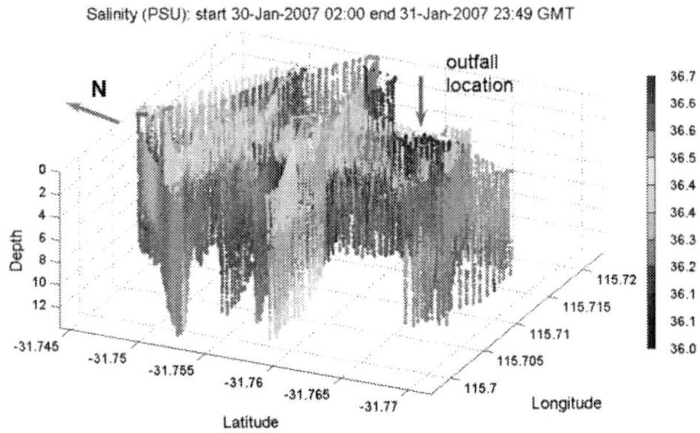

Figure 3 Three-dimensional plot of salinity over the two days of deployment.

The three-dimensional plot of salinity clearly shows the extent of the treated wastewater plume (Fig. 3) Because of the high accuracy of the temperature and conductivity sensors, the derived salinity profiles clearly showed the vertical and horizontal extent of the outlet waters even though the total salinity range during the mission was only ca. 0.5 PSU. The width of the effluent plume was approximately 1.5 km; a sharp salinity front to the south of the outlet highlights the effect of the southerly winds on northerly transport of the treated wastewater.

Figure 2 shows one of the south to north transects collected on January 30-31 between 23:44 and 02:42 Perth local time. The glider flew near the outlet, approximately 700-800 m from the start of the transect. The glider drifted at the surface while transmitting data from 750-900 m over which no vertical profiles were collected as shown in Fig. 5.

Two water bodies are clearly identifiable on the T–S diagram (Fig. 4). The salinity of the weakly stratified coastal waters averaged 36.6 PSU (blue crosses). In contrast the plume waters were confined to the upper 4 m with a salinity range of 36 to 36.5 PSU (red circles).

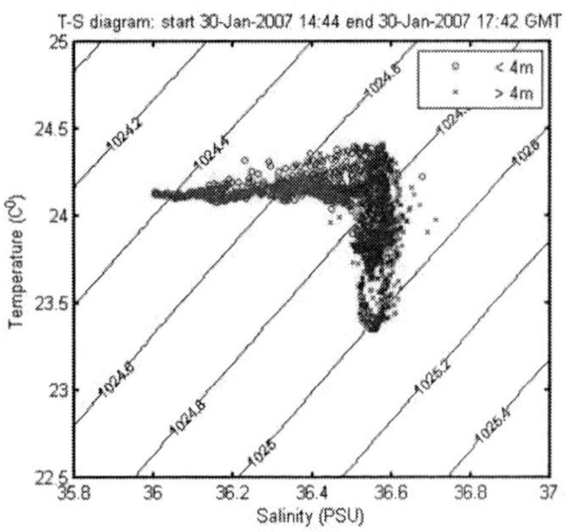

Figure 4 Temperature-salinity diagram for south to north transect with density contours in the background.

During this transect the combination of cooling of the surface waters along with stable salinity stratification led to a mid-depth temperature maxima between 2-6 m. (Fig. 5a). The outlet plume was generally confined to the upper 4 m of the water column as seen on salinity and density signal in Fig. 5b and Fig. 5c. Spatial variations in salinity (horizontally and vertically) decreased substantially with distance from the diffuser locations with only a 0.1 PSU decrease in the ambient salinity at a distance of several kilometres from the outlet (Fig. 5b).

A mid-depth chl-a maximum was recorded between 4-6 m just below the plume waters, although the highest chl-a levels were observed in the bottom waters between 8-12 m depth near the sediment-water interface (Fig. 5d). These cold waters had salinity signatures of the ambient marine receiving waters (i.e. 36.6 kg m^{-3}). Potential mechanisms of these elevated chl-a levels near the sediment-water interface include sedimentation of planktonic algae, transport from other regions and resuspension of benthic phytoplankton. The mid-depth chl-a maxima often recorded during strong stratification may have been caused by similar mechanisms. For example, the sedimentation of planktonic algal cells could accumulate at the density discontinuity (i.e. the thermocline, halocline, pycnocline) prior to settling to the bottom waters. Alternatively, resuspension of benthic phytoplankton in shallower waters and subsequent transport to deeper waters could also cause these observed patterns.

Turbidity levels were highest at 1 m above the sediment-water interface, and rapidly decreased to limit of detection levels several metres above the bottom (Fig. 5e). Presumably, resuspension of bottom sediments was the primary mechanism of this consistent feature. The entire water column was well oxygenated and did not exhibit substantial vertical or horizontal spatial variability; dissolved oxygen levels ranged from ca. 8-9 mg L^{-1} (Fig. 5f).

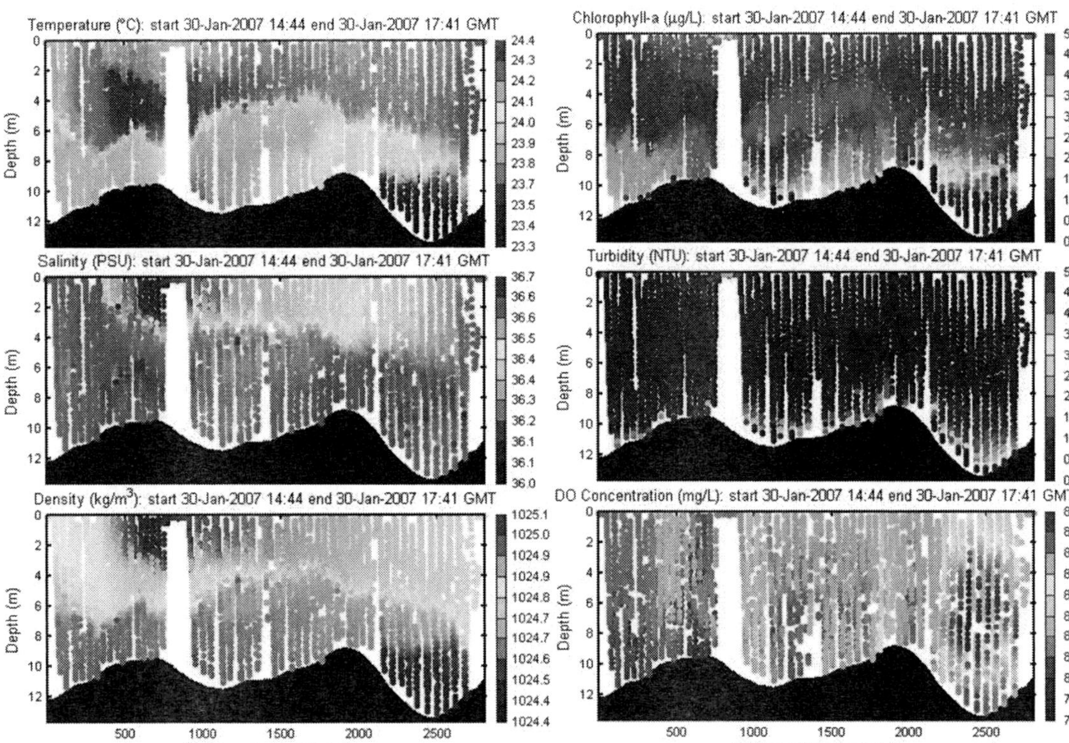

Figure 5 Two-dimensional transects of temperature, salinity and density (top to bottom, left plots) and chl-a fluorescence, turbidity and dissolved oxygen concentration, (right plots). Data are not gridded, individual point measurements are shown. The outfall is located approximately 700-800 m from the start of the transect.

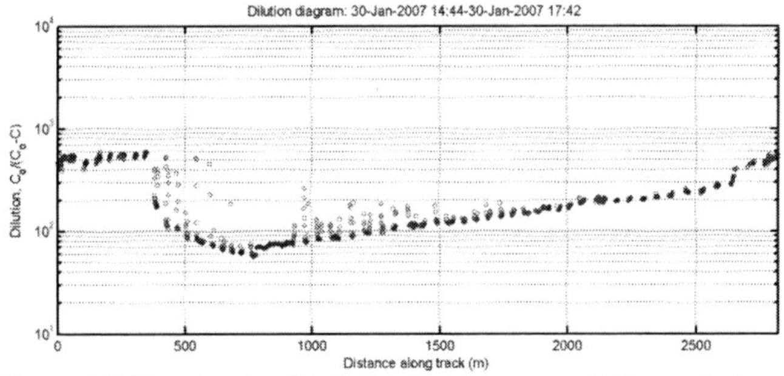

Figure 6 Effluent water dilution estimates versus distance for transect from shown in Fig. 2. The outfall is located approximately 700-800 m from the start of the transect.

Estimates of the dilution of the treated wastewater were computed for the surface waters from the glider data to characterise cross- and along-shore length scales over the trial as follows:

$$Dilution = \frac{C_0}{C_0 - C} \qquad (1)$$

where C_0 is the ambient salinity of the coastal waters during the trial (i.e. 36.6 PSU) and C is the average salinity of the upper 3 m of the water column. Dilution estimates were limited to the surface waters (i.e. upper 3 m) where the treated wastewater had a substantial effect on salinity. Dilution estimates greater than 500 were classified as background ambient salinity.

The dilution estimates indicate that approximately 2.5 km to the north of the outlet the plume had dispersed to values (i.e. 500) (Fig. 6). A sharp front to the south of the outlet is readily identifiable from the dilution estimates, which again highlights the effect of the southerly winds on northern transport over the monitoring period.

Monitoring continental shelf dynamics

On 17 January 2006 the glider was deployed off Perth's coast to examine three-dimensional flow patterns around Rottnest Island, Western Australia. Over 22 days the glider traversed a total of 485 km and collected over 2420 vertical profiles (Fig. 7). During the deployment the glider was remotely controlled from Rutgers University, USA.
Over the deployment period the tidal range was approximately 0.5 m with low tide in the early mornings and high tide in the early evenings. Meteorological conditions were characterised by summer air temperatures (22-33°C) and strong westerly sea breezes in late afternoons.

The full glider path with estimates of depth-averaged currents is shown in Fig. 7. Two regions are clearly identifiable from the velocity estimates. Shallow coastal waters to the east of Rottnest Island where wind-induced currents led to northerly transport under the influence of diurnal sea breezes. In contrast, the path to the west of Rottnest Island over the continental shelf captured the southerly Leeuwin Current.

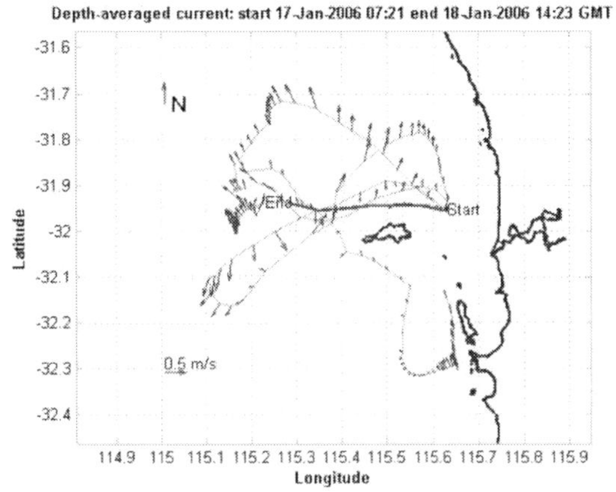

Figure 7 Location of the deployment site and glider track. Estimates of depth-averaged velocities are shown as green arrows.

The selected transect spans over 30 hours from afternoon of 17 January 2006 to late evening of 18 January 2006 (Fig. 7, red line). The glider went in a straight line over the continental shelf from 25 m of waters to a depth of 120 m to the north of Rottnest Island (Fig. 7). The instrument was equipped with CTD and optical ECO backscatter sensors; the glider's deployment depth was limited to 150 m.

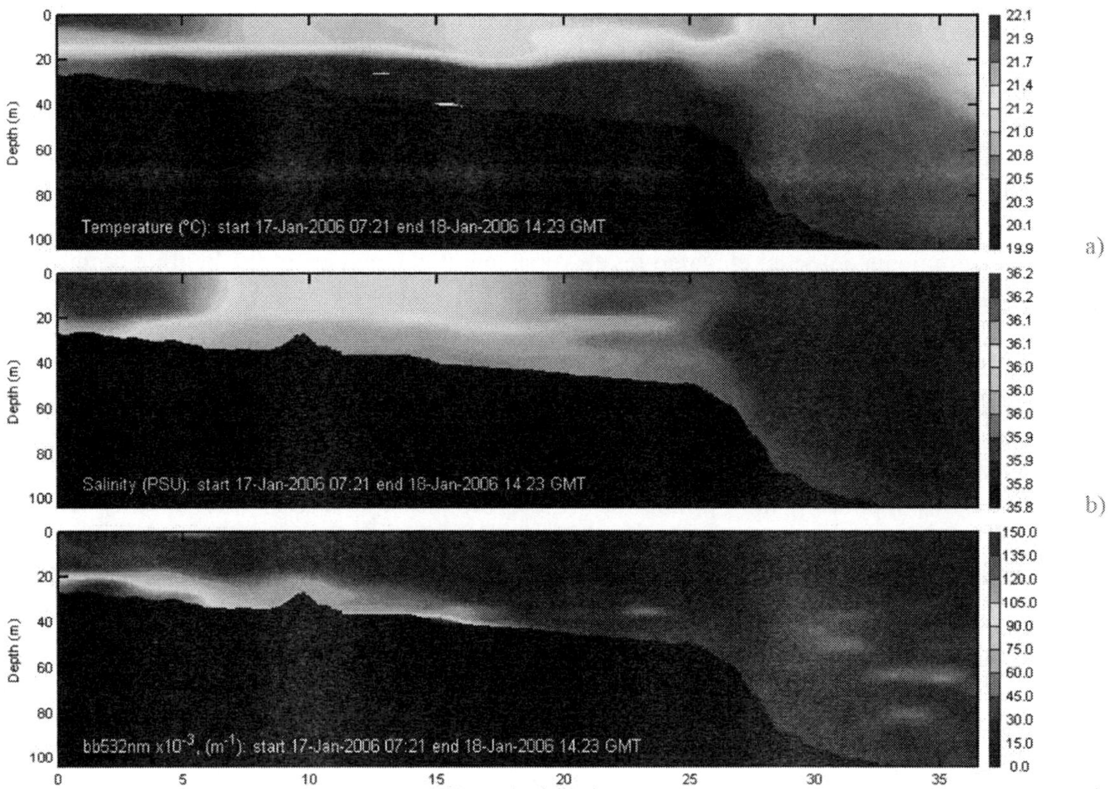

Figure 8 Two-dimensional fields of temperature, salinity, and optical backscatter (wavelength 532 nm) collected by the glider over 37 km E-W transect (Fig. 7).

The water was stratified with a sharp halocline between 15-18 m in the shallow regions and linear stratification over the deeper waters (Fig. 8a). Warmer saltier waters were observed in the near-shore region (Fig. 8b), possibly from elevated evaporation in shallow waters. The salinity profile clearly illustrates a density-driven underflow that cascades down the continental shelf. Further, internal waves are clear identified within the thermocline between 15-20 m depth. The warmer waters of the Leeuwin Current were identifiable on the western portion of the transect.

Elevated concentrations of suspended particles were detected by the backscatter sensors in the benthic boundary layer (Fig. 8c). Optical backscatter measurements can serve as a proxy for suspended particles (Sullivan et al 2004). The process causing the formation of turbid layer is unknown, but can be triggered by the resuspension of particles by wind-induced currents or near-bottom currents associated with internal wave activity.

CONCLUSION

Slocum glider is an autonomous underwater vehicle designed for the operation in coastal waters. Versatile and maneuverable the glider can be deployed over a depth range of 4-200 meters for 15 to 30 days with a 600 to 1500 km range. Its flexible payload allows it to carry customized sensors via a science bay with an exchangeable 5L payload capacity.

The glider was deployed in shallow waters (4-15 m) off Perth's coast to characterise a treated wastewater plume. The glider measured a 0.5 PSU decrease in the ambient salinity near the diffuser outlet, which rapidly decreased to a 0.1 PSU difference within several kilometres. Elevated chl-a and turbidity were detected near the bottom. Estimates of the dilution were estimated for the surface waters with the glider data to characterise the cross- and along-shore horizontal length scales of the treated wastewater plume.

Deep water glider deployment off Western Australian continental shelf identified two distinct regions: salty warm waters in the shallow coastal zone and the warm Leeuwin current over the continental shelf to the west of Rottnest Island. Optical backscatter measurements detected elevated levels of suspended particles in the benthic boundary layer likely from wind-induced currents or internal wave activity.

ACKNOWLEDGEMENTS

The authors thank Peter Kelly from AUV Pty Ltd for providing the data from Rottnest Island deployment. We also acknowledge Water Corporation, Perth, through Oceanica Pty Ltd, who supported the application of the Slocum glider for water quality monitoring. AUV Pty Ltd finance and develop the new Seaglider systems. GHD Pty Ltd market and apply the new Slocum glider technology on the Australian market.

REFERENCES

Alaee J.M., Pattiaratchi C.B. and Ivey G. (2007) "Numerical simulation of the wake of Rottnest Island, Western Australia". Dynamics of Atmospheres and Oceans. (in press)

Bachmayer, R., B. deYoung, C. Williams, C. Bishop, C. Knapp and J. Foley. (2006) "Development and Deployment of Ocean Gliders on the Newfoundland Shelf" Proceedings of Underwater Vehicle Symposium

Eriksen, C. C., T. J. Osse, R. D. Light, T. Wen, T. W. Lehman, P. L. Sabin, J. W. Ballard and A. M. Chiodi (2001) "Seaglider: A longrange autonomous underwater vehicle for oceanographic research" IEEE J Oceanic Eng. 26:424-436.

Fiorelli, E. Leonard, N.E. Bhatta, P. Paley, D. Bachmayer, R. Fratantoni, D.M. (2004) "Multi-AUV control and adaptive sampling in Monterey Bay". Proc. IEEE Autonomous Underwater Vehicles 2004: Workshop on Multiple AUV Operations (AUV04), June 2004

Graver J. G, Bachmayer R., Ehrich Leonard N., and Fratantoni D. M. (2003) "Underwater Glider Model Parameter Identification" Proc. 13th Int. Symp. on Unmanned Untethered Submersible Tech. (UUST), Durham NH August 2003

Jones, C., E. Creed, S. Glenn, J. Kerfoot, J. Kohut, C. Mudgal, O. Schofield (2005) "Slocum Gliders - A Component of Operational Oceanography". Proc. Autonomous Undersea Systems Institute Symposium 2005.

Oceanica (2006) "2006 Summer Water Quality Survey Compliance Report", internal report for Water Corporation, Perth.

Rudnick D. L., Davis R. E., Eriksen C. C., Fratantoni D. M. and Perry M. J., (2004) "Underwater Gliders for Ocean Research" Mar. Tech. Soc. Journal, 38 (1), 48–59

Sherman, J., R. E. Davis, W. B. Owens, and J. Valdes (2001) "The autonomous underwater glider 'Spray'." IEEE J Oceanic Eng. 26:437-446.

Stommel, H. (1989) "The Slocum mission" Oceanography, (2):22–25

Sullivan, J.M, Twardowski, M.S., Donaghay, P.L, and Freeman S.A. (2004) "Use of optical scattering to discriminate particle types in coastal waters" Appl. Optics, 44 (9), 1667-1680.

Webb D. C., Simonetti P. J. and Jones C. P. (2001) "SLOCUM: An underwater glider propelled by environmental energy" IEEE Journal of Oceanic Engineering 26 (4), 447–452

Coming to Terms with the New SOLAS Probabilistic Sub-Division and Damage Stability Provisions

Rob Gehling
Australian Maritime Safety Authority, Canberra, ACT, Australia
rob.gehling@amsa.gov.au

ABSTRACT

The amendments to Chapter II-1 of SOLAS that enter force on 1 January 2009 include harmonized requirements for the damage stability of both passenger ships and cargo ships based on probabilistic principles.

Although probabilistic damage stability has been reflected in some IMO instruments for many years, the finalization of these amendments marks a point in time from which naval architects and the general shipping and shipbuilding industries need to be familiar with the probabilistic method if they are to be involved with new ships covered by SOLAS standards. In particular, all concerned will need to be aware not only of the time and effort required to design a ship to meet these new SOLAS requirements but also of the flexibility afforded designers in comparison with the traditional sub-division by means of transverse bulkheads.

A brief history of the use of the probabilistic method in IMO instruments is given, including the process followed in the development of the subject SOLAS amendments and the associated explanatory notes.

The SOLAS probabilistic damage stability calculation method is described in broad terms, together with details of the limits of the application of the method to vessels covered by IMO safety instruments other than SOLAS.

INTRODUCTION

This paper has been prepared with the intention of focusing attention on the probabilistic approach to damage stability adopted by IMO and, in doing this also providing information on the considerations given to this subject over a long period of development. It will be seen that the latest SOLAS regulations that mandatorily apply probabilistic damage stability methods to both passenger ships and cargo ships have evolved through careful analysis of damage data and that they constitute a rational method for an overall assessment of the damage stability of a ship that allows the designer substantial flexibility in the ship's arrangement.

While many of the pertinent IMO documents are referenced, it would have been impractical in a paper of this length to provide a more complete list of documents considered by IMO during the development process. This paper should therefore be regarded as providing an overview of the process and any more detailed examination should refer to the individual documents considered at each IMO session.

In December 2005, following a development process that began in 1990, IMO adopted amendments to SOLAS Chapter II-1[1] through which a harmonized set of probabilistically-based damage stability rules were introduced to cover cargo ships of 80 metres length and over and all passenger ships to which SOLAS Chapter II-1 applies. These amendments, which have subsequently been re-adopted[2] to facilitate incorporation of other amendments, will apply to ships of which the keels are laid on or after 1 January 2009.

The process commenced in 1990 and completed in 2005 covered only the latest phase of this development work, which had its roots in the following recommendation 6 of the 1960 SOLAS Conference[3]:

> *The Conference has considered carefully the question of the watertight subdivision of passenger ships in light of the results achieved since the International Convention for the Safety of Life at Sea, 1948, came into force, and has agreed on certain additional requirements calculated to secure greater safety. It recognizes, however, that the questions of watertight subdivision and stability deserve further study which the limited time available to the present Conference did not permit, and accordingly recommends that the Organization (IMCO, renamed IMO in 1982) should at the earliest practicable date initiate further studies of watertight subdivision on the basis of proposals which any participating Government may submit, including proposals submitted to the Conference. The objective should be to review the existing criteria of subdivision, stability and damage and to consider the relative merits of these criteria in comparison with other possible criteria from the point of view of safety and practicability.*

Recommendation 8 of the Conference further stated, with regard to cargo ships:

> *The Conference, having considered proposals made by certain Governments to include in the present Convention regulations concerning the subdivision and damage stability of cargo ships, and having regard to studies which the Organization is already undertaking on the subject of tonnage measurement, recommends that the Organization should, at an early date, initiate studies on the extent to which it would be reasonable and practicable to apply subdivision*

and damage stability requirements to cargo ships, taking into consideration the proposals referred to above and any other material which interested Governments may make available, and having as its aim the formulation of such international standards as may appear necessary.

While the impetus to develop the probabilistic approach to subdivision and damage stability is generally attributed to Wendel's 1960 paper[4], examination of the IMO documents indicates that the development was conducted independently over many years as described below. In recounting this history, of necessity only the primary submissions and decisions are listed, as this paper would expand almost infinitely if a full account of the IMO work on the subject were to be attempted.

Outline of the new SOLAS Regulations

In essence the regulations establish for any ship design a required subdivision index "R" on the basis of length and passenger numbers (passenger ships only) that is to be met by the attained subdivision index "A" detailed in the regulations.

The attained index is weighted according to percentages 40, 20 and 40 respectively in relation to the deepest subdivision draught, light service draught and an intermediate "partial subdivision draught".

For each draught, all possible damage cases involving one, two, three or more compartments/zones are analysed and assigned probabilities of the damage being limited to those compartments (p) and of survival after flooding the compartments (s). These probabilities are multiplied together for each flooding case and the products obtained are summed to give an attained index for each draught. Figures 1 and 2 below, which are taken from the interim explanatory notes relating to the regulations, respectively illustrate the division of a ship's length into zones and the consideration of assessment of one or more zones in a damage case.

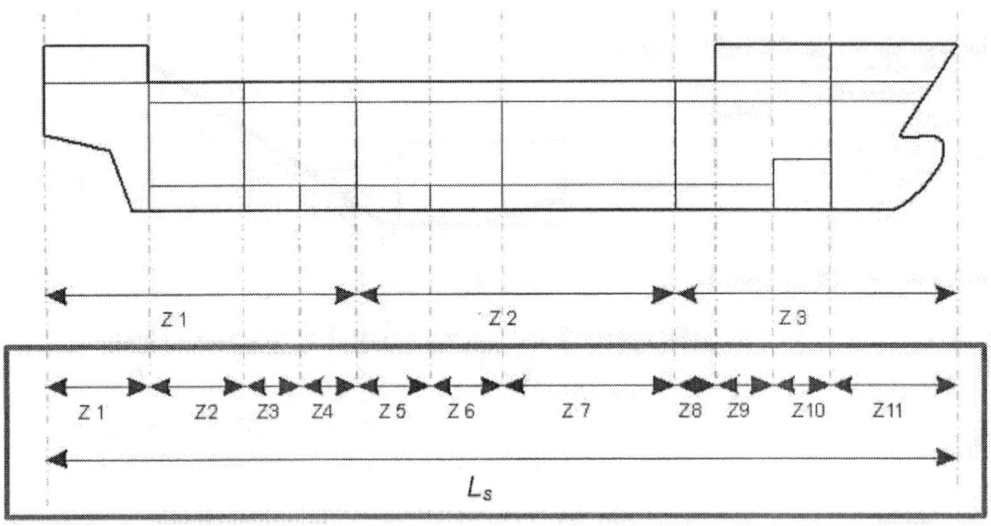

Figure 1 – Subdivision arrangement of example ship

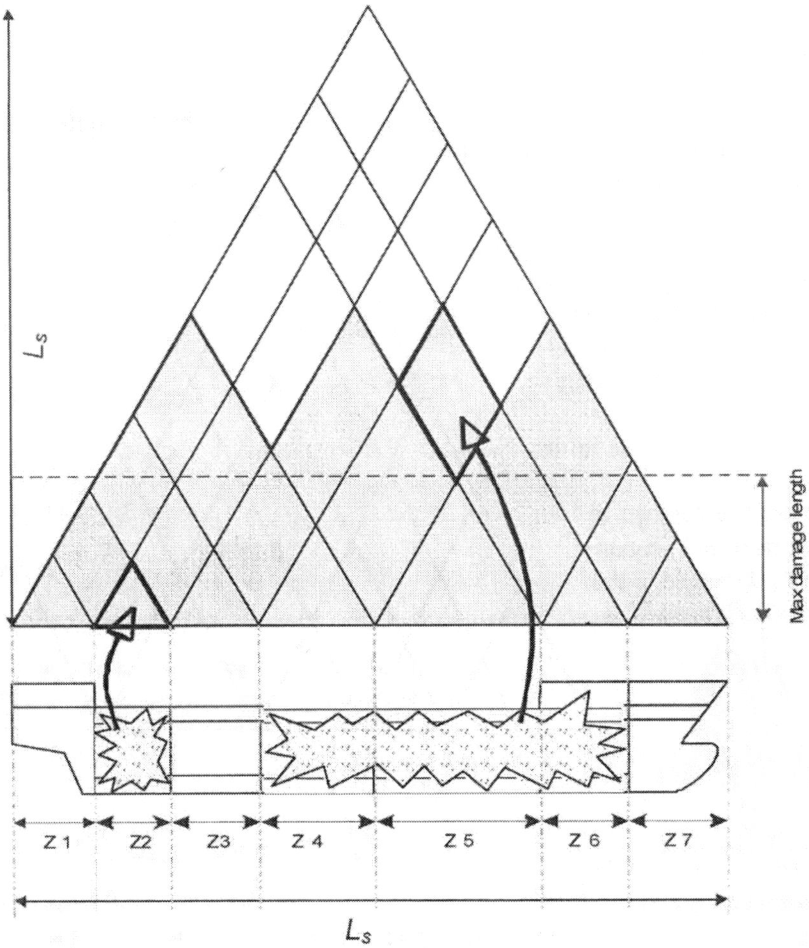

Figure 2 – Damage conditions to be taken into account

Although the calculation method is harmonized for passenger ships and cargo ships, the following differences in details of the analysis apply to cargo ships relative to passenger ships:

- Less onerous criteria for acceptance of partial attained indices at the different draughts (reg. II-1/6.1);
- Different scale for required subdivision index R (reg. II-1/6.2);
- Survival factor not applied to intermediate flooding and heeling moment cases (reg. II-1/7-2.2 and .4);
- Horizontal evacuation routes on bulkhead deck not subject to immersion criterion (reg. II-1/7-2.5.2.2); and
- Special passenger ship requirements not applicable (regs. II-1/8 and 8-1).

Accordingly the "R" and "A" values for a cargo ship are not directly comparable with those for a passenger ship.

PASSENGER SHIPS

Development of regulations for passenger ships reflecting the probabilistic damage stability concept

The SOLAS'60 Conference recommendations were taken up as major agenda items by the very first session of the Sub-Committee on Subdivision and Stability Problems (STAB) that was held in May 1962[5]. Concentrating on passenger ships as being the area of highest priority, a Working Group on Watertight Subdivision and Damage Stability of Passenger and Cargo Ships[6] met shortly before the Sub-Committee's second session in March 1963 and set about the collection of casualty data for use in developing probabilities of:

- size of damage in relation to the size of ship;
- location of damage in relation to length of ship; and
- prevailing weather conditions and sea location of the collision.

Together with parallel data to be collected on damage stability incidents, the above-mentioned data was to be analysed to determine, among other things, the probability of survival in a passenger ship damage stability incident.
Analysis of this damage information continued until the fourth meeting of the Working Group on Watertight Subdivision and Damage Stability of Passenger and Cargo Ships (SDS) in May of 1965, at which Germany presented an initial analysis[7] of damage cards submitted to date. Further detailed analysis continued after the session in parallel with consideration of the probability concept that might be based upon this data, constituting "improved subdivision rules of relatively simple form, but none the less taking account of significant physical parameters related to the effectiveness of subdivision and damage stability"[8].

It is noteworthy that the SDS working group, at its fourth session, received a report from the United States on analysis of wind and wave data reported on the damage cards submitted to IMO[9]. This investigation indicated that "the majority of collisions take place at a low sea state and that high waves are not very probable. On the other hand, high wind velocities appear to be somewhat more likely. Both of these circumstances are considered to be consistent with the fact that most collisions take place in harbour approach areas". The working group, and subsequently the STAB Sub-Committee[10], agreed that this analysis could be appropriately considered in subsequent probability studies.

A first draft of the new subdivision regulations that resulted from these studies was submitted in 1967 to the sixth session of the STAB Sub-Committee by Norway[11]. Comments made at that session resulted in improvements to that text being jointly submitted to the Sub-Committee's May 1968 session by Norway, the United States and the Soviet Union[12]. A further improved proposal, including draft explanatory notes, was submitted to STAB's ninth session in January 1969 by a group comprising the Federal Republic of Germany, Norway, the Soviet Union, United Kingdom and United States[13]. This text clearly outlines the fundamentals of the probabilistic method as being an "attained index" comprising the sum, for all possible extents of damage in defined loading conditions, of the product of probabilities for:

(a) damage to occur in a given longitudinal location (location);
(b) damage being confined to a particular compartment or group of compartments (extent); and
(c) the ship surviving that damage (stability).

Those fundamental principles have remained valid for all subsequent improvements. Following consideration of this joint proposal, draft replacement text for regulation 5 and other consequential amendments to SOLAS'60 Chapter II were appended to the report of that session[14]. Using the draft text, Sub-Committee members were requested to undertake calculations of the attained index A for consideration at the next session, particularly in relation to setting the required index R.

It is noteworthy, however, that the concept of subdivision provided by longitudinal bulkheads as raised in a Greek paper[15] was rejected by STAB IX as such designs "could not be considered as complying with the 1960 Safety Convention". This is illustrated by the fact that the draft text in IMCO 1969 contains in sub-regulation 5(d) the words "the attained subdivision index A shall be determined taking into account the placing of main transverse bulkheads, the floodable length" and so on.

A further improved draft[16] and associated explanatory notes were agreed by STAB X as a basis for test calculations and further development over the ensuing two sessions before being considered by the Maritime Safety Committee at its 71st session. It is noteworthy in relation to the above-mentioned point made by Greece that this improved text includes at sub-regulation 6(d) provision for longitudinal bulkheads to be taken into account.

But, far from STAB XI producing the desired confirmation of the text from the previous session, use of that text in test calculations resulted in a large number of papers to STAB XI, covering subjects including application of the proposed regulations, elimination of the link between subdivision and lifeboatage and longitudinal subdivision, among others. Disagreement with regard to the probability distribution functions (PDFs) for extent and stability is also apparent from the report[17], while the late submission of new proposals did not permit their consideration at the session. Discussion of these subjects was obviously rather highly-charged, as it appears to have resulted in the resignation of the Chairman towards the end of the meeting. In consequence, little progress was made.

Progress resumed with consideration of these issues continued at STAB XII[18], particularly in relation to the stability factor and defining the ships to be used for test calculations. However, despite the remaining differences of opinion regarding calculation of the stability factor, a revised draft set of SOLAS'60 amendments was produced for use in test calculations including a tentative formulation for the stability factor. It is of interest to note that these amendments provide for a residual GM requirement of 0.05 metre. This work continued inter-sessionally, at STAB XIII and through an *ad hoc* group that met in The Hague two months after STAB XIII.

The 26th session of the Maritime Safety Committee (MSC) expressed the view[19] that efforts should be made to finalise the work on this subject at STAB XIV. Notwithstanding the fact that the probabilistic regulations had until then been prepared as amendments to SOLAS'60, the Sub-Committee succeeded in completing them as instructed in the form of "an equivalent and total alternative to the requirements of Part B of Chapter II" of SOLAS'60 for passenger ships[20]. This text was submitted for consideration by the MSC at its 27th session in late March 1973 with a view to adoption by the 8th Assembly. In addition, as a guide to the innovative approach of the regulations and to assure their uniform implementation, STAB XIV agreed to develop explanatory notes which would be finalized by an ad hoc expert group meeting in Leningrad before the 8th Assembly.

Bearing in mind the previously-mentioned disagreement at STAB XI, the dissenting note expressed by Denmark at STAB XIV[21] is informative with regard to the circumstances in which these equivalent regulations were finalized. Denmark had submitted to STAB XIV its own alternative proposal for the regulations, but subsequently withdrew them while expressing concern that the draft developed by STAB did not "simplify the subdivision concept and tends to upgrade safety standards over and above what is required by the 1969 Safety Convention".

Notwithstanding the Danish position, the 8th session of the Assembly adopted the regulations[22] and issued the associated explanatory notes[23]. A footnote to Part B of Chapter II-1 in the new SOLAS Convention adopted in 1974[24] gives similar effect to the equivalent regulations under that Convention.

Some supplementary explanatory notes were subsequently agreed by STAB at its twenty-second session and issued by MSC 40[25].

The Sub-Committee on Stability and Load lines and on Fishing Vessels Safety (SLF), at its second session after being formed by amalgamation of STAB and the Sub-Committee on Safety of Fishing Vessels, commenced the task of reviewing experience with the equivalent regulations and reached the general conclusion[26] that the probabilistic standards were more severe than the SOLAS regulations to which they equated. Arising from this re-examination was consideration by SLF 30[27] and SLF 31[28] of the possible need to amend the equivalent regulations, resulting in the conclusion that amendment would be premature.

The "Herald of Free Enterprise" amendments to SOLAS

The sinking of the Herald of Free Enterprise on 6th March 1987 off Zeebrugge resulted in IMO's fifteenth Assembly resolved[29] as a matter of priority to develop SOLAS amendments to enhance the safety of ro-ro passenger ferries. It is noteworthy that, in addition to the amendments to SOLAS that are detailed below, this incident gave rise to what in time became the ISM Code[30].

Work on the task of improving SOLAS requirements for ro-ro passenger ferries commenced at an extended 55th session of MSC in April 1988.[31] This resulted in two sets of SOLAS amendments dealing with design and equipment issues but also including the so-called SOLAS 90 damage stability standards for passenger ships in the latter set of amendments, which entered into force on 29 April 1990[32,33].

The 58th session of MSC in May 1990 saw proposals regarding damage stability of existing ro-ro passenger ships starting to filter through as a result of a research programme by the United Kingdom, leading to a proposal to retrospectively apply SOLAS 90 standards to existing ro-ro passenger ships.[34] Consideration of the matter by the 35th session of SLF in February 1991 led to an alternative approach being developed, under which the subdivision of these ships would be assessed against a modified version of the original probabilistic standard, Res. A.265(VIII), using the so-called A/A_{max} criterion to determine which ships were in most urgent need of upgrading. This approach was endorsed by MSC 59 and further developed together with a phase-in scheme by SLF 36 in February 1992, resulting in its adoption by MSC 60 in April 1992[35] for entry into force on 1 October 1994 with phase-in dates extending up to 1 October 2005. The A/A_{max} standard itself is contained in an MSC Circular issued by MSC 59 in June 1991[36] and is subject to the interpretations issued by MSC in 1994[37].

Harmonised damage stability requirements – passenger ship perspective

At the start of this paper, mention was made of the most recent amendments to SOLAS (IMO 2005, IMO 2007), in which harmonised regulations for subdivision and damage stability of both cargo and passenger ships replaced the previous provisions of SOLAS Chapter II-1, following a development process that commenced in 1990. The superseded provisions included floodable length calculations. Since the new provisions were previously summarised there is no need to detail them here.

However it should be noted that the time taken to complete the development of these regulations was extended to allow for them to take account of the outcome of the "Harmonisation of Rules and Design Rationale" (HARDER) project, funded by the European Commission. Although conducted outside of IMO, the progress and outcomes of HARDER were reported to IMO in various papers. The first of these was a submission to MSC 72 in May 2000[38] advising of the start of the project and requesting corresponding amendment of the target completion dates for SLF's harmonisation work. An overview of the project was provided in a newsletter issued by the project coordinator in June 2004[39], while the outcome of the project was reported in a series of submissions to SLF 46[40] in September 2003.

In attempting to incorporate the outcome of HARDER, and bearing in mind that at the same session it was considering for the first time the improvement of safety of large passenger ships, SLF 46[41] found that a conflict existed between the improvement of safety of large passenger ships and its instructions to establish R values corresponding to existing SOLAS provisions. Specifically, current passenger ship designs exhibited a declining value of index A with increasing length, which it considered to be an unsatisfactory feature. This matter was corrected at SLF 47[42] and the resulting SOLAS amendments forwarded to MSC 79 for approval with a view to adoption, under protest by some interests who considered the provisions unduly onerous for cruise ships. To resolve these concerns, MSC 79 agreed to hold an inter-sessional working group meeting in February 2005 so that the group's recommendations could be taken into account when the amendments were adopted by MSC 80 in May 2005. In view of their effect upon ship design and construction and the need to resolve some peripheral issues with further SOLAS amendments, the subdivision amendments had attached to them a relatively long lead time before their entry into force on 1 January 2009.

In the author's view, the new SOLAS provisions reflect not only a rational and science-based approach to damage stability of passenger ships, but also a substantial improvement in safety over the regulations that they supersede.

This improvement is reflected in the fact that the regulations were being developed in parallel with consideration of broad improvements to the safety of "large passenger ships" (noting that the word "large" was dropped in the latter stages) by the various technical sub-committees of IMO. Bearing in mind the improved safety afforded by the probabilistic method and the fact that the required index "R" for larger ships had been set at a value greater than that corresponding to the trend line for existing designs, the resulting SOLAS amendments for improvement of passenger ship safety simply re-adopted the existing subdivision amendments.

Contemporary developments in passenger ship subdivision

In preparing a near-final text of the requirements of the amended SOLAS, SLF 47 produced text for the regulations[43] that does not include provision for accounting for water-on-deck, despite this feature being included in the text developed by the specialist working group at that session[44]. The author has since learned of significant concerns in this regard, particularly on the part of leading United Kingdom researchers into damage stability of ro-ro passenger ships. Whether these concerns lead to SOLAS amendments will no doubt be seen during 2008.

CARGO SHIPS

Development of regulations for cargo ships reflecting the probabilistic damage stability concept

With the adoption by the eighth Assembly of the equivalent regulations for passenger ships, the 16th session of the STAB Sub-Committee invited submissions[45] on extending to cargo ships the probabilistic approach to damage stability.

As an aside, students of IMO Conventions would be interested to know that STAB XIII[46] considered a possible basis for amalgamation of the 1960 SOLAS and 1966 Load Line Conventions in accordance with the recommendation of the latter Conference[47]. However, STAB XVIII concluded that coordination of such an amalgamation would not be a simple process. Noting that (deterministic) provisions for subdivision and damage stability had subsequently been included in the 1973 MARPOL Convention[48] and the Codes for Chemical Tankers[49] and Gas Carriers[50], the 33rd session of the Maritime Safety Committee decided that consideration of the harmonization of subdivision and stability requirements in the various conventions and codes should be done as a first step before commencing amalgamation of the conventions. Statements over the past year by the Chairman of IMO's Marine Environment Protection Committee give this proposal some contemporary currency but may need to be considered with the previous conclusion in mind. But I digress.

When considering the general question of possible harmonization of subdivision and damage stability requirements across various conventions and codes, the nineteenth session of STAB[51] concluded that harmonization does not necessarily mean unification of numerical values but rather that attention should be paid to the different objectives, approaches and related criteria associated with particular numerical values. However, it is apparent from the report of the STAB Sub-Committee's twenty-first session[52] that the harmonization being contemplated at that stage was, as suggested in a submission by the United States[53], deterministic rather than probabilistic and related to MARPOL and the Bulk Chemical and Liquefied Gas Carrier Codes.

STAB XXI also considered sub-division and damage stability of ro-ro ships, which were considered to be appropriately covered by the parallel discussion on sub-division and damage stability of dry cargo ships in general. In this regard STAB was presented with a submission by the USSR[54] suggesting that following the aforementioned deterministic work on MARPOL and related codes a second stage of standard development could be based on "probabilities of a damaged ship survival". The relevant working group[55] subsequently mentioned the USSR paper and its concept of the probability of survival as a prospect for future development as "the conventional type of survival requirements........would not necessarily be the most appropriate method of assessment" of survival capability of various types of dry cargo ships. MSC, at its 39th session, decided that development of subdivision and damage stability criteria for these ships should be prioritized. STAB XXII[56] then agreed to a general approach to development of the standards along the lines that had been proposed by the USSR, but deferred development until research had produced sufficient information for development of comparative assessment methods. But, pending the outcome of the necessary research, STAB turned its attention to providing to the master information relating to the damage stability characteristics of his/her ship, settling on a damage control plan for this purpose[57] at the last STAB meeting prior to the SLF amalgamation.

At its second session in its new guise, SLF 28 noted that "development of a probabilistic damage standard may take many years" and some delegations therefore advocated using a one-compartment deterministic standard as an optional standard for the interim. The Sub-Committee nonetheless agreed upon a proposed method for test calculations of assessment of dry cargo ships' damage stability[58], based upon the existing equivalent probabilistic regulations for passenger ships (IMCO 1973b). This test calculation method was revisited at SLF 31[59], with one of the significant amendments being to delete for the time being the formula for required index "R" with the intention of later re-inserting a criterion based on the analysis of attained indices "A" reflected in the test calculations using ship's length as the only variable.

While this work was going on, draft guidelines for on-board information on the effect of flooding on dry cargo ships were also agreed by SLF 30 and subsequently promulgated by the fifty-second session of MSC[60]. MSC 54[61] subsequently reaffirmed the decisions of its two preceding sessions that IMO should develop minimum subdivision and damage stability standards for all types of dry cargo ships within a year.

The text to be used for test calculations was again reviewed by SLF 31[62], resulting in a number of changes in matters of detail, although the main visible change was to expand the text to be self-standing rather than reliant on the equivalent regulations for passenger ships. One significant area of contention was whether the stability probability "s" should be a step factor having possible values of either 0 or 1 or whether it should be scaleable between those values. The latter view prevailed when the standards were completed and made applicable to ships of 100m length and above, by SLF 32[63] as a proposed non-mandatory circular in order to gain more experience before they could be made mandatory. The required subdivision index "R" remained to be finalized with the accumulation of experience, particularly in relation to ro-ro ships of less than 140m in length. MSC 55[64] issued the circular[65] for use as the basis of evaluation with the stated intention of finalizing mandatory requirements at MSC 57 in 1989 with a view to adoption by MSC 58, as occurred in May 1990[66] with application to cargo ships built on or after 1 February 1992. Explanatory notes were prepared in parallel with the process of finalising the regulations and were adopted by the 17th Assembly[67].

Having been tasked by the MSC with development of subdivision standards for cargo ships of less than 100m in length, the damage stability resources of SLF were taken up with the previously mentioned work on passenger ships until SLF 38 in March 1994. At that meeting and on the basis of analysis of smaller dry cargo ships, SLF agreed to apply the SOLAS calculation method for ships over 100m to ships in the range down to 80m, but that a new and less onerous curve for the R value would apply to these ships[68]. The relevant SOLAS amendments were approved by MSC 64 in December 1994 and adopted by MSC 66 in June 1996[69].

No new explanatory notes were developed in relation to these requirements, this task being left to be undertaken in relation to the harmonized SOLAS provisions relating to both passenger and cargo ships.

Harmonized damage stability requirements – cargo ship perspective

In addition to finding a problem with the R values determined for passenger ships, as outlined above, SLF 46 (IMO 2003) also unsurprisingly found that the HARDER statistics demonstrated a need for two standards of R for cargo ships, since car carriers and ro-ro ships demonstrated a lesser standard of damage stability performance than other cargo ships. MSC 78 subsequently agreed with the Sub-Committee that a single standard for R should be adopted, even if it meant upgrading the present requirements for car carriers and ro-ro ships[70]. The SOLAS requirements were finalized on this basis.

EXPLANATORY NOTES

Subsequent to finalizing the SOLAS harmonized damage stability requirements, SLF has been developing explanatory notes to ensure uniform international implementation of those requirements. These notes have been developed from the explanatory notes associated with the original "equivalent regulations" (IMCO 1973c). In their interim form, they were promulgated by MSC 82 as an MSC Circular[71]. It is anticipated that the full explanatory notes will be adopted by MSC 85 in December 2008 in time for the entry into force of the regulations on 1 January 2009.

CONCLUSION

The new harmonized SOLAS requirements for the probabilistic damage stability of both passenger ships and cargo ships have been developed by IMO over many years of painstaking work. They provide for the ship's actual subdivision to be taken into account, rather than only consider traditional transverse subdivision arrangements. The probabilistic element of the method applies mainly to how the outcome of the damage stability calculations are assessed, so the method should be readily implemented by any naval architect familiar with damage stability work.

REFERENCES

[1] IMO 2005, *Res. A.194(80) Amendments to the International Convention for the Safety of Life at Sea, as amended*, IMO, London, 2005, Annex 2

[2] IMO 2007, *Res. A.216(82) Amendments to the International Convention for the Safety of Life at Sea, as amended*, IMO, London, 2007, Annex 2

[3] IMCO 1960, *International Conference on Safety of Life at Sea, 1960 – Final Act of the Conference with Annexes including the text of the adopted Convention*, IMCO, London

[4] Wendel, K. (1960): *Die Wahrscheinlichkeit des Ueberstehens von Verletzungen, (The probability of surviving damages)*, Schiffstechnik, Vol. 7, No 36, pp. 47-61

[5] IMCO 1962, *STAB I/1 Sub-Committee on Subdivision and Stability Problems – Provisional agenda for the first session*, IMCO, London May 1962

[6] IMCO 1963, *SDS II/11 Working Group on Watertight Subdivision and Damage Stability of Passenger and Cargo Ships – Draft report*, IMCO, London, March 1963

[7] IMCO 1965, *SDS IV/13 Working Group on Watertight Subdivision and Damage Stability of Passenger and Cargo Ships, 4th session –Report*, IMCO, London, May 1965, paras 21-25.

[8] IMCO 1966, *SDS V/11 Working Group on Watertight Subdivision and Damage Stability of Passenger and Cargo Ships –Report to the Sub-Committee on Subdivision and Stability*, IMCO, London, September 1966, para 19

[9] IMCO 1965, para 15

[10] IMCO 1965a, *STAB IV/7 Sub-Committee on Subdivision and Stability Problems, 4th session – Report to the Maritime Safety Committee*, IMCO, London, December 1965, para 14

[11] Norway 1967, *STAB VI/3 Proposal for new subdivision and damage stability regulations*, IMCO, London, April 1967

[12] Norway, United States and USSR 1968, *STAB VIII/5 Proposal for new subdivision regulations – Joint submission*, IMCO, London, 1968

[13] Norway, United Kingdom, United States and USSR 1968a, *STAB IX/4 Proposal for new subdivision regulations*, IMCO, London, October 1968

[14] IMCO 1969, *STAB IX/29 Report to the Maritime Safety Committee*, IMCO, London, January 1969, paras 5 to 12 and Annex II

[15] Greece 1969, *STAB IX/19 Subdivision of Passenger Ships with Longitudinal Bulkheads*, IMCO, London, January 1969

[16] IMCO 1969a, *STAB X/13 Report to the Maritime Safety Committee*, IMCO, London, December 1969, Annex II

[17] IMCO 1970, *STAB XI/8 Report to the Maritime Safety Committee*, IMCO, London, October 1970, paras 7 to 16

[18] IMCO 1971, *STAB XII/8 Report to the Maritime Safety Committee*, IMCO, London, May 1971, paras 4 to 12 and Annex II

[19] IMCO 1972a, *MSC XXVI/19 Report of the Maritime Safety Committee on its twenty-sixth session*, IMCO, London, November 1972, para 43

[20] IMCO 1973a, *STAB XIV/10 Report to the Maritime Safety Committee*, IMCO, London, February 1973, paras 5 to 10 and Annexes III and IV

[21] IMCO 1970, Annex II

[22] IMCO 1973b, *Resolution A.265(VIII) Regulations on subdivision and stability of passenger ships as an equivalent to Part B of Chapter II of the International Convention for the Safety of Life at Sea, 1960*, IMCO, London, November 1973

[23] IMCO 1973c, MSC/Circ.153 *Regulations on subdivision and stability of passenger ships as an equivalent to Part B of Chapter II of the International Convention for the Safety of Life at Sea (Resolution A.265(VIII) – Explanatory notes to the Regulations*, IMCO, London, November 1973

[24] IMCO 1975, *International Convention for the Safety of Life at Sea, 1974*, IMCO, London, 1975

[25] IMCO 1979, *MSC/Circ.250 Regulations on subdivision and stability of passenger ships as an equivalent to Part B of Chapter II of the International Convention for the Safety of Life at Sea, 1960 – Additional explanatory notes to the regulations*, IMCO, London, May 1979

[26] IMCO 1984, *SLF 29/15 Report to the Maritime Safety Committee*, IMO, London, February 1984, para 12.2

[27] IMO 1985, *SLF 30/18 Report to the Maritime Safety Committee*, IMO, London, March 1985, section 11

[28] IMO 1986, *SLF 31/22 Report to the Maritime Safety Committee*, IMO, London, June 1986, section 9

[29] IMO 1987, Res. A.596(15) *Safety of ro-ro passenger ferries*, IMO, London, November 1987

[30] IMO 1993, *Res. A.741(18) International Management Code for the Safe Operation of Ships and for Pollution Prevention (ISM Code)*, IMO, London, December 1993 (subsequently amended)

[31] IMO 1988, *MSC 55/25 Report of the Maritime Safety Committee on its fifty-fifth session*, IMO, London, May 1988, section 23

[32] IMO 1988, para 23.28 and Annex 19

[33] IMO 1988a, *MSC 56/9 Report of the Maritime Safety Committee on its fifty-sixth session*, IMO, London, November 1988, section 2 and Annex 2 (res. MSC.12(56))

[34] United Kingdom, *MSC 58/3/8 Consideration and adoption of amendments to SOLAS 74 and related codes – Ro-ro safety research programme*, IMO, London, March 1990

[35] IMO 1992, *Res. MSC.26(60) Adoption of amendments to Chapter II-1 of the International Convention for the Safety of Life at Sea, 1974 – Existing ro-ro passenger ships*, IMO, London, April 1992 (Annex 2 to MSC 60/21)

[36] IMO 1991, *MSC/Circ.574 The calculation procedure to assess the survivability characteristics of existing ro-ro passenger ships when using a simplified method based upon Resolution A.265(VIII)*, IMO, London, June 1991

[37] IMO 1994, *MSC/Circ.649 International Convention for the Safety of Life at Sea, 1974, as amended – Interpretations of provisions of resolution MSC.26(60) and MSC/Circ.574*, IMO, London, June 1994

[38] Norway & United Kingdom 2000, *MSC 72/21/10 Proposed amendments to the work programme of the SLF Sub-Committee with respect to subdivision and damage stability harmonization*, IMO, London, March 2000

[39] Rusaas S, *Editorial – Project HARDER*, Thematic Network saferEUroro, June 2004, http//:www.safereuroro.org/SEII_Newsletter_Issue_2_June_2004_RE1.pdf

[40] Norway & United Kingdom 2003, *SLF 46/3/3 Development of the revised SOLAS Chapter II-1 Parts A, B and B-1 - Final recommendations from the research project HARDER*, (also SLF 46/INF.4, SLF 46/INF.5, SLF 46/INF.6, SLF 46/INF.7, SLF 46/INF.8 and SLF 46/INF.9), IMO, London, July 2003

[41] IMO 2003, *SLF 46/16 Report to the Maritime Safety Committee*, IMO, London, September 2003, section 3

[42] IMO 2004, *SLF 47/17 Report to the Maritime Safety Committee*, IMO, London, September 2004, section 3

[43] IMO 2004, Annex 1

[44] IMO 2004a, *SLF 47/WP.6/Add.1 Development of revised SOLAS chapter II-1 parts A, B and B-1 - Report of the SDS Working Group*, IMO, London, September 2004

[45] IMCO 1974, *STAB XVI/8 Report to the Maritime Safety Committee*, IMCO, London, July 1975, para 32

[46] IMCO 1972, *STAB XIII/13 Report to the Maritime Safety Committee*, IMCO, London, March 1972 1970, paras 38 and 39 and Annex V

[47] IMCO 1972a, *International Conference on Load Lines 1966 – Final Act of the Conference*, IMCO, London, 1972, recommendation 4

[48] IMCO 1973, *International Conference on Marine Pollution 1973 – Final Act of the Conference*, IMCO, London, 1973, regulations 22 to 25

[49] IMCO 1971a, Res. A.212(VII), *Code for the Construction and Equipment of Ships Carrying Dangerous Chemicals in Bulk*, IMCO, London, November 1971, Chapter 2 Part A

[50] IMCO 1975, *Res. A.328(IX) Code for the Construction and Equipment of Ships Carrying liquefied gases in Bulk*, IMCO, London, November 1975, Chapter II

[51] IMCO 1976, *STAB XIX/13 Report to the Maritime Safety Committee*, IMCO, London, July 1976, para 24

[52] IMCO 1977, *STAB XXI/13 Report to the Maritime Safety Committee*, IMCO, London, November 1977, paras 5 to 9

[53] United States 1976, *STAB XIX/7 Harmonization of subdivision and stability requirements in various conventions and codes*, IMCO, London, February 1976

[54] USSR 1977, *STAB XXI/3/1 Harmonization of subdivision and stability requirements in various conventions and codes*, IMCO, London, July 1977, para 4

[55] IMCO 1977, STAB XXI/WP.4/Add.1 *Harmonization of subdivision and stability requirements in various conventions and codes – second report of the ad hoc group*, IMCO, London, October 1977, paras 7 and 8

[56] IMCO 1978, *STAB XXII/14 Report to the Maritime Safety Committee*, IMCO, London, October 1978, section 10
[57] IMCO 1982, *STAB 27/13 Report to the Maritime Safety Committee*, IMCO, London, April 1982, para 6.4
[58] IMO 1984, para 3.10 and Annex 3
[59] IMO 1985, para 3.13 and Annex 3
[60] IMO 1986a, *MSC/Circ.434 Guidelines for the preparation of information on the effect of flooding to be provided to masters of dry cargo ships*, IMO, London, February 1986
[61] IMO 1987a, *MSC 54/23 Report of the Maritime Safety Committee on its fifty-fourth session*, IMO, London, May 1987, paras 20.18 and 20.19
[62] IMO 1986, section 3 and Annex 2
[63] IMO 1987b, *SLF 32/21 Report to the Maritime Safety Committee*, IMO, London, September 1987, para 3.10 and Annex 2
[64] IMO 1988, paras 6.1 to 6.15
[65] IMO 1988b, MSC/Circ.484, *Subdivision and damage stability of dry cargo ships including ro-ro ships*, IMO, London, May 1988
[66] IMO 1990, *MSC 58/25 Report of the Maritime Safety Committee on its fifty-eighth session*, IMO, London, June 1990, para 3.26 and Annex 2 (res. MSC.19(58))
[67] IMO 1991a, *Res. A.684(17) Explanatory notes to the SOLAS regulations on subdivision and damage stability of cargo ships of 100 metres in length and over*, IMO, London, November 1991
[68] IMO 1994a, *SLF 38/20 Report to the Maritime Safety Committee*, IMO, London, March 1994, paras 5.6-5.8 and Annex 5
[69] IMO 1996, *MSC 66/24 Report of the Maritime Safety Committee on its sixty-sixth session*, IMO, London, June 1996, Section 3 and Annex 2
[70] IMO 2004b, MSC 78/26 *Report of the Maritime Safety Committee on its seventy-eighth session*, IMO, London, May 2004, para 12.8
[71] IMO 2007, *MSC.1/Circ.1226 Interim explanatory notes to the SOLAS Chapter II-1 subdivision and damage stability regulations*, IMO, London, January 2007

Integrated Waste Management in the Maritime Environment

Kim Harvey
Weir Strachan & Henshaw Australia
kim.harvey@weirsandh.com

1 ABSTRACT

Shipborne waste systems have traditionally been seen as separate, unrelated entities, each having its own supplier and interface challenges. The various liquid and solid waste streams have no commonality and are discharged or stored without thought for cross-connection to other systems. Integration of these systems would minimise waste streams, decrease manual input and reduce port disposal needs and costs.

An Integrated Waste Management System (IWMS) allows the unification of the individual treatment technologies to create a single intelligent system which can monitor system status and optimise the operation of each piece of equipment, to ensure maximum efficiency with varying input requirements.

By employing an IWMS to manage the collection, transfer, treatment, stowage and disembarkation of both fluid and solid waste streams, the following key features are achievable:

- Full environmental compliance to current and considered future environmental legislation to facilitate unrestricted operation world-wide;
- Increased operational effectiveness by minimising onboard storage of waste, improving ease of maintenance and establishing a waste treatment and discharge methodology;
- Greater operational flexibility by reducing time spent offloading waste in port, minimising costly land-based treatment/disposal and minimising dependence on port reception facilities;
- Cost savings by reducing manpower requirements through system integration and automation, leading to substantial through-life cost savings; and
- Improved health and safety by minimising manual intervention in waste handling, and due to the safe inert nature of the reduced solid waste to be stored onboard.

This paper describes the flexible Weir Strachan & Henshaw Australia (Weir SHA)/QinetiQ IWMS solution, which achieves the above key features, and can be tailored to suit specific vessel and operational requirements, either for a new design or for retrofitting on an existing ship.

2 INTRODUCTION

Conventional waste management onboard both Navy and commercial vessels handles, processes and discharges or stores each waste stream independently, without consideration of what benefits may be possible through integration of the systems. This method is manpower-intensive, has health and safety implications, induces feelings of drudgery amongst the crew, is a poor use of the space, and will generate increasing port reception costs.

An integrated approach to dealing with the waste streams would ease operation and maintenance, reduce manpower requirements, minimise the onboard storage of waste, and reduce port disposal needs.

These benefits are achieved because integrated waste management:

- Allows full availability of the system;
- Ensures flexibility under varying demands;
- Takes products and automatically processes them further;
- Provides a largely autonomous processing capability;
- Produces less waste to store and put to shore; And
- Produces less to discharge overboard.

This would provide substantial through-life cost savings and also facilitate the world-wide operation of the vessel through its full compliance to current and future environmental legislation.

The Weir SHA/QinetiQ Integrated Waste Management System (IWMS) addresses the collection, transfer, treatment, stowage and disembarkation of both fluid and solid waste streams. The IWMS can handle metals, plastics, glass, bilge water (machine wastes, deck wash, aircraft wash and sea water), sanitary waste, clinical waste, food waste, black water (sewage and medical waste water), grey water (showers, basins, laundry and galley) and general trash (wood, paper etc).

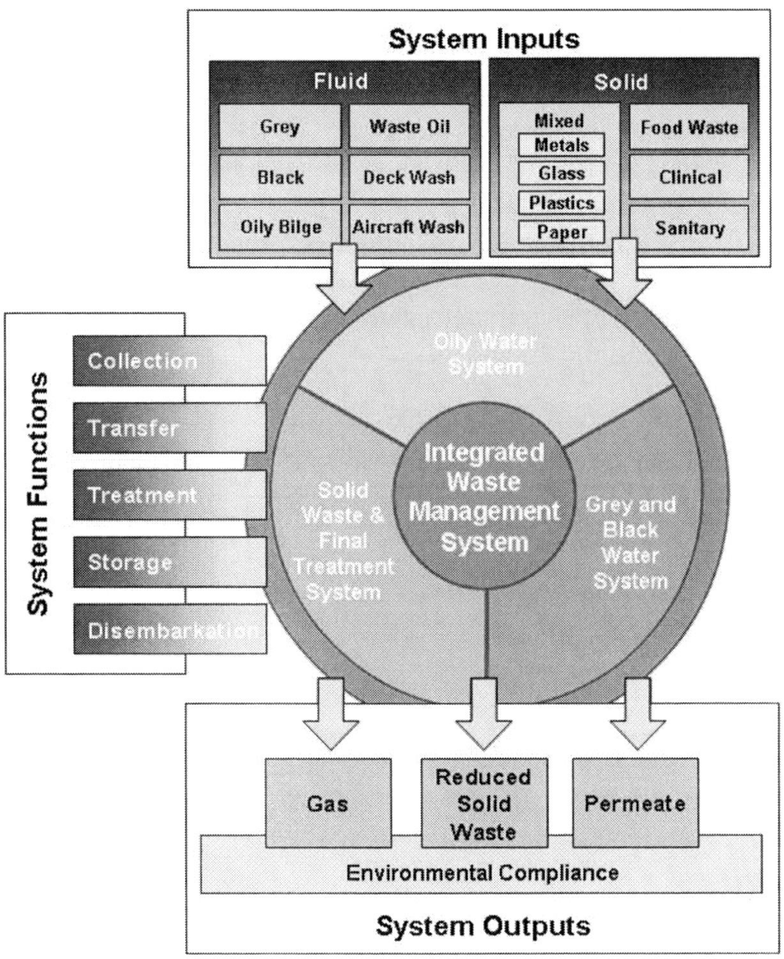

Figure 1: Schematic of the Integrated Waste Management System

These waste streams are handled by three sub-systems, which are the Grey and Black Water, the Oily Water and the Solid Waste Treatment systems. These sub-systems are unified by the IWMS, linking them with transfer and control systems to optimise the operation of each piece of equipment, and to ensure maximum efficiency with varying input requirements.

The outputs of the IWMS are:

- a clean permeate that meets MARPOL requirements for discharge to sea;
- MARPOL compliant gaseous emissions, which are vented to the atmosphere; and
- ash (with a 90% reduction in volume from the solid waste), which can be stored in containers until it can be disposed to shore.

The system configuration may be tailored depending upon vessel type and to meet operational and customer requirements.

3 KEY FEATURES OF INTEGRATED WASTE MANAGEMENT

3.1 Compliance to Environmental Legislation

International pollution regulations for ships are developed and maintained by the International Maritime Organisation (IMO) in the form of MARPOL 73/78 (the International Convention for the Prevention of Pollution from Ships). Its six annexes state the regulations for the prevention of pollution by: Oil, Noxious Liquid Substances, Harmful Substances in Package Form, Sewage from Ships, Garbage from Ships and Air Pollution from Ships.

Increasingly stringent environmental legislation, such as MARPOL, has led to the concept of an "Environmentally Compliant Warship" being considered and researched by a number of navies, as discussed in Reference 1. Environmental compliance is a consideration for all types of shipping, making integrated waste management an attractive proposition for both military and commercial vessels. IWMS design should ensure that as future legislation introduces tighter discharge standards, individual items of equipment can be readily upgraded or new equipment retrofitted to increase the capability of the system.

3.2 Meeting the Operational Requirements for a Warship

A warship must be self sufficient at sea for extended periods of time, so its design must consider a variety of operational requirements, many of which are different from those for a commercial vessel. Key design drivers for a warship are as follows:

- Crew size
- Waste generation
- Legislation
- Mission length
- Budget
- Available compartment space and weight constraints
- Build/ installation requirements (e.g. modular build and interfaces)
- Redundancy

Commercial-Off-The-Shelf (COTS) solutions intended for commercial vessels may not withstand the operating requirements expected from a naval vessel, such as mission length or the limited availability of space and weight.

3.3 Implementation of a Waste Treatment and Discharge Philosophy

To reduce the amount of waste created, and therefore the amount required to be processed, a Waste Treatment and Discharge Philosophy should be implemented which would involve the following, amongst other considerations:

- Increased crew awareness of the waste disposal regulations
- Point source reduction, e.g. remove excess packaging before taking onboard
- Recycling and re-use

3.4 Meeting Health and Safety Requirements

Historically the inclusion of waste processing equipment on ships caused an increase in manpower requirements, and also impacted on crew morale due to the tedium of operating the equipment. However, through system integration and automation, operation and maintenance can be simplified. Manual input is reduced and simplified, which improves health and safety and leads to substantial through-life cost savings.

3.5 Less Reliance on Port Facilities

Through a good Waste Treatment and Discharge Philosophy and also through the effectiveness of the Solid Waste and Final Treatment sub-system, the volume of waste to be offloaded at shore is minimised. This reduces time spent offloading waste in port, minimises costly land-based treatment/disposal, and minimises dependence on port reception facilities, giving faster turnaround and greater operational flexibility.

4 THE WEIR SHA/QINETIQ INTEGRATED WASTE MANAGEMENT SYSTEM

4.1 Top Level System

The IWMS unifies the Grey and Black Water, Oily Water and Solid Waste Treatment sub-systems, linking them with transfer and control systems to ensure optimum waste management is achieved. The routes of the individual waste streams and the interconnection of waste streams within the system are shown below.

In order to manage a variety of waste generation rates and achieve the required level of redundancy, the system configuration is tailored for each specific application, depending upon vessel type, operational requirements and customer requirements.

The overall IWMS incorporates the separate waste management systems, which are capable of operating independently of each other if required, with cross-connecting pipe work facilitating transfer of wastes between the systems, e.g. sludge from the MBR can be transferred to the Solid Waste and Final Treatment system for thermal destruction. The system design will ensure that the IWMS has the correct buffering capacity either within the equipment or by utilising storage tanks available within the vessel.

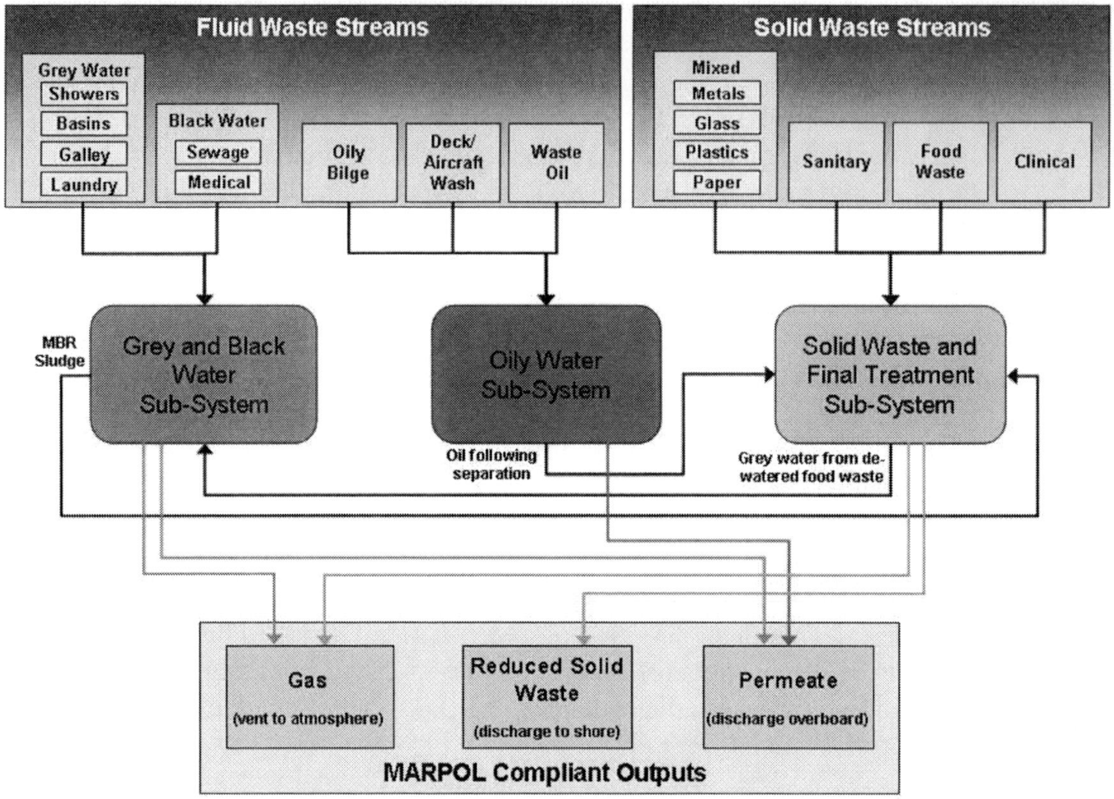

Figure 2: Integration of the Grey and Black Water, Oily Water System and Solid Waste and Final Treatment Sub-Systems.

The control of the entire process is achieved by the Waste Management Control System (WMCS), a top-level SCADA PC system capable of communication with the vessel's Integrated Platform Management System (IPMS), overseeing operation via module Programmable Logic Controllers (PLCs). These PLCs control the operation of the individual treatment equipment, pumping and transfer units, and the associated directional valves.

4.2 Grey and Black Water Sub-System – Membrane Bioreactor (MBR)

Black water from the heads and sick bay is vacuum collected and transferred directly to a Membrane Bio-Reactor (MBR), which handles the black water as it arises.

Grey water from the showers, basins, laundry and galley is gravity fed into a structural collection tank, sized to provide a sufficient buffer capacity; the transfer pumps then supply the grey water to the MBR on demand. Grey water from the galley is passed through grease separators prior to transfer to the collection tank to ensure the MBR membranes are not fouled by fats and greases.

The MBR is based on biological degradation and membrane separation. The MBR system produces a clean permeate that is MARPOL compliant. Other by-products are off-gas, mainly Carbon Dioxide which is vented to atmosphere, concentrated bio sludge and grease from the grease separators both of which can be treated in the Solid Waste System (Section 4.4).

4.3 Oily Water Sub-System (OWS)

Oily water from the bilges, aircraft and deck wash-down and waste oil is treated by an oily water separator.

The effluent produced easily meets the requirements of MARPOL Annex I, MEPC 107(49) by meeting the more onerous NATO MCG7 (Marine Capability Group) target of <5ppm oil content.

A concentrated oil sludge is also produced which can be stored in double bottom dirty oil tanks and then transferred to the Solid Waste System.

4.4 Solid Waste Sub-System

The Solid Waste system thermally treats mixed solid waste, which can include plastic, paper, cardboard, food, rags, workshop waste, glass, metal and sanitary waste with no need for segregation. Medical waste can also be treated providing that it is first autoclaved.

The system is also able to interface with the MBR and OWS to enable the final processing of waste oil, bio sludge and grease (via thermal destruction) in order to extend the operational duration of the vessel without need to off-load waste.

Garbage is manually loaded into a shredder, then screw fed into the thermal destruction system.

Food waste, collected from the galley and dining halls is manually transferred to the waste treatment compartment where it is passed through a grinder, dewatered, and transferred to a collection tank. The liquor produced from the dewatering process is transferred via a grease trap to the MBR for further processing. The grease from the traps can be collected and processed through the feed screw of the thermal destruction unit.

A metered daily fraction of the waste oil and sewage sludge is passed to a heated tank where it is mixed with the dewatered food waste and pumped directly into the first stage of the thermal destruction system. The heated tank is required to ensure that the mix remains in a state that is suitable for pumping.

The Solid Waste System thermally destroys the mixed solid waste to ash with a volume reduction of >90%. The ash is automatically discharged to the chosen containers for safe, inert storage until it can be disposed of to shore. The Solid Waste System also produces gaseous exhaust emissions that are compliant with MARPOL Annex VI, MEPC 76(40).

5 IWMS SOLUTIONS

There are two basic approaches to IWMS solutions:

- Bespoke: a tailored solution with equipment designed specifically for the requirements of the vessel; and
- Commercial Off The Shelf (COTS): utilises commercially available equipment that has been proven in a marine environment.

Bespoke equipment is designed specifically for the vessel, which allows complete flexibility. If the waste management system is to be fitted into an existing vessel a bespoke design may be the optimum solution. This is because the equipment can be designed to fit in the allowed space envelopes, considering maintenance access, any installation difficulties and existing services as well as the sheer volume of the kit.

COTS equipment enables a de-risked solution, as the equipment is already developed. COTS systems are most viable when waste management is considered at the design phase as the equipment will require a fixed space envelope.

Both the COTS and Bespoke solutions can be either integrated solutions or stand alone equipment. The IWMS solution could combine both COTS and Bespoke equipment.
The pros and cons of each of these solutions are listed in the table below.

Solution	Pros	Cons
COTS	No design or development required for the equipment Ability to select type-approved equipment Low risk	Future environmental compliance may be limited Difficult to retrofit due to space envelopes required
Bespoke	More suitable for small space envelopes More suitable for retrofitting Tailored to specific application Increased potential to meet future environmental discharge targets	Equipment type-approval is required, e.g. Pyrolysis Cost of design and development

6 SYSTEM OPTIMISATION

In order to manage a variety of waste generation rates and to achieve the required level of redundancy, the IWMS configuration is tailored for a specific vessel or class of vessels.

The equipment should be designed and selected with consideration of the system producibility and integration into the vessel, considering modularisation of the equipment.

Design and delivery of the IWMS in modular form allows:

- Ease of installation;
- Ease of maintenance;
- Improvement of performance characteristics; and
- Future improvement with the development of new technologies.

While elements of a waste processing system can be procured using requirements specifications, it is unlikely that any equipment will be totally compliant, and each variance from specification will impact on optimal system performance.

Weir SHA have significant experience applying discrete event process simulation to design and to model integrated systems for defence and nuclear based applications, and have developed a system process modelling tool which is able to simulate the interactions and statistical performance of highly complex systems, so that predicted outputs can be produced.
A process model of the IWMS can be used to investigate the system performance under a variety of different operational conditions. This modelling allows the system to be optimised for a particular application using real time simulation to generate "what if" scenarios to validate the system.

Using this technique allows Weir SHA to make the most appropriate equipment choices, and provide validated system performance data, assessing downtime, manning profiles, buffering capacities and waste consistency. Additionally, being able to determine accurately the levels of waste being produced at any time during operational cycles enables optimal sizing for tanks, pumps, valves and associated pipe work, leading to savings in weight and cost.

7 EQUIPMENT

As discussed there are many solutions available, which are dependent upon waste amounts, available compartment space, installation and operational requirements. This section provides some further information on the technologies available.

7.1 Grey and Black Water System

The ideal treatment solution for both the black and grey water waste streams is a Membrane Bio-Reactor. Various companies offer type approved, commercially available MBRs, and the number of systems available, combined with the maturity of the technology makes purchase of a commercially available system a low-risk option for waste water treatment. Difficulty may arise when attempting to retrofit the technology to any vessels already built. However if consideration is given to the integration of a COTS system during the ship design, this option is viable.

Based on experience with the RN Type 23 Frigate, QinetiQ are able to exploit existing holding tanks and facilities onboard a vessel, by retrofitting them into functional MBRs. This solution offers a less intrusive, more cost-effective means of fitting advanced technologies onto existing vessels or vessels with difficult space constraints.

Water generation on naval vessels operating in waters near to shore can become an issue due to sediment levels and blinding of the Reverse Osmosis membranes. QinetiQ has conducted cost benefit analysis studies into solutions for water production and are currently investigating stand alone applications for recycling of the grey water generated by RN vessels into potable water. Variations of micro and nano-filtration systems are being tested by QinetiQ as well as methods for treating the effluent from MBRs to meet the potable drinking water standards.

7.2 Oily Water System

Current MARPOL requirements are that a system must be in place capable of treating Oily Water waste to 15ppm or lower prior to discharge, including an in-line oil content monitor and an automated recirculation function linked to the discharge monitor.

Although current IMO regulations require 15ppm oil content or less in overboard discharge from ocean-going vessels, it is likely that in the near future certain regions will impose more stringent requirements on oil/water discharges. The NATO Marine Capability Group 7 has projected a target discharge level of 10ppm by 2015 and anticipates that some regions will implement further restrictions down to 5ppm. With this in mind, and in the interest of future-proofing, advanced oil/water separation technologies such as cross flow filtration should be utilised rather than coalescers and tortuous path separators.

Two technologies, ceramic and polymeric membrane cross flow filtration, are currently available from a range of manufacturers that can reliably offer <5ppm effluent quality with single-pass treatment. However the cost of these equipments is commensurately higher than more conventional technologies but will ensure future operations are not restricted within the life-time of the vessel.

7.3 Solid Waste System

The use of thermal destruction technology is recommended for the volume reduction of solid wastes either by advanced incineration, of which there are many companies that offer type approved systems in accordance with IMO regulations, or Pyrolysis which is a new technology for ship-board application.

There are more than twelve type-approved incinerators available to date, ranging in design from basic, single-staged batch burning systems to more advanced multi-stage incinerators with flue gas cooling and constant feed systems with integrated shredders.

QinetiQ's experience has shown that advanced incinerators can deal with not only general garbage, but also food waste, sewage sludge and waste oils. In order to apply these wastes directly to the combustion chamber of an incinerator, it is necessary to include certain ancillary equipment. As described in Section 4.4, the food waste, sewage sludge and waste oil are required to be mixed in a collection tank and then sprayed directly into the incinerator chamber using a pressurised lance system. A multi-stage incinerator, with greatly increased quality of the exhaust gases produced, would ensure the system is complaint with future legislation.

Pyrolysis is an advanced alternative to incineration, and involves the heating of waste in the absence of oxygen. The organics in the waste are thermally degraded to produce fuel gases which are then combusted with a small amount of air, approximately three times less than required for incineration. The combustion process is an exothermic reaction which produces the heat required to enable the system to become self sustaining, provided that waste with sufficient calorific value is being processed.

The mixed solid waste is shredded and compacted into the high temperature chamber where pyrolysis takes place. The ash produced from the process is typically only 5-10% of the volume of the waste input. This ash is inert and safe to store onboard for disposal to shore facilities.

8 FUTURE TECHNOLOGY

As a key player in waste management systems, the Weir SHA and QinetiQ team are always looking for new technologies to exploit, whether through in-house research and development such as our small Membrane Bio-Reactor, or through collaboration with other companies developing new technologies suitable for marine waste management. New systems such as those for thermal destruction, oily water separation and waste food drying are currently being investigated, with trade-off studies being carried out to gauge benefits and added value to the customer. Testing and trials programmes are also undertaken to ensure that the technology will perform as expected in service. By keeping one step ahead, we ensure that our systems are the best value solution for our customers and wherever possible, adaptable for future retrofit.

9 CONCLUSIONS

An Integrated Waste Management System that treats all major onboard waste streams for full environmental compliance and maximum operating potential is the future of waste management.

An integrated system is recommended rather than piecemeal solutions as this enables cost effective equipment and procedures to comply with present requirements and provide a smooth transition to compliance with future restrictions as they develop.

Integrated waste management can be retrofitted into an existing ship design, most likely through the use of bespoke equipment. COTS equipment is a lower cost option, but must be considered during the ship's design due to the larger space envelopes and to ensure the system is future proofed.

The Weir SHA/QinetiQ IWMS facilitates unrestricted operation world-wide through full compliance to current environmental legislation, and also allows for compliance to potential changes in legislation.

10 REFERENCES

1. Tsamenyi, M., Kaye, S., Rose, G., Snushall, B., Moore, C., Mfodwo, K., Munoz, L., (2005), "Future Environmental Policy Trends to 2020, Impact on Ship Design and Operation", Papers in Australian Maritime Affairs: No. 13.

11 BIBLIOGRAPHY

1. Committee on Shipboard Pollution Control, (1996), "Shipboard Pollution Control: U.S. Navy Compliance with MARPOL Annex V.
2. The Naval Architect January 2006 pages 27 & 28
3. "Evaluation Report'" The Navy Shipboard Pollution Control Equipment Program, Report No. D-2000-010, November 2000
4. http://www.nato.int/structur/AC/141/docs/D8%20for%20Publc%20Domain%20Web%20Site.doc - NATO Website.

Alternative Shock Trial Techniques for the RAN Air Warfare Destroyer

by Warren Reid, Leo Deyong and Paul Elischer

Defence Science & Technology Organisation (DSTO), Melbourne
Maritime Platforms Division
P.O. Box 4331 Melbourne VIC 3001
email: warren.reid@dsto.defence.gov.au
Ph 61 3 9626 8205

Abstract

Naval front-line combatant vessels by their very nature are expected to be able to survive the loadings of various above and below water weapon attacks and continue fighting. Near misses by underwater weapons such as mines and torpedos may not damage a vessel from the structural viewpoint, however equipment and systems may be seriously damaged, often to the extent of incapacitation of power, navigation, propulsion and combat systems.

Incapacitation is due to the resulting shock wave generated by an underwater explosion transmitted through the vessel structure, resulting in mechanical loadings and possible damage to equipment and systems. Vital equipment and systems are therefore usually shock tested prior to installation aboard a vessel and a shock trial on the First-of-Class vessel performed to ensure that, as a whole, systems continue to operate as installed on the vessel. A conventional shock trial requires an explosive charge to be detonated adjacent to the ship to simulate a weapon attack. First-of-Class ship shock trials can however be very expensive, time consuming and increasingly require extensive environmental considerations.

This paper outlines an alternative technique that could be used in a trial to shock test vital equipment and systems aboard the First-of-Class RAN Hobart Class Air Warfare Destroyer. The technique involves the underwater use of air-guns to produce a mechanical shock load upon a vessel and a similar loading on equipment and systems that would be produced by an underwater explosive. Potentially, air-guns may offer numerous advantages over explosives, including cost of testing, time, safety and a considerably lower environmental hazard during a shock trial. DSTO is investigating the use of air-guns in collaboration with the US Navy and private industry to further develop this technology.

INTRODUCTION

In order to provide an operational capability, naval vessels must be able to survive specified threats from underwater exploding weapons. Naval platforms are complex structures and it is difficult to predict the shock response or compliance against specified requirements, without the validation provided by First-of-Class full-scale shock tests on the vessel and/or equipment. These tests are supplemented by controlled small-scale testing of panels, equipment or full-scale test sections.

Vessels are designed to meet specified shock requirements which can be validated by exposure to repeated underwater shock loadings without significantly affecting their operational performance or creating a hazard to personnel. Onboard equipment, which includes commercial off the shelf (COTS) equipment, must survive specific shock loadings to ensure that the vessel is able to function should it come under attack.

Environmental pressures, availability of naval assets, costs and logistics are prime drivers for the need to identify and implement alternative shock testing methods to ensure performance requirements are met, that equipment used for upgrades and modification of systems meet appropriate functional specifications, and that design capability is not compromised.

In support of the need for the RAN to acquire a new class of destroyer, the Hobart Class, the Air Warfare Destroyer – AWD Project was launched and will be based upon the Navantia F100 class of vessel (Figure 1). The definition of the performance requirements of this new class of ship were detailed in the Functional Performance Specification (FPS) and, in a broader context, in the Operational Concept Document (OCD). Both these documents either explicitly or implicitly required the platform to survive a defined underwater shock level and refer to the requirements set out in the Materiel Requirements Set [1].

Figure 1. Navantia F100, basis for the RAN Hobart Class Air Warfare Destroyer.

This paper discusses the existing shock trial, testing and verification issues and the application of seismic air-guns in undertaking shock testing of naval assets such as the AWD.

WHY A SHOCK TRIAL?

Why does a surface naval vessel need to be subjected to a shock trial? A shock trial is conducted to evaluate the vulnerability of a vessel to underwater shock loads in three major areas – structural, human and equipment. Incapacitation in any or all of these three areas is due to the shock wave being transmitted through the structure of the ship, resulting in mechanical loadings and possible damage to equipment and systems or injury to crew.

Specifically, the main reasons for conducting an at sea shock trial are to:

1. prove and verify that the ship as a complete system, rather than individual items of equipment (that have been shock qualified in a laboratory), has achieved the specified shock requirements;
2. prove and verify the ship's structural response and resistance to underwater shock; including induced hull whipping;
3. obtain data to understand the vessel's response to shock loading and to aid in the verification of simulation models. The models may then be used to evaluate the safety of a vessel and its personnel against higher shock loads;

4. identify and determine the operational limits for a vessel when operating in a threat environment, and
5. identify problems or areas of the design and equipment that are susceptible to shock damage.

It is common for vessel classes that have had significant equipment or structural modification to be re-shock tested in order to validate the changes. A shock trial is generally performed up to a level where there is little danger to the crew and no danger of damaging critical equipment, although in some cases some structural damage such as frame tripping may occur. The United States Navy (USN) conducts shock trials using full-scale explosive charges (up to 5 tonnes) and record the response of all critical items of equipment. US shock trials are broad and include shock compliance inspections during ship construction, repairs and defect rectification, and modelling and simulation (M&S) activities.

SHOCK TRIAL ISSUES

The major issues that generally arise when a requirement to conduct a shock trial is proposed are the cost, the time and the environmental considerations. A USN shock trial may cost in the order of $US60 million [6] whilst a RAN shock 'threat representative test' may cost an order of magnitude less.

Another consideration is the environmental concerns of conducting an at-sea shock trial particularly those associated with the use of high explosive charges in a marine environment and adherence to relevant Commonwealth, State, Territory and Local Government environmental legislation. The Department of Defence requirements for environmental management are covered by Defence Environmental Instructions [3], and the RAN adheres strictly to these requirements and the principles espoused. For past RAN shock trials, the environmental requirements were an integral part of the trial. Also Environmental Management Plans (EMPs) have been developed for all major exercises and all RAN's Offshore Exercise Areas.

The cost of environmental compliance is difficult to accurately forecast but the cost of developing the EMP inclusive of the consultation processes with regulators and interested parties is significantly less than that for a US full-scale shock trial, and is about $200K. The cost of the marine mammal observation program is also likely to be about $200K.

SHOCK TRIAL TESTING

The issues outlined above have led to significant research over the last 10 years into alternative methods of verifying shock compliance. Initially the alternatives concentrated on the use of M&S, but US experience has shown that M&S has failed to predict a number of large problems that have occurred during at-sea shock trials. This situation has led to a variety of alternative techniques being examined.

The aim of a shock trial is to partially verify the structural integrity, but more importantly to verify that the equipment installation aboard a vessel is adequate to survive the specified vulnerability design limit. Equipment critical to the operational capability of the vessel is typically only laboratory shock tested to a limit prescribed in a shock standard. However, the equipment is usually tested as a sub-component or without ancillary equipment, and certainly not in the structural environment of a naval vessel. The performance of a critical system, as installed on a vessel, is therefore far from guaranteed.

By performing a shock trial, shortcomings with the installation of equipment and sub-systems are typically highlighted, along with structural responses that excite non-desirable resonant responses in equipment. An example of this is the ANZAC Modular Gun System. The original gun was a US design and in its US ship configuration satisfactorily passed its US shock standard test. When this system was mounted in a modular mounting system on the RAN ANZAC ships and re-tested against the US shock standard requirements, it failed. This was not a fault of the modular mounting system, but a result of the mismatch between testing techniques and differing resonant frequencies generated by the design and the test techniques. These same conditions can occur with laboratory shock tested equipment and its installation on a ship (or submarine).

Explosive Shock Testing

Underwater shock waves produce a complex excitation pattern within structures, typified by a primary shock wave with a fast, steep, rise time, followed by an exponential decay period. Other loadings may also result from a negative pressure environment due to surface cut-off, shock waves reflecting from the sea-bed, which are more significant in shallow water, and the effect of the bubble pulse due to expansion and contraction of the gas bubble. The significance of these parameters depends on the size of the detonation, explosive mass, water and detonation depth, and the distance of the vessel from the detonation. Deploying underwater pressure gauges at a known distance and depth from the vessel and recording the shock parameters quantifies the shock profile around the vessel. Analysis of the resultant pressure-time responses provides information on the arrival time of the shock wave and its magnitude in terms of shock loading via energy flux density calculations. Shock arrival times are used to confirm the exact position of the charge in terms of depth and position relative to the vessel at the time of detonation. This information is used to calculate the shock loading, validate that the correct standoff distances are achieved, and correlate the blast parameters with the shock-induced motion at specific locations in and around the vessel.

Onboard motion responses resulting from underwater detonations are also complex and varied, and can be described in terms of the time dependant history of the acceleration, velocity or displacement associated with the detonation interaction with equipment etc. All parameters can be derived from acceleration measurements by integration or by using either velocity or displacement sensors. Typically large displacements are associated with low frequencies and small displacements with high frequency responses. Accelerations are large at high frequencies and small at low frequencies. Generally the shock motion properties include amplitudes of acceleration, velocity or displacement, the frequencies involved, and the durations and information derived from shock spectra and Fourier analyses.

Full scale shock testing

The most common approach to shock testing naval vessels and hardware is to conduct a full-scale test using explosives or explosive ordnance, commensurate with the likely threat environment, to generate the shock loading or the adopted testing philosophy. DSTO has been involved in a number of First-of-Class shock tests; Minehunter Inshore, Minehunter Coastal, Collins Submarine, and a range of mine counter measures hardware. However it is becoming increasingly difficult to undertake these trials at sea due to the logistics involved, environmental constraints, cost and availability of naval assets. During full-scale shock tests, a vessel is exposed to a number of underwater detonations with the charges located at a predetermined distance from the vessel to deliver a shock loading. This is usually well below a vessel's shock design level. There are a number of variations to these tests depending on whether the tests are conducted in deep or shallow water and also the testing philosophies of particular navies. For example, the USN uses large charges which are detonated well away from the vessel and produce a planar shock wave along the length of the vessel. The advantage of this approach is that the shock responses and the propagation of the shock front throughout the vessel can be correlated with a single event, and provides more accurate information when scaling the shock loading to the vessel and the response of equipment to the respective design levels. Other navies, RAN included, use smaller charges in close proximity to the vessel to deliver the required shock loading. This approach does not subject the vessel to a single uniform shock loading but delivers a more focused loading to a smaller section of the vessel and subsequently to fitted equipment. To ensure that the full profile of the vessel has been subjected to the required loading, additional detonations are necessary at different locations along its length. Whilst still enabling the vessel to be shock qualified, unlike those obtained from the larger charges, the results are not easily scaled up to the design level as the scaling laws are not linear.

Irrespective of the charge/stand off distance configuration, the results from full-scale tests allow the vessel to be shock qualified and provide a valuable data base of shock data that can be used to validate M&S tools.

Scaled shock testing

Scaled shock testing in purpose built ponds or through the utilisation of flooded quarry sites is another option which is currently used. Like full-scale trials, items under test are subjected to underwater detonations with charges placed at a predetermined distance from the vessel to generate the required shock loading. Although the inherent issues associated with environmental considerations, logistics, safe handling, transport and initiation of explosives are applicable, such facilities provide a controlled environment that is regarded as a good supplement to full-scale testing. It is much more suited, depending on its size, for shock and specification compliance testing of smaller items of machinery and equipment through the utilisation of a calibrated floating shock platform. The performance of ordnance, explosives and testing of full-scale sections or panels of the platform can also be assessed. It provides a means for validating M&S tools and in this regard supplements data obtained during full-scale shock testing. The DSTO Underwater Explosive Test Facility (above) has supported experimental work to develop the RAN material requirements for shock tolerant RAN structures and systems, and shock testing to develop shock attenuation techniques to protect equipment and personnel.

Modelling and Simulation

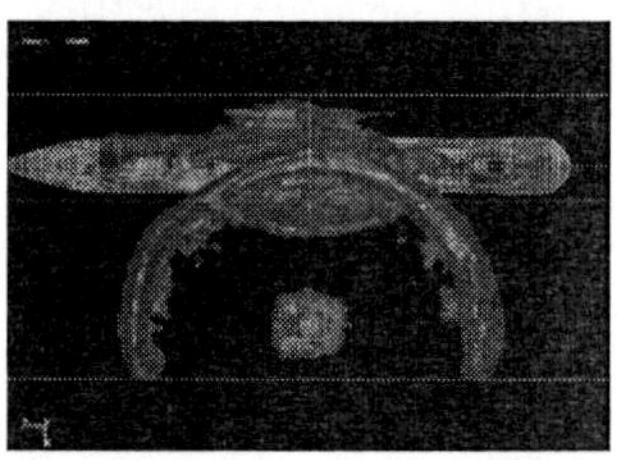

Due to escalating costs, explosives safety and environmental concerns, M&S technologies may offer a means in the future of replacing or off-setting some of the costs associated with full-scale shock trials. Full-scale shock testing of the AWD may not occur as it is a three ship class. It has been suggested that shock qualification will be by analysis, M&S, actual tests on specified equipment, by extension of previously qualified equipment onboard similar vessels or shock qualification using other established means, i.e. shock machine or a floating shock platform. DSTO research on structural response to underwater effects and underwater explosive performance assists the development of a M&S capability to predict underwater explosive behaviour and underwater weapon effects on target platforms. This includes the physics of underwater explosions and fluid/ structure interaction and structural response of naval vessels and fitted equipment. M&S tools still need to be validated against shock data derived from either full-scale or scaled shock trials as the prediction capability often only pertains to that specific ship design or class. The capability would need to be supported by detailed drawings of all ship components and fitted equipments to fully describe the design and/or equipment characteristics. Translating the M&S capability across various ship types is still in the development stage. The sophistication of M&S tools is improving rapidly and in the future may enable full-scale shock testing to address more specific testing issues related to more realistic underwater threat scenarios, but they still require validation against real data and are vessel specific.

Air Gun Technology

The oil and gas industry routinely use seismic air-guns in off-shore exploration of new reserves (Figure 2). The air-guns consist of compressed air within a chamber (at up to 210 bars/3000 psi) that is suddenly released, creating a high pressure gas bubble with an associated pressure pulse. The pressure pulse reflects off the sea-bed strata and time of arrival information at sensors is used to assess whether there is any likelihood of oil or gas reserves in the area. The air-guns are towed behind a ship, typically in arrays of 20 guns or more.

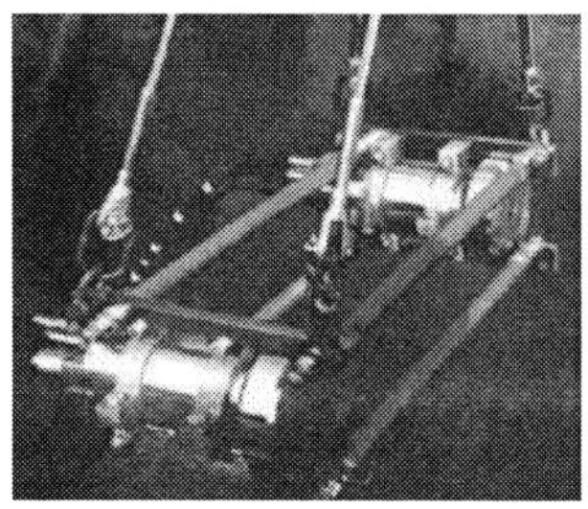

Figure 2 Seismic air gun and a pair of seismic air guns being deployed for oil exploration.

Air-guns are currently being investigated as an alternative shock testing (trialling) technique for naval vessels by the USN. Weidlinger UK is leading the development and application of this technology in shock loading of naval vessels and test barges and have taken out UK and US patents on the application of the technology [4]. A summary of the patent states that "by using a partially phased and partially synchronous array of seismic air guns of the type used in oil exploration, it is possible to simulate the explosive shock wave and provide a safe, repeatable and more environmentally benign test. The method can be used to provide shock excitation directly onto the keel of a vessel due to the relatively low energy of the air-gun discharge bubbles in addition to the more conventional off-beam scenarios. This novel method of shock testing can be applied to ships, submarines, underwater equipment and test models or specimens".

Tests on USN barges conducted by Weidlinger UK have so far concentrated on the feasibility of the process to produce similar explosive shock load tests on the US Floating Shock Platform [5]. The results of these tests are believed to be very successful and the next phase of any experimental program will require scaling up to large test platforms and or small ships.

In early 2007 DSTO began to develop an air-gun testing capability. This involved purchasing a number of air-guns to assess their potential, both individually and in arrays, for shock loading naval vessels and equipment shock test barges. The air-guns are an off-the-shelf item and therefore technically low risk. Part of the research program will also require manufacturing of structures to support arrays of air-guns, along with acquisition of the air supply components such as high-pressure compressor, hosing, firing units and ancillaries).

Potentially this technology will benefit a number of sections within the RAN, including DMO, DNPS, Capability Development, Dtrials, Training and Operations Divisions and numerous Defence Industry contractors. Ultimately this work could result in the development of a permanent shock testing facility. Naval ships could routinely undergo (low risk of damage) shock testing not only to ensure that they meet class-ship-shock-trial vulnerability requirements, but also for re-testing following equipment refits/structural modifications and for regular exposure/training of crews for 'fighting' the ship after an underwater explosive loading event. It is envisaged that M&S would then be used in conjunction with this technology to assess the vulnerability of the vessel at its design level.

The advantage of air-guns are that they are a directional device and therefore could be detonated in shallow water such as an inlet or river system, and they do not suffer many of the logistical problems associated with detonating explosive charges in blue water trials. The amount of energy emitted by a single air gun is not large, however when used as an array, they could produce sufficient energy for localised shock loads on sections of a vessel. Effectively this is the way the RAN submarine HMAS RANKIN was shock trialled, i.e. by moving a number of small explosive charges around to 'expose' different parts of the vessel to locally load the structure and equipment in specific areas.

Another major advantage with this system is that air-guns can be recharged and fired within a few seconds. This would enable shots to be fired without the dangerous, costly and time consuming procedures associated with explosive charges. The associated gas bubble and pressure wave would also be a valuable aid in vulnerability research for shock, blast (gas bubble) and hull whipping assessments. A potential danger with air-guns is the high pressure air component which needs to be considered due to the extremely high pressures. Even though the guns are rated at up to 3000 psi as seismic sources, they have been fired at up to 5000 psi in shock loading experiments.

Air Gun Shock Trials

Air-guns could be assembled onto a purpose built structure so that an array of the guns can be used to concentrate the air blast onto ship (or submarine) sections. Figure 3 is a schematic of the structure for shock testing a navy vessel.

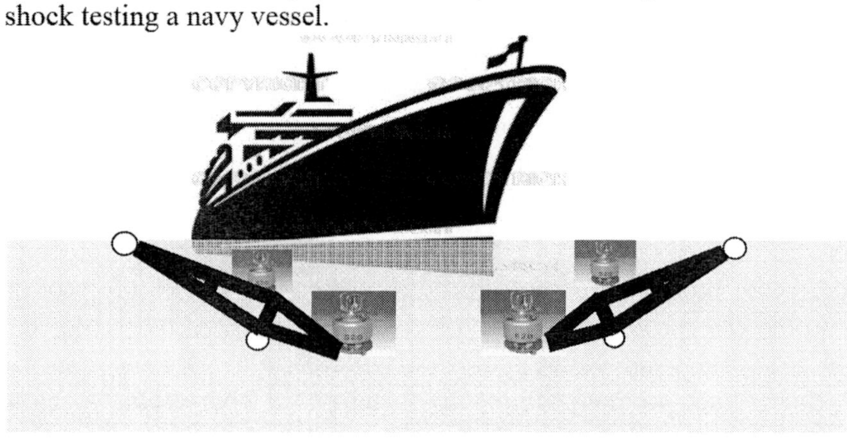

Figure 3. Schematic of air-gun array for shock testing a Navy vessel.

The array consists of multiple sections of the structure shown in Figure 3 supporting a number of guns (e.g. 20 to 30). The guns are charged with compressed air and fired electrically.

Since compressed air is available, buoyancy tanks could be used to raise and lower the guns into position, by emptying/filling with water via the compressed air supply.

DSTO is devising an experimental program to evaluate the performance of the guns at its Underwater Explosive Test Facility. Commissioning of the guns is initially planned for early 2008 with a small number of test firings to evaluate the deployment, firing and recovery of single guns in a target free environment. Once gun firing parameters have been established, the effect of single guns on a number of scaled hull forms will be assessed. This will then be followed by multiple gun firings with and without target vessels. Figure 4 shows a schematic of the gun, storage cylinders and compressor layout when fired as an array. The air-guns are SERCEL type 520 cubic inch (8.5 litres) capacity as shown in Figure 5, along with a typical pressure profile emitted from two of these guns when firing at a pressure of 3000 psi (5 metres depth) [6]. Each gun weighs 98 kg and measures 597 mm length by 287 mm diameter.

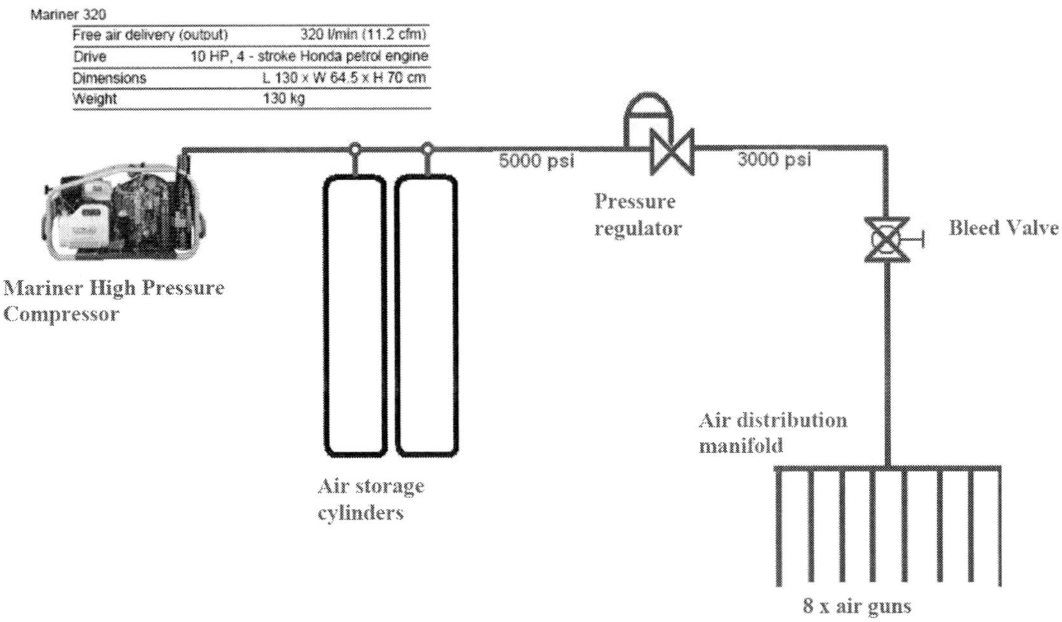

Figure 4. Schematic of air-gun array and air supply system for initial DSTO experiments.

The guns produce a peak pressure pulse considerably less than an explosive charge as shown in Figure 6, which shows the pressure pulse from 10 kg of Pentolite at a distance of 23 metres [7]. Even at the much greater stand-off distance, the explosive is almost an order of magnitude greater in the peak pressure output. An equivalent performance from the air guns, in terms of the shock loading, requires a number of guns at much closer stand-off distances. As discussed earlier, this will require using the guns as an array and even then such an array would only load a portion of a vessel at a time, e.g. a 10 to 15 metre length section for a large ship such as the AWD. However, the portability of an array system and the short time between firings mean that a whole ship could be readily shock loaded in a very short period of time. The limitations are repositioning of the array between successive firings and the around time to reset the instrumentation if required. Safety and margin for error resulting from incorrect placement of an explosive charge would also be reduced as the guns would most likely be placed very close to a vessels' hull.

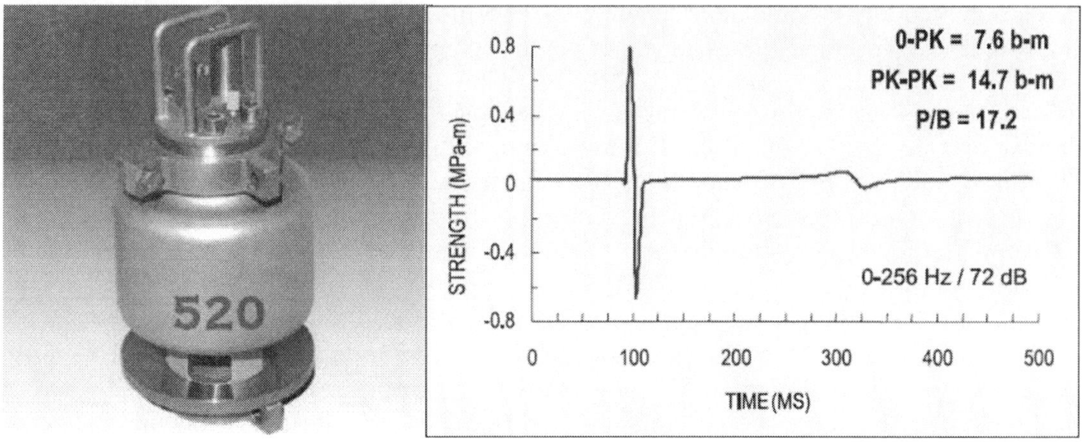

Figure 5. Sercel GI 520 air-gun along with typical pressure pulse.

Given that the actual pressure pulse produced by the guns is considerably different (in peak pressure and loading duration) to that of an explosive, consideration would have to be given to phasing their firing sequence. This is readily achievable given that each gun can be fired individually and in theory at varying pressures if air regulators are placed on the guns or each gun has a separate air supply line. In initial DSTO experiments it is planned to fire all guns at the same pressure.

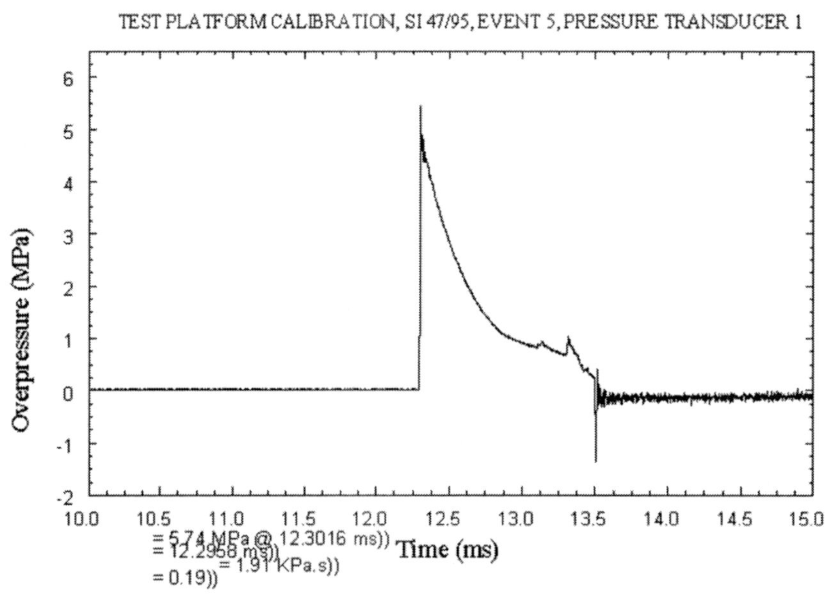

Figure 6. Pressure pulse from 10 kg Pentolite explosive at 23 metres stand-off (10 metres charge depth).

SUMMARY

This paper has detailed the basis for shock trialling naval vessels and the RAN Hobart Class Air Warfare Destroyer in particular. Conventional shock testing techniques have been discussed with their limitations and associated problems. An alternative shock trial technique involving air-guns has been discussed along with some of the potential advantages such as cost, logistics, environmental impact, safety, crew training and viable retesting opportunities.

DSTO is investigating air-guns as an alternative shock trialling technique in collaboration with foreign government and industry organisations. These investigations will involve a series of experiments at DSTOs Underwater Explosive Testing facility which if successful could then result in the establishment of a test/retest facility for RAN vessels and the shock trialling of the Air Warfare Destroyer.

REFERENCES

1. DEF(AUST)5000 Volume 2 Part 6 – Whole of Ship Requirements.
2. "Air Warfare Systems Centre, Shock Trial Considerations", draft White paper, AWD-SC-XXXX_01D01, September 2006.
3. http://defweb.cbr.defence.gov.au/home/documents/deptinstructs/environment.htm
4. http://www.freepatentsonline.com/6662624.htm.
5. NAVSEA, *Shock Tests (High Impact) Shipboard Machinery Equipment and Systems, Requirements for*. 1989, Department of the Navy (USA), Naval Sea Systems Command. (MIL-S-901D).
6. SERCEL "Marine Sources" Air gun Information PDF, http://www.sercel.com/en/Products/Seismic-Sources/Marine-Sources.php, September 2006.
7. Reid W and Burch I., "Calibration of a Shock Platform for Testing COTS Equipment", DSTO-TR-0776, February 1999.

A Trial Sensor Network for the Armidale Class Patrol Boat

C. P. Gardiner[1], P. Vincent[1], A. Wilson[1], D. Ellery[2] and T. Armstrong[2]
[1] Defence Science and Technology Organisation, Melbourne, Australia
[2] Austal Ships, Fremantle, Australia

ABSTRACT

A sensor network being trialled on a Royal Australian Navy (RAN) Armidale Class Patrol Boat (ACPB) in a collaborative project between the Defence Science and Technology Organistaion (DSTO) and Austal Ships is described. The network includes novel sensors that have not previously been used for ship hull monitoring. Potentially, they will provide versatile, small, energy-efficient options for strain, motion and corrosion sensing of hull structures. The evolution to a larger semi-planing hullform and an all aluminium construction provides an opportunity to deliver significant benefits with the installation of a hull-condition monitoring system. Improvements in assessing hull structure performance for the benefit of through-life fleet management and service life assessment can be realised if the system is complemented with strategic data analysis and modelling and these are also briefly discussed in the paper.

INTRODUCTION

Several new aspects of hull design and construction are introduced with the ACPB compared to previous RAN vessels. The hull is constructed with aluminium alloys rather than steel and is larger and faster than previous semi-planing designs such as the Fremantle Class Patrol Boat. This is consistent with international trends as other navies now begin using higher-speed designs for patrol and littoral duties. Although there is some experience in the commercial sector in managing higher-speed aluminium hullforms the relevance to naval application is limited. For example, passenger or freight vessels usually operate on well-defined routes and have significantly different operational patterns compared to a navy patrol vessel which is required to perform a range of tasks over wide and variable operational areas.

The construction and through-life maintenance of the ACPB fleet has been contracted by the RAN to the private sector through a leasing arrangement. Although this relieves the RAN from the day-to-day structural management issues it remains important to understand the issues and technology relevant to patrol boat capability development. For example, future marine patrol capability will depend on the consideration of platform upgrade, life-extension and fleet replacement options. The commercial sector can also benefit from an improved understanding of the through-life structural management of high-speed aluminium vessels operating beyond pre-defined routes.

Austal Ships, the vessel builder and the DSTO have undertaken a project to trial a novel hull-condition sensor network on an ACPB that incorporates a new range of sensors developed by DSTO. This paper briefly reviews the use of hull-condition monitoring systems and outlines the benefits of the existing trial for the RAN and wider applications. The sensor network is then described in detail including the sensors, network design, sensor locations, the data to be collected and how it can be used for future application.

PROJECT OBJECTIVES

The DSTO has prototyped low-powered strain and corrosion sensors to monitor airframe structures. They require less power, are generally smaller and cheaper than existing sensors and have been applied in an adaptable network. The first objective of this project is to determine if these sensors can be reliably proven in a robust marine application. The benefits of more affordable, versatile and energy efficient hull-condition monitoring systems will then be available for future surface fleet applications. In this regard, it is notable that the sensors and network being trialled are equally applicable to large surface vessels as to the ACPB.

The parameters to be monitored are strain, motions and internal corrosion. These are to be used to assess the incidence and effects of slamming loads, hull-girder loads and habitability. An important objective is for the hull-monitoring system to be easily adaptable to changing demands. This should be more achievable with the sensors investigated as adaptability has driven the detailed network design. Equally importantly by developing a suitably versatile networked application the project seeks to provide a tool to improve the understanding of the through-life structural response of high-speed aluminium platforms.

HULL MONITORING SYSTEMS

Background

Strain-gauge measurements have been undertaken for research applications since the 1960s (Smith, 1966). Initially the main principle purpose was to investigate hull-girder loads as a function of wave height, speed and heading by undertaking a trial in the vicinity of a wave-buoy. Evaluation of frequency response was also used to assess more dynamic transient loads such as springing and hull whipping. Other applications were developed for fatigue logging, ship motion measurement using accelerometers and slamming response using pressure transducers and strain gauges (Clarke, 1982; Hachmann, 1991).

The main uses and benefits of hull structure monitoring may be summarised as design load verification (for new or upgraded vessels), safety through load monitoring display and logging, contribution to through-life hull structure management and damage indication/detection through changes to the dynamic response of the hull. The last couple of decades in particular has seen a much wider application of commercial hull-stress monitoring systems on large merchant ships. Accordingly, the International Maritime Organisation released a guidance note on the application of hull-stress monitoring systems (IMO, 1994) and classification societies have also established guidelines.

As computer hardware has become more efficient and durable there has been a corresponding enhancement of networking to link structural response, motion and wave radar measurements with more common sensor information. Examples include engine data, ship motion, GPS and damage control monitoring data. Hull-monitoring information is now routinely displayed in real-time as well as stored for later analysis. Real-time display can be complemented with statistical trend data to indicate if the structural response is changing or increasing to critical levels, as indicated by classification design loads. Guidance on ship speed and heading can therefore be inferred and in some cases this is coupled with weather routing software.

In recent years there has been a shift towards using fibre-optic sensors for hull-structure monitoring. The main advantages are corrosion resistance, immunity from electromagnetic interference and reduced cabling requirements as multiple sensors can be "written" onto a single fibre (Wang et al., 2001).

Benefits for RAN Application

Currently very minimal quantified knowledge is available on the operation of aluminium deep-vee and high-speed monohulls over the course of a typical 15-20 year service life. In the context of RAN operation gaining such knowledge is vital given that

- the vessel construction and hull-form is significantly different to previous RAN vessels,
- routine patrol vessel operations and profiles are significantly different to most similar vessels in the commercial sector,
- the ACPB operational areas and strategic environment are unique compared with closely allied navies such as the Royal, United States or Royal Canadian Navies,
- the ACPB fleet is significantly the biggest in number where the RAN has used rotational crewing.

What is the risk of such factors to the RAN and the supporting commercial sector? It may be argued, for example that any risk to the RAN is mitigated by the private leasing arrangement for the build and maintenance of the ACPB fleet. However, an incomplete understanding of the magnitude and drivers of potential hull and system deterioration during operation will result in higher maintenance costs, reduced platform availability and resource for fulfilling patrol boat group commitments, as well as possibly limited re-sale or re-lease value. Crew confidence and morale can also be diminished in some cases. Equally importantly for the RAN is the legacy of poor hull structure knowledge and the limitations this could have for planning patrol boat group capability beyond the existing 15-year ACPB service contract. However, such outcomes be avoided with suitable hull-structural monitoring and focused analysis to track and manage hull-structure condition and the drivers thereof. For example, verifying design loads and generating information on the occurrence of bow slamming, hull fatigue and corrosion has several benefits

- reduction of risk of reduced ACPB fleet capability and subsequent effects,
- remaining ACPB service life options can be more accurately assessed,
- increased opportunity to optimally acquire similar or related vessels in the future,
- availability of diagnostic information for unforseen damage events, and
- improved safety and crew knowledge.

The fundamental goal of applying a structural hull-monitoring system is to facilitate effective future fleet management while maintaining high-levels of readiness and operability. Naturally this requires more than collection of hull-structure measurement data, nevertheless a structural hull-monitoring system is central to meeting such an objective.

TRIAL SENSOR NETWORK

Scope of Proposed Measurements

It is proposed to test the sensor network firstly with a dedicated trial and then longer term during normal RAN operation. Arrangements and approval for the latter are to be finalised.

The aim of the dedicated trial is to;

- prove initial sensor durability and data collection capability, and
- determine hull-girder loads and detect occurrences of bow slamming, in correlation with operational conditions such as vessel speed, heading and wave height.

The measured strain and motion data will also be used with the calculated hull loads to determine initial values to trigger data storage during the following ongoing trial.

Longer-term operational durability of the system will then be tested during normal RAN operation. This will also provide data on typical operational loads, stresses and cycles for the ACPB. It is expected that initial data analysis procedures and criteria for storage and download of data will require refinement.

The objectives of the continuous measurement trial are two-fold;

- implement a strategy for the collection, management and analysis of data to enhance the through-life management of the hull structure, and
- develop a real-time display mode (available to the Bridge).

The real-time display is to be developed to provide guidance to the operators for safe operation. This may involve displaying loads with respect to design values and trend analysis to indicate recent changes in loading.

Sensor Network System

The DSTO has developed a combined sensor interface and networking technology with the following key features relevant to the current application:

- low power sensor signal conditioning and network interface (RS485) in one small, light weight package,
- remote software upgrade capability,
- the ability to handle large networks of sensors,
- software and hardware adaptability to accommodate new sensors,
- a software interface that is flexible enough to accommodate single or multiple sensors,
- reduced network bus traffic by the use of local data storage and processing and local timing for data logging.

The original system was developed for low power, low data rate applications associated with corrosion monitoring. However, the system was versatile enough so that with a slight change in the philosophy of operation it can now support data rates up to 100Hz from 3 strain gauges or accelerometers simultaneously. This adaptability is largely due to the modular approach used to develop both the hardware and software. Figure 1 shows the sensor interface with the electronics and applications for specific sensors consisting of an invariant core communications and processing module interfaced to specific electronics and software for the sensors. The ACPB application required the development of a high speed device with significant local data storage for the signal conditioning and monitoring of strain gauges and accelerometers. Again, the modular approach and the method used for downloading the sensor software was vital to the rapid development and debugging of this software and hardware.

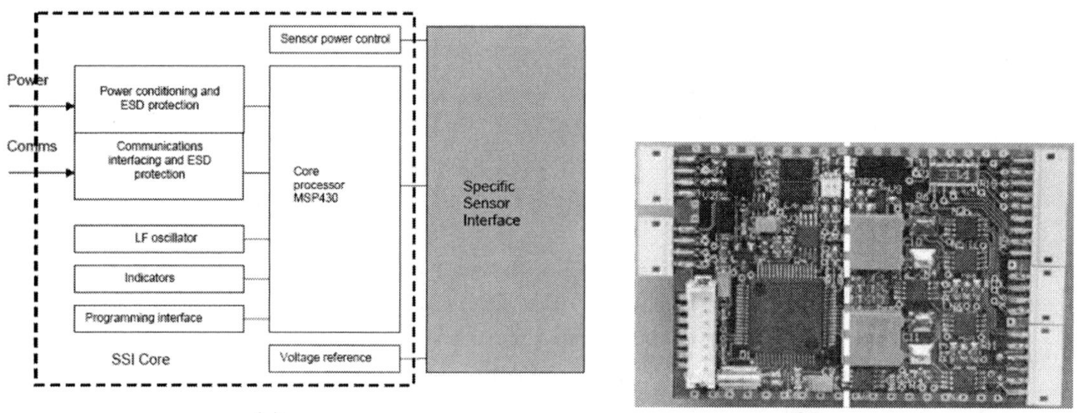

(a) (b)

Figure 1: Sensor interface is modular with the software and the hardware divided into a core communications and processing module (left-hand side) and software and hardware for specific sensors (right-hand side). b) shows the device used to monitor the corrosion and environmental sensors. This unit is 35mm wide and 56mm long.

The communications and core routines allow for the local sensors to be initialised and high speed data logging initiated at the sensor level. Data is then retrieved asynchronously in large blocks at somewhat longer time intervals than the maximum acquisition rate (e.g. blocks of 300 data points corresponding to 1 seconds worth of data from a strain rosette). This allows for easy mixing of sensors with different data acquisition rates on the one network. Time synchronisation is maintained by the transmission of time stamp identification codes to all devices on the network. These identification codes are then incorporated into the data stream stored on the sensor.

The sensors on the ACPB are grouped into a number of networks determined by the physical location of the sensors and the practicability of the cable runs. This also reduces the data rate required on any one network to achieve pseudo real-time data transfer. Each sensor on a network has a unique address and can be accessed individually. The networks are controlled by a single embedded PC operating at 1GHz with multiple RS485 communications at a data rate of 115200 Baud. Future implementations could consider higher speed (and higher power) networks. The embedded PC also stores the data from a Kongsberg MRU-6®, a commercial 6-axis motion sensor.

Choice of Sensors and Locations

The position of sensors is shown in Figure 2. They are used to measure corrosion and related environmental parameters, global and local motion and strain as follows:

1. Corrosion/Environment

The corrosion/environment sensors consist of a suite of 4 sensors comprising a DSTO developed 2-electrode corrosion sensor, time-of-wetness sensor and a commercial temperature and humidity sensor. They are located in areas judged to be potentially damp: under the chain locker, above the lowest point in the bilge, close to the sea chest and under the floor in the austere accommodation. The corrosion sensor suite will initially be monitored every 10 minutes and the frequency decreased for longer term monitoring once the system is initially tested and collected data is evaluated.

2. Global Motion

Global motion sensors consist of 4 low G, 3-axis accelerometers positioned port and starboard and fore and aft in the hood region in front of the bridge. The positions were chosen to give large lever-arms (about the centre of gravity) in reasonably accessible locations. The Kongsberg MRU-6® is located in the hood close to these sensors. A 3-axis accelerometer is also located at the top of the mast. These sensors will initially be monitored at up to 20Hz.

3. Global Strain

Global stress is to be measured using strain gauges at three longitudinal sections; forward, towards midships (Frame 27-28), and aft in the austere accommodation (Frame 42). The criteria governing the choice of strain gauge locations at each section was to select more highly stressed locations with a relatively uniform stress distribution and minimal influence from local loading. There is a trade-off however in practice due to structural detail, fit-out and accessibility. Stress magnitude and distribution was assessed using finite element analysis of the complete hull structure with the design load cases.

4. Local Motion and Strain: Slamming

Three-axis accelerometers and strain rosettes are located on the keel and several metres up from the keel, port and starboard, approximately 12.8m and 20.0m from the forward perpendicular to detect high acceleration and strain events due to slamming. These sensors can be monitored at up to 100Hz.

Future Sensor Development

The selection and layout of the existing sensors may be refined according to initial trial results. Following the dedicated trial to validate the system, analysis will be undertaken to determine suitable magnitudes for triggering permanent data storage consistent with significant events. It is planned to develop this further by localising this function to the sensors so that the local processing power of the microcontrollers on the sensors are used to only report on significant events. This will be particularly relevant for detection of slamming events.

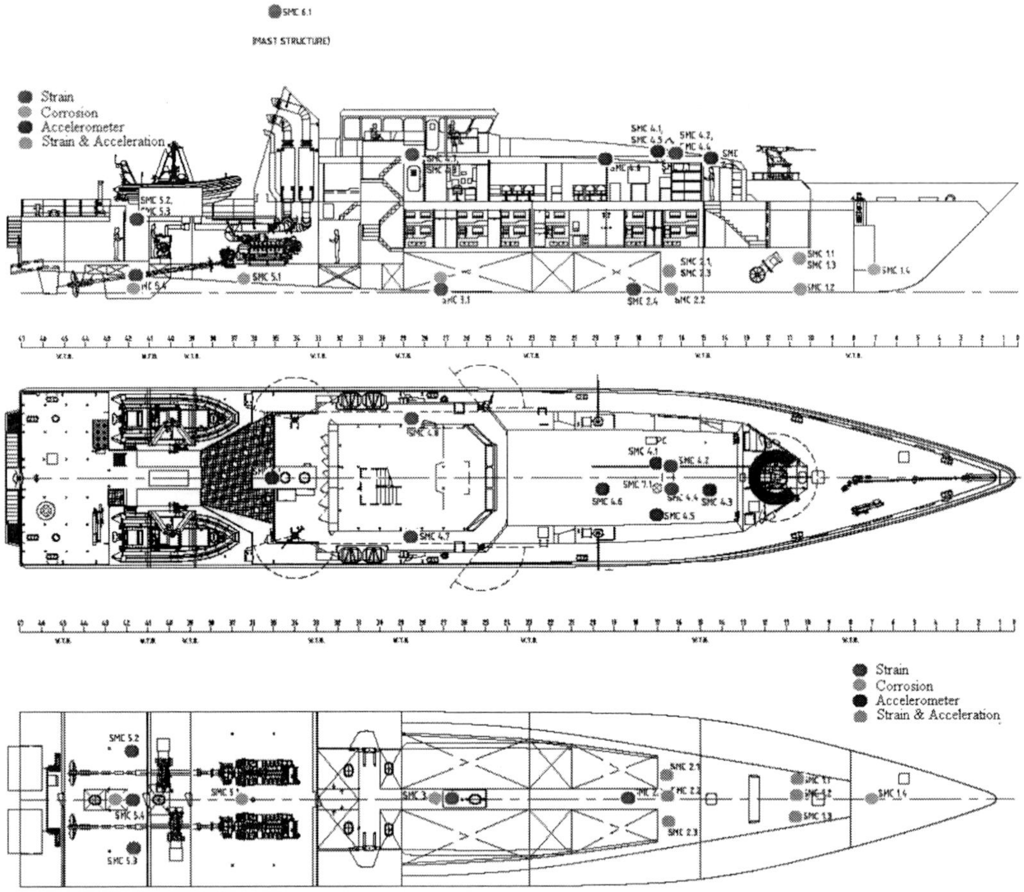

Figure 2: General arrangement drawing showing the sensor locations.

DATA ANALYSIS

Global Loading

The main global load to be determined at each cross-section is the wave-induced vertical bending moment (VBM). The wave-induced horizontal bending moment, torsional bending moment and vertical shear force will also be analysed. The sensor cross-section near midships is where the vertical bending moment is likely to be close to the maximum. The basis of the analysis is to use the measured time-series of strains and apply load-to-strain factors that are determined from FEA to calculate the relevant load.

Denoting the vertical bending moment by v, the lateral bending moment by l, the torsional moment by t, and the vertical shear force by s, then the strain component of interest at each gauge location is related to the relevant load by;

$$\begin{Bmatrix} \varepsilon_1 \\ \varepsilon_2 \\ \vdots \\ \varepsilon_n \end{Bmatrix} = \begin{bmatrix} R_{1v} & R_{1l} & R_{1t} & R_{1s} \\ R_{2v} & R_{2l} & R_{2t} & R_{2s} \\ \vdots & \vdots & \vdots & \vdots \\ R_{nv} & R_{nl} & R_{nt} & R_{ns} \end{bmatrix} \begin{Bmatrix} v \\ l \\ t \\ s \end{Bmatrix} \tag{1}$$

where n is the number of strains and $[R]$ is a matrix of load-to-strain factors that are determined from FEA. Given the measured strains $\{\varepsilon\}$ for a particular run and the matrix of load-to-strain factors $[R]$, the load time series that gives the best fit to the measured strains can be determined. If the error $\{\delta\}$ is defined as the difference between the measured strains $\{m\}$ and the strains $\{\varepsilon\}$ calculated using equation (1), then the sum of squares of errors, Ω can be defined as;

$$\Omega = \{\delta\}^T \{\delta\} = (\{m\} - \{\varepsilon\})^T (\{m\} - \{\varepsilon\}) \tag{2}$$

where the T superscript refers to the transpose. It can be shown that the loads which minimise the error Ω for a given set of measured strains $\{m\}$ is given by;

$$\begin{Bmatrix} v \\ l \\ t \\ s \end{Bmatrix} = \left[[R]^T [R]\right]^{-1} [R]^T \{m\} \tag{3}$$

Considering each cross-section where the strain gauges are located, there is one tri-axial strain gauge on the centreline near the keel and two tri-axial strain gauges higher up symmetrically either side of the centreline. It is proposed to use the longitudinal strain at each of the three locations to calculate the vertical and lateral bending moment. The 45° strain will be used to determine the torsion moment and the vertical shear force. Four strain components are therefore assessed to give n=4 in equation (1). The lateral and longitudinal forces may also be estimated.

The accuracy of the estimated loads from equation (3) can be assessed by substituting back into equation (1) and comparing the resulting time-series of calculated strains against the measured values. Higher frequency loading will be filtered initially to determine wave-induced loading prior to assessing the occurrence of higher frequencies for possible whipping loads.

The (wave) load-to-strain factors are determined using a static finite-element analysis by modelling an idealised load case separately for each load component. This involves applying a unit moment load at a position remote from the strain gauge location seperately in-turn for the vertical, lateral and torsional wave bending moments. The strain due to vertical shear force is determined by applying a unit vertical load away from the sensor location and then subtracting out the strain due to the bending moment, as shown by Jensen et al., (2001).

LONG-TERM HULL STRUCTURE ASSESSMENT

In addition to facilitating a more accurate evaluation of the existing condition of a hull structure, the installation of a hull structure monitoring system provides the means to make a more qualified prediction of the condition for future operation. A proposed conceptual outline of an evaluation process is shown in Figure 3. More detailed refinement is to occur as the project continues.

The initial input includes a definition of the structural configuration, material properties and the loads on the vessel. The latter is based on measurements from the sensor system. Analysis is then required at a component and system level. The failure modes, structural resistance models and limit states relevant to the overall hull structure system can be a unique combination of those that apply to each component of the system. Alternatively, component limit state may directly define the limit state for the entire system.

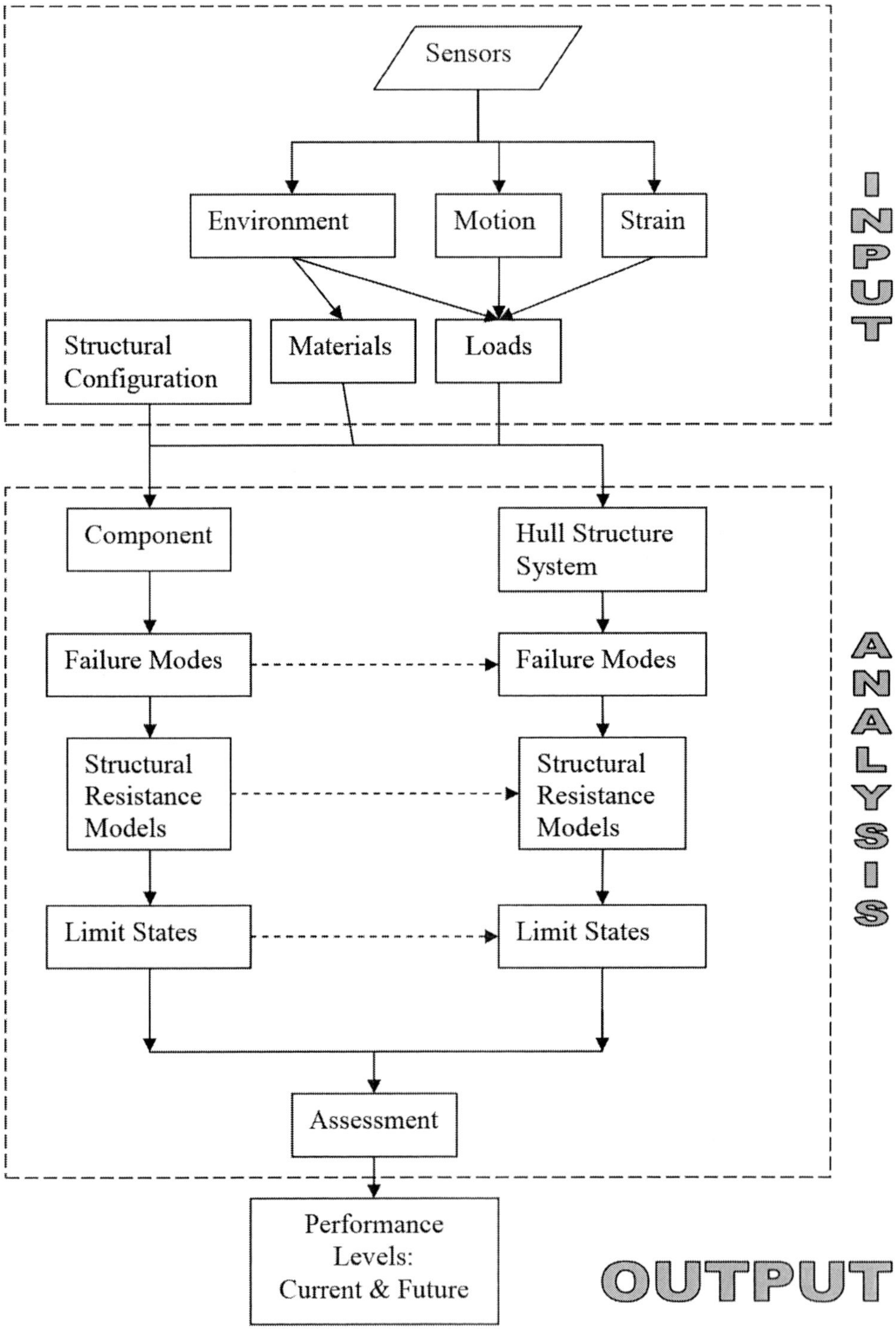

Figure 3: Through-life Hull Structure Performance Assessment.

Specific analysis of each component begins with an identification of how it may fail (failure modes) and the generation of a model defining the factors that contribute to the structural resistance to those failure modes. For example, local scantlings, and hence the structural resistance may decrease with time due to corrosion, or material stiffness may decrease due to fatigue. Both of these will be dependent on the operational environment and specific environmental factors.

For each failure mode there is a limit state to be defined outlining the condition for which the structural resistance of a component becomes unacceptable for the relevant load or action effects. It is important that each limit state is defined in terms of parameters that are available to define the failure mode. There will often be more than one failure mode for each component and it is therefore necessary to define a means to evaluate component performance based on the probabilities of failure for each failure mode.

Integrating the component performance levels in relation to the performance of the hull structure system provides the final output – an assessment of performance of the overall hull structure in the current condition, at service life completion or an intermediate time. The ship structure can then be assessed against other systems eg., electrical and mechanical systems, in terms of capability and availability. Such an assesment can then be used to provide information to support remaining life estimation, operational planning, maintenance scheduling and repair strategies.

SUMMARY

The ACPB fleet features the largest semi-planing hullform to be used to date by the RAN and the first all aluminium hull structure. The means to assess the future performance capability of the hull structure may offer an improved assurance level of fleet availability and assist planning ongoing patrol boat group capability. Additionally, the ACPB fleet provides an ideal application to develop knowledge on operational issues relating to aluminium hull structures that can be applied to the maintenance and scheduling of the ACPB fleet and the planning and design of future RAN vessels. The use of a hull-condition monitoring system is an essential part of establishing such a capability. The hull-condition monitoring system that has been developed and is being trialled is innovative in terms of sensor design and networking. The trials and analysis outlined aim to verify potential improvements in sensor technology such as useability, cost and power-requirements for both Defence and commercial marine applications.

ACKNOWLEDGEMENTS

This work is supported by the DSTO Initiative in Smart Materials and Navy Minors Project 1001, "Management of Marine Aluminium Alloy Hulls".

REFERENCES

Clarke, J.D (1982). "Measurement of hull stresses in two frigates during a severe weather trial." Transactions RINA, Vol. 124, pp. 63-83.

Hachmann, D (1991). "Calculation of pressures on a ship's hull in waves." Ship Technology Research, Vol. 38. pp. 111-132.

International Maritime Organisation (IMO). "Recommendations for the fitting of hull stress monitoring systems." MSC/Circ.646, London, June 1994.

Jensen, A. E., Taby, J., Pran, K., Sagvolden, G. and Wang, G. (2001). "Measurement of global loads on a full-scale SES vessel using networks of fiber optic sensors." Journal of Ship Research, 45(3), pp. 205-215.

Smith, C.S (1966). "Measurement of service stress in warships." Proc. Conf. Stresses in Service, Institution of Civil Engineers, London.

Vincent P. S, McMahon P. J, Muscat R. F, Zeve L, Wilson A. R (2006). "A small low-power networked and versatile sensor interface." SPIE International Symposium on Smart Materials, Nano-, and Micro-Smart Systems 2006: Smart Structures, Devices and Systems III, pp. 6414-6439.

Wang, G. Pran, K., Sagvolden, G., Havsgard, Jensen, A.E., Johnson, G.A. and Vohra, S.T. "Ship hull structure monitoring using fibre optic sensors." Smart Structures and Materials, 10(3), pp. 472-478, 2001.

Predicting reduction of tankers' Hull Girder Section Modulii

Ge Wang, Lyuben D Ivanov, Ah Kuan Seah
American Bureau of Shipping, Houston, USA

Summary

Periodical dry-docking and close-up inspection of tankers have long been an established regulatory requirement to monitor the health of the tanker's hull structure. While simple tool of extrapolation based on historical measured thickness data has been used by many to predict future corrosion and to assist in hull maintenance, pooling of data across tanker fleets has not been widely attempted. Also, hull girder section modulus (HGSM) has long been regarded as a more realistic representation of hull girder strength than mere thickness readings. This paper describes such an attempt by ABS using gauging records of several hundred tankers. The as-measured structural member thicknesses are used to calculate the HGSM loss. Statistically, the pooled data suggest that HGSM loss may best follow a Weibull distribution. Weighted regression analysis is used to obtain the mean values and standard deviations of the HGSM loss and the then the corresponding Weibull parameters are calculated. The result of this study could be used to investigate time-variant reliability of tanker hull structures.

Introduction

The industry has established a practice of using geometric properties of ships' hull structure to control the safety level of ships. This serves as a "representative" of ships' reliability although the embedded reliability level is not generally known (for a method to reveal how accurately the control of the geometric properties represents the control of the ship hull structure, see Ivanov et al, 2007). The objective of this paper is to calculate the probability that the Hull Girder Section Modulus (HGSM) will meet given renewal Criteria requirements (in the paper, the IMO Renewal Criteria for tankers, 2000) using the gauging records for around two hundred tankers.

Database of Hull Girder Section Modulus of tankers

The majority of publications are about individual structural members such as plates and stiffeners (TSCF 1992, Yamamoto and Ikegami 1998, Gardinar C P and Melchers R E 2001, Ivanov L D 2001, Harada S et al. 2001, Wang G et al. 2003a&b, Paik J K et al. 2003, Garbatov Y et al. 2005, Wang G et al. 2005, Thuanboon S et al. 2006). This is a necessary step before considering the probabilistic nature of the HGSM. As an example, the results of the statistical analysis of gauging records of bottom plating thickness of single hull tankers in the ABS fleet are shown in Figs. 1 and 2. Based on the gauging records for all structural components, the reduction of the HGSM is calculated and shown in Figs. 3 - 5. The data for the ships used in the analysis are given in Table 1. Figs. 1-5 reveal the large scatter of data for the means and standard deviations. One of the reasons for the scatter might be the insufficient statistical data for given ships' age. Although this data set is the largest ever reported in the public domain, it is apparently necessary to continuously add data to this database.

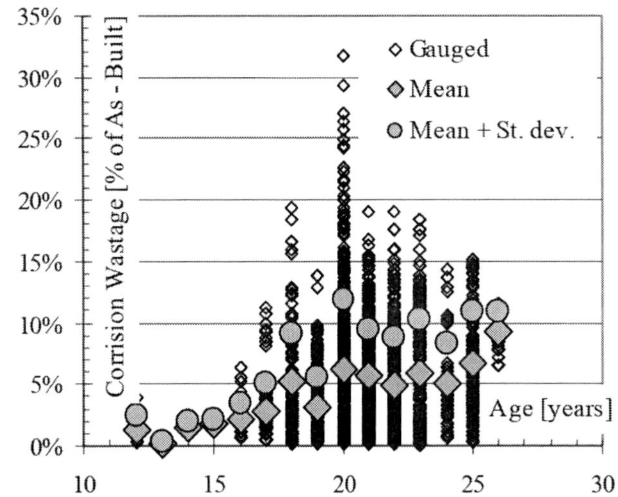

Fig. 1 Single Hull Tankers. Bottom Plate, Cargo Holds, 3580 readings

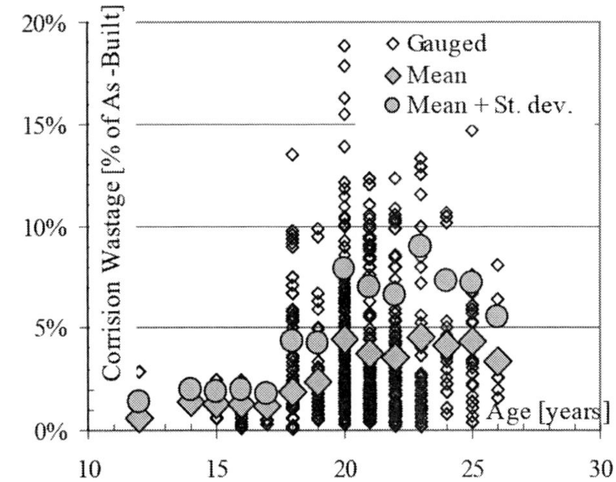

Fig. 2 Single Hull Tankers. Bottom Plate, Ballast Holds, 1016 Readings

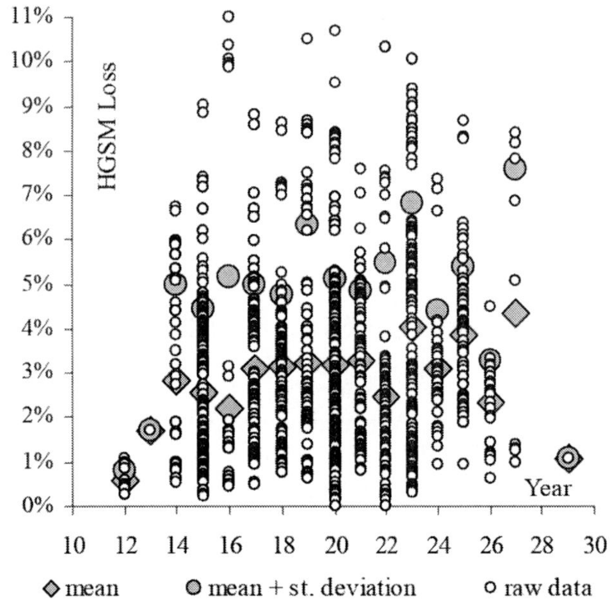

Fig. 3 HGSM loss collected from 2195 sections of 211 single hull tankers

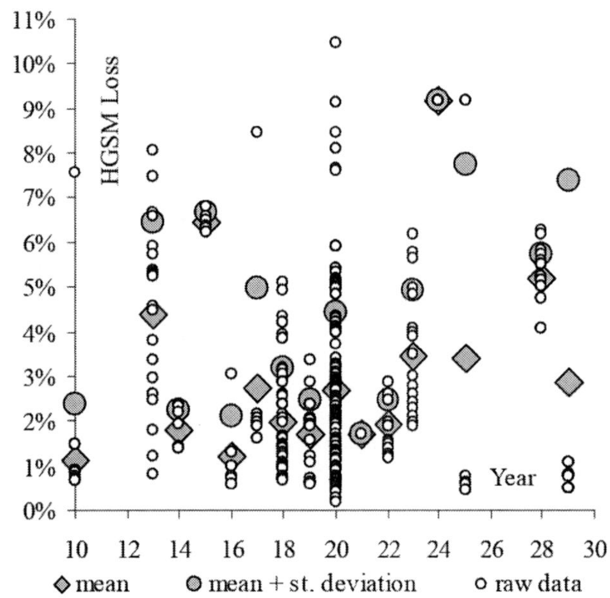

Fig. 4 HGSM loss collected from 573 sections of 51 double hull tankers

Statistical trends

A second reason for the large scatter is that the volume of data is not uniform at different ship age. Therefore, a weighted regression scheme was selected to account for the influences of the different number of data. The analysis was performed to obtain smooth curves for the mean values and standard deviations.

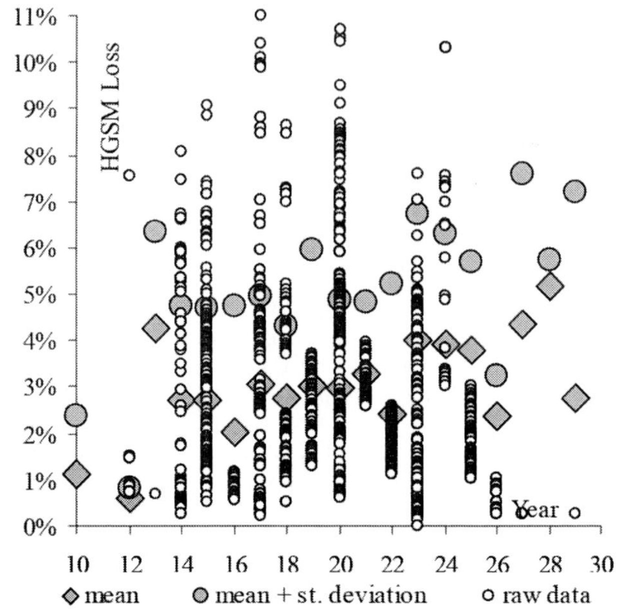

Fig. 5 HGSM loss for all tankers

Table 1. Data used in the analysis

Ship type	Tankers
Data sources	SafeHull Condition Assessment Program
Vessel length	100 – 400 meters long
Ship age	12 - 32 years old at the time of gauging
Class	5 IACS members
Year built	1970's and 1980's
Time at the time of gauging	1992 – 2004
Vessels and data points of HGSM	211 single hull tankers with 2195 transverse sections 51 double hull tankers with 573 transverse sections

This type of regression analysis takes into consideration the number of raw data for each ships' age and allows for obtaining more realistic curves for the mean values and standard deviations of the HGSM loss as a function of the ships' age. Once the curves (and the corresponding equations) for the mean values and standard deviations for the HGSM loss are available, one can build curves with different levels of confidence, e.g. curves for mean values plus one standard deviation, mean values plus two standard deviations, mean values plus three standard deviations as shown in Fig. 6.

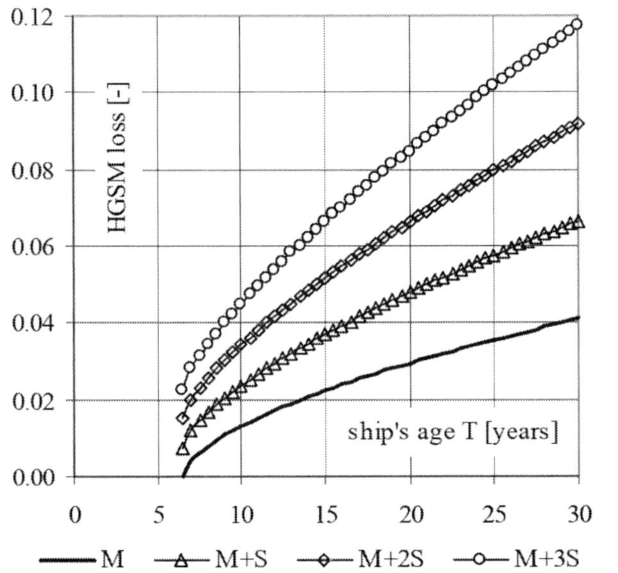

Fig. 6 HGSM loss at different levels of certainty (M = mean value, S = st. deviation)

Dozens of equations were tested for the mean value and standard deviation using Jandel Scientific Curve-fir program. The following equations were found to better fit the data:

$$M = 0.0062\,(T-6.5)^{0.60} \qquad S = 0.0017\,T^{0.80} \tag{1}$$

where M stands for the mean value and S stands for the standard deviation of the HGSM loss.

Time-varying probability density function

As to the probabilistic distribution of the HGSM loss within any given ships' age, calculations were performed to find the type of probabilistic distribution that best fits the sample of statistical data (an example for the histograms of the HGSM loss for ships' age of 25 years is shown in Fig. 7).

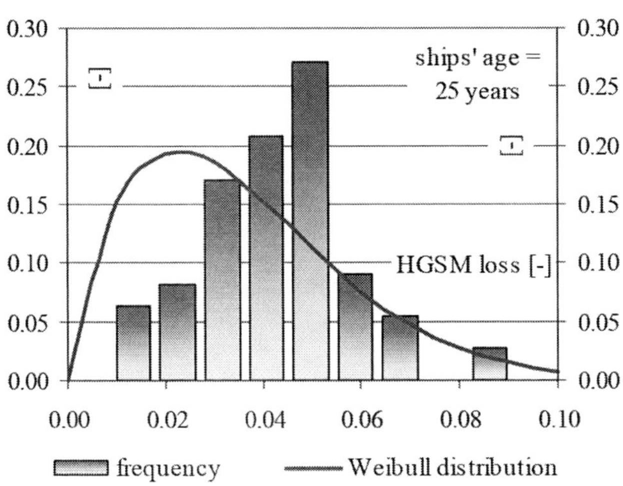

Fig. 7 Histogram of HGSM loss for 15 year old ships

It was found that the Weibull distribution best fits the database. The equation of the probability density function for the HGSM loss is (Ochi M, 1989):

$$f(x) = \frac{\lambda}{\alpha^{\lambda}} x^{\lambda-1} \exp\left[-\left(\frac{x}{\alpha}\right)^{\lambda}\right], \tag{2}$$

where "x" stands for the HGSM loss, λ = shape parameter, α = scale parameter.

When the shape and scale parameters of x are known, the mean value and variance of x can be calculated by the formula (Ochi M, 1989):

$$m_x = \alpha\Gamma\left(1+\frac{1}{\lambda}\right) \qquad D_x = \alpha^2\left\{\Gamma\left(1+\frac{2}{\lambda}\right) - \left[\Gamma\left(1+\frac{1}{\lambda}\right)\right]^2\right\} \tag{3}$$

where m_X = mean value of x and D_X = variance of x.

In the case, we know the mean values and standard deviations and have to determine the shape and scale parameters. Unfortunately, analytical solution of this problem is not available because of the complexity of the Gamma function (Fig. 8).

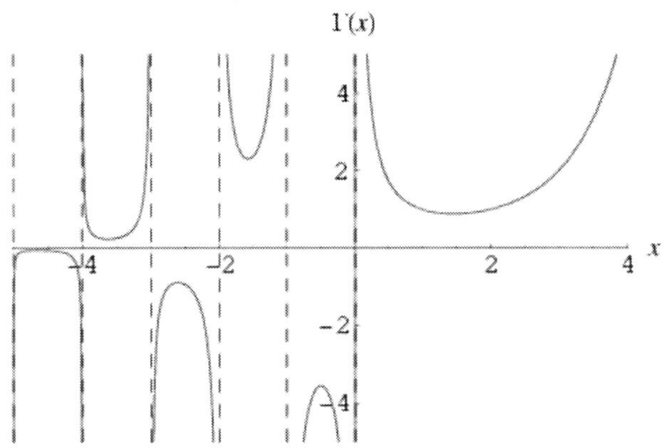

Fig. 8 Gamma function

However, we are dealing only with the part of the Gamma function where x > 0. For this case, the following numerical solution was developed to obtain the shape and scale parameters λ and α of the Weibull distribution:

- Based on Eq. (3), the scale parameter α is represented as:

$$\alpha = m_x / \Gamma\left(1+\frac{1}{\lambda}\right) \tag{4}$$

- Then, α is substituted in the second equation of formula (3):

$$\frac{D_x}{m_x^2} = \frac{\Gamma\left(1+\frac{2}{\lambda}\right)}{\left[\Gamma\left(1+\frac{1}{\lambda}\right)\right]^2} - 1 \tag{5}$$

- Bearing in mind that (Spiegel M R, 1968)

$$\Gamma(x) = (x-1)\Gamma(x-1), \tag{6}$$

one can derive the following condition for finding the unknown shape parameter λ:

$$\frac{D_x}{m_x^2} = 2\lambda \frac{\Gamma\left(\frac{2}{\lambda}\right)}{\left[\Gamma\left(\frac{1}{\lambda}\right)\right]^2} - 1 \tag{7}$$

If we build a graph for the expression right of the equality sign in Eq. (7), the crossing point of the graph with the horizontal line D_x / m_x^2 will determine the shape parameter λ (see Fig. 9).

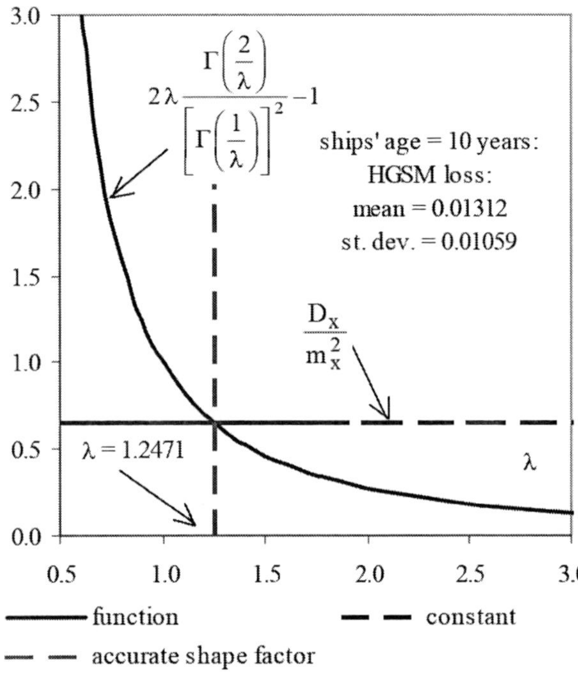

Fig. 9 Example for determining the Weibull shape parameter

- When the shape parameter λ is calculated, the scale parameter α is calculated by Eq. (4).

To make easier the application of the procedure, a parametric study was performed within wide range of λ (from 0.3 till 6) to find an approximate but sufficiently accurate formula for λ.

$$\lambda = 0.0068 + 1.0189 \frac{\text{Mean}}{\text{St. dev.}} \tag{8}$$

The maximum error imbedded in the approximate analytical method is within the range of minus 4.5% to plus 2.5% relative to the accurate numerical method for coefficients of variance (i.e., standard deviation divided by the mean value) above 0.7. For c.o.v. below 0.7, the error is smaller than 1%.

The parameters λ and α of the Weibull distribution of HGSM loss for each ships' age (they were calculated by the accurate numerical method) are given in Fig. 10.

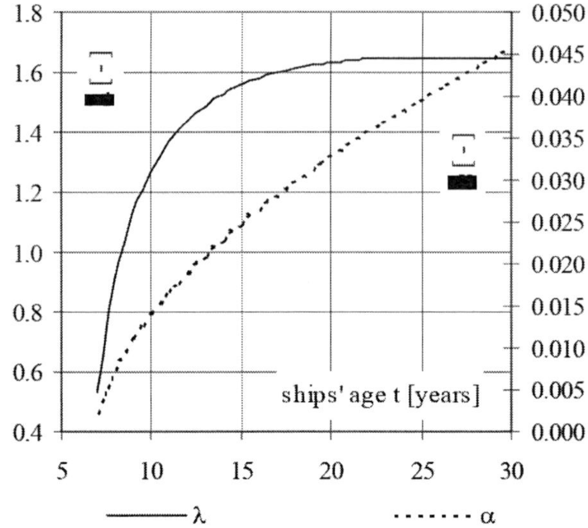

Fig. 10 Shape and scale parameters of the Weibull probabilistic distribution

Knowing these parameters, one can build the probability density functions or the cumulative distribution functions and probability of exceedence of HGSM loss for any given year (see Fig. 11). The graphs in Fig. 11 could be labeled graphs of the "annual" probabilistic distributions, i.e. they refer to ships with specific age.

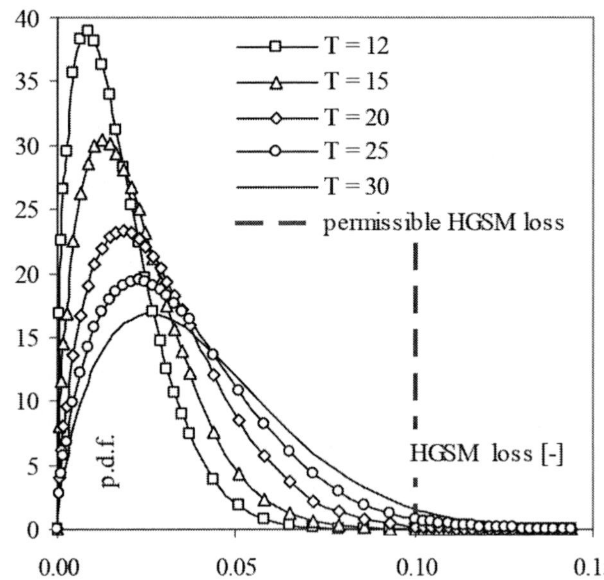

Fig. 11 Annual p.d.f. of HGSM loss

Probabilistic distributions of the HGSM loss for the whole ships' age span

To know the condition of hull girder strength of an entire

fleet, it is necessary to know the probability distribution of the HGSM loss of all vessels for the entire intended service time. This can be derived by following these steps:

- Build the envelope of the p.d.f. (Fig. 12). In this case, the volume below the envelope should be equal to unity.

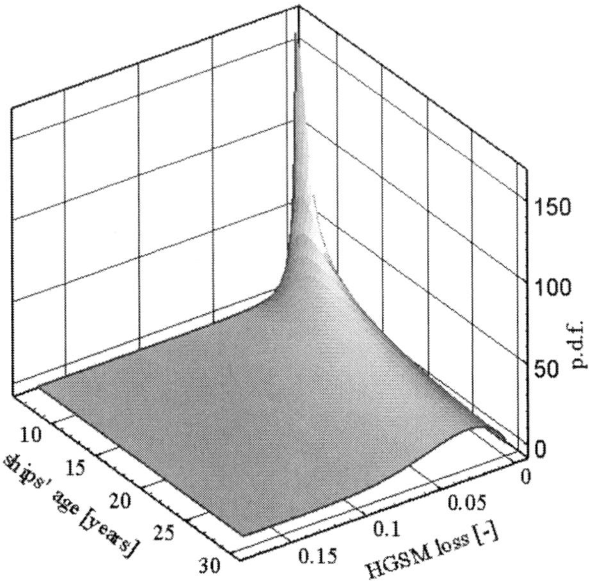

Fig. 12 Envelope of the annual p.d.f. of HGSM loss

- Specify the value of the HGSM loss beyond which the vessel is viewed not fit for the intended service. This specified value is the permissible value of HGSM loss.
- Calculate the volume below the envelope of the annual probability density functions between zero and the permissible value of the HGSM loss.
- Build the cumulative distribution function (cdf) for the whole ships' age span of the given fleet (see Fig. 13). Each ordinate of the c.d.f. that corresponds to any assumed value of the HGSM loss (say 50 assumed values of the HGSM loss between zero and 0.15 as in the paper) is calculated by division of the volume below the envelope of the annual probability density functions between zero and the permissible value of the HGSM loss divided by the total volume.
- Create the p.d.f. for the whole ships' age span by numerically differentiating the so-derived c.d.f. (see Fig. 13). In geometric terms, the value of the p.d.f. at any given HGSM loss is equal to the tangent of the angle between the tangent line at this HGSM loss and the horizontal axis. The accuracy of the calculated p.d.f. is also checked by integrating it to derive the area below the curve. This area should be equal to unity and this is the criteria used to assess the accuracy of the procedure. It has been found that the error resulting from application of the trapezoidal rule and the numeric differentiation of the c.d.f. is smaller than 1%.

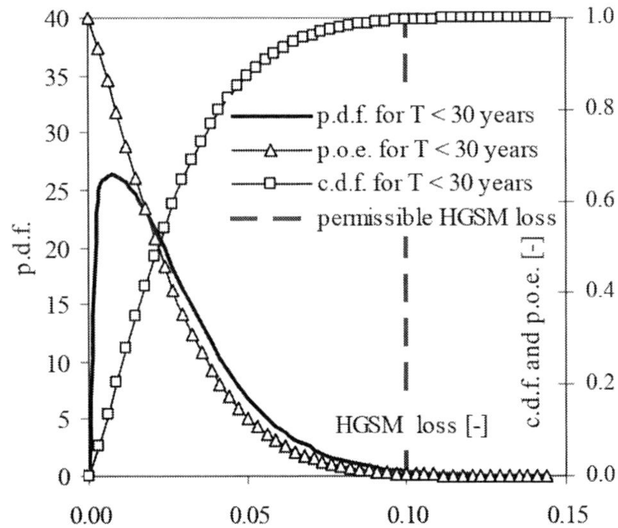

Fig. 13 P.d.f., c.d.f. and p.o.e. of the HGSM loss for all ships younger than 30 years

One should emphasize that the ordinates of the probabilistic distributions for the whole ships' age span are not equal to the algebraic average of the corresponding annual probabilistic distributions. As an example, Fig. 14 illustrates the difference between annual p.d.f. and p.d.f. for the whole ships' age span.

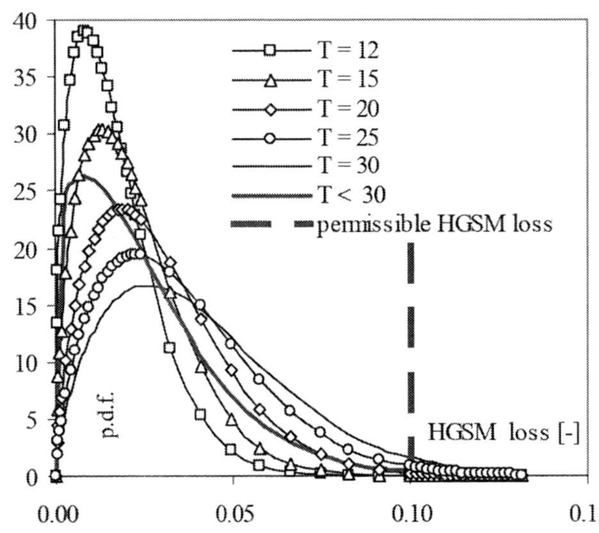

Fig. 14 Annual p.d.f. and p.d.f. of the HGSM loss for all ships younger than 30 years

Embedded safety of tankers' hull girder strength

Once the probabilistic distributions for the whole fleet (see Figs. 13 and 14) are available, one can calculate the embedded probability of meeting certain criteria of hull girder strength. As an example, the criterion formulated by the International Maritime Organization (IMO, 2000) for the permissible 10% reduction of the HGSM is used. Note that the 10% allowable reduction is against the requirement for new build. As most tanker's section modulus to the deck is close to this requirement, it is convenient to assume that the as-built section modulus to the deck equals the required. This assumption is conservative as it gives a lower estimate of the actual hull girder section modulus of tankers

The results are illustrated in Fig. 15 for ships of given age and for all ships in a fleet younger than any given age.

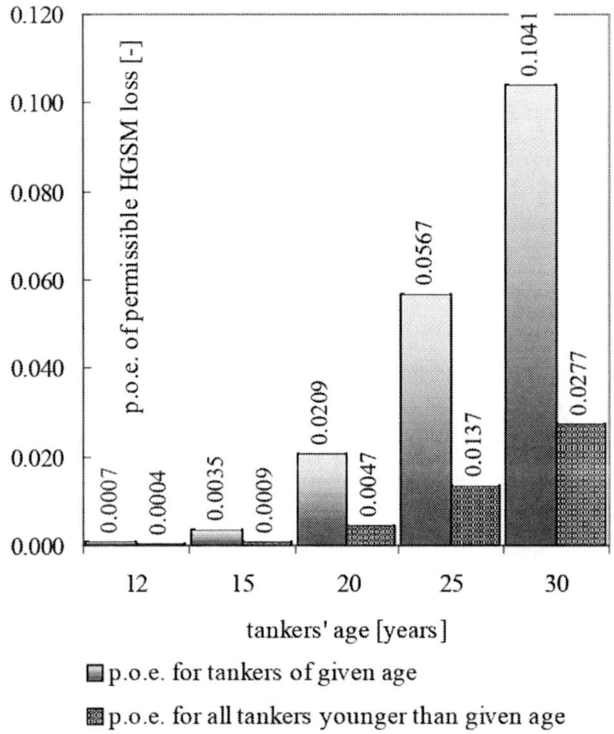

Fig. 15 Example for the probability of meeting the IMO permissible reduction of the HGSM

One can observe that the probability of exceeding the IMO permissible reduction of the HGSM is around 4.10^{-4} for tankers younger than 12 years, which may explain why some operators view this age of tankers as a "limit of caution". Analysis of this type allows for comparison of fleets of different ship type, owner, builder, class, etc., which provides valuable information to all parties engaged in ships' operation. Comparison based only on the average fleet's age is useful but not technically sound.

It is interesting to compare the results with the statistical data for the world tanker fleet in 2006 (Clarkson, 2007). The comparison is made in Table 2.

In some sense, the existing worldwide practice of controlling the geometric properties of ships' hull structure serves as a "representative" of ships' reliability. One example is the IMO requirement for aging ships to have a 10% maximum reduction of their hull girder section modulus (HGSM) relative to its required value (IMO, 2000). This practice is purely empirical but one should not automatically label it as wrong. It is based on data for structural failures collected and analyzed by the classification societies for dozens of years.

Table 2 Average tankers' age of the analyzed and world fleet (Clarkson, 2007)

Analyzed tanker fleet			World tanker fleet		
Ships' age span [years]	Average age [years]	p.o.e. permissible HGSM loss [-]	Ships' age span [years]	Average age [years]	p.o.e. permissible HGSM loss [-]
10-15	14.11	0.00093	0-15	7.40	0.00024
10-20	17.59	0.00465	0-20	9.15	0.00030
10-25	19.62	0.01368	0-25	10.66	0.00035
			0-30	11.84	0.00039

The table suggests that the probability of exceeding the IMO permissible HGSM reduction will be even smaller than that obtained for the analyzed tanker fleet (see Fig. 16).

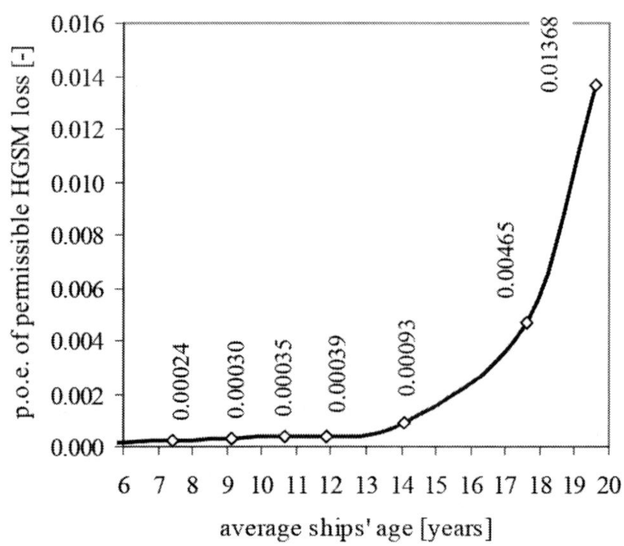

Fig. 16 Probability of exceeding the IMO permissible HGSM reduction for the world tanker fleet (Clarkson, 2007) and the analyzed tanker fleet

Determining the probability of meeting the IMO Renewal Criterion for the ships' HGSM can be used as a first approach toward assessment of the imbedded reliability, both in any given ship and in any given fleet. This approach is applied to a tanker fleet but it can be applied to any other type of fleet.

Although the probability of meeting the IMO Renewal

criteria represented in geometric terms (i.e., the control of the HGSM) does not directly represent the probability of failure, one can observe that the world tanker fleet at this moment is in good shape.

Conclusions

The following conclusions may be drawn from this paper:

- The mean value of the HGSM loss is below 5% from its required value over the whole ships' service life.
- Weibull distribution better fits the statistical data for the HGSM loss for any ships' age. Formulae are derived for calculation of the shape and scale parameters of the Weibull distribution.
- A method is proposed for calculating the embedded safety of hull girder strength of a given fleet. As an example, the probability for a tanker fleet to meet the IMO permissible 10% HGSM reduction was calculated.

References

Clarkson Research Studies – The tanker register 2007

Garbatov Y., Guedes Soares C., Wang G., (2005) Nonlinear time dependent corrosion wastage of deck plates of ballast and cargo tanks of tankers. OMAE 2005-67579, OMAE 25th International Conference on Offshore Mechanics and Arctic Engineering (OMAE'05), Halkidiki, Greece, 12-17 June 2005.

Gardiner C.P., Melchers R.E., 2001, "Bulk carrier corrosion modeling", International Offshore and Polar Engineering Conference, IV: 609-615, Stavanger, Norway.

Harada S., Yamamoto No., Magaino A., Sone H., 2001, "Corrosion analysis and determination of corrosion margin, Part 1&2", IACS discussion paper.

IMO Annex 13 to IMO Resolution MSC.105 (73) adopted on 5th of December, 2000

Ivanov L D, Serratella Ch., Wang G, Lee A.K – Development of Reliability-Based Renewal Criteria of Marine Structures, Marine Technology, October 2007

Ivanov L D – Probabilistic presentation of the geometric properties of shipbuilding structural profiles when assessing elastic bending strength, vol. 1 Theory, vol. 2 Numerical examples, ABS Technical Report RD-2001-19, December 2001

Ochi M - Applied Probability and Stochastic Processes in Engineering and Physical Sciences, John Wiley & Sons, New York, Chichester, Brisbane, Toronto, 1989

Paik J.K., Wang G., Thayamballi A.K., Lee JM., 2003, "Time-variant risk assessment of aging ships accounting for general / pit corrosion, fatigue cracking and local dent damage", SNAME annual meeting, San Francisco, CA.

Spiegel M R – Mathematical Handbook of Formulas and Tables, McGraw-Hill Book Company, Schaum's Outline Series, 1968

Thuanboon S., Tordonato D.S., Navidi W., Olson D.L., Mishra B., Wang G. (2006). A Statistical analysis of corrosion wastage of transverse members in single hull tankers, 25th International Conference on Offshore Mechanics and Arctic Engineering (OMAE'06), 4-9 June 2006, Hamburg, Germany.

TSCF (Tanker Structure Co-operative Forum), 1992, Condition evaluation and maintenance of tanker structures, Witherby & Co. Ltd, London.

Wang G., Spencer J., Elsayed T., (2003b), "Estimation of corrosion rates of oil tankers", 22nd International Conference on Offshore Mechanics and Arctic Engineering, Cancun, Mexico, 8-13 June 2002.

Wang G., Spencer J., Sun H.H., (2003a), "assessment of corrosion risks to aging oil tankers", 22nd International Conference on Offshore Mechanics and Arctic Engineering, Cancun, Mexico, 8-13 June 2002.

Wang G., Spencer S., Olson D., Mishra B., Saidarasamoot S., Thuanboon S. (2005) Chapter 25 Tanker corrosion. Handbook of Environmental Degradation of Materials, edited Kutz M., William Andrew Publishing, Norwich, NY.

Yamamoto N., Ikegami K., 1998, "A study on the degradation of coating and corrosion of ship's hull based on the probabilistic approach", Journal of Offshore Mechanics and Arctic Engineering, 120, 121-128.

Improvements and updates to Intact Stability Standards applied to domestic vessels in Australia

Mori Flapan
National Marine Safety Committee Inc
mflapan@nmsc.gov.au

ABSTRACT

The National Marine Safety Committee has recently revised two standards on intact stability, the first dealing with intact stability criteria, the second with stability tests and stability information. The revision incorporates a more performance-based structure with lessons learnt from the application of the previous standards over a 25 year period.

The standard on intact stability criteria has been formulated to widen options for compliance by distinguishing between so-called comprehensive and simplified criteria; and between criteria of general application and criteria for special operations. As an example of the last of these, there are criteria for vessels engaged in towing rather than being applicable to just tugs. Likewise there are criteria for vessels engaged in trawling rather than being limited to trawlers. Significant changes have been made to ensure simplified and comprehensive criteria provide similar safety outcomes; for example heeling moments are now directly comparable. A more performance based approach has been taken for determining the maximum allowable angle of heel resulting from a static heeling moment. A major change has been a significant increase in the minimum assumed mass of persons to take into account the changing demographic characteristics of the Australian population and serious incidents that occurred in the USA.

The standard for stability tests and stability information has also been significantly revised. It has replaced discretionary clauses with specified criteria for the application of inclining experiments, lightship measurements, simplified stability tests, draft markings, stability books, abbreviated stability calculation methods and other forms of stability information. Procedures for inclining experiments and other stability tests have been revised to comply with standard experimental procedure. Both standards have been prepared with significant input from both government and industry stakeholders.

THE NATIONAL STANDARD FOR COMMERCIAL VESSELS

Responsibility for the safety regulation of domestic commercial vessels in Australia is split between the Commonwealth government and the various State and Territory governments. The National Marine Safety Committee Inc (NMSC) was established to facilitate national consistency and mutual recognition between these jurisdictions. As part of its work, the NMSC is reviewing standards for domestic commercial vessels. The current standards for intact stability are contained in the Uniform Shipping Laws Code[1] (USL Code). A new standard called the National Standard for Commercial Vessels[2] (NSCV) is being developed to replace the USL Code.

Subsections 8A, 8B and 8C of the USL Code specify the current standard for intact stability. These standards were developed in the 1970s. Starting with what were largely the SOLAS Regulations and US Code of Federal Regulations of the day, the USL Code stability standards were developed incorporating modifications that were distinctly local.

The standards writing activities of the NMSC are governed by a Strategic Plan[3] that arose from a Council of Australian Governments agreement[4] and has been endorsed by the Australian Transport

Council. The Strategic Plan contains a number of principles that shape the revision of standards for commercial vessels, including standards that apply to intact stability. These include:

- Incorporate recognized and relevant national and international standards;
- Encourage professional competence;
- Incorporate a performance-based approach;
- Facilitate approval of new technologies;
- Incorporate OH&S principles;
- Encourage recognition of duty of care; and
- Develop the safety system based on sound information.

The current USL Code is largely prescriptive in its application. The strategic principles require the revised standards to incorporate a performance-based approach, with a greater awareness of the safety outcomes and improved flexibility of application; but with compatibility with other relevant standards and while maintaining national consistency and mutual recognition. This provides a challenge for a number of reasons that include:

1. The "science" to model and determine the adequacy of a vessel's intact stability taking into account the full range and probabilities of potential environmental and operational factors is still in its relative infancy.
2. Flexibility can sit awkwardly with the objective of national consistency and mutual recognition, unless the flexibility can be limited to the method of solution, while there remains consistency in the achieved safety outcomes.
3. Relevant national and international standards are themselves in a state of continual flux, and they do not necessarily sit consistently with one another.
4. The benefits of standardization have to be weighed against the costs associated with changing from the current system, including its impact on existing designs.

The process for developing standards requires input from key stakeholders at crucial stages in accordance with Council of Australian Governments' Guidelines for Standards Setting Bodies[5]. Stakeholders are involved in the drafting, public comment and review of public comment phases. A regulatory impact statement is prepared to identify the impacts and consider their benefits and costs. Stakeholder participation serves to focus the review on issues of major import and as a barometer to check that the proposed changes on the industry are viable and justifiable.

THE NATURE AND MAGNITUDE OF INTACT STABILITY RISK

Incidents involving inadequate intact stability are amongst the most unforgiving that can befall a vessel. This is because capsize can be very sudden giving little or no time to take reasoned steps to mitigate the consequences. With so little time to prepare, responses tend to be desperate and instinctive, and the chances of survival through the various stages of escape, evacuation, immersion and rescue are far from certain.

This is illustrated by the fact that capsize was the initial incident in 36% of fatal marine accidents (both commercial and fishing) in Australia from 1992 to 1998 (O'Connor 2004[6]).

Inadequate intact stability is a latent defect that just requires the right combination of factors (loading, wind, waves, heading, etc) to become manifest. The highly sporadic nature of occurrence (just the right combination of factors) plus the likely catastrophic nature of its consequences means that reliance on limited statistical data can be highly misleading and dangerous. This is the type of hazard that may not occur for decades within a region, but then happens with high mortality from a single incident. Before the incident, the statistics look favourable. After the incident, the data is often regarded as being skewed by the single occurrence – or was it? Under the HSC Code 2000[7] Annex 3, the tolerable probability of catastrophic loss is greater than 1×10^{-9} per hour of operation.

This equates to once in every 114,000 years[a] of operation assuming 24 hours operation per day. It would probably take six or more times this amount of operation before a reliable trend could be observed. Hence, there is a need to keep statistical data in perspective as a tool rather than as a proxy for the decision maker.

An analysis of 108 Australian commercial vessel incidents over the last 15 years is given in Figure 1 (Flapan 2007)[8]. Intact stability may have been a direct contributor to the fate of vessels that foundered and were lost or missing, amounting to 58% of the total. Intact stability may play a more indirect role in the loss of the further 29% of vessels that were wrecked or involved in collisions.

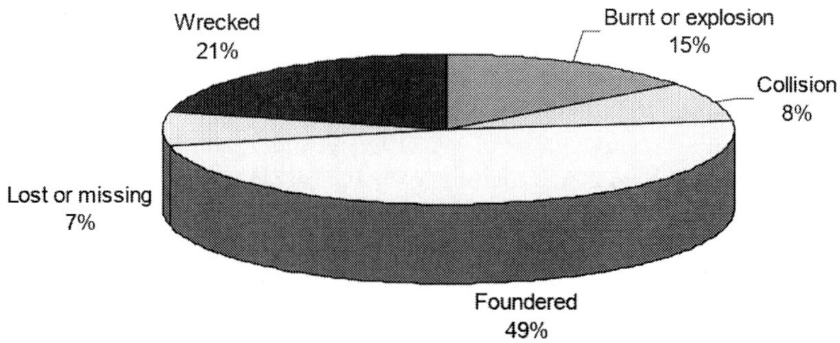

Figure 1—Analysis of 108 Australian vessel losses 1992-2007 (Flapan 2007)

An analysis of 1326 incident statistics for commercial vessels in Australian jurisdictions over the years 2005-2006 (NMSC 2007)[9] indicates that potentially intact stability related incidents (swamping, capsizing, loss of stability and loss or presumed loss of the vessel) accounted for only 5.1% of the total. However, the relatively low occurrence is to be expected given the catastrophic nature of the consequences discussed above, and should not be grounds for complacency.

Anecdotally, there have been concerns raised as to the intact stability performance of the domestic Australian fishing fleet. Deaths of commercial fishermen are the overwhelming majority of commercial fatalities (41 out of 49). Of these, about 40% of incidents were classified as the sort that may be associated with a stability problem, comprising capsize 18%, sinking 18% and loss of stability 5%. (O'Connor 2004).

An analysis of Australian and New Zealand commercial vessels lost by foundering (Flapan 2007) highlights the relatively high proportion of fishing vessel losses attributed to capsize. Of the total 32 fishing vessels that are recorded as having foundered since 1992, 43% were attributed to capsize. Of the remaining 21 vessels, all of which were not engaged in fishing, capsize accounted for 20%. In addition, 3 of 4 vessels that went missing during this period were fishing vessels.

Part of the explanation could be the relatively hazardous nature of fishing that increases the exposure of the vessel to the combination of critical factors. However, another more compelling reason is the relatively large proportion of the existing fishing fleet that still have not had stability characteristics verified against the current stability criteria. Prior to the introduction of the USL Code, very few fishing vessels were subjected to stability analysis. For historic and other reasons, the uptake of standards for fishing vessels has lagged behind those for other vessels. Thus, the relatively high stability losses in the fishing fleet are likely to be more a function of problems in implementation rather than the standards themselves.

[a] While the period of 114,000 years appears excessive, another perspective can be gained by comparing it against other everyday events such as the likelihood of winning the NSW Lotto game. Statistically, if a single game is played every hour, the player could expect to win the major prize once every 580,000 years. This means that the acceptable probability of being on a commercial vessel during a catastrophic incident (a fatality of loss of the vessel) caused by any one hazard if five times greater than winning the major prize in Lotto.

THE REVISED STANDARD HAS A MORE USER-FRIENDLY FORMAT

The review of the intact stability standards has involved the complete restructuring of the format. The stability subsections that comprise NSCV Part C Section 6—Stability are arranged as shown in Table 1. Intact stability criteria have been separated from the requirements for stability tests and stability information. The stability tests are considered to be procedures for input, while the stability information provisions are specifications for output. Subsections 6A and 6C have been completed and are the subject of this paper. Work on Subsection 6B Buoyancy and Stability after Flooding is currently underway. Subsection 6C Stability Tests and Stability Information will be expanded to accommodate specific damaged stability information once Subsection 6B has been finalised.

Table 1 – Structure and contents of NSCV Part C Section 6 Stability

Subsection 6A	Subsection 6B	Subsection 6C
Intact stability criteria	Buoyancy and stability after flooding	Stability tests and stability information
1 Preliminary 2 Intact stability outcomes and solutions 3 Determining the applicable intact stability criteria 4 Maximum displacement and longitudinal stability criteria 5 Comprehensive stability criteria of general application 6 Additional comprehensive stability criteria for special operations 7 Simplified criteria of general application 8 Simplified criteria for vessels engaged in special operations	Under development	1 Preliminary 2 Required outcomes 3 Methods for establishing and verifying lightship particulars 4 Methods for conducting simplified stability tests 5 Presentation of Stability Information 6 Arrangements for determining draft

Key changes in the format include the following:

- The incorporation of objectives and required outcomes in accordance with the performance-based approach being adopted for the NSCV[10].
- A clear differentiation between so-called comprehensive and simplified criteria, comprehensive criteria being defined as those for which a righting lever (G_fZ) curve is derived.
- Clarification as to the application and options available for criteria
- Separation of prerequisites for the application of criteria as compared to criteria themselves.
- Presentation of criteria in a tabular format with numbering to reduce errors and omissions.
- The aggregation of methods of calculation as Appendices.
- Increased use of illustrations to facilitate understanding and reduce the need for interpretations.

SPECIFIC ISSUES ADDRESSED BY THE REVISED STANDARDS

Authority discretion

Many of the USL Code provisions were expressed in a way that provided for or even required surveyor discretion in order that the provision is applied. This resulted in inconsistent application of the standards. While still available in the assessment of an equivalent solution, surveyor discretion has been largely eliminated for the application of the deemed-to-satisfy solutions. This has been achieved by identifying the factors upon which discretion would have been exercised and incorporating that information as part of the standard.

Removal of anomalies between simplified and comprehensive criteria

The USL Code Subsection 8C[11] contains both comprehensive and simplified stability criteria. There are significant differences in the requirements and costs of the two, with the simplified criteria being much cheaper to verify. The anomaly arises because the simplified criteria also specify a lower threshold for compliance than if the same vessel were measured against the comprehensive criteria, notwithstanding that the methodology is more approximate and subject to error. Figure 2 illustrates the righting lever curve of a vessel that would meet the simplified Category S criteria from the USL Code, but would be unable to comply with the comprehensive Category P criteria. This means that vessel designers and builders argue strongly for the use of the simplified criteria, even when it is clear that the vessel form and/or operations are not properly suited to the application of simplified stability testing methods.

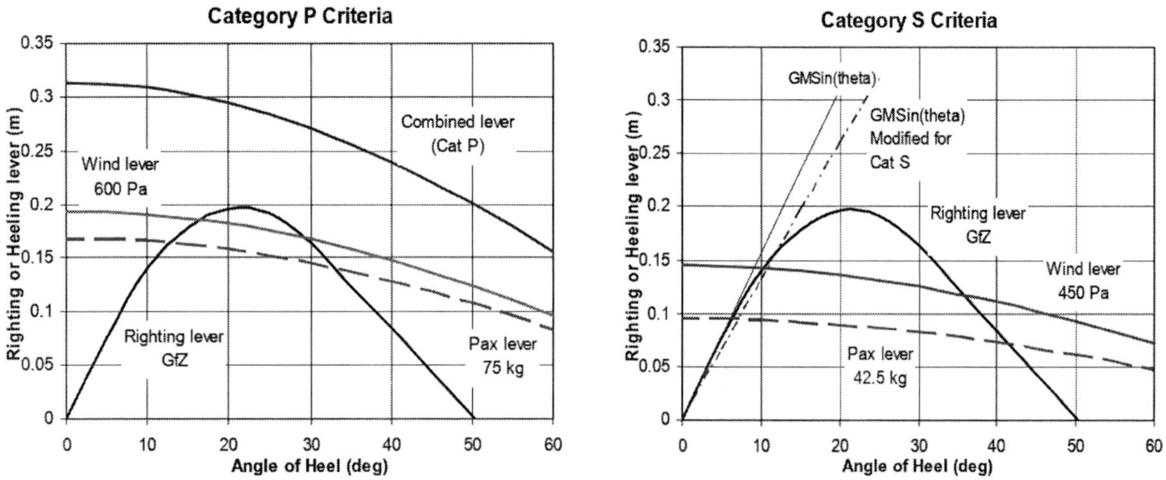

Figure 2—Comparison of Category P versus Category S criteria

The anomaly has been addressed by modifying both the comprehensive and simplified criteria so that they are consistent. The simplified criteria have been modified by:

- adjusting the passenger heeling moment to align with that used for comprehensive criteria;
- providing for testing in near laden and near light conditions of loading; and
- applying criteria for turning moments.
- allowing comparable angles of heel from single heeling influences

To help compensate for the increased threshold for stability, a safety margin that was previously added to the Australian simplified criteria has been removed, aligning the criteria to the original source criteria[12]. Likewise, the comprehensive criteria have also been modified to better align with the simplified criteria by:

- allowing for comparable wind heeling moments for sheltered waters and restricted offshore operations;
- removing the combined lever requirement for vessels of size and area of operation similar to that for which the simplified criteria could apply.

Rationalization of the allowable angle of heel criteria

The USL Code criteria specify maximum allowable angles of heel that vary between 5 degrees for certain crane barges to 15 degrees for combined heeling moments. The requirements were inconsistent in that certain vessels applying Category S or T (simplified criteria) could apply up to

14 degrees for a single heeling moment, while the same vessel applying Category P, Q or R (comprehensive criteria) could only apply up to 10 degrees. The anomaly was sometimes overcome administratively by the Authorities having discretion to increase the allowable angle to greater than 10 degrees. However, this concession was exercised inconsistently and without guidance. The approach taken has been to specify performance-based maximum angle criteria within the standard itself. The maximum allowable angle of heel is determined by the risk characteristics of the vessel and its operation, see Table 2. The result is that the criteria can accommodate the different characteristics of a wide variety of vessels including most monohull sailing vessels that are arranged to sail at relatively large angles of heel.

Table 2 — Maximum allowable angles of static heel

Heel conseq-ence level	Allowable maximum angle of static heel for heeling moment(s)		Conditions of application
	Single θ_s degrees	Combined θ_c (A) degrees	
1. High	5	5	No specified conditions of application – applicable to any vessel that is unsuited to the application of large values of heel.
2. Moderate	10	15	θ_s or θ_c (if combined lever criteria are applied) may exceed 5 degrees where— 1. if the vessel is fitted with a slewing crane that is subject to the lifting criteria, the crane is capable of safe operation at angles of heel up to at least θ_s, and 2. if the vessel is carrying unsecured deck cargo, the deck cargo shall either— i) comprise vehicles having rubber tyres; or ii) have a maximum potential moment from cargo shifting that does not exceed 20 per cent of the greatest value of M_P, M_W or M_T.
3. Low	14	18	θ_s may exceed 10 degrees or θ_c (if combined lever criteria are applied) may exceed 15 degrees where— 1. all cargo including deck cargo is secured against shifting; 2. seating is provided for all persons; 3. furniture is fixed when in use and/or when stowed; 4. sufficient grab rails are provided in spaces that normally contain persons; and 5. decks and deck surfaces are arranged to reduce slipping hazards.

KEY:

(A) Combined moments apply only where combined levers are specified in the criteria.

Clarifying the application of wind heeling moments

The Category P, Q and R criteria USL Code do not specify the nature of the wind heeling lever; i.e., whether it should be applied straight line, to the cosine of the heel angle, or to the square of the cosine of the heel angle, see Figure 3. As a result, different approaches have been adopted by the various jurisdictions. The new standard specifies that the wind heeling lever should reduce by a factor of Cos θ for criteria of general application. This takes into account the resolution of wind force that is normal to the projected surfaces but does not take into account any reduction in area because, as a three dimensional object, the projected area tends not to reduce at heel angles less than 40 degrees. The more conservative straight line wind lever adopted by IMO was considered but not adopted because it would be inconsistent with the Cos θ factor used for the simplified criteria (see above). Furthermore, the drafting group considered that the costs of applying the more conservative IMO approach could not justify the benefits; especially given that the larger passenger vessels have additional criteria applicable to the combined lever. It was noted that the wind moment calculation was, at best, a broad approximation with no consideration being given to wind gradient effects or drag coefficients.

The wind heeling lever applied to vessels that carry sail was also revised. The USL Code criteria had been based on the US CFR46 criteria[13] that applied $\text{Cos}^2\theta$. However, subsequent investigations by the Wolfson Unit in the UK[14] based on wind tunnel testing indicated that a more appropriate factor for sailing vessels was $\text{Cos}^{1.3}\theta$. A subsequent amendment to the USL Code[15] allowed the alternative application of criteria based on the MCA Standard[16] that applies $\text{Cos}^{1.3}\theta$. The new standard retains both sets of criteria applicable to vessels that set sail. However, the different methods of calculation were inconsistent with a performance based approach. This was resolved by applying the more soundly based $\text{Cos}^{1.3}\theta$ factor to both sets of criteria, see Figure 3.

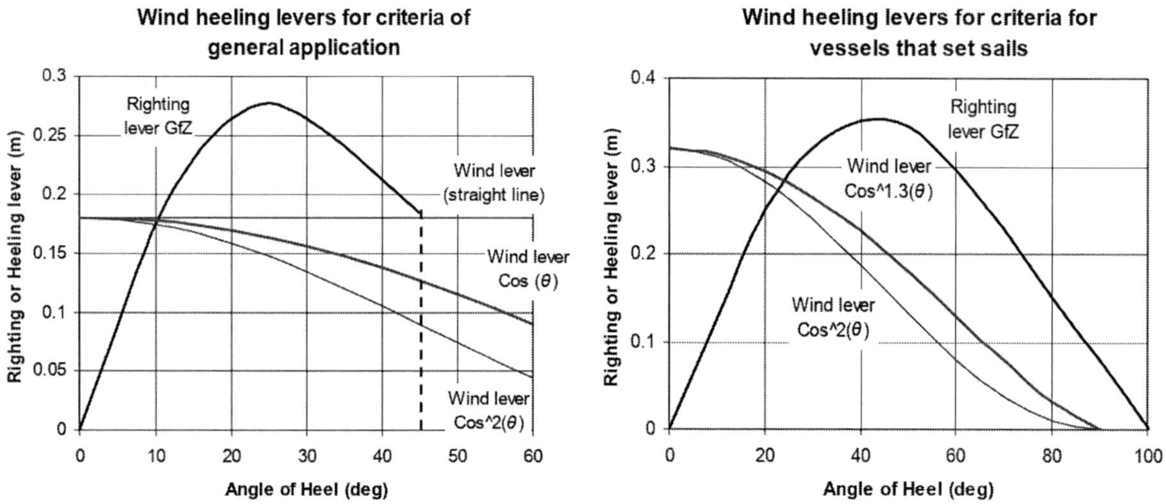

Figure 3—Comparison of wind heeling lever factors

Maximising the choice of applicable criteria

The USL Code specifies a single set of criteria applicable to a particular vessel. It then says where a vessel fails to meet the relevant criteria then the marine authority may apply discretion as to the application of alternative criteria. This leads to variations in the application of stability criteria. The revised NSCV standard is arranged so to maximise the choice of deemed-to-satisfy criteria available without having to go to the Authority for a ruling. Criteria that are better suited to catamaran vessels have been added based on HSC Code 2000 Annex 7. Criteria have also been provided for collared vessels (including RIBs). Standards that were applicable only to Hire and Drive vessels are now available for other vessel types where they have a similar area of operation and configuration. While all vessels can apply at least one set of comprehensive criteria, certain vessels can also choose between one or more sets of simplified criteria.

Criteria apply to special operations rather than particular vessel types

The USL Code contains criteria applicable to particular vessel types such as sailing vessels, tugs, trawlers, crane barges. The revised standard specifies 'additional' criteria applicable to vessels when engaged in special operations such as carrying sail, towing, trawling or lifting. This has two benefits. The first is that the widest choice of criteria of general application is maintained. The second is that a vessel may engage in more than one type of operation. Thus, a police boat can apply the towing criteria even though it is not a tug, or a research vessel can apply the trawling criteria even though it is not a trawler. Again, the criteria for special operations have been presented to maximise the choice available.

More realistic assumptions for the mass of persons

The USL Code assumes 75 kg per person for seagoing vessels and 65 kg per person for sheltered water vessels. These are clearly lower than the average masses in the community. Recent incidents in the USA[17] have highlighted the dangers of overloading on passenger vessels with recommendations that the assumed mass of passengers be reviewed. The latest comprehensive data for Australia[18] (2004-05) on average mass are 84.0 kg for men and 68.1 kg for women, giving an overall average in the order of 76 kg (ABS 2006)[b]. In 1995, the figures were 80.2 kg for men and 64.5 kg for women (ABS 1998)[19], indicating an increase in the average mass of the population of about 3.7 kg or 5 per cent in a 10 year period. The minimum assumed mass of persons in the revised standard as given in Table 3. The proposed average mass of 80 kg is in excess of the average 2004-05 figure of 76 kg. The figure of 80 kg was adopted because the average weight figures are likely to underestimate the reality for two main reasons. Firstly, they are calculated from self-reported figures. ABS (1998) reports that the 1995 self-reported average mass is 1.8 kg lower than the measured average mass for males and 2.5 kg for females. Secondly, the masses quoted are likely to be for people when naked; i.e., without taking into account clothing, shoes, keys, wallets, mobile phones, hand bags or other hand luggage carried by day passengers.

Table 3 also specifies the minimum assumed mass for divers. The additional 53 kg for divers includes the mass of diving masses, diving suit, flippers, mask, wet suit, regulator and 2 filled gas bottles. Of the total additional mass, 36 kg is included in the minimum assumed mass of the diver and the other 17 kg is the minimum allowance for the second gas bottle as additional diving equipment. The effect on a vessel of the different minimum values for mass of persons and their effects is illustrated in Figure 4. The change effects both the loading of the vessel (and thus the shape of the G_fZ curve) and the person heeling lever. Note that normally, the number of divers would be reduced to result in a mass comparable to that of passengers.

Table 3 — Minimum values for mass of persons and their effects

Person type	Minimum assumed mass per person Kg per person (w)	Minimum allowance for baggage Kg per person	Minimum allowance for additional diving equipment Kg per person
Passenger or crew day only	80	Nil	Nil
Passenger or crew overnight	80	15	Nil
Diver day only	116	Nil	17
Diver overnight	116	15	17

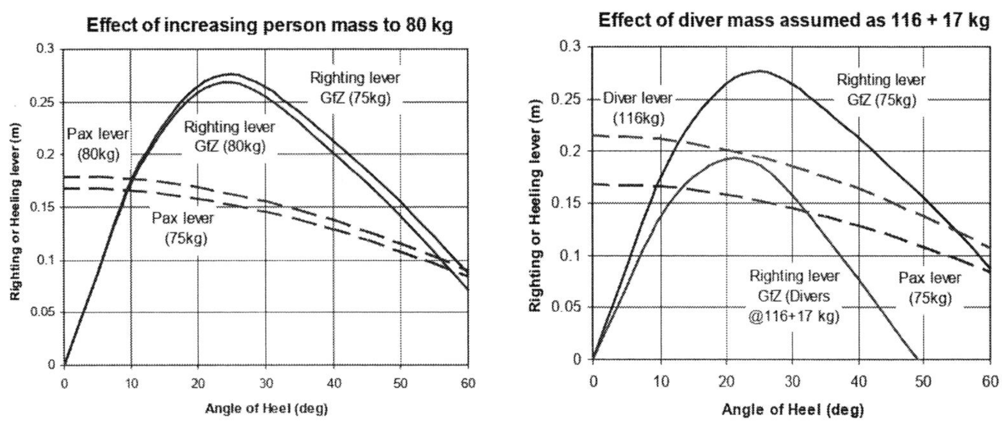

Figure 4—Illustration of the effect on a vessel of assuming 75 kg, 80 kg and diver mass for same total number of persons carried

[b] As these figures relate to people 18 years of age and over, they would overestimate the average weight of a person when children are carried and underestimate it when more males than females are carried.

Revised approach for catamarans that set sail

An amendment to the sailing catamaran criteria of the USL Code in 1997[20] treats catamarans that set sail in a manner similar to sailing monohulls, applying a criterion that relied on dynamic balance. There were two issues with this approach. The first was that the increased windage effect of the bridge deck as the windward hull emerged was ignored. As well, the assumption that such vessel (not being an off-the-beach sailing catamaran) would still be under control and would properly recover from an incident at a large angle of heel with one hull a-flying was unsound. A criterion developed by Lock Crowther, one of the leading pioneers of sailing catamaran development, for use on his own designs has been adopted. Lock Crowther's approach was to avoid the windward hull emerging in all circumstances by adopting a principle that the draft on the windward hull should not reduce by more than 25% in average conditions. This then corresponds to approximately 50% draft loss during a gust. The approach has been interpreted by specifying a criterion that the heeling lever calculated from the mean wind pressure shall not intersect the righting lever (G_fZ) curve at a value not greater than 25% of the maximum righting lever (G_fZ_{max}). Because the heel angles are relatively small, there is no need to analyse the increased windage of the bridging deck at large angles of heel.

Stability standards applicable to small fishing vessels

The USL Code does not specify stability criteria for small fishing vessels less than 7.5 metres in length. At the time when the USL Code was first drafted, there were no stability standards applicable to recreational craft in Australia. In recent years, the Australian Builders Plate standard[21] has required that a large proportion of recreational craft be fitted with a plate declaring that the vessel is capable of carrying a given load and number of persons in accordance with a relevant national or international standard. Maintaining the status quo for small fishing vessels would be inconsistent with the broader policy objectives of the NMSC to incorporate Occupational Health & Safety (OH&S) principles into the standards for the design, construction and operation of vessels and to encourage vessel operators to recognise their duty of care to employees and passengers. It would also be inconsistent with the requirements for recreational vessels.

The revised NSCV stability standard now specifies a range of sets of criteria applicable to small fishing vessels that include the AS1799 standard for recreational vessels[22]. While the vast majority of these small fishing vessels are not required to carry a certificate of survey under legislation, the criteria will still provide a benchmark for operators having to comply with their Occupational Health and Safety obligations.

Determination of lightship particulars on vessels with large stability

Designers and builders of catamarans have often objected to undertaking an inclining experiment to determine the lightship *VCG* on the basis that the quantity of inclining masses required to heel the vessel is excessive. The result is high costs to source, verify and transport the masses and difficulties locating, supporting and moving the masses during the experiment. Exemptions from this requirement have been granted subject to a detailed weight estimate being verified by a lightship measurement and the application of a margin of safety in the assumed *VCG*. Such exemptions are discretionary and have sometimes been applied subjectively and inconsistently. The revised standard allows for the determination of lightship *VCG* by means of detailed weight calculation with an added safety margin calculated as the lesser of 10% of *KG* or 0.5 metres, subject to verification of the calculated displacement and *LCG* by a lightship measurement. The approach is limited to vessels having a ratio of lightship GM_O to *KG* of more than 2.

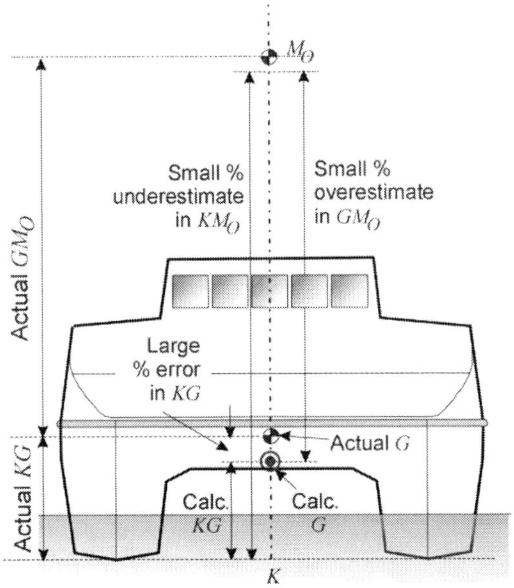

Figure 5—Cumulative effect of errors on vessels with large ratio of GMO to KG

The rationale is illustrated in Figure 5. As the ratio of GM_O to KG increases, errors in the determination of KG will be magnified by any errors in the determination of GM_O and KM_O. For example for a ratio of GM_O to KG of 3, a 2% overestimate in GM_O plus a 1% underestimate in KM_O could result in an underestimate of 2% x 3 + 1% x 4 = 10% in the determination of KG. Because of the generally favourable stability characteristics of most catamarans, such errors are often of little significance for intact stability. However, catamarans frequently must also be designed to withstand unsymmetrical flooding. The error in KG may invalidate what would otherwise be a compliant damaged stability analysis, resulting in a latent defect that only becomes apparent at time of crisis. The standard acknowledges the limitations in application of an inclining experiment on vessels having a high ratio of GM_O to KG and substitutes a conservative calculation, improving the reliability while saving unnecessary cost.

CONCLUSION

The revision of the USL Code to become the new NSCV intact stability standards have been a process of evolution rather than revolution. The new standard contains many of the same criteria as the previous USL Code. However, these have been updated to reflect the experience of over 25 years of application of the previous standards. The most significant modification has been the increase in minimum assumed person mass, a change which, though unwelcome, was considered necessary if safety standards are to be maintained. The potential impact of greater assumed persons mass could be in some cases an increase in the size of the vessel needed to carry the same number of persons. But other reforms within the new standard should help to offset the additional burden by eliminating inconsistencies, increasing flexibility of application, avoiding the need for interpretation and reducing the likelihood of stability reports having to be reworked[23]. The new standards establish the framework for performance-based solutions as an alternative to deemed-to-satisfy solutions. Whilst not providing the performance-based solutions themselves, the standard provides an environment that can accommodate their ongoing development and innovation.

ACKNOWLEDGEMENTS

The author gratefully acknowledges the contributions made to the development of the standard by members of the intact stability reference group and the many stakeholders who provided public comment[24]. Special mention should be made for the efforts of Graham Taylor, Terry Ehret and Robin Gehling.

REFERENCES

---

[1] Australian Transport Council. *Uniform Shipping Laws Code.* 1991. 1993. Australian Government Printing Service. Canberra. http://www.nmsc.gov.au/uslcode_1.html

[2] Australian Transport Council. *National Standard for Commercial Vessels* 2005. 2006. 2007. National Marine Safety Committee. Sydney. http://www.nmsc.gov.au/nscv.html

[3] Australian Transport Council. *National Marine Safety Strategy.* August 1998. National Marine Safety Committee Inc. Sydney. http://www.nmsc.gov.au/documents/strategy.pdf

[4] Council of Australian Goverments (COAG). *Agreement Establishing a National Marine Safety Regulatory Regime.* November 1997. http://www.nmsc.gov.au/documents/agreement.pdf

[5] Council of Australian Governments. *Principles and Guidelines for National Standard Setting and Regulatory Action by Ministerial Councils and Standard-Setting Bodies.* COAG. 2004. http://www.coag.gov.au/meetings/250604/coagpg04.pdf

[6] O'Connor, P. (2004) *National Assessment of Boating Fatalities in Australia 1992-1998* Flinders Consulting Pty Ltd for the NMSC, March 2004. http://www.nmsc.gov.au/documents/Fatal%20Report_screen.pdf

[7] International Maritime Organization. *International Code of Safety for High Speed Craft 2000 with Resolution MSC.222(82) amendments.* IMO. 2000. 2004.

[8] Flapan, Mori. *Register of Australian and New Zealand Ships and Boats.* Unpublished. 2007. Sydney. www.boatregister.net

[9] National Marine Safety Committee Inc. *Incident Data Analysis 2005-2006.* NMSC Inc. 2007. Sydney. http://www.nmsc.gov.au/researchdata_ibt.html

[10] Preamble to the NSCV. http://www.nmsc.gov.au/documents/NSCV/PARTB.PDF NSCV Part B. http://www.nmsc.gov.au/documents/NSCV/PARTB.PDF

[11] Uniform Shipping Laws Code. *op. cit.* Subsection 8C. Stability Criteria. http://www.nmsc.gov.au/documents/USL/usl-section8-sub-sectionC.pdf

[12] United States Government. *US Code of Federal Regulations 46—Shipping.* 2007. Chapter I—Coast Guard, Department of Homeland Security: Part 170—Stability requirements for all inspected vessels, clause 170.170 Calculations required and also Part 171—Special Rules Pertaining To Vessels Carrying Passengers, 171.050 Intact stability requirements for a mechanically propelled or a non self-propelled vessel. http://www.access.gpo.gov/nara/cfr/waisidx_05/46cfr170_05.html and http://www.access.gpo.gov/nara/cfr/waisidx_05/46cfr171_05.html

[13] US Code of Federal Regulations 46. *op. cit.* Part 171—Special Rules Pertaining To Vessels Carrying Passengers, 171.055 Intact stability requirements for a monohull sailing vessel or a monohull auxiliary sailing vessel.

[14] Deakin, B. *The development of stability standards for U.K. sailing vessels.* Royal Institution of Naval Architects. Spring Meetings 1990.

[15] Marine and Ports Group. *Uniform Shipping Laws Code Amendment List 4.* 4 March 1997. http://www.nmsc.gov.au/documents/USL/uslc-amendment%20list%20no%204.pdf

[16] Maritime and Coastguard Agency. *The Small Commercial Vessel and Pilot Boat Code of Practice.* 2004. MGN 280(M). http://www.mcga.gov.uk/c4mca/mgn_280-2.pdf

[17] National Transportation Safety Board. *Capsizing of U.S. Small Passenger Vessel Lady D Northwest Harbor, Baltimore, Maryland March 6, 2004.* Washington D.C. 2006. NTSB/MAR-06/01

[18] Australian Bureau of Statistics (ABS) (2006) *National Health Survey: Summary of Results 2004-05.* Catalogue Number 4364.0, issued 27 February 2006.

[19] Australian Bureau of Statistics (ABS) (1998). How *Australians Measure Up 1995.* Catalogue Number 4359.0, issued 6 October 1998.

[20] Marine and Ports Group. *Amendment list 4. op. cit.*

[21] Australian Transport Council. *National Standard for The Australian Builders Plate for Recreational Boats. NMSC Inc. 2005.* Sydney. http://www.nmsc.gov.au/documents/ABP_Standard_Edition_3_Nov_2005.pdf

[22] Standards Australia. *AS 1799.1—1992 Small Pleasure Boats Code Part 1: General requirements for power boats.* Sydney. 1992

[23] Margaret Starrs for National Marine Safety Committee Inc. *Regulatory Impact Statement for NSCV Part C Subsections 6A and 6B.* NMSC Inc. Sydney. 2007

[24] National Marine Safety Committee Inc. *Review of Public Comment - NSCV Part C Subsection 6A; Intact Stability Requirements.* NMSC Inc. Sydney. 2007. http://www.nmsc.gov.au/documents/NSCV%20C6A%20RG%20PC%20outcomes.pdf

Integrated Ship Survivability for the Royal Australian Navy

Leo de Yong
Maritime Platforms Division,
Defence Science and Technology Organisation,
Melbourne, Australia
leo.deyong@dsto.defence.gov.au

ABSTRACT

Survivability of a modern warship is dependant on several fundamental factors – its susceptibility, its vulnerability and its recoverability. These three factors are widely seen as the key elements in the equation to describe survivability. For many years, research programs have provided advice to many navies on the individual survivability factors but largely in isolation to each other. This is now changing with an emphasis on the provision of advice on a ship's overall survivability or the relationship between susceptibility, vulnerability and recoverability and the "trade space" between them.
This paper discusses the fundamental elements of survivability, their relationship to each other and how various methods of survivability estimation have been developed. It also discusses the areas that need to be further explored to evaluate ship survivability and the balance of investment and mission effectiveness.

INTRODUCTION

The nature of warfare at sea has changed significantly in the last fifty plus years since WWII. Naval platforms have become increasingly network enabled with other military platforms, operations are more focused on brown and green water regions and crew sizes are decreasing. The maritime environment is increasingly more complex as are the weapons deployed against naval ships. In this context, naval ships are required to undertake operations where they consciously put themselves into harms way. The Royal Australian Navy (RAN), like any navy, needs to be able to undertake operations anywhere in the world, to arrive ready to carry out its mission and complete it with minimal losses.

Navies today also generally have fewer ships than they used to and the overall cost of a modern ship is a significant investment of dollars. The political consequences of losing such valuable capital assets let alone the potential loss of crew in the event of conflict would be significant. Thus, the goal of mission success, with minimal loss of either personnel or assets, is the fundamental basis for survivability. From this, one may choose to define survivability as the ability to complete a mission successfully in the face of a hostile environment with minimal losses.

Mission success may be defined in relation to a very clearly stated operational profile or to the capabilities required to achieve mission success. These include such capabilities as conduct anti-air warfare, conduct surface warfare and conduct under-sea warfare and others. Whatever the approach taken, the fundamental element is that the ship should have the capability to avoid or to withstand threats whether man made or not, to continue on its mission and to achieve success in the execution of that mission with minimal losses. Often the threats are stated as military loads (e.g. missile attack) or non-military loads (e.g. grounding) and ship survivability is affected by one or the other.

There are three key elements to survivability and these are defined as susceptibility, vulnerability and recoverability. These elements may also be stated as damage avoidance, damage tolerance and damage repair. The fundamental problem today is that many navies, including the RAN, are provided with high quality scientific advice on one or more of these elements individually, but the combination of them into an overall evaluation of survivability is not often done to any significant degree. As isolated elements, it is difficult to determine the overall ship survivability or the trade off between the individual elements. For example, how much is survivability increased by halving susceptibility, or given that it may be cheaper to halve the vulnerability, does that halve the survivability? Perhaps a more important question today is what is the survivability of the ship given that it is more than likely to have been built and contain equipment that is qualified to commercial standards?

This paper discusses the fundamental elements of survivability, their relationship to each other and how various methods of survivability estimation have been developed. It also outlines some of the unanswered questions and the areas that need to be further explored to answer some of the questions on ship survivability posed above.

CURRENT SURVIVABILITY APPROACH

DSTO provides "frank and fearless" S&T advice to the RAN in a wide range of scientific fields which includes advice to solve current fleet problems, future fleet requirements, and upgrades of existing RAN assets and the procurement of new assets. It is usually in this latter area that advice on the trade off between the elements of ship survivability is asked for. These include what is the effect on the survivability of the ship if the designer trades off signature reduction (which may be expensive) with reduced vulnerability through thicker steel (which is cheap) in critical locations? If own ship weapons are effective in destroying threats, then what level of signature reduction is needed? If the ship will be hit (due to the signature being too high, and decoys being ineffective), what level of damage control and recoverability will be required to retain a defined level of a specific capability? All these questions require an understanding of the individual elements of survivability, the relationship between each element and the trading space between them.

There is also a high degree of emphasis today on the reduction in the susceptibility of the platform, especially in relation to hard kill and soft kill solutions. This approach tends to be based on the assumption that if signatures are adequately addressed in the initial design, then the probability of a hit is negligible and vulnerability reduction is a lower priority.

Survivability is critical to the ship mission and avoiding a hit is a central element of survivability. But, there is increasing evidence that indicates that susceptibility cannot be reduced to zero, there is always a probability of a hit and the ability to withstand the hit and recover some degree of capability is fundamentally important. Thus all the elements in platform survivability need to be considered and the trade off between them when determining the overall survivability required.

Survivability advice is also usually provided at the individual platform level. But when under hostile threat, most ships are operating as part of a task group and a broader analysis is required. Also, at the task group level, the development of tactics for the individual platforms feeds into the task group requirements for mission success. However, analysis at the task level is extremely difficult as it depends on many variables such as the task group composition, the different levels of survivability of the individual task group elements and the interaction between the task group elements etc.

SURVIVABILITY DEFINITIONS

The definition of the elements of survivability in the naval environment need to be spelt out as the terminology is frequently very loosely and inaccurately used for a variety of applications. As stated above, there are three basic elements to survivability and they are shown in Figure 1 along with the major components of each element.

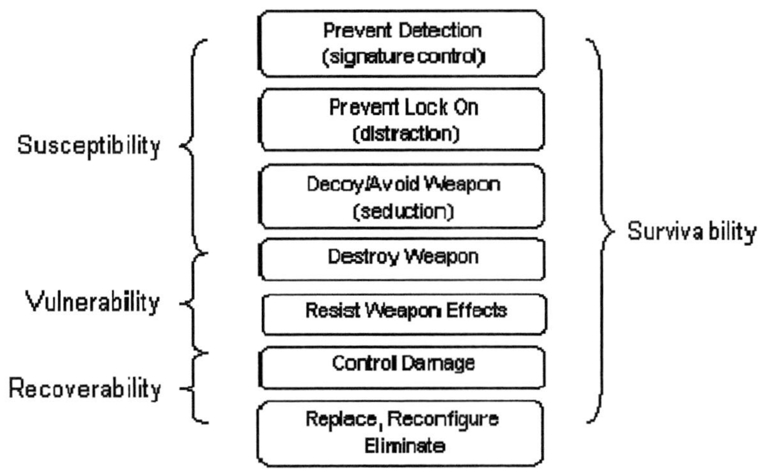

Figure 1: Definition of ship survivability showing the individual components that comprise the three main elements

1. Susceptibility: this is defined as the likelihood of the platform being hit by a threat weapon - does the ship get hit, or the likelihood of "engaging" with a non-military threat. Susceptibility is related to the platform being detected, being classified or being targeted by a threat. The susceptibility of a platform to a military threat is a function of its signatures (infrared, radar, acoustic, magnetic etc.), the performance of its decoy systems and its hard kill systems and the capability/sophistication of the threat. These systems assist in preventing detection, preventing classification, preventing lock on and in destroying or decoying the threat.
2. Vulnerability: this is defined as the likelihood of the platform being disabled or damaged after being hit - how much damage does the ship suffer? This is a very broad definition and it may be narrowed down to the extent of degradation of a system after having been subject to a military or non-military load. The vulnerability of a platform is a function of its design (zoning, separation and

redundancy), equipment (blast doors/bulkheads etc.) and the materials used (armour, hardening etc.). These systems are designed to resist or to mitigate the effects of weapons or non-military threats, which may be primary effects (blast, shock, fragmentation, injury) or may be secondary effects (fire, flooding, toxic gases).

3. Recoverability: this is defined as the likelihood of recovering the capability after having been hit and encompasses the damage control and repair/restoration of systems and capabilities on the platform – if the ship is damaged, how much capability can be recovered and how long will it take? Note that the degree of capability recovered may be less that the initial level of capability before the hit. There are numerous systems involved in the recoverability phase ranging from the fixed and portable fire fighting systems, the damage control reporting and monitoring systems (situational awareness), flood mitigation, toxic gas removal including ventilation etc. However, one major factor that is paramount to this part of survivability is the human factors dimension. Recoverability involves the actions of the crew, which are difficult to predict under stressful conditions. Also, with reduced crew sizes, the integration of systems and humans is becoming more widespread and is exceedingly complex.

Several approaches to survivability analysis include recoverability as one of the elements of vulnerability rather than separating recoverability out from the vulnerability element. DSTO's approach is to separate out the recoverability element as it is seen as a fundamental activity on its own. The UK MOD defines three types of survivability [1]:

- Functional survivability – the probability that a specific function (e.g. air warfare, propulsion etc) will survive a damaging hit. This is equal to (1 – Vulnerability) and is also called the probability of escaping critical damage.
- Basic survivability – the probability that the vessel will retain some level of function after damage (e.g. that the ship will still float and be stable).
- Overall survivability – this includes all three elements of susceptibility, vulnerability and recoverability.

The first two are simply subsets of the overall survivability and if the overall survivability is determined in an appropriate manner, the other two should be able to be determined.
The intent of any survivability analysis should be to ultimately evaluate the overall survivability.

Figure 2 shows the elements of survivability in a temporal representation of the capability of the ship, which is degraded after a damage event. After the damage event, there is a period of instantaneous damage where a significant reduction in capability occurs very rapidly. This is followed by a longer-term progressive damage phase which reflects the secondary damage caused by the initial event. This includes flooding, loss of sea-keeping/stability, progression of fire etc. The capability of the platform continues to rapidly decrease. Damage control processes are initiated during the progressive damage phase and when they have achieved some degree of control over the damage, some restoration of capability commences. This phase also includes the recovery of vital systems and capabilities. The diagram shown in Figure 2 may be replicated for all the three basic functions (float, move, fight), for individual capabilities (self defence, air warfare, propulsion, etc.) or for the total ship survivability.

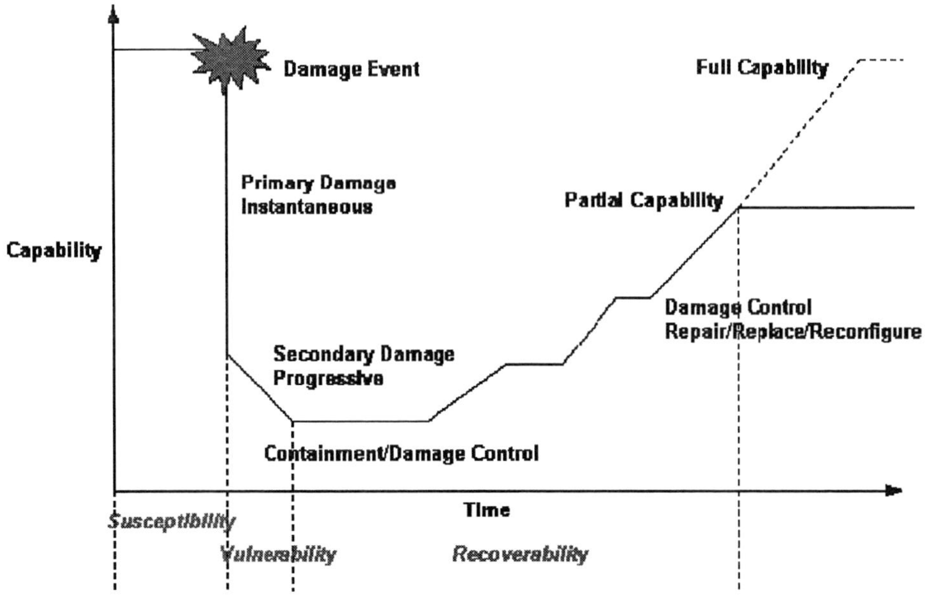

Figure 2: Temporal representation of the damage process from being detected, through the damage event until final capability recovery

The aim of a total ship survivability research program is to enable the elements of survivability to be modelled in such a way as to enable the overall survivability of the ship to be determined, to evaluate the trade off (e.g. cost Vs capability, cost Vs schedule for new ship construction) between the elements of survivability, the benefits of technical upgrades to survivability, to enable setting of minimum survivability requirements, to identify survivability capability shortfalls, to evaluate new technologies or systems and to evaluate bottlenecks in operations such as damage control. It may also be used as a tool to evaluate new operational constraints such as reduced crewing levels and their impact on survivability.

CURRENT APPROACH

Many navies are provided with advice on the susceptibility, the vulnerability and to some degree, the recoverability of its ships. But integration of the three elements to give a single statement or overall evaluation of ship survivability is not often provided. Complicating the process is the fact that the evaluation of the three individual elements of survivability is usually done at different levels of maturity and fidelity.

The evaluation of the aspects of susceptibility is usually carried out using complex and sophisticated computer models of the ship, defensive weapon systems and countermeasures, which evaluate the interaction between the signature elements and the ship. Many such models exist that detail the behaviour and performance of missile seekers, decoys, hard-kill weapons, electronic warfare systems and ship signatures. There are many signature models (visual, IR, RF, acoustic etc.) but their fidelity differs and many naval countermeasure models are limited in their capability. Many models detailing the performance of threats (e.g. missiles) have been developed and attempts to validate them in field trials or hardware-in-the-loop simulations has had mixed success.

The next element in the survivability equation is the vulnerability of the ship and again this is determined largely by advanced computer modelling techniques or by an analysis of the ship design and equipment. The modelling is generally based on small scale experimental evaluation of the vulnerability of various ship design elements such the construction materials, the design of bulkheads, the type of blast doors, etc. which provide the damage algorithms that are used by the models. Vulnerability evaluation also relies on an examination of the design in terms of location of critical equipment and the issues of separation, zoning and redundancy. The vulnerability is then often related back to various capabilities (e.g. propulsion, air warfare, self defence etc.) and how the threat may affect that capability post event. Vulnerability evaluation generally does not examine the effect of doctrine or operability as it focuses purely on the design elements in the ship. Also, the threat scenarios tend to focus on military threats as opposed to non-military threats.

The final element of survivability is recoverability, which includes damage control and the repair, restoration or replacement processes. Damage control aims to prevent, control and to limit the effects of fire, flooding and contamination and to maintain the combat capability and personnel safety. Recoverability also includes the ability to repair and recover vital systems such as power, chilled water and data transport. In global survivability analysis, often the damage control and recoverability factors are not considered. It is sometimes claimed that recoverability is an operational factor that relies on the training of the crew and by inference cannot be adequately evaluated [2].

Much research has been conducted detailing the performance of fire fighting systems, evaluated damage control tools and techniques, fire and smoke progression in structures and flooding progression and structural consequences. However, the development of models to evaluate many of these aspects of recoverability is limited. The increasing importance on the ability to recover vital capabilities coupled with reduced crewing, will lead to more attention to automation. This also requires that operational procedures and system design will have to be matched to each other.

In terms of the modelling of the overall ship survivability it should be noted that the UK have embarked on an extensive research program to develop a ship survivability modelling capability [3, 4].

MODELLING SURVIVABILITY

There are a number of mathematical approaches to evaluating survivability and some are outlined below.

The simplest approach to survivability evaluation is a probabilistic method where the probability of survival is calculated as:

$$P_{survive} = 1 - (S \times V \times (1-R)) \qquad [1]$$

where S = the probability of the ship being hit
V = the probability of the ship being damaged given that it has been hit
R = the probability of recovering the ship after being damaged after being hit

One problem with this approach is that S, V and R are not independent as has been assumed. They are conditional variables and cannot be simply combined to give an overall equation for survivability.
Many approaches to survivability analysis also dismiss the recoverability component as an operational aspect of the platform and as such, is neglected in the analysis [2]. For such approaches, the survivability is usually calculated as:

$$P_{survive} = 1 - (P_H \times P_{K/H}) \qquad [2]$$

where P_H is the probability of being hit
$P_{K/H}$ is the probability of being killed or suffering serious damage when hit

This equation is exactly the same as Equation 1 with the recoverability component removed. Another approach [3] which dismisses the recover component expands on the equation above:

$$P_{survive} = 1 - (P_{ec} \times P_{en} \times P_h \times P_k) \qquad [3]$$

where P_{ec} = the probability of being encountered
P_{en} = the probability of being engaged
P_h = the probability of being hit
P_k = the probability of being kill

Equations 2 and 3 are essentially identical with $P_H = P_{ec} \times P_{en} \times P_h$.

A recent approach to quantify capability recoverability after damage was based on the three options of replacement, elimination or reconfigure [5]. These were deemed as the three recover options available for the components that made up the capability. A recovery tree was developed for each of the major capabilities, the major components defined and each was assigned a vulnerability factor (V_f) and a probability of being able to replace, eliminate or reconfigure that component. These were summed together to provide an overall probability of recovering the capability according to:

$$RF_s = \left\{ \frac{1}{n} \sum_{x=1}^{n} \left(1 - RF_{c_x}\right)^3 \right\} \qquad [4]$$

where n = the number of components making up the system
x = the component number
RF_s = the system recoverability factor
RF_{cx} = the component x recoverability factor

The value of RF_c for each component x was calculated from:

$$RF_c = \frac{1}{3}\left(1 - VF\right)\left(R_1 + R_2 + E\right) + VF \qquad [5]$$

where, VF = the component vulnerability factor
R_1 = the probability of recovering the component through repair
R_2 = the probability of recovering the component through reconfiguring
E = the probability of recovering the component through elimination

This approach examines the recoverability purely in terms of repairing, replacing or eliminating components of the capability.

All these equations simply attempt to produce a number that is related to the survivability of the component, the system, the platform or the task group. However, the simplistic nature of the equations belies the complex nature of the elements that make up survivability. Not the least of which are the interrelationship between the variables, that a platform should be studied as part of a task group rather than as a single element, that each of the variables depends on the mission, the focus on a single encounter when most encounters involve multiple threats, the lack of data for recoverability and the fact that both susceptibility and vulnerability are not fixed as they will almost certainly change during the course of the encounter.

THE WAY AHEAD

The future direction in this area would be to develop a capability to answer some of the questions posed in the introduction by developing a capability to model ship survivability. The question is how may this to be achieved? There are a number of fundamental steps to this that may be coarsely broken up into the following:

1. Identify what current core models that would be the foundations for a modelling capability. Define the fidelity and the capability of these models to provide the information and identify the gaps in the existing models;
2. From the gap analysis, decide whether a new model is required to fill the gap or whether the current model can be expanded to fill the gap;
3. Build a platform level framework of survivability based on "mind maps" of the top level and lower levels influences on susceptibility, vulnerability and recoverability;
4. Develop a capability to integrate or link the individual survivability models together; and
5. Expand the platform level framework to a mission level framework.

As noted above, there are models available to evaluate the susceptibility and the vulnerability of naval ships and the performance of a range of damage control technologies. These would be classed as the "core" models that would be used. However, the two areas where there are knowledge gaps are in the modelling of recoverability and the integration of the "core" models into a platform level and mission level framework.

RECOVERABILITY MODELLING

Many of the individual elements of damage control such as fire fighting and flood mitigation have been studied scientifically but there have been limited attempts to model these operations and to validate the modelling approaches. Some of the areas that would need to be modelled include:

1. The behaviour of fire and its effects on the critical functions/capabilities of the ship, the effect on personnel and equipment and the effect on the structural strength of the ship;
2. The effectiveness of fixed and portable fire protection systems;
3. The effects of flooding on the critical functions/capabilities of the ship, the effect on personnel and equipment and the effect on the structural strength of the ship;
4. The effectiveness of fixed and portable flood protection systems including shoring;
5. The behaviour and effectiveness of damage control processes (e.g. DC parties) including human factors and human systems integration issues; and

6. The behaviour and effectiveness of repair, restoration or replacement processes.

There are some existing models of damage control processes including fire fighting and flooding [6,7] but the validation and the fidelity need further analysis. The approaches are also generally relatively simplistic and there is a need to capture more detail of the DC processes. This increases the complexity of the modelling with the attendant problems of run times and cost. Another fundamental problem with recoverability modelling is the lack of data on the processes involved and their relationship to capabilities. To enable this to be achieved the following are required:

1. The definition of the recoverability measure and the actions that need to occur to reach this value;
2. A method of measuring capability and defining the capability-time behaviour; Figure 2 is a useful visualization but is not the model;
3. Identify the more difficult factors of recoverability such as repair priorities, damage reporting; and
4. Identify the human aspects of recoverability; crew movement data is available but there is a heavy emphasis on statistics; the crew may be well trained but can behave very differently to real situations.

One of the major problems with modelling recoverability is the need to accurately (or inaccurately) model human behaviour. This is not only difficult to do but may be affected by human behaviour that is unexpected or irrational. For example, it is almost impossible to model the behaviour of the "heroic" person that has a major impact on the outcome of a damage control operation.

INTEGRATION

Integration of the many "core" models into a single survivability model is also difficult and presents significant challenges. These include:

1. The design of the platform and mission level frameworks; this includes the definition of the task group and the elements that make up the task group;
2. The design of the engagement scenario to include multiple encounters;
3. The management of variation in one or more aspects of survivability during any military and non military encounters;
4. The definition and the acquisition of the requisite fidelity for all the core models;
5. The identification of the linkages between the individual survivability models being cognizant that there may be multiple linkages to consider;
6. The incorporation of tactics, doctrine, concepts and training into an integrated philosophy; and
7. The decisions regarding the weightings to be assigned to each of the survivability core models when integrating them into the total survivability.

FUNDAMENTAL QUESTIONS

Perhaps the one basic issue that has not been discussed above is exactly what does a survivability of 70% or 80% or even 90% actually mean in practical terms to the navy decision makers? Or what does it mean to the captain of the ship? If we use a statement of 90% survivable in a performance specification for a new ship, how is it to be tested to ensure compliance? And should not the survivability be related to the primary mission of the ship? If the prime function of the ship is AAW, then if this capability is recovered 100% after an attack, but the total survivability is 70%, is this acceptable?

CONCLUSION

Given the increasing emphasis on the survivability of naval platforms, there is a need to develop a capability to evaluate integrated survivability for naval platforms. The fundamental characteristics of survivability and how they may be mathematically combined to give a numerical measure of ship survivability are easily described. However, the framework of an integrated survivability modelling tool combines the core computer models and develops those areas where models do not currently exist. The major focus for this approach is in the development of models of recoverability and in the process of integrating the core models. The integration needs to start at the platform level and then progress to the mission or task group level.

REFERENCES

1. Manley, D., The Development of "Smart" Requirements for Ship Survivability, NATO Research and Technology Organisation, Proc. RTO AVT Symposium AVT-087/RSY-012 on Combat Survivability of Air, Sea and Land Vehicles, Aalborg, Denmark, September 2002
2. Boulougouris, E. and Papanikoloaou, A., Optimization of the Survivability of Naval Ships by Generic Algorithms, Proc. 3rd Int. Conference on Computer Applications and Information Technologies in the Maritime Industries, COMIT'04, Siguenza, Spain, May 2004
3. Wickes, J., Integrated Mission Survivability, J. Def. Science, 10, 2, R48, 2005
4. Thornton, J., Surface Platform Survivability Analysis, Proc. 2nd European Workshop on Survivability, Noordwijkerhout, The Netherlands, March 2004
5. SEA 4000 Air Warfare Destroyer Vulnerability and Recoverability Assessment, Directorate of Naval Platform Systems, Navy Systems Command, Royal Australian Navy, 2005
6. Smit, C., Ship Survivability through Optimization of Fire Fighting Systems and Damage Control, Proc. RTO AVT Symposium AVT-087/RSY-012 on Combat Survivability of Air, Sea and Land Vehicles, Aalborg, Denmark, September 2002
7. Gillis, M., Keijer, W. and Smit, C., Advanced Concept for Damage Control, Naval Forces, 1, 6-10, 2003

SEA4000: The Australian Air Warfare Destroyer Program – Combat Systems Challenges and Opportunities

John Davis
Air Warfare Destroyer Alliance, Australia

The Air Warfare Destroyer Program was established in response to the Defence White Paper 2000, which stated that the Australian Defence Force would replace the Navy's FFGs with a class of new Air Warfare Destroyer (AWD). The AWD Program is to ensure timely and efficient delivery of the new affordable, effective, flexible and sustainable AWD capability for the security of Australia.

The purpose of the new AWD is to achieve sea control in conjunction with Australian Defence Force or coalition maritime, air and land forces. The AWD is to support the full spectrum of operations from peacetime to war, and will be capable of supporting a number of primary missions, including control of the air environment, control of the surface and subsurface environments and tactical land strike. The AWD will operate in tropical to sub-Antarctic conditions and will be adept in both the littoral and open ocean environments. The AWD will be equipped with a highly-capable combat system founded on the Aegis Weapon System, which incorporates the state-of-the-art phased array radar, AN/SPY-1D(V).

The Commonwealth is using an alliance contracting strategy for the Build phase of the project. Under the alliance strategy, the DMO, together with ASC AWD Shipbuilder Pty Ltd and Raytheon Australia Pty Ltd, is working in an integrated team environment to achieve project objectives. However, opportunities for wider-industry participation remain and are crucial if the Program objectives are to be met.

MAN Diesel High Power Density Engines and After-Sales Support in Naval and Commercial Applications

By Ralf Losch
Head of Sales - Naval, Yachts and Fast Ferries

MAN Diesel SE
MSE6
Stadtbachstr. 1
86153 Augsburg

ralf.losch@man.eu
www.mandiesel.com

MAN Diesel is the world's leading designer and manufacturer of low and medium speed engines – an estimated 50% of world marine trade is powered by engines from MAN Diesel.

The company develops two-stroke and four-stroke engines, auxiliary engines, turbochargers and propulsion packages that are manufactured both by the MAN Diesel Group and its licensees.

As well as pure Diesel engines, the MAN Diesel engine range comprises two and four-stroke gas engines with dual-fuel, spark-ignition and MAN Diesel's unique "PGI" ignition technology. The engines cover a power range from 450 kW and 97.3 MW and are built at works in Germany, Denmark and France and India, as well as at a global network of licencees. In the near future MAN Diesel's own engine building capacity will concentrate on four-strokes with all two strokes built by licencees.

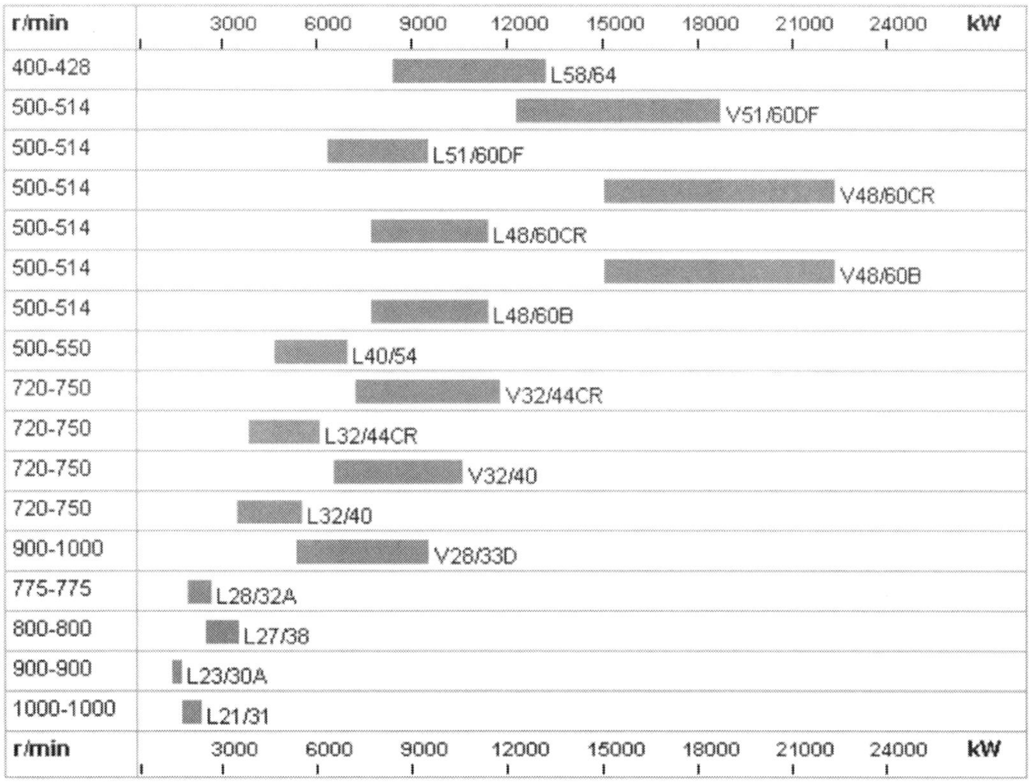

Figure 1: MAN Diesel four-stroke marine engine programme

With this mix of engines, MAN Diesel's major markets are marine propulsion, marine generator sets and stationary electrical power generation.

On the marine propulsion side the product range also includes CPP-propellers, reduction gears, propulsion controls and complete propulsion packages. Exhaust-gas turbochargers are built at the works in Augsburg, the Czech Republic and China (components and assembly).

Figure 2: CPP Propeller

Figure 3: TCA high efficiency axial turbocharger

As a consequence of this combining of forces, all engines and systems for naval applications will henceforth be sold under the “MAN Diesel” brand. Likewise, all sales and engineering activities are now concentrated into a new team drawn together from specialists throughout MAN Diesel’s expanded European organisation. Its members are all specialists with proven industry experience in naval applications from all our manufacturing locations

MAN Diesel in Australia

MAN Diesel Australia Pty Ltd occupies extensive office and workshops facilities located in N. Ryde, New South Wales and covers sales, distribution and service (MAN Diesel PrimeServ hub) of MAN Diesel engines, turbochargers, propellers, gears and complete propulsion system.

Engine sales emphases are stationary power generation and marine applications, including a strong position in the market for fast multi-hull ferries, where MAN is a preferred supplier to catamaran specialist INCAT of Tasmania.

On the PrimeServ after-sales side, MAN Diesel Australia Pty Ltd is contracted with the maintenance, repair and overhaul of a number of ships of the Royal Australian Navy.

Figure 4: MAN Diesel Australia supplied and commissioned diesel/electric systems for two hydrographic ships of the Royal Australian Navy.

Naval and Governmental Vessels

With the 2006 acquisition of the remaining shares in French engine builder S.E.M.T. Pielstick – now MAN Diesel SAS – the company added a huge fund of naval propulsion knowledge to the considerable experience and expertise inherited from UK subsidiary MAN Diesel Ltd and the longstanding connections the Danish works with the navies of World.

The combined customer base includes over 60 Navies around the world and one of the most extensive reference lists in the industry.

Figure 5: Sigma Corvettes for Indonesian Navy.

It covers over 2,500 propulsion engines with a combined output of over 8,300 MW and full spectrum of applications including OPVs, corvettes, frigates, destroyers, aircraft carriers, submarines and many types of logistical support vessel.

With a market share of 44%, MAN Diesel is the market leader in all types of naval vessels with prime movers in the power segment 3,000 to 10,000 kW per engine.

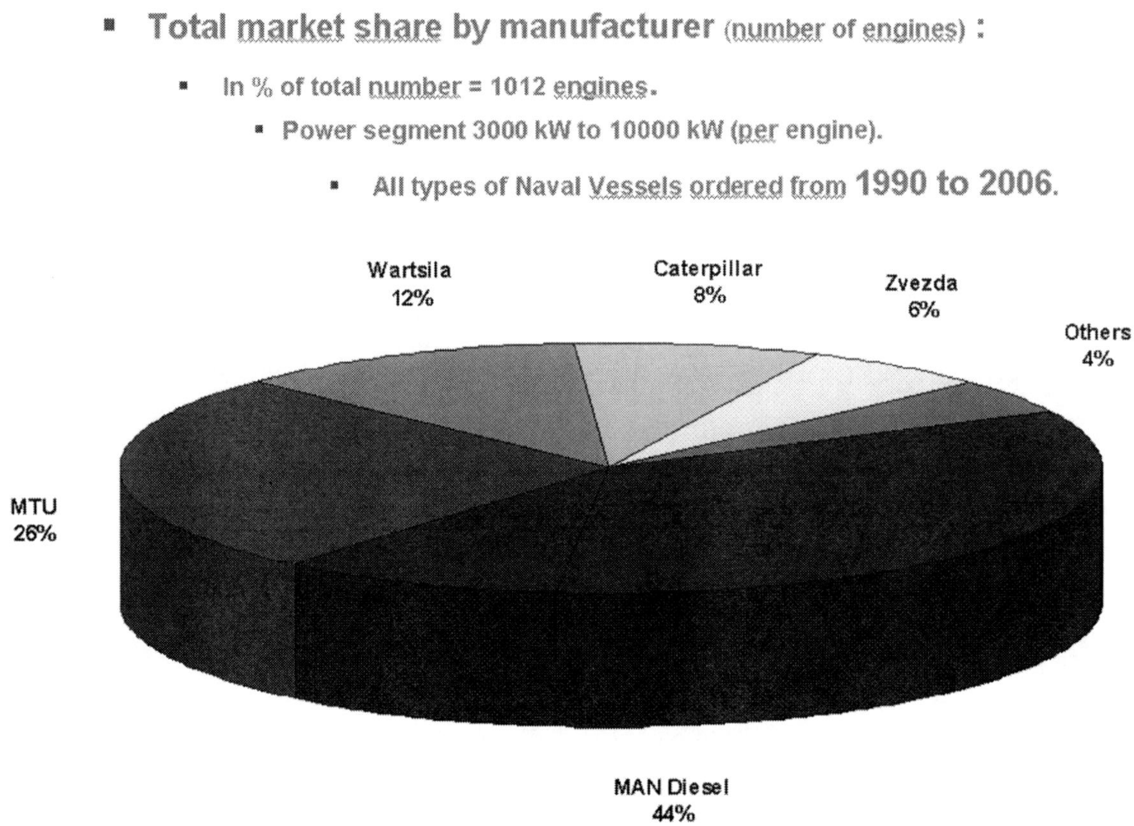

Figure 6: Market Share, Naval Vessels

Application expertise and experience encompass conventional reduction gear and propeller propulsion, fast craft with water jet propulsion, diesel electric systems with sophisticated load sharing and multi prime-mover propulsion systems like CODAD, CODAG and CODOG.

Submarines

MAN Diesel SAS in France can point to uninterrupted success with one of the most widely used submarine diesels in existence. With some 120 submarines equipped and 230 engines with 160 MW sold, the compact PA4 engine and generator-set package is a power source of choice for conventional submarines.

Of special interest is the PA4-200 SMDS, which is used in French "Scorpène" conventional submarines. To achieve best-in-class power output and absolute safety while snorkelling regardless of surface sea swell, the PA4-200 SMDS features a unique supercharging system which pairs an engine driven supercharger with an exhaust gas turbocharger.

Figure 7: French Marine Nationale Scorpène submarine

OPVs

In total, MAN Diesel has supplied over 650 engines with a combined output approaching 1,900 MW for OPVs of various types and sizes and employing both mechanical and diesel-electric propulsion systems.

Figure 8: Tenix / RNZN OPV

Fighting Ships

Since the early 1970's MAN Diesel has supplied over 840 propulsion engines totalling over 2,400 MW to frigate, corvettes and destroyers applications. The 265 vessels supplied feature both single engine-single reduction gear and combined prime mover systems like combined diesel and diesel (CODAD) and combined diesel or gas turbine (CODOG) - for example the propulsion system of the Franco-Italian designed Horizon/Orrizonte frigates employing MAN Diesel 280 mm bore engines with STC sequential turbocharging. Moreover, over the same period, the company has delivered numerous generator sets for fighting ships.

Figure 9: OPVs for Trinidad & Tobago Coastguard from VT Shipbuilding

Logistic Support Vessels

In total MAN Diesel has supplied almost 4,000 MW of propulsion and onboard generator power involving the full spectrum of its medium speed engines and some 820 ships covering a full spectrum of vessel types: support tankers, replenishment tankers, fleet tankers, transport oilers, submarine support vessels, LSD's and LST's, dock landing ships amphibious transport tankers, auxiliary oilers, fast sealift ships. icebreaking salvage vessels, helicopter carriers and water tankers.

Figure 10: HMAS Success

MAN Diesel Technology

The foundation of MAN Diesel's success in all applications, naval or merchant marine is its unparalleled understanding of large engine technology. A vital part of our technological leadership stems from a firm belief in the "total systems approach" to engine building.

MAN Diesel is the only medium speed engine manufacturer having all the strategic elements of engine development in-house.

As well as the design, development and manufacture of complete engines, MAN Diesel is also the market leader in:

- large exhaust turbochargers,
- injection equipment for heavy fuel
- advanced engine management software
- on-engine electronic hardware for control, monitoring and fault diagnosis on large engines and their driven equipment.

Moreover, as well as an impressive array of computer-aided design tools, one of the engine industry's largest, best-equipped foundries for high quality, high duty engine components allow a decisive shortening of product development, further development and application engineering times.

Advanced Fuel Injection

MAN diesel's latest fuel injection technology is its in-house developed common rail equipment for heavy fuel engines. Unlike common rail systems from other manufacturers, the MAN Diesel system employs pressure control of conventional injectors with solenoid valves on the system's segmented pressure accumulators, instead of injectors with integral magnetic valves. Electronic components are thus protected from conducted heat emanating from the combustion chamber. The system's high pressure pumps are installed in the same bores used for conventional unit pumps in a conventional injection system and the pressure accumulators (common rails) are segmented to make best used of available space on the engine between the injection pumps and the injectors. This construction and the use of standard, pressure controlled injectors are intended to facilitate common rail system retrofits.

Common rail fuel injection allows free selection of the injection pressure and injection timing parameters and enables precise injection rate shaping. As well as benefits in fuel consumption and exhaust emissions at part load – MAN Diesel engines with common rail injection feature smokeless exhaust gases under all load conditions – the power output of MAN Diesel common rail engines, like the all-new 32/44CR and upgrade 48/60CR, can be precisely matched to the needs of the application at all speeds and under all load conditions with resulting increases in fuel efficiency emissions.

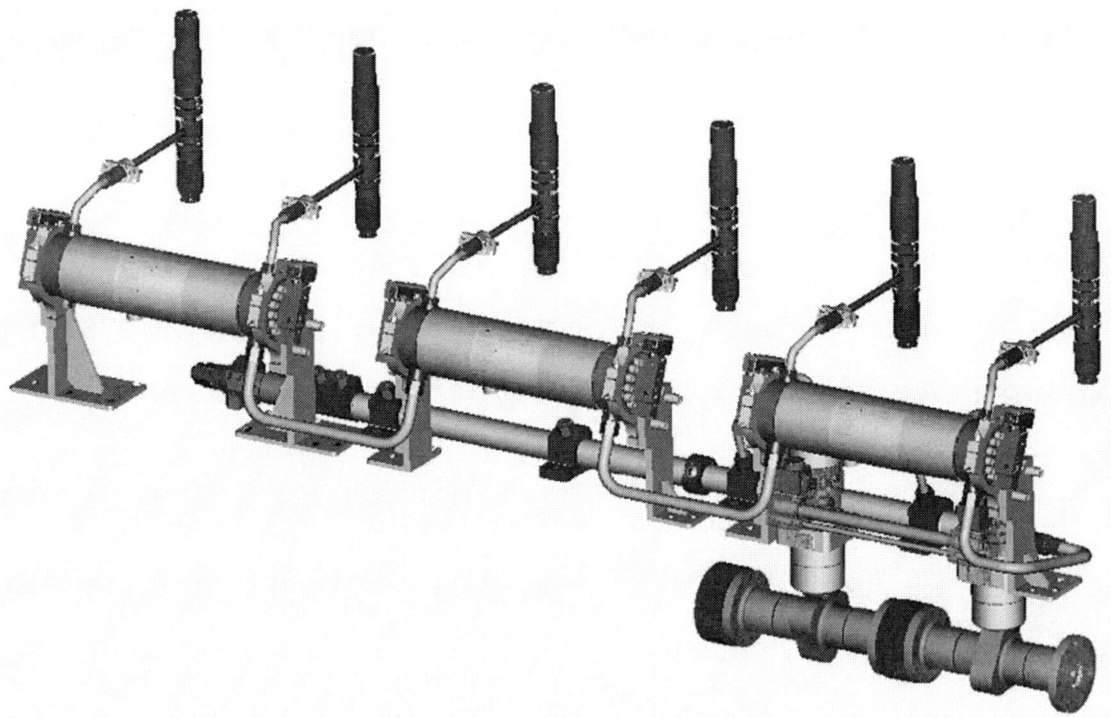

Figure 11: MAN Diesel Common Rail Injection System

Advanced Air Management

Paralleling the fuel management flexibility achievable with common rail injection is the air management flexibility offered by advanced turbocharging techniques. MAN Diesel recently announced its "VTA" (Variable Turbine Area) turbochargers and is currently in the process of developing the STC (Sequential Turbocharging System) for the high power density 28/33D engine, initially for application in naval vessels. Moreover, as part of the European Union sponsored, engine industry-wide "Hercules" engine development programme, early experience has been gained with two-stage, high pressure turbocharging.

The VTA system consists of a nozzle ring, equipped with adjustable vanes which replaces the fixed-vane rings used in MAN Diesel's standard TCA and TCR turbochargers. Adjusting vane pitch regulates the pressure of the exhaust gases impinging on the turbine to vary compressor output. The quantity of charge air can be more precisely matched to the quantity of injected fuel, resulting in reduced specific fuel consumption and emissions, in combination with improved dynamic behaviour of the engine-turbocharger system.

Figure 12: VTA Turbocharger showing adjustable vane nozzle ring.

Variable Valve Timing

A further "variable" which will be applied on MAN Diesel engines to the benefit of emissions and efficiency is variable valve timing, which enables efficiency-boosting, NOx-reducing techniques like a variable Miller process. The first engine to feature variable valve timing will be the vee configuration versions of the new 32/44 CR common rail engine.

Advanced Electronics

As the underlying technology of all these new engine control possibilities, MAN Diesel develops its own electronic engine management systems, encompassing engine control, monitoring and diagnostics. The company is autonomous in both software and hardware development, having initiated its own advanced hardware platform in 2005.

The latest development is $SaCoS_{one}$ .The abbreviation stands for "Safety and Control System" and the "one" suffix for "on engine, since the system is very compact and mounted entirely on the engine. In terms of its functionalities, $SaCoS_{one}$ combines enhanced mapping of all engine control parameters with improved monitoring and diagnostics, including remote "online" monitoring via telephone connections or the internet. To enhance operational safety, a variant of the SaCoS system with two electronic control units offers complete redundancy.

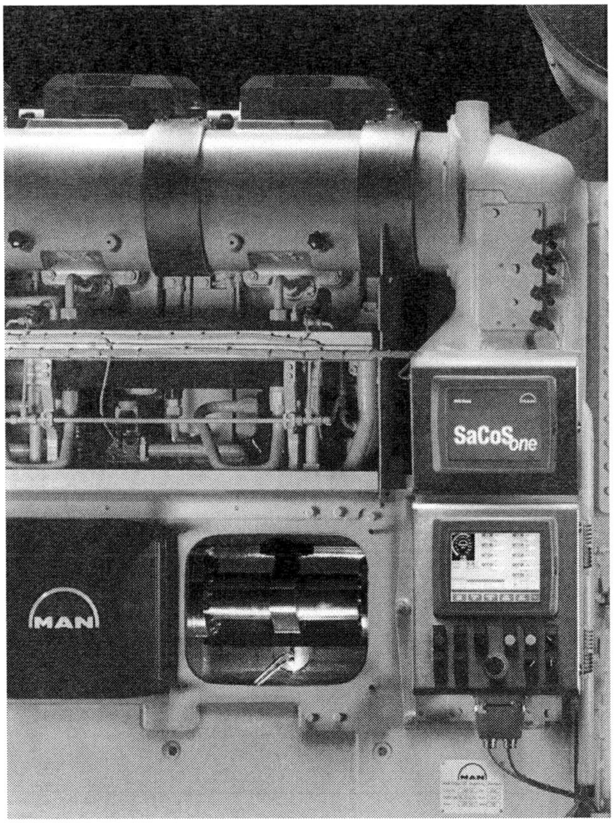

Figure 13: $SaCoS_{one}$ on-engine controller module.

Application Engineering

Within our overall product objectives for new and further developed medium speed diesels from MAN Diesel, the needs of naval ship and patrol boat applications are always given full consideration, including both maintainability and survivability. Their rationale may be different, but naval applications and commercial applications have many common requirements which MAN Diesel's total systems approach is ideally suited to meet. For example, efficient combustion leads to compliance with exhaust emission legislation and low fuel consumption, but also benefits a vessel's thermal signatures and its radius of operation.

As a precondition for low wear and long life through low deformation, robust construction is the basis of high availability but also the firm foundation of excellent shock resistance and survivability. Modular construction and ready accessibility ensure maintainability to the benefit of operating and service costs and high availability in all applications.

Our works are expert in closely adapting our engines and generator-sets to the needs of the specific application. In this area our flexible, electronically controlled technologies, detailed above, facilitate precise matching of engine output at all points on its load profile.

On the naval side, equipment at our works in Saint-Nazaire, France, includes test and simulation equipment for designing, testing and perfecting engine shock resistance and thermal and acoustic emissions to the standards of both individual navies and military alliances like NATO etc. Our experience in shock attenuating technologies, including the resilient mounting of engines and driven and auxiliary equipment, enable us to produce systems that can withstand loadings as high as 10G and more. Another highly specialised, longstanding activity is the preparation of engines for submarine applications, with their vital tolerance of high exhaust backpressure and intake air resistance.

Figure 14: The PA4-200 SMDS from MAN Diesel with its unique combination of supercharging and turbocharging.

After-Sales

MAN Diesel's global after-sales organisation, MAN Diesel PrimeServ, comprises an expanding network of company-owned service centres and authorised partners at strategic locations around the world.

MAN Diesel PrimeServ unites all the after-sales activities for all the products of the MAN Diesel Group under a single, high profile global brand.

Its products and services are conceived to complement the designed-in user-friendliness and maintainability of MAN Diesel's engines and propulsion equipment.

Integrated Logistical Support

In accepting naval contracts, MAN Diesel can be expected to commit to keeping a fighting vessel operational for several decades. With its constantly developing after-sales infrastructure, MAN Diesel PrimeServ is well equipped to meet the very special requirements of naval customers regarding comprehensive documentation, the transfer of know-how in the form of personnel training and local content agreements, and in fulfilling our undertakings to ensure high, long term engine availability, including guaranteed, full life-cycle replacement parts supply and the substitution of obsolete technologies.

These obligations can include air-freighting spare parts around the world at short notice, inventing or configuring new assemblies and systems to replace obsolescent/ obsolete technology and providing specialist assistance, likewise at short notice and anywhere in the world.

With more than 100 PrimeServ service stations and authorised service partners currently in operation around the world, MAN Diesel is making a long term investment to guarantee maximum vessel readiness via the high availability of the propulsion plant.

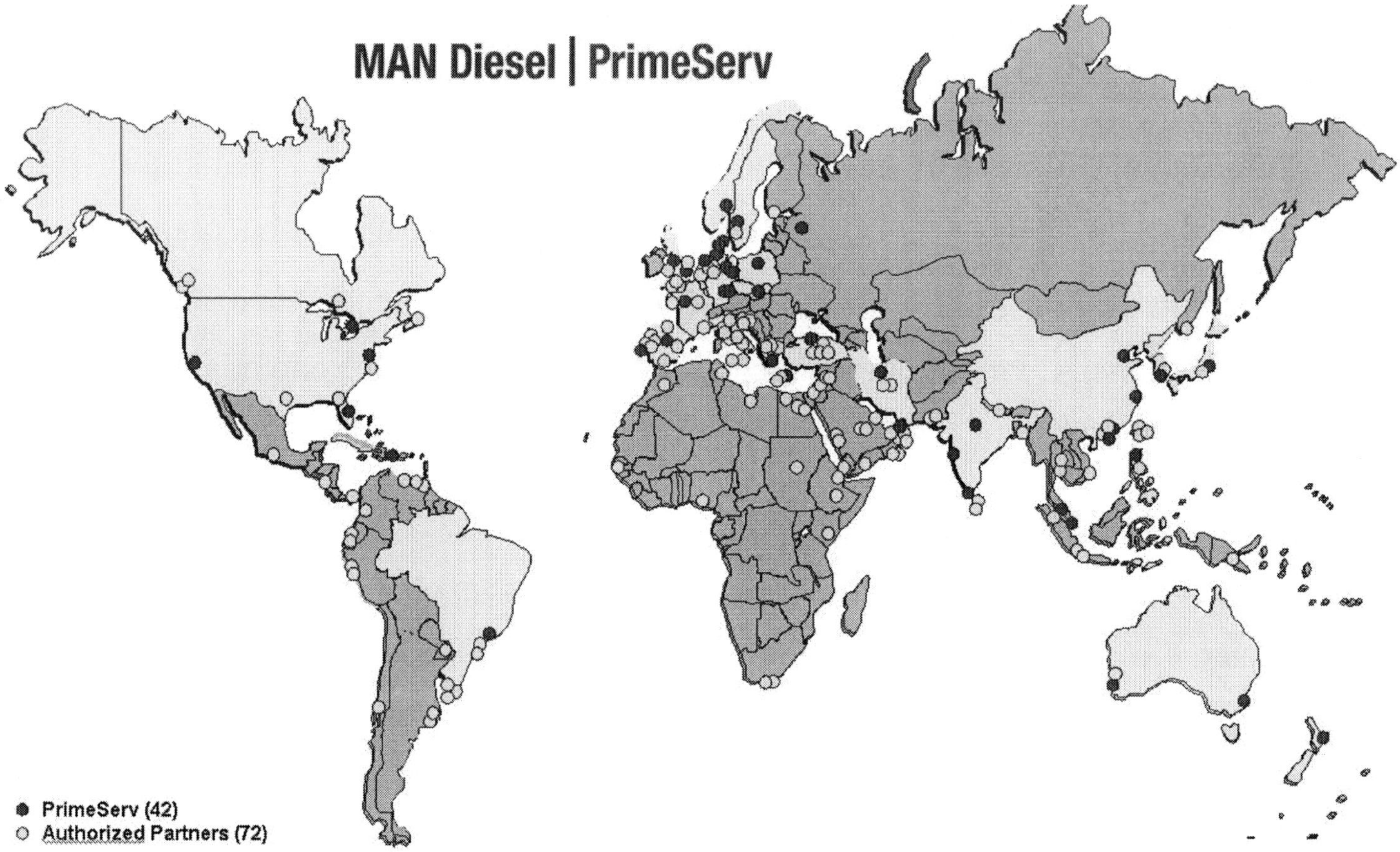

Figure 15: MAN Diesel PrimeServ's Local Presence

The PrimeServ organisation sets out to:

- Provide prompt, OEM-standard service based on experience and feedback from an immense population of engines and equipment in the field.
- Advise and qualify customers in order to improve the availability and viability of their installation over its entire service life, from commissioning to de-commissioning.

- Supply rapidly and all around the world, genuine, quality-assured MAN Diesel spare parts or replacing obsolescent components, assemblies and systems by new designs.

Since its 2006 launch, the PrimeServ organisation has made rapid advances in all areas:

- Geographic presence - in 2006 alone MAN Diesel PrimeServ established 11 major hubs and in 2007 and 2008 up to 20 further locations will be equipped and qualified.
- Spares Logistics - MAN Diesel PrimeServ has optimised and globalised its spares inventories. Stocks now more precisely reflect real ordering patterns. Inventories are being set up at strategic locations around the globe.
- Spares Kits - MAN Diesel PrimeServ has introduced spares kits comprising all the parts needed for a specific service or repair job. They simplify ordering and ensure that components are not omitted.

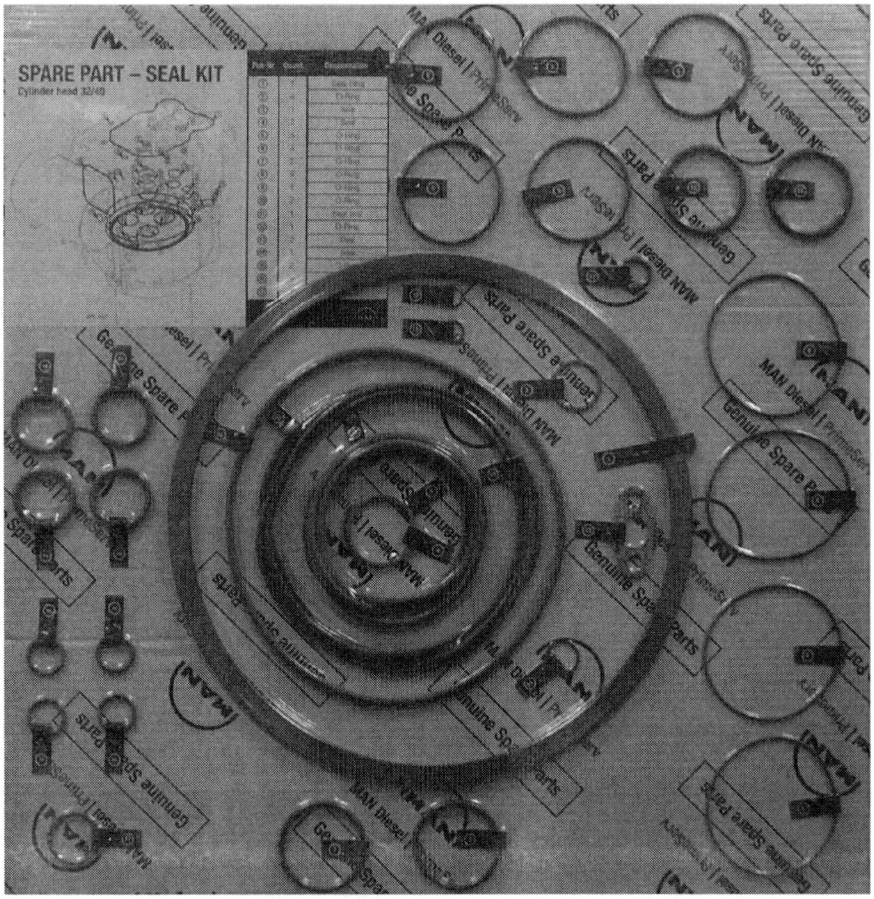

Figure 16: PrimeServ comprehensive spares kits

- Service Contracts - the new EMC "Engine Management Concept" from MAN Diesel PrimeServ comprises range of delegated after-sales service and maintenance agreements. Contracts are offered for both new and existing marine engines with a choice of 5, 10, or 15 year durations - or the life of the installation – and charged at a fixed monthly rate.
- Qualification: At the growing network of "PrimeServ Academies" MAN Diesel provides efficient, high-grade product training and regular retraining.

To facilitate both theoretical and hands-on learning, all Academies are equipped with a comprehensive range of modern teaching aids, including computer simulators and an array of engines, engine systems, sub-systems, assemblies and sub-assemblies. Courses range from tuition in basic service and repair skills to levels where naval technicians can acquire the qualifications they need to completely take over of the operation, maintenance, repair and overhaul of MAN Diesel products.

Figure 17: High level of teaching equipment at PrimeServ Academy, Augsburg

MAN Diesel Engine Types 28/33D and 28/33D STC

The newest and most versatile of MAN Diesel's engines for fast vessels, offshore patrol boats and naval and governmental vessels are the vee configuration (52° vee angle), type 28/33D with conventional turbocharging and the 28/33D STC with sequential turbocharging. They represent a logical evolution from their well proven predecessors including the MAN Diesel Pielstick PA6, engine with which they share key features like their 280 x 330 mm bore and stroke dimensions, and their direct progenitor the

MAN Diesel Ruston RK270 – the engine which pioneered the market for 1000 rpm medium speed engines in fast multi-hull craft.

Figure 18: Cutaway view of MAN Diesel type 28/33D Diesel

The new engines can point to a prominent pedigree. On the one hand they enjoy a long-standing dominance in the market for high speed, multi-hull ferries - including craft which have more than once taken the Hales Trophy (Blue Riband) for the fastest Atlantic crossing by a passenger vessel. On the other hand they look back on a long history of propulsion and power generation tasks in OPVs, classic surface fighting ships like corvettes, frigates, destroyers and aircraft carriers and a leading position in submarine applications. As well as its commercial applications, this engine has been involved in pioneering fast military catamaran projects, including a number of Australian built multi-hull patrol, fighting and logistic support craft.

The 28/33D is currently offered in 12V, 16V and 20V versions in a power range from 5400 to 9000 kW. As such it is the world's highest powered engine with 1000 rpm rated speed and the global engine of choice for fast multi-hull craft.

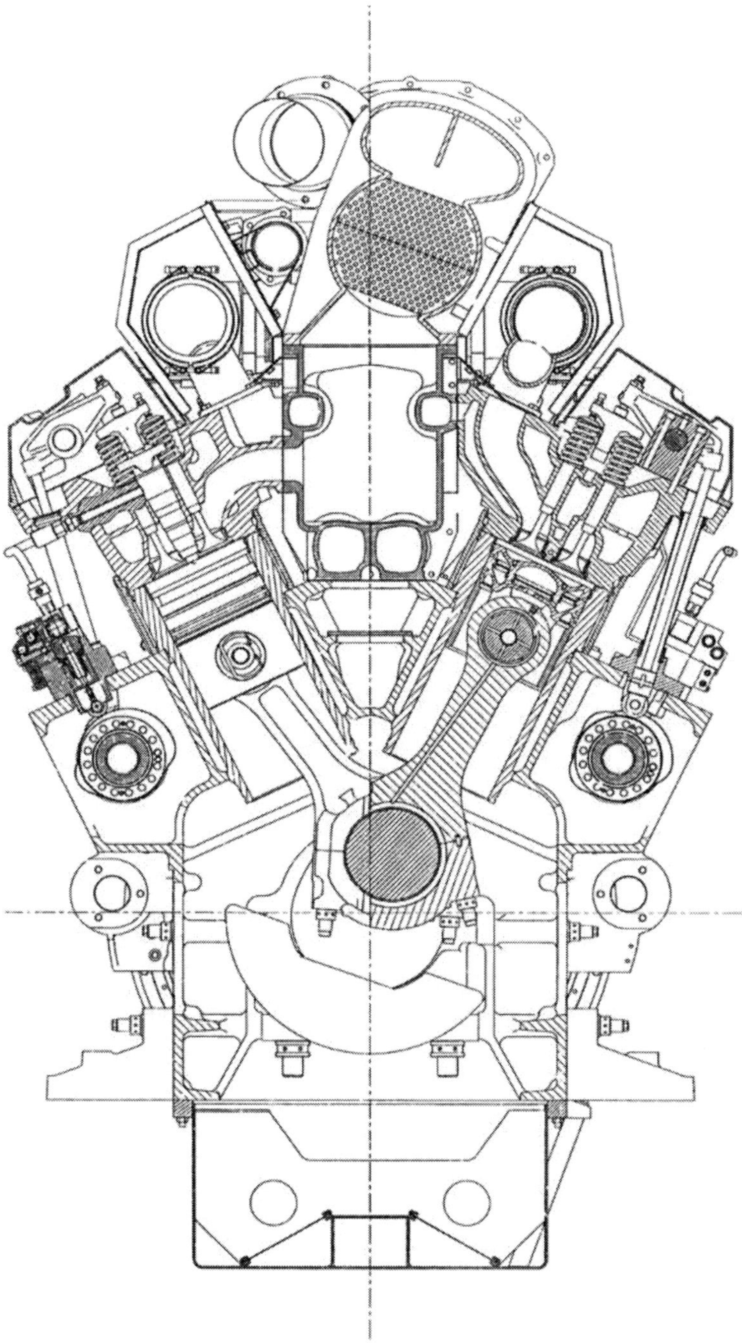

Figure 19: Cross sectional drawing of MAN Diesel 28/33D

At 20.32 litres per cylinder, the 28/33 engine platform combines the highest unitary displacement in its class with the one of the most compact dimensional envelopes. Its excellent ratios of power-to-weight and power-to-installation space render the 28/33D equally suited to fast mono- and multi-hull vessels, OPVs with conventional, single engine-gear-propeller systems or corvettes, frigates, and destroyers with combined prime mover systems like CODAD, CODOG, CODAG etc or as a power unit for generator sets.

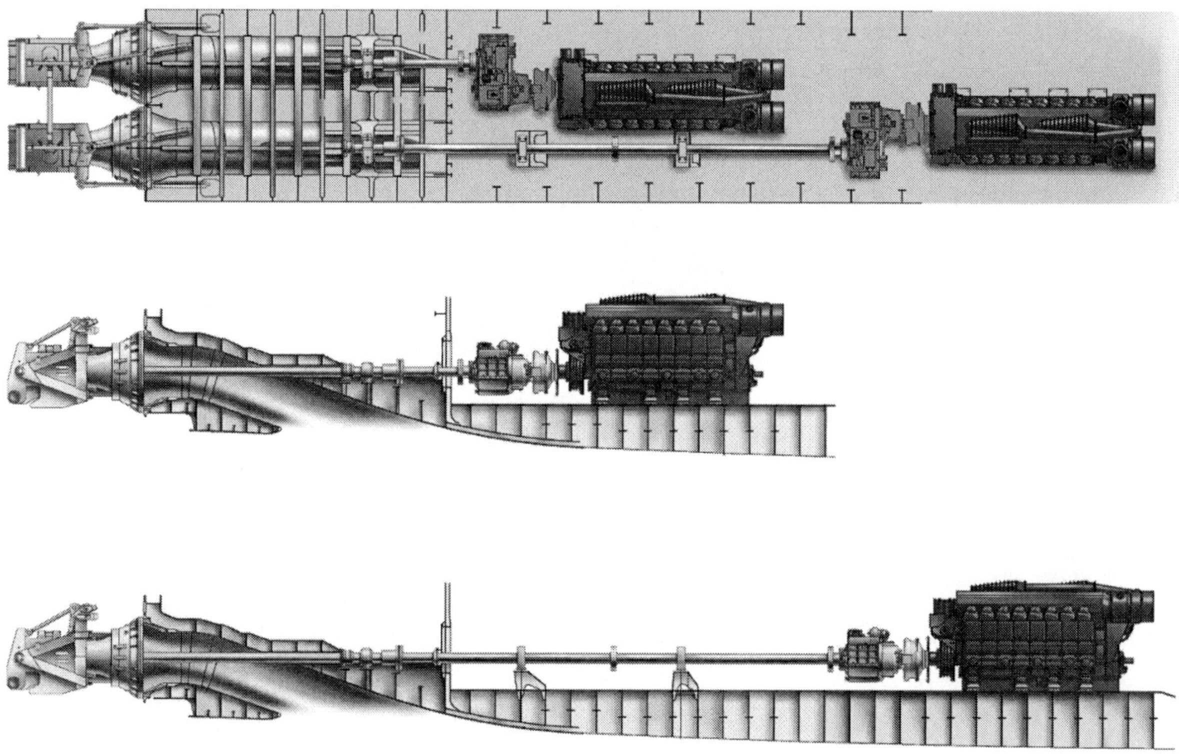

Figure 20: Typical layout of MAN Diesel type 28/33 engines in fast multi-hull ferries with water jet propulsion driven via reduction gears with horizontal shaft offset.

However, in spite of attaining its market-leading performance data at the very start of its product life cycle, the 28/33 platform nonetheless offers wide potential for future upgrades beyond its current 450 kW/cylinder rating. Indeed, a version with 10 MW rating (20V, 500 kW per cylinder) is already projected.

	12 cylinder 'vee'	16 cylinder 'vee'	20 cylinder 'vee'
Bore - mm	280	280	280
Stroke - mm	330	330	330
Power output - kWb	5400	7200	9000
Rating speed - r/min	1000	1000	1000
Fuel consumption - g/kWh*	190	190	190
BMEP - bar	26.5	26.5	26.5
Weight - tonnes	30	37	46
Length - mm	5490	6410	7330
Height - mm	3180	3180	3180
Width - mm	2100	2100	2100

*** includes all ancillary drives running and achieved at full power and 1000 RPM.**

Figure 21: Main Characteristics and Power Range 28/33D

As well as high power density, the 28/33D has been designed for high reliability and ease of maintenance as a route to low life cycle costs and low total cost of ownership. Both service engineers and shipbuilder representatives gave vital input throughout the development process. Significant results are the high power density and compact design of the engine (e.g. 52° vee angle, multifunctional camshaft/drive gear, hollow centre of camshaft used as a clean oil gallery, charging system incorporating compact manifold sections joined by expansion bellows.

Further features of the 28/33D and 28/33D STC engines include and underslung dry crankcase, a high performance cylinder head with intensively cooled flame face, cylindrical two-stage, high efficiency intercoolers, the manifold system is a quasi-constant pressure compact arrangement. The lubrication system is designed as an engine mounted module including duplex filters, cooler and thermostat, exchangeable cartridge filters. The lube oil sump features and integral oil suction pipe. All ancillary equipment is engine mounted with all pumps driven from PTOs at the free end of the crankshaft.

In standard versions two high performance axial turbochargers are employed, featuring an air by-pass arrangement to eliminate high temperatures at part load in variable speed applications and to maintain the surge margin.

STC + EFI

Among the clearest advances on the 28/33D and 28/33D STC is their full-authority electronic fuel injection technology (EFI). As well as favourable values for fuel consumption and exhaust emissions over the entire operating map, electronic control makes the 28/33D well suited to integration into overriding vessel management and automation systems.

On the 28/33D STC, microprocessor-managed, solenoid valve-controlled fuel injection technology meets its perfect complement – electronically managed sequential turbocharging.

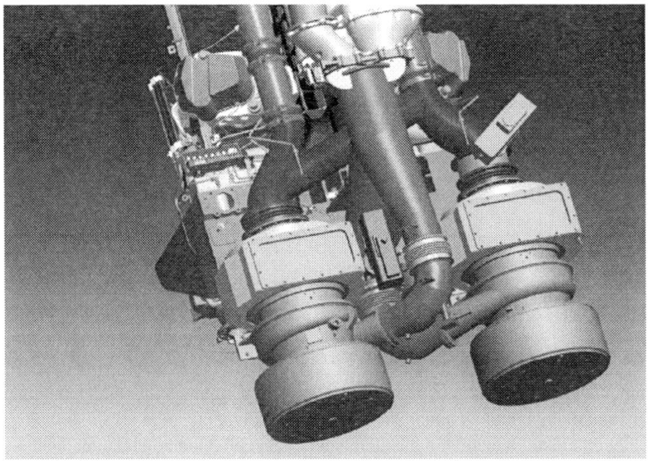

Figure 22: Sequential Turbocharging System

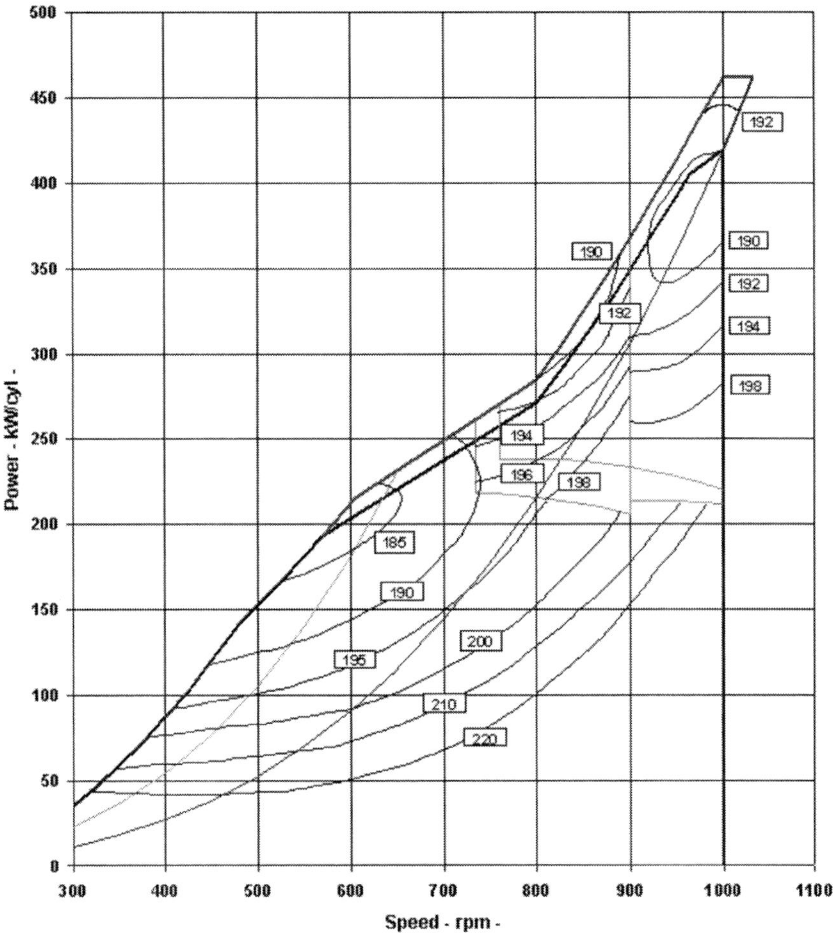

Figure 23: V28/33D STC, Specific Fuel Oil Consumption (g/kW.h, ISO 3046/1 conditions, Tolerance 5%)

MAN Diesel's STC system is derived from well proven equipment used on the PA6 engine and greatly increases the supply of charge air at low to medium engine speeds. However, in contrast to other sequential turbocharging systems on the market, the MAN Diesel STC system is essentially - and intentionally - simple, consisting of two identical, standard turbochargers, one providing copious charge-air at low and medium speeds with the second cutting in at higher speeds.

STC Flexibility and Stealth

In combination, the two technologies STC and EFI substantially extend the torque envelope of the 28/33D in its mid speed and load range. Improved torque at low and medium operating speeds confers a range of operating benefits encompassing excellent engine response and vessel acceleration, improved reliability due to reduced thermal stresses and enhanced manoeuvrability due to faster load acceptance. Likewise, economical operating modes are possible, like the operation of a single engine operating at twice the propeller law in twin engined propulsion systems with twin input / single output gears or in combined prime mover propulsion systems (CODAD, CODAG etc.).

Due to its enhanced fuel and air management, the 28/33D STC displays a low native thermal signature and reduced smoke emissions at part load. Moreover, the high torque reserves of the STC engine enable operating modes with extremely low noise emissions. These include cruising with a controllable pitch propeller (CPP) fixed at its optimum pitch for quiet running while still retaining the capability to accelerate rapidly.

Current Development Targets, MAN Diesel 28/33D / 28/33D STC

The ongoing development programme of the 28/33D includes:

- Adoption of SaCoSone engine control system
- FAT-US Naval Qualification
- Shock qualification
- Application engineering for GenSet applications
- Adaptation to operation on jet fuel
- Certification to IMO Tier 2
- 500 kW/cyl. versions

Medium term development aims also include the introduction of a further developed electronic fuel injection system with greater flexibility for injection pressure and rate shaping control. Shock calculations using DEF STAN 00-35 shock loading are already available

Applications

- **Delivered engines:**

▪ Fincantieri / Gotland	4 x 20V28/33D	Fast Ferry	In service since 07/2006
▪ Incat 062 / Trasmed	4 x 16V28/33D	Fast Ferry	in Service since 08/2006
▪ Tenix / NZ Navy	2 x 12V28/33D	OPV	Installing in vessel
▪ HDW /Sigma	2 x 20V28/33D	Super yacht	Installing in vessel
▪ Tenix / NZ Navy	2 x 12V28/33D	OPV	Delivered
▪ Incat 064	4 x 20V28/33D	Fast Ferry	Delivered

- **Manufacturing:**

▪ Incat 065	4 x 20V28/33D	Fast Ferry	Manufacturing in Augsburg
▪ Incat 066	4 x 20V28/33D	Fast Ferry	Order received 05/2007
▪ Incat 067	4 x 20V28/33D	Fast Ferry	Order received 05/2007
▪ Vosper : Trinidad Navy 1	2 x 16V28/33D	OPV	Notice to proceed 06/2007
▪ Vosper : Trinidad Navy 2	2 x 16V28/33D	OPV	Notice to proceed 06/2007
▪ Vosper : Trinidad Navy 3	2 x 16V28/33D	OPV	Notice to proceed 06/2007
▪ Schelde : RNLNavy 1	2 x 12V28/33D	Naval Patrol Vessel	Notice to proceed 08/2007
▪ Schelde : RNLNavy 2	2 x 12V28/33D	Naval Patrol Vessel	Notice to proceed 08/2007
▪ Schelde : RNLNavy 3	2 x 12V28/33D	Naval Patrol Vessel	Notice to proceed 08/2007
▪ Schelde : RNLNavy 4	2 x 12V28/33D	Naval Patrol Vessel	Notice to proceed 08/2007

- **18 engines already delivered, 26 more engines on order at time of writing.**

Figure 24: 28/33D Sales References

Commercial

Commercial MAN Diesel recently announced a contract for eight engines to power the largest catamarans ever built in Australia, the acknowledged centre of excellence for fast multi-hull craft.

The 112 metres overall length, 30 metres beam catamarans are built by INCAT of Tasmania and the contract involves the first 28/33D diesel engines built at MAN Diesel's engine works in Augsburg, following the engine's transfer from its UK facilities.

Figure 25: INCAT 112 metre Catamaran Natchan Rera

In detail, the contract covers eight of the largest, 20 cylinder versions each rated 9,000 kW at 1000 rpm and represents a seamless transition from UK-built engines to German-built engines in INCAT vessels. In the 112 metre catamarans two of the 20 cylinder engines are installed in each of the twin aluminium hulls where they drive high capacity reversible water-jets via reduction gears with horizontal offset input and output shafts. The total installed power of 36 MW gives the catamarans a loaded speed of 40 knots, INACAT states.

Naval and Governmental

The 28/33D and 28/33D STC engines are already making inroads into the naval market with their combination of high power, long endurance, economical, low-signature operation and, especially in the case of the sequentially turbocharged STC versions, rapid engine response.

Recent highlights include adoption as the prime mover in the mechanical propulsion systems of three OPVs to be built by VT Shipbuilding of the UK for the Trinidad and Tobago Coastguard (2 x 16V28/33D), engines for two Tenix OPVs of the Royal New Zealand Navy (2 x 12V28&33D) and for four naval patrol vessels being built for the Royal Netherlands Navy by Schelde Shipbuilding (2 x 12V28/33D).

Figure 26: The MAN Diesel type 28/33 diesel has been chosen for the mechanical propulsion systems of three OPVs to be built by VT Shipbuilding for the Trinidad and Tobago Coastguard.

KSS – EFFICIENT CATAMARAN BUILD PROCESS IN COMPOSITES.

By Derek Kelsall.

INTRODUCTION.

The twin hulled Catamaran configuration and composite construction are two areas of boating which have seen spectacular expansion over the past 20 years. This applies to craft for both pleasure and commercial use. The sizes of craft to which both are being applied has steadily increased throughout that time. My forty + year career has been based on catamarans and on the development of custom building techniques in composites, particularly sandwich of fiberglass skins on foam core. We have been successful in most areas of pleasure sailing. Our recent developments make our build techniques particularly applicable to larger and commercial craft.

BACKGROUND.

In 1964 I entered the Single Haned Trans Atlantic Race in a trimaran. My yacht was the first to race without lead ballast. Trimarans have totally dominated this race over the past 35 years.

In 1965, I got the chance to build my first own design. I chose to use PVC foam sandwich with fiberglass skins. Foam as the basis of a build method was virtually unknown at that time. I entered the 1966 Round Britain Race and collected all the main trophies. It was a designed, built, raced and won situation, with a first design, in a new style and build method.

Over the next few years we produced winners of the 68 solo Atlantic Race and the first Whitbread around the World race (both were the largest fiberglass sailing yachts at the time) amongst many others. Within a dozen years the Kelsall style of racing multihull and one-off foam construction became the norm for all offshore racing.

For about ten years Foam was my province alone. I continued to develop and refine the ways we used the foam sandwich as both boat builder and designer and continue to do so today. For the typical catamaran, the basic materials specification I chose 43 years ago has not changed. Ie. 80 kg. PVC foam, unidirectional glass and polyester resin. In the search for more efficient building methods, the way we handle those materials has changed totally.

In 1973, we built the first full size mould table. It immediately proved to be a huge time saver. We moved more and more of our work onto the table and away from the traditional approach. The table lead directly to KSS (Kelsall Swiftbuild Sandwich) which is the term we have used for our build method since. All parts of the boat structure start on the table.

Important advances have been made to KSS over the last six years, particularly since resin infusion became standard. We refined the infusion process for our use on the KSS table, where it is very much simpler to apply and more reliable than it is in production moulds. The materials are laid out onto the table dry. Vacuum spreads the resin. The end result is best quality. The whole task is neater, cleaner and smell free.

World wide, custom boat building is a huge industry, consisting of thousands of small companies. Many are moving away from wood and metal into composites. However, the standard composite methods, based on the traditional approach are woefully inefficient and very expensive. Some, for example, build expensive moulds for each part even when building a one-off. Others spend literally thousands of hours applying a finish to the outside. There is no vehicle for R and D within the industry. KSS is the only effort to ever address the efficiency problem.

KSS has been the topic of regular R and D through-out my career. It has benefited from a hundred + opportunities to trial ideas on full size projects. Regular workshops which we hold in order to demonstrate KSS, also provide opportunities for trials. In any comparison with alternatives, the savings in build time, the improvement in the conditions for the boat builders and the improved quality results are nothing short of spectacular.

EXPANDING USE OF KSS.

Our work is best known in the pleasure boating field, but with catamarans and composites expanding into commercial vessels and larger sizes we are now promoting KSS for all sectors of the marine industry.

THE MATERIALS.

The sandwich cores we specify are PVC foam or Corecell foam. These are an excellent match with glass and either vinylester or polyester resins as the skins of the sandwich. Sandwich panels are ideal for boat structures due to the panel stiffness of a sandwich. As the core has good thermal insulation properties, additional insulation is not usually required.

Our 40 yo. boats are still passing survey. For more extreme race boats we do use epoxy. For some of the recent boats we have specified vinylester resin or a combination of vinylester and polyester. Both are simpler to handle than epoxy, less cost and less toxic.

The glass we specify for most projects is E glass in knitted, unidirectional form. This allows the use of mostly one weight of material and to tune the fiber directions and number of layers for the best structure.

Gel coats can be the final finish on commercial projects avoiding the need for a full paint system to be applied later. Joins are painted after usual smoothing operation. Where the perfect yacht finish is required a paint system is applied over the gel coat.

As a modern example of current composite practice, I refer to the World's largest sloop. The 250ft. yacht, Mirabella launched about 2 years ago is an example of the current use of the materials and the build method in typical fashion. The specification for the materials is the same as I used in 1965 for my first design. A full female mould was built, at great expense, in which to mould the hull. A well publicized picture during build showed a double-decker London bus looking very small inside the hull.

THE MOULD TABLE.

The table is usually the length of the boat and as wide as practical. If a full size table is not practical, joining of panels is not a major issue. There are many forms of table in use. The basic requirement is for the surface to be smooth and give a release for the finished part. There is no limit to the length or width of panel that can be made using resin infusion.

THE KIT OF PARTS.

Each part as it comes off the table is a basic flat panel. As designers we provide precise drawings for each part so that it can be produced as a kit. It is easily handled, transported and stored if necessary. Each part is custom made for the purpose, with additions as below. The foam can be CNC cut, which then also provides the pattern for each part. The boat builder can choose to make the kit or to buy from a few specialist suppliers.

THE LAMINATING METHOD.

Resin infusion is the process used for all work on the table. The power of vacuum spreads the resin into the fibers to create the laminate. It is a simple set up. All the materials are laid out dry and remain in place on the table. The vacuum ensures a high quality end product at least weight. The panel as it comes off the table can be either the full sandwich structure or it can be one skin and the foam core, with the second skin to be added after shaping. A major advantage compared to the typical boat building environment for laminating is that the whole process is neat, clean and smell free. The conventional approach to building boats is for the glass skins to be applied by hand to both sides of the core, which requires getting the material to difficult to reach areas and applying liquid resin to angled surfaces.

KSS HULL SHAPING PROCESSES.

The ideal offshore catamaran hull style for most applications has slim, round bilge hulls which are displacement hulls. A smooth ride, low drag and good speeds come from this hull style. A round bilge KSS catamaran hull is normally made from two panels. Part of the control of the final shape is built into the panels. The typical hull will have what we call developed panel topside area, which is the full structure with gel coat finish. Below the topsides is an area of partly one skin and foam and partly foam only. This part is formed into the round bilge area below the water line. A simple timber frame is used to set up the panels and to provide some shape control. To achieve compound shapes, the area of foam without skins is subject to pressure, which stretches and compresses the foam to the final shape needed. Additional glass is applied to complete the structure, setting the shape. The pressure normally comes from an air bag set between the hull panels. Large hulls in this style are complete in just one or two weeks.

THE ADDITIONS TO THE PANELS WHILE PRODUCING.

Edge treatment, provision for joining, recesses for fittings such as windows, capping of openings and some inside finishing can be applied to each panel while making the panel is on the table. This not only saves time, by doing the job when it is most efficiently done, it saves materials and finishing time.

BONDED ASSEMBLY.

By pre-planning the bonding edge treatment, most of the usual laminating on the boat associated with composite construction, is eliminated. This also means less finishing. The bonding process is very much quicker than making laminated joins. For example, where applicable we design a radius for the sheer line of the hulls. This can be either a solid composite arc in section or a full tube to carry wiring etc., which is made as part of the panel. A matching radius is applied to the deck panels. The decks are bonded to the hulls after fit-out. Reliable bonding materials are readily available.

MODULAR ASSEMBLY.

The KSS designs for modular assembly allow for three or more entirely separate units to be produced. These are taken to near completion before there is need for them to be assembled into one unit. This allows the boat builders ready access to all parts of the boat, without climbing steps or working off scaffolding. The particular designs are normally based on hulls which have a relatively low sheer line. Ie the hull are relatively small in section and less work is involved in building them. This arrangement allows for the engines, tanks and all the main systems to be within the hulls and installed before the decks are fitted. The machinery is away from the main part of the boat, where there is plenty of space and no requirement for plumbing or electrics to be hidden. This efficiency applies to both the structure and the fit-out. Better then halving the build time involved is typical.

COMPARISON TO METAL.

Metal requires closely spaced stiffeners and the addition of insulation. Both add a lot of extra cost and extra build time. In effect, another whole surface has to be added to the inside where accommodation is involved Foam sandwich is less weight for the same strength and stiffness as aluminium. It does not require extra insulation and a simple paint finish inside is often sufficient.

Steel is seen by many commercial operators as the material which withstands the most use or misuse. The truth is that at the sizes we are considering here – 30 – 75m composites compare well in all factors and it is easy to make provision for extra protection in vulnerable areas.

DURABILITY AND MAINTENANCE.

Recent Surveys of Kelsall craft built in the 60's are proving that the materials, with relatively little maintenance have a very long working life. 50 years would be a conservative estimate.

FIT OUT.

Most KSS designs use the same basic materials in their fit out. This is chosen for least weight and for using the same personnel as for the structure. By bonding to the structure, stiffness is added.

FINISHING.
If a particularly high level of finish is required a paint system will be applied over the gel coat. Many owners choose to stay with the gel finish and only paint the join areas. A paint or lined interior finish is usual. The inside surface is free of stiffeners making it easy to do.

SMALL PRODUCTION RUNS.
It is entirely practical to combine work on the KSS table with production style, part moulds for the more complex parts when more than one boat is planned. Operating with full female moulds presents other problems as well as being very expensive to make. We can quote one instance where a builder has chosen to make a full hull mould for the second hull of a 60 foot power catamaran design, only to find that it took longer to produce the hull from the mould than from panels made on the table. Above a certain size, unless there are a lot of boats to be produced and good facilities are in place, there is no gain from an investment in full moulds compared to KSS and part moulds.

MULTIPURPOSE CATAMARAN DESIGNS.
There are many advantages for the boat builder. He can offer a wide range of designs, all of which use the same technique with no extra investment in tooling for each model. Another advantage is that one hull design, which could be set up for part moulding, can be utilized in several different ways by making the necessary changes to the central accommodation unit but retaining the same hulls.

FAST FERRIES.
Catamarans as fast ferries have been extremely successful and warrant special mention. The majority are built of aluminium, which is chosen for it's low weight and for the material which most commercial operators are familiar with. Composites however are gaining ground in this field. KSS build would offer further weight saving, hence lower operating costs as well as lower initial costs. The later cost saving comes from the efficiency of KSS building of the structure and fit out.

LOCAL BUILDING USING KSS.
We find that boat builders quickly learn the skills involved in KSS composite construction. One experienced individual, which we can normally arrange, can train a crew and carry a project through to completion of the structure. This opens up the option of building the boats local to the operators. Building in low cost locations also favors KSS, where the necessary expertise can be arranged. We have a number of KSS projects where woman work the table very effectively.

SHARING OF INFORMATION AMONGST KSS BUILDERS.

KSS is a complete build system. However, refinements have been added on a regular basis and we expect this to continue. Most of our clients are willing to share information where they have found improvements in technique or in the materials being used.

LATEST INNOVATIONS FOR KSS.

KSS Pressure Shaping is a recent addition to KSS. It utilizes the plastic property of foam core in a way that has not been done before. A sheet of foam is deformed under pressure to assume a compound curvature shape. The pressure is usually applied via an air bag, but water can also be used. At the present time the process is used to achieve the round bilge area of the KSS catamaran hulls (see above). The process can be applied to anywhere that compound curvature is needed. At this early stage in the development, a number of trials are planned to determine what is needed to apply this process to other boat parts.

OBJECTIVES.

We are seeking partners to assist in this effort to bring the efficiency of KSS to the commercial and larger catamaran yacht markets. We invite designers, boat builders and operators in these fields to contact us.

derek@kelsall.com

Meeting Future Military Capability with a Network Centric Architectural Framework

Alex Cameron

EDS, Australia

Peter Williams

EDS, United Kingdom

ABSTRACT

Our armed forces unquestionably need rapid dissemination of information and knowledge superiority when engaging the enemy. This information is becoming less formally consolidated as modern systems become more modular, self contained and highly decentralised. A wave of miniaturisation is producing finer granularity of data from sensors in robots and even from soldier mounted technology. As technology moves closer to the "edge" not only must platforms and systems support loose coupling and decentralisation but the underlying architecture must also be able to orchestrate and control the data in this new information centric world. The military response to these trends is an overarching hypothesis that future Battlespace information systems that are interconnected, interdependent and networked, will lead to enhanced Shared Situational Awareness. Such an architecture (properly designed and implemented) will overcome many of the limitations of today's "platform centric" approach to acquisition towards a steady evolution of capability.

"Network Centric" means achieving military effect through the networking of existing and future military capabilities. It allows for a higher operational tempo through efficient utilisation and deployment of assets by supporting higher levels of awareness and decision making. A similar need and scenario is playing out in industry where competitive forces in business are driving the development of agile, interoperable Information Systems. It is therefore inevitable that more and more military capability will be inextricably intertwined with industry through a common technology foundation driven by the need to share information. Greater interoperability between industry and defence will be far reaching in its implications. However before this can be achieved there will need to be an order of magnitude improvement in leverage over what is presently achievable.

Today, the reality is that both military and industry suffer from having Information Systems that have become rigid with respect to modern needs and suffer from technical issues that inhibit agility and timely access to and sharing of vital information. Therefore, the purpose of this paper is to help answer the questions as to how this vision can be realised:

- Is there an Imperative for a Network Enabled Capability?
- What are the Technology and Business Objectives of NEC?
- What are the Challenges of Implementing a Network Centric Architecture?

INTRODUCTION

Network Enabled Capability (NEC) as it is known in the UK and Network Centric Warfare (NCW) as it is known in the US and Australia relies on the ability to collect, fuse and analyse data in near real-time within increasingly interconnected, interdependent and networked military systems. While the difference in terminology between the UK and Australia identifies differences in emphasis between centralisation and enablement, the desired objectives are identical. However the language of NCW is important because it shapes the way military leaders and planners think about what is possible[1]. NEC and NCW will be used interchangeably throughout this paper.

The intent of NEC is to enable for rapid decision making and the rapid delivery of the most appropriate military force to achieve the desired effect, but also to achieve an order of magnitude increase in leverage of existing and future assets. In addition to the provision of a digital secure communications network itself, it is also important to ensure that the net centric components, with the appropriate level of granularity, are available to gather and process information. This is to ensure that our forces have the appropriate reach and 'deployability' to achieve rapid effect. The ability to respond quickly and decisively will act as a force multiplier, allowing the same military effect to be achieved with fewer resources.

The key business objectives of the vision of NEC have been stated as:

- The timely provision of information and intelligence drawn from a broad range of sources;
- The integration of sensors, decision makers, weapon systems and support capabilities to better enable commanders to synchronise effects;
- Increased interoperability between the armed services and other government organisations, both nationally and internationally;
- Improve the synchronisation of logistic, medical and personnel information in order to optimise the use of available resources;
- Enable effective homeland defence through inter-agency operational support to security services, police, customs, etc.
- Improve the level of support to deployed forces through the optimised sharing of relevant information with Defence Industry.

Realisation of the vision will require three key challenges to be addressed:

- To describe a future information systems architecture in which the aspirations of NEC can be delivered in their fullest sense. This means current information and operational silos will no longer constrain the way operations are conducted to effectively utalise intelligence gathered.
- A means of incremental transition from the current state towards the future vision. It is inconceivable that NEC can be achieved in a single step and the only practical approach is incremental using existing technology investments to minimise costs and avoid high risk step changes;
- To define and implement a practical governance framework under which the transformation to the future NEC architecture will be achieved.

THE NET CENTRIC IMPERATIVE AND FUTURE DEFENCE LANDSCAPE

Technology is an enabler and is exploited by the prevailing environmental factors such as operational tempo and political will to invest in Defence capability. These factors are broad and multi-faceted, but technologies that offer maximum effectiveness and preserve existing investments will be preferenced.

Political backdrop:
- The defence capability gaps between the major military powers will narrow, producing increased innovation as leaders seek to gain and maintain superiority.
- Competition for natural resources will increase with population growth and economic expansion in China and India.
- Resource-rich nations are potential flashpoints as other nations vie for scarce resources.
- Oil rich countries in the Middle East will continue to invest in defence.
- Counter-terrorism will not be the only call upon military resources.

Battlefield developments:
- Increasing use of remote sensors (e.g. UAV's) producing increased data volumes.
- Rapid developments in soldier-mounted technology - enhanced vision and information display, health monitoring, communications. The man will genuinely become a node of the network as the technologies become more practical.
- Information operations will become more important (the use of information to win hearts and minds as opposed to winning by force).
- NEC advances from information aggregation and display, to analysis, advice and hypothetical scenario support.
- Allies share data seamlessly and securely, enabling greater integration of military capabilities (UK soldier designates a target using imagery from an Italian UAV which is hit by a US missile and industry replenish).

Business developments
- Increasing use of telematics and supply chain automation to maximise cost-effective equipment availability. Engineering improvements impregnate the supply chain via shared data environments, and spares holdings are optimised through holdings-wide RFID-enabled reliable visibility.
- Output driven defence enterprises operate across nations on major procurement projects.

Organisational Developments:
- Delineation continues to blur between Defence/Intelligence and Civil Government services as countries look to utilise all assets in the fight against global terrorism;
- Border control becomes significantly linked to Intelligence agencies;
- Blue light services command and control becomes increasingly linked to military equivalents to provide 'joined up' domestic capability supported by underlying 'joined up' information services;
- Virtual teams, staffed from across government departments/organisations and tasked with addressing new threats will become increasingly common.

THE TECHNOLOGY AND BUSINESS OBJECTIVES OF NEC

Overview of NEC

NEC is a military response to the premise that technology can deliver high levels of discoverability, integration and synchronisation of resources, capability and data. In that sense, it is not new, as many existing capability elements are networked. However technology now provides the means for Information to be assimilated at all levels with the goal of providing 'Shared Situational Awareness'; the art of providing each element in the command chain with accurate and timely - but most of all relevant - information to meet the demands of their respective roles. Effective Shared Situational Awareness will improve self-synchronisation, and will enhance command speed. Through NEC the command structure will improve its responsiveness to events on the ground, and have the flexibility to respond in near real-time to fleeting targets, even where higher level decision making is required prior to engagement.

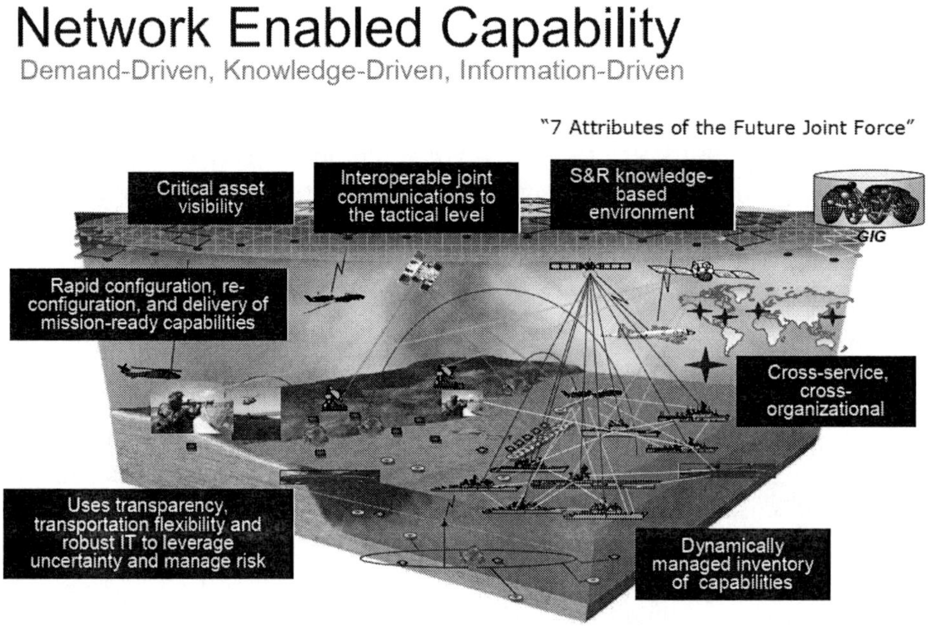

Source: OSD, Office of Force Transformation

In the short term NEC will focus on improving the implementation of current doctrine, while in the future it will enable the application of new solutions to emerging challenges. NEC has the potential to fundamentally influence doctrine; i.e. not just to help to 'do things better', but also to enable organisations to 'do better things'. Technology enabled leaps in capability are not without precedent; the advent of radar in World War II is an early example of a Network Enabled Capability which, by separating sensor and weapon, changed the RAF's modus operandi for defending the UK coastline. Considering the concept of continual separation of functionality leads to a concept of loosely coupled elements orchestrated or choreographed at the force or capability element level or higher.

NEC Benefits within Business and the Battlespace

The core theme behind the NEC objectives is the timely provision of relevant and accurate information applicable across the whole of Defence and extends beyond into industry for supply chain and into other government organisations for purposes such as intelligence exchange and Command & Control. In short then, the aim is to provide each key decision maker with superior information to his/her enemy counterpart; in this context 'superior' critically means more than accurate and timely, it is also the breadth of pertinent information sources which really enriches the quality of information available to the decision maker.

It follows that an IS architecture which enables the NEC objectives to be realised must be enterprise wide, capable of dynamically accessing the very latest information from a broad and changing list of sources and must also be highly flexible to enable rapid (re)configuration to implement new and innovative ideas.

THE CHALLENGES OF IMPLEMENTING A NETWORK CENTRIC ARCHITECTURE

Many existing military Information Systems environments comprise a series of disparate solutions which have been procured to meet the existing project/programme requirements. It follows that these solutions rarely integrate easily. There is often a lack of a common data architecture (or data model) across such solutions, and so attempts to integrate them not only face technical protocol and programming language challenges, but there is often semantic incompatibility which can indicate deeper problems. Worse still it is often difficult to justify the Return On Investment to integrate past systems when in competition with new acquisitions and so in such cases previous investment may be lost.

In many cases the advantage can come not from new platforms or technology, but rather through being able to transform or leverage the existing investment. Therefore, a sustained and systematic approach to Enterprise Architecture can identify the natural integration points within legacy systems and future acquisitions, ultimately leading to a reduction in overall Total Cost of Ownership (TCO). The four steps outlined in the next paragraphs outline a proven approach to developing Enterprise Architectures.

Establish Enterprise Architecture and Governance Programme

A key stage in the transition to NEC then is to establish an enterprise architecture (EA) for the organisation; that is to say ONE architecture. The Defence EA must incorporate key design principles which must be agreed and enforced across the organisation – the objective here being to start to turn all new and, where possible, current projects in the same direction – to the unified Defence architecture. The definition and implementation of the Enterprise Architecture strategy is absolutely fundamental to realising the benefits of NEC. Governance, whilst a key plank, must socialise the framework through participation of key architects and should be concerned with the integrity and evolution of the framework, not the actual implementation of the system components but the interfaces between them.

Establish Enterprise Architecture Framework

Defence have adopted and developed tools for capturing the architecture framework (see figure), such as DAF/MODAF/DODAF. The Defence Architectural Framework provides the basis for describing the various views of the overall architecture supporting the defence capability. The DAF typically provides a hierarchy of view, such as Strategic Views (StV), Common View (CV), Operational View (OV), Technical View (TV) and System View (SV). Combined with clear taxonomies and semantic clarity these views provide different perspectives necessary to enhance the interoperability of platforms and systems. Like most frameworks, the declared views are a necessary first step in communicating common architectural components for the purpose of governance and compliance, but they do not necessary assist in developing systems that satisfy the tenets of NEC.

While a necessary step in the overall journey, the Enterprise Architecture Framework in itself does not guarantee that an agile architecture will be developed within the framework. To achieve this, we require the development of an architectural style that, when instantiated, is agile and capable of supporting the NEC vision. This step is rarely well understood or well executed, and is the downfall of many "Enterprise Architecture Planning" initiatives[3].

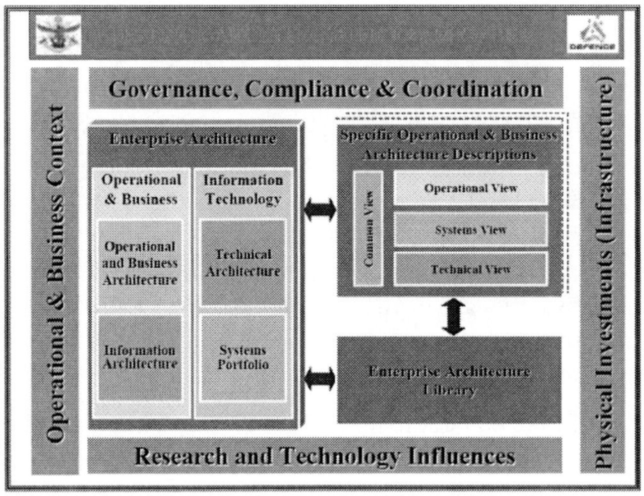

Establish Enterprise Architecture Approach

Service Oriented Architecture

Analysis of the NEC business objectives leads to a conclusion that the future Defence IS architecture must comprise IS components that are loosely coupled and engineered for interoperability – components must interact with one another only via strictly defined 'service' interfaces which prevent one component depending on the specific implementation characteristics of another.

Service Oriented Architecture (SOA) is a current architectural style that embodies these key design principles and this approach has several benefits, both in business and technical terms and offers the potential of meeting the design challenges outlined above:

- There will be little technical dependency between components/applications which means that they can be changed quickly and with minimal business risk. In a Defence sector hampered by a complex legacy of poorly integrated applications, this becomes a transforming advantage;
- It will be possible to dynamically discover and invoke services based on key service attributes. Consider an Intelligence service which defaults to accessing a local Operational Data Store but that this service is superseded as and when suitably communications equipped aircraft (e.g. AWACS) are within electronic reach;
- Further, with the key functions of each component/application exposed to the network as services, it is then possible to combine such services to perform higher level units of work and ultimately to drive business processes; and

- These too will be made available as new services to be called upon when required. Later it is also possible, by combining these business level services, new solutions will be conceived to as yet unconsidered business problems.

The adoption of NEC will be incremental, enabling the organisation to absorb the costs and training burden of transformation. The transformation roadmap, will involve the progressive integration of applications into the architecture but the actual sequence of this integration will be dictated by the business value of the associated business processes; in this way savings from a specific transformation increment can help to fund the next. The opportunities to exploit legacy applications and, crucially, existing data help to smooth the architectural transformation and to minimise costs.

The SOA philosophy lifts the focus away from the complexity of applications to relevant information: this enables business processes to prioritise the outputs of NEC – delivering effectiveness to the front line, as well as efficiency. For example, SOA will enable the sharing of accurate and timely Supply Chain information to Front Line Commands in order to implement the Joint Logistics Picture (JLP).

Understanding the Influence of SOA on Re-Use

A key benefit of SOA is the ability to re-use existing services rather than create all of a new solution's functionality from scratch. Re-use is not a new concept, has been the software development holy grail for many years and has been a singular failure, often because the effort was misplaced. Service reuse within a SOA paradigm offers visibility into the benefits to be realised through workflow and Battlespace Process Modelling that can be readily understood by the analysts and decision makers. The following diagram shows the evolution of software component re-use and helps to position the re-use capability of SOA against the more traditional approaches such as object orientation, and the role of orchestration and choreography as the key mechanisms towards achieving dynamic (re)configuration. It also shows that a very little return or understanding derives from investment at the code and object level.

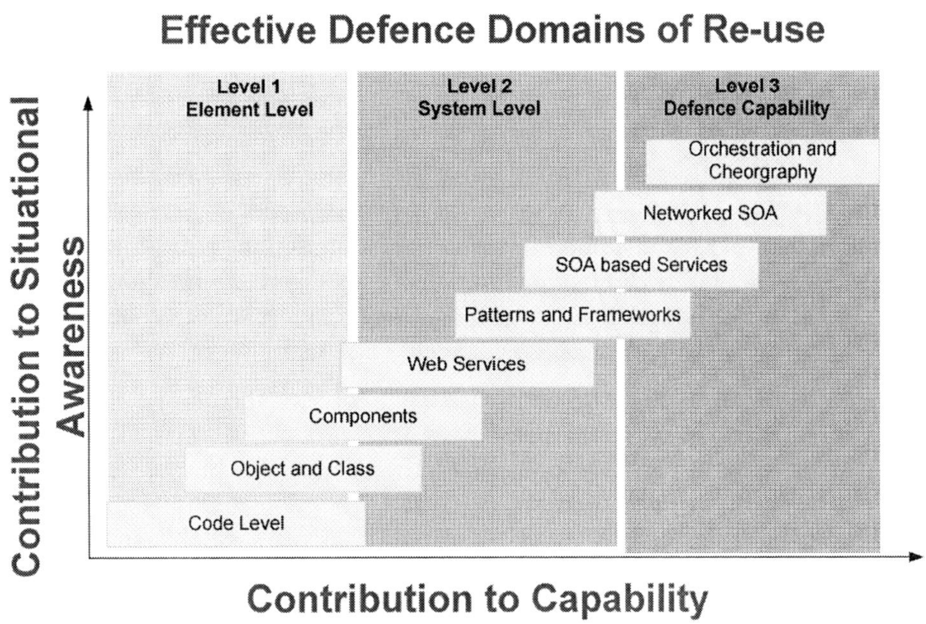

Develop, Implement and Maintain the Architecture

A Defence Enterprise Architecture, underpinned by SOA principles, is therefore the basis for achieving a Network Enabled Capability because all applications and other software components become accessible to both one another and higher level business services. Further, existing applications can be enhanced by accessing additional information and new applications can be created which draw upon information provided by any service based application/component in the architecture.

The first step is to define a Reference Architecture – a logical, layered, framework that provides technical and conceptual context and cohesion for all solutions with the Defence organisation. It also allows for the classification of services, a vital component in developing a sturdy and maintainable architecture. The following graphic shows the various core services/capabilities which form the SOA Reference Architecture; note that in addition to the run-time production services the framework all supports both development and operation of solutions.

Defence Agile Application Architecture

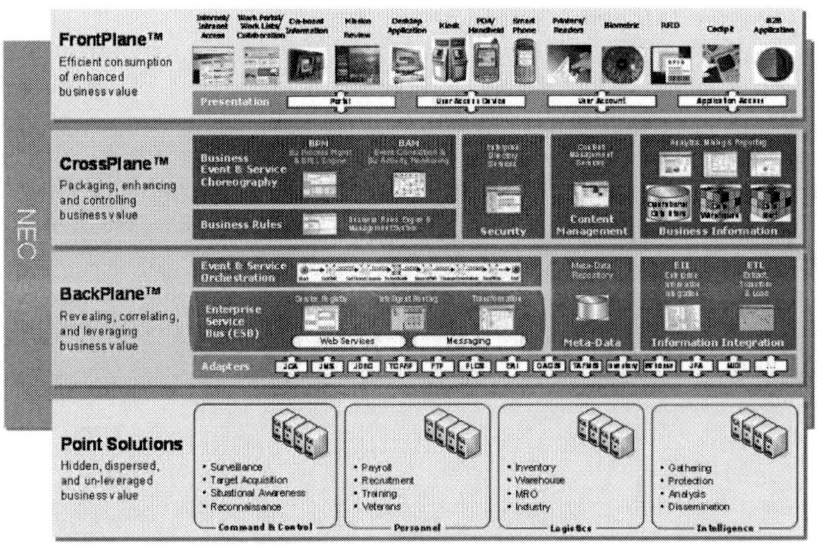

In the future COTS applications will be provided from vendors in an 'SOA ready' representation; that is to say that their key functions/transactions will be accessible to the SOA infrastructure which binds architecture components. In the near term though, in the early steps in a transformation to a Service Oriented Architecture a major task is to enable existing applications to present 'service interfaces'. Such a task comprises three key steps:

Step 1: Establish Basic Core Services: This step involves integrating current applications via Enterprise Service Bus adapters. In many cases SOA product vendors will provide adapters for the more pervasive applications but older and/or more specialist bespoke applications may require the development of bespoke adapters; a further option is to introduce a new, modern, component into the architecture which sits in front of the old specialist application, integrates to it and presents key functions to the network as services. Service enablement may also involve the building of minimum functionality services that can exist as facades within the CrossPlane. These facades provide the intermediate point to abstract deeper functionality and aid the development of an agile development. They can at a later stage be removed as the architecture matures.

Examples of 'Basic Services' are:

- Querying a Logistics application to find an item in a warehouse;
- Querying a C2 application for the current status of an aircraft within a mission.

Step 2: Introduce Service Orchestration: This step involves the combining of basic services to produce new, higher level, services which perform 'units of work' within the business. These new 'orchestrated' services also include logic which controls the sequence in which Basic Services are invoked and the data passed/received with them. It is essential that orchestration through code be avoided and that Business Process Model be developed to generate the necessary Business Process Execution Language (BPEL) constructs in order to create the composite solutions.

Examples of Orchestrated Services are:

- Querying a 'Supply Chain' service to find an item in the entire supply chain - invoking Basic Services many applications to find a solution to the item request;
- Querying an 'Air Mission Status' service which accesses various C2 applications to return an aggregated summary of all aircraft involved in a mission.

Step 3: Evolve to Service Choreography: This step introduces yet higher levels of services. These new services also comprise logic but at a business process level. It is permissible for the Choreographed services to access Basic Services directly but the general intention is to access Orchestrated services (which in turn then access the Basic Services). Thus a hierarchy of services is formed with clear levels of abstraction between the layers – strong governance is critical to the successful implementation of the service hierarchy.

Examples of Choreographed Services are:

- A 'Supply Chain' service to manage the progression of sourcing, ordering and delivering an item. This service would invoke Orchestrated Services which implement inventory, warehouse and order management functions (in turn these services will eventually access Basic Services which access the applications);
- An 'Air Mission Planning' service which manages the flow of building up a force for a specific mission by accessing Orchestrated Services which implement HR (find a pilot), Engineering Management (confirm configuration of aircraft) and C2 (review intelligence and plan route) applications.

AN EXAMPLE DEFENCE A3 IMPLEMENTATION

Overview

The solution to NEC is based on the Defence Agile Applications Architecture (DA3) is presented here and shown in the figure below. However, before discussing the specifics of the architecture it is appropriate to explain a few of the more significant technical building blocks:

- Enterprise Service Bus: ESB is a relatively new concept which builds upon the principles of Enterprise Application Integration (EAI). EAI addressed the issues of multiple applications needing to exchange information in often complex sequences without creating a tangled mess of point-to-point links. The ESB goes further and applies this concept to any software component wishing to access any other and the ESB is therefore the foundation of a Service Oriented Architecture (SOA). The concept here then is that a given Service X will invoke another Service Y but X is not only unaware of Y's network location but also has no knowledge of how Y is logically and technically implemented. This means enhanced technical and business flexibility as changes can be made quickly with significantly constrained impact on the rest of the IS architecture. As with EAI, the ESB still needs to access applications and this is achieved via specialist software components known as adapters;

- Business Process Orchestration/Choreography: When interactions between applications/components over the ESB are quite complex then the architecture requires a new capability to control, or 'Orchestrate', the flow of information. In its simplest form this could just mean that a call from a service X to two other services Y and Z requires that the data from their response be combined before being passed back to X; or perhaps the call to Y must happen before the call to Z so the sequence must be enforced. In its grandest form the flow of interactions will represent a multi-step business process. Processes run in a Process Execution Engine which is connected to the ESB;
- Business Activity Monitoring: With business processes executing in a Business Process Execution Engine it is possible to monitor them and capture information about them in order to determine, for example, whether KPIs are being achieved. The architecture component focused on this area is known as Business Activity Monitoring (BAM). Significantly BAM means that the business processes can be monitored on-line. This could therefore enable a current business KPI event stream to be passed into a Business Intelligence component for later analysis or into the Service Management component for correlation of IT and business events.
- Composite Application: A composite application is a new type of application which is formed partially by traditional means but also partly by accessing existing services within the architecture. The benefit is one of re-use – only new functions be developed and this has benefits both in terms of cost and change management should business rules need to be modified.

The following diagram shows the logical architecture for a proposed NEC solution that would be mapped to the reference architecture to guide development:

The main components in the architecture diagram are:

- Defence ESB: The ESB used for communication between 'domains' (examples of domains being Logistics, C2, Personnel and industry);
- Defence Process Orchestration: Business process models with a pan-domain context;
- Defence Business Activity Monitoring: The monitoring of the pan-domain business processes in order to capture high level business metrics;
- ESB Gateway: A component which controls the flow of information between Ebbs;
- Domain ESB: This is the ESB used for communication between applications/components within a domain;
- Applications: These are the current Defence applications;
- Domain Process Orchestration: Business process models with a domain specific context;
- Domain Business Activity Monitoring: The monitoring of the domain specific business processes in order to capture domain specific business metrics;
- Composite Application Portal: This is a UI function which is used to access information either within or across domains via the ESB. The composite application may utilise process models in order to orchestrate the more complex interactions;

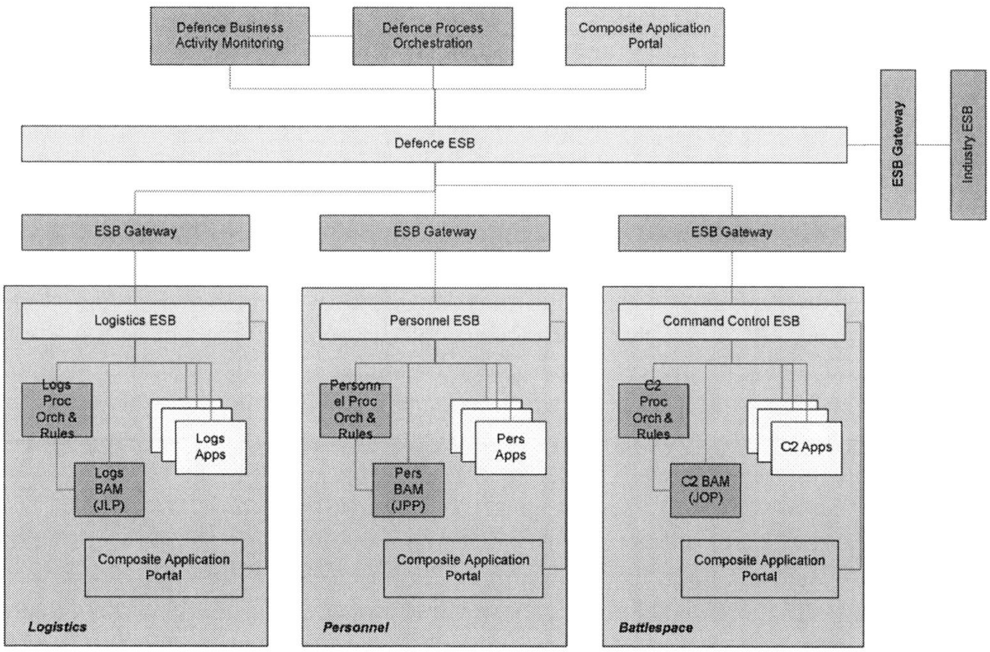

The architecture has some key points that need to be drawn out and point the way to future enhancements:

- Design Time Discovery: Future solutions will introduce new services but must also seek to exploit existing services; to that end designers and developers will access a UDDI Service Registry which contains WSDL definitions and human friendly descriptions of existing services – hence enabling designers to identify the 'delta' between their new requirements and the functionality provided by existing services. Designers will be able to discover services from other domains as well as those more locally available;
- Application Integration: Within each domain all applications will be integrated with the ESB which will provide high integrity transactional messaging; the ESB will not only decouple applications in terms of communication but will also break any data dependencies between the applications by utilising a standards based canonical data model. All 'basic' application services will be discoverable via the UDDI Service Registry;
- Business Process Orchestration and Activity Monitoring: Within each domain the core business processes will be modelled and loaded into a BPEL process orchestration engine. Process models are also services and so will also be discoverable via the UDDI Service Registry. Key process points will be monitored by BAM which will provide both KPI monitoring and business event processing capabilities – the querying of both will be available as further services which are also discoverable within the UDDI Service Registry;
- Dynamic Discovery and Invocation: All services will be dynamically discoverable via the UDDI Service Registry at run-time. Sophisticated client (consumer) architecture components are capable of dynamically finding new services, evaluating their suitability and then auto-configuring and executing a service invocation;

- Cross-Domain Service Invocation: Service invocations are controlled by the ESB; a service invocation across domains must pass out of the originating domain, traverse the Defence ESB and be routed to the target domain where the invocation will be finally routed to the target service. This approach enables the organisation to control and, as appropriate, constrain both the visibility and accessibility of services within each domain and so helps to reduce the magnitude of the task of governing the architecture.

SUMMARY AND CONCLUSIONS

In this paper we have presented an architectural perspective of NEC. The framework presented has the potential to unlock existing capability within Logistics, Personnel and Battlespace domains. The paper also advocates a Service Oriented Approach that can be implemented incrementally, thus keeping projects small and agile, thereby greatly reducing programme risk. It is clear that the vision of NCW/NEC will not be achieved overnight, nor will it be achieved at a programme level. It will be achieved through a gradual and systematic approach that adheres to standards and an architectural style always reflecting the needs of the military. The collective wisdom today (based on lessons from the past) is that an architecture that is loosely coupled and based on services of appropriate granularity is the next evolutionary step towards achieving the sought after Information Superiority. It is argued that these capability improvements will be gained through small incremental changes, over time and as such it is very likely that NEC/NCW doctrine as it is now will change. Adopting an architecture that can support this incremental approach means that changes can be continually incorporated without a massive haemorrhage of funds and confidence.

References:

1. David Schmidtchen, *The Rise of the Strategic Private: Technology, Control and Change in a Networked-Enabled Military,* Land Warfare Studies Centre, Canberra, 2006.
2. Richard Dawkins, *Climbing Mount Improbable*, Penguin, London, 2006.
3. Steven H. Spewak, Enterprise Architecture Planning, Wiley, New York, 1993.
4. Carlo Kopp, 'Understanding Network Centric Warfare', *Australian Aviation,* Jan./Feb. 2005.

Optimal Naval Warship Design For Fabrication And Maintenance

G W Sloan
ASC Pty Ltd, Australia
Glen.Sloan@asc.com.au

SUMMARY

The optimal warship design is no longer merely the one that has the greatest fire power, endurance and detection systems for its class. The Navies of the world are now including the additional requirements of reduced acquisition costs and through-life costs into the equation and are then expecting an optimum balance of all these issues in the end product. This paper explores the impacts this has on the design of the product for both fabrication and maintenance with respect to welding and structural design to reduce fabrication and through-life maintenance costs as well as the skill set required by the designer and fabricator.

1. INTRODUCTION

The owners of naval warships, either surface vessels or submarines, have to operate the vessels to fill a vast range of mission requirements within an era of reduced defence naval budgets. They are therefore very conscious of the need to procure cost effective vessels which have advanced capability, multi-role functionality and therefore complexity of systems, reduced and minimised manufacturing costs and more importantly reduced through-life costs. It is not just cost that they are interested in. One of their main requirements is to maximise sea days and if possible have guaranteed availability. An expensive or cheap naval platform is of no value to the operator if the platform is not out there fulfilling its mission roles. These requirements are related to each other and are best expressed in the form of Total Through Life Cost per Sea Day with the Total through Life Cost being the summation of the Design and Build Cost plus Total maintenance Cost plus Operating cost.

Reducing overall costs for the naval vessel is not just a matter of fabricating the hull as cheap as possible using the cheapest general metal work subcontractor. They tend to erect sections using basic welding techniques, to deposit welds of poor profile and quality, with no systematic or extensive non-destructive testing or with vary broad acceptance criteria, and then having the main assembly yard smash, bash and force the sections together. This can result in a very unreliable hull. Similarly reducing in-service costs is not just achieved by designing and installing more rugged reliable equipment with generous maintenance envelopes. If you have to rip out extensive equipment at refits or major maintenance periods to get at badly corroding hull areas or sections of pipe systems, then the duration of the maintenance periods will be increased above those originally envisioned in the operating and maintenance profile. There is not just the direct cost of the corrosion repair to consider. There is also the increase in total accumulated overhead costs due to the time taken to do the repairs and replace the equipment. The performance of the hull and its initial quality are frequently forgotten when considering ways to reduce total life of vessels costs.

Reducing hull and pipe system fabrication, assembly, and installation costs is best achieved by minimising fabrication and assembly time as this reduces the cumulative indirect costs. Throwing huge levels of labour resources at the workload is not the way to achieve this as it is very inefficient. It gets to the point where production personnel are actually interfering with each others work and, therefore, there is extensive lost or idle time or rework from the interference. The most effective method to reduce the fabrication time is to

- minimise or eliminate errors in design and production and therefore reduce rework;
- match the quality requirements to the service requirements, to the inspection requirements and these to the design of the structure, and;
- minimise the use of inspection or testing processes which are not compatible with other work activities.

The first point is achieved by ensuring that the structure and pipe systems have been designed to match the fabrication, welding and installation processes used by the builder and their subcontractors and that both the design and fabrication are aimed at the minimisation of access restrictions, weld defects, excessive distortion, dimensional error. This is also achieved by ensuring the welders and boilermakers are adequately trained to produce the required dimensional accuracy and quality of welds with each material and process used.

The second objective is achieved by using designers who have the right knowledge, experience and skills who, through extensive pre-design or production testing, understand the service performance of the materials and welds. This needs to be coupled with training of the designers on the access restrictions for each NDT method so that the individual welds can actually be tested by the selected methods.

The third point is achieved through research and testing to validate that the testing methods achieved will detect the defects of concern This is coupled with training of the designers and the customer on the capability of the chosen methods.

With regard to through-life costs, it must not be forgotten that the naval vessel has to operate in one of the most corrosive and erosive environments for, on average, a design life of 30 years. The hulls of naval vessels will be subjected to galvanic corrosion and attack from a variety of microbiological mechanisms which can result in severe structural degradation. Similarly the pipe systems will also be subject to these corrosion mechanisms plus flow erosion. The actual operating environment for the hull and systems will vary with change of operational location and within the systems or hull itself. These environmental conditions may also be outside those factors with which the designer has had previous experience. Therefore if costs are to be reduced, the designer must understand the mechanisms applicable to the customer's operating profile and then must either try to eliminate the corrosion or mitigate it in such a way that there is a low probability of occurrence. The worst case event for maintenance costs is to have to cut away or move extensive equipments and pipe systems in order to reach the underlying structure to address the corrosion issue as this has the potential to adversely impact the duration of the maintenance availability. The designer must consider that if they can not eliminate the corrosion erosion issues entirely, then they should ensure that the removal of the pipe systems and equipment is as easy as possible. They must also ensure that this activity does not itself cause future degradation. which would impact on costs and vessel availability.

It must also be appreciated that the structure, and therefore the welds, will also be subject to severe fatigue loadings. A surface vessel, for example, will, throughout its service life, be subjected to varying levels of hogging, sagging, slamming, racking and twisting as a result of the wave profiles it operates in and yet its structural scantlings have invariably been optimised to reduce the weight of hull structure to maximise its ability to carry equipment, systems, fuel and personnel within a given displacement. However, if the hull and systems are to have low maintenance costs through-life they must be designed to ensure there is low risk of fatigue. This requires avoidance of crack initiators by design, as well as by fabrication, and in addition in such a way that the non-destructive testing methods used to validate the absence of fatigue cracking or stress corrosion cracking can be undertaken using methods which are relatively non intrusive.

From the points raised above, it is obvious that the best strategy to reduce fabrication costs and reduce through-life costs from a fabrication and welding perspective and to reduce the fabrication and maintenance time is to apply the principal of optimum engineering. A level of realism is also necessary as the ultimate design for maintenance access, be it to structure or equipment, would result in a very large naval platform of degraded operational performance. These issues will be discussed in more detail in the rest of this paper..

2. WHAT ARE THE WELD DESIGN CONSIDERATIONS FOR OPTIMUM ENGINEERING?

There are many opinions on what constitutes an optimum weld in the design/shipbuilding field. To some the optimum weld is one with huge factors of safety on the stress levels from the applied loads, thereby presumably ensuring that the weld will never fail. To others it is designing the smallest possible weld to accommodate the applied load. Some will contend that it is a weld that meets both stress and fatigue requirements while to others the designer does not even come into it, as in their opinion optimum engineering is purely being able to throw the weld down as quickly as possible, using code pre-qualified procedures and undertaking no Non-Destructive Testing (NDT) other than a quick visual inspection. There is also the designer who considers that optimum engineering for weld design is defining extensive controlling property or process requirements on the design to achieve high performance regardless of whether they are consistently achievable in the production environment. Very rarely do the initial designer and manufacturer consider how to repair the welds or undertake new welds when the component or fabrication is in the service environment.

An optimised weld design involves a combination or balance of both production/process considerations and performance considerations. It is one which:

- is easily produced within the space restrictions of the structure by the processes identified;
- can be easily repaired by the same welding process or alternate welding process;
- the required mechanical properties for the welds are readily achievable by the available processes;
- can be validated to meet acceptance criteria with respect to weld defects;
- has acceptance criteria for weld defects which are related to criticality of weld or high probability of structural failure;
- can be produced in a cost effective manner and so achieves a balance between weld deposition rate, defect rate, size of weld, extent of NDT and the time to do the NDT, and infrastructure to undertake the weld;
- can be deposited within the skill level of the available workforce or, skill levels after realistic amounts of training;
- can operate in the customer's defined environment and therefore does not suffer from fatigue, corrosion fatigue, stress corrosion cracking or higher rate of galvanic corrosion; and
- does not require introduction of other processes to recover weld shrinkage and excessive distortion.

While traditionally it was considered that the majority of these factors or issues were left up to the production personnel to sort out, (with a heavy burden for this falling on the welding engineer, the welding foreman and the NDT technicians), it is essential in the current naval economic climate for the design engineer to accept a large part of the burden and optimise the design of the welds. If the design of the welds is not optimised then either profit will drop, production schedule increase, or product reliability decline. The last two points greatly impact on the cost effectiveness of the platform to the customer. Detailed in the following pages is discussion on how these aspects affect weld design and are frequently inter-related, and how good practice can result in an optimised cost effective engineering solution and where deficiencies lie in designer education.

3. ARE THESE CONSIDERATIONS IMPLEMENTED IN REAL DESIGNS AND FABRICATIONS?

One of the most frequently encountered issues with the design of welds is that they are grossly oversized or undersized. The undersized weld occurs when the designer attempts to optimise the design of the structure against weight and distortion. However, the designer frequently fails to address whether this smaller weld can physically be deposited using the processes available in the fabrication shop. Many of us have seen 2mm fillet welds specified on a structure which is to be fabricated by the Manual Metal Arc Welding (MMAW) process. If the welder attempts to achieve this size weld then the weld deposit is well out of normal specification parameters and will be hard and brittle due to a low heat input. Hence, there is a higher risk of weld failure. Generally the welder will deposit a weld which is larger than that specified, particularly when the weld is deemed undersized for the process. When this occurs, both of the justifications for the design optimisation of the weld have been defeated as the larger weld will result in greater distortion and increased weight. For successful design minimisation of weld size it is important that the designer understands the deposit limitations of each welding process that is available, preferably those used in his company's own fabrication shop.

The oversized weld by design usually occurs on the mistaken belief that this ensures safety either from failure by overload or by fatigue growth of a weld defect. In many cases the weld is not sized by analysis but by rules of thumb which are inherently conservative. An oversized weld has the detrimental effects of increasing the distortion level resulting from the weld shrinkage, thereby also increasing the likelihood for rework to be undertaken to obtain required alignment when joining subsequent components or structures. An oversized weld also increases the production costs due to the direct cost (of depositing the extra weld metal) and increasing the indirect costs (as a result of the increased overheads due to the impact of the production time on overall schedule). While it is important that safety is put first, it is also essential that the designer achieves this a) via stress analysis of the structure to determine loads and stresses in welds and b) by applying a logical factor of safety structure developed considering all the varying strength aspects of the weld and realistic probability of defect occurrence.

Comprehensive knowledge of fracture mechanics, fatigue and corrosion fatigue performance, of both the weldments and the base materials, can be used to assess the impact of weld defects and, in particular, determine what is the critical defect size which can lead to a catastrophic failure of the weld within its expected service life. Most fracture mechanics calculations are done after the event when a problem has occurred in-service or during build. This is the wrong approach. They should be undertaken before production specifications are issued based on the results of 'First Article Testing', either industry developed (and therefore available in codes), or in-house for the specific project, to determine these key properties of the weld.

By knowing the primary stresses in the welds based on analysis and the general consequences of failure of the weld on the structure it is joining and, if possible, combined with basic fracture mechanics calculations, the designer should be able to:

- specify the appropriate level of NDT based on a criticality of the weld or probability of failure due to stress and not just blindly invoke standards;
- determine the correct acceptance criteria to be applied both for surface profile, surface breaking and subsurface defects;
- assign the right NDT method to detect the critical defects identified as important for through life maintenance structural integrity;
- design the structure and place the welds so that the correct NDT can be undertaken with minimum test restrictions to find the critical defects; and
- minimise the number of weld repairs required.

In other words the designer should strive to apply optimum engineering practices to maximise the assurance of the welds while minimising the production time and cost by appropriately designing the weld.

In practice, it is frequently the case that the design structural engineer or draft person has very little knowledge of NDT methods and fracture mechanics. They also have very little knowledge of the ability of each NDT method to detect sub-surface crack-like indications. They - at best - follow codes blindly with the net result that NDT is over-specified or incorrectly applied. Similarly there is a tendency to assign the higher weld quality acceptance criteria to welds which, if they fail through thickness, will have minimum safety or operational impacts.

A frequent deficiency in designs of welded structures is that the specified NDT can not be adequately undertaken or is inappropriate. This is particularly the case with Ultrasonic Testing (UT). Welds or structural parts are frequently placed close together and do not allow for traversing of the ultrasonic probe to cover the full length and thickness of the weld. Applying high acceptance criteria to this situation is pointless as it only wastes money. Similarly, specifying radiography and UT of fillet welds is pointless, but it happens. If the designer is worried about a subsurface defect in the fillet weld or its fusion face with the base material propagating into the base material, then the optimum engineering solution is to design the weld as a full penetration weld using plate thicknesses greater than 6mm so that the joint can be adequately ultrasonically tested.

Many fabrication specifications for merchant ships, naval surface vessels and submarines mandate a level of non-destructive testing of the welds to be undertaken by radiography in addition to other surface and subsurface methods such as UT. This approach applies to both the hull welds and also for the welds of pipe systems. The percentage of inspection by this method will vary from contract to contract depending on the codes invoked. Generally as a minimum the codes invoke a requirement for 10% of welds to be subjected to this method of inspection and also all weld repairs 100% inspected by radiography. This mandating of radiography is done without any consideration of whether the other NDT methods are more than adequate to detect the critical defects. Radiography is one of the most intrusive processes to the fabrication and maintenance of ships and submarines due to the need to isolate, for Occupational Health and Safety issues, a large area around the structure to be tested. This results in considerable disruption to the progress of work and effectively increases the total fabrication or refit time. As stated earlier, any increase to the construction time is very costly due to the accumulation of the daily overhead costs for the shipyard or maintenance facility and also impacts the available sea days. Therefore, the use of radiography must be fully justified by the designer/ customer before invocation in codes and specifications for initial fabrication or maintenance work.

There have been considerable advancements in UT techniques, particularly in respect to the Phased Array and Time of Flight Techniques as well as what is termed Conventional UT. These techniques, either individually or in combination, have the capability or potential to detect all of the significant defects which affect the integrity of the structure provided the designer has given adequate access on the surfaces adjacent to the weld. Radiography is only suitable for detection of voluminous defects and porosity. It has very little value in detection of lack of fusion type defects or cracks as it is dependent on the plane of the defect being parallel to the incidence of the radiation. The various ultrasonic methods are more than capable of detecting these types of defects as well as the voluminous defects. A permanent record is also maintained by these methods as the trace is digitally recorded giving the added advantage of post testing processing or evaluation.

Ultrasonics are also suitable for most materials used in naval vessels with the exception of stainless steels. Therefore, one of the most appropriate methods to reduce the cost of warship construction and maintenance is to not mandate radiography but leave it up to the signatory for the NDT to invoke it as required when limitations exist for implementing one of the UT techniques. This approach, if adopted will be a positive step towards reducing costs. So why is it not implemented? Generally this is as a result of lack of knowledge or experience of the customer. This can be overcome by implementing a testing program using test plates with known defects to validate that UT can match or is superior in detection capability to radiography and presenting the results to the controlling authority. A program of this nature is currently in progress in Australia involving the naval warship construction and maintenance yards and the Welding Technology Institute of Australia, (WTIA).

The design factors of safety or margins for the weld must take into consideration the variation in mechanical properties due to parameter change. Within the weld procedure qualification structure there are defined essential variables with specified allowable variation of these. In the real environment, particularly in extreme access conditions for MMAW and Gas Metal Arc Welding (GMAW), it is unrealistic to expect that the operator can consistently maintain the correct travel speed, and therefore the correct heat input. Therefore, there may be variation in strength and toughness. Most designers just use the properties defined in the applicable welding code or specification and are blissfully unaware that a variation can exist and its potential impacts. A good designer should establish, via a testing regime, the impacts of large variations in heat input on the key mechanical properties needed to support the design in service and then consider this in the factor of safety structure.

The variation in heat input and the associated variation in mechanical properties may result in the weldments being susceptible to Stress Corrosion Cracking (SCC). High hardness and high strength due to low heat input are some of the key factors which cause this problem. For optimum service performance it is essential that the designer is aware of this problem. While it may seem to some that this is beyond the control of the designer, this position is not correct. For starters, the designer chooses the material. If they choose to use a high strength steel then they must take responsibility for establishing that the material or the welds deposited by the applicable processes are not susceptible to SCC. Secondly, the designer is the one who sets the strength requirements for the weldments. As identified in References [1] and [2], the designer can place the welds in low stress areas of the structure. Therefore it is not always essential to match or over-match the weld metal strength to the plate material. Hence a lower strength consumable may be able to be used quite safely with adequate factors of safety and therefore reduce the risk of SCC.

As also identified in References [1] and [2] the lower strength weld deposits may be associated with an increase in toughness and therefore greater resistance to crack growth. Why is avoidance of SCC so important? Put simply, if the structure or component experiences this type of cracking or there is a high probability that it is susceptible, it may have to be subject to a vast regime of NDT to establish if it has occurred, and if detected, withdrawn from service for extensive repairs - or in the worst case, scrapped following considerable downtime. The net result is a structure which is not an example of optimum engineering, represents a considerable cost to the end user and is a cost impact that is not predicted for and included in long term maintenance budget strategies. Another option is to not incorporate into the design materials which have any susceptibility to SCC or similarly not use any welding process that can result in a susceptibility of the welds. If there is no alternative and the design performance requirements necessitate the use of processes with a slight possibility of initiation of SCC if welding is undertaken outside of approved parameters then the designer must also take steps to provide mitigation by ensuring that the design of the structure and the welds, independent of welding position, can be undertaken with optimum access thereby minimizing the risk of the welder depositing grossly out of parameter welds from this cause. This mitigation strategy also has the advantage that it is one of the most effective methods to minimise the occurrence of other weld defect types.

The designer can use optimum engineering practices by undertaking extensive 'First Article Testing' on the material and its weldments early in or prior to the concept design stage to establish the limiting factors for avoidance or to vastly reduce the probability of occurrence. Regretfully, most structural designers are blissfully unaware of this problem and many design and fabrication codes (with the exception of those for the gas pipeline industry and the submarine design industry), do not insist on including testing to detect susceptibility to SCC. With the more extensive use of higher strength steels in the design of structures, to minimise weight and optimise dimensions, it is anticipated that the instance of failure by this mechanism will increase. This can be avoided by a structured approach of training the designer and also changing the codes to insist on adequate 'First Article Testing'.

4. THE ROLE OF THE DESIGNER IN THE OVERALL FABRICATION PROCESS

The role of the designer for a fabrication or structure is not simply limited to the stress analysis, qualification of the materials, and defining the NDT. The designer plays an even more significant role in the total fabrication process and the ability to optimise the engineering task and reduce costs by ensuring that the design can accommodate the welding processes of all stages of fabrication and reflects the accuracy control requirements. Both are essential to minimise rework.

The best example of this is with respect to the current practices in European and American shipyards [2].In order to compete against lower manpower costs in Asian shipyards, while living with a reduction in skilled worker levels, it has been essential for these shipyards that the end-to-end process is as cost effective as possible throughout the hull fabrication, outfitting and test and trials phase. For the hull and superstructure fabrication this has been achieved by the designer and fabricator working in conjunction to optimise the design and fabrication processes. In sizing the scantlings and designing the welds the designer has taken account of the access restrictions for each process to be used- not just those for welding but also those for cutting, edge preparation and aligning. For the welding processes the access limitations of both the robotic and manual welding processes have been considered so that when one process cannot complete the task, another can. This has also been achieved taking account of the lack of skilled tradespeople by using processes such as GMAW that require less skill.

The designer's input does not just end with considering access limitations. One of the key cost saving areas has been the elimination or minimisation of extra process steps such as cutting off excess material on parts and assemblies just prior to integration together. This has involved the designer accurately incorporating allowances for cutting kurf, the required joint gap, and weld shrinkage in the original part sizing. The weld shrinkage is based on that applicable for the welding process used, ie: heat input, and the material thickness. Translational or interpretation errors have been eliminated by the direct transfer of the cutting information from the 3D CAD drawings to the cutting equipment. The same is applied to all the forming and bending equipment as well as automatic welding equipment. The weld preparations on all plates are milled and not flame cut so that there was a high level of accuracy on joint fit-up provided the designer 'got it right'. Similarly, all profiles were cut to length and weld preparations incorporated using oxy cutting undertaken using robots and automated indexing tables. Again, if the designer got it wrong, the impact was huge.

The interesting point is that both these approaches work successfully at many shipyards independent of the level of automation used and the welding processes. To achieve this required the designer to have a high level of understanding of the fabrication processes. The designer had been a major participant in optimising the engineering and fabrication process. For this to be achieved consistently in the future it will require a change in approach and re-training of design personnel to increase their knowledge of fabrication processes.

5. CORROSION PREVENTION

One of the most effective methods to reduce the in-service cost of a naval vessel and to maximize sea days and therefore make it more affordable is to minimise the incidence of corrosion or erosion damage and the extent of that damage. The key areas that need to be addressed for the hull and pipe systems are;

- Material selection
- Implementation of corrosion protection layers
- Operational environment
- Variation in chemical composition

Singularly, or in combination, these can result in rapid unacceptable degradation. It is essential that the designer fully understands the influences of these factors on the performance of the structure or pipe system. The designer should not assume that they understand fully the materials behaviour in the intended application. For example nickel performs well as a protective coating in cool flowing sea water. In warm stagnant sea water its performance dramatically deteriorates. In fact, this only occurs in a narrow temperature band between 20 degrees and 50 degrees. Above this temperature an oxide layer forms which decreases the corrosion rate. Many metals have good corrosion resistance in flowing sea water but perform badly in stagnant sea water found in dead legs of pipe systems. The designer must understand this and take mitigation steps to minimise the impact. Some of this may be as simple as specifying through ILS documentation that cross connection valves between pumps be regularly cycled to allow exchange of the stagnant water normally trapped in the dead leg. The other major area frequently forgotten is the performance of metals with respect to crevice corrosion, an example here being the frequent use of 316 stainless steels for potable water tanks resulting in failure of flange joints or even the tank walls themselves by crevice corrosion.

Similarly it is essential that the designer understands the impact of slight chemical variations. This is particularly important for the design of corrosion protection layers deposited by welding. The level of iron dilution can have a significant effect on the corrosion performance and ductility and resistance to hot cracking of corrosion prevention weld deposits of nickel alloys or monel. Monel is particularly susceptible to pitting and crevice corrosion when the iron dilution level is high for the exposed final surface. It is also very susceptible to hot cracking when there is high iron dilution and the weld joint is tight or contains stress raisers or notches. Nickel alloy weld deposits are also very brittle if the iron content on the surface is measured to be above 6%. This can result in through cracking of the corrosion protection layer when subject to fatigue loading exposing the anodic substrate to very rapid corrosion. Therefore, all weld deposited corrosion protection coatings must be designed as a two layer system and deposited by processes which use low heat input. The designer must therefore consider the access restrictions for when the weld is deposited as these have a potential to cause out of parameter weld deposits and therefore non optimum chemical composition.

A modern naval vessel contains many different materials particularly for pipe systems. These range form carbon steel, Copper Nickel 90/10 and 70/30, stainless steels, nickel aluminium bronze and copper for domestic piping. These pipes have to penetrate through bulkheads and structures and therefore there are many dissimilar joints. Similarly the valves are either made from Nickel Aluminium Bronze, Monel or a variety of stainless steels and these also have to be mounted to the hull or bulkhead structure resulting in more dissimilar metal joints.

The designer must ensure that some fundamental principals are followed in designing dissimilar metal joints. These are;

- The interfaces between dissimilar metals must never hidden behind flanges or internal to pipe systems. This is particularly the case for metals where one is highly anodic to the other in the presence of an electrolytic fluid.
- The interface is designed so that the application of corrosion of protection schemes such as paint can be undertaken with minimum risk of breakdown. This is ensured by avoiding rough profile of the joint and also by extended overlap.
- If the joints involve clad welded deposits understand the implications of chemical composition on performance.
- Similarly if the installation welds are dissimilar, fully validate the performance of the weld via the weld procedure qualification testing and include in this corrosion testing.
- If there is a combination of several metals in the system, fully test the performance via corrosion testing, and in particular, simulate the mass effects of the metals in the system.
- Avoid using studs of highly noble bolt materials into carbon steel or low alloy steel structures without incorporating a means of excluding electrolytic fluid from the interface.

By careful design of the dissimilar metal joints or interfaces, the risk of structural failure by corrosion can be minimised and therefore maintenance costs greatly reduce making the vessel more affordable to the customer.

The other major area of corrosion induced structural degradation or failure in the marine environment is caused by microbiological induced corrosion. This failure mode affects both structure and pipe systems. It is particularly insidious in pipe systems as it usually results in inside to out failure which is difficult to predict prior to failure. Key areas or systems for attack are bilge areas and the suction pipes form the bilge. For prevention of corrosion of the later it is essential that the designer selects the correct material and does not include low lying segments in the pipe system which can be a catchment point for stagnant foul fluid. Similarly, the designer should ensure that mitigation strategies such as regular flushing with fresh water are institutionalised in the ILS documentation and made visible in the training documents for the crew.

Corrosion of the structure forming bilge areas of surface ships and submarines is frequently influenced by microbiological attack. While it is not possible to eliminate the microbes themselves - as these are always present in sea water and are regularly replaced or refreshed by the addition of water into the bilge there are some practical mitigations which can be introduced The first and most obvious is to get rid of the oil films and other matter upon which the microbes feed and also contribute to the creation of the anaerobic conditions that the microbes thrive in. This can be achieved by frequent pumping of the bilges and regular steam cleaning. This may require larger oily water and oily bilge holding tanks than have been traditionally used and this may impact on the vessels overall dimensions and initial costs. However, this is more cost effective than leaving it to the refit yard to repair the corrosion damage.

This approach will not solve the problem on its own. While dosing regimes can be adopted to kill the microbes using chemicals, the environmental laws are becoming more restrictive on what can be pumped overboard. Therefore, the best additional mitigation strategy is to improve the anti corrosive coating applied to the structure. The coatings applied in the bilge areas are frequently subjected to impact damage and chipping from dropped tools, etc in addition to the usual problem of pin holes in the paint schemes. Therefore the designer should consider the application of impact resistant coatings in the bilge areas and to ensure that the integrity of the coating is ensured on application by testing for pin holes. The scheme used must be fully validated by a rigorous test program before incorporation into the naval platform.

In reality it would be naïve to assume that the occurrence of corrosion can be designed out. However, by the designer not assuming that they know everything about the corrosion performance of the intended platform and materials in the operational environment and implementing a validation testing program during the concept design stage, or preferably before, to explore the influence of variations in composition, temperature, microbiological fouling and implement the simple mitigation strategies mentioned above the incidence and extent of corrosion damage can be reduced. The maintenance costs will also be reduced, both by direct reduction of the work scope and by reduction in the overhead cost All the mitigation strategies mentioned previously are relatively simple to adopt but are frequently forgotten or not understood.

6. IN-SERVICE REPAIR OF PIPES AND STRUCTURES

For some industries consideration of optimum engineering for the design of welds and the fabricated product ends with the initial fabrication of the product. This is because their prime business is initial sales, and not sales plus extended maintenance and repair. For example the whitegoods industry does not optimise its product for long life and ease of maintenance and repair. For other industries, in particular the defence industry, the opposite applies. Major defence hardware acquisitions are usually intended for long life and frequent upgrades. However, these upgrades must be undertaken within a minimum time scale, so as to maximise force availability, and within tight budgets. The designer is aware of this upfront through the original purchase order and should, therefore, design the structure and systems to minimise maintenance. The designer should also ensure that maintenance, when needed, can be undertaken achieving the same quality requirements as at build.

For example, it is important that piping systems in ships and submarines are designed to minimise flow erosion which is an insidious failure from inside to out. Several factors influence the occurrence of this including:

- fluid velocity above erosion inception speed;
- vortex generation from excessive weld bead penetration for butt welds;
- vortex generation from welded pipe joint misalignment; and
- vortex generations due to interruptions to the internal surface of the pipe due to things such as flanged joints or other forms of mechanical joint or marine growth.

As can be seen from the above, the welding process used to join segments of pipe has an influence on the incidence of flow erosion. Excessive weld bead penetration can be avoided by designing the joints for welding by the Gas Tungsten Arc Welding (GTAW) process, either manual or automatic, which gives a smooth and small root bead, provided sufficient access surrounding the joint is available. If access is poor and there is no clear space around the joint, then the automatic processes such as Orbital GTAW cannot be used. Therefore, manual GTAW is used, which, in difficult access conditions, may result in non-optimum internal profile. It is the designer who usually determines where the weld joints will be, noting that this is important to maximise pre-manufacture of parts and minimise assembly time. So, it is the designer's responsibility to position the welds so that the optimum process can be used. Noting that pipe segments are pre-cut to length and shaped using 3D CAD information a large portion of the responsibility for joint fit up now falls on the designer. If they have not allowed for adequate weld shrinkage and pipe bending spring back during computer controlled 3D pipe forming, then the joints will have to be forced together and again the optimum Orbital GTAW can not be used. This again results in an increased probability of flow erosion.

To put things in perspective there are many kilometres of pipes of all sizes in a modern naval surface vessel or submarine. Whether it is a surface vessel or a submarine, optimisation of the use of the available space is a major design requirement. To fit the pipes in the available space, or conversely to minimise the size of the ship, welded joints are preferred over flange joints. There is also the advantage that welded joints minimise the number of potential leak sources. To further maximise space usage, pipes are frequently stacked in relatively tight bundles to maximise the space for equipment. Therefore, even for initial build, the designer has to have considerable skill and knowledge to obtain an optimised solution for weld joint position against the multitude of constraints.

Regretfully, those constraints due to welding may not be foremost in their minds depending on their skill level, or also may be difficult to achieve because of the complexity of competing requirements both in the macro and micro perspective. In the macro perspective, the designer is trying to limit the size of the vessel while optimizing the warfare capability of the vessel and physically accommodating all the equipment, piping systems and habitation requirements. The designer has a very difficult task.

It is even more of a problem in the maintenance periods for the vessels where segments of pipes may have to be cut out of the way for removal of equipment or to gain access to the structure. In this scenario it is often not possible to re-weld the pipe segments into place using the preferred butt welds, as the position of the welds cannot necessarily be optimised for access. This is because the original pipe installation at build is invariably undertaken with the vessel relatively empty of equipment. The designer does not consider the implications of latter repair work, and is more interested in getting the build drawing packages out on time. In the maintenance scenario, the opposite is the case with respect to access restrictions due to equipment. Therefore repair/replacement welds are invariably sleeve joints for small bore pipe which use fillet welds and therefore have a lower level of assurance of integrity as only surface inspection techniques are applicable. For larger bore pipes butt welds still have to be used. These are undertaken using manual GTAW. As a result of the access restrictions the root bead profile is not as controlled and the profile is lumpier and higher. This can initiate vortex generation and down stream erosion. The repairer has little control or mitigation possibilities other than taking due care. The net result is that, by the time of the second major maintenance activity, insidious inside to out failure of pipe systems may become prevalent therefore increasing the cost and duration of maintaining the in-service vessel.

Consideration of the implications of maintenance and repair activities by the designers at the concept and detailed design phase may lead to an optimised solution for both minimum cost of build and repair. The key is providing adequate space envelopes around every weld joint to be undertaken so that in the event of repair of the joint or pipe or temporary removal of the pipe the optimum weld process can be used. While this is not feasible for every joint in the pipe systems (as this will grossly impact the size of the vessel itself and therefore the cost), there is value in adopting it for areas where there is a high probability of requiring access to the hull to repair corrosion damage. The obvious areas that fall into this category are the bilge areas of submarines and the bilge areas and wind and waterline of surface ships. Optimising the separation between pipes in these areas will inevitably increase the size of the vessel. However, by limiting the optimisation to these locations the increase will not be excessive and the benefit for through life costs will be greater.

Why is this not usually done? The issue is that the designer is not usually aware of these maintenance impacts due to a lack of visibility of shop floor work, or being from a specialised design house, or a lack of training.

7. HOW TO IMPROVE DESIGNER CAPABILITY

It would be naïve to think that there is a simple solution to improving designer knowledge of how to optimise the structures they design with respect to welding and fabrication and of how to reduce maintenance requirements of the hull and pipe systems. In reality, it will involve approaching the problem from many angles.

The first logical area that it is worth tackling is at the university level. Universities should be encouraged to increase emphasis on training and teaching of weld design and fabrication technology for structural, civil, automotive, aeronautical and marine engineers and practical methods to avoid corrosion and erosion. The current material science and manufacturing process subjects can be expanded to include topics such as NDT, greater description of process physical limitations, consequences of welding with respect to defect types and initiation, and impact on the life of welds. Greater emphasis should also be placed on the basic design methodology subjects - optimising designs for fabrication and being able to access and adequately undertake repairs to meet the same performance requirements as the original weld, are a fundamental aspect for consideration in the initial design phase, even as early as the concept design.

Not everything should be left up to the universities or technical institutions as they will argue that there is so much to squeeze in to the four year or five year initial degrees that it is not possible to cover every topic in depth. A large part of the responsibility falls on the employer of the young engineer. Employers, and in particular the senior engineers of the employer, must make every effort to continually train the young design engineers and drafts people on how to design welds for optimum structural performance and in-service performance and how to avoid or mitigate corrosion and erosion. This can be achieved by using short courses - internal and external and direct on-the-job training including exposure to the maintenance environment - with the senior engineers making the effort to train the junior. This may mean that there is initially a net reduction in direct hours available for design work and generating profit, but the long term benefits will be a worthwhile return on investment both for the design house, shipyard and more importantly the customer. It is also important that the design house takes the initiative and takes steps to educate the customer on the advantages of adopting the methodologies described above. It is only through a knowledgeable customer that an optimisation solution can be implemented.

Organisations such as WTIA and The Welding Institute (TWI) can also play a large part in designer education through a change in strategy and approach to their educational technical documents and workshops. The good news is that this approach is being implemented. The WTIA Smart Group for the Defence sector is currently putting together six projects for inclusion in a proposal for funding under the ICIP program. The six projects include:

- conformance to specifications and standards;
- welding design for optimum fabrication, construction and repair of naval platforms;
- fabrication and repair of in-service piping on naval vessels;
- renewal of worn or damaged components;
- inspection techniques; and
- OH and S for ship builders.

The Expert Technology Tools which are the output from these projects will assist in improving the knowledge of designers. However, they will only be of benefit if the industry gives the designers the time and the structured education programme to learn how to optimise the design of welds.

Similar education programs should also be implemented with respect to optimisation of the design of naval vessels to avoid detrimental corrosion.

8. CONCLUDING REMARKS

The key issue for making affordable warships with respect to hull fabrication costs and through-life maintenance of the platform is that they must do it better and more efficiently than in the past. Development of designs optimised to reduce cost of fabrication, in service maintenance direct costs, and minimized number of days required to undertake the maintenance play a major part in achieving this objective. To this end the skill of the designer and the fabricator is paramount. We must make sure that the designer's skills can meet the task and that the fabricators chosen have the right balance between advantages of implementation of technology, skill levels of the production personnel, cost of introduction of technology, and the ability to fabricate within a reduced time scale.

9. REFERENCES

[1] SLOAN, G W 'Structural Integrity of Submarine Pressure Hull Welds', *ASM 7th International Conference on Welding Research.*

[2] SLOAN, G W 'New Technologies In Ship and Submarine Fabrication', *Pacific 2006 International Maritime Conference.*

10. ACKNOWLEDGEMENTS

This paper was first published in the Proceedings of the "Affordable Warship Conference" Bath, UK, 2007

10. AUTHOR BIOGRAPHY

Glen Sloan holds the current position of Principal Submarine Technical Authority at ASC Pty Ltd, South Australia, Australia. He is responsible for Pressure Hull Design, Material Selection and

Qualification, Naval Architecture and Welding. His previous experience includes 27 years in the design, construction and maintenance of submarines.

Flammability and Fire Resistance Issues of Composite Structures for Ships

Z Mathys, I Grabovac, N St John and S Russo
Defence Science and Technology Organisation
PO Box 4331, Melbourne, Victoria 3001, Australia.

ABSTRACT

Glass reinforced polymer (GRP) composite materials are increasingly being introduced in marine applications both as structural and non-structural components. Composite are combustible materials due to their polymer matrix and therefore are required to meet stringent safety requirements governed by the International Maritime Organisation (IMO) and the High Speed Craft Code (HSCC) to be accepted on commercial vessels. This paper will discuss the requirements composite structures must meet to be accepted on commercial vessels and some examples of their thermal properties and their ability to meeting these requirements will be presented. The properties that have been evaluated at the Defence Science and Technology Organisation (DSTO) and which are discussed hereunder include: (i) fire reaction, (ii) fire resistance and (iii) structural integrity during fire. This work showed that the use of composites can meet the IMO criteria particularly when thermal protection is applied.

1. INTRODUCTION

Glass reinforced polymer (GRP) composite materials are increasingly being introduced in marine applications both as structural and non-structural components. This trend is likely to continue because composites are lightweight and offer good mechanical properties including high resistance to corrosion and chemicals. Another useful property is their low thermal conductivity. However composites are combustible materials due to their polymer matrix, but if adequately engineered, composites can offer superior service properties over metal counterparts.

In marine applications composites are suitable candidates for the replacement of some aluminium and steel structures on ships. Composite structures are known to possess enhanced fire containment properties due to their low thermal conductivity and thus reduce the risk of ignition in adjoining compartments in comparison to metallic structures [1, 2]. The use of thermal barriers on composite structures would result in improved fire resistance boundaries and reduce the risk of fire therefore increasing the time available for escape and fire fighting [3-5]. However, the combustible nature of composite materials due to their organic matrix is still a significant issue especially on board ships [6-8].

In the view of various ship classification societies, composites structures can be used on ships but stringent safety requirements have to be adhered to which are governed by the IMO and the HSC Code [9-11]. These regulations stipulate certain fire performance criteria that composites must meet to be considered safe to use in marine structures.

In this paper an overview of composite thermal properties such as fire reaction, fire resistance and structural integrity during fire will be presented and issues related to meeting the stringent IMO requirements is discussed.

2. FIRE REACTION

One of the most significant issues related to the use of composites in marine structures in both structural and non structural applications is the susceptibility to combustion of the organic matrix component of the composite materials. Concerns regarding the spread of flame, smoke and toxicity of the organic matrix have led to stringent safety requirements being implemented by the IMO. The IMO HSC Code, Resolution MSC.40 (64) classes composites as "combustible materials' and sets out stringent fire requirements to enable the use of composites on commercial passenger ships [11]. The fire performance criteria within the HSC Code require fire testing of the "as installed composite material" within a room corner fire test, ISO 9705 [12]. The main fire reaction properties required to be meet by the HSCC are shown in Table 1.

Table 1. IMO Performance Criteria for Fire Restricting Materials

Quantity	IMO Requirement
Time average Rate of Heat Release (RHR), excluding heat release from ignition source	Max 100 kW
Peak RHR excluding heat release from ignition source, averaged over any 30 s period	Max 500 kW
Time average smoke production	Max 1.4 m^2/s
Peak smoke production rate, averaged over any 60 s period	Max 8.3 m^2/s
Flame spread down walls, over 1.2 m away from ignition source corner	Min 0.5m from floor
Flaming drops or debris	Nil

2.1. Testing for Fire Reaction

Testing of materials in a large-scale corner room burn according to the ISO 9705, as required by the IMO, is very expensive. Cone calorimetry is a method that could be used to screen materials prior to conducting large-scale fire tests. This small scale method is capable of testing the fire reaction properties which can give an indication of the flammability, toxicity, smoke and spread of flame characteristics. Ostman [13] have used cone calorimetry to predict the flashover in a corner room burn for wood products. In this paper, DSTO has used cone calorimetry to determine the flammability properties of several commonly used polymer matrices that are used in the manufacture of composites structures or components.

The rate of heat release (RHR) during combustion is generally considered to be one of the most important parameters in characterizing material combustion behaviour [14, 15] and is related to the spread of flame and the propensity to flashover in the event of fire. Figure 1 shows the RHR vs time profiles of the polymer matrix resins at a heat flux of 50 kW/m^2. The polyester and vinyl ester resins show similar RHR values, whereas the phenolic matrix resins shows a significantly reduced RHR than the other systems whilst the epoxy resin system shows significantly more flammable characteristics. From Figure 1, it possible to observe the times at which ignition of the matrix resins occurs. The epoxy, polyester and vinyl ester matrix resins have similar ignition times whereas the ignition time for the phenolic matrix resin is significantly longer.

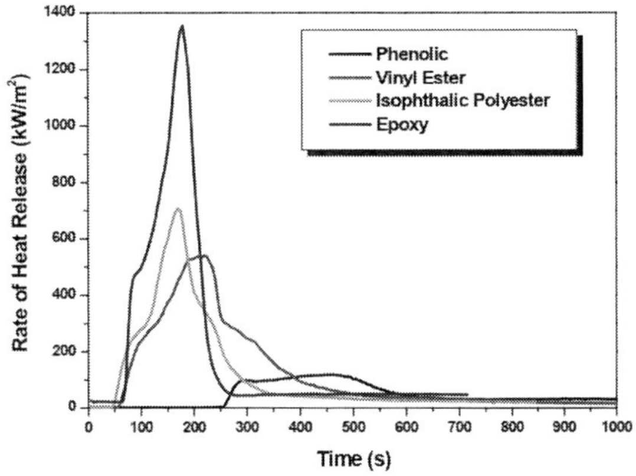

Figure 1. Rate of heat release (RHR) versus time profiles for matrix resin castings at a heat flux of 50 kW/m^2[6].

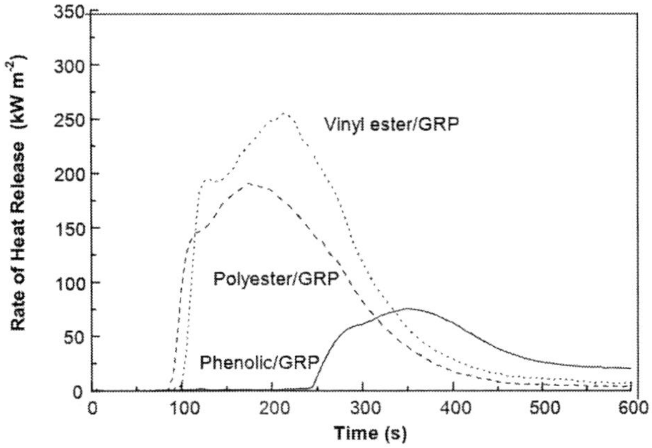

Figure 2. Rate of heat release (RHR) versus time profiles for woven roving vinyl ester (dots), polyester (dash) and phenolic (line) GRP laminates at a heat flux of 50 kW/m^2 [6].

Figure 2 shows the RHR vs time profiles of polyester-, vinyl ester- and a phenolic-glass laminate at a heat flux of 50 kW/m^2. This graph indicates the superior fire retardancy of the phenolic resin can be clearly translated to the combustion properties of laminates with reductions in the rate of heat release and ignition times. The longer ignition times and the use of less flammable matrix resin systems are important if improvement in the safety and survivability of navy vessels is to be achieved. This reduction in flammability enables an increase in the time available for escape and the activation of the fire fighting procedures.

2.2. GRP sandwich panels

Cone calorimetry rate of heat release results of a phenolic/PVC foam core sandwich panel with thick and thin phenolic skins are shown in Figure 3. The results show that the thick skin phenolic/PVC sandwich panel performs slightly better than the thin skin phenolic sandwich panel. This is due to the thicker (4mm) skins providing additional protection and thus delaying the decomposition of the foam core which is a combustible material. The shorter ignition times observed by the phenolic sandwich materials in comparison to the phenolic matrix resin alone (Figure 1) can be attributed to the decomposition of the foam core. Hence thicker skins or the implementation of thermal barriers would improve the flammability of these materials to ensure greater safety and the survivability of vessels if these materials were employed.

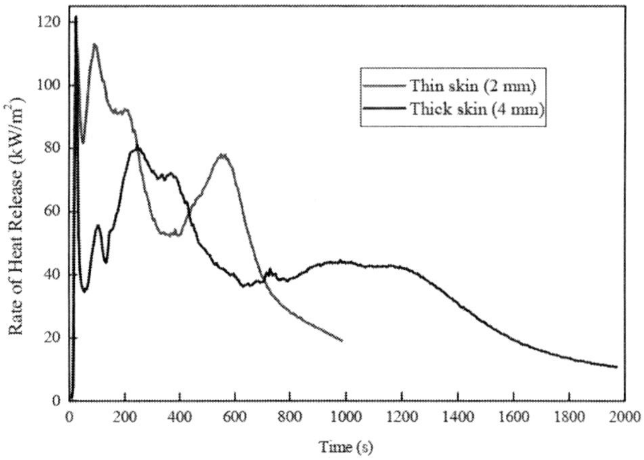

Figure 3. Rate of heat release profiles of a 2 and 4mm phenolic skin sandwich panel with a PVC core at a heat flux of 50 kW/m^2 [16].

3. FIRE RESISTANCE

Fire resistance is the ability of a structure to prevent a fully developed fire spreading from one compartment to a neighbouring compartment within a specified time. Typical fire resistance properties are often assessed by furnace tests with the focus on *Fire insulation, Flame and smoke containment and Structural integrity*. To rate the fire-resistance divisions, the IMO has developed Resolution A.754(18) "Standard fire test" requirements as part of the International Convention for the Safety of Life at Sea (SOLAS) [9]. There are many specifications, all of which have accepted the standard SOLAS time-temperature profile for qualifying materials according to their fire protection. The equivalent standard used in Australia is the Australian Standard AS 1530 Part 4 [17], *"Fire-Resistance Test of Structures"* that closely follows the time-temperature curve as specified by IMO [9]. For the material to pass the test it is required that ..." *the average temperature rise of the unexposed surface of the specimen shall not increase by more than 140°C above the initial temperature, and the temperature of the unexposed face shall not increase at any point by more than 180°C or reach a value higher than 221°C, whichever is the lesser value, for any specified period in the fire resistance test".*

3.1. Fire Resisting Divisions

The HSC Code also introduced the SOLAS principle of subdividing the craft by "Fire Resisting Division" (FRD) primarily to reduce the spread of fire from one compartment to another for a specified time. This is an important part of the passive fire protection, which makes it easier to fight a fire spread from its place of origin. The HSC Code divides spaces on high-speed craft into 6 groups based on fire risk, Table 2.

Table 2. Fire risk division of High Speed Craft

CLASS DIVISION	FIRE RISK
A	Areas of major fire hazard
B	Areas of moderate fire hazard
C	Areas of minor fire hazard
D	Control stations
E	Evacuation stations and external escape
F	Open spaces

Fire resisting divisions on SOLAS ships are to be tested in accordance with IMO Resolution A.754(18) *"Recommendation on fire resistance tests for "A", "B" and "F" class divisions"*. SOLAS also describes the time-temperature curve for the "*Standard fire test*" to which such divisions are to be tested. For high speed craft the standard fire test applies only to "A" and "B" class division.

3.2. Small Scale Fire Test

The fire tests required to meet the IMO Resolution A.754(18) to enable composites or metallic structures to the meet these regulations is conducted using gas-fired furnaces for which the operating conditions are specified. The furnace is versatile industrial scale equipment capable of testing various sizes of test panels in either vertical (bulkhead) or horizontal (deckhead) orientation. To meet the Resolution A754(18) the panel sizes should be 2440 mm x 3040 mm for decks and 2440 mm x 2500 mm for bulkheads. However our research focuses on conducting fire tests using smaller panels to obtain indicative results for fire resistance performance is possible.

3.2.1. Polyester and Phenolic glass Panels

Results from furnace testing in the vertical orientation of a polyester (9.5 mm) and phenolic (10.5 mm) laminate (300 mm x 300 mm) are shown Figure 4. An interesting observation from the temperature vs time profiles is that the backface temperatures are similar for both the phenolic and polyester glass laminates even though phenolic laminates are significantly less flammable than the polyester matrix resin system as was discussed in Section 3.1. This is an important observation with respect to fire resistance and meeting the IMO criteria.

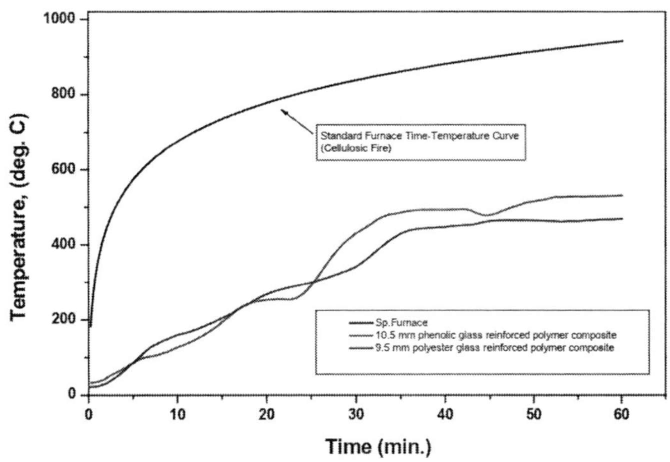

Figure 4. Furnace Time-temperature plot for unprotected polyester and phenolic GRP panels

3.2.2. Polyester/PVC Foam core sandwich panel

The fire performance of a polyester/PVC foam core sandwich panel (1200 mm x 1200 mm) under typical cellulosic fire conditions is illustrated in Figure 5. The graph shows a comparison of the temperature increase on the panel's exposed and unexposed surface. It is evident that while there is a rapid increase of surface temperature on the exposed panel side, the heat progression through the panel is considerably delayed resulting in a monotonic temperature rise on the unexposed side for about 12 minutes into the test. The panel failed in relation to the IMO fire resistance requirements after 22 minutes when the average temperature of the five thermocouples exceeded 158°C (140 °C plus room temperature). Note, at this time there was no evidence of a burn-through effect on the panel for the 30 minutes duration of the fire test. Thus for this type of composite structure, passive thermal barriers would need to be considered to enable this material to meet the IMO criteria for fire resistance.

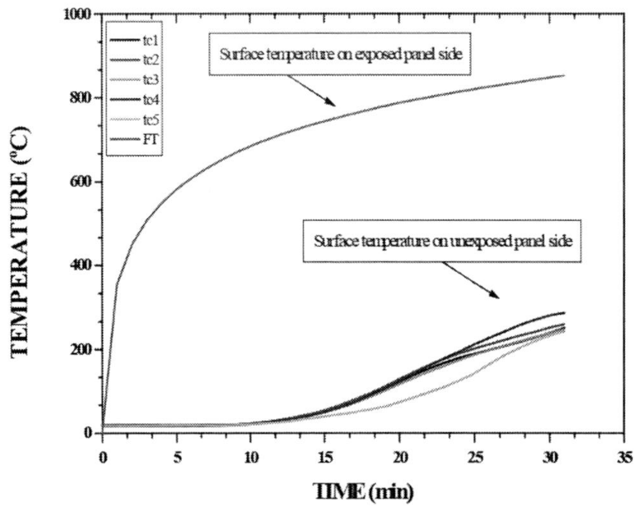

Figure 5. Thermocouple data for Polyester/PVC sandwich panel - Cellulosic fire test

4. THERMAL BARRIERS

Protective coatings or passive thermal barriers are used to enhance the fire performance of materials and/or structures. Their use is to minimise the spread of flame, reduce toxic by-products and prevent flashover of a material or structure. Protective coatings or fire hardening materials can also provide thermal insulation to reduce the rate of heat transmission, protecting the underlying material from heat and fire damage. The use of protective coatings or thermal barriers allows more combustible materials to be used in applications where combustible materials do not currently meet the required fire performance. There are a variety of different types of protective coatings or thermal barriers. These include inorganic, organic, passive and active insulative materials. Combinations of these can be used to achieve one or more of the fire performance capabilities required.

4.1. Phenolic/PVC Sandwich Panels coated with intumescent mat and coating

The effectiveness of a thermal barrier applied to two thicknesses of phenolic sandwich panels with a PVC core (300 mm x 300 mm) is shown in Figure 6. The two phenolic sandwich panels tested had different skin thicknesses (2.5 and 5.6 mm). The other two panels were protected with a thermal barrier coating consisting of an intumescent mat with a intumescent paint. The result of the cellulosic type fire test that was conducted simultaneously is given below.

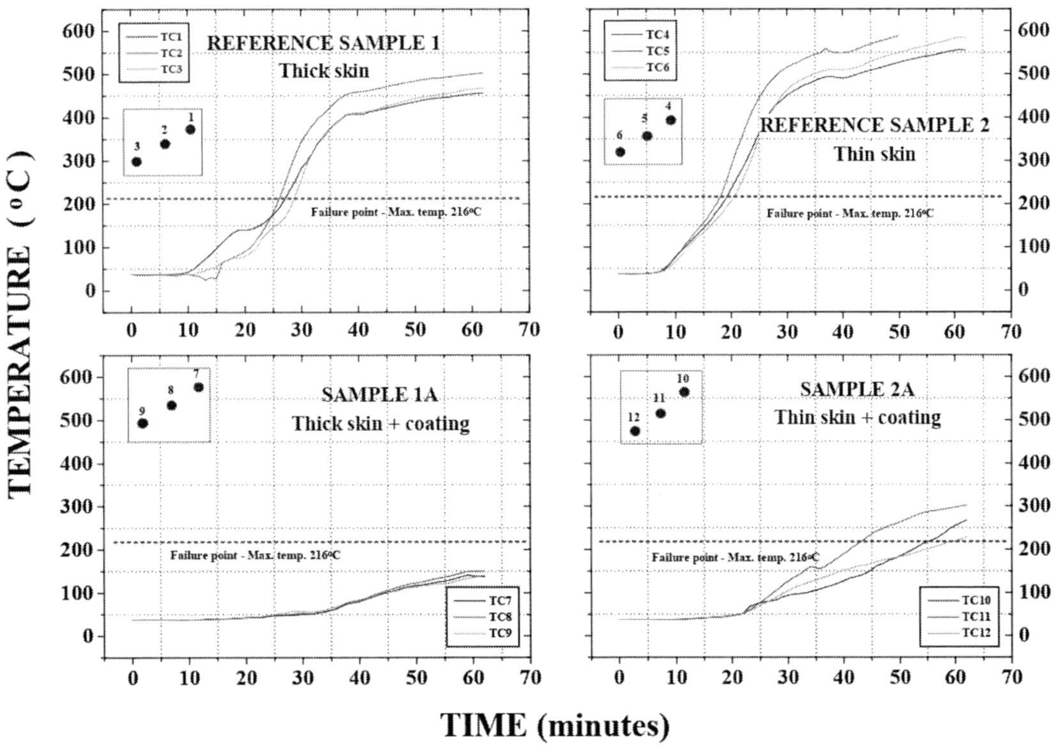

Figure 6. Effect of thermal barrier materials on fire performance of GRP composites

The panels failed in the order of increasing skin thickness and surface protection i.e. reference panel thin skin (18 minutes), the reference panel thick skin (27 minutes), protected panel with the thin skin (43 minutes) and finally the fourth panel, thick skin with the thermal barrier protection which passed a specification of achieving an A60 rating. The test pass mark for this indicative test was based on the individual thermocouple reading that exceeded 180°C above the ambient (i.e. 216°C). The panels with the thermal barrier coating on the fire exposed skin showed a large expansion of both the intumescent mat and paint. The results clearly indicate that the use of thermal barriers enhances the composite panels to meet the IMO criteria for fire resistance.

4.2. Vinylester/Glass Composite Panel fitted with Microtherm insulation

Indicative furnace testing of a vinyl ester/glass laminate with and without two layers of a Microtherm® insulation (Figure 7) showed that the backface temperatures achieved for panels with insulation had a significant reduction in temperature when exposed to the IMO cellulosic fire curve.

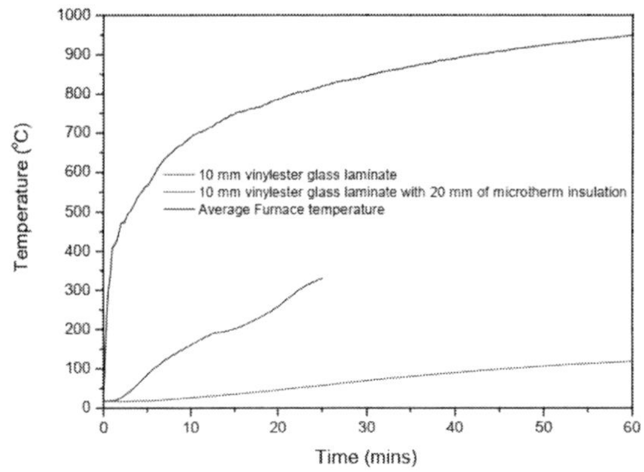

Figure 7. Effect of Microtherm passive insulation on fire performance of GRP composites.

5. STRUCTURAL INTEGRITY

Structural integrity of fire resisting division is an important property that structures on marine vessels are required to meet during a fire event to be considered safe. Steel structures are tested only against the IMO Resolution A754 (18) as steel is considered to remain structural during a fire event. Aluminium structures on the other hand, are required to additionally meet a load bearing fire-resisting divisions criteria as aluminium loses strength during elevated temperatures. The load-bearing requirement for aluminium structures, as specified in the IMO Resolution MSC. 45(65) [10] for "A-class divisions", is to meet a "core temperature rise of not more than 200° C above the initial temperature at any time within the classification period". This is in addition to the above requirements of the IMO Resolution A754 (18).

Composites structures, like aluminium structures, lose strength at elevated temperatures. Hence composite are evaluated against thermal and load bearing ability with a static load when exposed to a cellulosic fire. Such a test is commonly performed on test panel representing either; a ship deck (2440 mm x 3040 mm) or a ship bulkhead (2440 mm x 2500 mm) using the purpose-built furnace .

A composite panel's performance is thus assessed for the thermal and mechanical requirements as specified by IMO Resolution A.754(18) [9] and IMO Resolution MSC. 45(65) [10], respectively. These are summarized as follows:

- The thermal performance criterion which must be met by the panel under test was stated earlier in Section 4. Essentially, the rate of heat progression through the thickness of a composite structure should not cause the average temperature rise on the unexposed panel side to exceed 140° C above ambient.
- The mechanical performance criterion requires the load bearing fire-resisting panel (eg. ship deck or bulkhead panel) to meet (i) the limiting deflection and (ii) the limiting rate of deflection. The level of static loading applied to a deck is $3.5 kN/m^2$ and for a bulkhead 7.0 kN/m of the width. The deflection limits are calculated as per equations below:

  (i) Limiting deflection = $\left(L^2/400d\right)$ mm and

  (ii) Limiting rate of deflection = $\left(L^2/9000d\right)$ mm/min

where:

L = the clear span of the specimen
d = the distance from the extreme fibre of the design compression zone to the extreme fibre of the design tension zone of the structural section

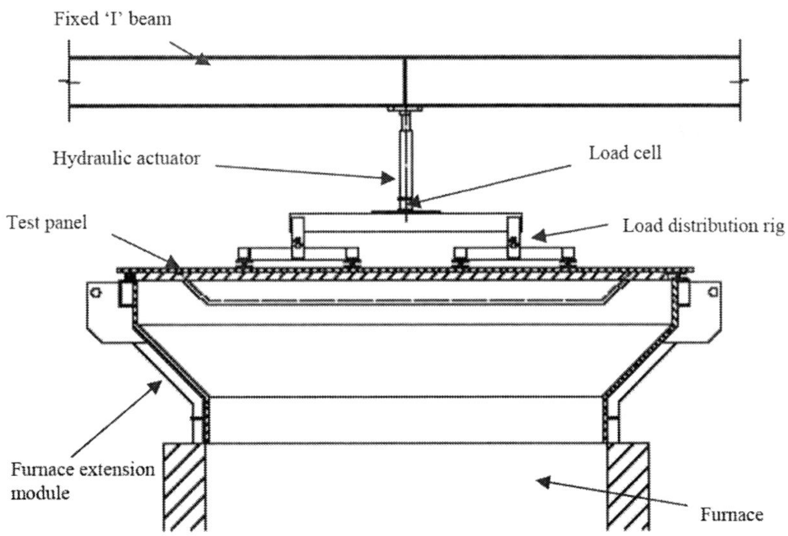

Figure 8. Test set-up for thermo-mechanical evaluation of composite structure.

A typical set-up of a structural fire test is shown in Figure 8. In recent tests DSTO has successfully conducted a thermo-mechanical evaluation of a full-scale naval composite structure using a test set-up similar to that shown in Figure 8. Both, thermal and mechanical requirements were met with minimal thermal protection.

6. CONCLUSION

Some examples of research performed by the DSTO in the field of composite reaction and resistance to fire including the structural integrity of some composite material systems have been described. This work shows that the use of effective and suitable thermal protection materials for composites structures is a means of meeting the stringent IMO standards. Thus composite materials can be used safely and hence are a viable option for ship structures.

7. REFERENCES

1. Pegg, R.L., Reyes, H., *Progress in Naval Composites*, in *Advanced materials and Processes inc. Metal Progress*. 1987. p. 35-38.
2. Grabovac, I., Turley, D.M., Townsend, C.R. *Fire Testing of a Composite Bulkhead in a Navy Ship*. in *The Second International Conference on Composite Engineering (ICCE/2)*. 1995. University of New Orleans, New Orleans USA, August 21-24.
3. Hoyning, B., Taby, J., *Fire Protection of Composite Vessels: Fire Protection and Structural Integrity - An Integrated Approach*, , in *Fourth International Conference on Fast Sea Transportation, July 21-23*. 1997: Sydney Australia

4. Caplan, I.L., *Marine Composites- The US navy experience, Lessons learned along the way* in *Composites Materials for Offshore Operations: Proceedings of the First International Workshop*. 1993, NIST Special Publication 887, pp. 91-114: Houston, Texas, October 26-28.

5. Fischer, K.J., *Is Fire a Barrier to Shipboard Composites*. Advanced Composites, 1993. May / June: p. 20-28.

6. Brown, J.R., Mathys, Z., *Reinforcement and Matrix Effects on Cone Calorimeter Fire Properties of Glass Reinforced Polymer Composites*. Composites, Part A, 1997. 28A: p. 675-681.

7. Sorathia, U., Rollhauser, C.M., Hughes, W.A., *Improved Fire Safety of Composites for Naval Applications*. Fire and Materials, 1992. **16**: p. 119.

8. Sorathia, U., Beck, C., *Fire Protection of Glass/Vinyl Ester Composites for Structural Applications*, in *Int. SAMPE Symp., 41, p. 687*. 1996.

9. IMO, *Recommendation on fire resistance tests for "A", "B", and "F" class divisions*. Resolution A.754(18), International Maritime Organisation.

10. IMO, *Test procedures for fire-resisting divisions on high-speed craft*. Resolution MSC. 45(65), International Maritime Organisation.

11. IMO, *Standards for qualifying marine materials for high-speed craft as fire restricting materials*. Resolution MSC. 40(64), International Maritime Organisation.

12. ISO, *Fire tests - Full scale room test for surface products, ISO 9705*. 1993, International Organisation for Standardization.

13. Ostman, B.A.-L., Tsantaridis, L. D., *Heat Release and Classification of Fire Retardant Wood Products*. Fire and Materials, 1995. **19**: p. 253-258.

14. Babrauskas, V., Peacock, R.D., *Analysis of large-scale fire tests data*. Fire Safety J., 1991. **17**: p. 387.

15. Redfern, J.P., *Rate of heat release measurement using the cone calorimeter*. J. Theraml Anal., 1989. **35**: p. 1681.

16. Grabovac, I., Mathys, Z., *Fire Performance Of Marine Composites*, in *The U.S. Pacific Rim Workshop on Composite Materials for Ship and Offshore Structures*. 1998: Honolulu, Hawaii, 7 - 9 April.

17. AS-1530, *Fire-Resistance Test of Structures, AS 1530, Part 4-1975*. (1976), Standards Association of Australia, Sydney N.S.W., Australia.

SEA4000: The Australian Air Warfare Destroyer Program – Shipbuilding Challenges and Opportunities

Jim Carr
Air Warfare Destroyer Alliance, Australia

ASC Shipbuilder Pty Ltd was selected by the Federal Government as shipbuilder for the HOBART Class Air Warfare Destroyers in May 2005.

In June 2007, after a successful Second Pass evaluation process the Federal Government chose the Spanish F100 platform design for the basis of developing the AWD.

This paper describes the major effort that will be required to execute the F100 Class design build process for the AWD shipbuilding program. It will highlight the opportunities for collaborating design build process with Navantia. The paper will summarize the key challenges associated with modifying the build process from an inclined ship way within a single organisation to a land level construction supported by significant subcontractors. The paper will also discuss the challenges for work force recruitment, facility development, and subcontract management that the Shipbuilding team within the Alliance management team anticipates.

The Application of Classification to High Performance Craft

by Tom Ingram and Derek Novak
American Bureau of Shipping

TIngram@eagle.org, Dnovak @eagle.org

SUMMARY

The American Bureau of Shipping and the U.S. Navy have been working together over the last several years in a measured initiative to develop and apply classification Rules for naval vessels. The first complete set of Rules resulting from this effort was embodied in the ABS Guide for Building and Classing High Speed Naval Craft. It laid the foundation for the subsequent development of the ABS Rules for Building and Classing Naval Vessels. In addition, in its own right, it has been used in the classification of the U.S. Navy SEA FIGHTER, SWIFT, Torpedo Recovery/Security Craft, U.S. Naval Academy Training Boats, the Egyptian Fast Missile Craft, Egyptian Fast Patrol Boat, Omani Patrol Boats and Canadian Navy Patrol Boats. In addition, the High Speed Naval Craft Rules were used as a core of the ABS classification requirements for the U.S. Navy's Littoral Combat Ships, both the monohull and trimaran versions. Finally, this Rule set will be used to class the U.S. Navy Joint High Speed Vessel and the U.S. Coast Guard Deepwater Fast Response Cutters.

This presentation will summarize the approach to applying this tool on naval craft and will address how the requirements fit together. In addition, valuable lessons learned from the ongoing applications will be presented.

INTRODUCTION

In the pursuit of vessel design, the naval architect will invariably rely on established processes that provide reliable and tested approaches to design convergence. These processes can be found in many forms including design tools, "rules of thumb", statutory criteria, established standards and classification Rules. These serve two purposes – to put the designer in a part of the design space which promises convergence and to ensure that the design is based on a set of bounding criteria which assure the vessel will be fit for its intended purpose. Once in this region, first principle approaches are often used to refine and optimize the design. Classification, as a process, provides such criteria with the addition of an independent, un-biased check for compliance which establishes credibility with the prospective owner, underwriting bodies, regulatory bodies and the public with whom we share a social contract for responsible design. After the design is complete and approved, the shipyard can then rely on the Classification Society surveyor team to work with them to ensure that the vessel is built in accordance with the approved plans and, thus, is in compliance with the established criteria set upon which the owner has established his requirements. Navies are coming to realize that this process holds value for the range of projects they initiate and have begun to require it as a part of their acquisitions.

CERTIFICATION AS A PROCESS

Verifying that a complex system such as a naval vessel meets the expectations of those that must use it is central to both the acquisition and through-life maintenance processes. With respect to any given acquisition and at any point in the life of a vessel, the cognizant Naval Administration must be assured that key technical and production issues are being addressed correctly and adequately. Critical to gaining such assurance is the ability to credibly verify and document proven engineering and production practices as verified by the designated technical authorities. In pursuit of this goal it must be recognized that there are several competing dynamics at work. The acquisition manager, the designer, the shipyard, the vendor, the owner, the operator and the general public each have separate and sometimes conflicting objective functions related to the vessel. Expectations and the constraints within which they are to be fulfilled must be clearly defined. Experience has demonstrated that when the process of assessing satisfaction of these expectations is placed in the hands of an independent, trusted agent, disappointments and disputes can be minimized and quickly resolved. It is best if this agent does not have a vested interest; that is, its sole job is to assess satisfaction of agreed-upon criteria. When this can be done, conflicts of interest can be eliminated and both the procurer and the provider can rest assured that decisions relating to certification are being made independent of economic, political or organizational influences. This is best achieved through the development and implementation of a comprehensive master certification plan. This plan identifies the necessary system, subsystem, and component certifications; identifies the relevant technical criteria and standards; identifies the designated certifying agent; and summarizes this information in a Certification Matrix. The result is a simple yet comprehensive means for documenting ship certification requirements, and their implementation and management throughout naval vessel design, construction, and life cycle support activities.

In this approach, the Naval Administration, the Naval Technical Authorities, the American Bureau of Shipping and industry design and life cycle support team members work together to identify the most reasonable and economical approaches to certifying a ship and its systems. Such a plan would notionally include certifications performed both during design and construction and throughout the life of the vessel, and would incorporate application of "Naval Rules" as the core of the criteria for system certification. Naval Rules are comprised of the ABS High Speed Naval Craft Guide and the ABS Naval Vessel Rules (NVR) which address the bulk of hull, mechanical, electrical, environmental and safety related criteria. In addition to these must be added those criteria retained and maintained by the Navy for the military unique aspects of combatants (including mission systems) for which classification society Rules are inappropriate.

ABS HIGH SPEED NAVAL CRAFT GUIDE

The *ABS Guide for Building and Classing High Speed Naval Craft* (aka ABS High Speed Naval Craft Guide (HSNC Guide)) has been developed in cooperation with the U.S. Naval Surface Warfare Center - Carderock Division - Combatant Craft Department, which represents the technical authority for the U.S. Navy's high speed craft. This is a set of standards that was developed for high speed craft that will operate in a variety of missions in environments that range from riverine to the open sea, and may be called upon to go into harm's way.

The High Speed Naval Craft Guide is divided into parts, chapters, and sections that reflect the current numbering system of the American Bureau of Shipping. The division of the Guide is as follows:

> *Part 1 General Provisions:* Administrative notes and Classification details that apply to the entire craft.
>
> *Part 2 Requirements for Materials and Welding:* These requirements are applicable to all craft that are being Classed and include requirements for steel, aluminum, and fiber reinforced plastics. The High Speed Naval Craft Guide contains the requirements for aluminum and fiber reinforced plastics only. The requirements for steel are part of the Steel Vessel Rules. ABS has also developed a supplement to this book specifically for Naval Vessels.
>
> *Part 3 Hull and Structures:* General requirements for all craft, structures, stability, and fire protection.
>
> *Part 4 Machinery and Equipment:* General requirements for all craft, prime movers, propulsion and maneuvering, pressure vessels, deck machinery, piping systems, fire safety systems, electrical systems, and remote propulsion control and automation.
>
> *Part 6 Survey After Construction:* Requirements for maintaining vessel in Class

The High Speed Naval Craft Guide incorporates the following items as part of the Classification process:

- Hull, mechanical and electrical (HM&E) systems
- Subdivision and stability
- Mission system interface with vessel and HM&E systems (but not the mission systems themselves)

The obvious difference between a naval craft and a commercial craft is that a naval craft is designed and built to go into harm's way. The Naval Administration determines, on a case basis, the levels of threat from enemy aggressions (weapons and weapons effects) that the vessel will face, the manner in which the vessel will avoid them (e.g., by signature reduction), and the level of survivability against those threats. However, naval craft also encounter hazards at sea during their normal operations, due to the environment, accidents or the inherent risks involved in training for war, and must be able to survive them. Therefore, the requirements for Classification under the High Speed Naval Craft Guide distinguish between "inherent" survivability requirements against fire, collisions, flooding and the operation of own-ship weapons (e.g., gun blast), and requirements imposed by the Naval Administration on survivability and signature reduction against enemy threats. These distinctions are as follows:

- Mandatory for Class:
    - Flooding and damaged stability
    - Fire protection
    - Damage control
    - Protection against own-vessel weapons effects
    - Structural integrity and residual hull girder strength
    - Propulsion system redundancy
    - Electrical system redundancy

- Guidance for supplemental requirements determined by Naval Administration
    - Protection against chemical, biological and radiation
    - Protection against underwater threats, e.g., shock and whipping
    - Protection against above-water threats, e.g., blast, fragmentation, electromagnetic pulse
    - Signature reduction

As stated above, the High Speed Naval Craft Guide provides requirements intended to protect the craft during their normal operations, due to the environment, accidents or the inherent risks involved in training for war. The Guide and the philosophy of ABS does not specify materials for certain systems unless we find them to be unfit for use, nor does the Guide specify a specific shape or arrangement of the craft, or any of it's appendices, unless it will interfere with the overall safety of the craft. If there are any specific preferences by the Naval Administration regarding the material selection or geometry of the craft or any of its appendices, it should be specified in the specifications for the craft.

Additionally, there are many areas in the High Speed Naval Craft Guide that allow for relaxation or increases in the requirements based on the needs of the Naval Administration and the concurrence of the Naval Technical Authority. If the Naval Administration would like to exercise any of these options, it should be indicated in the specifications for the craft and clear written technical justification produced showing why deviations were made from the Rules and the technical justification for such.

ABS NAVAL VESSEL RULES

Building on the established partnership between ABS and the United States Navy, the ABS Guide for Building and Classing Naval Vessels (Naval Vessel Rules) has been developed to address larger combatant classification.

The Naval Vessel Rules scope evolved during development. The Navy partners recognized that the Rules would be an extremely convenient tool with which to capture a wider range of design and construction criteria than those traditionally addressed by classification society Rules. Subsequently, the development effort expanded to include such areas as seakeeping, maneuvering, human systems integration, ship design margins, habitability and environmental protection systems. The makeup of the Naval Vessel Rules took on a modified version of the NAVSEA Ship Work Breakdown Structure:

Part 0 - General Provisions
Chapter 1 General Application of Naval Vessel Rules
Chapter 2 Classifications
Chapter 3 General Technical Requirements
Chapter 4 Approvals, Surveys and Requirements During Design and Construction
Chapter 5 Subdivision and Stability
Chapter 6 Reliability and Maintainability
Chapter 7 Human Systems Integration
Chapter 8 System Safety
Chapter 9 General Arrangements
Chapter 10 Design and Build Margins

Part 1- Hull and Structure
Chapter 1 General Provisions
Chapter 2 Structural Fire Protection
Chapter 3 Hull Structural Requirements
Chapter 4 Composite Hull Construction
Chapter 5 Structural and Non Structural Closures
Chapter 6 Hull Testing

Part 2 Propulsion and Maneuvering Systems
Chapter 1 General Provisions (including machinery arrangements)
Chapter 2 Diesel Engines
Chapter 3 Gas Turbines
Chapter 4 Transmission Systems (Gears, Shafting and Propellers)
Chapter 5 Waterjets and Thrusters
Chapter 6 Propulsion Redundancy
Chapter 7 Boilers, Pressure Vessels and Fired Equipment

Part 3 Electrical Systems
Chapter 1 General Provisions
Chapter 2 Systems design
Chapter 3 Electrical Equipment

Chapter 4 Shipboard Installation and Tests
Chapter 5 High Voltage, Electric Propulsion and DC Systems

Part 4 Control and Navigation Systems
Chapter 1 General Provisions
Chapter 2 Ship and Machinery Control and Automation Systems
Chapter 3 Navigation

Part 5 Auxiliary Machinery Systems
Chapter 1 General provisions
Chapter 2 Piping Systems
Chapter 3 Heating, Ventilation and Air Conditioning Systems
Chapter 4 Firefighting Systems and Damage Control
Chapter 5 Deck Systems
Chapter 6 Underway Replenishment Systems
Chapter 7 Environmental Protection Systems

Part 6 Habitability and Outfit
Chapter 1 General Provisions
Chapter 2 Accommodations and Workspaces
Chapter 3 Whole-Body Vibration
Chapter 4 Airborne Noise
Chapter 5 Indoor Climate
Chapter 6 Lighting

Part 7 Military Environment
Chapter 1 General Provisions
Chapter 2 Weapons Effects Design Considerations
Chapter 3 Signature Reduction Design Considerations

Part 8 Materials and Welding
Chapter 1 Material for Hull Construction
Chapter 2 Materials for Machinery, Boilers, Pressure Vessels and Piping
Chapter 3 Welding and Fabrication
Chapter 4 Coatings and Corrosion

NAVAL VESSEL CLASSIFICATION – APPLYING THE ABS HIGH SPEED NAVAL CRAFT GUIDE

In the normal course of Classification to the High Speed Naval Craft Guide, the Bureau would be contracted to provide a Class certificate as required by the Naval Administration. In this scenario, the Naval Administration takes the decision to build its vessel under Class, where ABS acts as the classification agent for the vessel and certification agent for HM&E systems; i.e., those items covered by the Rules. In this scenario, ABS is normally contracted directly by the shipyard and delivers its plan approval and certifications to them. However, it is possible that Classification can be carried out with ABS directly under contract to the Naval Administration. Thus, the Naval Administration has the flexibility to maintain an oversight role while concentrating its resources on more critical technical and programmatic issues, or leave the matter entirely in the hands of the shibuilder, as has been the case on some auxiliary ship and service craft acquisitions.

During the initial steps of the Concept/Design Phase, the Naval Administration generally concentrates on strategic planning, requirements definition and budgeting, and less on developing a technical framework for execution. ABS involvement will typically begin during early stage design by advising the Naval Administration and Naval Technical Authority on the Rules and Classification. ABS may assist in developing early contract requirements to invoke the High Speed Naval Craft Guide in a Statement of Work that may require a Class certificate from the shipyard as part of the deliverables. ABS may also be tasked to guide the process of developing a total ship certification plan including all elements of the vessel, as well as certifications of voluntary compliance with international conventions. ABS may also assist the Naval Technical Authority and/or Design Agent in preparing the ship specification, which naturally involves tailoring the specific application of the HSNC Guide.

During the Detail Design portion of the Build Phase, ABS works with the shipyard, and HM&E system vendors / integrators to ensure that the vessel, and its HM&E systems and mission system interfaces conform to the Rules and any other requested certification standards, such as SOLAS or MARPOL. The final product to guide Detail Design and construction will be the approved Ship Specification. ABS will issue approvals of design plans or CAD models to the shipyard. ABS will also establish a technical file for the ship including pertinent documents and drawings. The Naval Administration and Naval Technical Authority will accept the design and authorize construction, based in part on the shipyard's receipt of ABS and other approvals.

During the construction portion, ABS will carry out survey, inspection and testing of the vessel, HM&E systems and mission system interfaces to ensure compliance with approved drawings and the Rules. At the end of the Build phase, ABS issues a Class Certificate, along with other certifications as requested by the Naval Administration. These certificates are delivered to the Naval Administration as part of the overall package required for acceptance into fleet service.

After delivery, if the vessel is maintained in Class during the service life of the vessel, ABS will conduct both annual and periodic special inspections and surveys, reporting the suitability of the vessel to remain in class directly to the Naval Administration. The technical database for the vessel will be maintained to reflect the material condition of the vessel and HM&E systems. During maintenance and modifications, ABS will be contracted through the shipyard to conduct engineering plan approvals, survey, tests and certification for the vessel and HM&E systems during overhauls, conversions, repairs and mission system upgrades. These certificates are delivered to the Naval Administration as part of the overall package required for acceptance and return to fleet service.

The Office of Naval Research vessel X-Craft (now called SEA FIGHTER (FSF-1)) was the first application of this process using the ABS High Speed Naval Craft Guide. This application, as affirmed by the Chief of Naval Research, has been highly successful and the classification process introduced a disciplined structure to the enforcement of baseline criteria which added value to the final product. In addition, the SWIFT, U.S. Navy Torpedo Recovery/Security Craft, U.S. Naval Academy Training Boats, the Egyptian Fast Missile Craft, Egyptian Fast Patrol Boat, Omani Patrol Boats and Canadian Navy Patrol Boats have all been or are in the process of being classed by ABS under the High Speed Naval Craft Guide. The Lessons Learned from these applications will be discussed in a later paragraph along with those coming to light from the application of the Naval Vessel Rules to the LCS and DDG-1000.

The SEA FIGHTER, SWIFT and Canadian Navy vessels will be maintained in class as will the LCS and DDG-1000.

NAVAL VESSEL CLASSIFICATION – APPLYING THE ABS NAVAL VESSEL RULES

In parallel with the above effort, the Littoral Combat Ship (LCS) and DDG-1000 Land Attack Destroyer Programs began the effort to include the requirement to class to the ABS Naval Vessel Rules. Because of the legacy processes and associated thinking from prior acquisition approaches, this has been an evolving effort with a gradual recognition of the value of the classification process and an increasing integration of ABS into the acquisition structure. The SEA FIGHTER experience provided an experiential basis for the shaping of certain processes and the classing process was shaped for each individual program by the molding of ship specific Rule sets as represented by the invoked Rules and associated Justifications for Technical Determination (JTDs). These JTDs are the documentation for tailoring the generic Rules, including the rational consideration of requested equivalencies, deviations, and waivers of prescriptive Rule cites which must be modified to some degree due to the nature of the specific platform or mission. This is where the flexibility of the process allows an accepted criteria baseline to be shaped to accurately fit Program needs while providing discipline to the decision process. Criteria cannot just be ignored; there must be a basis for the decision which addresses any risk inherent in deviating.

LESSONS LEARNED

ABS has gathered significant experience now in the application of the High Speed Naval Craft Guide and Naval Vessel Rules on several programs in various stages of completion and has issued the first class certificate for a naval vessel designed and constructed to the Guide in the process described above. Classing the U.S. Navy X-Craft has provided valuable insight into the structure and nature of the relationships which are necessary in classing a naval vessel. The US Navy vessel SWIFT (HSV-2) was brought into ABS class under the HSNC Guide after it was chartered to the U.S. Navy and this experience is providing day-to-day lessons regarding keeping a naval vessel in class. The U.S. Navy's new Littoral Combat Ship and DDG-1000 are being classed by ABS. In addition, ABS is now applying the HSNC Guide in classification and certification efforts for the Canadian Navy, Egyptian Navy, the Government of Oman and the Mexican Navy.

In light of these experiences to date it would be useful to note several observations which might foster discussion that could result in improvement of the application.

1. It is imperative that the criteria against which certification is to be evaluated are clearly stated and that objective thresholds be established to the maximum extent possible. Subjective criteria can result in a range of expectations amongst the various parties involved in the process and invariably lead to disagreement. It will be necessary to retain subjective criteria to some degree and, as such, the importance of the next observation is underscored. It is important to realize that no single, static set of criteria (Rules) can be expected to adequately address the certification needs of any specific naval platform and that Rule sets must be shaped for each application to ensure that they are relevant to the program and embrace all current applied technology. The JTD process allows the shaping of the Rule set within the constraints of a disciplined risk assessment environment resulting in a set of governing criteria which are both relevant and comprehensive.

2. The roles of each involved party should be documented and clearly understood. The specification or contract must clearly empower the certification agent and make clear the process by which interpretation of the meaning of criteria is established and the process through which disputes are handled. Of necessity, ABS and the Navy have agreed to share the role of interpreting any requirement in the Rules during their application. In the case of a naval vessel, the final decision for interpretation, equivalency or waiver must remain with the naval technical authority but the class society should be included in all discussion and technical adjudication that leads to the final decision. In that manner, the class society will be equipped to enforce the criteria with a full knowledge of the intent and desire of the naval technical authority. Roles and relationships should be addressed in pre-contract discussions and again at a post-award kickoff conference. The objective of this initiative is to develop the procedure whereby the Navy can rely upon ABS as a classification and certification agent on vessels and associated systems, equipment and components which are more militarily unique than those for which the Navy has engaged ABS in the past (navy auxiliaries and service craft). One goal is that this should be done in such a manner as to allow the Navy to focus its in-house resources on work which could not be delegated. As such, it is imperative that redundant review, inspection and survey be avoided as much as possible. Having the Navy redo all that ABS does would defeat the intent of the practice. Thus, it is not expected that the Navy would analyze each part of the certification process undertaken by ABS, but rather would selectively employ monitoring of the process to ensure that systems certified in vessels built and classed by ABS are designed, built and installed to a level of acceptability equivalent to those certified using the Navy's traditional approach. This again underscores the importance of establishing at the outset a clear

understanding of the roles and responsibilities of each party. Organizations which provide navy or government representation to the designer or at the yard during construction must engage with the classification society very early to establish coordinated and complementary working processes to avoid confusion and misunderstanding. In the current programs, those decisions affecting Rules are documented by JTDs while integrated roles during plan review and survey are addressed in Cooperative Agreements between the Navy and ABS.

3. Open communication throughout the entire process is a key to success. It must be understood from the beginning that there will be difficult aspects to resolve enroute, but each party has the same overall objective - to deliver a vessel which meets or exceeds specified expectations. Thus, it is absolutely necessary to avoid selective "information management" in order to try to hide or otherwise misrepresent factual information. The contract should contain language which facilitates free and open communication between all parties insofar as the acceptability of the systems are concerned. It must also establish a clear adjudication process whereby disagreements or questions of proper procedure may be quickly addressed with all affected parties involved.

4. It is inherent in the nature of the new ship design process for a naval application that every effort will be made to incorporate state-of-the-art technology wherever possible. The reasons for this are obvious. The challenge to those who must establish technical design and certification criteria is equally obvious – how does one establish criteria for developmental technologies? Oftentimes such certification criteria are based upon experiential feedback and this is non-existent for some of what is desired. Examples include unique hull forms and alternative materials and there are a number of instances where this has occurred in the programs cited above. The solution is to ensure that a disciplined and documented risk-based analysis approach is applied to ensure that all possible outcomes of such an application are identified and mitigated to an acceptable level. The entity to make this decision is necessarily the Naval Technical Authority. The strong partnership and close working relationship between NAVSEA and ABS has been the key to ensuring that this takes place for applications of new technology on US Navy platforms. The requirements for the establishment of Safe Operating Envelopes, associated Operator Guidance and supporting ship instrumentation are examples of innovative solutions applied in such cases on the X-Craft and LCS.

5. As a corollary to the first lesson learned discussed above, it is extremely important to ensure that the Rule set is stable and well-defined prior to the initiation of design approval. Although it is imperative that there be a process for addressing interpretations, equivalency determinations, deviations and waivers, these should be the exception rather than the rule. Using these to justify design or production errors should be avoided as much as possible as this will lead to compromise of the original intent of the owner in choosing the criteria basis he did. Flowing from this, it is important to strive to have review and approval of the design submittals completed prior to initiation of construction. This would seem to be obvious but, under the time

and cost pressures which confront an acquisition team, it is surprising how often common sense is abandoned. Concurrent design and construction will more often than not lead to increased costs and impact on the building schedule from rework.

6. Finally, it is extremely important to realize that an inherent part of the classification process is the certification of key components including machinery and mechanical systems at their point of manufacture by the classification society. Items such as main propulsion components, shafting, switchboards and the like must be built by the vendors to approved plans and must be certified by the classification society to be compliant with the established Rule set. It is the responsibility of the builder to ensure that his procurement documents relate this requirement to the relevant vendors and it is the vendor's responsibility to comply and deliver the resulting certification documentation with the equipment. If this is not done, it is extremely difficult to do it after the fact. The owner is then put in the awkward position of having to accept key elements in his vessel of which the acceptability is uncertain.

CONCLUSION

In light of the reduction of retained resources within most governments and navies, the ship classification approach outlined above provides a mechanism for the continued accomplishment of the core navy technical authority role of certification. Its success lies in clear alignment of the navy technical authority with their designated ship classification and certification agent along with the need for open, continuous communication amongst all parties regarding the engineering aspects of certification. The goal is the design, construction and installation of systems, equipment and components for which there is a high degree of confidence that the stated performance requirements will be safely met. The service to be delivered is certification to clearly defined criteria, and the product is a certificate of compliance which the vendor and builder can use to demonstrate to the navy that the requirement has been satisfied.

Residual Strength Variation of a Welded Structural Component due to a Pore

A.Ghosh
ASC Pty Ltd., Adelaide, S.A., Australia
Abir.Ghosh@asc.com.au

ABSTRACT
In general, flaws in welded components in Collins Class submarines are repaired during maintenance periods. However, it is not always feasible to conduct such repairs in areas that are considered non-critical or areas that are inaccessible. Therefore, it is helpful to analyse for fatigue in welded components in submarines. Welded components subjected to repeated loading are normally analysed using S/N curves to find the end of the fatigue life and Palmgren-Miner's law to determine the damage at any intermediate period of fatigue life. If cracks are evident then fracture mechanics is used to determine the remaining lives of components. A method of analysis is being developed that gives the variation of the residual strength of the component during the fatigue life and therefore allows for more comprehensive analysis of the component at any intermediate period of the fatigue life. In this method, the residual strength variation is determined due to a pore in the weld of a structural component in a submarine. Thus, as the pore develops with fatigue loading, the variation in strength of the component during the entire fatigue life is determined. The proposed method will permit estimation of the maximum allowable stresses with inherent flaws in the welded structure and allow further analysis of such a weld at any intermediate period of its fatigue life.

1 INTRODUCTION

At present, welded components are designed for static strength and fatigue life (1-4). It is assumed that the strength of a structure does not change during the fatigue life (Fig.1). There are no procedures to assess the variation in strength during the life of the structure. Palmgren-Miner's law is used to give fatigue damage caused to the structure (2, 5). This is given as a ratio and gives only a preliminary idea of the condition of the structure. It does not give any idea of the change in the strength of the structure. If cracks are detected then fracture mechanics is used to find the strength of the component and crack propagation laws are used to determine the remaining life (6, 7). This only gives the strength of the cracked component and does not give the variation of strength during its remaining life.

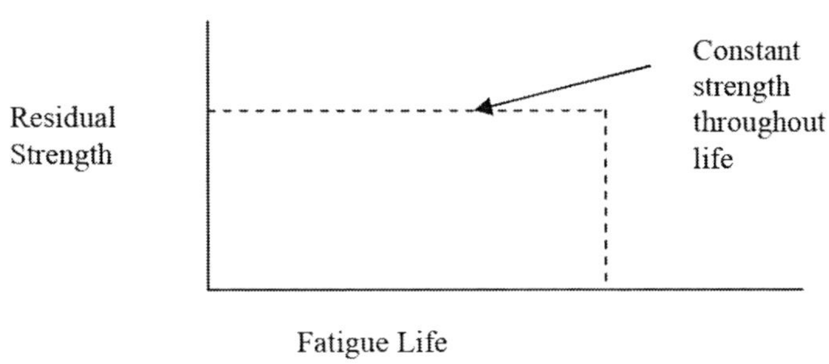

Figure 1 (8)

A component subjected to fatigue undergoes i) fatigue hardening and softening ii) crack initiation iii) crack propagation and iv) failure during its lifetime. Therefore, the component would have undergone the first three processes prior to failure. Fatigue hardening and crack propagation particularly affects the strength of the component, so the strength of the component will change during its lifetime.

Components can fail by unstable crack propagation failure and plastic deformation failure. A procedure for developing the residual strength variation of a welded component failing by unstable crack propagation and the residual strength variation for a component failing by plastic deformation has already been determined. The residual strength variation due to unstable crack propagation failure is compared with the residual strength due to plastic deformation failure to give the actual residual strength of a component.

The procedure of development of residual strength variation can be used to determine the variation in strength of a component during its fatigue life. The fatigue life in such cases is found from S/N curves. The procedure of development of the residual strength curve can also be used to find the variation of strength of a component with a crack. In both cases this allows assessment of the strength of the structure and the maximum overload that can be applied.

The above procedure of assessment can be used to determine the residual strength variation of welded components for different types of structures. The procedure can also be used to determine the residual strength variation of welds in submarines. Here the procedure of determining the residual strength variation due to the growth of a pore in a weld is discussed. The procedure is illustrated by arbitrarily considering a material and a loading pattern. Thus, the procedure is demonstrated without divulging the load pattern and materials used in the submarine. The process of developing the residual strength variation is explained first. The specific case of determining the residual strength variation due to the growth of a pore is discussed later.

2 RESIDUAL STRENGTH VARIATION

The residual strength variation of components is determined by comparing the residual strength due to unstable crack propagation and residual strength due to plastic deformation.

2.1 Residual Strength due to Unstable Crack Propagation

The unstable crack propagation strength of component is found using fracture mechanics. The stress intensity of a component with a crack is given by the equation (9)

$$K = \sigma M \sqrt{\pi a} \qquad (1)$$

where σ is the stress applied, a is the crack length and M is the magnification factor. The stress intensity equation at failure is given as

$$K_c = \sigma M \sqrt{\pi a} \qquad (2)$$

where K_c is the critical stress intensity factor or the fracture toughness. For a given component the above equation can be considered to give a relation between strength and crack length. The number of cycles required for a crack to propagate from crack length a_1 to another crack length a_2 has been given by Paris (10) as

$$N = \int_{a_1}^{a_2} C(\Delta K)^m \tag{3}$$

where N is the number of cycles required for a crack to propagate from a_1 to a_2, ΔK is the range of stress intensity factor and C and m are material constants. For a given component, the above equation can be considered to give a relation between number of cycles and crack length.

The stress intensity equation at failure can therefore be considered as an equation that, for a given component, gives a relation between strength and crack length while the crack propagation equation can be considered to give a relation between crack length and number of cycles for a given component (11). Combining these two equations provides a relation between strength and number of cycles for a given component. A closed form solution can be obtained if the value of the magnification factor M is considered to be constant (12). For common steel the variation becomes linear and is given by the equation

$$N_x = -G_1\sigma_x + H_1 \tag{4}$$

where

$$G_1 = \frac{\sqrt{2}}{\pi^{3/2} C M^3 (\Delta\sigma)^3 K_c} \quad \text{and} \quad H_1 = \frac{2}{C\left(M\Delta\sigma\sqrt{\pi}\right)^3 \sqrt{a_i}}$$

σ_x being the strength at N_x cycles with a_i being the initial crack length. However, only idealized components such as an infinite plate with a centre crack has a constant value of M. For real cases, the value of M will vary with the crack length and the residual strength variation for failure by unstable crack propagation has to be obtained by numerical integration. The residual strength variation due to unstable crack propagation for such components can be found (see Fig.2) by taking small increments of crack length.

2.2 Residual Strength due to Plastic Deformation

Welded components do not fail by unstable crack propagation alone. Components can also fail by plastic deformation failure (13). The plastic deformation strength of a component with a flaw has been given by McClintock (14) by the equation

$$P_{mp} = f_u\left(1 - \frac{a}{t}\right) \tag{5}$$

according to Tresca's theory where P_{mp} is the plastic deformation strength, t is the thickness and f_u is the tensile strength of the material. According to Von-Mises criteria the strength of the component is given by

$$P_{mp} = \frac{2}{\sqrt{3}} f_u \left(1 - \frac{a}{t}\right) \quad (6)$$

However equations 5 and 6 do not give the strength of a cracked component subjected to fatigue as the strength of fatigued component may not be given by the tensile strength f_u of the material (15-17). This is because of fatigue hardening or softening. When fatigue loading is applied to metallic material, the first process it undergoes is fatigue hardening or softening. Fatigue hardening/softening involves rearrangement of the molecular structure of the entire material being subjected to fatigue loading. When loading initially commences, the molecular structure of the material alters with repeated loading until the molecular structure enters into a stable state and the material is known to be saturated. As further loading is applied, the molecular structure remains in this stabilised state. If a material is subjected to controlled stress or strain amplitude loading, fatigue hardening/softening causes the hysteresis stress-strain loops of the material to vary with such loading as the molecular structure rearranges. With the application of fatigue loading, as the molecular structure stabilises the hysteresis loops also become stable. A plot through the peaks of stable hysteresis loops of a fatigue hardened or softened material at different controlled strains or stresses is known as a cyclic stress strain curve. The cyclic stress strain curve is often considered a property of the material under cyclic loading in the manner in which the tensile plot of a material is considered the property of the material under monotonic loading.

Therefore, the strength of a component subjected to fatigue loading is not the same as the strength of a component subjected to monotonic loading. Fatigue hardening/softening occurs early in the fatigue life of the component. The strength of a component that has undergone saturation will be given by the cyclic stress strain curve which is different from the monotonic stress strain curve. If the maximum strength of the material of the component as given by the cyclic stress strain curve is given as f_c then the plastic deformation strength of a saturated component will be given as

$$P_{mp} = f_c \left(1 - \frac{a}{t}\right) \quad (7)$$

according to Tresca's criteria and

$$P_{mp} = \frac{2}{\sqrt{3}} f_c \left(1 - \frac{a}{t}\right) \quad (8)$$

according to Von-Mises criteria.

Equations 7 and 8 need to be combined with equation 3 to give the residual strength variation due to plastic deformation of a component that has been subjected to fatigue and has undergone saturation. The residual strength variation of a component failing by plastic deformation in the same manner as the residual strength variation of a component failing by unstable crack propagation was determined. Thus, small increments of crack lengths need to be taken to determine the residual strength variation for the entire life (see Fig.2).

2.3 Actual Residual Strength Curve

The residual strength of a component can be determined by proper positioning of the residual strength due to unstable crack propagation with the residual strength variation due to plastic deformation. The curves are placed as shown in Fig.4. The curve which gives the lower values gives the actual strength of the component and the mode of failure (18, 19). The estimation of the residual strength curve can be improved by using more accurate plastic deformation and crack propagation laws, provided relevant material properties are available.

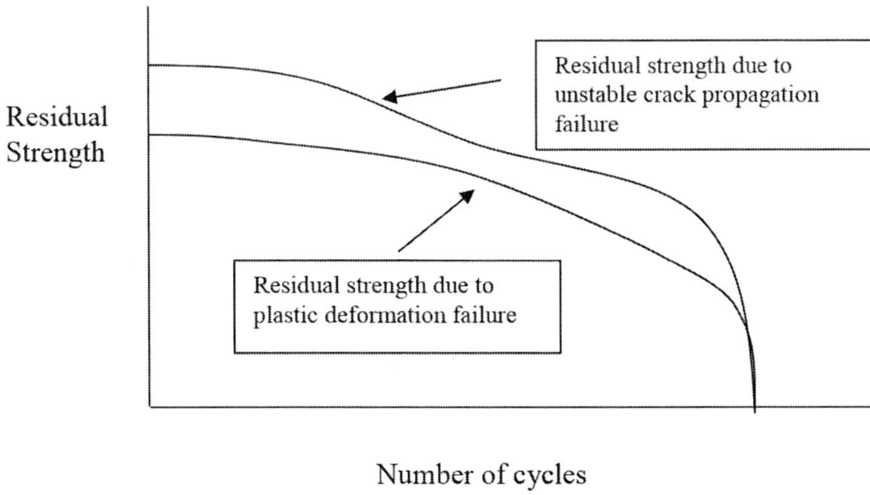

Figure 2

3 RESIDUAL STRENGTH VARIATION DUE TO GROWTH OF A PORE

A procedure of determining the fracture mechanics properties of a crack emanating from a pore has been developed by Albrect and Yamada (20). It has been reported that approximately 80% of cracks in plain welded beams originate from porosities caused by gas entrapment (21). The pores can be considered as rounded cavities on the fracture surface. It has been shown that the leading edge of a disk-like crack that originated from a pore was almost circular in shape (21). This circular crack shape was found at various stages of growth until the crack has reached the farthest fibres of the flange.

Based on this information, Albrect and Yamada have modelled it as a disk-like crack emanating from a spherical porosity. Hence, they have determined the magnification factor M used in Eqn.1. The magnification factor M is given as

$$M = F_E F_s F_W F_G \qquad (9)$$

where F_E accounts for the crack front, F_s accounts for effects of a free surface, F_w is a factor accounting for finite width and F_G is the factor accounting for geometry.

The procedure has been illustrated for a pore of 0.5 mm radius in a weld. The attachment with the weld is shown in Fig.3. The corresponding magnification factor obtained is also shown in Fig.3.

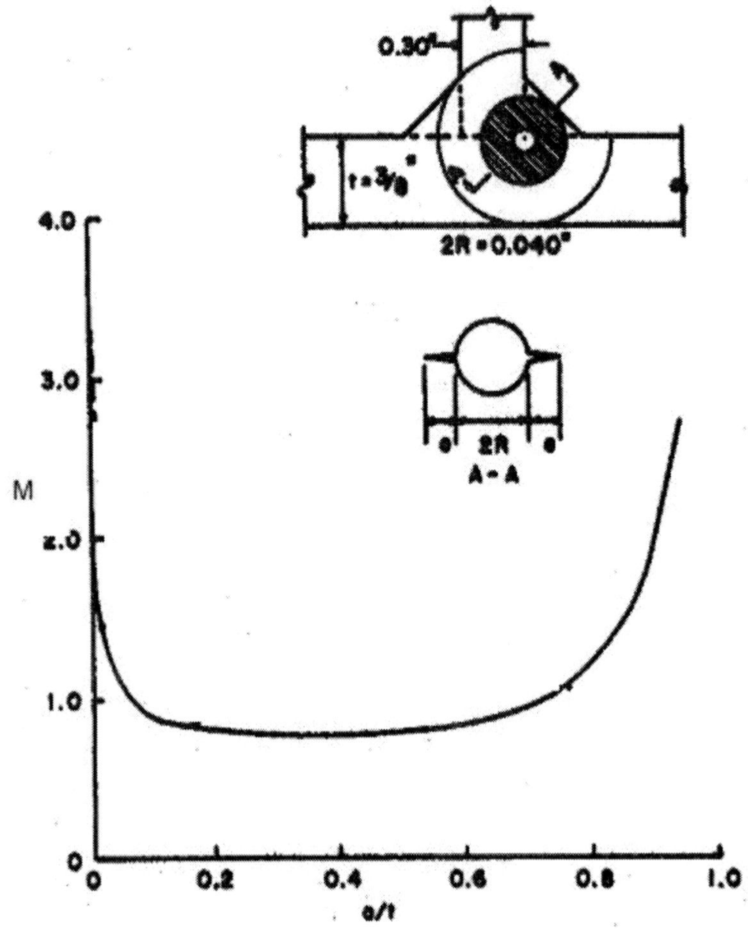

Figure 3 (20)

The information can be used to determine the residual strength variation of the welded component. The residual strength variation can be used to assess the component at any intermediate period of its life.

Consider a component with crack a=0.24 mm. Let us consider the throat thickness of the weld to be 6 mm. The material considered is 4340 steel which has a high yield and tensile strength. Consider the application of a constant stress amplitude of 70 MPa.

The fracture toughness of 4340 steel is taken as 1470 $N/mm^{3/2}$. The tensile strength is taken as 1310 MPa and the strength after saturation as 970 Mpa (17). The value of C in Paris' equation is 2×10^{-10} and the value of m is 2.25 for martensitic steel.

The residual strength variation is given in Fig.4. The strength is given for a portion of the life by the plastic deformation curve which considers that the component is saturated. For a portion of the life the strength is given by the unstable crack propagation curve.

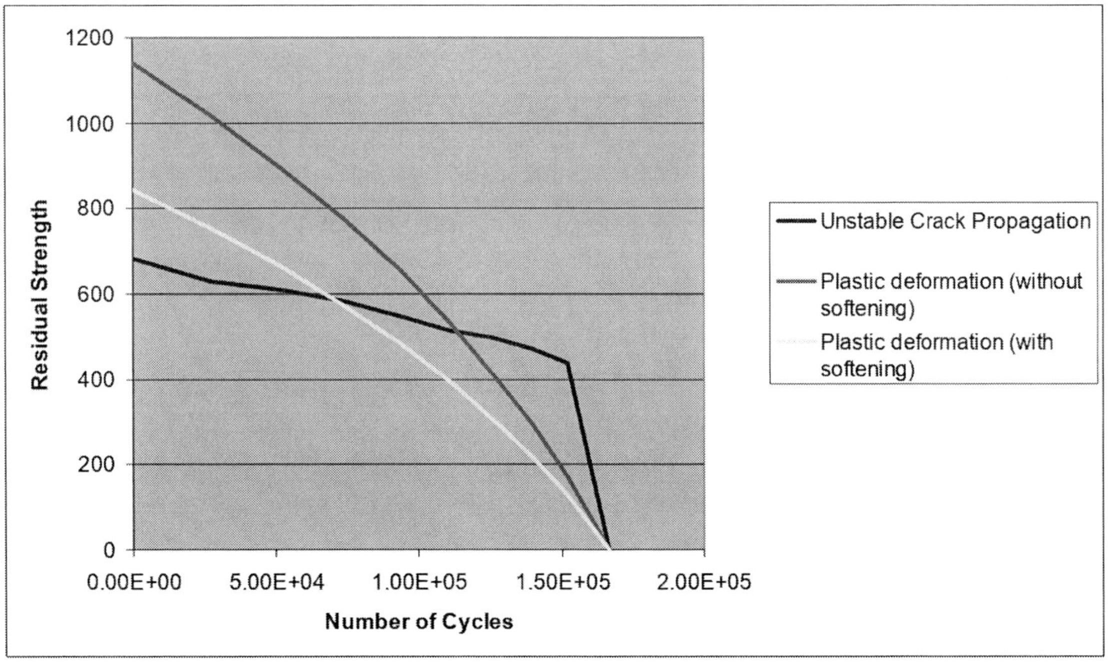

Figure 4

The strength of the component after 5 x 10^4 cycles is given by the unstable crack propagation curve as 610 MPa, and after 1 x 10^5 cycles it is given by the plastic deformation curve which considers softening as 513 MPa.

4 CONCLUSIONS

Welded components subjected to fatigue are designed for static strength and fatigue life. There are no procedures of assessing the strength and condition of such components. Palmgren-Miner's law gives only a ratio of the fatigue damage. If cracks are found then the strength is determined and remaining life calculated. There are no procedures to find the strength of such a cracked component at any intermediate period of its remaining life. A procedure of assessing based on residual strength at any intermediate period of the life of a component is being developed. The procedure has been used to determine the residual strength variation of a welded component with a pore. It has been reported that approximately 80% of cracks in welds are due to pores. The procedure showed here can be used to assess the strength of welded components in submarines suspected to have a growing pore.

5. REFERENCES

1. American Association of State Highway and Transportation Officials (AASHTO). (1989). *Guide specifications for fatigue design of steel bridges*, Washington, D.C.

2. British Standards Institution (1980). "Part 10: Code of practice for fatigue". *BS 5400, Steel, concrete and composite bridges*, London, England.

3. Commission of the European Communities. (1992). "Common unified rules for steel structures". *Eurocode No.3,* Luxembourg.

4. Standards Association of Australia. (1990). "Steel Structures" *Australian Standard AS 4100*, Standard House, North Sydney, NSW. Australia.

5. Ghosh, A (2002). "A residual strength model for Palmgren-Miner's law." *Australian Journal of Structural Engineering,* 4 (2), 135-142.

6. Rolfe and Barsom (1979). *Fracture and Fatigue Control in Structures*. Prentice Hall.

7. British Standards Institution (1999). "Guidance on methods for assessing the acceptability of flaws in fusion welded structures". *BS 7910*. London, England

8. Oehlers, D. J. (1990). "Deterioration in strength of stud shear connectors in composite bridge beams". *J. Struct. Eng., ASCE,* 116 (12), 3417-3431.

9. Irwin, G.R. (1962). " The crack extension force for a part through crack in a plate." *Trans.ASME, J.Appl.Mech.,* 1962, pp.651-654.

10. Paris, P.C. and Erdogan, F. (1963). "A critical analysis of crack propagation laws" *J.Bas.Engng, Transactions, ASME,* 85, No.4528, 1963.

11. Ghosh, A., Oehlers, D.J. and Wahab, M.A. (1993). "A New Fatigue Design and Analysis Approach Based on Residual Strength" *Research report no. R108*, October, Department of Civil and Environmental Engineering, University of Adelaide.

12. Ghosh, A., Oehlers, D.J., Wahab, M.A. (1996). "Linear fracture envelopes for fatigue assessment of welds in bridges." *Structural Engineering and Mechanics Journal*, July, 1996:347-364.

13. Cox, E.P. (1987). "A methodology for characterising weldment fracture behaviour". *Welding Journal*, pp 39s-49s.

14. McClintock, F.A. (1961). "On notch sensitivity". *Welding research supplement*, pp.202s-209s, May.

15. Ghosh, A (1999). "Modifying the residual strength model of welds to allow for plastic deformation failure." *Australian Journal of Structural Engineering,* 2(1), 1-9.

16. Klesnil, M. and Lukas, P. (1980). *Fatigue of metallic materials*. Elsevier scientific publishing company.

17. Landgraf, R.W., Morrow, J., and Endo, T. (1969). "Determination of the Cyclic Stress Strain Curve". *Journal of Materials*, Vol.4, No.1, pp. 176-188.

18. Ghosh, A. (1997). A residual strength approach for the fatigue analysis of welded components." Phd Thesis, University of Adelaide, Adelaide.

19. Ghosh A.(2004). "Modern approach to design of welded components subjected to fatigue." *J. Struct. Engrg., ASCE*, 130(5), 812-821, 2004

20. Albrecht, P., and Yamada, K. (1977). "Rapid calculation of stress intensity factors." *J.Strut.Engrg., ASCE* ,(2), 377-389,1977.

21. Hirt, M.A. and Fisher, J.W. (1973). "Fatigue crack growth in welded beams", *Journal of Engineering Fracture Mechanics* Vol 5, No 2.

Classification of Naval Support Ships

Bob Simpson, Lloyd's Register

Lloyd's Register's naval ship classification regime was developed primarily for naval ships that have or will be designed/constructed in accordance with standards applicable to frontline naval ships and that will be operated in a naval environment. The regime covers a range of ships such as destroyers, aircraft carriers and frigates, and includes vessels with offensive and defensive roles. The advantages of ships operating within the naval ship classification regime are recognised by different navies. Changes to Lloyd's Regulations for the Classification of Naval Ships now allow for auxiliary naval ships used for the support of civil and naval operations to be brought within the scope of the naval ship classification regime.

Naval auxiliary support ships may have a variety of roles including the movement of military and other personnel, ammunition, vehicles, stores and fuels and their transfer other naval ships. They do not have a defined offensive role but may have limited self defensive capability. In general, the ships will also need to comply with other Lloyd's Register classification rule requirements of the International Convention standards for arrangements and equipment which are applicable to the ship type.

This paper will describe the requirements and processes for the classification of auxiliary naval support ships within the scope of Lloyd's Register's Naval Ship classification regime. It will include the types of ships that can be brought within the regime and the standards that need to be addressed, as well as the need for the submission of a design and operating scenario statement and additional assessments of the hull structure and machinery system arrangements.

The key differences in the classification requirements relating to ships that have been designed in accordance with other Lloyd's Register classification Rules and the requirements that have been developed specifically for frontline naval ships will also be highlighted.

1 INTRODUCTION

Long before Lloyd's Register agreed to develop a set of Rules for the Classification of Naval Ships, it had been involved in the classification of naval support vessels for the UK MoD, the Royal Australian Navy and the Royal New Zealand Navy, and had built to and maintained in Class many other naval and paramilitary vessels for other countries.

Since 1999, when the draft Rules and Regulations for the Classification of Naval Ships (the Naval Ship Rules) were initially released,, having been formally agreed by the Lloyd's Register Naval Ship Technical Committee in November 1999 and released publicly in January 2000, the Rules have been applied to many naval projects as a more appropriate suite of standards that reflect the operational requirements of the vessels. While the Rules were originally developed for ships with a clear front line purpose, many naval vessels currently being designed or under construction have a support or logistics role which makes them similar to various merchant ship types such as RoRos, tankers and passenger ships.

The more comprehensive package of operational requirements presented by such ships, coupled with experience gained in dealing with naval ship projects, has prompted Lloyd's Register to make significant changes to the Naval Ship Rules, both in the way that they are applied and in their scope. One of the most important changes has been the introduction of the Naval Ship Auxiliary (NSA) notation, which is intended to specifically deal with the challenges that support ships present.

2 CURRENT PROJECT PORTFOLIO

The following list of projects gives a summary of Lloyd's Register's current position in the naval market:

Project	**Phase**
UK	
Type 45 Destroyer	Construction
MARS project	Design
Future Aircraft Carrier (CVF)	Design
NATO Submarine Rescue System (NSRS)	Construction
PASCAT (Partially Air Supported Catamaran)	Design
Future Surface Combatant	Concept
Australia	
Landing Helicopter Dock	Design
Canada	
Joint Support Ship	Design
New Zealand	
Protector OPV	Construction
Protector IPV	Construction
UAE	
Baynunah	Construction
Indonesia	

Project	Phase
Sigma Class Corvettes	Construction
Spain	
BPE	Construction
BAC	Construction
Chile	
OPV	Construction
Oman	
OPV	Construction
Trinidad & Tobago	
OPV	Construction
Iceland	
OPV	Construction

Other naval vessels built to Lloyd's Register class, in service since 1995, include:

Galicia/ Castilla (Spain)
HMS Ocean (UK)
HMS Albion and *Bulwark* (UK)
HMS Scott (UK)
HMS Echo & *HMS Enterprise* (UK)
HMNZNS Canterbury (NZ)
HMAS Sirius (Australia)

3 CLASSIFICATION OF NAVAL SHIPS

3.1 GENERAL

There have been a number of papers [2], [3] and [4] on the Classification of Naval Ships The ideas espoused in these papers are summarised below:

- classification is an assurance process showing that a ship complies with a suite of technical standards appropriate to its function
- assurance is given by an independent third party audit of procurement processes, particularly in the supply chain (where it is based on a 'policing' of supply chain activities but does not necessarily include complete duplication of a manufacturers' inspection obligations)
- technical standards provide a readily accessible body of corporate 'lessons learnt' that provide significant input into formal safety management techniques which can be inadvertently ignored if the expert panel is too small.
- class societies need ships to be maintained in Class to close the knowledge management loop
- basic Classification is only a minimum standard and navies will need to define their own levels of safety that reflect their operational and assurance requirements

Of these, the one that has emerged as being particularly useful to naval projects are the embedded 'lessons learnt and hazards identified' that exist in technical standards, whether they are Classification Rules, IMO Conventions and Resolutions or other national or international requirements. This was the subject of Ref [7], which identified that for machinery and electrical systems alone, some 1,300 hazards are not only identified in Lloyd's Register's Naval Ship Rules, but by designing and building to Class and then maintaining ships in Class, hazards could be actively managed to at least an As Low As Reasonably Practicable (ALARP) level, and in the majority on qualitative basis, to a BROADLY ACCEPTABLE level.

This was shown by the following diagram (Fig 1).

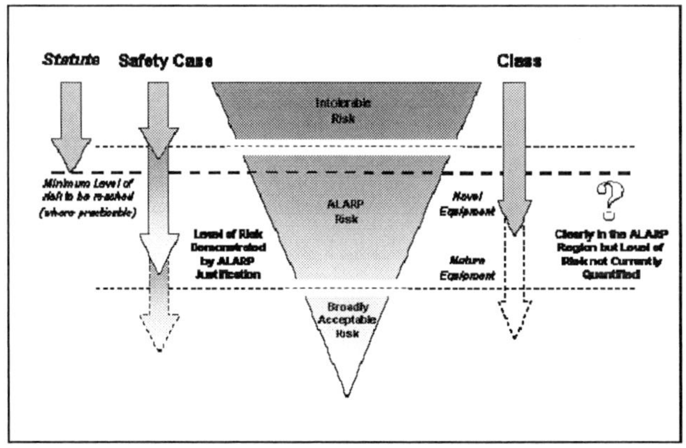

Fig 1 - Comparison of Risk Levels given by implementing UK MoD Safety Case and Classification

3.2 SELECTING THE CLASS SOCIETY

The challenge that is frequently faced by navies or designers when given a free hand in selecting both the class society, or the Rule set to be applied, is understanding the issues that need to be considered. For example, we have a number of examples of navies issuing ship specifications that leave the choice of class society entirely to the designer without guidance as to any preference that the navy has in the application of Class. Of note is the comment in Ref [9] that the Canadian Navy tried to act as class society neutral for the Joint Support Ship but found this impossible to achieve.

In the extreme we can easily end up with a similar situation to that encountered by the US Coastguard (USCG) 'Deepwater' project. At the US Senate enquiry following the problems and major cost overruns experienced, the testimony of Richard L. Skinner (Inspector General US. Department of Homeland Security) [8] noted that the contract had allowed the contractor to select standards without comparing them to those which the USCG was familiar with and then allowed them to concess against these standards when the contractor required, without reference to the USCG.

This situation also arises in the merchant shipping business and while owners will not always specify Class, major ship owners frequently do and will pay a premium on ship costs to have their preferred class society involved.

3.3 DIFFERENCES BETWEEN MERCHANT AND NAVAL CLASS RULES

In order to assist a navy in understanding the differences between Rules for merchant and naval Ships, Lloyd's Register has produced a comparative report that identifies the differences and categorises them into major, intermediate and minor impact.

Differences with a major impact include:

- a Concept of Operations statement is required for various machinery systems
- systems are categorised as being either Mobility, Ship Type or Ancillary, with varying requirements for Classification
- mobility and certain Ship Type systems will require formal Failure Mode and Effects Analysis to be undertaken.
- identification of transverse issues specified for a project by the Navy for shock, EMC, Vibration, ambient conditions, *etcetera*
- additional smoke and toxicity test and verification requirements for cables and piping not in enclosures
- additional tests to demonstrate suitability for environment in which equipment is installed
- possible changes to IP rating of equipment
- additional requirements for prime movers, including short term high power operation and effective means of adjusting shaft rotation speed
- equipment operation at a static heel angle of 30deg inclination
- increased scope of Lloyd's Register verification due to ship type and ancillary categories, requires Lloyd's Register certification for systems not normally certified.

The need to determine the acceptance criteria that requirements will be judged against is therefore an important step in defining the procurement process, which needs to be taken during the concept phase of a naval project. Those involved in the project therefore need to actively engage with a number of potential class societies early on in the process. By understanding the expertise and resource class societies can bring to a project, one that will meet both the technical assurance and political requirements of the project can be selected.

4 CHALLENGES ADDRESSED THROUGH CLASS

In undertaking classification activities for these projects some very clear issues have been exposed. These include :

- the Classification Rules are not the only standards that merchant ships comply with. Others include IMO conventions, resolutions and guidance; flag state requirements; and IACS Unified Requirements
- Commercial off the Shelf (COTS) equipment for ships is not supplied solely with QA type certification
- Classification has a set of operational assumptions that can be inappropriate for naval operations
- navies, even when directed to, cannot fully comply with SOLAS
- statutory legislation is applicable to naval ships in some parts of the world
- Naval Authority requirements need to be laid over any other technical standards adopted
- civil liability exposes navies to a duty of care that needs to be demonstrated through an auditable process
- Classification is a holistic process that needs input from other duty holders where the class society is not directly involved with their domain activities (such as magazine safety and damage control)
- how levels of safety achieved by compliance with technical standards are defined in terms recognisable to a Safety Case philosophy
- cor certain naval aspects, Naval Engineering Standards will need to be applied directly.

Looking in more detail at some of these points, we can see some examples of issues arising that need to be addressed.

4.1 OPERATIONAL ASSUMPTIONS

It is quite rightly commented that, upon leaving port, merchant ships normally accelerate to a particular speed, at which they remain for the majority of the voyage. However, naval ships will vary speed frequently – sometimes almost continuously – in order to match operational requirements on passage. These changes mean that naval ships' machinery must be able to accommodate these load changes, while remaining reliable. Equipment is therefore likely to have very different usage profiles and the safety aspects addressed by Class will need to recognise that duplication of equipment may not be adequate to give the desired levels of simple redundancy. To identify these issues a formalised Failure Modes Effects Analysis (FMEA) is required for essential systems (in Lloyd's Register's Rules Mobility and some Ship Type systems).

4.2 OTHER STATUTORY COMPLIANCE REQUIREMENTS

Other than the well recognised requirements documented in Classification Rules or in SOLAS and MARPOL, others are embedded in either a class society's guidance or IACS's Unified Requirements (IACS URs). IMO's Resolutions, such as access arrangements or ballast water management, may also be relevant. Which requirements are applicable, whether they are to be applied and if only partial compliance is required, needs to be identified early in concept development if implications for a vessels' General Arrangement are to be understood. In the UK MoD's Acquisition Management System (AMS), there is a clearly defined need for a project to undertake a Strategic Standardisation Implementation Plan (SSIP). Thus process was the subject of reference [1]

4.3 INTEGRATION OF OTHER STANDARDS

In order to assess the applicability of standards within any naval project, a process map is useful . The one shown in Fig 2 has been used by Lloyd's Register for a number of years with significant success. Fundamentally, any design aspect of a naval vessel has to satisfy two criteria; it must meet its design objective and be demonstrably safe. For design aspects, the design objective sits somewhere along a spectrum related to Primary Capability. For those aspects at the lower end of this spectrum, compliance through commercial standards such as Classification Rules, SOLAS or other Statutory Standards, IEC or ISO requirements will be satisfactory.

For aspects that are distinctly naval, such as weapon integration or magazines, then naval engineering standards are most likely to be the single source of safety requirements.

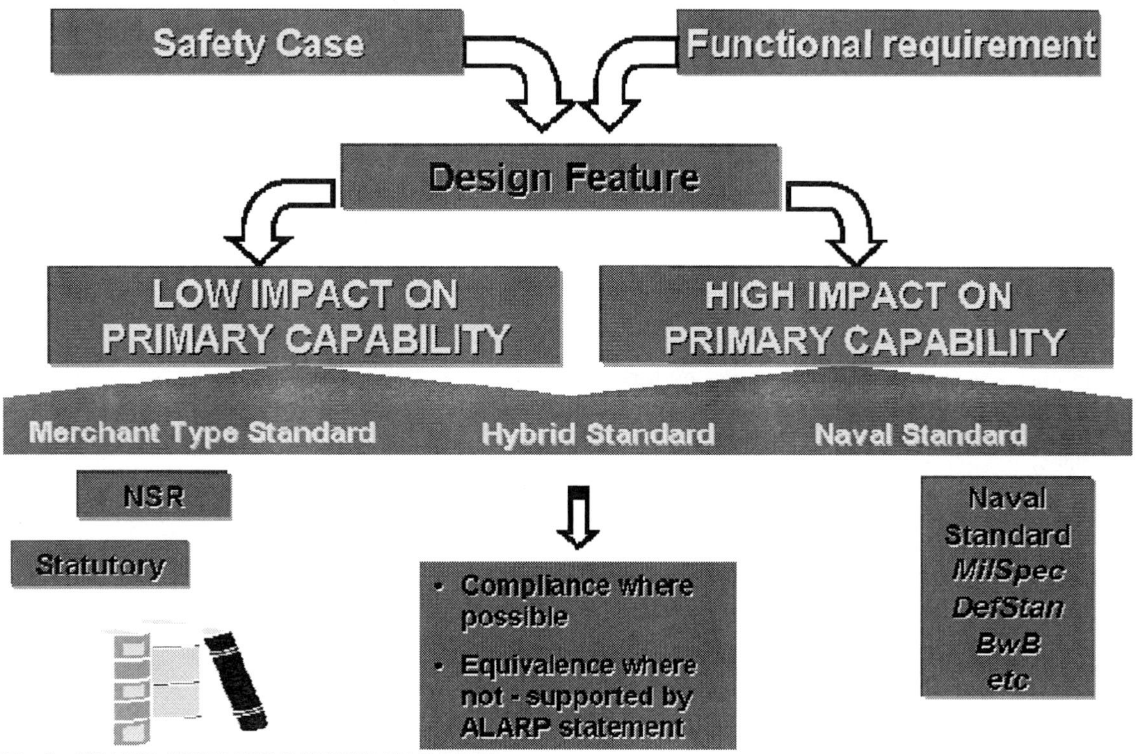

Fig 2 - STANDARDS SELECTION PROCESS

Irrespective of the source of the adopted technical standards, simple compliance with the corporate knowledge they embody will be the most cost effective approach. However, all standards incorporate a set of operational assumptions and if these are incorrect for the Concept of Operations for the vessel, then it becomes necessary to demonstrate compliance through equivalence or by engineering judgement as part of the design Safety Case.

If this initial assessment of the applicability of standards is done during the concept stage of the project, then instructions for the design process in terms of a 'Tailoring Document' can be developed. This document will also allow for different ambient conditions and both external factors and internal control limits to be identified, as well as the design aspects that they will impact on. A typical section from a tailoring document is included in Annex 2.

4.4 NAVAL AUTHORITY COMPLIANCE/ NAVAL SHIP CODE

Full compliance with statutory requirements is generally not possible for navies, especially with regard to external control of safety management systems. In addition, as has already been noted, conventions such as SOLAS and MARPOL have been developed based on assumptions that apply to merchant ships.

To make SOLAS work for a naval ship, other than accepting the various non-compliances and simply stating them not to be applicable, the only route so far has been to use the Alternative Design Methodology [10] introduced in the 2001 edition of SOLAS and demonstrating that the 'Goals' in the various sections have been achieved. This process is the basis of the Safety Equipment and Arrangement notations in the Lloyd's Register Naval Ship Rules.

In response to this challenge, NATO navies decided to develop a Naval Ship Code in 2004, which emulated SOLAS, but did it in such a way that the relevant Goals that a naval ship should meet are identified and a process of verification is set out. This 'Naval SOLAS' is now nearing completion of the major sections (structure; stability; machinery; electrical systems; fire; and escape, evacuation and rescue). The current draft of this can be found at www.inso.co.uk or www.nakmo.co.uk, under 'Library'.

A number of navies have indicated their intention to adopt it and to add country specific issues in the same way that flag states can have additional requirements to SOLAS.

This alternative approach will now offer a means to those navies that have a Naval Authority process to use the Naval Ship Code and country specific requirements, while other navies can select the scope of statutory compliance using Lloyd's Register's Safety Equipment and Arrangements notations.

These currently address:

- Fire (FIRE)
- Lifesaving Arrangements and Equipment (LSAE)
- Escape and Rescue (ESC)
- Safety of Navigation and Communications (SNC)
- Pollution (POL)

In addition where FIRE, LSAE and ESC have been integrated such that common criteria have been applied, the Escape, Evacuation and Rescue (EER) notation can be achieved. Finally where a navy needs to demonstrate that a vessel complies with its own Flag State requirements, such as the UK Maritime and Coastguard Agency (MCA) or Transport Canada, the * can be applied as a suffix to the notations.

Further development of the Naval Ship Code will complete the work and once ratified by sufficient NATO navies, it can be adopted as a STANAG rather than remaining as an ANEP, supported by certain navies and a number of classification societies.

5 SUPPORT SHIP CLASSIFICATION

Using the ideas indicated above, it is worthwhile then to consider the process of selecting a suitable suite of standards for a naval support ship and to consider what decisions a navy needs to make early on so that designers can work efficiently.

5.1 CONCEPT OF OPERATIONS

All following activities need the navy to develop and agree the ConOps statement. The NSC has a very good template to use and is included in Annex 2.

5.2 CLASSIFICATION NOTATIONS

There will be a large number of permutations of possible notations depending on the ConOps statement. Ones related to areas of operation (Sea Area, Ice, Winterisation, *etcetera*) will be straightforward to select; criteria for operational capability (RAS, LMA, Redundancy, DP, *etcetera*) can be addressed if a clear statement of performance is defined, bt it needs to be noted that Class will not set the performance; and demonstration with statutory compliance (SOLAS and MARPOL aspects) can be selected from the Safety Equipment and Arrangement options.

At a ship type level, the NS1, NS2 and NS3 notations are intended for frontline vessels, whilst the NSA notation recognises that support vessels may operate in close support to frontline vessels and be in 'harm's way', but that they are not a combatant.

For fundamental aspects of a support ship, determining if the NSA notation is needed is more complex. A decision process diagram in Fig 3 below shows the major questions to be considered

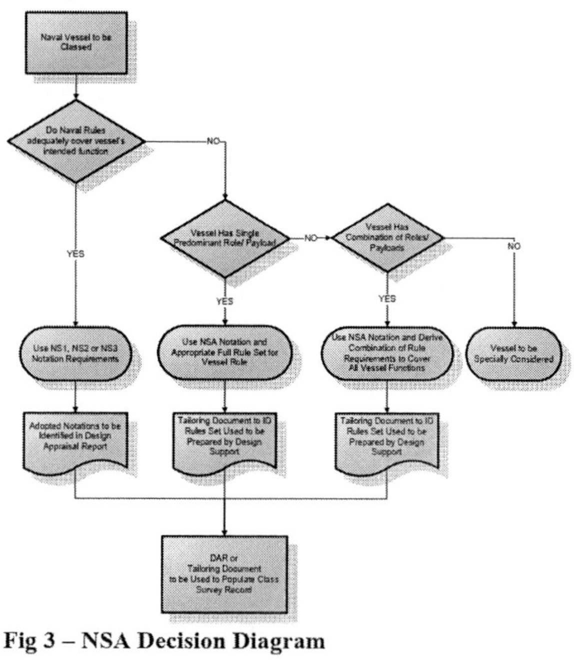

Fig 3 – NSA Decision Diagram

If the NSA notation is appropriate for the vessel, then the associated requirements from other Classification Rules will need to be selected and recorded in the tailoring document. Guidance for this is shown in Fig 4.

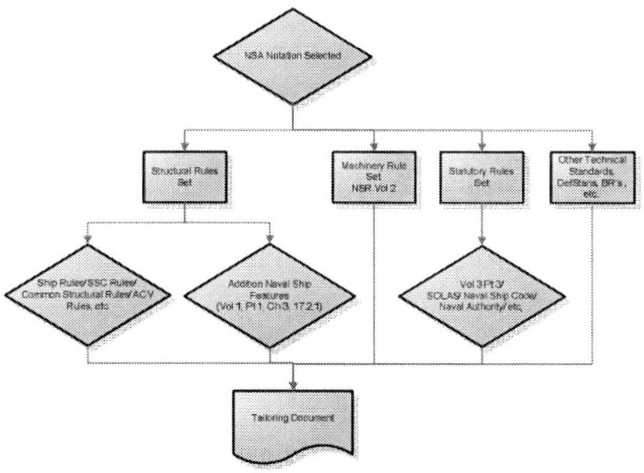

Fig 4 - Selection of Rules for Tailoring Document

Issues that arise at this stage include the need to understand the applicability of the external Rule requirements identified. For example, a medium sized fleet tanker, if likely to be in excess of 150 metres, would need to adopt the Common Structural Rules (CSR) [5] for the cargo tank section. However, if the vessel is of RoRo configuration for the carriage of vehicles as well, the CSR are not applicable as they only deal with traditional tanker arrangements. In this latter case, suitable rules for the tanks that will contain the 'cargo oil' need to be identified and agreed. These will include considerations of how suitable cofferdams will be allowed for, arrangements for double hulling and issues such as tank access for rescue of injured personnel [6].

5.3 USE OF COMMERCIAL OFF THE SHELF EQUIPMENT

Whilst the use of Commercial of the Shelf (COTS) equipment is a clearly avowed intent of navies, and the Naval Ship Rules also aspire to this objective, there are a number of aspects of operational requirements and naval practice that need careful management if undesirable end results are to be avoided.

This is most clearly seen by some of the differences in ambient conditions that are defined in the Naval Ship Rules for aspects such as angles of inclination, IP ratings, operating temperature ranges to deal with close down scenarios and overpower operation. These are generally referred to as 'transverse' issues and these need to be lowed down to equipment suppliers if equipment is to operate in these conditions.

In order to manage this, at an early stage of the project when equipment procurement is being initiated, a certification matrix that captures the transverse issues, applicable acceptance standards and means of demonstrating compliance need to be set up.

An extract from a generic certification matrix is included in Annex 3.

The overall costs of the certification should not be more than a commercial ship incurs, as processes are the same and it is only the criteria that are different.

6 DEALING WITH NON COMPLIANCES

As identified previously, one of the main arguments for not using 'prescriptive standards' is that they constrain innovative design. If this were true then many of the innovations that are seen regularly in merchant shipping, such as the size and configuration of passenger ships, wing-in-ground effect (WIG) craft, marine evacuation systems, alternative propulsion methods (water jets/ gas fuelled ships), *etcetera*, could not have been introduced.

The means by which these alternative solutions have been accepted by class societies and flag states is by application of a rigorous safety engineering judgement, normally including aspects of a Safety Case. As previously mentioned, for SOLAS, Regulation 17 in Chapter II-2 provides specifically for Alternative Design Justification [10]

This sets out a clear mechanism whereby a designer can demonstrate that their design meets the 'Goals and Objectives' of SOLAS.

For the Naval Ship Rules a similar arrangement exists, which allows further alternatives to be applied, to the extent where, if Lloyd's Register is still unconvinced of suitability but the navy will take on the risk, it can be ring-fenced and exempted for Class. The four options are:

- interpretation
- exception
- alternative/ additional standard
- exemption.

They are defined as:

- **interpretations** are applied where strict compliance with a Rule requirement cannot be achieved but the proposed arrangement clearly satisfies the intent of the Rules.
- **exceptions** are non compliances with rule requirements where Lloyd's Register, with the agreement of the owner, naval authority and prime contractor are satisfied that the arrangements are an acceptable alternative. Exceptions arise at both design and construction

stages; they are a formal record of the agreement and any mitigation which might be in place and which will form part of future surveys. At the design stage, these may be captured in a 'tailoring' document.

- Where more major **'alternatives'** or '**additions**' are proposed, such as the use of an alternative standard to replace a LR Rule chapter or the use of an standard in addition to LR Rule requirements, an (NS) character is appended to the notation. Lloyd's Register, with the agreement of the project and Naval Authority, are to be satisfied that the arrangements are an acceptable technical alternative or addition. A tailoring document or design statement is required to define the alternative proposal in advance of design appraisal.
- Where a departure from Rules is not agreed by Lloyd's Register and an **exemption** is required, by the Naval Authority senior management or personnel at Board level must confirm this.

These mechanisms then provide a clearly transparent process where the inherent risk management provided by the technical standards has been followed through and the appropriate management level made aware of the resulting decisions. This information can then be reflected in operational practice to ensure that an agreed level of safety performance is achieved throughout the ship's life.

Examples where these processes have been used include the use of naval pattern lifesaving equipment where the SOLAS pattern is not practical; a corvette where significant redundancy in electrical generating equipment has negated the rule requirement for an emergency generator; non-fire proof vent trunking accepted as crossing the fire boundary; and where undersize air pipes on ballast tanks have been fitted. Each of these examples demonstrate a compromise between the recognised solution that commercial standards would apply against and the operational impact that their imposition would cause with vessel design. Many more exist and have been managed by Lloyd's Register in dealing with the vessels it has or is classing and the process has been found to achieve the required degree of robustness in assuring safety without imposing unacceptable prescriptive solutions.

7 FUTURE DEVELOPMENTS

Having clearly identified and developed a need for independent third party assurance showing that naval vessels comply with an appropriate and practical suite of technical standards, the robustness of the regulatory processes within navies now needs to evolve such that it becomes the normal way of working, rather than a project-specific decision. This will help avoid the current problems associated with each project trying to reinvent the process to its own definitions. When this is in place we will start to see the real cost benefits of having a 'standard' process, rather than one being customised each time.

With the introduction of the Naval Ship Code, we have this opportunity, but navies will need to put some country specific perspectives to one side and cooperate and agree minimum levels of safety for naval ships, and accept that safety management is a discipline rather than a tradeable option. However, one risk with the Naval Ship Code is ensuring that it remains firmly linked to SOLAS and is not seen as means by which navies perpetuate naval standards but under a 'commercial standard' guise.

8 SUMMARY

In the years since Lloyd's Register first introduced its Rules and Regulations for the Classification of Naval Ships, a good understanding now of how safety assurance practices from the commercial environment can be applied to naval ships.

It is clear that through the application of technical standards, irrespective of their origin, there are important lessons to be learned, that are ignored at a project's peril. Expressing these lessons as hazards, highlights the need for them to be addressed and not ignored on a low risk justification; safety lapses later in a project's life due to misinformed judgements will incur disproportionate cost and disruption in the navy.

Application of the Classification process provides a cost effective means of engaging effectively with a peer review process that is not swayed by commercial factors, but does understand the need for pragmatic solutions to be devised.

Continued benchmarking of process against commercial practice and ongoing monitoring of changes to safety practice will ensure that navies achieve the objectives of being at least as good as statutory legislation, but not constrained as such that they cannot meet their primary objective of delivering their commitments to a nation's defence policy.

9 REFERENCES

1. Selection and use of standards for naval ships, Norman Rattenbury, INEC 2004
2. Lloyd's Register's approach to naval ship classification, R V Pomeroy,Lloyd's Register of Shipping, London, RINA 2002
3. Application of Classification Rules to Combat Ships: The Italian Experience, Boccalatte/ Simone/ Dogliani/ Vaccarezza, INEC 2004
4. Classification of Naval Ships – ANZAC Class, L Petersen, INEC 2004
5. Common Structural Rules for Double Hull Oil Tankers, April 2006
6. Tank Access Arrangements (IMO)

7. The Assurance Afforded To A Risk-Based Naval Authority Certification Regime By Commercial And Naval Class, Guy Roberts & Andy Smaller, BMT Isis Ltd,, Robert Simpson, Lloyd's Register, UK, Mike Plaskitt MOD, RINA 2005
8. Statement of Richard L. Skinner Inspector General U.S. Department of Homeland Security before the Sub-Committee on Oceans, Atmosphere, Fisheries and Coast Guard Committee on Commerce, Science, and Transportation, UNITED STATES Senate, The United States Coast Guard's Deepwater Program, February 14, 2007
9. Canadian Forces Joint Support Ship – Systems Requirements Definition and Validation, David Morris & Andrew Curran RINA 2007,
10 IMO Convention for the Safety of Life at Sea, Chapter II-2, Regulation 17

10 AUTHORS BIOGRAPHY

Robert M. Simpson, CEng, MBA, FIMarEST. - A former seagoing engineer, presently Naval Business Development Manager for Lloyd's Register (LR), responsible for the development of Lloyd's Register services to naval clients. He is currently involved with all of the UK MoD's major naval projects along with various MoD Naval Authorities as they develop a robust internal regulatory process modelled on international and national statutory legislation. Also working with a number of other navies as they adopt Naval Classification as their assurance methodology

7 Escape, Evacuation and Rescue				
8 Radiocommunications				
9 Safety of Navigation				
Chapter 10 Carriage of Dangerous Goods				

Survey, Maintenance and Disposal philosophy

Survey philosophy	*(overview of survey and inspection philosophy)*
Survey schedule	*(survey cycle and scope of survey if different from that laid down in the present Naval Ship Code)*
Maintenance philosophy	*(overview of survey and inspection philosophy)*
Maintenance schedule	*(maintenance cycles and depth of planned maintenance)*
Disposal philosophy	*(overview of survey and inspection philosophy)*

<Official seal of the Naval Administration>	**CONCEPT OF OPERATIONS STATEMENT**	Lloyd's Register

Particulars Of The Ship

Ship Name — *(pennant number and name)*

Class — *(name)*

Date last updated — *(date)*

The Owner

defines the ship details, role and extreme threat survivability and agrees the Ship Type, foreseeable damage survivability, maintenance philosophy, principal standards and environmental conditions.

Signed

Name

Position

Address

Date of Signature

Official Seal

The Naval Administration

agrees the ship details, role and extreme threat survivability and defines the Ship Type, foreseeable damage survivability, maintenance philosophy, principal standards and environmental conditions.

Signed

Name

Position

Address

Date of Signature

Official Seal

Project Name Concept of Operations Statement

Primary and secondary roles

Primary Roles	*(high level overview of primary role in sufficient detail for standards to be selected and the design completed)*
Secondary Roles	*(high level overview of primary role in sufficient detail for standards to be selected and the design completed)*

Ship Attributes

Ship Design Life	*(years)*
Length Overall	*(m)*
Breadth Overall	*(m)*
Displacement (Light)	*(te)*
Displacement (Deep)	*(te)*
Speed (maximum)	*(knots)*
Cruise speed	*(knots)*
Range at cruise	*(nautical miles)*
Endurance	*(mission length in days)*
Cargo / Payload	*(weights, volumes and locations)*
Emergency Loading	*(weights, volumes and locations)*
Embarked personnel	Crew: Embarked forces: Special personnel: Casualty allowance: Passenger allowance: Persons carried in an emergency: *(numbers by type, persons carried in an emergency is an estimated practical limit)*
Accommodation	*(space, access, facilities)*

Project Name

Concept of Operations Statement

Survivability

Scenario reference number	Damage Extent DCA Limited DCB Moderate DCC Severe	Damage Location DLI Internal DLSI Specific Internal DLE External DLS Side DLOB Outer Bottom	Vulnerability VB Basic VM Moderate VN Naval	Post-damage Capability PC1 Safe Abandonment PC2 Float and Move PC3 Operational	Recovery Philosophy RPB Basic RPI Intermediate RPA Advanced	Supplementary Notes Relevant additional notes for clarification
1						
2						
3						
4						
5						
..						
n						
Additional notes	*(CBRN requirements, damage control philosophy)*					

Environment

A - Meteorology and climatology (above surface)	
Wind	*(maximum Beaufort Force or speed for operation and for survival)*
Precipitation	*(if specifically required, eg Tropical Storm)*
Air temperature - high	*(specify eg. mean daily max)*
Air temperature - low	*(specify eg. mean daily min)*
Air humidity	*(if not 100% relative humidity at all air temps)*
Visibility	*(if specifically required, eg. night operations)*
Atmospheric pressure	*(if specifically required)*
Solar radiation	*(if specifically required, eg equatorial)*
Electro-magnetic discharge	*(if specifically required)*
Air quality	*(if specifically required, eg operations in coastal waters near deserts)*

Flora and fauna	*(if specifically required, eg in waters of know high activity)*
B - Sea surface (interface)	
Waves	*(Sea State, significant wave height, maximum wave height)*
Waves - other situations	*(if specifically required, eg operations in surf, tidal bore)*
Tide	*(range (height) and maximum speed (relevant to berthing))*
Green seas and spray	*(area affected, frequency)*
Ice navigation	*(if specifically required, eg icebreaking)*
Sea surface quality (floating objects, pollution)	*(if specifically required, eg operations in estuaries)*
Ship motions	Maximum from equilibrium / Period Roll degrees seconds Pitch degrees seconds Yaw degrees seconds Heave metres seconds Surge metres seconds Sway metres seconds *(design values for deviations from the static position)*
Vibration	*(motion induced and wave induced)*
C - Bathymetry and oceanography (below surface)	
Pressure (depth)	*(for specific features in head of sea water)*
Ocean currents	*(if specifically required, eg drift)*
Water quality	*(if specifically required, eg operations in estuaries)*
Sea temperature	
Flora and fauna	*(if specifically required, eg in waters of know high activity)*
D – Geotechnical	
Bottom/Ground conditions	*(if specifically required)*
Banks (inc. canals)	*(dimensions, bottom conditions if specifically required)*
E – Human Caused Environment	
Berthing	*(maximum speed of contact)*
Beaching	*(bottom conditions if specifically required, eg landing craft)*
Docking	*(docking arrangements assumed for design purposes covering intentional docking and emergency docking)*
Towing and salvage	*(bottom conditions if specifically required, eg landing craft)*
Acoustic fields	*(if specifically required)*
Electro-magnetic fields	*(if specifically required)*
Launching	*(assumptions for build)*
Noise and vibration	*(if additional to statutory limits for accommodation and working spaces)*

Project Name Concept of Operations Statement

Operating philosophy

Naval or civil manning	*(type and level of expertise, eg training craft)*
Restrictions and limitations	*(including all restrictions and limitations that are acceptable under the role of the ship)*
Management of hull strength	*(approach to management of structure, eg loading tool)*
Buoyancy and stability	*(approach to management of stability, eg loading tool)*
Machinery	*(watch patterns, operating envelopes, profiles)*
Electrical supply	*(approach to generation and distribution)*
Fire safety	*(including number of safety zones, number of fires to be considered at any one time, the use of boundary cooling, re-entry of spaces and smoke clearance)*
Escape, evacuation, rescue	*(pertinent aspects including issue of personnel equipment, communication, mustering)*
Radiocommunications	*(requirements in additional to statutory requirements eg secure communications)*
Carriage of dangerous goods	*(requirements in additional to statutory requirements eg. embarking at sea, movement on board, breaking out)*

Ship Type, principal standards and authorities

Ship Type	**A, B, B(H), C, C(H), D, D(H)** *(delete as applicable referring to Chapter 1 Regulation 2)*			
NSC Chapter	Principal standards *(and modifications if required)*	Naval Administration	Recognised Organisation	Recognised Organisation delegation *(Full, Partial, Limited)*
2 Structure				
3 Buoyancy and Stability				
4 Machinery Installations				
5 Electrical Installations				
6 Fire Safety				

Introductory paragraph stating basic standard is Lloyds… or other if substituted wholesale… justification for substitute st[d]

3.14 Gas Turbine Alternators (GTAs)

48. The GTAs are being designed and procured to meet Lloyd's Rules and Regulations for Classification of Naval Ships for design, functionality and operational performance, design appraisal and type approval, including construction under survey.

only those standards which are not referred to in the Lloyds rules… and clarify if only partial application. Identify Lloyds involvement eg. plan approval, survey etc

3.14.1 Additional Standards

49. Alternators are to be designed and constructed to appropriate IEC standards.

50. Frequency to STANAG1008 to meet all user requirements.

additional requ's… not standards… Identify Lloyds involvement eg. plan approval, survey etc

3.14.2 Additional Navy/Client Requirements

51. NR Part 2, Ch 2, 4.2.2 – Means is to be Provided to Shutdown the GT before an over critical situation occurs - Vibration detection and warning will be fitted but without an auto shut-down.

3.14.3 Additional Project Requirements

52. None

significant issues (not minor issues for which no written justification is needed) where the rules are not met… justification in section if brief or reference to justification doc[t]/e-mail etc if longer

3.14.4 Exceptions/Non-Compliances

53. NR Part 1, Ch 2, 5.11 - Provision of Suitable Closing Devices on Enclos… enclosure ventilation will have a shut-off device on the supply trunking to the enclosure but not on the exhaust trunking as this air is fed directly into the engine exhaust.

54. NR Part 2, Ch 2, 4.4.2 – Detection and Prevention of Icing Up of Intakes – Detection will be fitted but the inclusion of anti-icing devices is under discussion with the Client.

55. NR Part 2, Ch 2, Table 2.7.1 – Alarms and Shutdowns – The full list of alarms and shutdown will be agreed with RR and the Client taking into account personnel and equipment safety and the RN's fight through capability requirements.

56. NR Part 2, Ch 2, Table 2.8.1 – Condition Monitoring Techniques – CM techniques will be agreed with RR and the Client taking into account equipment maintenance requirements and standard RN practice.

57. NR Part 1, Ch 2, 5.11.4 – Prevention of Contamination of the Machinery Spaces in a NBCD Environment – The GT has a positive pressure ventilation system and contamination of the machinery space is likely if the GTA is run in an NBCD environment.

58. NR Part 1, Ch 2, 5.11.9 - Enclosures are to be Provided with a Means to Enable Gas Purging with a Portable Fan in Addition to the Fixed Ventilation System – This requirement is under review as the likely enclosure firefighting system is water mist.

59. NR Part 1, Ch 2, 5.11.10 – Means is to be Provided to Monit… is not intended to fit such a device.

clarifications of wording/meaning etc… not issues where rules are not met

3.14.5 Interpretations

60. Full Lloyds Acceptance will be during SATs, not during FAT, as it is not intended to FAT the completed GTA.

Details on the procurement strategy where is differs from the basic compliance with NSR Requirements

3.14.6 Procurement strategy

61. The GTA set will be procured in accordance with LR Naval Ship Rules 2005.

62. Additional test requirements will be imposed for;
    Operation following 30eg inclination.
    Cold weather temperature in accordance with environment specification.
    Endurance testing in accordance with ABS requirements

The Tenix Marine LHD Ship Build Program

Keith Malpress
LHD Program Manager
Email: keith.malpress@tenix.com

INTRODUCTION

Some military and industry personnel and observers mistakenly refer to the Landing Helicopter Dock (LHD) ship as a large amphibious truck designed to move troops and equipment in and out of operational areas. While an LHD can be employed in such a role, the observation only serves to illuminate the degree of capability ignorance that surrounds an LHD platform.

The origin of such a perception in Australia may be embedded within the historical role played by the converted aircraft carrier, HMAS Sydney which was widely known for transporting troops and equipment between Australia and South Vietnam during the Vietnam War. This transport perception continued with the introduction into service of the Landing Ship Heavy (HMAS Tobruk) and the two Landing Platform Auxiliary (LPA) ships during the 1980s and 1990s, when they deployed troops on exercises around the coast of Australia because it was often cheaper than alternative methods. Whatever the reasons, RAN ships with amphibious potential were largely employed for over 30 years in sea lift and sea transport tasks and this has served to shape the Australian perception of amphibious capability.

LHD Functionality

The two ship Tenix-Navantia LHD mission system has the potential to profoundly change the face of the ADF and its concept of operations. An immense capability and potential complexity is resident in these 4th generation ships. Each LHD will be capable of accommodating, transporting, launching, commanding, protecting, hospitalising, resting and sustaining a large military force for a significant period in a hostile, uncertain or permissive environment. Indeed, the LHD will provide the means for the ADF to exploit technological advantages and apply them in the difficult littoral interface of sea, land and air. The LHD mission system will empower an ADF Task Force to apply unified effects at a time and place of political choice, entirely independent of shore infrastructure.

The Tenix Navantia LHD

JP2048 Phase 4A/4B - the LHD component of the Amphibious Deployment and Sustainment capability build is a demanding program that is being managed by Tenix Marine in close partnership with DMO and the Spanish designer and shipbuilder, Navantia. The LHD program will further establish Tenix Marine as a leader in naval systems integration as we move into building amphibious platforms and capability, following on from our successful ANZAC Frigate Program.

Aim

The aim of this brief paper is to provide an overview of the LHD Ship Build Program.

THE TEAM

Tenix Marine

Tenix Marine is a prime ship systems integrator in the Asia Pacific region. Our focus is on the delivery of cost-effective solutions to complex Defence and commercial projects for new builds and associated support activities. With shipyards in Australia at Williamstown (Victoria) and Henderson (WA) and in New Zealand at Whangarei, Tenix Marine has a strong capability for new construction, repairs and maintenance and the ability to perform in-service support.

Tenix Marine has a strong project focus demonstrated by its heritage in major projects such as the $1b Frigate Project and the $7b ANZAC Ship Project. Tenix Marine is now in the final stages of delivering one Multi Role Vessel, two Offshore Patrol Vessels and four Inshore Patrol Vessels under Project Protector to the RNZN.

Our Integrated Materiel Support capabilities centre on the supportability and life cycle management of any asset, Defence or commercial, by positively influencing design and build.

Tenix Marine was recently awarded the Victorian Industry Capability Network achievement award for our support and development of Australian industry.

Navantia

Navantia is one of the world's leading ship designers and shipbuilders with 250 years of shipbuilding experience. As well as its shipbuilding capability, in which it has provided new modular construction techniques, Navantia has advanced skills in the full range of engineering required to bring all the elements of a naval mission system together. Navantia is particularly blessed in that it has both the Ferrol and Fene shipyards across the harbour from one another, thereby providing inherent flexibility to Navantia's ship building programs.

Tenix Navantia Teaming

The Tenix-Navantia team formed approximately two years ago after both companies carefully viewed the DMO options for Project JP2048. Tenix uses similar shipbuilding techniques to Navantia, meaning that our existing project management and systems integration approach is easily adapted to this program to offer a low risk solution. Teaming was a natural follow-on as both companies complement each other's strengths and weaknesses. Both companies have exchange staff in place with protocols and mature systems for communication. Importantly, both Tenix and Navantia are committed to the long term support of these ships in service.

Saab and L3

Saab and L3 officially joined the team on 19 June 2007 when the Commonwealth selected Tenix Defence as the Prime Contractor for Project JP2048 Phases 4A/4B. Saab joins the team as the combat systems integration arm and L3 is the communications system team provider.

THE AUSTRALIAN LHD DESIGN

Little reference has been made to this point in time to the 'Australianisation' of the two Spanish LHDs. Australian design standards call for the Spanish Navy BPE to be adapted in order to meet RAN classification and certification requirements. Both LHDs will be built in accordance with Lloyd's naval ship rules and Australian Defence Force military standards. Indeed, there is a heavy emphasis on reducing the vulnerability and increasing the survivability of the platforms in the event that they are employed in a hostile environment.

Significant changes have been made to the Spanish BPE in order to meet higher ADF operational and safety requirements. The following 12 major category changes demonstrate the difference between the Australianised LHD and the Spanish Navy BPE:

- Lloyd's register Naval rules and owner requirements
- Combat system changes to include self-defence, ballistic protection and aviation changes
- New armouries and changes to magazines
- Increased endurance by increasing spaces for food storages and spares
- Environmental (MARPOL) improvements to reflect compliance with fuel tank protection, Marine Green Passport (MGP) and oily water separation
- Habitability improvements to accommodation, hospital facilities, galley, and ablution spaces
- Electrical conversions
- Safety improvements to fire detection, fire fighting and protection systems
- Refrigeration upgrade including a 30% increase to chilled water plants
- Paint adaption to RAN standards
- Tactical lighting in the dock and on the deck which is Night Vision Device (NVD) friendly
- Lifesaving improvements including 100% capability on both sides of the ships

PROGRAM MANAGEMENT OVERVIEW

To achieve the contracted cost and schedule imperatives with the Tenix team dispersed around the globe, three Integrated Product Teams have been established to manage the lines of business. These are:

- Combat and Communication
- Platform
- Acceptance

These projects are in turn supported by functional teams from Mission System Integration Engineering, Commercial, Finance and Project Controls and ILS. The Program Office includes the functions of Operations, Mobilisation and Risk Control.

As Tenix is managing all prime contractor responsibilities a Project Management Office is established in Williamstown, Victoria. DMO, Navantia, Saab and L3 are invited to collocate staff in the PMO. An on-site office is also established within the Ferrol shipyard where a Tenix team will work closely with Navantia. The Commonwealth has observers in this office.

Systems Engineering Overview

Tenix has overall responsibility for integration of the Mission System, acceptance (including overall Verification and Validation), Set to Work & sell off of the whole ship.

In their role as Platform System Designer, Navantia's responsibilities include the integration of the combat and communications equipments into the design of the platform, including 'hotel services'. Verification is to be performed on an "as early as possible" basis, whereby the functionality of platform systems is verified to the maximum possible extent in Spain, with Tenix performing HAT and SAT level verifications on the complete LHD in Australia. Similarly, an Australian Land Based Test Site will be used to pre-integrate and test the combat and communications systems elements, with platform-dependent testing being performed during HAT and SAT.

THE LHD BUILD

Build Strategy Objectives

The ship build strategy for both LHDs is a joint DMO and Tenix decision in response to a raft of competing Government and commercial requirements, not the least of which is value for the taxpayers' dollar. The main objectives of the LHD ship build strategy include:

- An optimum build cost
- A low risk design and build schedule for 2013 and 2015 completion of LHD 01 and LHD 02 respectively
- A limited Australian infrastructure development program
- Low risk to product performance, quality, schedule and build cost
- Sufficient capacity for Tenix and Australian industry to support concurrent module construction of the AWD Project.

Low risk and on time delivery are paramount amongst these objectives. A key enabler of these objectives is the absence of a break in Navantia's production line, as construction of the Australian LHDs will immediately follow the roll-off of the Spanish Navy BPE.

Work Share Interface

The optimal work share interface for the build was determined to fall between the construction of the hull and flight deck (individually referred to a Zones 1 to 7) on one hand and the superstructure (or Zone 8) on the other. The work share interface calls for Navantia to build Zones 1 – 7 in Ferrol and Fene, while Tenix builds Zone 8 in Williamstown.

While the Zone 8 superstructure of each LHD may not appear to be overly significant to some, it should be noted that the LHD superstructure houses the vast majority of the high end technology resident in the ship. This includes the combat management system and communications systems whilst land, sea and air operations are directed via the ships operations room, two embarked forces operations rooms, intelligence facilities and the air traffic control facilities, which are all resident in the superstructure. Indeed, construction of Zone 8 in Australia brings with it those skills in which future upgrades of combat and communications systems are most likely to be needed.

Production Flow

Tenix and Navantia have established a three phase production process for each LHD. Within each ship build a concurrent production flow will exist within and between Spain and Australia. The most evident example will be the Tenix construction of the Zone 8 blocks during Navantia's construction and outfitting production cycle.

Schedule

The build schedule for both ships is similarly concurrent in nature to facilitate maximum effectiveness of the Ferrol, Fene and Williamstown shipyards. For example, in Spain the consolidation of blocks 1 – 7 of LHD 01 will be immediately followed by the consolidation of blocks 1 – 7 of LHD 02. In a similar manner the delivery of the LHD 02 hull to Williamstown, will immediately follow the scheduled acceptance of LHD 01.

Most significantly, Navantia is scheduled to commence cutting steel cutting on LHD 01 on 23 September this year. Tenix is scheduled to commence cutting Zone 8 steel on 16 July 2010. LHD 01 hull arrives in Williamstown on 23 July 2012 and the ship is scheduled to be accepted on 19 January 2014. LHD 02 is scheduled to be accepted on 19 September 2015.

Heavy Lift Vessel Transportation

The transportation of both LHD hulls from Spain to Australia, a distance of 12,500 nautical miles, is a massive logistic exercise. A range of options was examined, however the most cost effective and efficient manner is to transport the 19,000 tonne hull to Australia on the back of a heavy lift vessel (HLV). Our task is comparable to the 59,000 ton Thunderhorse Offshore Unit which was transported from South Korea to Texas via the Cape of Good Hope in 62 days in 2004.

In comparison, the Navantia LHD hulls will move via the Suez Canal and Australian east coast to Melbourne where they will be offloaded in Port Philip Bay – which incidentally is deep enough to provide the minimum 21.5 metres of water depth for the docking down procedure.

Williamstown Facilities

The most immediate challenge for Tenix Marine centres on a requirement to retain sufficient work in the Williamstown dockyard in the two years between completing work on Project Protector and commencing steel cutting for LHD 01 Zone 8 block construction on 16 July 2010. A number of prospects are currently under consideration to ensure production capabilities and resources are available for LHD cut steel date.

All LHD Zone 8 blocks will be built within the existing Williamstown facilities. Consolidation will then occur alongside Shenandoah Wharf with the blocks being lifted by a crane similar to that used on the ANZAC project.

Williamstown will also be the site for LHD crew training, training development and the Combat System Land Based Test Facility.

When structurally complete the LHDs will be moved to Nelson Pier for final outfitting, final testing and Set to Work evolutions before undertaking sea trials off the east coast of Australia.

INFRASTRUCTURE

The current division of work between Tenix and Navantia provides the Commonwealth with price and schedule advantages. It is also significant in that no new infrastructure is required in Spain and only minimal infrastructure investment is necessary in Williamstown, eg:

- Demolition of Reid St and Dockyard Pier
- Installation of an automated Profile Cutting Line
- An extension module will need to be applied to the Unit Blast Facility (UBF) door in order to accommodate larger blocks
- A crane will be required to lift the larger than normal LHD blocks in Zone 8.

CONCLUSION

In conclusion, the LHD program is an extremely ambitious and demanding program that will deliver to the RAN a capability likely to redefine the face of, and concept for ADF operations. Tenix is confident that the very capable team we have assembled, working closely with DMO, will assist us to deliver this capability and program on time, and also on budget.

ABSTRACT

JP2048 Phase 4A/4B – the Landing Helicopter Dock (LHD) ship component of the Amphibious Deployment and Sustainment capability will profoundly change the face of the ADF and its concept of operations in the next decade. This paper examines the demanding two ship build program that is being managed by Tenix Marine in close partnership with DMO and the Spanish designer and shipbuilder, Navantia. The combat systems integration and communications system team expertise is being provided by Saab and L3 respectively.

Innovative Gear Solutions in Naval Propulsion Systems

Franz Hoppe
RENK Aktiengesellschaft, Augsburg, Germany
franz.hoppe@renk.biz

1. INTRODUCTION

What could innovation in the context of marine gear technology mean? Viewing at traditional propulsion of larger vessels such as corvettes, frigates, anti submarine warfare ships or air defense combatants, an answer becomes fragile because main propulsion gears are supposed to be highly developed since an extended time. There seems no room for further improvements.

However, such statement is invalid as main gearing is to be steadily oriented to demands imposed by enhanced vessel designs impacting continuously propulsion configurations for utmost flexible and redundant operation. The propulsion system aboard a naval vessel has to include features to cope with requirements of state of the art, followed by incremental innovational designs steps which need to be introduced to modern vessel propulsion concepts.

Naval projects for new building vessels are normally asking for the latest technology state. Today, combining propulsion systems are the trend, whether in form of CODAG as mechanical solution or in form of CODELAG as electro-mechanic propulsion. In all cases, innovation with marine gearing has to be translated into trustworthy and reliable designs ever meeting the specifications of newly developed ship technologies.

In addition, damages to main gearing components are being considered as hazardous to the propulsion train because of their vital importance to reliable vessel propulsion performance. By all means, operational failures need to be minimized to practically zero events on all levels. This is only possible by respecting all interfaces in mechanical and electrical ways matching with the environment aboard the vessel. For example, the main reduction gear operates in a vibratory active surrounding, and has to cope with sometimes indefinable external forces.

Overall, the experience in design and production of marine gearings, encountered by numerous applications throughout the world's Navies, provides the fundament for further enhancement of propulsion systems. Below given aspects and examples view into the spread of available gearing configurations and their critical components.

2. DIESEL ENGINE PROPULSION

The propulsion by solely diesel engines may be considered as a traditional configuration for surface combatants such as corvettes or frigates in the 2000 ton class. A number of variants are known, whether straight forward one diesel engine to one propeller via a single stage reduction gear, or twin diesel engine propulsion combining to one propeller shaft.

Twin input to single output main gearings are available in a broad variety including different arrangements of the diesel engines versus propeller shaft lines, see Fig. 1. Structural or redundancy aspect requirements may ask for configurations in T-, U-, or Z designs showing the flexibility of a gear manufacturer in adjustment to the shipyard's prime needs such as engine room layouts, center distances between diesel engines, propeller shaft lines, or diesel engine rotational directions. In addition, all configurations are only limited by the main driver capabilities regarding power and speeds. Thus, up to 16 MW per shaft line are realised with as many as 300 applications world wide.

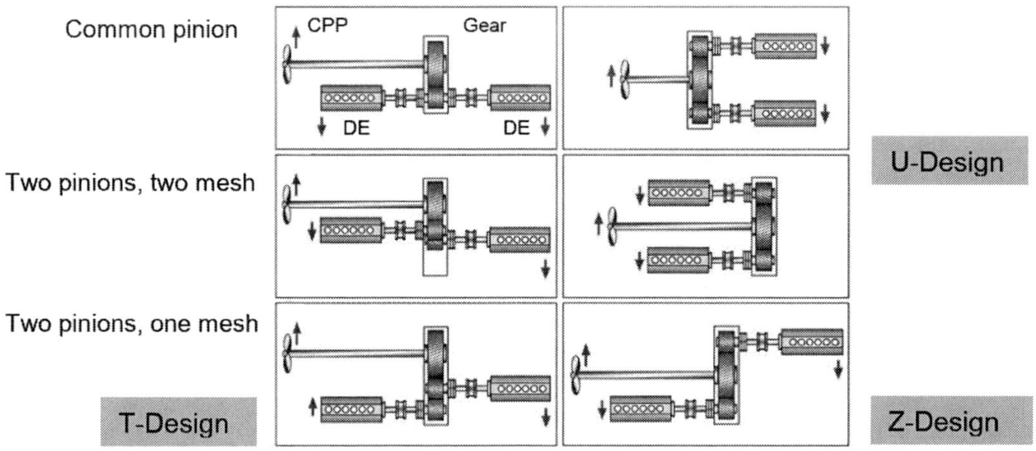

Fig 1: Various designs of diesel engine propulsion systems as to match with different rotations at input vs. output.

With the basic designs as shown, fundamental features are always in common, irrespective of the single arrangement. This is represented with double helical toothed gears for utmost load carrying capacity and lowest radiated noise, combined with integral clutches in fluid coupling or multi disc clutch designs. Finally, sleeve bearings are being used throughout all applications serving for smooth running shafts totally resistant against vibrations and structure borne noise transfer. Compared to alternative anti-friction bearings, these features play an extraordinary role in achievement of reliable and low noise gears.

Fig. 2 shows a typical diesel engine combining gear (CODAD) with fluid couplings at both DE inputs, which include the comfort of smooth power transition between diesel engine and shaft line whilst change over modes from one DE to full power and vice versa. Due to specific engine room space constraints aboard, the couplings are arranged aft of the gear, where the pinion shafts are hollow bored and the diesel engines are connected via quill shafts penetrating the pinions to the couplings. The relatively long and thin inner shafts were designed considering torsional vibrations and their natural frequency behavior. Such investigations need to include the mass/stiffness distribution of the entire propulsion train.

Alternatively, instead of fluid couplings, hydraulically operated multi disc clutches could be applied, see Fig. 3, which provide by means of electro-hydraulic actuation utmost flexibility for quick operation changes between the diesel engines in a broad speed range. However, with this design, the torsional vibrations induced by diesel engine are to be absorbed by additional torsional damping couplings arranged external of the gear train.

Fig 2: 2* 7,5 MW, 1350/250 rpm twin diesel engine gear with fluid couplings connected to torsional input shafts.

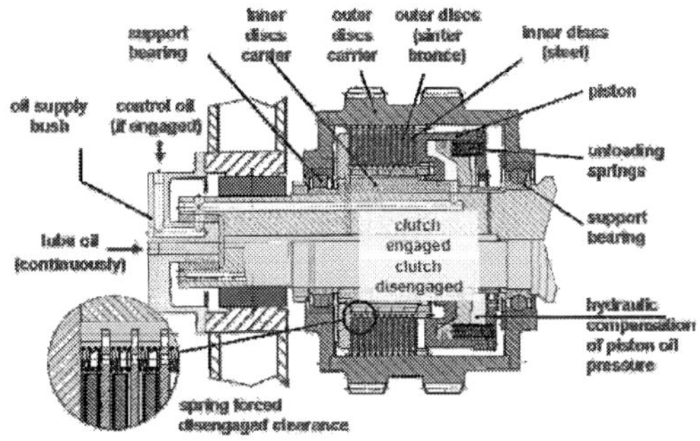

Fig 3: Multi-disc clutch in special RENK design for reduced wear resistance.

3. MECHANICAL COMBINING PROPULSION SOLUTIONS

Further advanced propulsion systems provide arrangements where diesel engines and gas turbines are being combined to a variety of mechanical drive train solutions. Gas turbines are being supplemented mainly to manifest high speed performance in the area of slightly below 30 knots. Today, CODOG (**Co**mbined **D**iesel **O**r **G**asturbine) or CODAG (**Co**mbined **D**iesel **A**nd **G**asturbine) systems are available in different configurations regarding driving engine arrangements and practical operational modes.

The Classic CODOG

The introduction of modern gas turbine propulsion required a two shaft line gear system, with gear trains specifically designed to transmit the gas turbine power at fairly low weights. It had also to transmit power alternatively from diesel engines at lower (cruise) speeds. Thus, the CODOG was borne.

As such, frigate combatants up to 5000 tons displacement require two separate propulsion trains, with one gas turbine each for sprint speed providing around 20 Megawatt, and two diesel engines at four to six Megawatt each for cruise propulsion. Fig. 4 shows a typical gear train schematic of this type. The power is transmitted via double reduction gears to controllable pitch propellers. For most weight efficient gears, so-called "locked train gears" are applied, where the main reduction gear is designed in the split power principle including intermediate torsional quill shafts between the first and second gear stage for load transmission at practically equal load share. By this, one tooth mesh transmits only half of the nominal power, thus providing limited dimensions of the gearbox.

Separate pre-stage gears transmit cruise speed power from the normally inboard to the shaft lines arranged diesel engines, being connected to the main gear train. This causes entrained no load rotation of the gas turbine locked train in the DE mode. Appropriate means have to be taken to avoid excessive noise in this particular mode. At the diesel inputs, fluid couplings are used as vibration dampers. Overrunning SSS clutches, see Fig. 5, provide automatic shift from gas turbine to diesel mode and backwards. For the avoidance of elevated no load speeds to the diesel engine input in gas turbine mode, the related SSS clutch disconnects automatically once the gas turbine is gaining speed.

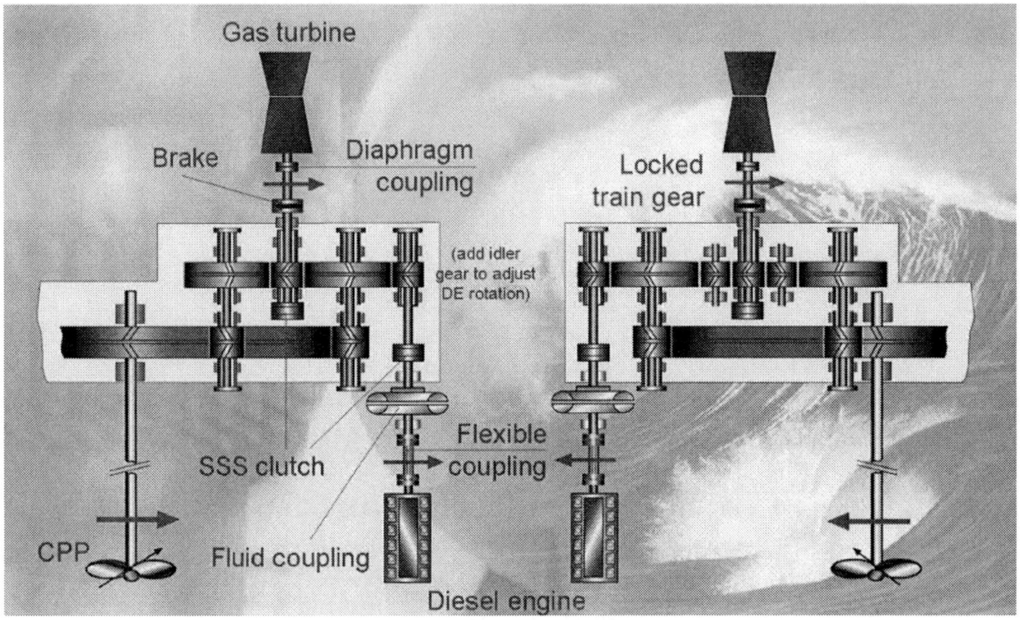

Fig 4: CODOG gear schematic for 22 MW GT and 3,8 MW DE power per shaft line. Starboard with additional intermediate gears for inverse propeller rotation.

RENK realised CODOG propulsion systems in the eighties and nineties in many NATO vessels, still operating today at no lack in performance. The gas turbine locked train gear has to be carefully designed to ensure equal load sharing between the intermediate shafts. Primary gears and secondary pinions are connected via specially configured diaphragm couplings, where the circumferential position between gear and pinion should be adjusted and fixed at high resolution to avoid any incremental load share deviation. Additionally, all gears have to be manufactured by use of most accurate grinding tools. At no doubt, in this very special configuration, tooth accuracies of ISO grade 1 to 2 including appropriate tooth flank corrections are compulsory for safe operation. The noise performance could be improved by application of sophisticated elastic supports to the foundation, where distinctively the natural frequency behaviour is to be considered.

Apart from above aspects, the twin shaft CODOG requires two gas turbines of the 20 MW class for only 3 to 5 percent operational time, followed by high investment and maintenance costs for these driving engines. Always two main drivers are to be operated in parallel to both shaft lines, either two diesel engines for cruise or two gas turbines for sprint speed. In diesel engine modes, in particular, the gear efficiency is considerably reduced to approx. 93 %, because the entire gear train rotates at only 30 percent of nominal speed, and the main gear train is being operated at relatively low load. In any event, and in respect of all features as mentioned, the classic CODOG may be regarded as viable option for surface combatants even today.

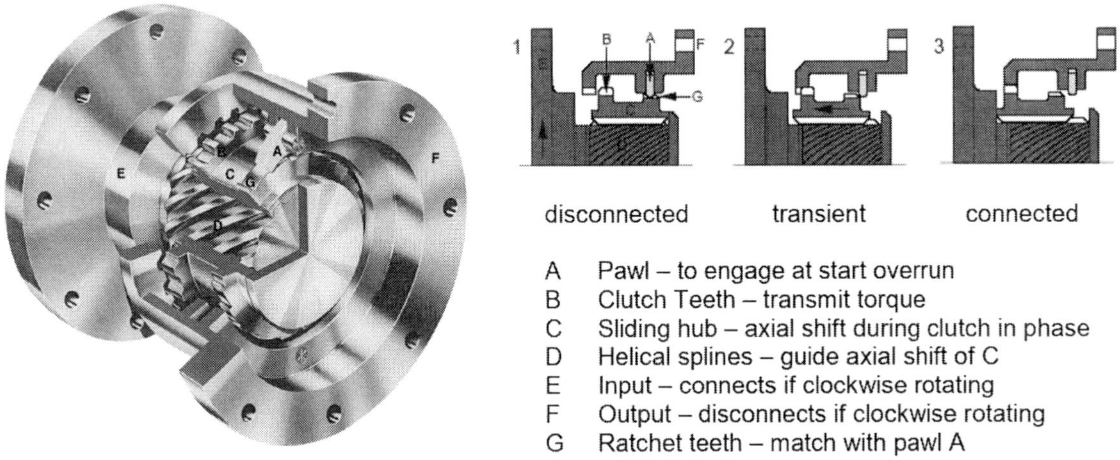

Fig. 5: Working principle of SSS clutch (**S**elf **S**hifting **S**ynchronising, SSS Gears Ltd).

CODAG with Cross Connect Gear

The significant improvement regarding flexibility and power efficiency is to use more than one driver for sprint speed, and, alternatively, only one diesel engine driving both propeller shaft lines at loiter or slow cruise. Necessarily, this leads to a CODAG gear system of today's state of technology, combining one gas turbine and two diesel engines, with an advanced gear train featuring an interconnection between the two propeller shaft lines.

Fig. 6 shows the gear rotor schematic, as it is introduced with the highly sophisticated German Sachsen class frigate propulsion. Due to its convincing technology showing some 15000 operational hours since 2003 aboard the three vessels of this kind, and versatile operability in propulsion modes, it was selected as prime propulsion concept also for the Deepwater acquisition program's National Security Cutter (NSC) of the US Coast Guard.

The two controllable pitch propellers (CPP) are being driven by one central arranged gas turbine and/or one or two diesel engines, reserving some remarkable advantages compared to conventional trains. With four modes of operation defined, the total range from slow to sprint speed is covered at highest economy, effected by strategically distributed multi-disc clutches in the gear train (see Fig. 3). The basic main features are:

1. Three drivers (gas turbine or diesel engines) are to being connected per automatic bridge control actuation to one or both propeller shaft lines.
2. Maximum speed is achieved at simultaneous power input of three drivers (CODAG).
3. Switch over between all operational modes is realised within a wide spread without vessel propulsion interruption.

The cross connecting gear as link between the two main gears distributes the gas turbine power to two intermediate shafts. It does also serve as transmission gear from one diesel engine to the opposite shaft once the „one diesel engine mode" is selected. By this, the most advantageous propulsion with only one diesel engine to two propellers is enabled, covering up to 18 knots speed of the vessel, and allowing for switch over between the both diesel engines to minimise maintenance costs. The diesel engines are arranged inboard of the shaft lines, connected via torsional vibration damping fluid couplings and highly elastic couplings with the main gears.

For optimised propeller efficiency, both main gears include a two stage gear at the diesel engine inputs, where reduced speeds are generated for diesel modes, and at elevated speeds the diesel engine power is adjusted to the gas turbine connected (CODAG mode). Hydraulically operated multi disc clutches in all intermediate shafts guarantee for smooth switch over between the modes.

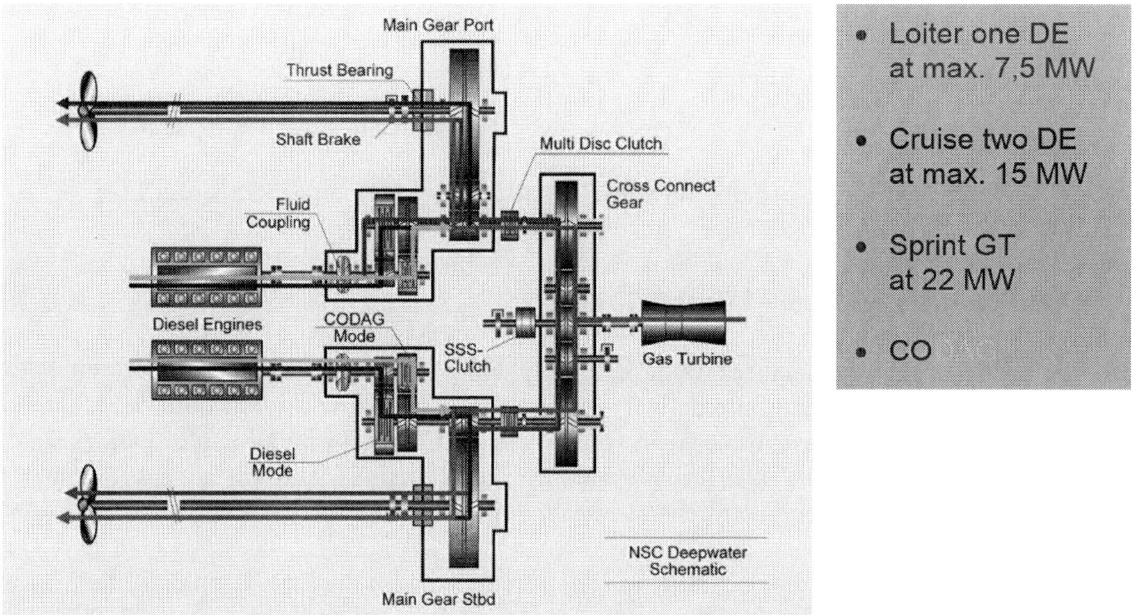

Fig. 6: Schematic of CODAG propulsion train aboard USCG "Bertholf".

Against CODOG gears, the cross connect CODAG arrangement is meeting even improved noise signatures. This is achieved by gears loaded under torque transmission in all modes – an important aspect regarding noise radiation performance compared to the twin shaft CODOG installation as discussed earlier, where in diesel engine mode the gas turbine locked train gears are rotating load free, thus generating unpredictable additional noise by tooth hammering. In general, Fig. 8 shows the development of noise signatures at gear mesh frequencies over the past decades depending on achieved technology enhancements and high accuracy gear grinding to be involved.

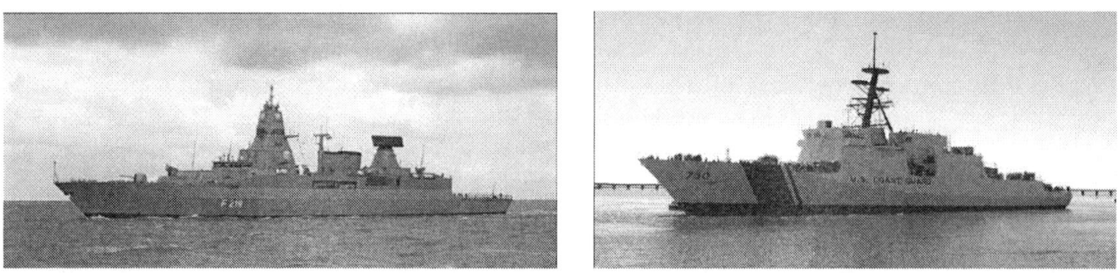

Fig. 7: Frigate „Sachsen“ (left) and NSC „USCGC Bertholf“ (right) with CODAG aboard.

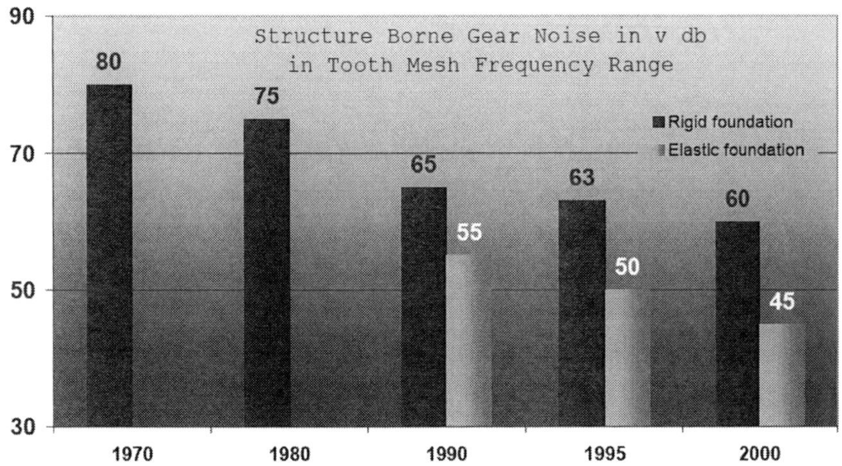

Fig. 8: Gear structure borne noise of RENK gears over three decades.

In particular with the CODAG propulsion principle, new steps had to be taken regarding the adjustment of the relatively continuous gas turbine torque output compared to the diesel engine power at highly vibratory amplitudes. In this context, RENK developed gear controls matching with the digital BUS ship control system. At immediate reaction times to speed variations, operational flexibility is guaranteed. Gear controls comprise locally arranged PLCs with touch screens, combined with the selection of a multitude of input parameters.

4. AN EXCURSION TO EXOTICS

As a matter of fact, gas turbines to water jet drives could develop fantastic speeds. With catamarans, this performance is being supported, in addition, with their low water resistance. In case of surface effect ships (SES), where the ship's superstructure is lifted relative to the water line at high speeds by vertically active air fans between the semi-hulls, the water resistance is brought to minimum. All these design features are combined within the Norwegian Navy Skjold program fast patrol craft of only 55 m in length, comprising two COGAG propulsion modules of 6 MW in each semi-hull, serving for a vessel's top speed close to 60 knots.

The gear arrangement is specifically fit into the environment aboard the vessel, respecting narrow space requirements. As shown with Fig. 9, the two gas turbines are cantilevered one over the other, with the smaller 2 MW engine on top, and the water jet shaft underneath the lower gas turbine module. In a first stage, a planetary gear reduces speed from 20000 to 8000 rpm, to enable the speed limited SSS overrunning clutch switching between drive modes. The 4 MW engine provides power to the gear train directly via its SSS clutch to a pinion in a hollow shaft arrangement, where the torque from the upper engine arrives at the water jet in COGAG mode simultaneously.

From there, the design power totalling 6 MW is transmitted via two gear stages to the output shaft. The entire gear train includes also double helicals as with all sophisticated marine gears. In this respect, the lay out shows advantages, in particular, because the lack of axial forces allow the casing structure to be at lowest weight possible. The high speed shaft bearings of the input trains are of tilting pad type for utmost vibration stability. Both input shafts are linked to the gas turbines by use of diaphragm couplings, to compensate for eventual misalignments, although these are attached directly in most rigid ways to the gear casing.

In summary, the gear systems between the gas turbines and the water jets have multiple demands such as:

- Reduce inputs speeds 20000 rpm and 16000 rpm, respectively, to 1000 rpm water jet output
- Single or twin gas turbine operation
- anti shock resistance up to 5 g above elastic supports
- optimized efficiency grade
- gas turbines cantilevered to gear
- fit into tight semi hull structures
- ultra light weight to meet 3700 kg limit

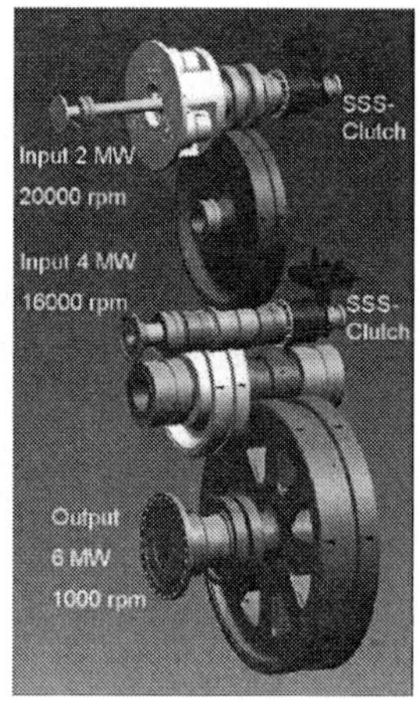

Fig. 9: The COGAG marine gears for Skjold: right rotor lay out, mid arrangement on test bed, left SES catamaran during trials.

The propulsion gears are being operated in single GT mode, or by use of both GT in parallel, respectively, which is called the COGAG (**Co**mbined **G**as **A**nd **G**as) operation. The SSS clutches allow for automatic switch over between the modes, as they engage at overrun and disengage at lowering input speeds. Thanks to the gas turbine's capability serving for a constant torque over a wide speed range, still at reasonable fuel consumption, secondary gear stages to adapt for lower output speeds in single GT modes are avoided.

High speed gears are normally critical with their efficiency grade, because power losses generated by tooth mesh friction and rotational ventilation sum up to extreme values. By use of thin air flow shields strategically placed inside the gear casing, in addition to energy loss optimised tooth geometries, and minimised oil circulation as synthetic oil is applied, it is regarded as success to diminish the full load power loss to 240 kW per gear unit only.

5. HYBRIDS INTRODUCED TO MAIN GEARINGS

Notwithstanding the successful CODAG propulsion newly introduced, the market searches still for alternative propulsion systems. As a result, and involving electric propulsion as part of the "all electric ship" philosophy, some projects are being elaborated today, where frigate propellers are driven by so-called "hybrid" propulsion configurations. Under this title it is understood that the diesel engines are replaced with electric motors whilst maintaining one center arranged gas turbine as prime maximum speed mover. Correspondingly, the CODAG is transferred to a CODELAG system.

From the gear manufacturer point of view, the electro-mechanic ("hybrid") propulsion is regarded as modified version of a cross-connecting gear system. Commonly, the electric motors are arranged in line with the propeller shafts, thus avoiding additional gear sets for speed reduction, and saving noise generating tooth meshes. This pre-condition reserves the gear train to be solely designed for gas turbine power, still encountering a cross connecting gear for the central arranged gas turbine, connected to two main gears providing the secondary speed reduction to the two propellers.

Fig. 10 shows the principle layout as generated for the Italian version of the franko-italian frigate program FREMM.

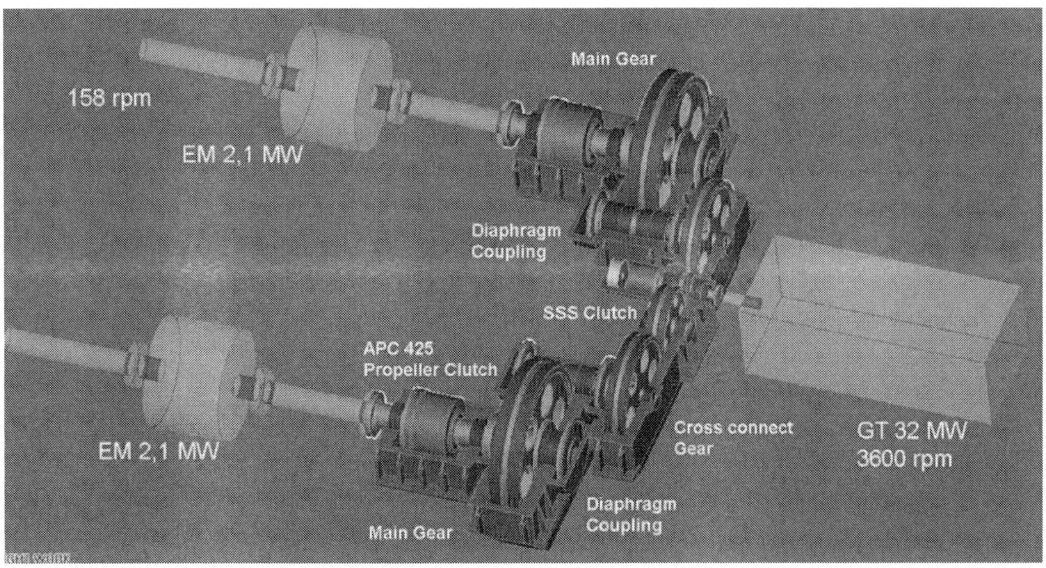

Fig 10: CODELAG gear arrangement for the Italian FREMM project.

The RENK „Advanced Propeller Clutch" (Type APC 425), filed as pending patent, is arranged at the main gear outputs and transmits half the gas turbine power onto the shaft train. The electric motors arranged astern supply the power either individually (cruising speed) with disconnected APC clutch or in connection with the gas turbine (high-speed), the APC clutches being connected.

The APC clutch was especially developed by RENK for this application. It is placed on a foundation separated from the gear system in order to separate vibrations from the gear units during connection and disconnection procedures. The APC clutches are wet running multi-disc clutches, size wise extended by the torque requirement of the slowly running propeller shafts.

The control is working fully automatically and is adapted to the control system of the entire propulsion system.

The CODELAG type propulsion experiences increased attraction, obviously because it shows the ideal combination of cruising with electric power gained from the ship's electric generator system, and high speed propulsion by adding one 30 MW class gas turbine. For this, it has been also selected for the new F125 class frigates for the German Navy, however, in different load share between gas turbine and electric motors.

6. CONCLUSION

Developments in gear technology continue to orientate at increasingly sophisticated propulsion systems. Starting from classic CODOG arrangements initiated in the early eighties, enhanced systems are available. In today's cost and efficiency sensitive times, alternative naval propulsion concepts have been developed, such as CODAG, supported by high speed controls, leading to the CODELAG, the "hybrid" in naval propulsion. RENK main propulsion gears follow the technical needs, based on renowned experience with more than 400 installations aboard fleets of 30 Navies. Extremely compact gears such as high speed COGAG systems are required for specialised vessel designs. The trend with all applications are highly efficient compact solutions at low weights, including low noise signatures, even extinguishing tooth mesh frequencies in audible and structure borne noise. To combine all these factors to a reliable gear system is the challenge in the yet again enlarged market place for surface combatant propulsion gears.

7. REFERENCES

1 Hoppe, Franz; Pinnekamp, Burkhard: Gear Noise – Challenge and Success based on Optimized Gear Geometries. Transaction of AGMA Fall Technical Meeting, Milwaukee, Oct. 2004

2 Weiß, Toni; Hoppe, Franz: High Speed Gears for Extreme Applications in Industrial and Marine Fields. Proceeding Conference Verein Deutscher Ingenieure (VDI), Sept 2005

3 Oster, Peter: Influence of overlap ratio on noise exciting level. Research program of FVA 58/T252, 1985

4 Hoppe, Franz: Innovative Gear Systems for Naval Propulsion. Proceeding of Conference Mecon, Hamburg, 2006

DAD Helicopter-Ship Aid for Maximising Helicopter and Ship Operational Capability, Improving Safety and Usage Monitoring

David Howe
Advanced VTOL Technologies Pty Ltd (AVT)

Abstract

Navy ships need to have the flexibility to manoeuvre to avoid threats and provide expansive operating envelopes for helicopter operations to maximise operational effectiveness. Safe operating limits are required to enable the flight crew and deck crew to perform particular on-deck operations. These include landing and takeoff, traversing the helicopter from the hangar to the flight deck, or use of a particular lashing scheme or restraint system to secure the helicopter. Traditionally RAN limits have been based on simulation for a wide range of conditions and have been conservatively defined using just pitch and roll at worst case wave encounter frequencies. Little account has been made for the effects of ship manoeuvring and list.

AVT is developing a ship based helicopter aid, termed the Deck Availability Designator (DAD). The system couples the Australian Defence Science and Technology Organisation (DSTO) Ondeck helicopter-ship dynamic simulation code with an Inertial Motion Unit (IMU) to provide a real time predictive capability of the helicopters response to actual ship motion as well as an estimate of the loads transferred through the helicopter. By using real-time ship motion data and direct measures of limit exceedence like wheel sliding, wheel lift off or lashing load levels the DAD system will provide an optimal solution for maximising the operational envelope of both the helicopter and ship and will improve flight crew and deck crew safety as well as provide a means of helicopter usage monitoring.

Advanced VTOL Technologies in collaboration with the Australian Defence Science and DSTO have been modelling helicopters on ships for many years. This paper discusses the attributes and benefits of the proposed system.

1. Introduction

In this paper the current Australian Defence Force (ADF) methodologies as well as the proposed Deck Availability Designator (DAD) real-time system is discussed. Work towards this goal is being undertaken by AVT with the objective improving the manner in which helicopter operations are conducted.

1.1 Background

The Air Operations Division (AOD) of the Defence Science and Technology Organisation (DSTO) has been developing dynamic models of helicopters on ship decks using their Ondeck simulation code for this purpose and for analysing other particular on-deck interface problems since 1987. Advanced VTOL Technologies (AVT)[1] has collaborated with DSTO on this problem since 1998.

The Ondeck code can be driven by real or simulated time-domain ship motion data. The motion of the helicopter on the ship deck is determined by integrating the standard equations of motion with the helicopter fuselage modelled as a rigid body. Multi degree of freedom non-linear undercarriage models and complex tyre models are used to account for spring and damping effects in the undercarriage, tyre forces and tyre adhesion and sliding. Main and tail rotor thrust and gyroscopic moments can be accounted for, together with wind on the fuselage. Lashings used to secure the helicopter as well as Decklock and RAST restraint systems can also be modelled. A wide range of helicopter types have been modelled, including Seahawk, Black Hawk, Sea King, Squirrel, Chinook, Seasprite and Lynx[2], MRH 90 and Tiger helicopter models.

Figure 1 Seasprite Ondeck Model

1.2 Overview

Optimising the combined effectiveness of both the ship and helicopter is of paramount importance in hazardous conditions or high threat environments. The ability to take off and land a helicopter in high states to perform a rescue or military operation as well as having the flexibility to rapidly manoeuvre the ship to avoid subsurface or surface threats is very important.

The motion of the flight deck in high sea states, turbulent high winds, and waves breaking over the deck create an extremely hazardous environment when operating helicopters from ships. The launch and recovery of helicopters can at times be very demanding and at times pose a significant risk to flight and deck crew safety. The high loads experienced during landing and the cyclic loading of a helicopters undercarriage, tie down points and airframe due to ship motion can result in airworthiness issues and operational uncertainty. In addition, the environmental conditions can create a number of problems for helicopters when they are on the ship deck. Safe operating limits are required to enable the deck crew to know when it is safe to perform particular on-deck operations. These may include traversing the helicopter from the hangar to the flight deck, engaging the rotors for take-off, or the use of a particular lashing scheme or restraint system to secure the helicopter.

1.3 Traditional Limits

The accurate prediction of safe Ship Helicopter Operating Limits (SHOL's) is essential to maximise safety and operational capability as well as ensure airframe availability. Traditionally RAN limits have been expressed in terms of roll and pitch (Ref. 1) since these are measures that have been widely used by Navies throughout the world and are readily available from the ships gyros. In more recent times attempts at using alternative combinations of parameters such as flight deck accelerations (Ref. 2) have been made. These limits are generally arrived at by conducting simulation for a wide range of operating conditions and are generally conservatively defined to account for worst case likely conditions and the effects of ship manoeuvring and list. Operational limits are based on the occurrence of wheel sliding or lift-off, or in the case of a restrained helicopter loads applied to the fuselage tie-down points, ship tie down points, lashings or restraint systems such as the Indal Recovery Assist and Traverse System (RAST).

Existing on-deck limits defined only in terms of roll and pitch or a limited combination of state variables fall short of providing an optimal indicator of limit proximity due to the assumption that they will adequately define the complex non-linear response of the helicopter on the ship. The highly non-linear response of the ship in 6 degrees of freedom is dependent on the sea state, the number of wave fronts, the relative wave heading, ship speed, ship loading condition, wind speed and direction, stabilisation system and whether the ship is manoeuvring.

The helicopters response is also highly non-linear and the state of the models is defined not only by the rigid body motion of the airframe but the degrees of freedom associated with the undercarriage components. Apart from the complex motion the deck the helicopter is also influenced by the wind, the deck surface condition, helicopter loading, rotor engagement and restraint systems.

Given this complexity, methods employing just a few parameters to define the response of the helicopter must be applied with great caution. Significant effort is required to calibrate such reduced order systems to ensure that they do indeed provide a conservative estimate of the full spectrum of possibilities. Therein exists the reduction in accuracy and operational capability.

Systems that rely on a conservative limit database for guidance are limited in usefulness since the ships Relative Wave heading (RWH) cannot be easily measured and the fact that the same encounter frequency can be generated for a wide range of RWH's, wave spectrums and ship speeds yet yield a completely different ship motion response.

1.4 The DAD System

The proposed DAD aid overcomes the above deficiencies by coupling the Ondeck simulation code with an IMU to provide a real time estimation of the helicopters response to actual ship motion. The system will reside on the ship and the IMU will be placed on the flight deck. This approach negates the requirement to use alternative less accurate measures such as pitch and roll that need to be conservatively applied to cover the myriad of possible combinations of helicopter response due to the motion of the ship and other environmental factors.

The benefit of the DAD system is that it will use the actual ship motion data provided by the IMU in real time to simulate the response of the helicopter using more accurate measures of limit proximity such as:

- Tyre lift-off proximity
- Tyre sliding
- Lashing load

- Restraint system loads (RAST / DeckLock)

More accurate knowledge of limit proximity will allow deck crew to more safely conduct their tasks and provide a better indication of the level of helicopter restraint required. Logging of the system limit data against time will provide the observing officer with improved situational awareness and an improved basis for predicting the response of the helicopter and ship.

The capability to predict the loads transferred through the helicopter means that the system can also be used for usage monitoring and airworthiness inspection guidance. This capability does not currently exist and cannot be provided by alternate systems not employing simulation.

The system will also enable the Commanding Officer more freedom to manoeuvre the ship in adverse conditions or minimise the ships exposure to surface, subsurface or air attack. By assessing the limit measure provided by the DAD system the ship will be able to be manoeuvred accurately within helicopter limits at the maximum permissible rate depending on the value of the asset.

In addition to features described above, an improved means of providing wind envelope limit guidance is proposed so as to provide a complete ship helicopter operating limit (SHOL) monitoring tool. The system shall include a wind envelope monitor based on existing flight trail limit databases. Recommendations provided by the flight operations observer in addition to a traffic light type system arranged on the flight deck will give pilots a better indication of when it is safe to takeoff, land, shut down and start-up particularly in poor meteorological conditions (PMC). Further enhancements are planned to provide guidance on whether the helicopter can land safely within dynamic response and structural limitations.

2. Current Methodologies for Determining Ondeck Limits

Two procedures are currently used by the ADF for determining on-deck operating limits. Both define limits in terms of pitch and roll amplitude, measures that have been widely used by Navies throughout the world as the ships pitch and roll gyros have typically been the only source of ship motion data aboard ships. Both methods are based on the reasonable premise (when conservatively applied) that pitch and roll can be used to define the helicopters limits.

The first traditional approach relies on conservatively extracting the most likely critical pitch and roll amplitude and frequency data combinations for a range of ship speeds, significant wave heights (SWH) and relative wave headings (RWH). This data may be derived through spectral analysis of actual experimental data, tank test data or deduced from ship motion simulation codes. Sinusoidal motion is then used to drive the Ondeck simulation code for each roll and pitch frequency and amplitude combination to determine those combinations that are most likely to adversely excite the response of the helicopter.

The sinusoidal limit analysis involves performing pure and combined roll and pitch scans with increasing amplitude with different phase angle differences between pitch and roll[3] to determine the most adverse pitch and roll combinations. The limit points are then plotted on an x-y chart of pitch versus roll. Pitch and roll limit boxes are then conservatively chosen within the envelope based upon suitable combinations that fit beneath the curve, as illustrated in Figure 2 so that a tabled set of limits can be produced. Primary and alternate limit boxes are sometimes chosen in an attempt to maximise the operational envelope. The placement of the box limits is also governed by the ship motion characteristics that are actually experienced operationally during a FOCFT.

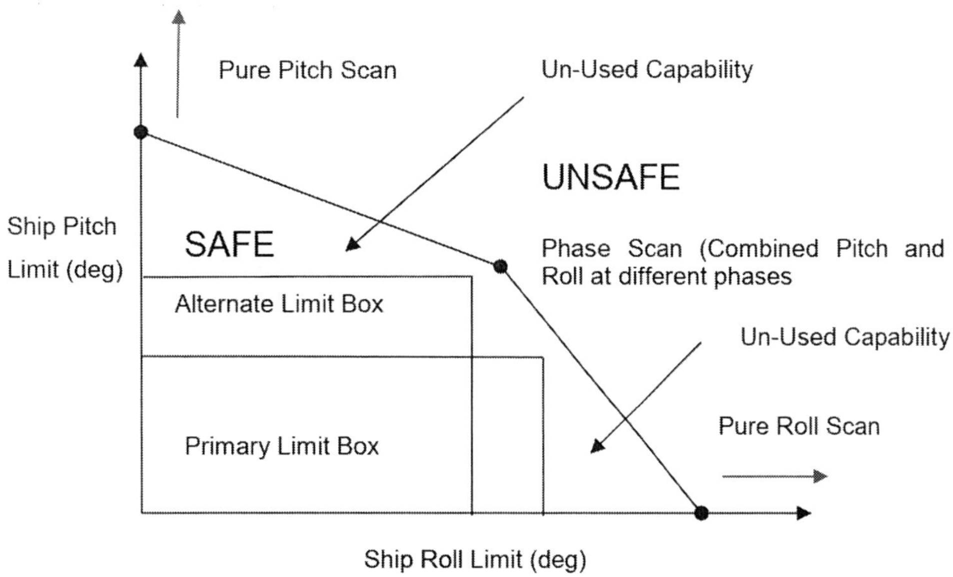

Figure 2 Representation of Pitch and Roll Limits

Despite the apparent success of the first method, reservations have remained over the approach, because of the large degree of conservatism that needs to be applied to account for the superposition and phasing of other degrees of freedom such as heave, sway and yaw and the effects of ship list or the response of the ship during manoeuvring.

A second approach was recently developed by AVT and DSTO (Ref. 3), that utilises frequency response data from the ship motion simulation package, SHIPMO, developed by Defence Research & Development Canada (DRDC) – Atlantic. This software can calculate the frequency response of the ship for a wide range of ship speeds, RWH's, and wave frequencies. Statistical wave information relating wave amplitude or sea state to wave period is then used to realistically determine the likely wave frequency and hence ship encounter frequency for each Significant Wave Height (SWH), RWH and ship speed (SP) combination.

In this method the SWH is increased for different combinations of RWH at a fixed ship speed until a limiting condition is reached. Here regular, constant amplitude uni-directional waves are assumed. For each limiting condition pitch and roll amplitude is extracted and plotted as before on a pitch versus roll graph. Figure 3 shows a set of limits derived from this approach as well as limits derived using the existing methodology. The negative pitch axis is used to denote following seas and the positive axes for waves from ahead. Unlike the existing methodology each pitch-roll limit combination encompasses the effects of all 6 ship motion degrees of freedom, and so is more likely to capture critical limit conditions resulting from all degrees freedom of ship motion.

Depending on the ship speed and encounter frequency the limits obtained when compared to the sinusoidal approach and box limits, demonstrates that significant potential exists for expanded operational capability if ship speed and RWH can be taken into account.

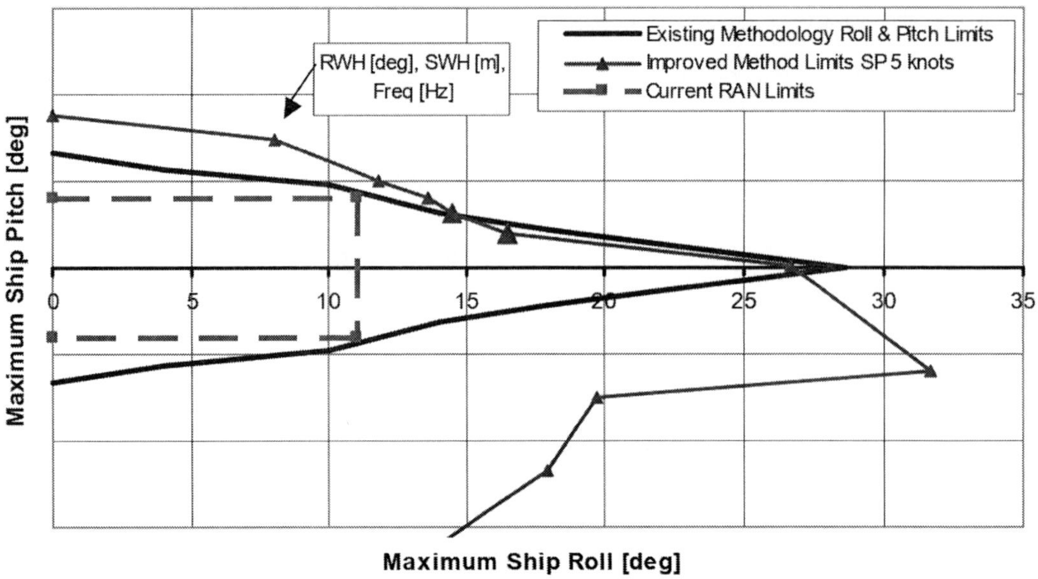

Figure 3 Typical Roll versus Pitch Ship Motion Limits

3. Operational Use of Current Methodology Limits

ADF helicopter operations on ships are governed by Ship Helicopter Operating Limits (SHOL's) defined in the Navy operations document ABR5419. On-deck limits are expressed in terms of roll and pitch attitude because they are typically the only measures of limit proximity available on the ship. Flying limits are defined by a wind envelope whose boundaries are expressed in terms of wind speed and direction. Traditionally, these limits are first estimated using analytical means and previous flying experience and then revised during First of Class Flight Trials (FOCFT's) to account for operational factors, helicopter performance, control margins, handling qualities, and pilot workload. Since 1988 the Ondeck software has been used to provide the initial FOCFT on-deck engineering ship limit estimates.

The observer tasked with guiding the operational use of the helicopter on the ship continually monitors the ship pitch and roll gyros as well as the relative wind speed and direction to ensure that on-deck and flying operations are being conducted within limits. The observer is required to track all these limits and mentally log the ships response and the sea state trends and during landing phases in PMC help provide an indication of when a quiescent period may occur. The Figures 4, 5 and 6 show typical SHOL limits as well as the instrumentation that is typically available on ADF ships.

Condition	Lashing[4]	Limits	
		Pitch	Roll
Land-on	Nil	3	7
Prepare to Shutdown	Short Term	5	10
Prepare to Shutdown	Intermediate Term	6	12
Secure	Long Term	10	25

Figure 4 Traditional Roll and Pitch Limits

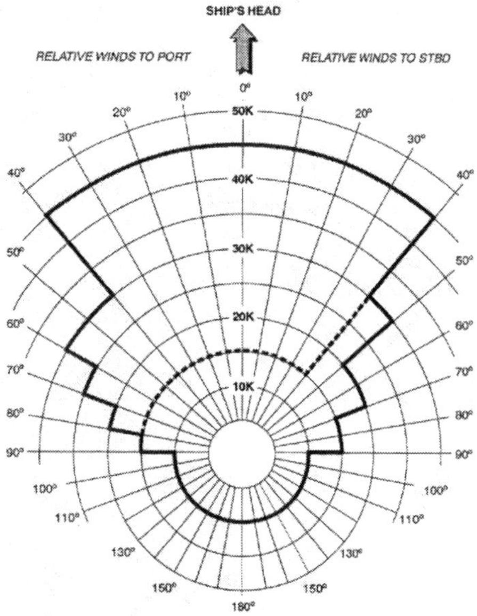

Figure 5 Wind Speed and Direction Limits

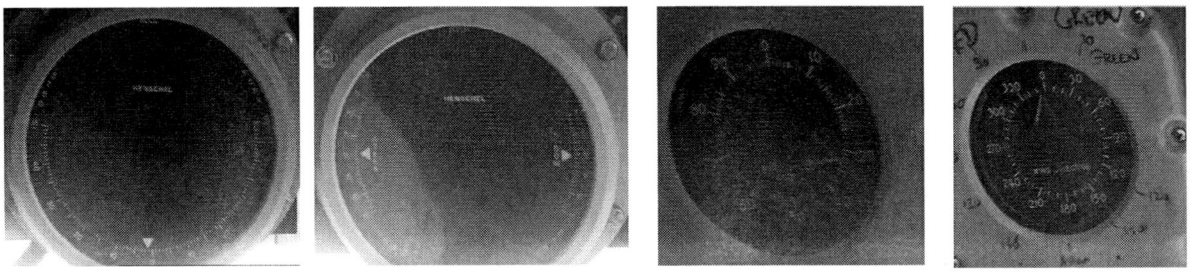

Roll Gyro **Pitch Gyro** **Wind Speed** **Wind Vector**

Figure 6 Typical Ship Instruments

The tabled limits employed vary depending on the operation being performed. In the case of on-deck operations a helicopter may need to be traversed from the hangar to the flight deck using running lashings or a deck handling system. Once traversed, additional lashings, may need to be applied to the helicopter if the ship has to be manoeuvred to get within the required wind envelope. Before takeoff, the lashings and restraint systems must be removed and at the same time the rotor engaged. The observer is required to ensure that appropriate limits are used and must be able to quickly provide advice to the flight crew, deck crew and the person steering the boat on how to minimise the risk of exceeding limiting conditions.

4. Deck Availability Designator System

The DAD system was first proposed by AVT in early 2004. It is proposed that the ship mounted system couples an IMU with the Ondeck code to give the simulated response of the helicopter to actual ship motion in real time. The IMU would be mounted close to the flight deck and a small LCD touch screen for the observer would provide the interface. The system inputs would include

[4] Lashing configurations are generally defined as 'short' (usually around 4 lashings and used only for short periods of time on the flight deck), 'intermediate' (usually around 8 lashings), and 'long-term' (a significant number of lashings) used for prolonged periods of time on the flight deck or hangar.

IMU data for all ship motion degrees of freedom, the wind speed and direction, the helicopter configuration, whether the rotors are spinning, the type of lashing scheme being utilised and whether or not a restraint system like RAST was being employed. Figure 7 below illustrates the concept.

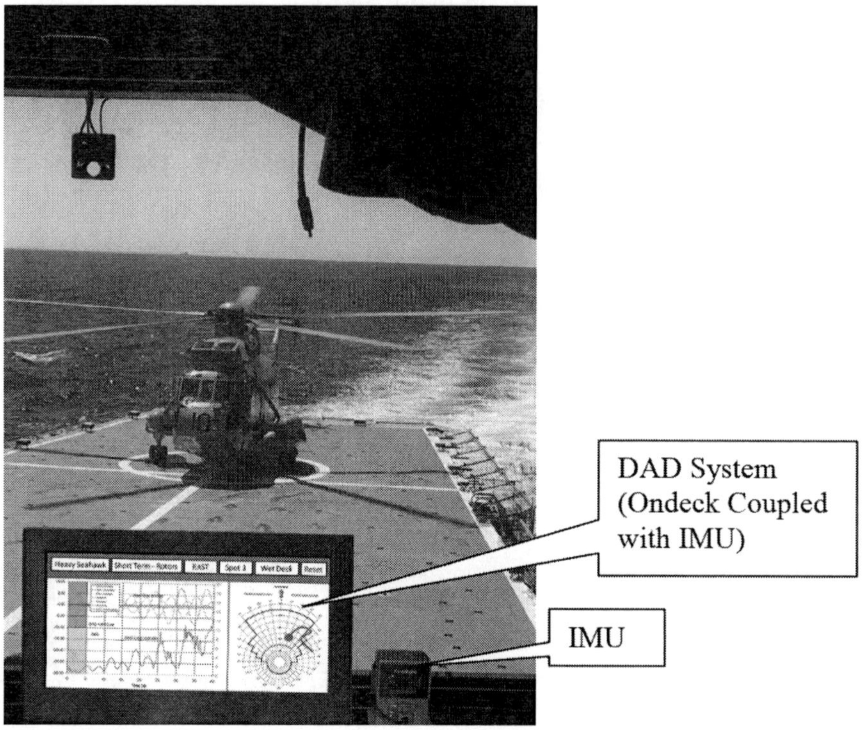

Figure 7 Illustration of the Real Time DAD Simulation System

The mock up in Figure 8 below illustrates (potentially) what type of data will be displayed. The left portion of the display shows the limit indicator which is represented by the red line. The IMU ship motion data is shown at the top of the plot. The colour bands provide an indication of risk. The time scale may be tuned (stretched / compressed) depending on the sea state and it may be possible that extrapolation could be used to provide a limited extrapolated capability depending on the regularity of the sea state. The limit indicator is simply based on either wheel lift off, degree of sliding, lashing loads or restraint system loads, whichever is most critical. Ship motion limits may also be defined to minimise the likelihood of helicopter or ship structural damage during landing phases. On the right hand side is a wind envelope limit diagram with moving marker driven by wind speed and direction data from a reliable ship anemometer. A trail trace could be provided to provide the observer with an indication of wind variability and an indication of which direction it is best to steer based on logged data.

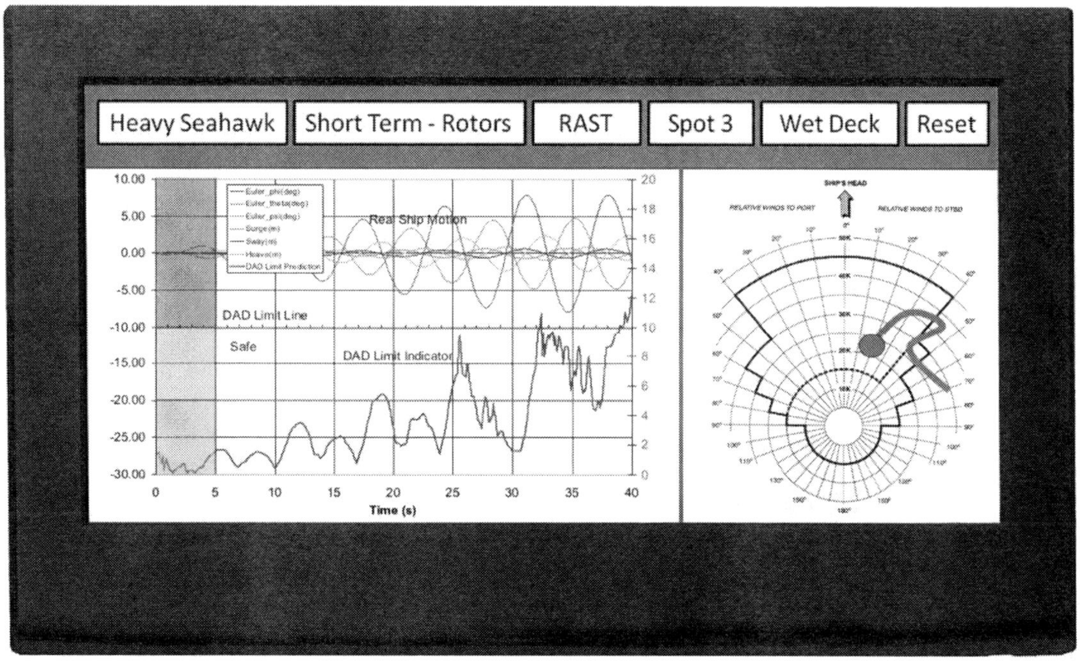

Figure 8 DAD System Mock Up Showing the On-deck Limit Indicator and Wind Envelope

Figure 8 shows a 40 second window of the ship motion data that is ramping up in amplitude as well as the corresponding DAD limit indicator (red trace) for an unrestrained helicopter. The plot shows a limit exceedence (DAD value >10) occurring at approximately 32 seconds and again at 38 seconds. The green, orange and red colour bands indicate safe, approaching limits and unsafe regions respectively. The weighting between the green and orange bands may be tuned based on operational experience with each helicopter and ship type. The duration of the time history window may also be varied to adequately capture a given sea state's variability to provide additional guidance on when it may be safe or unsafe. The DAD limit line clearly shows the cyclic build up to limiting conditions and shows short quiescent periods where it may have been possible to conduct certain on-deck or flying operations.

4.1 Derivation of the DAD Limit Indicator

The DAD limit indicator is partially a weighted sum of the proximity to all possible limiting factors. It utilises the tyre normal force (proximity to wheel lift-off), lateral and longitudinal tyre forces (proximity to sliding), the degree of tyre sliding, lashing loads (proximity to exceeding a load limit at the airframe tie-down point or in the lashing itself) and any restraint loads. A short description of these parameters follows.

Tyre Normal Force Ratio

The wheel lift off limit indicator is given by the following expression where as the tyre normal force tends to zero the limit indicator approaches one.

$$TyreNormForceRatio = 1.0 - MIN\ (\ TyreNormForce\ /\ StaticGearLoad\ ,\ 1.0\ )$$

where *TyreNormForce* is the tyre normal force and *StaticGearLoad* is the tyre static load at simulation initialisation.

Friction Load Ratio

The friction load ratio is given by the maximum ratio of the tyre force divided by the limiting friction load (μN).

$$FriLoadRatio = Ftyre / (\mu N)$$

where *Ftyre* is the total magnitude of the tyre load and μN is the limiting friction load in the direction of the applied load.

Slide Ratio

Each tyre is permitted to slide over a small area before it is considered a limiting condition. The slide ratio is the maximum ratio of tyre displacement from the initialised position over the allowable sliding envelope for each tyre. Figure 9 illustrates the slide ratio limit measure.

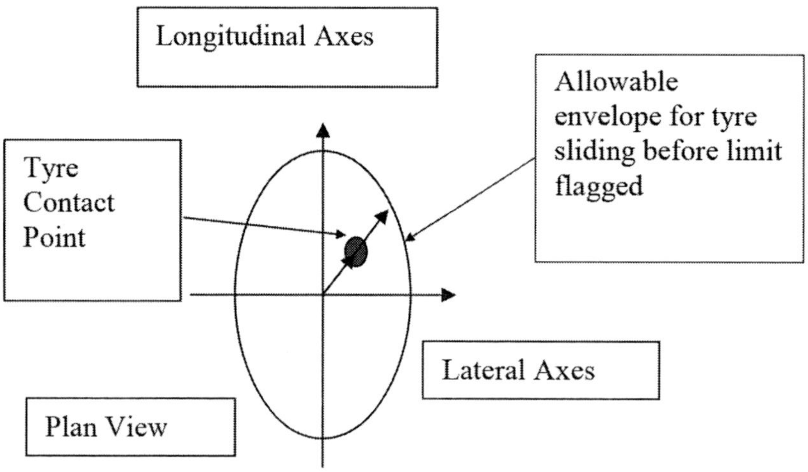

Figure 9 Slide Ratio Derivation

For the unrestrained case the DAD proximity limit value is simply based on the weighted sum of the above parameters for all tyres.

$$maxtyrenorm = MAXVAL(TyreNormForceRatio) \times 10.0$$

$$maxfrictionload = MAXVAL(FriLoadRatio) \times 8.0$$

$$maxslide = MAXVAL(SlideRatio) \times 2.0$$

where *maxtyrenorm* is the maximum tyre normal force ratio of all the tyres, *maxfrictionload* is the peak friction load ratio of all the tyres and *maxslide* is the maximum ratio of tyre sliding of all the tyres.

The weighting between the limit friction load (weighting of 8) and the degree of sliding (weighting of 2) within the allowable slide envelope may be adjusted and tunned. A high friction load ratio is probably a better indicator of overall closeness to sliding at any given instant.

The final DAD (combined) proximity limit value is given by

$$maxlimit = MAX(maxfrictionload + maxlslide , maxtyrenorm)$$

Unrestrained limit exceedance is deemed to have occurred if the *maxlimit* is greater than 10. In all cases, tyre lift-off results in a limit exceedance.

In some instances a *maxlimit* value of 10 may occur when the tyre is outside the allowable slide patch because the tyre has already partially slid within the patch prior to a high friction load ratio building up. In the case of a lashed helicopter the ratio of lashing load to limit load would be used instead.

5. Concluding Remarks

The attributes and limitations of the current methodologies used by the RAN and other Navies for the attainment of safe helicopter-ship on-deck operating limits have been discussed. An alternative system developed by AVT based on the coupling between the DSTO Ondeck code and an IMU to provide real time limit prediction has been proposed.

By using real-time simulation and measures of limit proximity relating to the response of the helicopter the DAD system offers the potential of expanded helicopter and ship operational capability, a means of structural load monitoring and airworthiness inspection guidance as well as significantly improving safety.

Advanced VTOL Technologies in conjunction with DSTO have been assessing helicopter-ship operating limits as well as other dynamic interface issues since 1998. A wide range of helicopter types have been developed, including Seahawk, Black Hawk, Sea King, Squirrel, Chinook and Seasprite, Lynx, MRH 90 and Tiger helicopter models.

REFERENCES

1. ABR 5419 – Vol 1, Revision 2: Ship Helicopter Operations Manual, September 2004.
2. Tadros, R. A., Langlois G., R., LaRosa, M., "Investigation of Ship Flight Deck Motion Limits for Safe Embarked Helicopter Operations", IPLC Conference, Virginia, October 30th –November 1st, 2000.
3. Howe, D, McNeice, B, and Blackwell, J, "Development of an Improved Methodology for Determining Helicopter On-Deck Limits Using simulated ship Motion," DSTO Technical Report 1696, Defence Science and technology Organisation, March 2005.
4. McTaggart, K.A., "SHIPMO7: An Updated Strip Theory Program for Predicting ship Motions and Sea Loads in Waves," DREA Technical Memorandum 96/243, March 1997.
5. Ferrier, B., Baitis, A., Manning, A., "Evolution of the Landing Period Designator (LPD) for Shipboard Air Operations." American Society of Naval Engineers. Naval Engineers Journal. Vol. 112, Number 4. Arlington. July 2000.
6. Colwell, J.L. "Flight Deck Motion System (FDMS): Operating Concepts and System Description," Defence R&D Canada, Technical Memorandum DRDC Atlantic TM 2004-003, January 2004.
7. Howe, D., & Blackwell, J., & "Development of a Dynamic On-Deck Model for a Super Seasprite Helicopter," AOD Technical Report DSTO-TR-1385, Defence Science and Technology Organisation, Department of Defence, Australia, January 2003.
8. Blackwell, J. & Howe, D., " Development of a Dynamic On-Deck Model for a Black Hawk Helicopter," AOD Technical Report 1223, Defence Science and Technology Organisation, Department of Defence, Australia, October 2001.

Advances in Gyro-Stabilisation of Vessel Roll Motion

Tristan Perez[1] and Paul Steinmann[2]

[1]Centre for Complex Dynamic Systems and Control (CDSC), The University of Newcastle, Australia.

[2]Halcyon International Pty. Ltd., Western Australia.

tristan.perez@newcastle.edu.au, paul.steinmann@halcyon.net.au

Abstract

The use of gyro-dynamic forces to counteract the wave-induced roll motion of marine vessels in a seaway was proposed over 100 years ago. These early systems showed a remarkable performance, reporting roll reductions of up to 95% in some sailing conditions. Despite this success, further developments were not pursued since the systems were unable to provide acceptable performance over an extended envelope of sailing and environmental conditions, and the invention of fin roll stabilisers provided a satisfactory alternative. This has been attributed to simplistic controls, heavy drive systems, and large structural mass required to withstand the loads given the low strength materials available at the time. Today, advances in material strength, bearings, motor technology and mechanical design methods, together with powerful signal processing algorithms, has resulted in a revitalized interest in gyro-stabilisers for ships. Advanced control systems have enabled optimisation of restoring torques across a range of wave environments and sailing conditions through adaptive control system design. All of these improvements have resulted in increased spinning speed, lower mass, and dramatically increased stabilising performance. This brief paper provides an overview of recent developments in the design and control of gyro-stabilisers of ship roll motion. In particular, the novel Halcyon Gyro-Stabilisers are introduced, and their performance is illustrated based on a simulation case study for a naval patrol vessel. Given the growing national and global interest in small combatants and patrol vessels, modern gyro-stabilisers may offer a significant technological contribution to the age old problem of comfort and mission operability on small ships, especially at loiter speeds.

INTRODUCTION

The earliest appreciation of gyroscopic effects can be dated back to the times of Newton and his study of the motion of the Earth, but the mathematical foundations describing its working principles has been attributed to Euler. The word 'gyroscope' seems to have introduced by J.B. Foucault in 1852 (Arnold and Maunder, 1961).

Otto Schlick (1904) was the first to propose use of the gyroscopic effects of large rotating flywheels as a ship roll reduction system. White (1907) described, on behalf of Schlick, the installation and testing of a system on the ex-German torpedo-boat destroyer See-Bar. The Schlick gyroscope presented some problems in adjusting the precession moment according to the magnitude of the waves, and although it worked well for the See-Bar, it did not perform as expected in other vessels—see Chalmers (1931) for details.

The American company Sperry then developed a system that addressed the problem of the Schlick gyroscope by using an electric motor and a brake to control the precession of the gyroscope. The motor was commanded by switches which responded to feedback from a small measurement gyroscope. In this arrangement, the velocity of precession was made proportional to the roll rate of the vessel. Further developments involved considering twin gyros with precession in opposite directions to cancel out extraneous cross-torques that may be generated by a single gyro during manoeuvres or due to pitching motion. The use of geared twin gyros was proposed by the Sperry company. This approach was also followed by Fieux in France, who introduced horizontal spinning axis gyro-stabilisers. For a thorough review of the early developments of these type stabilisers, see, for example, Chalmers (1931).

Although the performance of these early gyro-stabilisation systems was remarkable, reported to be up to 95% roll reduction, their high cost, the increase in vessel displacement, and the invention of roll fins which operate well at speed prevented further developments. In addition, the lack of advanced control techniques prevented the early development of gyro stabilisers from achieving high performance over an extended envelope of sailing and environmental conditions. Today, over 100 years later, advances in materials, bearings, mechanical design methods and digital control systems have revitalized the interest in gyro stabilisers. Indeed, improved bearing materials and custom electric motors allow higher spinning rates and the use of smaller gyros to produce high roll reduction moments. Also, the use of digital control systems imbedded in compact, reliable hardware, allows adapting the dynamic characteristics of the precession control to maximize performance in changing sailing and environmental conditions. Therefore, the time for further development of gyro-stabiliser technology may be propitious. This is reinforced by current trends towards reduced manning of marine operations and the increasing attraction of small and medium sized combatants and civil policing vessels. Demand for stabilisation at zero vessel speed for luxury motor yachts and recreational craft is creating a similar need for gyro-stabilisation in non-military roles.

In this brief paper, we provide an overview of recent developments in the design and control of gyro-stabilisers for ship roll motion reduction. In particular, we introduce the novel Halcyon Gyro-Stabilisers and illustrate their performance based on a simulation case study for a naval patrol vessel.

HALCYON GYRO-STABILISERS

In recent years Halcyon has committed to ongoing research and development aimed at modernising the previously investigated concept of using gyros to control rolling motions of ships. The HGS units have now been developed in a 'Defence' range suitable for patrol boats and paramilitary vessels, a 'Yacht' range suitable for luxury motor yachts, and a 'Sport' range suitable for recreational and fishing vessels.

Figure 1, presents a rendered image of one of the novel Halcyon Gyro Stabiliser units (HGS). The unit illustrated has two flywheels in an integrated bearing mounting. These units have the following features:

- Robust mechanical design and judicious use of advanced high strength materials
- Adaptive control with a minimum number of sensors to ensure high reliability
- Failsafe to passive operation in case of sensor or controller fault
- Quiet, efficient, long life custom designed electric motors
- Condition monitoring systems
- Self aligning, oil lubricated bearings

- Flexible mounting arrangements

These systems require minimum maintenance and can either be incorporated at initial stages of vessel design, or retro-fitted to existing vessels.

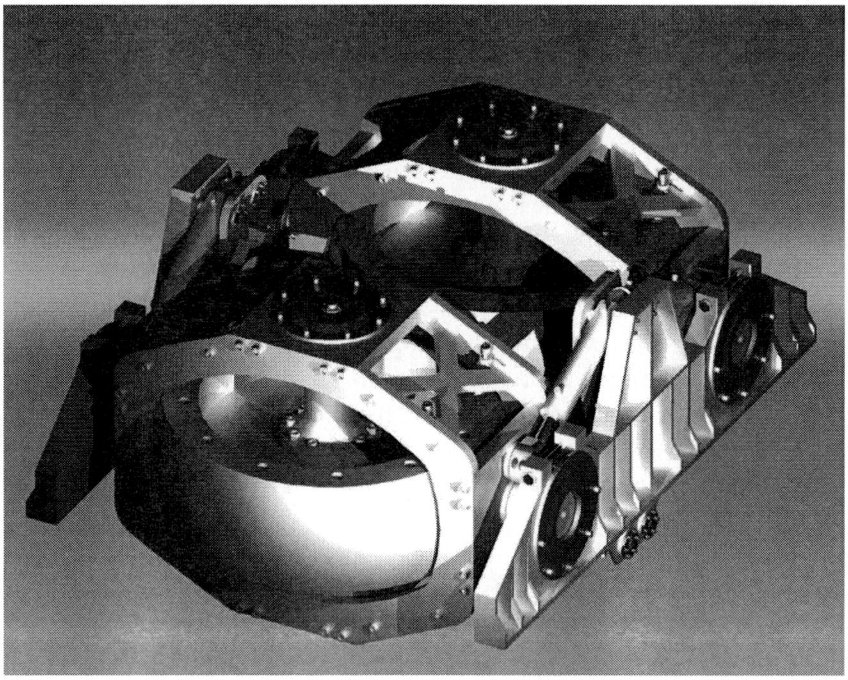

Figure 1 A twin flywheel HGS unit.

Mechanical Design Characteristics

The mechanical modernisation of the gyro as a ship roll motion reduction device has involved increasing performance dramatically in a number of areas. This has been done through the adoption and application of proven advances in structural material properties and the performance of system components like bearings, fastenings and electric motors. A conservative design approach remains essential for producing robust practical product designs.

The combination of oscillating wave induced forces and the high spinning rate of the HGS flywheels leads to very high cycle loading on critical components of the gyros. In order to produce a safe product, the design of the flywheels and supporting structure must be based on utilising the mechanical strength properties of the materials to well below the fatigue endurance for these cycles. Modern high strength steels offer a real advantage here as their fatigue endurance limits generally increase with increasing tensile strength, so long as the material is not overly heated, and the surface is not excessively damaged with discontinuities.

Bearing performance is crucial to a modern gyro-stabiliser as this is the primary limitation on the speed at which the flywheel may be spun, and this defines the torque density of the units. Where required, the HGS units feature circulating oil lubrication and water cooling systems to maintain bearings at safe temperatures which maximise the speed attained and prolong the life of the bearings. Each bearing is also fitted with a bearing condition monitoring system comprising an accelerometer and temperature sensor which is constantly monitored by the PLC based main controller. Exceeding safe levels in either of these measured data streams triggers a non-critical alarm at the touch-screen control panel(s) and may result in a reduction in spinning speed until safe levels are achieved. If a bearing is determined to be significantly damaged the unit will initiate a safe shut down sequence.

Hydraulic linear actuators have been used to provide a resistive control torque on the precession axis. The cylinders apply a braking torque to the precession axis and as a result, no external power is required beyond that to control hydraulic valves. No electric hydraulic power packs or PTO power are required to run a HGS unit.

Control Systems

Figure 2 shows a block diagram representative of the vessel an *n*-number of identical gyro-stabiliser units. The coupled equations of motion of the vessel and the gyro can be represented by

$$I_{44}\ddot{\phi} + B_{44}\dot{\phi} + C_{44}\phi = \tau_{\omega} - nK_g\dot{\alpha}\cos\alpha \qquad (1)$$

$$I_g\ddot{\alpha} + B_g\dot{\alpha} + C_g\sin = K_g\dot{\phi}\cos\alpha + \tau_g \qquad (2)$$

where the (1) represents the roll dynamics and (2) the dynamics of the gyro units in the precession axis. The roll stabilising moment provided by the gyros is given by the last term on the right-hand side of (1). This term is proportional to the precession rate of the gyro, with a proportional coefficient Kg, which amounts to the product of the inertia about the spinning axis and the angular velocity.

For precession angles up to 70deg, a linear analysis gives reliable insight about the system behaviour. The controller implements a torque on the precession axis of the gyro using hydraulic actuators. This torque is represented by the last term on the right hand-side of (2). The linear analysis provides Halcyon with a tool to assess initial gyro control system design.

Then GyroSim, a comprehensive proprietary simulation package with high-fidelity mathematical models is used to appropriately size gyro units for given applications, to refine the designs, and to establish the initial settings for the adaptive control characteristics. This high-fidelity model allows the simulation of various well-known wave spectrum formulations and sinusoidal excitation, and includes a non-linear integrated vessel-gyro dynamic model in 2 and 3 degrees of freedom.

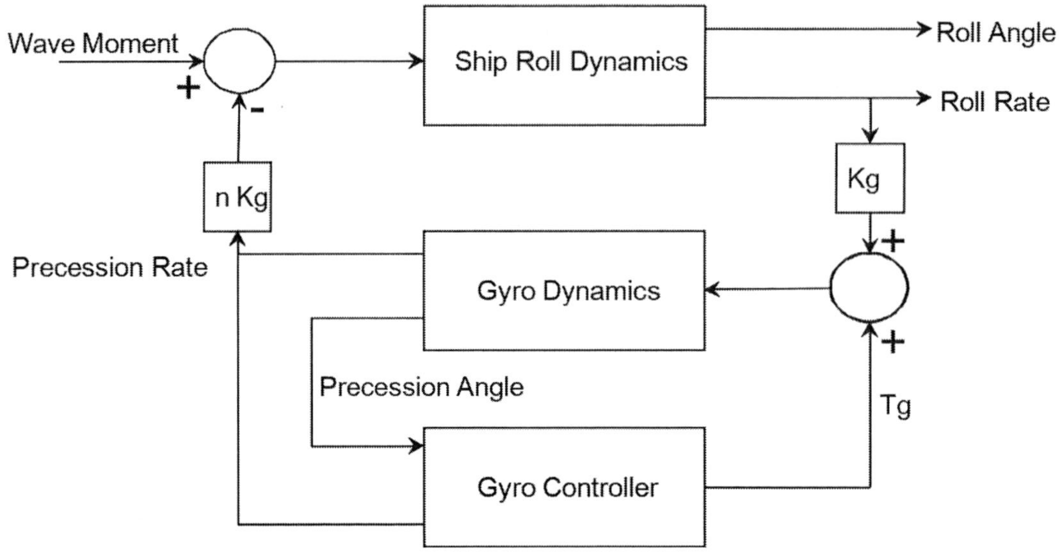

Figure 2 Block Diagram of a Gyro-stabilisation unit.

Using a dynamic controller, which takes only measurements of precession angle and rate, the adaptive precession rate controller enables significant roll reductions to be achieved for varying wave environments and sailing conditions. Figure 3 shows a typical frequency response (from roll exciting moment to roll angle) obtained for a 364tonne naval patrol vessel with the characteristics given in the Appendix, fitted with a Halcyon Gyro-stabiliser. In this example, the responses due to three different controllers are presented in Figure 3. The controller that produces the lowest reduction in the response is intended for use in severe sea states---in this instance the gains of the controller are limited to prevent damage of the unit. The second controller is intended for use in intermediate sea states. The third controller produces the maximum reduction at frequencies around the natural frequency of the vessel; however if this controller is used in long-crested seas of low wave frequencies, it can amplify the motion at those low frequencies with respect to the non-stabilised vessel.

In order to maximise the performance of the gyro in varying conditions, the controller implemented on the HGS units is adaptive. This ensures both optimum performance and integrity of control of the units, by applying appropriate controller tuning as the wave environment or sailing condition changes. The control system also performs monitoring tasks and can trigger safe-fail mode in case of a fault.

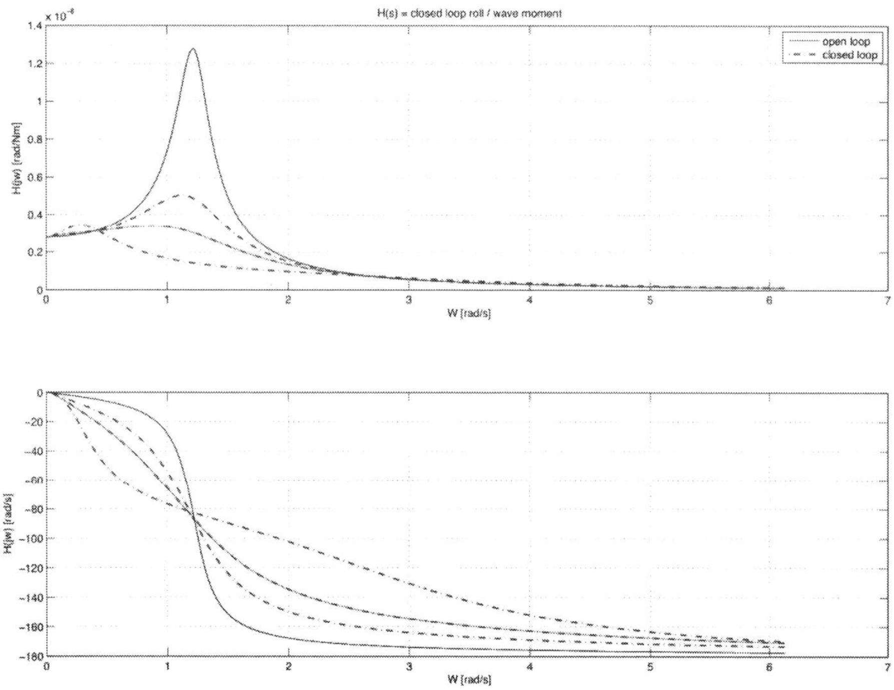

Figure 3 Typical frequency responses from roll excitation moment to roll angle of a naval patrol vessel. Open loop—no gyro stabiliser; Closed-loop—active gyro stabiliser with three different control tunings (highest line – lowest gain, middle line – medium gain, lower line – highest gain)

CASE STUDY: NAVAL PATROL BOAT

In order to illustrate the performance of different stabilisers, we consider three potential designs for a twin HGS unit (shown in Figure 1). The most relevant parameters of the HGS are summarised in Table 1.

Table 1 Naval Patrol Vessel Twin HGS particulars.

ID	**Total Unit Mass [tonne]**	**Kg [kN m.s]**
HGS 1	9.2	22.5
HGS 2	13.6	33.7
HGS 3	16.8	45.0

As a testing condition, we limit the example to a sea state described by the JONSWAP spectrum with significant wave height of 1.5m and modal period of 7s. The sailing conditions are given by zero forward speed and beam seas. **Figure 4** shows the results of the GyroSim simulation package corresponding to the vessel, sea state and sailing conditions described above.

Table 2 summarises the results of numerical GyroSim simulations for the three designs. The data in this table corresponds to 3 particular realizations of the sea state, and are representative of the expected performance of the system. The corresponding time series of roll motion, rate and accelerations are shown in Figure **5**, Figure **6**, and Figure **7**. These results correspond to the system operation after a short period of adaptation, and illustrate the performance of the system.

Table 2 Naval Patrol Vessel roll motions with and without the HGS units operational, in Hs=1.5m, Tp=7sec irregular waves in beam seas at zero speed.

ID	**Roll RMS Open loop [deg]**	**Roll RMS with HGS [deg]**	**RR %**
HGS 1	6.5	1.66	74.3
HGS 2	6.61	1.06	83.9
HGS 3	5.99	0.77	87.2

RR means roll reduction.

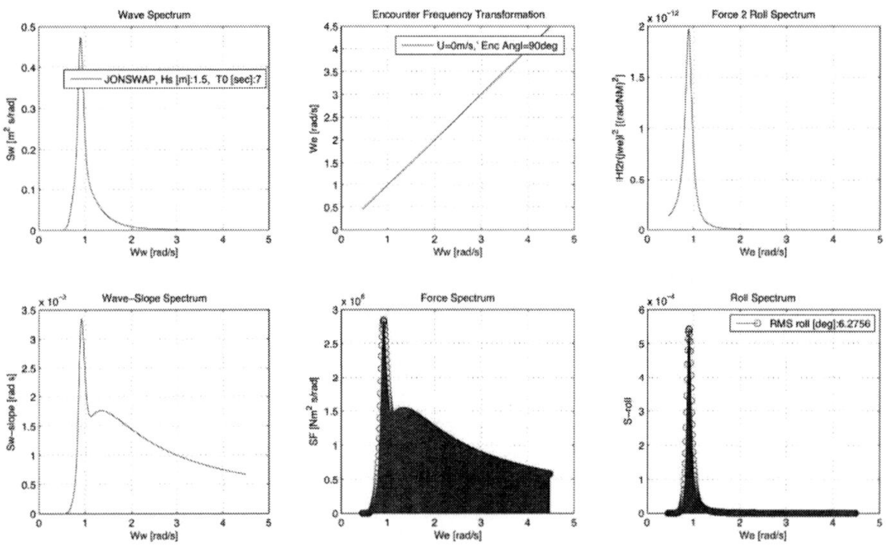

Figure 4. Sea state, roll response operator and power spectrum for the tested conditions.

CONCLUSIONS

In this brief paper, we have provided an overview of recent developments in the design and control of gyro-stabilisers for ship roll motion reduction. In particular, the novel Halcyon Gyro-Stabilisers (HGS) were introduced, and their performance was illustrated based on a simulation case study for a naval patrol vessel. Reductions in RMS rolling motion were found to be in the range of 75% to 87% for the considered designs. The performance indicated by the simulations described in this paper suggest that modern gyro-stabilisers such as the Halcyon Gyro-Stabilisers may offer a very attractive and competitive alternative to other types of stabilisers; either as a replacement for those systems or in combination with them, for operations at both zero and forward speed.

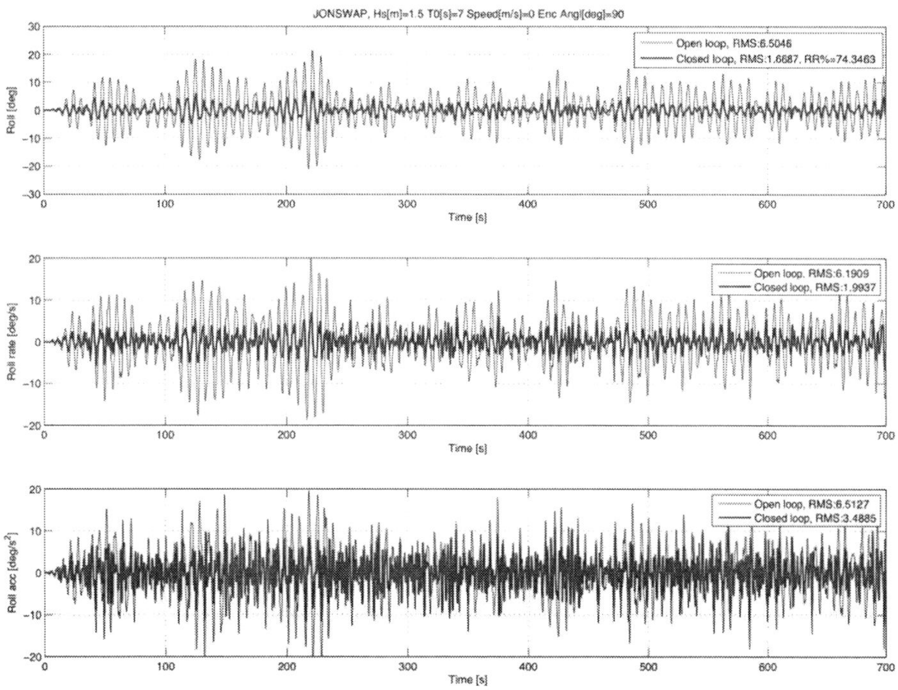

Figure 5. Roll time series of predicted performance corresponding to HGS 1. RMS roll angle reduction of 75%.

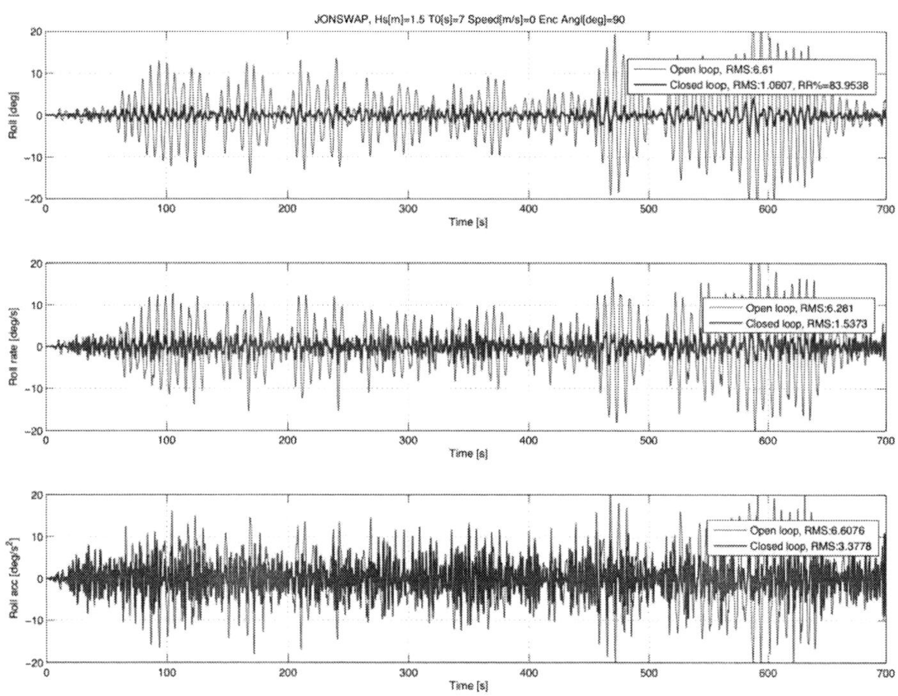

Figure 6. Roll time series of predicted performance corresponding to HGS 2. RMS roll angle reduction of 84%.

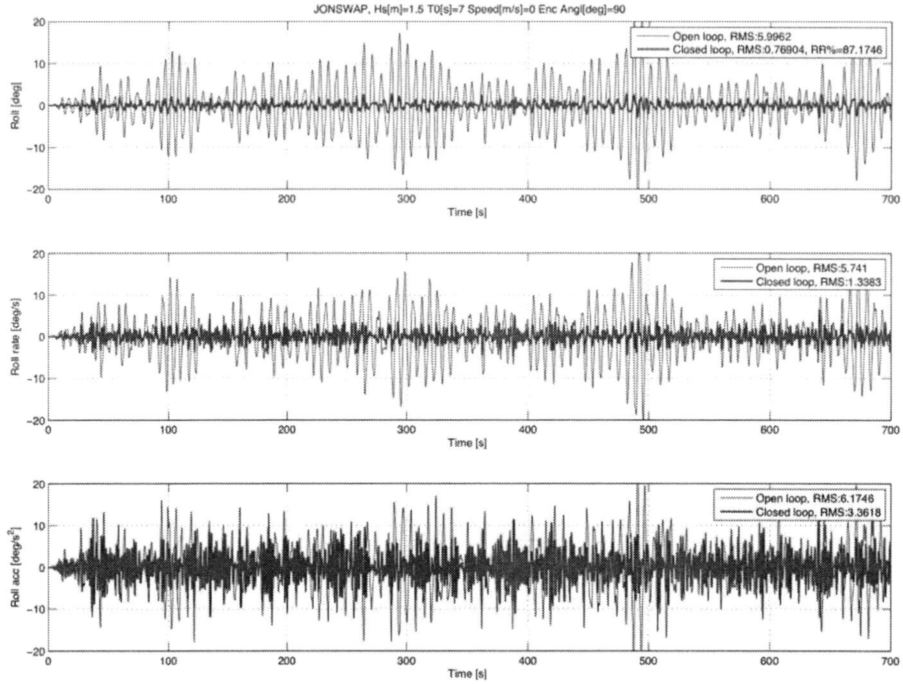

Figure 7. Roll time series of predicted performance corresponding to HGS 3. RMS roll angle reduction of 87%.

REFERECES

- Arnold, R.N. and L. Maunder (1961) "Gyrodynamics and its Engineering Applications" Academic Press, New York and London.

- Chalmers, T. W., (1931) "The Automatic Stabilisation of Ships" Chapman and Hall, London.

- Perez, T. (2005) "Ship Motion Control" Springer-Verlag London. ISBN-10: 1852339594.

- Schlick, O. (1904). Gyroscopic effects of flying wheels on board ships. Transactions of The Institution of Naval Architects, INA.

- White, W. (Sir) (1907) Experiments with Dr. Schlicks's gyroscopic apparatus for steadying ships. Transactions of The Institution of Naval Architects, INA.

APPENDIX: Naval Vessel

The naval vessel used in this paper was introduced by Perez (2005) as a benchmark example for roll stabilisation. This vessel has the size and represents a typical design for a naval patrol vessel. The main particulars are summarised in Table 3.

Table 3 Naval Patrol Vessel main particulars\. Adapted from Perez (2005).

Quantity	**Value**	**Units**
Length between perpendiculars	51.5	m
Displacement mass	364,000	Kg
Transverse metacentric height (GMt)	1.0	m
Roll radius of gyration about CG	3.05	m
Non-dim. roll damping (from model decaying test)	0.114	-
Roll natural period	6.76	s

ABOUT THE AUTHORS

Dr Tristan Perez completed his Electronic Engineering degree at the National University of Rosario, Argentina, in 1999. He then moved to Australia and completed his doctoral degree at The University of Newcastle in 2003, where the focus of his work was on performance limitations and motion control of marine vessels. In 2002, he was also employed as a technical officer in naval architecture for ADI-Limited Australia where he worked in the area of seakeeping. From Nov 2003 until May 2004, he was a research fellow at the Mechatronics Research Centre, University of Wales, United Kingdom; where he worked on control and fault detection of thrusters for underwater vehicles. From Jun 2004 to Jul 2007 he was a senior scientist at the Centre for Ships and Ocean Structures (CeSOS), Trondheim, Norway, and a lecturer in marine control systems and computer control systems at the Norwegian University of Science and Technology (NTNU). Since Oct 2007, he has been with the Centre for Complex Dynamic Systems and Control (CDSC) at The University of Newcastle, Australia. Dr Perez is author of the book "Ship Motion Control" (Springer, 2005); he is a member of Technical Committee on Marine Control Systems of the International Federation of Automatic Control. He has also provided consultancy services in modelling, identification and control of marine systems for the industry in Europe, Australia and New Zealand. Further information can be found at http://cdsc.org.au/TP/Site/TPwb.html

Paul Steinmann completed his Mechanical Engineering degree with honors at The University of Melbourne in 1991. During his university career he completed three Sydney to Hobart sailing races and sailed vigorously in the Melbourne area. His engineering career has been strongly biased towards innovative design covering a diverse range of marine projects including: design of motion stabilising systems for high-speed craft, paramilitary vessels and luxury motor yachts; mooring system design and optimization for FPSO; design and project engineering of high speed aluminium fast ferries and offshore oil and gas field developments; and design and testing of keel, bulb and winglet designs for Americas Cup yachts. Paul has extensive sea trials experience on a wide range of vessel types ranging from sailing yachts, commercial ferries and paramilitary vessels, through to luxury motor yachts. He has also spent a number of years providing deck officer training, commissioning and servicing, for ship stabilisation systems around the world. Paul is the managing director at Halcyon International. He completes his MBA at The University of Western Australia in February 2008.

Delegate List

29 -31 January 2008

Sydney Convention & Exhibition Centre, Sydney, AUSTRALIA

Name	Organisation	Email
AUSTRALIA		
Abraham-James, CMDR Derek	Department Of Defence	derek.abraham-james@defence.gov.au
Ackroyd, Mr Bretton	Defence Materiel Organisation - Amphibious Deployment And Sustainment	corinne.fitzgerald@defence.gov.au
Adams, Mr. Keith	The Royal Institution Of Naval Architects	kadams@zeta.org.au
Aiyelananthan, Chelliah	Directorate Of Naval Sensor & Intelligence	aiyel.chelliah@defence.gov.au
Allerhand, Mr Peter	Centre for Maritime Engineering, Department of Defence	peter.allerhand@defence.gov.au
Andrewartha, Mr Tristan	Defence Science And Technology Organisation	tristan.andrewartha@dsto.defence.gov.au
Anokhin, Mr Vadim	GHD	vadim.anokhin@ghd.com.au
Ashfield, Mr Steve	EDS Defence	steve.ashfield@eds.com
Barton, Mr James	Tenix Marine	bartonj@tenix.com
Beale, CAPT Steven	A.I.M.S.	MIPEC@bigpond.net.au
Beer, Mr Jason	Tenix Marine	jason.beer@tenix.com
Behnke, Mr Kurt	Tropical Reef Shipyard	john.kormany@trshipyard.com.au
Bendall, Mr David	Nautical Institute	david@maritrade.com.au
Benjamin, Mr John	Thales Australia	john.benjamin@thalesgroup.com.au
Best, Mr Roger	Tenix Marine	bestr@tenix.com
Bixley, Mr Bill	Bixley Marine	wbixley@bigpond.net.au
Bonner, Mr David	Weir Strachan And Henshaw Australia	dave.bonner@weirsandh.com
Bose, Prof Neil	Australian Maritime College	n.bose@amc.edu.au
Bower, LCDR Brett	Defence Materiel Organisation - Amphibious Deployment And Sustainment	corinne.fitzgerald@defence.gov.au
Bradley, Mr Rodney	ANZAC SPO	rod.bradley@anzacspo.gov.au
Brennan, Mr Rod	Defence Materiel Organisation	rod.brennan@defence.gov.au
Broadbent, Mr Adrian	Lloyds Register Asia	adrian.broadbent@lr.org
Brown, Mr Allen		allen.brown@dsto.defence.gov.au
Bryen, Mr Noel	Tenix Marine	bryenn@tenix.com
Budd, Mr Alan	Mercy Ships Australia	alan.budd@mercyships.org
Burke, Mr Darren James	Navy Certification	darren.bruke@defence.gov.au
Burrowes, Mr Darren	ATSA Defence Services	darren@atsa.com.au
Cameron, Mr Alex		alex.cameron@eds.com
Campbell, Prof Peter	Defence And Systems Institute	peter.campbell@unisa.edu.au
Cannon, Dr Stuart	Defence Science And Technology Organisation	stuart.cannon@dsto.defence.gov.au
Capizzi, Mr Vincent	Tenix Marine	capizziv@tenix.com
Carr, Mr Simon		jkay5@csc.com
Chalak, Mr Saeb	Centre for Maritime Engineering, Department of Defence	saeb.chalak@defence.gov.au
Chowdhury, Mr Maghfar	Centre For Maritime Engineering - Department Of Defence	Maghfar.chowdhury@defence.gov.au
Clifford, Mr Robert	Incat Australia	incat@incat.com.au
Cobb, Mr Glenn	Lloyd's Register Asia	glenn.cobb@lr.org
Coleman, Mr Ben	Department of Defence	ben.coleman@defence.gov.au
Colquhoun, Mr John	Department Of Defence	john.colquhoun@defence.gov.au
Consiglio, Mr Paul	Tenix Marine	consiglp@tenix.com
Crumlin, Mr Paddy	Maritime Union of Australia	paddycrumlin@mua.org.au
Curtis, Mr Dan	Navy Platform Systems	dan.curtis@defence.gov.au
Cussen, Mr Dermot	ICN South Australia	dermot.cussen@state.sa.gov.au
Dane, Dr Robert	Advanced Technology Watercraft Pty Ltd	robert@solarsailor.com.au
Davies, Mr Paul	Hub Consulting Pty Ltd	paul.davies@hubconsulting.com.au

Delegate List

29 -31 January 2008

Sydney Convention & Exhibition Centre, Sydney, AUSTRALIA

Name	Organisation	Email
AUSTRALIA		
Davis, Mr John	Raytheon	John.davis@ausawd.com
De Yong, Mr Leo	Defence Science And Technology Organisation	leo.deyong@dsto.defence.gov.au
Di Lorenzo, Mr Maurizio	DMO	maurizio.dilorenzo@defence.gov.au
Downing, Mr Robert	Tropical Reef Shipyard	john.kormany@trshipyard.com.au
Duff, Mr Tony	Thales Australia	anthony.duff@thalesgroup.com.au
Duggan, Mr Andrew	Boeing Australia Limited	andrew.j.duggan@boeing.com
Dylejko, Mr Paul	Defence Science & Technology Organisation	paul.dylejko@dsto.defence.gov.au
Edwards, Mr Bill	Defence Material Organisation	corinne.fitzgerald@defence.gov.au
Elischer, Mr Paul	DSTO	paul.elischer@dsto.defence.gov.au
Elliott, Mr Andrew	Tenix Marine	elliotta@tenix.com
Elms, Mr Tony	Ship Dynamics	tony.elms@shipdynamics.com
Emerton, Mr Graeme	Centre For Maritime Engineering - Department Of Defence	Graeme.Emerton@defence.gov.au
Emmett, Mr Lindsay	Australian Maritime Safety Authority	lve@amsa.gov.au
Equid, Mr Rod	AWD Alliance	rod.equid@ausawd.com
Flapan, Mori	National Marine Safety Committee Inc	mflapan@nmsc.gov.au
Fletcher, Mr Andrew	Defence SA	andrew.fletcher@defencesa.com
Foley, Mr Peter	Australian Transport Safety Bureau	Peter.Foley@atsb.gov.au
Fox, Mr E.C George	Cubic Transportation Systems Australia Pty Ltd	
Gabrovska, Ms Violeta	Tenix Marine	gabrovsv@tenix.com
Gallacher, Mr John	AWD Alliance	john.gallacher@ausawd.com
Gillies, Mr Donald	IMarEst	dongillies@idx.com.au
Goswell, Mr Stephen	DMO ANZAC SPO	stephen.goswell@defence.gov.au
Granger, Mr R. Stuart	Tenix Marine	
Gravenall, Mr David	Tenix Marine	gravenalld@tenix.com
Gray, Mr Norman	Thales Australia	jude.bavcar-skubla@thalesgroup.com.au
Grech, Dr Michelle	Defence Science And Technology Organisation	michelle.grech@dsto.defence.gov.au
Groot, Mr Olav	Sydney Ferries Corporation	Olave.Groot!@sydneyferries.nsw.gov.au
Hallett, Mr Mark	DSTO	mark.hallett@dsto.defence.gov.au
Harding, Mr Nathan	BMT Defence Services (Australia)	nharding@bmtdsl.com.au
Hardy, Prof Tom	Australian Maritime College	t.hardy@amc.edu.au
Harvey, Kim	Weir Strachan & Henshaw Australia	kim.harvey@weirsandh.com
Harzmeyer, Ms Heather	AWD Alliance	heather.harzmeyer@ausawd.com
Haynes, Mr Wayne	Thales Australia	wayne.haynes@thalesgroup.com.au
Helmore, Mr Phillip	The University Of New South Wales	p.helmore@unsw.edu.au
Henry, Mr Peter	Centre for Maritime Engineering, Department of Defence	peter.henry1@defence.gov.au
Hogan, Mr Jim	ANZAC SPO	jim.hogan@anzacspo.gov.au
Hughes, Mr Craig	ABS Pacific	crhughes@eagle.org
Jaffrey, Mr Mark	DSTO	mark.jaffrey@defence.gov.au
Jaksic, Mr Zoran	Thales Australia	zoran.jaksic@thalesgroup.com.au
James, Mr Roger	Government of Victoria	roger.james@diird.vic.gov.au
Jarvis, Mr Robert	DSTO	hugh.torresan@dsto.defence.gov.au
Jeremy, Mr John	Royal Institution of Navel Architects	jcjeremy@ozemail.com.au
Johnson, Mr Colin	Tenix Marine	johnsonc@tenix.com
Jones, Mr Ian	Australian Customs Service	ian.jones@customs.gov.au
Kaper, Mr Roald	Dockwise Shipping Australia	roald.kaper@dockwise.com
King, Mr Clive	DMO - AWD	clive.king@ausawd.com
Kingston, Mr David	Defence Materiel Organisation - Amphibious	corinne.fitzgerald@defence.gov.au

Name	Organisation	Email
AUSTRALIA		
	Deployment And Sustainment	
Knox, Mrs Jennifer	Lighting Naval Architecture	navlight@bigpond.com
Kordic, Mr Phil	MTU Detroit Diesel Australia	christina.yianni@mtudda.com.au
Kormany, Mr John	Tropical Reef Shipyard	john.kormany@trshipyard.com.au
Laverock, Mr Ian	Australian Customs Service	ian.laverock@customs.gov.au
MacDonald, Mr Gordon	BMT Defence Services (Australia)	gmacdonald@bmtdsl.com.au
Macfarlane, Mr Gregor	Australian Maritime College	G.Macfarlane@amc.edu.au
Magill, Mr Dave		davewma@aol.com
Mai, Mr Tu	Tenix Marine	maitu@tenix.com
Makin, Mr Murray	Thales Australia	Murray.Makin@thalesgroup.com.au
Malpress, Mr Keith	Tenix Marine	keith.malpress@tenix.com
Marshall, CDRE Peter	Department Of Defence - NAVY	peter.marshall6@defence.gov.au
Martin, Mr Tony	Defence SA	tony.martin@defencesa.com
Mathys, Ms Zenka	Defence Science Technology Organisation	zenka.mathys@dsto.defence.gov.au
May, Mr Peter		
McCleery, Mr Kim	Defence SA	kim.mccleery@defencesa.com
McConachie, Mr Robert	Centre For Maritime Engineering - Department Of Defence	robert.mcconachie@defence.gov.au
McGloin, Ms Kelly	Defence SA	kelly.mcgloin@defencesa.com
McLachlan, Mr Richard	Defence SA	richard.mclachlan@defencesa.com
McLean, Mr Andrew	DSTO	andrew.mclean@dsto.defence.gov.au
Mechanicos, Mr Michael	Germanischer Lloyd Australia Pty Ltd	michael.mechanicos@gl-group.com
Meldrum, Mr John	ANZAC SPO - DMO	john.meldrum@anzacspo.gov.au
Menderes, Mr Mike	Defence Meteriel Organisation	mike.kenderes@defence.gov.au
Mensforth, Mr Adrian	Tenix Marine	mensfora@tenix.com
Milne, Mr Richard	Thales Australia	Richard.Milne@thalesgroup.com.au
Moore, Mr Sean	Tenix Marine	moorese@tenix.com
Morris, Mr Brett	DSTO	brett.morris@dsto.defence.gov.au
Mroz, Mr Alek	DNSIE	alek.mroz@defence.gov.au
Nazor, Mr Tony	CSC Australia	jkay5@csc.com
Neill, Dr Roger	Defence Science And Technology Organisation	roger.neill@dsto.defence.gov.au
O'Donnell, Mr Gary	RAN	gary.odonnell@defence.gov.au
Ojeda, Mr Roberto	Australian Maritime College	r.ojeda@amc.edu.au
Paglia, Mr Mario	AVEVA Pty Ltd	kristy.faulkner@aveva.com
Paliaga, Mr Dave	Defence Materiel Organisation - Amphibious Deployment And Sustainment	corinne.fitzgerald@defence.gov.au
Perez, Dr Tristan	CDSC, The University of Newcastle	tristan.perez@newcastle.edu.au
Pierce, Mr David	Defence Materiel Organisation - Amphibious Deployment And Sustainment	corinne.fitzgerald@defence.gov.au
Pisaniello, Mr Victor	DMO	andrew.gates@defence.gov.au
Pourzanjani, Prof Malek	Australian Maritime College	M.Pourzanjani@amc.edu.au
Prandolini, Laurie	The Royal Institute Of Marine Engineering Science And Technology	
Pregnalato, Dr Craig	BMT WBM Pty Ltd	cpregnalato@wbmpl.com.au
Priestnall, Mr Graham	SyntheSys Systems Engineers Ltd	graham_priestnall@synthesys.co.uk
Rahman, Mr Tauhid	ABS Pacific	trahman@eagle.org
Ramsey, Mr Roger	Tenix Marine	ramseyr@tenix.com
Ranmuthugala, Dr Dev	Australian Maritime College	D.Ranmuthugala@amc.edu.au
Rashid, Mr Aminur	DMO - AWD	aminur.rashid@defence.gov.au

Delegate List

29 -31 January 2008

Sydney Convention & Exhibition Centre, Sydney, AUSTRALIA

Name	Organisation	Email
AUSTRALIA		
Ray, Dr Tapabrata	UNSW@ADFA	t.ray@adfa.edu.au
Reid, Mr Warren	Defence Science & Technology Organisation	warren.reid@dsto.defence.gov.au
Rekkas, Mr George	DMO	george.rekkas@defence.gov.au
Relf, Dr Phillip	DSTO	Phillip.Relf@DSTO.Defence.gov.au
Renilson, Prof Martin	Australian Maritime College	m.renilson@amc.edu.au
Richards, Mr Marc	Aimtek Pty Ltd	admin@aimtek.com.au
Roshan-Zamir Malayer, Dr Saeed	Tenix Marine	roshanz@tenix.com
Sadovnikov, Dr Dmitry	Austal	dmitrys@austal.com
Sanford, Mr Darren	Defence Science And Technology Organisation	darren.sanford@dsto.defence.gov.au
Sare, Dr Ian	Department of Defence	ian.sare@dsto.defence.gov.au
Schmidt, Mr Frank	Priority Engineering Services	fschmidt@p-e-s.com.au
Shah Khan, Dr Zafar	DSTO	zafar.shahkhan@dsto.defence.gov.au
Sirutis, Mr John	AWD Alliance	john.sirutis@ausawd.com
Slater, Mr Karl	Tenix Marine	slaterk@tenix.com
Smith, Mr Jonathan	Australian Marine Complex (WA) Management Pty Ltd	deidre.watson@amccuf.com.au
Smith, Dr Warren	ACME, UNSW and ADFA	w.smith@adfa.edu.au
Smith, Mr Wayne	AWD Alliance	wayne.smith@ausawd.com
Sothirajah, Mr Sothy	Centre for Maritime Engineering, Department of Defence	thiyagarajah.sothirajah@defence.gov.au
Squires, Mr Michael	Australian Transport Safety Bureau	michael.squires@atsb.gov.au
Stein, CAPT Frederick R.	Marine National Facility, CSIRO	Fred.Stein@csiro.au
Steinmann, Mr Paul	Halcyon International Pty Ltd	paul.steinmann@halcyon.net.au
Stevelt, Mr Hans	Lightnight Naval Architecture	navlight@bigpond.com
Steward, Mr Jason	Jacobs Australia	fiona.brayshaw@jacobs.com.au
Swain, Mr Peter	Thales Australia	peter.swain@thalesgroup.com.au
Swann, Mr Michael	Defence Materiel Organisation Air Warfare Destroyer Alliance	michael.swann@ausawd.com
Sykes, Mr Wayne	Tenix Marine	sykesw@tenix.com
Tait, Ms Sam	BMT Defence Services (Australia)	stait@bmtdsl.com.au
Taylor, Mr Graham	Taylortech Pty Limited	graham.taylortech@gmail.com
Thomson, Mr Andrew	CSC Australia	jkay5@csc.com
Thornton, Mr John	Defence And Systems Institute	john.w.thornton@unisa.edu.au
Tildesley, Mr Mathew	Centre for Maritime Engineering, Department of Defence	mathew.tildesley@defence.gov.au
Tink, Mr Rowan	Tenix Marine	rowan.tink@tenix.com
Toffoletto, Mr Robert	DSTO	robert.toffoletto@dsto.defence.gov.au
Toman, Mr Raymond	DMO - Air Warefare Destroyer Systems Centre	Raymond.Toman@ausawd.com
Torresan, Mr Hugh	DSTO	hugh.torresan@dsto.defence.gov.au
Town, CAPT Rick	Canadian High Commission	andrew.harrison@international.gc.ca
Tully, Mr Michael	Advanced Technology Training	michael.tully@attraining.com.au
Vale, Mr Evan	Directorate Of Naval Certification	evan.vale@defence.gov.au
Valentinis, Dr Francis	DSTO	francis.valentinis@dsto.defence.gov.au
Van Bronswijk, CAPT Ted	Company Of Master Mariners	TedVanBronswijk@iprimus.com.au
Vaughan, Mr Andrew	Tenix Marine	vaughana@tenix.com
Walker, Mr Phillip	Thales Australia	phil.walker@thalesgroup.com.au
Walsh, Mr Jeff	GHD Pty Ltd	jeff_walsh@ghd.com.au
Walters, Mr Dale	Tenix Marine	Dale.Walters@Tenix.com
Ward, Mr Greg	Defence SA	greg.ward@defencesa.com
Watson, Mr Graham	DMO	graham.watson3@defence.gov.au

Delegate List

29 -31 January 2008

Sydney Convention & Exhibition Centre, Sydney, AUSTRALIA

Name	Organisation	Email
AUSTRALIA		
Watson, Dr Nigel	Ship Dynamics	nigel.watson@shipdynamics.com
Whittaker, Mr Andrew	AWD Alliance	andrew.whittaker@ausawd.com
Wilhelm, Mr Geoffrey	Centre For Maritime Engineering - Department Of Defence	Geoff.Wilhelm@defence.gov.au
Williams, Mr Martin	Thales Australia	martin.williams@thalesgroup.com.au
Wilson, Mr Bruce	Australian Customs Service	bruce.wilson@customs.gov.au
York, Mr David	Centre for Maritime Engineering- Department of Defence	David.York1@defence.gov.au
Young, Mr Kevin	APC Technology	kyoung@apctechnology.com.au
Zbikowski, Mr Andrew	Centre for Maritime Engineering - Department of Defence	Andrew.Zbikowski@defence.gov.au
AUSTRIA		
Middleton, Mr Phil	Tenix Marine	phil.middleton@tenix.com
CANADA		
Bramwell, CAPT Eric	National Defence Of Canada	bramwell.eg@forces.gc.ca
Burd, Mr Douglas	Ultra Electronics Maritime Systems	doug.burd@ultra-uems.com
Colwell, Mr James	DRDC Atlantic	jim.colwell@drdc-rddc.gc.ca
Cullen, Ms Patricia	National Defence	cullen.pac@forces.gc.ca
MacKinnon, Dr Scott	Memorial University of Newfoundland	smackinn@mun.ca
Neri, John	National Defence	neri.jp@forces.gc.ca
Ross, Mr Dan	National Defence	Ross.D@forces.gc.ca
GERMANY		
Fach, Mr Karsten	Germanischer LLoyd AG	Karsten.Fach@gl-group.com
Fenske, Mr Christoph	MTU Friedrichshafen GmbH	christoph.fenske@mtu-online.com
Hoppe, Dr Franz	Renk AG	constanze.wunderlich@renk.biz
Losch, Mr Ralf	MAN Diesel SE	ralf.losch@man.eu
Pranger, Mr Heiner	MTU Friedrishafen GmbH	heiner.pranger@mtu-online.com
INDONESIA		
Hartawan, Mr Yoyok	ICMI Indonesia	icmi-indonesia@telkom.net
JAPAN		
Hashimura, Prof Ryusuke	Sojo University	ryusuke@eco.sojo-u.ac.jp
Kido, Dr Hachiro	Marine Technical College Japan	kido@mail.mtc.ac.jp
Thanh Thuy, Ms Nguyen	Kobe University, Japan	ngthanhthuy@gmail.com
NEW ZEALAND		
Fillmore, Ms Jessie	Maritime New Zealand	jessie.fillmore@maritimenz.govt.nz
Johnson, Mr Alexander Tulloch		marinecon@clear.net.nz
Neville, Mr Bruce	Beca Applied Technologies Ltd	bruce.neville@beca.com
Wishart, CMDR Kelvin	Royal New Zealand Navy	kelvin.wishart@nzdf.mil.nz
P.R. CHINA		
Teng, Mr Yanli	Shandong Ketong Maritime Equipment Co.,Ltd.	exhibiton_show@hotmail.com
SINGAPORE		
Choo, Prof Yoo Sang	National University Of Singapore	cvecys@nus.edu.sg
Davison, Ms Pelin	DNV	Pelin.Davison@dnv.com
Tan, Mr Thomas	American Bureau of Shipping	ttan@eagle.org

Delegate List

29 -31 January 2008

Sydney Convention & Exhibition Centre, Sydney, AUSTRALIA

Name	Organisation	Email
SOUTH AFRICA		
Akpata, Mr Paul	Ordec Communications And Engineering	pabak2003@yahoo.com
SWEDEN		
Ullman, Dr Johan	Ullman Dynamics	johan.ullman@ullmans.com
THE NETHERLANDS		
Bos, Dr Jelte	TNO Human Factors	Jelte.Bos@tno.nl
Engelbracht, Mr Dick	NLDA	DJ.Engelbracht@nlda.nl
Keuning, Dr Lex	Delft University Of Technology	j.a.keuning@tudelft.nl
Van Ginkel, Ms Hilde	NLDA	hta.v.ginkel@nlda.nl
UNITED KINGDOM		
Bricknell, Mr David	Rolls-Royce	david.bricknell@rolls-royce.com
Brink, CAPT Allen	International Institute Of Marine Surveying	iimsceo@netscape.net
Buckingham, Mr John	BMT Defence Services Limited	johnb@bmtdsl.co.uk
Buckley, Mr James	BMT Defence Services Ltd, UK	James.Buckley@nsass.org.uk
Dobbins, Dr Trevor	Human Sciences & Engineering Ltd	trevor.dobbins@humansci-eng.com
Holder, CAPT Leonard	Nautical Institute	lenholder@lenholder.freeserve.co.uk
Kimber, Mr Andy	BMT Defence Services Limited	akimber@bmtdsl.co.uk
Maclean, Mr David	BMT Group Ltd	dmaclean@bmtdefence.co.uk
Myers, Mr Stephen	University Of Chichester	s.myers@chi.ac.uk
Owen, Mr Jeff	Weir Strachan & Henshaw	Jeff.Owen@weirsandh.com
Pattison, Mr Denis	UK MOD	denis.pattison789@mod.uk
Quilliam, Mr Roy	BMT Defence Services Ltd	Heyles@bmtdsl.co.uk
Simpson, Mr Robert	Lloyd's Register EMEA	glenn.cobb@lr.org
Warren, Mr Steve	BP Marine Fuels Ltd	steve.warren@uk.bp.com
USA		
Agbolosu, Mr Georgeson		sea2land_2003@yahoo.co.uk
Assumeng Ankrah, Mr Henry	Regional Maritime Academy	assumengankrah@yahoo.com
Benedict, RADM Terry	Program Executive Officer For Integrated Warfare Systems	terry.benedict@navy.mil
Blankenship, Mr Jeff	Naval Surface Warfare Center Panama City Division	jeffrey.w.blankenshi@navy.mil
Carlin, Mr Joe	Argon ST	joe.carlin@argonst.com
Cumm, Mr Eugene	Northrop Grumman Corporation	gene.cumm@ngc.com
Doughty, Mr Alan	Bath Iron Works	alan.doughty@biw.com
Fumo, Mr John	Argon ST	john.fumo@argonst.com
Gallagher, Mr Herbert	Computer Sciences Corporation	hgallagh@csc.com
Glass, Mr Barry	The Boeing Company	barry.n.glass@boeing.com
Goehrig, Mr William	CSC	wgoehrig@csc.com
Hicks, RADM Alan	Missile Defense Agency	alan.hicks@mda.mil
Holt, Mrs Rhonda	Computer Sciences Corporation	ccarr20@csc.com
Hulse, LEUT Marion	Missile Defense Agency	katie.hulse@mda.mil
Ingram, Mr Thomas	American Bureau of Shipping	tingram@eagle.org
LaBrecque, Mr Joseph	NSWC Panama City Division	joseph.labrecque@navy.mil
MacQuarrie, Mr Jeff	Argon ST	jeff.macquarrie@argonst.com
Moe, Mr Christopher	Raytheon Company	mike_moe@raytheon.com
Naqvi, Dr Waseem	Raytheon	Waseem_Naqvi@raytheon.com
Patamia, Mr Kingsford Yao	REGIONAL MARITIME ACADEMY	kingptm@aol.com
Pierce, Mr Eric	Naval Surface Warfare Center Panama City Division	eric.pierce@navy.mil
Price, Mr Brian	Naval Surface Warfare Center Panama City	brian.l.price@navy.mil
Rispin, Dr Paul	Office Of Naval Research	paul.rispin@navy.mil

Delegate List

29 -31 January 2008

Sydney Convention & Exhibition Centre, Sydney, AUSTRALIA

Name	Organisation	Email
USA		
Schneidt, Ms Ashley	Aegis Ballistic Missile Defense	ashley.schneidt.ctr@mda.mil
Shampine, Ms Lauree	Raytheon Company	Fawn_J_Fox@raytheon.com
Stark, Dr Julie	U.S. Navy	julie.stark.ctr@navy.mil
Sweet, Mr John	Argon ST	jack.sweet@argonst.com
Vardakas, George	Raytheon	george_m_vardakas@raytheon.com
Whiting, Mr Geoffrey	Raytheon Company - Integrated Defense Systems	Geoffrey_A_Whiting@raytheon.com
Woodward, Mr David	Raytheon IDS	David_W_Woodward@raytheon.com

AUTHOR INDEX

Adams, S. 408
Allsop, J. 94
Anderson, J. 408
Andrewartha, T. 53
Anokhin, V. V. 460
Armstrong, T. 504
Bendall, D. 36
Bertram, V. 215
Bose, N. 408
Bricknell, D. J. 22, 74, 84
Brink, A. R. 104
Brown, P. 451
Buckingham, J. 315
Buckland, M. 350
Buckley, J. 306
Budd, A. 65
Cameron, A. 583
Cannon, S. 53
Carr, J. 619
Chan, S. 1
Chang, M. 144
Clifford, C. 198
Colwell, J. 379
Cox, G. 285
Crumlin, P. 169
Cuckson, B. 306
Davies, A. 360
Davis, J. 543
Davison, P. 32
De Yong, L. 533
Deyong, L. 494
El-Atm, J. 360
Elischer, P. 43, 494
Ellery, D. 504
Emmett, L. 235
Fach, K. 215
Fjeldstad, C. 84
Flapan, M. 522
Foley, P. 389
Gardiner, C. P. 504
Gehlin, R. 470
Ghosh, A. 631
Giles, W. 329
Grabovac, I. 609
Grech, M. R. 43
Hardy, T. 315
Hardy, T. A. 415
Harvey, K. 484
Helmore, P. J. 425, 437
Hodge, C. 315
Holder, L. A. 36
Hoppe, F. 661
Horstmann, P. 112
Howe, D. 671
Ingram, T. 620
Ivanov, L. D. 515
James, P. 306
John, N. S. 609
Kelsall, D. 577
Kido, H. 259
Kilbourn, P. 339
Kimber, A. 329
King, W. 233
Kucharzewski, H. 144
Lawrence, N. 275
Lee, K. H. 370
Lewis, R. 408
Losch, R. 544
Macdonald, G. 112
Macfarlane, G. J. 285
Macgillivray, I. 204
Malpress, K. 655

AUTHOR INDEX

Marco, J. 204
Marshall, S. 306
Mason, L. 415
Mathys, Z. 609
McConochie, J. 415
Milne, R. S. 425
Morrisseau, P. 153
Novak, D. 620
Ojeda, R. E. 275
Partridge, R. 22
Perez, T. 682
Pregnalato, C. 370
Prusty, B. G. 249, 275
Ranmuthugala, D. 360
Ray, T. 223, 249, 265
Reid, W. 494
Renilson, M. 163
Romero, J. R. 460
Russo, S. 609
Sadovnikov, D. 296
Sajdak, J. A. W. 153
Sakaguchi, K. 259
Scardino, A. 397
Seah, A. K. 515
Sidek, A. R. M. 249
Simpson, B. 640
Sloan, G. W. 595
Smith, W. 223, 265
Steedman, R. 460
Steinmann, P. 682
Stredulinsky, D. 379
Thomas, G. 53
Tso, Y. 360
Vedlog, P. E. 74
Vincent, P. 504
Wang, G. 515
Wiedemann, R. 144
Williams, P. 583
Wilson, A. 504